ELECTROMAGNETIC ENERGY TRANSMISSION AND RADIATION

TO E. A. G.

FOREWORD

This book is one of several resulting from a recent revision of the Electrical Engineering Course at The Massachusetts Institute of Technology. The books have the general format of texts and are being used as such. However, they might well be described as reports on a research program aimed at the evolution of an undergraduate core curriculum in Electrical Engineering that will form a basis for a continuing career in a field that is ever-changing.

The development of an educational program in Electrical Engineering to keep pace with the changes in technology is not a new endeavor at The Massachusetts Institute of Technology. In the early 1930's, the Faculty of the Department undertook a major review and reassessment of its program. By 1940, a series of new courses had been evolved, and resulted in the publication of four related books.

The new technology that appeared during World War II brought great change to the field of Electrical Engineering. In recognition of this fact, the Faculty of the Department undertook another reassessment of its program. By about 1952, a pattern for a curriculum had been evolved and its implementation was initiated with a high degree of enthusiasm and vigor.

The new curriculum subordinates option structures built around areas of industrial practice in favor of a common core that provides a broad base for the engineering applications of the sciences. This core structure includes a newly developed laboratory program which stresses the role of experimentation and its relation to theoretical model-making in the solution of engineering problems. Faced with the time limitation of a four-year program for the Bachelor's degree, the entire core curriculum gives priority to basic principles and methods of analysis rather than to the presentation of current technology.

J. A. STRATTON

PREFACE

The addition of a new work to the multitude of excellent textbooks based on classical electromagnetic theory can be justified only insofar as the book represents an effort to solve a new educational problem. Such a new problem has arisen from the postwar growth of the Electrical Engineering field as a whole. This growth has led the Department of Electrical Engineering at the Massachusetts Institute of Technology to develop during the last six years an undergraduate "core curriculum" program comprised of eight individual subjects. These are taken by *all* of its students, regardless of their personal goals for ultimate specialization.

Quite early in our considerations, it was agreed that one of these subjects should deal with the dynamics of distributed systems and the related field theory. A much longer time was involved in defining more precisely the appropriate subject matter, deciding its position in the sequence of others, and, finally, in preparing the material for the unspecialized background and interests of the students. As our ideas crystallized, it became clear that, in spite of our earlier hopes to the contrary, it would be wise to limit the considerations to purely electromagnetic phenomena, and to place the study beyond the junior level. Accordingly, a first-term senior subject was developed, designed to provide for all electrical engineering students the very minimum of contact with electric waves and oscillations that we regarded as acceptable for anyone who would bear the title "Electrical Engineer" some ten or twenty years from now. The present book contains a rather extended form of the material covered in this subject.

We were fortunate in being permitted by the concurrent development of other subjects to assume that the students involved would have these prerequisites: two and a half years of calculus; more than a year of college-level experience with electromagnetic fields, including one term of quasi-statics based completely upon Maxwell's equations in vector-analytical form;[1] and two full years comprising various aspects of linear-system theory, involving considerable exposure to superposition principles, Fourier series and integrals, and matrix notation. We could not, however, count upon any previous experience with partial differential equations.

It has, broadly speaking, been our effort to gather up the important linear-system concepts of the time and frequency domains, and the energy-power relations from the theories of lumped circuits and quasi-static fields, and extend them to cover "distributed" situations, in which *space* assumes an importance equal to that of the other independent variables. Our objective in this connection has been the formulation of a complete four-way viewpoint—time, frequency, energy (or power), and space.

Nevertheless, the traditional emphasis on the solution of formal boundary-value problems in a variety of coordinate systems would be inconsistent with the balance we wished to achieve between physical understanding, pure technique, and limited specialization of interest. We have, instead, adopted the approach of presenting first the solutions of field or wave problems in the source-free, unbounded space, and then examining the properties of these solutions in considerable detail, with a view to discovering the kinds of boundary conditions that can be met by combinations of a few of them. Moreover, only the simplest of the infinite set of solutions are considered carefully, because it is our conviction that the most important situations that arise in practice under the heading of "boundary-value problems" are those requiring either a sound qualitative understanding of what can (or cannot) happen under the given circumstances, based on very general principles, or, in more particular quantitative cases, the design of a set of *boundaries* that will yield some rather simply stated field configuration within a certain region of space.

The method of presentation described above reflects our commitment to organize the material around principles of lasting value, using examples of the most elementary character which will still illustrate the points involved. In this context, practical devices are often replaced

[1] The subject referred to is based upon portions of Robert M. Fano, Lan Jen Chu, Richard B. Adler, *Electromagnetic Fields, Energy, and Forces*, John Wiley and Sons, New York, 1960.

by reminiscent configurations of boundaries, requiring only very limited combinations of elementary solutions.

The foregoing philosophy of teaching field theory may make more difficult the use of this book as a professional technical reference work; if so, it represents one result of the difficult conscious compromises made necessary by the nature of our educational objectives.

In any effort to develop a "core-curriculum" subject, ruthless discarding of nonessential material is mandatory so as not to overstep the boundaries of time. Correspondingly, a very clear-cut concept of the principal objectives of the work must be borne in mind continuously. For those students described above, we believe we have established such a teachable core-curriculum subject in the following portions of this book: Practically all of Chapters 1–5; approximately two-thirds of Chapter 6; selected sections on *uniform* plane waves and skin effect from Chapters 7 and 8; sections on the *two-conductor* lossless line from Chapter 9; and practically all of Chapter 10. The classroom part of this subject is built upon the following general ideas:

1. The role of stored energy, dissipation, and energy flux in unifying field problems.
2. The circumstances under which voltage, current, and impedance can be defined, and the importance of the energies in linking the circuit and field descriptions of physical systems.
3. The description of waves and oscillations from multiple points of view; time, space, and (complex) frequency domains; the corresponding analytical techniques and physical interpretation of the mathematics; the behavior of the energies.
4. The role of boundary conditions in dividing a complicated physical problem into simpler tractable pieces, and the ideas relating the forms of the elementary solutions to the appropriate boundary conditions.

Of course it has not been possible to give generalized demonstrations of all the items mentioned in the foregoing list, because most of them represent not definite theorems, but rather viewpoints of great utility in the engineering applications of electromagnetic phenomena. Instead, the simplest possible analytical models, suggestive of practical devices, were chosen in each case to exhibit clearly the concept in question. Conversely, we have tried, where possible, to treat examples from several different points of view to develop flexibility with the concepts and depth of understanding of the examples. This whole procedure turns out to contrast strongly with the more conventional one of organizing the subject matter around "real" devices, the analysis of which often mixes several difficult conceptual and mathematical matters together in such a way as to yield the most "expeditious" treatment of the particular case in question.

However desirable we may believe is the inclusion of a third contact with electromagnetic phenomena at the undergraduate level, we recognize that situations will arise in which this contact must be postponed until a student's early graduate career. By this time, although the student's needs may be essentially the same as those we have described previously, his interest in specialization may well have increased greatly. Moreover, his over-all maturity and mathematical facility are expected to be considerably improved. Similar comments can be made about the exceptionally able undergraduate student who may progress relatively rapidly through the material described above as constituting the core subject. It is essentially for these reasons that we have added to this book the sections on coupled resonators in Chapter 6; the rather lengthy discussions of nonuniform plane waves and guided waves in Chapters 7 and 8; and the section on multiconductor lines in Chapter 9. Some justifications of approximation techniques are also given in much greater detail than is suitable for a basic subject. We have marked with an asterisk (*) any section a major part of which is considered to be beyond the "core" subject, and we have occasionally added footnotes concerning elaborate individual developments within a section that can either be omitted or simplified greatly by reducing the generality. We hope these additions have increased the flexibility of the text for application over a wide range of levels from undergraduate to graduate study.

In a work of this kind, it is impossible to acknowledge the many sources, written and personal, from which the authors have drawn their backgrounds for the task. But we cannot refrain from mentioning specifically the tremendous impact made upon us by our long association with the stimulating environment of the M.I.T. Research Laboratory of Electronics, under the able direction of Prof. Jerome B. Wiesner. This association has made natural for us the blurring of boundaries between education and research considered so important to the educational developments underlying this book.

Similarly, the fact that major portions of the book have been taught for about five years makes it difficult to assess accurately the debt we owe to the staff and students whose points of view have been reflected in the final work. Foremost among these, however, with regard to contributions in the form of early text material, constructive criticism of notes, proofreading of preliminary versions of notes, and extensive laboratory developments, are respectively Professors Herman A. Haus and Edward I. Hawthorne, and Messrs. Frederick Hennie, John Blair, and Ronald Massa.

For the critical reading of the final manuscript, numerous extremely helpful criticisms and suggestions for improvement, and the provision, organization, and solution of a great many problems for the text, we are most deeply grateful to Drs. John Granlund and Morton Loewenthal.

We wish also to thank those who helped by preparing typewritten copy and by performing all the secretarial duties that accompany the development of notes into a text; in particular: Mrs. Bertha Hornby, Miss Dorothea Scanlon, Mrs. Margaret Park, Mrs. Carolyn Bennett, and Miss Nancy Rhodenizer.

This work could not have been undertaken or carried to completion without the encouragement, support, and imaginative leadership of Dr. Gordon S. Brown, whose dynamic efforts and great investment of energy, as Head of the Electrical Engineering Department and our own personal Maxwell demon, made possible the entire core-curriculum program, of which this book represents a tiny fraction.

Richard B. Adler
Lan Jen Chu
Robert M. Fano

Cambridge, Mass.
August, 1959

C O N T E N T S

SECTIONS THAT MAY BE OMITTED FROM A MINIMUM UNDERGRADUATE SUBJECT

C H A P T E R O N E

Lumped-Circuit and Field Concepts

It is often necessary in physical situations to deal with the relations between lumped circuit and field concepts. Indeed, the existence of a voltage-current characteristic describing any lumped-circuit element arises from the validity of a particular kind of approximation to the field distribution in the space occupied by and surrounding the corresponding physical device. It is helpful, therefore, to be able to express both lumped-circuit and field behavior in similar language. The concept of energy in its various forms, the principle of its conservation, and the conservation of charge are important aids for achieving this goal.

This chapter is intended to supply a rapid review [1] of familiar circuit and field principles, presented in a form especially suitable for our subsequent use.

1.1 Lumped Electric Circuits

1.1.1 Time Domain

To obtain the dynamic-equilibrium equations for a lumped electric circuit, we use essentially three laws:

(a) Conservation of charge (Kirchhoff's current law).
(b) Single-valuedness of voltage drop (Kirchhoff's voltage law).

[1] For a more thorough treatment, see Robert M. Fano, Lan Jen Chu, Richard B. Adler, *Electromagnetic Fields, Energy, and Forces*, John Wiley and Sons, New York, 1960.

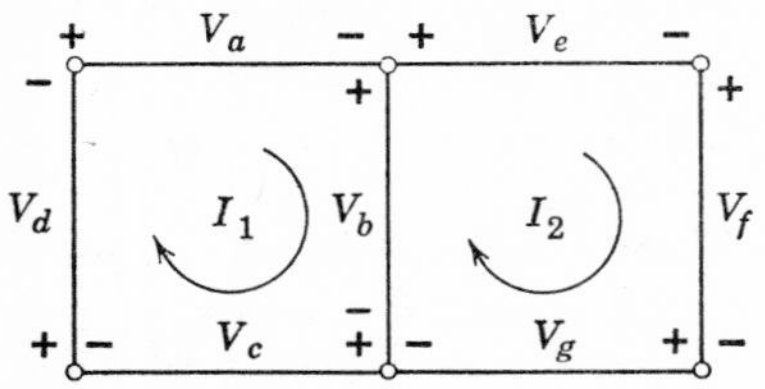

Fig. 1.1. A simple circuit geometry.

(c) Voltage-current relations for the lumped elements:
1. Capacitance.
2. Inductance (and mutual inductance).
3. Resistance.
4. Voltage or current sources.

Charge may be taken to be a fundamental concept that appears in electric systems, as distinct from mechanical ones. Current is the rate of flow of charge along a clear-cut conduction path. In the general case of time-varying charge or current, voltage is a somewhat subtle concept. For our immediate purposes, we wish to emphasize its property of being the instantaneous power divided by the instantaneous current.

With these ideas in mind, it is easy to show that the Kirchhoff laws imply conservation of energy for the whole network. Consider the schematic circuit geometry of Fig. 1.1, in which the detailed character of the branches is not shown, but it is presumed each contains only one of the lumped elements mentioned above. Analysis [1] in terms of loop currents I_1 and I_2 automatically satisfies Kirchhoff's current law. The voltage law yields

$$\begin{aligned} &\text{(a)} \qquad V_a + V_d + V_c + V_b = 0 \\ &\text{(b)} \qquad V_e + V_f + V_g - V_b = 0 \end{aligned} \tag{1.1}$$

[1] Notation conventions in the text are: italic lightface (A, a) for real scalars; roman lightface (A, a) for complex scalars (phasors); italic boldface ($\boldsymbol{A}$, $\boldsymbol{a}$) for real space vectors; roman boldface (**A**, **a**) for complex space vectors; and script ($\mathscr{A}$, $\mathscr{a}$) for matrices, with the lower case script ($\mathscr{a}$) reserved for column matrices. The distinctions between real and complex character (italic and roman respectively) and scalar and space-vector character (lightface and boldface respectively) in all four cross combinations are so important to a clear quantitative understanding of the subject matter of this book that they must be carried over carefully into blackboard and problem-solution techniques. To do this most conveniently by hand, we recommend use of underscores and overscores to indicate complex character and real space-vector character respectively, with both together being used for complex space vectors.

After multiplying Eq. 1.1a by I_1, Eq. 1.1b by I_2, and adding the results, we find

$$(V_a + V_d + V_c)I_1 + V_b(I_1 - I_2) + (V_e + V_f + V_g)I_2 = 0 \quad (1.2)$$

Equation 1.2 says that the net power delivered to the network is always zero. Obviously the demonstration is readily extended to an arbitrary network geometry.

The passive lumped elements (C, L, R) either store energy (C, L) or dissipate it (R). From the point of view of lumped circuits, in the strictest sense, we can tell which ones store and which dissipate—but there is no basis for identifying the specific forms of energy involved. Thus, a linear, time-invariant capacitance is defined by the voltage-current relation

$$I = C\frac{dV}{dt} \quad (1.3)$$

where positive I flows in the direction of voltage drop V through the element. Multiplying Eq. 1.3 by V, we find

$$VI = \frac{d}{dt}\left(\frac{1}{2}CV^2\right) \quad (1.4)$$

Equation 1.4 has a left side which is the power delivered to the element. Therefore the quantity $\frac{1}{2}CV^2$ must be energy, and it has proper dimensions. Moreover, it is stored energy because the right side of Eq. 1.4 may be either positive or negative, even though CV^2 is always positive. Without further assumptions, however, we have no right to call it stored electric energy; this nomenclature would invoke a field concept which is beyond the domain of "pure" circuit theory. Nevertheless, let us denote instantaneous energy stored in capacitors by W_e.

A similar derivation would lead to energy W_m stored in linear, time-invariant inductance, of the form $\frac{1}{2}LI^2$. For mutually coupled coils (L_1, L_2, M) carrying currents I_1 and I_2, the stored energy is

$$\tfrac{1}{2}L_1I_1^2 + \tfrac{1}{2}L_2I_2^2 + MI_1I_2$$

which we include in W_m. The extension to more mutually coupled coils is clear.

The case of resistance is different from that of capacitance or inductance, since its volt-ampere relation is $V = RI$. Thus $VI = RI^2$, and if $R \geq 0$ it follows that $RI^2 \geq 0$. Power is always delivered to a positive resistor—at every instant of time. None ever comes from it. The energy delivered to it has gone down the drain as far as the circuit is concerned. Where has it gone? Again, the answer to this question depends upon other than lumped-circuit concepts. We shall simply call instantaneous power put into (dissipated in) resistors P_d.

Finally, the instantaneous power delivered *by* sources in the network we shall call P.

According to Eqs. 1.2 and 1.4 and the preceding remarks, if for a linear, time-invariant network we include all resistors, capacitors, inductors (including mutual inductance), and sources in the respective energy or power terms, we have

$$P = P_d + \frac{d}{dt}\left(W_e + W_m\right) \tag{1.5}$$

Equation 1.5 expresses the conservation of energy in the network at any instant. It is to be noted that W_e and W_m, being stored energies, are never negative. P_d is positive by virtue of its algebraic form as a sum of RI^2 terms, with all R positive.

1.1.2 Frequency Domain

It might be expected that for a network under any special conditions, e.g., in the sinusoidal steady state, a full discussion of all important energetic matters would stem from appropriate substitutions into Eq. 1.5. This is not so. The reason is that, although Kirchhoff's two laws imply conservation of energy for the whole network, the law of conservation of energy for the whole network, taken with either one of Kirchhoff's laws alone, is not equivalent to the other Kirchhoff law. Reference to Eqs. 1.1 and 1.2 will clarify this fact. Equation 1.2 may be regarded as the result of applying both conservation of energy and conservation of charge (Kirchhoff's current law) to the network of Fig. 1.1. It is only one equation containing two unknown currents. If there were more loops, Eq. 1.2 would still only be one equation, but there would be even more unknown currents in it. Without introducing new assumptions into Eq. 1.2, there is no way of recovering the voltage law relations (Eq. 1.1). The two Kirchhoff laws must therefore say more about a network than is contained in either one of them plus the law of conservation of energy. Hence it may well be possible to state in terms of energy concepts a good deal more about networks than is contained in Eq. 1.5; e.g., though we shall not pursue them here, the Lagrangian and Hamiltonian formulations of network equilibrium have been derived.[1]

[1] Ernst A. Guillemin, *Introductory Circuit Theory*, John Wiley and Sons, New York, 1953, Ch. 10; and David C. White and Herbert H. Woodson, *Electromechanical Energy Conversion*, John Wiley and Sons, New York, 1959, Ch. 1.

Our immediate interest is in the sinusoidal steady-state behavior of a lumped, linear, time-invariant network. If we represent the currents and voltages in Fig. 1.1 by their complex amplitudes V and I, the Kirchhoff voltage law reads:

$$\text{(a)} \qquad \mathrm{V}_a + \mathrm{V}_d + \mathrm{V}_c + \mathrm{V}_b = 0$$

$$\text{(b)} \qquad \mathrm{V}_e + \mathrm{V}_f + \mathrm{V}_g - \mathrm{V}_b = 0 \tag{1.6}$$

Now, defining complex power P as $\frac{1}{2}\mathrm{VI}^*$, which differs from the definition $\frac{1}{2}\mathrm{V}^*\mathrm{I}$ sometimes used, we are led to multiply Eq. 1.6*a* by I_1^* and Eq. 1.6*b* by I_2^*, and to add the results. We find

$$\tfrac{1}{2}(\mathrm{V}_a + \mathrm{V}_d + \mathrm{V}_c)\mathrm{I}_1^* + \tfrac{1}{2}\mathrm{V}_b(\mathrm{I}_1 - \mathrm{I}_2)^* + \tfrac{1}{2}(\mathrm{V}_e + \mathrm{V}_f + \mathrm{V}_g)I_2^* = 0 \tag{1.7}$$

which expresses a new law of conservation of complex power in the network as a whole. Note that Eq. 1.7 is *not* just a special case of Eq. 1.2.

The complex power delivered to any passive branch of the circuit is easily related to the energy stored or dissipated in it—in particular the time-average value thereof. For example, in the sinusoidal steady state the volt-ampere relation of a constant capacitor is

$$\mathrm{I} = j\omega C\mathrm{V} \tag{1.8}$$

making

$$\frac{1}{2}\mathrm{VI}^* = -j\frac{\omega C}{2}|\mathrm{V}|^2 \tag{1.9}$$

But $|\mathrm{V}|^2/2$ is just the time-average value of $V^2(t)$ when $V(t)$ is sinusoidal. Therefore if we denote by a special bracket $\langle\ \rangle$ the time-average value of the enclosed symbol, we have for the time-average value of the instantaneous stored energy W_e given in connection with Eq. 1.4,

$$\langle W_e\rangle = \tfrac{1}{2}C\langle V^2(t)\rangle = \tfrac{1}{4}C|\mathrm{V}|^2 \tag{1.10}$$

In view of Eq. 1.10, Eq. 1.9 becomes

$$\tfrac{1}{2}\mathrm{VI}^* = -2j\omega\langle W_e\rangle \tag{1.11}$$

Similarly, for pure inductance the complex power is $+2j\omega\langle W_m\rangle$, where

$$\langle W_m\rangle = \tfrac{1}{4}L|\mathrm{I}|^2$$

in the sinusoidal steady state. If two coils are coupled (L_1, L_2, M), the time-average energy stored is

$$\langle W_m\rangle = \tfrac{1}{4}L_1|\mathrm{I}_1|^2 + \tfrac{1}{4}L_2|\mathrm{I}_2|^2 + \tfrac{1}{2}M\,\mathrm{Re}\,(\mathrm{I}_1\mathrm{I}_2^*)$$

and again the complex power is simply $+2j\omega\langle W_m\rangle$.

For resistors, the complex power is the time-average power dissipated, $\frac{1}{2}R|\mathrm{I}|^2$, which we express as $\langle P_d \rangle$.

Thus, writing the complex power *produced* by *all* sources in the network as $\langle P \rangle + jQ$, we recast Eq. 1.7 to read

$$\langle P \rangle + jQ = \langle P_d \rangle + 2j\omega \langle W_m - W_e \rangle \tag{1.12}$$

in which the average power and energy terms on the right must of course include sums over all passive elements in the circuit.

In connection with Eq. 1.12, an interesting result arises when only one source is driving the network. Let V_s and I_s be the complex source voltage and current respectively, such that $(\mathrm{V}_s/\mathrm{I}_s)$ is the impedance Z presented to the source by the network. Then we have

$$\tfrac{1}{2}\mathrm{V}_s\mathrm{I}_s^* = \langle P_d \rangle + 2j\omega \langle W_m - W_e \rangle \tag{1.13}$$

which, upon division by $\mathrm{I}_s\mathrm{I}_s^* = |\mathrm{I}_s|^2$, becomes

$$\frac{\mathrm{V}_s}{\mathrm{I}_s} = \mathrm{Z} = \frac{2\langle P_d \rangle}{|\mathrm{I}_s|^2} + j4\omega \left(\frac{\langle W_m \rangle}{|\mathrm{I}_s|^2} - \frac{\langle W_e \rangle}{|\mathrm{I}_s|^2} \right) \tag{1.14}$$

In other words, the real part of the input impedance is twice the average power dissipated in the network for an ampere of input current, whereas the reactance is proportional to the difference between the average inductive and average capacitive energies stored by an input ampere. It is to be observed that, with a fixed input current $\mathrm{I}_s = 1$ amp, $\langle P_d \rangle$, $\langle W_m \rangle$, and $\langle W_e \rangle$ will in general be complicated functions of frequency according to the redistribution of branch currents and voltages throughout the network. One should therefore not be misled by the fact that ω enters Eq. 1.14 explicitly in only a simple way.

1.2 Electromagnetic Fields

1.2.1 Time Domain

Having reviewed briefly purely lumped circuits, we turn now to review principles of the electromagnetic field in stationary media. At the outset we have six dependent field variables, all of which are in general functions of independent variables denoting space and time. These field variables are:

1. Free-charge density ρ (scalar), coulombs/m^3.
2. Current density $\boldsymbol{J}$ (vector), amp/m^2.
3. Electric field intensity $\boldsymbol{E}$ (vector), v/m.
4. Magnetic flux density $\boldsymbol{B}$ (vector), webers/m^2 = v-sec/m^2.

5. Electric flux density $\boldsymbol{D}$ (vector), coulombs/m^2.
6. Magnetic field intensity $\boldsymbol{H}$ (vector), amp/m.

We observe that the last five are vector functions of position and time, while the first (ρ) is a scalar function. The relations among these variables are fundamentally expressed in integral forms with which we are presumably already familiar. Without writing them, therefore, we merely remind ourselves that they express:

(a) Conservation of free charge (Gauss' law).
(b) Faraday's induction law.
(c) Ampere's circuital law, including displacement current.
(d) Absence of soluble magnetic charge.

By using the conventional techniques of vector analysis, the fundamental integral laws stated above can be converted to differential form. These are Maxwell's equations (in the order above):

$$\begin{aligned} &\text{(a)} && \nabla\cdot\boldsymbol{D} = \rho \\ &\text{(b)} && \nabla\times\boldsymbol{E} = -\partial\boldsymbol{B}/\partial t \\ &\text{(c)} && \nabla\times\boldsymbol{H} = \boldsymbol{J} + \partial\boldsymbol{D}/\partial t \\ &\text{(d)} && \nabla\cdot\boldsymbol{B} = 0 \end{aligned} \tag{1.15}$$

There are three other relations connected with Eqs. 1.15 that enhance our understanding of electromagnetic fields. If we take the divergence of Eq. 1.15*c*, we find that, since $\nabla\cdot(\nabla\times\boldsymbol{H}) \equiv 0$,

$$\nabla\cdot\left(\boldsymbol{J} + \frac{\partial\boldsymbol{D}}{\partial t}\right) = 0 \tag{1.16}$$

Recalling that $\partial\boldsymbol{D}/\partial t$ is a displacement current, we see in the above result that *total* current is continuous—it flows in closed paths. Where conduction current stops in space, displacement current takes over to "complete the circuit." Equation 1.15*a*, however, when differentiated partially with respect to time, yields

$$\nabla\cdot\frac{\partial\boldsymbol{D}}{\partial t} = \frac{\partial\rho}{\partial t} \tag{1.17}$$

Thus, from Eqs. 1.16 and 1.17, we discover that

$$\nabla\cdot\boldsymbol{J} = -\frac{\partial\rho}{\partial t} \tag{1.18}$$

which expresses directly, in a familiar form, the law of conservation of *free* charge. We see that it was contained completely in Eqs. 1.15, and is, therefore, not independent of them. Together with Eq. 1.16, Eq. 1.18 also tells us that a changing free-charge density forms the connecting link in the conversion from the flow of conduction current to the (equal) flow of displacement current.

Finally, we take the divergence of Eq. 1.15*b* and use the fact that $\nabla \cdot (\nabla \times \boldsymbol{E}) \equiv 0$. Then

$$\nabla \cdot \frac{\partial \boldsymbol{B}}{\partial t} = \frac{\partial}{\partial t} (\nabla \cdot \boldsymbol{B}) = 0 \tag{1.19}$$

Now Eq. 1.19 does not prove that $\nabla \cdot \boldsymbol{B} = 0$, but it does insist that $\nabla \cdot \boldsymbol{B}$ is not allowed to change with time. If, on the other hand, $\nabla \cdot \boldsymbol{B}$ had some constant value other than zero at any point in space, we would have to admit the presence of a net isolable magnetic "charge" at that point *forever.* Space would be dotted here and there with isolable magnetic charge that could not budge! The fact that not even one such charge has ever been found leads us to reject this possibility on physical grounds. Thus we conclude from Eq. 1.19, plus experiments, that

$$\nabla \cdot \boldsymbol{B} = 0$$

Therefore Eq. 1.15*d* is not quite contained in Eq. 1.15*b*.

To complete Eqs. 1.15 for the determination of all six field variables, we need some field relationships analogous to voltage-current relations, which involve the detailed properties of the medium at each point. The simplest such connections are:

$$\text{(a)} \qquad \boldsymbol{D} = \epsilon \boldsymbol{E}$$

$$\text{(b)} \qquad \boldsymbol{B} = \mu \boldsymbol{H} \tag{1.20}$$

$$\text{(c)} \qquad \boldsymbol{J}_c = \sigma \boldsymbol{E}$$

in which ϵ is the dielectric permittivity (f/m), μ is the magnetic permeability (h/m), σ is the conductivity (mhos/m) of the medium, and $\boldsymbol{J}_c$ is that part of the total current density $\boldsymbol{J}$ which varies with the field.

We have in mind, in connection with Eqs. 1.15*c* and 1.20*c*, that part of the total current density $\boldsymbol{J}$ may be considered as a "current density source" $\boldsymbol{J}_0$, *independent of the field.* This "source" may, of course, really be a current prescribed by an external agency; but more often it is simply one which we know from measurements before we start to find the fields. Alternatively, $\boldsymbol{J}_0$ is often a current we *imagine* we know, on a surface or in a region, and to which we will relate the fields in order to arrive finally at an equation determining the current. This use of $\boldsymbol{J}_0$ is merely a device for solving a problem; it has no special physical significance. Thus in Eq. 1.15*c*, we visualize

$$\boldsymbol{J} = \boldsymbol{J}_0 + \boldsymbol{J}_c \tag{1.21}$$

Equation 1.21 adds to the "voltage-current" characteristics of Eqs. 1.20 a possible source in addition to the passive elements. The analogy with lumped circuits may be completed by considering a "voltage" source in the form of a prescribed portion of the electric field, although we shall not need it at the moment.

By writing Eqs. 1.20, we are assuming that the medium is *isotropic* (i.e., has the same parameters at a given point, regardless of the direction of the field at that point). It is not, however, necessary in these equations to assume that ϵ, μ, and σ have the same values at every point in space (i.e., the medium need not be *homogeneous*), nor that they are necessarily independent of field strength (i.e., the medium need not be *linear*). In all of our work here, however, the medium will be considered to be linear unless otherwise stated specifically.

The question of homogeneity is a little more involved because we shall be interested in some problems concerning the boundaries between two different conductors, two different dielectrics, a conductor and a dielectric, etc. If, however, we imagine these boundaries to be abrupt, the parameters ϵ, μ, and σ will be invariant with position at all points of space, except right on these boundaries themselves. Then the steps leading from the integral form of the field equations to the differential form cannot be carried out directly at points right *on* the boundary surface. Since some of the field components may be discontinuous at such points, the very existence of the derivatives in Eqs. 1.15 may be denied. By methods which are presumably familiar, however, the fundamental integral laws stated in items (*b*) to (*d*) (preceding Eqs. 1.15) can be applied specifically to such sharp boundaries between media with the result that, if neither medium has $\sigma = \infty$, those components of $\boldsymbol{E}$ and $\boldsymbol{H}$ tangent to, and of $\boldsymbol{B}$ normal to, the interface surface must be continuous across that surface. In terms of Fig. 1.2, showing a surface S separating two different media 1 and 2, and possessing a normal unit vector $\boldsymbol{n}$, these conditions become simply

$$\begin{aligned}
&\text{(a)} \qquad \boldsymbol{n} \times (\boldsymbol{E}_2 - \boldsymbol{E}_1) = 0 \qquad \text{on } S \text{ if } \sigma_{1,2} \neq \infty \\
&\text{(b)} \qquad \boldsymbol{n} \times (\boldsymbol{H}_2 - \boldsymbol{H}_1) = 0 \qquad \text{on } S \text{ if } \sigma_{1,2} \neq \infty \\
&\text{(c)} \qquad \boldsymbol{n} \cdot (\boldsymbol{B}_2 - \boldsymbol{B}_1) = 0 \qquad \text{on } S \text{ if } \sigma_{1,2} \neq \infty
\end{aligned} \tag{1.22}$$

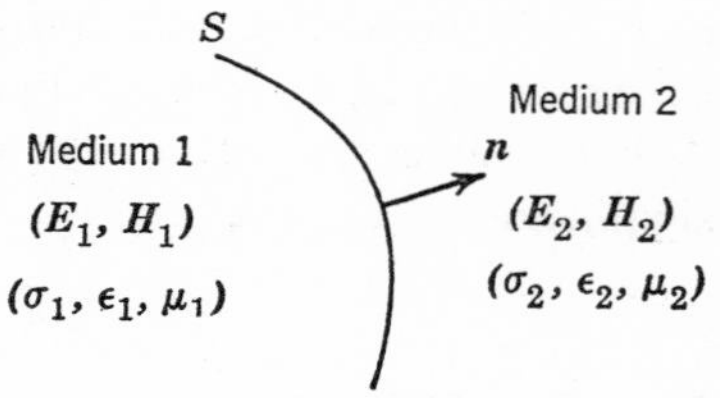

Fig. 1.2. A boundary between two media.

Equations 1.22 are to be regarded merely as those limiting forms of Eqs. 1.15*b* to 1.15*d* that govern the behavior of the field just at a boundary, where the differential equations simply do not apply. In other words, a correct field solution must not only satisfy Eqs. 1.15*b* to 1.15*d* *within* medium 1 and *within* medium 2, but it must also satisfy their limiting forms (Eqs. 1.22) right *on* the boundary. If it does not meet all these conditions, then it will not satisfy the basic physical laws (*b*) to (*d*) preceding Eqs. 1.15 in *all* regions of space (which must, of course, include regions containing the interface).

With respect to the law of conservation of charge [(a) preceding Eqs. 1.15] at interfaces, where Eq. 1.15*a* breaks down, there are several steps involved in reaching an answer. First, we note from Eq. 1.21 that part of $\boldsymbol{J}$ is $\boldsymbol{J}_0$, a current which we have regarded as a source of the field, known in advance. The rest of the current is $\boldsymbol{J}_c$, related directly to $\boldsymbol{E}$ by Ohm's law (Eq. 1.20*c*). If, then, we put $\boldsymbol{J} = \boldsymbol{J}_0 + \sigma\boldsymbol{E}$ in Eq. 1.15*c*, it will contain as *unknowns* only $\boldsymbol{H}$, $\boldsymbol{D}$, and $\boldsymbol{E}$, while Eq. 1.15*b* contains only $\boldsymbol{E}$ and $\boldsymbol{B}$. Now, however, using Eqs. 1.20*a* and 1.20*b*, we can eliminate from Eqs. 1.15*b* and 1.15*c* $\boldsymbol{B}$ and $\boldsymbol{D}$. Thus we have two first-order vector partial-differential equations relating $\boldsymbol{E}$, $\boldsymbol{H}$, their *curls*, and time derivatives. With these go their limiting forms, Eqs. 1.22*a* and *b*, relating *tangential* components across the interface. Similarly, Eqs. 1.15*d* and 1.22*c* govern $\nabla\cdot\boldsymbol{H}$ and the transition of *normal* components $\boldsymbol{n}\cdot\boldsymbol{H}$ across the boundary. Next, using Eqs. 1.16, 1.20, and 1.21, we find

$$\nabla\cdot\boldsymbol{J}_0 + \sigma\nabla\cdot\boldsymbol{E} + \epsilon\partial(\nabla\cdot\boldsymbol{E})/\partial t = 0 \tag{1.23}$$

for $\nabla\cdot\boldsymbol{E}$, and in the limit, for the change of $\boldsymbol{n}\cdot\boldsymbol{E}$ across S,

$$\boldsymbol{n}\cdot[\boldsymbol{J}_{02} - \boldsymbol{J}_{01} + \sigma_2\boldsymbol{E}_2 - \sigma_1\boldsymbol{E}_1 + \partial(\epsilon_2\boldsymbol{E}_2 - \epsilon_1\boldsymbol{E}_1)/\partial t] = 0 \qquad \text{on } S \text{ if } \sigma_{1,2} \neq \infty \tag{1.24a}$$

With other physical boundary conditions, Eqs. 1.15*b* through 1.15*d* and Eqs. 1.20 through 1.24*a* are sufficient to determine the entire field. Hence, except in some electrostatic problems, Eq. 1.15*a* and its limiting form at the boundary.

$$\boldsymbol{n} \cdot (\boldsymbol{D}_2 - \boldsymbol{D}_1) = \rho_s \qquad \text{on } S \text{ if } \sigma_{1,2} \neq \infty \tag{1.24b}$$

serve directly only to find the unknown free charge densities in the volume (ρ) or on the interface surface (ρ_s coulombs/m^2).

It will be noticed that in Eqs. 1.22 and 1.24 the condition is imposed that neither medium shall be a perfect conductor (for which $\sigma = \infty$). Suppose next that medium 1 (Fig. 1.2) is such a perfect conductor, so that $\sigma_1 = \infty$. This is, of course, an idealization, but we often use it to simplify problems in which a very good conductor appears. Then Ohm's law (Eq. 1.20*c*) requires that $\boldsymbol{E}_1 \equiv 0$, unless we are willing to let $\boldsymbol{J}_c$ become infinite *everywhere in medium 1*. Now, we reject the latter thought because the *total* current through any *finite* area (however small) of the medium would also be infinite, and no physical source could possibly supply it.

Observe, however, that we do not reject, as such, the idea that $\boldsymbol{J}_c$ may become infinite; indeed, if $\boldsymbol{J}_c$ becomes infinite over a zero area, we may well end up with a finite total current. Such a result, like the unit impulse of electric circuit theory, is a perfectly acceptable idealization. In static field theory, too, similar idealizations are very familiar—the most common one being the "point" charge (infinite charge density ρ over a zero volume, yielding a finite total charge).

Nonetheless, inside medium 1 we must put $\boldsymbol{E}_1 \equiv 0$, for the specific reason discussed above. Therefore $\boldsymbol{D}_1 = 0$ (Eq. 1.20*a*) and $\rho_1 = 0$ (Eq. 1.15*a*). Also $\partial \boldsymbol{B}_1/\partial t = 0$, by Eq. 1.15*b*, and of course $\partial \boldsymbol{D}_1/\partial t = 0$. There remains only the possibility of a *static* magnetic field, from Eqs. 1.15*c* and 1.15*d*. Even this, if it exists at all, must be determined by sources outside medium 1. We say this because there is no hope of impressing a known current density $\boldsymbol{J}_0$ throughout a material of zero resistance, and because $\boldsymbol{J}_c$ is completely indeterminate from the product of $\sigma_1 \equiv \infty$ and $\boldsymbol{E}_1 \equiv 0$; i.e., $\boldsymbol{J}$ in Eq. 1.15*c* is completely indeterminate. In fact, although a static magnetic field looks formally possible, it could never be set up; or, if once set up, it could never be destroyed ($\partial \boldsymbol{H}/\partial t = 0$)!

We are forced to conclude that all time-varying fields inside a perfect conductor are certainly zero, and that even a static electric field cannot exist in it. The question of static magnetic fields depends upon the idealization involved in any particular case, and need not concern us at all in our study of time-varying fields. Thus, instead of Eq. 1.22*a*, we find from the integral laws,

$$\boldsymbol{n} \times \boldsymbol{E}_2 = 0 \qquad \text{on } S \text{ if } \sigma_1 = \infty, \quad \sigma_2 \neq \infty \tag{1.25}$$

This is the only condition we really need, because now we know the field solution inside medium 1 (namely, zero field) and have only to find it in medium 2. As in electrostatics, it turns out to be sufficient to know the condition given by Eq. 1.25 on perfect conductors to determine field solutions outside them.

We may, however, be interested later in what happens to Eq. 1.22*b*. The limiting form must in that case be carried out with care, since we know that $\boldsymbol{J} = \infty$ *over a zero area* is possible. We must admit this possibility when $\sigma_1 = \infty$, and allow for a "surface current density," $\boldsymbol{K}_s$ (amp/m), which may be running along just under S. We then find, since $\boldsymbol{H}_1 \equiv 0$ (except for a trivial static possibility mentioned above),

$$\boldsymbol{n} \times \boldsymbol{H}_2 = \boldsymbol{K}_s \qquad \text{on } S \text{ if } \sigma_1 = \infty, \quad \sigma_2 \neq \infty \tag{1.26}$$

This equation really serves only to determine $\boldsymbol{K}_s$, after $\boldsymbol{E}_2$ and $\boldsymbol{H}_2$ have been found. In a similar way, Eq. 1.24*b* becomes

$$\boldsymbol{n} \cdot \boldsymbol{D}_2 = \rho_s \qquad \text{on } S \text{ if } \sigma_1 = \infty, \quad \sigma_2 \neq \infty \tag{1.27}$$

which determines ρ_s. There will come later examples to clarify the limiting and boundary conditions discussed above. Further consideration here is unnecessary.

We come next to the question of energy, and ask whether its conservation is in any sense guaranteed by Eqs. 1.15. From static field theory we know that the stored energy w_e per unit volume in an *electrostatic* field changes by

$$dw_e = \boldsymbol{E} \cdot d\boldsymbol{D} \tag{1.28}$$

as a result of a change in $\boldsymbol{D}$ of $d\boldsymbol{D}$. If we *assume* that the same expression remains valid for time-varying fields, then

$$\begin{aligned} \frac{\partial w_e}{\partial t} &= \boldsymbol{E} \cdot \frac{\partial \boldsymbol{D}}{\partial t} \\ &= \boldsymbol{E} \cdot \frac{\partial \boldsymbol{D}}{\partial t} \end{aligned} \tag{1.29}$$

Similarly for the magnetic energy w_m stored per unit volume, we would *assume* that the static relation $dw_m = \boldsymbol{H} \cdot d\boldsymbol{B}$ becomes in the time-varying case

$$\frac{\partial w_m}{\partial t} = \boldsymbol{H} \cdot \frac{\partial \boldsymbol{B}}{\partial t} \tag{1.30}$$

Thus we can calculate the rate of change of total stored energy per unit volume by using Eqs. 1.30, 1.29, 1.15*b*, and 1.15*c* as follows:

$$\begin{aligned} \frac{\partial}{\partial t}(w_e + w_m) &= \boldsymbol{E} \cdot \frac{\partial \boldsymbol{D}}{\partial t} + \boldsymbol{H} \cdot \frac{\partial \boldsymbol{B}}{\partial t} \\ &= \boldsymbol{E} \cdot (\nabla \times \boldsymbol{H}) - \boldsymbol{H} \cdot (\nabla \times \boldsymbol{E}) - \boldsymbol{J} \cdot \boldsymbol{E} \end{aligned} \tag{1.31}$$

But

$$\boldsymbol{E}\cdot(\nabla \times \boldsymbol{H}) - \boldsymbol{H}\cdot(\nabla \times \boldsymbol{E}) = -\nabla\cdot(\boldsymbol{E} \times \boldsymbol{H}) \tag{1.32}$$

So Eq. 1.31 becomes

$$-\nabla\cdot(\boldsymbol{E} \times \boldsymbol{H}) = \boldsymbol{J}\cdot\boldsymbol{E} + \frac{\partial}{\partial t}(w_e + w_m) \tag{1.33}$$

Using Eqs. 1.21 and 1.20*c*, however, Eq. 1.33 becomes

$$-\boldsymbol{J}_0\cdot\boldsymbol{E} = \nabla\cdot(\boldsymbol{E} \times \boldsymbol{H}) + \sigma|\boldsymbol{E}|^2 + \frac{\partial}{\partial t}(w_e + w_m) \tag{1.34}$$

where $|\boldsymbol{E}|^2 = \boldsymbol{E}\cdot\boldsymbol{E}$. We would interpret Eq. 1.34 on a per unit volume basis as follows:

$\sigma|\boldsymbol{E}|^2$ is power loss to the medium in joule heating;
$-\boldsymbol{J}_0\cdot\boldsymbol{E}$ is power fed *from* the source *into* the field;
$\frac{\partial}{\partial t}(w_e + w_m)$ is the rate of increase of total stored energy in the field.

Then, *if conservation of energy is to hold*, we must assign $\nabla\cdot(\boldsymbol{E} \times \boldsymbol{H})$ as the rate at which power leaves each unit volume by electromagnetic means.

Matters are clarified somewhat if we integrate Eq. 1.34 over some finite volume V bounded by a closed surface S (Fig. 1.3). We find

$$-\int_V \boldsymbol{J}_0\cdot\boldsymbol{E}\,dv = \int_V \nabla\cdot(\boldsymbol{E} \times \boldsymbol{H})\,dv + \int_V \sigma|\boldsymbol{E}|^2\,dv + \frac{\partial}{\partial t}\int_V (w_e + w_m)\,dv \tag{1.35}$$

But, by Gauss' theorem,

$$\int_V \nabla\cdot(\boldsymbol{E} \times \boldsymbol{H})\,dv = \int_S \boldsymbol{n}\cdot(\boldsymbol{E} \times \boldsymbol{H})\,ds$$

and we can rewrite Eq. 1.35 in the form

$$\underbrace{-\int_V \boldsymbol{J}_0\cdot\boldsymbol{E}\,dv}_{P} = \underbrace{\int_V \sigma|\boldsymbol{E}|^2\,dv}_{P_d} + \frac{\partial}{\partial t}\underbrace{\int_V (w_e + w_m)\,dv}_{W_e+W_m} + \int_S \boldsymbol{n}\cdot(\boldsymbol{E} \times \boldsymbol{H})\,ds \tag{1.36}$$

Equation 1.36 says that the power P delivered *to* the volume V by the current source $\boldsymbol{J}_0$ is accounted for by that dissipated in resistance, P_d, plus the rate at which energy is stored in the field, $(\partial/\partial t)(W_e + W_m)$, plus another term which we must interpret as the electromagnetic power *leaving* V via its surface S, if we are to have a conservation of energy at all.

It is convenient to give the vector $\boldsymbol{E} \times \boldsymbol{H}$ a special symbol $\boldsymbol{S}$,

$$\boldsymbol{S} \equiv \boldsymbol{E} \times \boldsymbol{H} \tag{1.37}$$

called the Poynting vector. The direction of $\boldsymbol{S}$ at any point in space is interpreted as the direction of power flow at that point. The magnitude of $\boldsymbol{S}$ is interpreted as the value of the power flowing per unit area across a surface normal to the direction of $\boldsymbol{S}$. Thus the dimensions of $\boldsymbol{S}$ are w/m^2. This interpretation is consistent with Eq. 1.36, although the latter concerns only the *integrated* effect over a closed surface. One

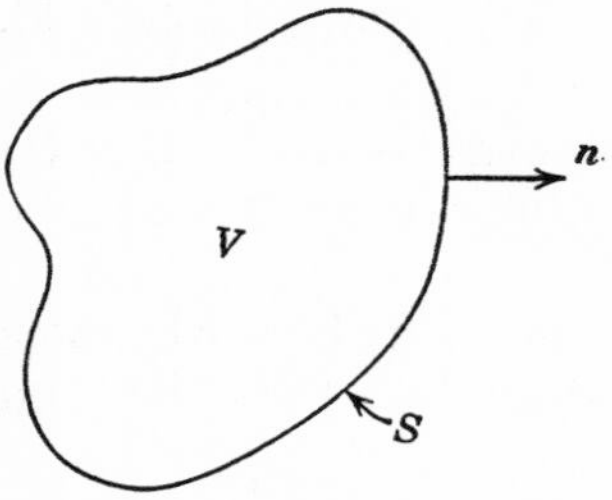

Fig. 1.3. A closed region of space.

therefore cannot prove from Eq. 1.36 that the foregoing physical interpretation of $\boldsymbol{S}$ at a point in space is correct. Any divergenceless function of the fields could be added to $\boldsymbol{S}$ without changing Eqs. 1.34 or 1.36. Usually, however, we are interested in computing the total power leaving a closed surface; so we eventually do perform an integral like that in Eq. 1.36. Thus the previous physical interpretation of $\boldsymbol{S}$ at any point in space is in most cases both convenient and safe.

In simplified notation, what we have learned from Eq. 1.36 may be written as

$$P = P_d + \frac{\partial}{\partial t}(W_e + W_m) + \int_S \boldsymbol{n} \cdot \boldsymbol{S}\, ds \tag{1.38}$$

which is often called Poynting's theorem. Equation 1.38 should be compared with Eq. 1.5. It is, of course, on account of the last term in Eq. 1.38 that electromagnetic energy transmission through space is possible.

1.2.2 Frequency Domain

1.2.2.1 Complex Notation. To discuss fields in the sinusoidal steady state, it is convenient to employ complex notation. Most of our fields are space vectors, however, so we do not know at once how to write them as sinusoidal functions of time, much less how to do so in complex form. Therefore we start with what we do know, namely, how to write a scalar sinusoidal function of time in complex form.

Charge density $\rho(x, y, z, t)$ is a scalar function of position and time. What we mean when we say that ρ undergoes sinusoidal time variation is simply that, at some arbitrary point (x_0, y_0, z_0) of space, ρ varies sinusoidally with time:

$$\rho(x_0, y_0, z_0, t) = \rho_0(x_0, y_0, z_0) \cos [\omega t + \psi(x_0, y_0, z_0)] \tag{1.39}$$

The only novel thing about Eq. 1.39 is that both the *amplitude* ρ_0 and *phase* ψ of the sinusoid may not have the same values at point (x_0, y_0, z_0) as they have at some other space point (x, y, z). The frequency ω, however, is the same everywhere. Thus, in complex notation, we would write for Eq. 1.39

$$\rho(x, y, z, t) = \text{Re}\,[\rho_0(x, y, z)e^{j\psi(x,y,z)}e^{j\omega t}] \tag{1.40}$$

or, more simply,

$$\rho = \text{Re}\,(\rho_0 e^{j\psi}e^{j\omega t}) = \text{Re}\,(\bar{\rho}e^{j\omega t}) \tag{1.41}$$

where we agree to remember that the amplitude ρ_0 and the phase ψ of the complex number[1] $\bar{\rho}$ may vary with position but not with time.

Now, to handle a space vector $\boldsymbol{E}(x, y, z, t)$ in the sinusoidal steady state, we first imagine it taken apart into its *space* components, which, for the present discussion, we may choose as rectangular components without loss of generality:

$$\boldsymbol{E} = \boldsymbol{a}_x E_x + \boldsymbol{a}_y E_y + \boldsymbol{a}_z E_z \tag{1.42}$$

Then any component—E_x, for example—is a scalar function of position and time, which we shall now agree is to be a sinusoidal function of time. By a space vector with sinusoidal time dependence we, therefore, actually mean *a space vector whose three space components are (scalar) sinusoidal functions of time.* The magnitudes and phases of these three components need not, of course, be the same at any space position (x, y, z). Also they will, in general, vary differently with space position.

[1] Since neither the distinction between italic and roman type, nor underscoring, is available in the Greek letters, complex character will be indicated for these symbols by a bar over the top. Thus $\bar{\gamma}$ is a complex scalar, $\bar{\boldsymbol{\gamma}}$ a complex vector.

Writing the sinusoidal space components in the form of Eq. 1.41, we find, for example,

$$E_x = \text{Re}\,(E_{x0}e^{j\psi_x}e^{j\omega t})$$

or

$$E_x = \text{Re}\,(\mathrm{E}_x e^{j\omega t}) \tag{1.43}$$

Again, in Eq. 1.43, we must remember that the amplitude $E_{x0}(x, y, z)$ and the phase $\psi_x(x, y, z)$ of the complex number E_x may vary with position. Similarly, we put

$$\begin{aligned} E_y &= \text{Re}\,(\mathrm{E}_y e^{j\omega t}) \\ E_z &= \text{Re}\,(\mathrm{E}_z e^{j\omega t}) \end{aligned} \tag{1.44}$$

where now the complex numbers E_y and E_z need not have the same amplitude or phase as E_x at any point, and need not vary with position (x, y, z) in the same way; but the frequency ω is the same for all space components at all positions. Collecting Eqs. 1.43 and 1.44 into 1.42, we have

$$\begin{aligned} \text{(a)} \qquad \boldsymbol{E} &= \boldsymbol{a}_x E_{x0} \cos(\omega t + \psi_x) + \boldsymbol{a}_y E_{y0} \cos(\omega t + \psi_y) \\ &\quad + \boldsymbol{a}_z E_{z0} \cos(\omega t + \psi_z) \\ &= \text{Re}\,[(\boldsymbol{a}_x \mathrm{E}_x + \boldsymbol{a}_y \mathrm{E}_y + \boldsymbol{a}_z \mathrm{E}_z)e^{j\omega t}] = \text{Re}\,(\mathbf{E}e^{j\omega t}) \\ \text{(b)} \qquad \mathbf{E} &\equiv \boldsymbol{a}_x \mathrm{E}_x + \boldsymbol{a}_y \mathrm{E}_y + \boldsymbol{a}_z \mathrm{E}_z \end{aligned} \tag{1.45}$$

in which $\mathbf{E}$ is called a "complex vector." It is a rather curious *space vector whose components are complex numbers.*

We have seen that what we mean by a space vector which has sinusoidal time variation is one whose *components* are sinusoidal in time, all with the same frequency ω. The amplitudes and phases of the components need not be the same at any point in space, and they may vary quite differently with position. The complex vector $\mathbf{E}$, in Eq. 1.45, is simply a shorthand (nonphysical) symbol which contains all this information about the amplitudes and phases of the three space components of $\boldsymbol{E}$. Since these amplitudes and phases depend only on position, not time, the complex vector $\mathbf{E}$ is a function of position only.

There is a way of writing a complex vector $\mathbf{E}$ which is alternative to, and often more useful than, the form given in Eq. 1.45*b*. We can arrive at the new expression by noting that the space components E_x, E_y, and E_z of $\mathbf{E}$ are simply complex numbers, which can therefore be written in rectangular form instead of the polar form we used in Eqs. 1.43 and 1.41. Thus, instead of putting $\mathrm{E}_x = E_{x0}e^{j\psi_x}$, we can as well put $\mathrm{E}_x = E_{xr} + jE_{xi}$, where now E_{xr} and E_{xi} are the ordinary real numbers giving, respectively, the real and imaginary parts of the complex number E_x. In the time domain, this means that, instead of writing the sinusoid $E_x = E_{x0} \cos (\omega t + \psi_x)$, we write $E_x = E_{xr} \cos \omega t - E_{xi} \sin \omega t$, which is a completely equivalent form. If we do this for all three components of **E**, Eq. 1.45*b* becomes

$$\mathbf{E} = (\boldsymbol{a}_x E_{xr} + \boldsymbol{a}_y E_{yr} + \boldsymbol{a}_z E_{zr}) + j(\boldsymbol{a}_x E_{xi} + \boldsymbol{a}_y E_{yi} + \boldsymbol{a}_z E_{zi})$$
$$= \boldsymbol{E}_r + j\boldsymbol{E}_i \qquad (1.46a)$$

in which now $\boldsymbol{E}_r$ and $\boldsymbol{E}_i$ are real space vectors. Note that $\boldsymbol{E}_r$ and $\boldsymbol{E}_i$ are functions of space coordinates (x, y, z) only; they are *not* functions of time. The result of Eq. 1.46*a* tells us that a complex vector **E** may now be thought of as a curious kind of *complex number whose real and imaginary parts are space vectors.*

This is an interesting alternative viewpoint to that originally given after Eq. 1.45, and is neither more, nor less, fundamental. It is somewhat more convenient, however, because, in terms of the representation in Eq. 1.46*a* for **E**, the actual sinusoidal electric field $\boldsymbol{E}$ can be expressed in the simple form

$$\boldsymbol{E}(x, y, z, t) = \mathrm{Re}\,[\mathbf{E}(x, y, z)e^{j\omega t}]$$
$$= \mathrm{Re}\,\{[\boldsymbol{E}_r(x, y, z) + j\boldsymbol{E}_i(x, y, z)]e^{j\omega t}\}$$
$$= \boldsymbol{E}_r(x, y, z) \cos \omega t - \boldsymbol{E}_i(x, y, z) \sin \omega t \qquad (1.46b)$$

Since $\boldsymbol{E}_r$ and $\boldsymbol{E}_i$ are any two ordinary (real) vectors in space, they must determine a plane. According to Eq. 1.46*b*, therefore, the actual electric field $\boldsymbol{E}(x, y, z, t)$ at the point (x, y, z) always lies in the plane determined by $\boldsymbol{E}_r$ and $\boldsymbol{E}_i$ at that point. Such a conclusion is not so easy to draw from Eq. 1.45.

It is now possible to visualize the most general behavior with time of a sinusoidal vector at one point in space. In terms of Eq. 1.46*b*, we may construct the two time-independent vectors $\boldsymbol{E}_r$ and $\boldsymbol{E}_i$, including an angle ζ between their positive directions. An example appears in Fig. 1.4. The vector $\boldsymbol{E}(t)$ is then the sum of two sinusoidal vectors, $\boldsymbol{E}_r \cos \omega t$ and $-\boldsymbol{E}_i \sin \omega t$, which always lie along $\pm\boldsymbol{E}_r$ and $\pm\boldsymbol{E}_i$ respectively, but oscillate back and forth 90° out of time phase with each other.

We know that, if $\boldsymbol{E}_r$ and $\boldsymbol{E}_i$ were mutually perpendicular ($\zeta = \pi/2$), the resultant $\boldsymbol{E}(t)$ would trace an ellipse whose major and minor axes were $\boldsymbol{E}_r$ and $\boldsymbol{E}_i$. This is, after all, what happens to an oscilloscope spot when different sinusoidal signals of the same frequency and 90° out of time phase are applied to the plates. If we were to look at the screen

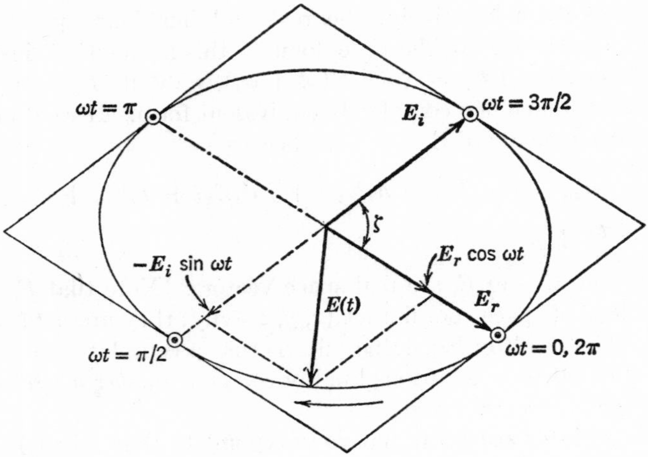

Fig. 1.4. Polarization ellipse for a vector with sinusoidal time dependence.

obliquely, the figure would still be an ellipse, and this oblique view is equivalent to looking directly at the screen when the plates (axes) are skewed. In other words, $\boldsymbol{E}(t)$ traces an ellipse even when $\zeta \neq \pi/2$.

Viewed in another way, we could resolve vectors $\boldsymbol{E}_i$ and $\boldsymbol{E}_r$ in their own plane along a pair of mutually perpendicular axes (ξ, η). If $\boldsymbol{a}_\xi$ and $\boldsymbol{a}_\eta$ are unit vectors along these new axes, the resolution reads

$$\boldsymbol{E}_r = \boldsymbol{a}_\xi E_{r\xi} + \boldsymbol{a}_\eta E_{r\eta}$$

$$\boldsymbol{E}_i = \boldsymbol{a}_\xi E_{i\xi} + \boldsymbol{a}_\eta E_{i\eta}$$

and, therefore,

$$\mathbf{E} = \boldsymbol{E}_r + j\boldsymbol{E}_i = \boldsymbol{a}_\xi(E_{r\xi} + jE_{i\xi}) + \boldsymbol{a}_\eta(E_{r\eta} + jE_{i\eta})$$

But

$$E_{r\xi} + jE_{i\xi} = E_{\xi 0}e^{j\psi_\xi}$$

$$E_{r\eta} + jE_{i\eta} = E_{\eta 0}e^{j\psi_\eta}$$

so

$$\boldsymbol{E} = \boldsymbol{a}_\xi E_{\xi 0} \cos(\omega t + \psi_\xi) + \boldsymbol{a}_\eta E_{\eta 0} \cos(\omega t + \psi_\eta)$$

which represents the sum of two mutually perpendicular, sinusoidally oscillating vectors with arbitrary relative time phase. Again, we know from the oscilloscope example that the resultant is an ellipse, but the ξ- and η-component vectors are not in general the principal axes.

Although it is pleasant to describe the behavior of $\boldsymbol{E}(t)$ in terms of the principal axes of the ellipse which its tip traces with time, and it is important to know how to determine this description, it is not always necessary to go to the trouble involved in doing so. The vectors $\boldsymbol{E}_r$ and $\boldsymbol{E}_i$ themselves, obtained by simply finding the real and imaginary parts of $\mathbf{E}$, tell us almost all we need to know.

Thus the rhombus drawn in Fig. 1.4, composed of lines parallel to $\boldsymbol{E}_i$ drawn through the tips of $\boldsymbol{E}_r$ and $-\boldsymbol{E}_r$, and lines parallel to $\boldsymbol{E}_r$ drawn through the tips of $\boldsymbol{E}_i$ and $-\boldsymbol{E}_i$, must be tangent to the ellipse at the tips in question. This fact follows from Eq. 1.46*b* by differentiation partially with respect to time,

$$\frac{\partial \boldsymbol{E}}{\partial t} = -\omega(\boldsymbol{E}_r \sin \omega t + \boldsymbol{E}_i \cos \omega t) \tag{1.47}$$

making clear that, when $\boldsymbol{E} = \boldsymbol{E}_r$, for example, so that $\sin \omega t = 0$, then $\partial \boldsymbol{E}/\partial t = -\omega \boldsymbol{E}_i$. In other words, $\partial \boldsymbol{E}/\partial t$ (which is by definition tangent to the ellipse) is parallel to $\boldsymbol{E}_i$ at $\boldsymbol{E} = \boldsymbol{E}_r$. This feature, applied similarly at the four points mentioned, makes it possible to sketch the form of the ellipse quite accurately.

Moreover, $\boldsymbol{E} = \boldsymbol{E}_r$ at $t = 0$, and moves toward $-\boldsymbol{E}_i$ as time advances. Hence, we know the "phasing" and the direction of rotation of the tip of $\boldsymbol{E}$ in its motion on the orbit.

Form of elliptic orbit traced by the tip, phasing in the orbit, and rotation direction, together, convey to us the most important information about the general behavior of a space vector with sinusoidal time variation. It is obvious why this behavior is referred to as *elliptical polarization* and why Fig. 1.4 is called the *polarization ellipse*. It is also obvious that two special cases may be of particular interest:

1. *Linear polarization*, in which $\boldsymbol{E}(t)$ does not rotate direction with time, but oscillates back and forth with a sinusoidally varying magnitude. The vector $\boldsymbol{E}_r$ is parallel or antiparallel to $\boldsymbol{E}_i$ ($\zeta = 0$ or π), or one of them is zero. This case is also quite easy to detect in the original form Eq. 1.45, because it requires that all the components oscillate in phase ($\psi_x = \psi_y = \psi_z$).

2. *Circular polarization*, in which the magnitude of $\boldsymbol{E}(t)$ does not vary with time, but the vector rotates at constant angular speed ω. This polarization occurs when $\boldsymbol{E}_r$ is perpendicular to $\boldsymbol{E}_i$ and is of equal magnitude ($|\boldsymbol{E}_r| = |\boldsymbol{E}_i|$ and $\zeta = \pi/2$). It is not so easy to detect circular polarization from Eq. 1.45.

It is worth while to ask whether the "rectangular" form Eq. 1.46*a* for a complex vector $\mathbf{E}$, which expresses it as a sum of its imaginary and real parts, implies the existence of a corresponding "polar" form, as would be true for complex numbers. Interestingly enough, the answer is that *there is, in general, no polar form for a complex vector*. The reason is that the real and imaginary parts $\boldsymbol{E}_r$ and $\boldsymbol{E}_i$ are space vectors, and their ratio (which would define essentially the "polar-form angle") is meaningless except in one case: Suppose $\boldsymbol{E}_r$ and $\boldsymbol{E}_i$ differ only by a real constant multiplier

$$\boldsymbol{E}_i = a\boldsymbol{E}_r \qquad \text{or} \qquad \boldsymbol{E}_r = b\boldsymbol{E}_i$$

Then $\boldsymbol{E}_r$ and $\boldsymbol{E}_i$ are either parallel or antiparallel, or one is zero. The vector $\boldsymbol{E}(t)$ is linearly polarized, and

$$\mathbf{E} = \boldsymbol{E}_r(1 + ja) \qquad \text{or} \qquad \mathbf{E} = \boldsymbol{E}_i(b + j) \tag{1.48a}$$

for linear polarization.

This means that $\mathbf{E}$ is of the form

$$\mathbf{E} = \boldsymbol{E}(x, y, z)e^{j\psi(x,y,z)} \tag{1.48b}$$

for linear polarization, where $\boldsymbol{E}$ is a *real* space vector. One could reasonably call the last result a "polar" form for the complex vector, but clearly it applies only to cases in which the complex vector represents a linearly polarized real vector. Indeed, it is probably wiser to summarize Eqs. 1.48 by saying that, in the special case of linear polarization, the complex vector degenerates to a real vector multiplied by a complex number, rather than making any reference to "polar" forms.

When we must deal with the general case of elliptical polarization, it is sometimes necessary to determine the directions and magnitudes of the principal axes of the polarization ellipse at the space point under consideration. There are many ways to do this, only one of which we shall carry out here. We use the fact that, when $\boldsymbol{E}(t)$ in Fig. 1.4 occupies a position along one of the principal axes, it is perpendicular to the tangent to the ellipse drawn at its tip, that is

$$\boldsymbol{E} \cdot \frac{\partial \boldsymbol{E}}{\partial t} = 0$$

along a principal axis.

From Eqs. 1.46*b* and 1.47 this condition becomes

$$\tfrac{1}{2}(E_{r0}{}^2 - E_{i0}{}^2) \sin 2\omega t_p + E_{r0}E_{i0} \cos \zeta \cos 2\omega t_p = 0$$

or

$$\tan 2\omega t_p = \frac{2 \cos \zeta}{[(E_{i0}/E_{r0}) - (E_{r0}/E_{i0})]} \tag{1.49}$$

where $E_{r0} = |\boldsymbol{E}_r|$, $E_{i0} = |\boldsymbol{E}_i|$, and t_p is a time when $\boldsymbol{E}(t)$ lies along any principal axis. Evidently in the range $0 \leq \omega t_p \leq \pi$, there are two solutions to Eq. 1.49, giving two *perpendicular* principal directions.

Substituting these values of ωt_p back into Eq. 1.46*b* locates the axes, and $|\boldsymbol{E}(t_p)|$ gives their lengths. Needless to say, these lengths are the largest and smallest values of $|\boldsymbol{E}(t)|$ that occur in the orbit.

One last feature of the elliptical polarization may well be of interest for our general understanding, although it is hardly an important consideration in practice. This feature concerns the angular speed of rotation of $\boldsymbol{E}(t)$ when its tip is at various points on the orbit. It certainly completes a revolution every $2\pi/\omega$ seconds. What this fact implies is that the *average* angular speed is ω, but *not* that its angular speed is ω at all times. To understand what does happen, refer to $\partial\boldsymbol{E}/\partial t$ in Eq. 1.47 and compare it with $\boldsymbol{E}$ in Eq. 1.46*b*. Clearly $\partial\boldsymbol{E}/\partial t$ is a vector whose tip traces an ellipse exactly similar to the polarization ellipse of $\boldsymbol{E}$ (Fig. 1.4). There are only two differences: the "polarization ellipse" of $\partial\boldsymbol{E}/\partial t$ is larger in all dimensions by the factor ω; $\partial\boldsymbol{E}/\partial t$ traces its polarization ellipse in a different "phase" than $\boldsymbol{E}(t)$ traces its own. Specifically $\partial\boldsymbol{E}/\partial t$ starts from essentially $-\boldsymbol{E}_i$ at $\omega t = 0$ and proceeds to $-\boldsymbol{E}_r$ at $\omega t = \pi/2$, etc. Therefore the principal axes of the $\partial\boldsymbol{E}/\partial t$ ellipse lie along the same directions as those of the ellipse of $\boldsymbol{E}(t)$ and, of course, are in the same ratio of lengths. Moreover

$$\frac{\partial^2\boldsymbol{E}}{\partial t^2} = -\omega^2\boldsymbol{E}(t)$$

so the condition

$$\frac{\partial\boldsymbol{E}}{\partial t}\cdot\frac{\partial^2\boldsymbol{E}}{\partial t^2} = -\omega^2\frac{\partial\boldsymbol{E}}{\partial t}\cdot\boldsymbol{E} = 0$$

defining the times at which $|\partial\boldsymbol{E}/\partial t|$ is maximum and minimum, yields the result of Eq. 1.49 for t_p again. Accordingly, it is clear from Fig. 1.4 and the phasing discussion given above that $|\partial\boldsymbol{E}/\partial t|$ is *largest* when $|\boldsymbol{E}(t)|$ is *smallest* and vice-versa. Since the two vectors are perpendicular at these times t_p, it follows that the angular speed is given by $|\partial\boldsymbol{E}/\partial t|/|\boldsymbol{E}|$ at these times. Therefore $\boldsymbol{E}(t)$ certainly spins fastest when its magnitude is smallest, and slowest when its magnitude is largest. Indeed, the fastest angular speed is ωg, and the slowest ω/g, with $g > 1$ being the ratio of major to minor axis of the polarization ellipse for $\boldsymbol{E}(t)$.

1.2.2.2 Complex Form of Maxwell's Equations. We are now ready to write the sinusoidal steady-state form of Maxwell's Eqs. 1.15 in complex notation. Since all the operations involved in these equations are linear ones (i.e., addition and differentiation), we can use complex numbers directly in them *with the understanding that we will always eventually take the real part of our complex answers to get the instantaneous physical ones.* Thus, with $\mathbf{E}e^{j\omega t}$ for $\boldsymbol{E}$, $\mathbf{B}e^{j\omega t}$ for $\boldsymbol{B}$, etc., we find for Eqs. 1.15 (Eqs. 1.15*d* and 1.19 are now redundant)

$$\text{(a)} \qquad \nabla \cdot \mathbf{D} = \bar{\rho}$$

$$\text{(b)} \qquad \nabla \times \mathbf{E} = -j\omega \mathbf{B} \qquad (1.50)$$

$$\text{(c)} \qquad \nabla \times \mathbf{H} = \mathbf{J} + j\omega \mathbf{D}$$

Note that $e^{j\omega t}$ cancels out; *so, if we find a complex solution* **E** *to some problem, we must remember that the physical field is* $\boldsymbol{E} = \text{Re}\,(\mathbf{E}e^{j\omega t})$, *etc.*

Similarly, Eqs. 1.20 and 1.21 become

$$\text{(a)} \qquad \mathbf{D} = \epsilon \mathbf{E}$$

$$\text{(b)} \qquad \mathbf{B} = \mu \mathbf{H} \qquad (1.51)$$

$$\text{(c)} \qquad \mathbf{J} = \mathbf{J}_0 + \sigma \mathbf{E}$$

Here however, it is well to observe that Eqs. 1.51 are somewhat more general than Eqs. 1.20 and 1.21. In particular, in Eqs. 1.20 and 1.21 the parameters ϵ, μ, σ must be *strictly constant* at any point in space, whereas in Eqs. 1.51 we may permit them to be functions of frequency $\epsilon(\omega)$, $\mu(\omega)$, and $\sigma(\omega)$. In other words, Eqs. 1.51 apply in some circumstances where Eqs. 1.20 and 1.21 would be invalid.

From the preceding results we can discuss "complex power" in the field. It is not difficult to show directly, by steps similar to those employed in circuit theory, that the time-averaged value of the Poynting vector becomes

$$\langle \boldsymbol{S} \rangle \equiv \langle \boldsymbol{E} \times \boldsymbol{H} \rangle = \tfrac{1}{2}\,\text{Re}\,(\mathbf{E} \times \mathbf{H}^*) \qquad (1.52)$$

Therefore we define a complex Poynting vector **S** (analogous to complex power in circuits) by the relation

$$\mathbf{S} \equiv \tfrac{1}{2}\mathbf{E} \times \mathbf{H}^* \qquad (1.53)$$

It is important that **S** is a space vector with complex components, i.e., a complex vector. We can learn something more about it by treating Eqs. 1.50*b* and 1.50*c* very much as we did in the circuits: namely, dot-multiply Eq. 1.50*b* by $\mathbf{H}^*$; then take the conjugate of Eq. 1.50*c* and dot-multiply it by **E**; finally subtract the two results. We find

$$\mathbf{H}^* \cdot \nabla \times \mathbf{E} - \mathbf{E} \cdot \nabla \times \mathbf{H}^* = -\mathbf{E} \cdot \mathbf{J}^* + j\omega(\mathbf{E} \cdot \mathbf{D}^* - \mathbf{B} \cdot \mathbf{H}^*) \qquad (1.54)$$

Using the same vector identity on the left as we did in Eq. 1.32, we find that Eq. 1.54 becomes

$$\nabla \cdot (\mathbf{E} \times \mathbf{H}^*) = -\mathbf{E} \cdot \mathbf{J}^* + j\omega(\mathbf{E} \cdot \mathbf{D}^* - \mathbf{B} \cdot \mathbf{H}^*) \qquad (1.55)$$

Then Eqs. 1.51 can be used to eliminate **D** and **B** from Eq. 1.55:

$$-\nabla\cdot(\mathbf{E}\times\mathbf{H}^*) = \sigma\mathbf{E}\cdot\mathbf{E}^* + j\omega(\mu\mathbf{H}\cdot\mathbf{H}^* - \epsilon\mathbf{E}\cdot\mathbf{E}^*) + \mathbf{E}\cdot\mathbf{J}_0^* \quad (1.56)$$

In terms of **S** (Eq. 1.53), we have

$$-\frac{1}{2}\mathbf{E}\cdot\mathbf{J}_0^* = \frac{1}{2}\sigma\mathbf{E}\cdot\mathbf{E}^* + j2\omega\left(\frac{\mu\mathbf{H}\cdot\mathbf{H}^*}{4} - \frac{\epsilon\mathbf{E}\cdot\mathbf{E}^*}{4}\right) + \nabla\cdot\mathbf{S} \quad (1.57)$$

But on a per-unit-volume basis:

$$\begin{aligned} \tfrac{1}{2}\sigma\mathbf{E}\cdot\mathbf{E}^* &= \sigma\langle|E|^2\rangle = \text{time-average power dissipated} \\ \tfrac{1}{4}\mu\mathbf{H}\cdot\mathbf{H}^* &= \tfrac{1}{2}\mu\langle|H|^2\rangle = \text{time-average magnetic stored energy}^1 \\ \tfrac{1}{4}\epsilon\mathbf{E}\cdot\mathbf{E}^* &= \tfrac{1}{2}\epsilon\langle|E|^2\rangle = \text{time-average electric stored energy}^1 \\ -\tfrac{1}{2}\mathbf{E}\cdot\mathbf{J}_0^* &= \text{complex power delivered } by \text{ source } \mathbf{J}_0 \text{ } to \text{ the field} \end{aligned} \quad (1.58)$$

Hence, if we integrate Eq. 1.57 over the volume of Fig. 1.3, and use Gauss' theorem on the last term on the right-hand side, we find:

$$\underbrace{\int_V \left(-\tfrac{1}{2}\mathbf{E}\cdot\mathbf{J}_0^*\right) dv}_{\langle P_{\text{in}}\rangle + jQ_{\text{in}}} = \underbrace{\int_V \tfrac{1}{2}(\sigma\mathbf{E}\cdot\mathbf{E}^*)\, dv}_{\langle P_d\rangle} + 2j\omega\underbrace{\int_V \tfrac{1}{4}(\mu\mathbf{H}\cdot\mathbf{H}^* - \epsilon\mathbf{E}\cdot\mathbf{E}^*)\, dv}_{\langle W_m - W_e\rangle} + \underbrace{\int_S \mathbf{S}\cdot\boldsymbol{n}\, ds}_{\langle P_{\text{out}}\rangle + jQ_{\text{out}}} \quad (1.59)$$

$\langle P_{\text{in}}\rangle + jQ_{\text{in}}$ represents the complex power provided to the field by all sources $\mathbf{J}_0$ located *inside* V. $\langle P_d\rangle$ is the time-average power dissipated as heat inside V, and $\langle W_m - W_e\rangle$ is the difference between time-average magnetic and electric energies stored within V. The term $\langle P_{\text{out}}\rangle + jQ_{\text{out}}$ represents complex power which leaves V through its bounding surface S. If we transpose this term to the left side, Eq. 1.59 says that the *net* complex power supplied to the volume V comprises that delivered by sources inside ($\langle P_{\text{in}}\rangle + jQ_{\text{in}}$) *plus* that delivered *to* V by sources outside ($-\langle P_{\text{out}}\rangle - jQ_{\text{out}}$); or, alternatively, that delivered by sources inside ($\langle P_{\text{in}}\rangle + jQ_{\text{in}}$) *less* that which passes *out* of V through $S(\langle P_{\text{out}}\rangle + jQ_{\text{out}})$. In any case, this *net* complex power supplied to volume V is accounted for by the time-average or "active" power

[1] The stated interpretations apply only when σ, μ, ϵ are constants. If these parameters are functions of frequency, the stored energies mentioned are more difficult to define and are of doubtful utility. See Probs. 1.6–1.9, 2.12, and 6.28.

$\langle P_d \rangle$ dissipated in it as heat, and the "reactive" power proportional to the difference between time-average magnetic and electric energies stored in V. Equation 1.59 is known as the *complex Poynting theorem.* It should be compared with Eq. 1.12 for lumped circuits.

Having reviewed how to formulate mathematically in both the time and frequency domains the lumped-circuit and field relations and all the energy balances, we are prepared to enter the new subject matter of our study.

PROBLEMS

Problem 1.1. The result Eq. 1.2 follows from applying both the Kirchhoff current law and the law of conservation of energy to the lumped network of Fig. 1.1. Keeping in mind that loop currents I_1 and I_2 are dynamically independent variables for the network, determine a reasonable and simple additional assumption which will lead back from Eq. 1.2 to the Kirchhoff voltage law statements in Eq. 1.1. (*Hint:* Consider the Lagrange or Hamilton formulations of network equilibrium. See, for example, Ernst A. Guillemin, *Introductory Circuit Theory,* John Wiley and Sons, New York, 1953, Chapter 10; or David C. White and Herbert H. Woodson, *Electromechanical Energy Conversion,* John Wiley and Sons, New York, 1959, Chapter 1.)

Problem 1.2. Lumped electric networks and mechanical systems with only one dimension of motion are analogous. What electrical principles correspond to: (a) Newton's first law? (b) second law? (c) third law? (d) the law that, in a mechanical system, the sum of all *relative* velocities of points on a closed path is zero?

Problem 1.3. Assuming we understand current, the problem of defining voltage in a truly *lumped-circuit* environment must, strictly speaking, be solved without recourse to "field" ideas like line integrals, etc. This problem is similar to the one of defining force and mass independently in classical mechanics, assuming we understand velocity. Develop the corresponding theory for the electrical case, stating all the electrical laws you use and their mechanical counterparts.

Problem 1.4. A linear time-varying perfect capacitor $C(t)$ is defined by the volt-coulomb relation $q(t) = C(t)V(t)$: (a) Is the electric stored energy still given by $W_e = CV^2/2 = q^2/2C$? Give an explanation based specifically upon the relations between lumped circuits and fields. (b) An open-circuited perfect linear capacitor C_0 carries charge q_0. The plates are pulled apart, reducing the capacitance to αC_0. How much work is done?

Problem 1.5. A short-circuited perfect linear inductor L_0 carries a circulating direct current I_0. Without breaking the circuit, several turns are unwound by straightening them out. The reduced inductance is αL_0. How much mechanical work is done?

Problem 1.6. If at a point in space the parameters $\sigma(\omega) + j\omega\epsilon(\omega)$ and $j\omega\mu(\omega)$ which appear in Eq. 1.56 are rational functions of complex frequency $s = j\omega$, but the medium is still linear and isotropic, write the most general form of the time-domain relations that characterize the medium and must replace Eq. 1.20.

Problem 1.7. (a) By Fourier integral or infinite superposition methods, we may generally write

$$\boldsymbol{E}(\boldsymbol{r}, t) = \int_{-\infty}^{\infty} \mathbf{E}(\boldsymbol{r}, \omega)e^{j\omega t}d\omega$$

where $\mathbf{E}(\boldsymbol{r}, \omega)$ is the complex transform of $\boldsymbol{E}(\boldsymbol{r}, t)$ with respect to time. Show that $\mathbf{E}(\boldsymbol{r}, -\omega) = \mathbf{E}^*(\boldsymbol{r}, \omega)$. (b) In a linear, isotropic medium we may have $\sigma(\boldsymbol{r}, \omega)$ and $\epsilon(\boldsymbol{r}, \omega)$, such that

$$\boldsymbol{J}(\boldsymbol{r}, t) + \partial \boldsymbol{D}(\boldsymbol{r}, t)/\partial t = \int_{-\infty}^{\infty} [\sigma(\boldsymbol{r}, \omega) + j\omega\epsilon(\boldsymbol{r}, \omega)]\mathbf{E}(\boldsymbol{r}, \omega)e^{j\omega t}d\omega$$

Show that $\sigma(\boldsymbol{r}, -\omega) = \sigma(\boldsymbol{r}, \omega)$; $\epsilon(\boldsymbol{r}, -\omega) = \epsilon(\boldsymbol{r}, \omega)$. In other words, the real part of the "admittance per meter" $(\sigma + j\omega\epsilon)$ is an *even* function of frequency while the imaginary part is an *odd* function of frequency.

Problem 1.8. From the time-domain form of Poynting's theorem, and from quasi-static energy conservation considerations, we have been led to *assume* that in general $\partial w_e/\partial t = \boldsymbol{E}\cdot(\partial \boldsymbol{D}/\partial t)$ where w_e is the instantaneous stored electric energy per unit volume. Suppose that $\boldsymbol{E}$ has sinusoidal time variation, at frequency ω, and that the dielectric constant of the isotropic linear medium is a function of frequency $\epsilon(\omega)$. (a) Determine $\partial w_e/\partial t$. (b) What can you say about w_e? (c) What can you say about $\langle w_e \rangle$? (d) In the simple case where $\boldsymbol{D} = \epsilon \boldsymbol{E}$, review very carefully the steps which suggest that, for *any* time variation, $w_e = \frac{1}{2}\epsilon|\boldsymbol{E}|^2$. Discuss the validity and uniqueness of this expression for w_e in that case.

Problem 1.9. With reference to Prob. 1.8, let

$$\boldsymbol{E} = \text{Re}\,[\mathbf{E}_1 e^{j\omega_1 t} + \mathbf{E}_2 e^{j\omega_2 t}]$$

and let $\epsilon(\omega)$ be a function of frequency. (a) Find $\partial w_e/\partial t$. (b) What can you say about w_e? (c) What can you say about $\langle w_e \rangle$? (d) What can you say about $\langle w_e \rangle$ as $\omega_1 \to \omega_2$, if in the process: (i) $\mathbf{E}_1 \neq \mathbf{E}_2 \neq 0$? (ii) $\mathbf{E}_1$ or $\mathbf{E}_2 \to 0$? (iii) $\mathbf{E}_1 \to \mathbf{E}_2$? (e) Choose an expression for $\langle w_e \rangle$, in the sinusoidal steady state at one frequency, which meets the following conditions: (i) It checks with all the results of (d) above; (ii) it does not conflict with the results of Prob. 1.8; (iii) it checks with the results assumed previously when ϵ is independent of frequency. See Prob. 6.28 in this connection.

Problem 1.10. Determine the most general form of the relations that must replace Eqs. 1.20 to describe a linear *anisotropic* medium with constant parameters.

Problem 1.11. Consider an enclosed simply connected region of space V filled with a linear, isotropic, time-invariant medium, and possibly some known currents $\boldsymbol{J}_0$. Let the boundary conditions specify completely tangential component of *either* $\boldsymbol{E}$ or $\boldsymbol{H}$ at each point on the bounding surface S. Suppose there are two different fields $(\boldsymbol{E}_1, \boldsymbol{H}_1)$ and $(\boldsymbol{E}_2, \boldsymbol{H}_2)$ which are solutions to Maxwell's equations in V, and meet the given boundary conditions on S. (a) What equations and boundary conditions are satisfied at points in V and on S by the new field $\boldsymbol{E}_3 = \boldsymbol{E}_2 - \boldsymbol{E}_1$, $\boldsymbol{H}_3 = \boldsymbol{H}_2 - \boldsymbol{H}_1$? (b) Apply Poynting's theorem to $(\boldsymbol{E}_3, \boldsymbol{H}_3)$, and show that, if $\sigma \neq 0$, $\boldsymbol{E}_3 \equiv \boldsymbol{H}_3 \equiv 0$ everywhere in V. This means $(\boldsymbol{E}_1, \boldsymbol{H}_1)$ is a *unique* solution, given $\boldsymbol{J}_0$ and the boundary conditions described above. (c) If $\sigma \equiv 0$ everywhere in V, interpret the possible nonzero solutions ($\boldsymbol{E}_3$, $\boldsymbol{H}_3$). In particular, does their existence mean that lossless systems do not have a unique response to driving forces in them or on their boundaries? (d) Set up and discuss the lumped-network analog of this whole problem. It may help, especially with (c). If not, review (c) and its network analog after studying Chapter 6.

Problem 1.12. (a) Repeat Prob. 1.11 in the frequency domain for the sinusoidal steady state, and generalize it to include the case in which ϵ, μ, and σ are functions of frequency. (b) Show that the same uniqueness theorem applies if the boundary conditions demand either

$$\mathrm{E}_i = \sum_{j=1}^{2} \mathrm{Z}_{ij}\mathrm{H}_j + \mathrm{E}_{0i} \qquad i = 1, 2$$

or

$$\mathrm{H}_i = \sum_{j=1}^{2} \mathrm{Y}_{ij}\mathrm{E}_j + \mathrm{H}_{0i} \qquad i = 1, 2$$

at each point on S, where rectangular axes 1 and 2 are tangent to the surface, E_{0i} and H_{0i} are specified functions of position on the surface, and the Z_{ij} are independent of field strength (but possibly functions of frequency and position on S).

Problem 1.13. Extend Probs. 1.11 and 1.12 to include linear *anisotropic* media.

Problem 1.14. Does the uniqueness theorem proved in Probs. 1.11 through 1.13 apply to nonlinear media? Explain.

Problem 1.15. Given an electric field $\boldsymbol{E} = \mathrm{Re}\,[\mathbf{E}e^{j\omega t}]$ and a magnetic field $\boldsymbol{H} = \mathrm{Re}\,[\mathbf{H}e^{j\omega t}]$, show that: (a) $\langle|\boldsymbol{E}|^2\rangle = \frac{1}{2}\mathbf{E}\cdot\mathbf{E}^*$; (b) $\langle\boldsymbol{E}\times\boldsymbol{H}\rangle = \frac{1}{2}\,\mathrm{Re}\,(\mathbf{E}\times\mathbf{H}^*)$; (c) putting $\mathbf{E} = \boldsymbol{E}_r + j\boldsymbol{E}_i$, show that $|\boldsymbol{E}|^2$ can be written in the general form $|\boldsymbol{E}|^2 = A + B\cos(2\omega t + \varphi)$, where

$$A^2 = \tfrac{1}{4}(|\boldsymbol{E}_r|^2 + |\boldsymbol{E}_i|^2)^2$$

$$B^2 = \tfrac{1}{4}[(|\boldsymbol{E}_r|^2 + |\boldsymbol{E}_i|^2)^2 - 4|\boldsymbol{E}_r|^2|\boldsymbol{E}_i|^2 \sin^2 \zeta]$$

$$\tan\varphi = \frac{2\cos\zeta}{(|\boldsymbol{E}_r|/|\boldsymbol{E}_i|) - (|\boldsymbol{E}_i|/|\boldsymbol{E}_r|)}$$

and ζ is the space angle between $\boldsymbol{E}_r$ and $\boldsymbol{E}_i$. (d) For what values of $|\boldsymbol{E}_r|/|\boldsymbol{E}_i|$ and ζ is the ratio $(B/A)^2$ of (c) largest? Smallest? What kind of polarization does $\boldsymbol{E}$ have in the two cases (show sketches)?

Problem 1.16. Can the state of polarization of a time-varying electric field be different at different points in space? How about the plane of the polarization ellipse?

Problem 1.17. (a) With reference to Prob. 1.15, determine the conditions upon the polarization at each point of an electric field $\boldsymbol{E}(x, y, z, t)$ in the sinusoidal steady state, that will make the electric energy $W_e = \int_V \frac{1}{2}\epsilon|\boldsymbol{E}|^2\,dv$ stored in a volume V of space go to zero periodically. (b) Show that the conditions found in (a) *can* always be met by a field solution of Maxwell's equations inside a lossless, source-free, region of space bounded completely by a perfect conductor.

Problem 1.18. Show that a change of time-phase reference can always be made to transform the given $\boldsymbol{E}_r$ and $\boldsymbol{E}_i$ of a polarization ellipse into its principal axes.

Problem 1.19. Find the principal axes of a polarization ellipse by considering the conditions for maximizing or minimizing $|E(t)|^2$.

Problem 1.20. Let $\boldsymbol{E}(t) = \text{Re}\,(\mathbf{E}e^{j\omega t})$ and $\mathbf{E} = \boldsymbol{E}_r + j\boldsymbol{E}_i = \boldsymbol{a}_r E_{r0} + j\boldsymbol{a}_i E_{i0}$, where $\boldsymbol{a}_r$ and $\boldsymbol{a}_i$ are (generally nonorthogonal) unit vectors along $\boldsymbol{E}_r$ and $\boldsymbol{E}_i$ respectively, and $|\boldsymbol{E}_r| = E_{r0}$, $|\boldsymbol{E}_i| = E_{i0}$. Then $\boldsymbol{E}(t) = \boldsymbol{a}_r E_r(t) + \boldsymbol{a}_i E_i(t)$. Show that the equation of the polarization ellipse in the (nonorthogonal) coordinates E_r and E_i is in "normal form." Show also that all chords of the ellipse parallel to $\boldsymbol{E}_i$ are bisected by $\boldsymbol{E}_r$, and all chords parallel to $\boldsymbol{E}_r$ are bisected by $\boldsymbol{E}_i$.

Problem 1.21. Find the principal axes of a polarization ellipse by transforming to a new *orthogonal and rotated* set of axes the expression for the ellipse in the skew axes $\boldsymbol{a}_r$, $\boldsymbol{a}_i$ of Prob. 1.20.

Problem 1.22. Interpret in the time domain for $\boldsymbol{E}(t) = \text{Re}\,(\mathbf{E}e^{j\omega t})$ the conditions: (a) $\mathbf{E}\cdot\mathbf{E} = 0$; (b) $\mathbf{E}\cdot\mathbf{E}^* = 0$; (c) $\mathbf{E}\times\mathbf{E} = 0$; (d) $\mathbf{E}\times\mathbf{E}^* = 0$.

Problem 1.23. If $\boldsymbol{E} = \text{Re}\,(\mathbf{E}e^{j\omega t})$ and

$$\mathbf{E} = \left[\frac{\sqrt{2}+1+j(\sqrt{2}-1)}{\sqrt{3}}\right]\boldsymbol{a}_x + \left[\frac{(2\sqrt{2}-1)+j(2\sqrt{2}+1)}{2\sqrt{3}}\right](\boldsymbol{a}_y + \boldsymbol{a}_z)$$

(a) Describe completely the orientation of the plane of $\boldsymbol{E}(t)$ with respect to the (x, y, z) axes by giving, for example, the direction cosines of the normal to this plane. (b) Make a scale drawing in its own plane of $\boldsymbol{E}(t)$ versus time, including the exact size of the principal axes of the polarization ellipse, and the positions of $\boldsymbol{E}(t)$ at $\omega t = 0, \pi/2, \pi, 3\pi/2, 2\pi$. (c) Determine the direction cosines with respect to the (x, y, z) axes of the principal axes of the polarization ellipse.

Problem 1.24. The propagation of *small amplitude* sound in a nondissipative gas involves three field quantities: the sound-pressure increment $p(x, y, z, t)$, measuring deviation of the total pressure P from the atmospheric pressure P_0; the small velocity $\boldsymbol{u}(x, y, z, t)$ of a differential element of gas at each point; and the incremental density departure $\rho(x, y, z, t)$ from the normal value ρ_0. The one vector equation and two scalar equations needed to relate these variables uniquely are:

$$\nabla p = -\rho_0 \frac{\partial \boldsymbol{u}}{\partial t} \qquad \text{Newton's second law}$$

$$\nabla\cdot\boldsymbol{u} = -\frac{1}{\rho_0}\frac{\partial \rho}{\partial t} \qquad \text{conservation of mass}$$

$$\frac{\rho}{\rho_0} = Kp \qquad \text{property of the medium}$$

where $K = -V^{-1}(dV/dP)$, the volume compressibility of the gas. (a) Derive the above equations on the basis of small-signal approximations. (b) What boundary conditions constitute the limiting forms of the above equations at an interface between two different media? (c) Develop a Poynting theorem for acoustic waves and identify the "acoustic Poynting vector."

Problem 1.25. (a) Repeat Prob. 1.24 in the frequency domain for the sinusoidal steady state. The equivalent of (c) in Prob. 1.24 will deal now with a *complex* Poynting theorem and Poynting vector. (b) Discuss the polarization possibilities for $\boldsymbol{u}(t)$ in the sinusoidal steady state, and suggest how one might arrange matters physically to obtain each type.

CHAPTER TWO

Quasi-Static Fields and Distributed Circuits

To understand a lumped-circuit element, we must consider the fields surrounding the physical structure. A thorough review of this matter will remind us of the meaning of quasi-static fields, and lead naturally onward to the notion of a "distributed circuit" or continuous transmission line.

2.1 The Dilemma of Lumped Circuits

If a direct voltage V is applied to a pair of infinitesimally thin, perfectly conducting plates, open-circuited as shown in Fig. 2.1*a*, and if fringing effects at the edges are neglected, the electrostatic field will be uniform:

$$\boldsymbol{E} = \boldsymbol{a}_x \frac{V}{b} \tag{2.1}$$

and the electrostatic "capacitance" will be

$$\left(\frac{+\text{Charge on plate 1}}{V}\right) = C = \left(\frac{\epsilon a}{b}\right) l \tag{2.2}$$

where ϵ is the permittivity of the homogeneous medium between the plates.

If the plates are short-circuited, and a total current I is impressed with a uniform distribution along y, as shown in Fig. 2.1*b*, then, with fringing effects again neglected, the magnetostatic field will be uniform

$$\boldsymbol{H} = \boldsymbol{a}_y \frac{I}{a} \tag{2.3}$$

and the magnetostatic "inductance" will be

$$\frac{\text{Flux in the } +y \text{ direction}}{I} = L = \left(\frac{\mu b}{a}\right) l \tag{2.4}$$

where μ is the permeability of the homogeneous medium between the plates.

In Fig. 2.1a no current will be drawn from the source of the static voltage V. Inasmuch as $dV/dt = 0$ and the current I drawn from the source is also zero, it is not at all obvious that the capacitance C of Eq.

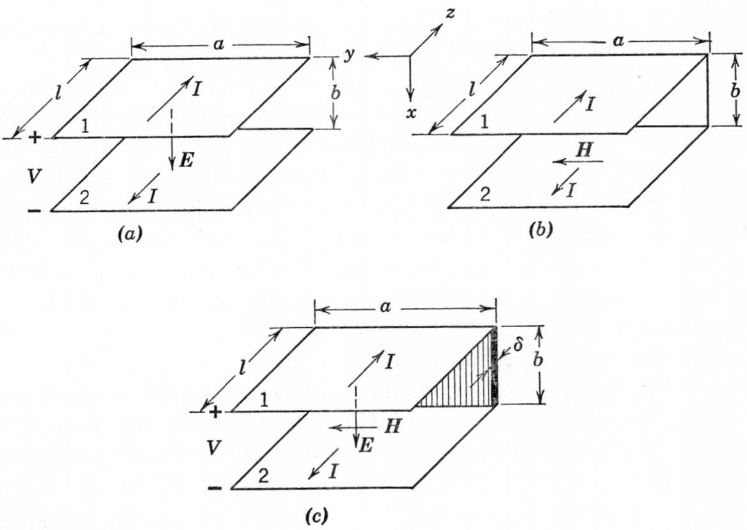

Fig. 2.1. Three simple structures.

2.2 bears any particular relation to the capacitance of lumped-circuit theory, which is actually defined by the relation

$$I = C\frac{dV}{dt} \tag{2.5}$$

Yet we usually assume that the two capacitance definitions are equivalent.

In the second case (Fig. 2.1b), there will be no voltage between the plates. Since $dI/dt = 0$ and the voltage V between the plates is also zero, the connection between L of Eq. 2.4 and the inductance defined in lumped-circuit theory by the relation

$$V = L\frac{dI}{dt} \tag{2.6}$$

is by no means evident. Still, we usually assume that the two inductance definitions are equivalent.

The real extent of our difficulties is perhaps not entirely clear from these two familiar examples, just because they are so familiar. Everyone recognizes a parallel-plate capacitor and a one-turn inductor when he sees them! Is it necessary to worry about mere problems of definition in such limiting cases? That it is, can be appreciated more fully in terms of a somewhat less conventional example.

What happens to our lumped-circuit concepts if the short circuit of Fig. 2.1b is replaced by a very thin sheet of carbon, leaving plates 1 and 2 perfectly conducting, as shown in Fig. 2.1c? For direct current, it is not difficult to see that the carbon sheet constitutes a resistance in the sense of lumped-circuit theory. We are more interested, however, in the a-c behavior of the device. Clearly, if the resistance of the carbon sheet were "very great," we should expect a behavior like that of Fig. 2.1a, and, accordingly, we would undoubtedly say that the device was almost an open circuit for direct current and a capacitor for alternating current. We would probably go even further and, in spite of the difficulties mentioned earlier, would insist that for alternating current the capacitance would be given by Eq. 2.2. If, on the other hand, the resistance of the carbon sheet were "very small," we would anticipate behavior like that of Fig. 2.1b, and, accordingly, we would say that the device was almost a short circuit for direct current and an inductance for alternating current. In fact, we would probably assign Eq. 2.4 as the value of the inductance for alternating current.

But, following the same line of reasoning, what would we expect for a-c behavior if the carbon sheet had only a "moderate" resistance? Surely we would still expect some resistance; but how about L or C? In short, having concluded (somewhat hastily) from static considerations alone, that for alternating current the device of Fig. 2.1a is to be treated as a lumped capacitance (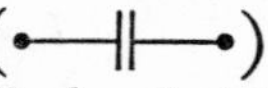) in any network to which it may be connected and, similarly, that the device of Fig. 2.1b is to be treated as a lumped inductance (), we now ask the simple question: How shall we draw a lumped circuit to represent the a-c behavior of the plates with a "moderate" resistance attached in the form of a carbon sheet?

Let us try to proceed along the same lines we followed in connection with Figs. 2.1*a* and 2.1*b*. Assume that a direct voltage V is applied to the plates (we could equally well start with an applied current I). If, again, we neglect fringing, this voltage will produce a total current in the $+x$ direction through the carbon sheet, and this current will be uniformly distributed in the y and z directions. The conductance of the sheet to such a current flow is given by Ohm's law

$$\frac{I}{V} = G_c = \frac{\sigma_c \delta a}{b} \tag{2.7}$$

where σ_c is the conductivity of the carbon. If we now suppose, for definiteness, that the voltage V is actually applied over the entire front edge of the plates, the current I will flow uniformly over the top and bottom plates as shown. The static magnetic field between the plates will then be given by

$$\boldsymbol{H} = \boldsymbol{a}_y \frac{I}{a} \tag{2.8}$$

while the static electric field will be

$$\boldsymbol{E} = \boldsymbol{a}_x \frac{V}{b} = \boldsymbol{a}_x \frac{I}{G_c b} \tag{2.9}$$

Evidently, as far as either electrostatics or magnetostatics is concerned, we can still define exactly the same L and C in this case as we did in Eqs. 2.2 and 2.4, since the above expressions for $\boldsymbol{H}$ in terms of I, and $\boldsymbol{E}$ in terms of V, are precisely the same as Eqs. 2.1 and 2.3. Of course we now have G_c as well, and the problem is: Can we put together the "static" G_c, L, and C of Eqs. 2.7, 2.4, and 2.2 in a lumped circuit in such a way that its a-c behavior will nevertheless represent properly the a-c relation between V and I in the physical structure of Fig. 2.1*c*?

At this point the question looks quite ridiculous. How can we expect some quantities like G_c, L, and C, which have been defined quite arbitrarily from a completely *static* situation, to have any meaning at all when the situation is no longer static? The objection raised here is valid. Unfortunately, however, our position now is extremely weak, because in connection with Figs. 2.1*a* and 2.1*b* we did suggest that the "static" C and L were, in fact, meaningful in a-c situations as well. Since Figs. 2.1*a* and 2.1*b* represent merely two limiting cases of Fig. 2.1*c* ("low" G_c and "high" G_c respectively), we must either raise the same objection in every case or else produce forthwith a lumped circuit valid for alternating current in the case of Fig. 2.1*c*, using *only* the static G_c, L, and C of Eqs. 2.7, 2.4, and 2.2. Faced with this embarrassing choice, we are forced to admit that we have not understood well enough the a-c behavior of *any* of the structures in Fig. 2.1.

2.2 Approximate Solutions

In spite of the dilemma into which Fig. 2.1*c* has thrown us, we know very well from experience that, for the physical configurations of Figs. 2.1*a* and 2.1*b*, we have obtained substantially correct answers (Eqs. 2.2 and 2.4) in some sense or other. Perhaps slightly different views of the matter will make it clearer in what sense. We can re-examine Fig. 2.1*c* again later.

2.2.1 An Energy Point of View

Let us admit, then, that we wish to consider a-c behavior: in particular, the sinusoidal steady-state at a frequency ω. Consider Fig. 2.1*b*, and let I_0 be the peak complex value of a sinusoidal current $I(t)$. Now we wish to compute the impedance of the structure in terms of the field configuration in it. Inasmuch as we have shown in Chapter 1 that both conservation of energy and conservation of complex power hold equally well for lumped circuits and fields separately, we shall take the reasonable position that the energy stored or dissipated in a lumped-circuit element is, in fact, stored or dissipated via its field. On this basis, Eq. 1.14 contains information which is of the greatest importance. If we apply it to the present problem, $\langle P_d \rangle \equiv 0$ because there are no losses in the structure. If $\omega = 0$ and $\mathrm{I}_s = \mathrm{I}_0 = 1$ amp, the equation yields $\mathrm{Z} = 0$, unless we allow $\langle W_e \rangle$ or $\langle W_m \rangle$ to become infinite—which we reject on physical grounds. Evidently this result ($\mathrm{Z} = 0$) is quite correct for Fig. 2.1*b* at $\omega = 0$.

Next we ask how Z behaves when ω is *very small, but not zero*. To be precise, realizing that in general any impedance is an analytic function of $\mathrm{s} = j\omega$ over most of the s-plane, we ask for the behavior of Z correct to the *first power* of s (or ω). Oddly enough, Eq. 1.14 tells us that for this purpose we need only compute $\langle W_e \rangle$ and $\langle W_m \rangle$ correctly to the *zero*th power in ω! At an exceedingly low frequency, however, the field configuration differs negligibly from the static field solution in Eq. 2.3, except that we have, nevertheless, a (slow) sinusoidal variation of I and $\boldsymbol{H}$ with time. To the zero order in ω, we may therefore conclude from essentially static considerations that

(a) $$\mathbf{E} \cong 0$$

(b) $$\mathbf{H} \cong \boldsymbol{a}_y \frac{\mathrm{I}_0}{a}$$

leading to

(a) $$\langle W_e \rangle \cong 0$$

(b) $$\langle W_m \rangle \cong \frac{1}{4} \int_V \mu \mathbf{H} \cdot \mathbf{H}^* \, dv \cong \frac{1}{4} \mu \frac{|\mathrm{I}_0|^2 bl}{a}$$

Thus, from Eq. 1.14,

$$\mathbf{Z} \cong \frac{4j\omega \langle W_m \rangle}{|\mathrm{I}_0|^2} = j\omega \left(\frac{\mu b}{a} l \right) \tag{2.10}$$

which should be valid for "low-enough" frequencies. In this sense, then, the answer obtained for L in Eq. 2.4 is correct because it agrees with Eq. 2.10. A similar (in fact, a "dual") analysis can be carried out with respect to Fig. 2.1*a* and the validity of Eq. 2.2.

From the foregoing energy point of view, the difficulties presented by Fig. 2.1*c* may be pin-pointed. Clearly, the device is not representable by a *single* lumped element because it contains energy dissipation, electric stored energy, and magnetic stored energy—all in the same physical space. This, of course, violates all the rules of lumped circuit elements, in which the whole idea is to isolate completely the three separate energetic processes, one in each of the three lumped elements. But this mixture of energies is really not unique to the structure of Fig. 2.1*c*. Even in Fig. 2.1*b*, for example, we know that at a high enough frequency the electric field, produced by Faraday's law from the changing magnetic field, will become appreciable. Then that structure will contain two kinds of energy in the same space and will not be representable by a *single* lumped element. A dual argument applies to Fig. 2.1*a*. The really curious feature of Fig. 2.1*c* is that it contains all three energetic forms *at arbitrarily low frequencies.* Thus it forces at once upon our attention a difficulty which is in fact universal, but which we may otherwise overlook by tacitly considering only low-frequency behavior: *No physical device is truly a single "lumped" circuit element, although it may behave approximately as such in certain circumstances.*

2.2.2 Quasi-Statics and Beyond

We may strengthen our understanding of this last point by continuing the examination of Fig. 2.1*b* in a direction which will render quantitative our remarks above regarding behavior of physical devices at higher frequencies. Thus we again apply a current of the form $I(t) = \text{Re}\,(\text{I}_0 e^{j\omega t})$. We wish to determine the field inside the structure (Fig. 2.2) at the nonzero frequency ω, but we insist upon finding the particular solution that approaches the proper static fields as $\omega \to 0$. For this purpose we shall use a method of successive approximations. Fringing will be neglected, and the fields will be taken in the form $\mathbf{E}e^{j\omega t}$ and $\mathbf{H}e^{j\omega t}$.

If the frequency were very low, our first guess would be the zero approximation used before:

$$\text{(a)} \qquad \mathbf{E}_0 = 0 \tag{2.11}$$

$$\text{(b)} \qquad \mathbf{H}_0 = \boldsymbol{a}_y \text{K}_0 = \boldsymbol{a}_y \frac{\text{I}_0}{a} = \boldsymbol{a}_y \text{H}_{y0}$$

which again amounts to saying that the field has the same space variation as it would have for direct current, but that it varies sinusoidally in time ($e^{j\omega t}$).

Evidently this is incorrect, because we know from Maxwell's equations that a time variation of $\boldsymbol{H}$ causes an electric field (Faraday's law). The *simplest* electric field that can account for the time-varying magnetic field of Eq. 2.11*b* is one in the x direction, $\mathbf{E}_1 = \boldsymbol{a}_x \text{E}_{x1}$, which does not vary with either x or y, but does depend upon z. The z dependence may be determined from path P_1 in Fig. 2.2, according to Faraday's law. Noting the boundary condition that $\text{E}_{x1} = 0$ on the

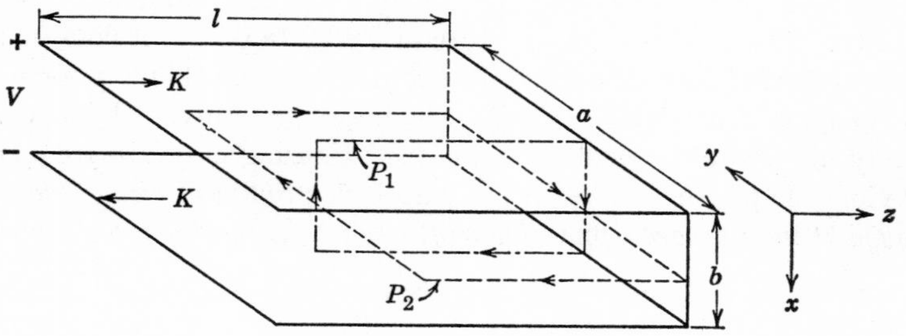

Fig. 2.2. For analysis by successive approximations.

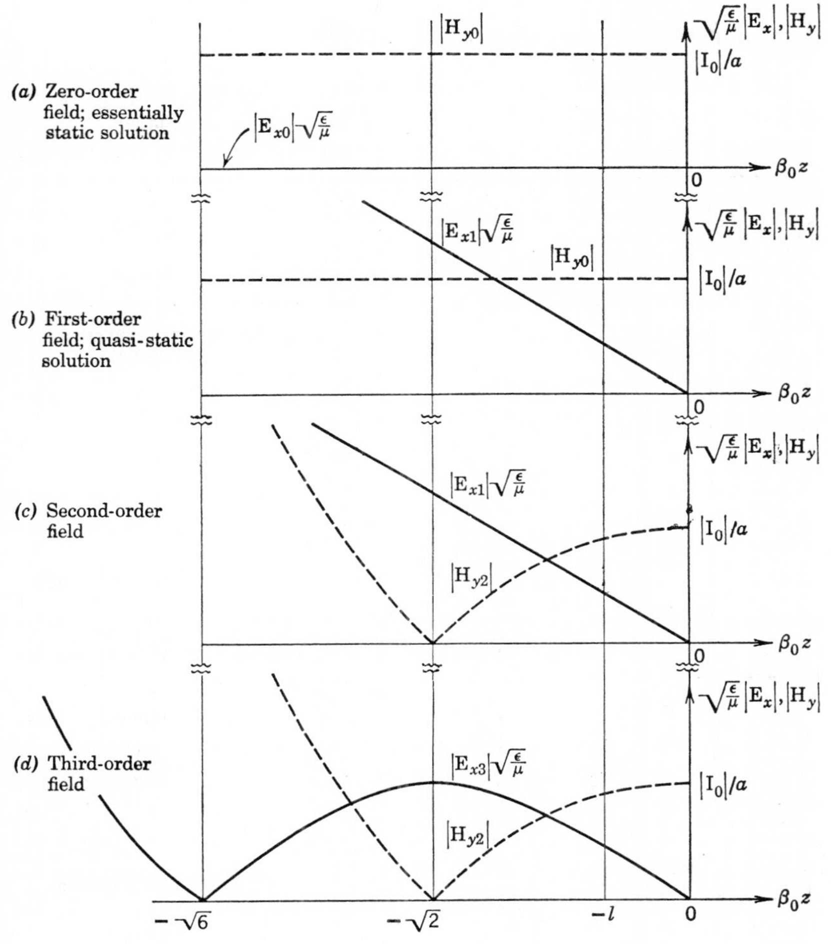

Fig. 2.3. Successive field approximations for the "coil" of Fig. 2.2.

short-circuiting plane taken to be at $z = 0$, we have

$$-bE_{x1}(z) = -j\omega\mu H_{y0}b(-z)$$

inasmuch as $z < 0$ on the left of the short-circuiting plane. Therefore the first approximation for **E** is

$$\mathbf{E}_1 = \boldsymbol{a}_x E_{x1} = -\boldsymbol{a}_x \frac{j\omega\mu I_0}{a} z \tag{2.12}$$

The *magnitudes* of the fields in Eqs. 2.11 and 2.12 are plotted versus

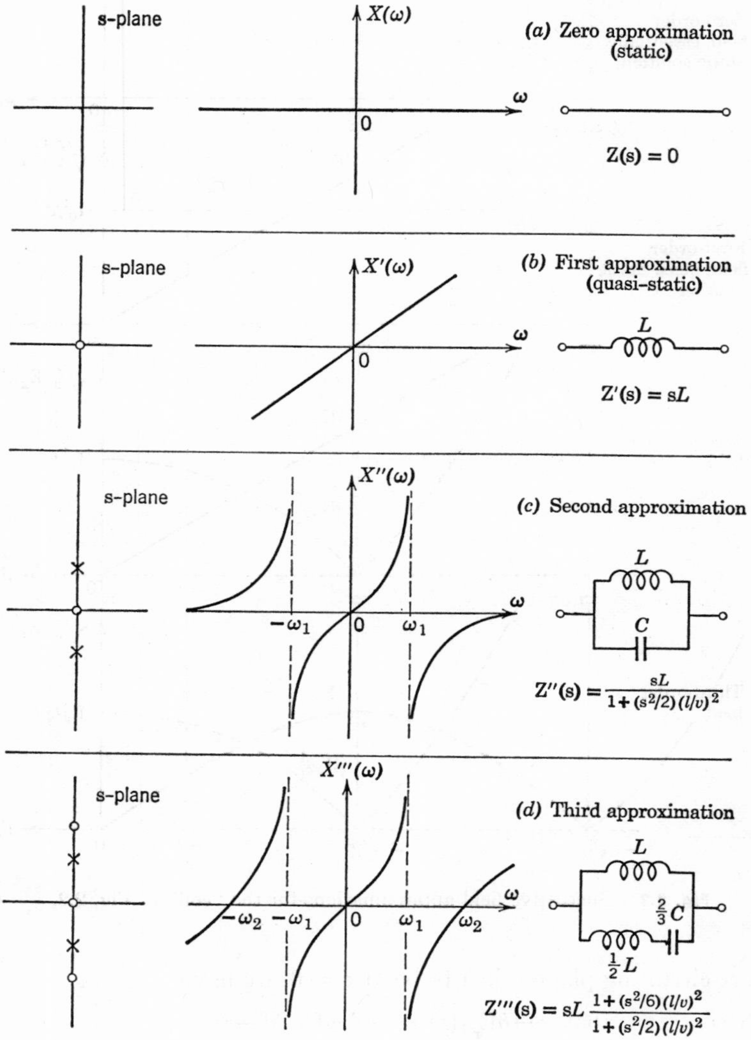

Fig. 2.4. Successive approximations to the input impedance Z(s) of the "coil" of Fig. 2.2. Note that $Z(j\omega) = jX(\omega)$, $Z'(j\omega) = jX'(\omega)$, $Z''(j\omega) = jX''(\omega)$, etc.

the convenient normalized variable $\beta_0 z \equiv \omega\sqrt{\epsilon\mu}\, z$ in Figs. 2.3*a* and 2.3*b*.

The input impedance $Z = jX$ (at $z = -l$) can now be calculated on the basis of a definition of the voltage V as

$$V = bE_x \tag{2.13}$$

We shall consider this definition more closely in Chapters 8 and 9. For the moment, let us accept it merely as a reasonable extension from the static case. Consequently, the zero and first approximations for Z (see Figs. 2.4*a* and 2.4*b*) are respectively,

$$\text{(a)}\quad Z = \frac{V(-l)}{I(-l)} = 0 \qquad \text{zero approximation}$$

$$\text{(b)}\quad Z' = \frac{V(-l)}{I(-l)} = \frac{bE_{x1}(-l)}{I_0} = j\omega\left(\frac{\mu bl}{a}\right) = sL \qquad \text{first approximation} \tag{2.14}$$

We should observe two important features of our results so far:

(a) The *time* variation of $\boldsymbol{H}$ must be accompanied by a *space* variation of $\boldsymbol{E}$ (in addition to its time variation, of course).

(b) The "static" $\boldsymbol{H}(H_{y0})$ and the first order $\boldsymbol{E}(E_{x1})$ together define what is called a *quasi-static* solution for the field in this case. According to Eq. 2.14*b*, this solution is the one that leads to the conventional idea of a "lumped" inductance. *The electric field that produces the "voltage" across the "inductor" actually exists in space and time* (Fig. 2.3*b*); yet it *cannot* be determined from purely static field considerations alone.

There is nothing magic about the quasi-static solution that makes it necessary to stop the calculation at this point. We can go on. Now we have a time-varying electric field $\boldsymbol{E}_1$. This produces a displacement current across the gap which we omitted from our first calculation of $\boldsymbol{H}$. In terms of the path P_2 of Fig. 2.2, Ampere's law (including displacement current) tells us that $\boldsymbol{H}$ must actually vary with z! This, in turn, requires that the current on the top and bottom plates also vary with z, a condition that can alternatively be phrased in terms of the continuity of total current (Eq. 1.16). It follows that we must be more careful about specifying the driving current for the system since the current depends upon z. We could, of course, imagine the input current (at $z = -l$) fixed at I_0. Although this procedure is perfectly reasonable, and not really difficult to manage, the analysis is a little easier if, instead, we regard the current in the short circuit at $z = 0$ as fixed at I_0. Since the system is linear, no generality is lost by this procedure. On this basis, the second approximation for **H** has on it the boundary condition

$$H_{y2}\Big|_{z=0} = \frac{I_0}{a} \tag{2.15}$$

From the path P_2 (Fig. 2.2) we then have, according to Ampere's law, the simplest **H** field solution in the form

$$a[H_{y2}(z) - H_{y2}(0)] = j\omega a \int_z^0 \epsilon E_{x1}\, dz$$

or

$$H_{y2}(z) = \frac{I_0}{a} + j\omega\epsilon(-j\omega\mu)\frac{I_0}{a}\int_z^0 z\, dz$$

$$= \frac{I_0}{a} - \frac{I_0}{2a}(\omega^2\epsilon\mu)z^2 = \frac{I_0}{a}\left[1 + \frac{1}{2}(j\beta_0 z)^2\right] \tag{2.16}$$

We notice from Eq. 2.16 that the first term is the zero-order solution, whereas the second term arising from displacement current involves the square of the frequency. Compare this with Eq. 2.12, where the first term (zero) is the zero-order solution, and the second arising from flux change involves the first power of ω. The sketch of $|H_{y2}|$ in Fig. 2.3c shows that when $l \ll \sqrt{2}/(\omega\sqrt{\epsilon\mu})$, the magnetic field in the structure has almost the static distribution in space. It also shows that this is no longer true for either large l or high frequency.

The impedance at $z = -l$ stems from the definition of the current as that in the top plate at $z = -l$,

$$I(-l) = aH_y(-l) \tag{2.17}$$

and the definition of the voltage in Eq. 2.13 applied at $z = -l$. There results the second approximation

$$Z'' = \frac{V(-l)}{I(-l)} = \frac{bE_{x1}(-l)}{aH_{y2}(-l)} = \frac{j\omega[(\mu bl)/a]}{1 + (j\omega)^2[(l^2\epsilon\mu)/2]}$$

$$= \frac{sL}{1 + \frac{1}{2}s^2(l/v)^2} \tag{2.18}$$

where we have introduced v $(\equiv 1/\sqrt{\epsilon\mu})$ for the velocity of light in the medium filling the space between the plates. At frequencies $\omega \ll \sqrt{2}\, v/l$, the impedance is just that of an inductance L. At higher frequencies, however, Eq. 2.18 and Fig. 2.4c show a "parallel" resonance from the increased stored electric energy in the structure. This resonance occurs at the frequency for which the zero of $|H_{y2}|$ in Fig. 2.3c falls just at $z = -l$. The interpretation of both events is simple. The ratio $l/v = l\sqrt{\epsilon\mu}$ is the time T that light (or electromagnetic waves) would take to traverse distance l. The period τ corresponding to radian frequency ω is $\tau = 2\pi/\omega$; so the important condition $\omega = \sqrt{2}\, v/l$ becomes $2\pi/\tau = \sqrt{2}/T$ or $T = \tau/\sqrt{2}\,\pi \approx \frac{1}{4}\tau$. In other words, serious departure from quasi-static (or, in this case, lumped-circuit) behavior occurs when the transit time of electromagnetic waves along a significant dimension of the structure becomes comparable to the period of excitation.

Another description of the above situation is often made by attributing the first parallel resonance of our elementary one-turn inductor "coil" to the "distributed capacitance" between the plates. Such an idea is sometimes useful, but is really even less precise than the simple statement of increasing electric stored energy made previously. Moreover, it avoids the very important fact that the *space* distribution of the magnetic field at such a frequency differs radically from its quasi-static form.

It is worth while to go one step further with our solution to establish more clearly our viewpoint. The third approximation for **E** follows from Ampere's law applied to path P_1 in Fig. 2.2, using the second approximation for **H** (Eq. 2.16). Thus

$$-bE_{x3} = -j\omega\mu b \int_z^0 H_{y2}\, dz = -j\omega\mu b \left(\frac{I_0}{a}\right) \int_z^0 \left[1 + \frac{1}{2}(j\beta_0 z)^2\right] dz$$

$$= \frac{j\omega\mu I_0 b}{a} z - \frac{j\omega\mu\beta_0{}^2 I_0}{3\cdot 2a} bz^3$$

or

$$E_{x3} = -\frac{I_0}{a}\sqrt{\frac{\mu}{\epsilon}}\left[j\beta_0 z + \frac{(j\beta_0 z)^3}{3!}\right] \tag{2.19}$$

The first term in Eq. 2.19 is the quasi-static solution obtained previously in Eq. 2.12. The second term is a correction which depends upon the third power of the frequency ω. Thus Eqs. 2.19 and 2.16 constitute a third-order approximation to the field (Fig. 2.3*d*).

It should be clear now that continuation of this successive approximation or iterative process will add terms containing higher powers of ω to the expressions for **E** and **H**, only odd powers of ω appearing in **E** and only even powers of ω in **H**. *We have succeeded in developing* **E** *and* **H** *in power series of ω, i.e., in Taylor series expansions about zero frequency. In this light, the so-called quasi-static field is merely a power* *series solution of Maxwell's equations, correct to the first power of ω. Indeed, this last description is actually the best definition of the name "quasi-static field"!*

If, for a given physical structure, we wish to consider a wide frequency range, then it is not sufficient to stop the series after the quasi-static or first-power term. Correspondingly, it is not sufficient to treat the physical structure as a single lumped circuit element either. As suggested by Figs. 2.4*b* and 2.4*c*, inclusion of additional terms shows up the more complicated aspects of the frequency dependence of the impedance over the extended frequency range. For example, the third approximation in the present problem yields an input impedance

$$Z''' = \frac{b\mathrm{E}_{x3}(-l)}{a\mathrm{H}_{y2}(-l)} = \frac{sL[1 + \frac{1}{6}(l/v)^2s^2]}{1 + \frac{1}{2}(l/v)^2s^2} \tag{2.20}$$

illustrated in Fig. 2.4*d*. It is not so easy to charge off the new "series" resonance at ω_2 to "stray" or "distributed" capacitance or inductance in a vague sort of way. The limitations of such a weak point of view have already become extremely obvious. However, it is easy to see that succeeding steps in the approximation process will introduce alternately new poles and zeros into the input impedance, etc., *ad infinitum*. Indeed, a little thought will indicate that the various series, if continued further by our methods, become expansions for

$$\text{(a)} \qquad \mathrm{H}_y(z) = \frac{\mathrm{I}_0}{a} \cos \beta_0 z$$

$$\text{(b)} \qquad \mathrm{E}_x(z) = -j\sqrt{\frac{\mu}{\epsilon}}\frac{\mathrm{I}_0}{a} \sin \beta_0 z \tag{2.21}$$

$$\text{(c)} \qquad Z(-l) = j\sqrt{\frac{L}{C}} \tan \beta_0 z$$

the first two of which are reminiscent of standing waves.

2.2.3 Energy Again

Before pursuing these notions in any greater detail, however, we can shed even more light upon the quasi-static situation by treating the bothersome example of Fig. 2.1*c*. This we shall do first by using the method which so successfully led to Eq. 2.10 for the structure of Fig. 2.1*b*.

Thus, from Eqs. 2.7, 2.8, and 2.9,

$$\text{(a)} \qquad \langle P_d \rangle \cong \frac{1}{2} \frac{|I_0|^2}{G_c} = \frac{1}{2} |I_0|^2 R_c$$

$$\text{(b)} \qquad \langle W_m \rangle \cong \frac{1}{4} \left(\frac{\mu l b}{a} \right) |I_0|^2 = \frac{1}{4} L |I_0|^2$$

$$\text{(c)} \qquad \langle W_e \rangle = \frac{1}{4} \int_V \epsilon \mathbf{E} \cdot \mathbf{E}^* \, dv \cong \frac{lab}{4} \epsilon \frac{|I_0|^2}{b^2 G_c^2} = \frac{1}{4} C R_c^2 |I_0|^2$$

where we have used Eqs. 2.2 and 2.4 to define C and L. Equation 1.14 then yields

$$Z \cong R_c + j\omega(L - CR_c^2) \tag{2.22}$$

which is represented by the circuit of Fig. 2.5*a*.

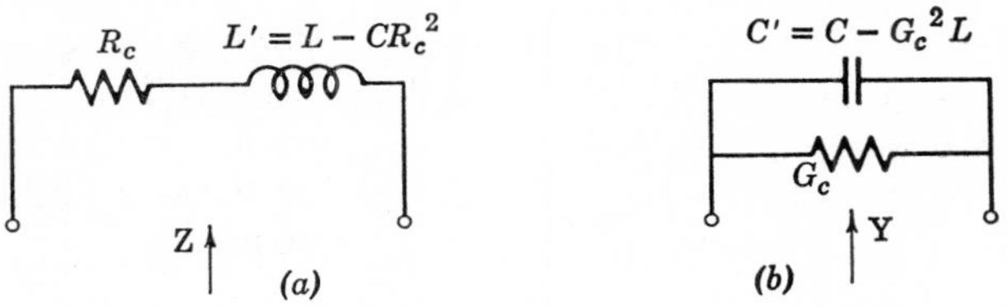

Fig. 2.5. Approximate low-frequency equivalent circuits for the structure of Fig. 2.1*c*.

This result shows that as $R_c \to 0$, the low-frequency behavior of the structure approaches that of an inductance L—as indeed it should. Also, if $L/C > R_c^2$, the resistance of the carbon sheet is "small" and the device appears from the terminals to be an R-L' circuit (at low frequencies). The average stored magnetic energy exceeds the average stored electric energy. The "net" stored energy $\langle W_m - W_e \rangle$ is therefore magnetic, making the system appear inductive with $L' < L$. Of course, the loss in the carbon sheet leads to the resistance.

On the contrary, if $L/C < R_c^2$, the resistance of the sheet is "large" and the average electric stored energy evidently dominates. We then expect the structure to look capacitive, but in such a manner that its impedance becomes R_c for direct current. Figure 2.5*a* actually shows this behavior correctly because $L' < 0$, but the circuit looks nonphysical on that account. A more reasonable looking circuit, just as correct *to the first order in* ω, stems from the admittance expression

$$Y = \frac{1}{Z} = \frac{1}{R_c}\left\{\frac{1}{1 + j\omega[(L/R_c) - CR_c]}\right\} \cong G_c\left[1 - j\omega\left(\frac{L}{R_c} - CR_c\right)\right]$$
$$= G_c + j\omega(C - LG_c^2) \tag{2.23}$$

which behaves properly like $j\omega C$ as $G_c \to 0$ ($R_c \to \infty$). The new equivalent circuit appears in Fig. 2.5*b*, with $C' > 0$ when $R_c^2 > L/C$.

2.2.3.1 Some Analytical Difficulties.*[1] There is one feature of Eqs. 2.22 and 2.23 (or Figs. 2.5*a* and 2.5*b*) which is disturbing at first glance. At any fixed *nonzero* frequency ω, Eq. 2.22 predicts $Z \to \infty$ as $R_c \to \infty$, while Eq. 2.23 predicts $Y \to j\omega C$ or $Z \to 1/(j\omega C)$ as $R_c \to \infty$. Similarly, Eq. 2.22 predicts $Z \to j\omega L$ as $R_c \to 0$, while Eq. 2.23 predicts $Z \to 0$ as $R_c \to 0$. The two equations simply do not agree about what happens when $R_c \to \infty$ or $R_c \to 0$, except in the d-c case $\omega \equiv 0$.

In one sense this should be no surprise, because the binomial series approximation used in Eq. 2.23 is valid only when $\omega(L/R_c - CR_c) \ll 1$. For a given nonzero frequency, this condition is violated whenever R_c is *either* very small or very large. Thus, *with a fixed* ω, we should not expect Eqs. 2.22 and 2.23 to agree for either very small or very large R_c. Nevertheless, given any finite but nonzero value of R_c, the condition for the validity of Eq. 2.23 is always met *when* ω *is small enough*. The *range* of ω for which the two equations will agree is simply reduced when R_c is either very large or very small.

A more troublesome appearing point about the above situation now comes to light. Which expression should be taken as correct for $R_c = 0$ and for $R_c = \infty$, if $\omega \neq 0$? The physical situation of Fig. 2.1*c* and our previous analysis of Fig. 2.1*b* leave no doubt that only Eq. 2.22 is correct when $R_c = 0$ and only Eq. 2.23 is correct when $R_c = \infty$. The reason why our present analysis does not show up this point lies in our somewhat hasty use of Eq. 1.14.

We computed $\langle P_d\rangle/|I_0|^2$, $\langle W_m\rangle/|I_0|^2$, and $\langle W_e\rangle/|I_0|^2$ *at* $\omega = 0$ (i.e., for direct current). If $R_c = \infty$, it is actually impossible to apply a direct current I_0 without getting an infinite electric field and an infinite voltage. Thus both the power loss and the electric stored energy *for one ampere of current applied* will become infinite. The implication is that when $R_c = \infty$, the true impedance Z has a pole at $s = j\omega = 0$. We know in general that no power series expansion of the form $A +$

[1] Sections marked with an asterisk (*), and all subsections under them, may be omitted from a minimum undergraduate subject.

$Bs + \cdots$ can correctly represent Z(s) *at* a pole; so we realize that Eq. 2.22 is invalid when $R_c = \infty$. Viewed in another way, when $R_c = \infty$, we can have $\langle P_d \rangle$ and $\langle W_e \rangle$ finite only by applying a voltage source. This results in $|I_0| = 0$ for direct current, invalidating the division by $|I_0|^2$ used in deriving Eq. 1.14. Therefore we reach the same conclusion as before.

On the other hand, we could instead have divided Eq. 1.13 by $V_s V_s^* = |V_s|^2$ and derived the admittance of the system:

$$\frac{I_s}{V_s} = Y = \frac{2\langle P_d \rangle}{|V_s|^2} + j4\omega\left(\frac{\langle W_e \rangle}{|V_s|^2} - \frac{\langle W_m \rangle}{|V_s|^2}\right) \tag{2.24}$$

Application of this formulation to our problem of Fig. 2.1*c* leads directly to Eq. 2.23, since

$$\langle P_d \rangle = \tfrac{1}{2}|V_0|^2 G_c$$

$$\langle W_e \rangle = \tfrac{1}{4}C|V_0|^2$$

and

$$\langle W_m \rangle = \tfrac{1}{4}L|I_0|^2 = \tfrac{1}{4}LG_c^2|V_0|^2$$

There is no difficulty in applying the method even when $G_c = 0$ ($R_c = \infty$), inasmuch as we have seen that a direct voltage can still be impressed without the danger of any energies becoming infinite. Of course, there would be trouble in applying Eq. 2.24 to a case where $R_c = 0$, when the voltage V_0 could not be applied at zero frequency (the true admittance then has a pole at $\omega = 0$). Thus we see that Eq. 2.23 is valid only when $0 < R_c \leq \infty$, whereas Eq. 2.22 is valid only when $0 \leq R_c < \infty$.

We can now find another interesting interpretation of the fact that Eqs. 2.22 and 2.23 agree with each other only over a very small range of $\omega \geq 0$ whenever R_c is either extremely large or extremely small (compared to $\sqrt{L/C}$). Suppose that R_c is very small, for example. Then, since Eq. 2.22 is correct when $0 \leq R_c < \infty$ for *some* range of $\omega \geq 0$, the principle of analytic continuation allows us to say that $Z(s) \approx R_c + sL'$ for values of $|s|$ which are not too large. This expression predicts a *zero* of Z(s) at $s_0 = -R_c/(L - CR_c^2)$; and if R_c is small enough, this root is very near the point $s = 0$, as shown in Fig. 2.6*a*. The smaller R_c is, the smaller is s_0, and the more valid becomes our calculation of s_0 from only the first two terms of the Taylor expansion of Z(s). More precisely, a full Taylor expansion of the form $Z(s) = A + Bs + Cs^2 + \cdots$ would converge in a circle extending out to the nearest singularity of Z(s). This singularity corresponds physically to the first shunt resonance of the structure, which, accord-

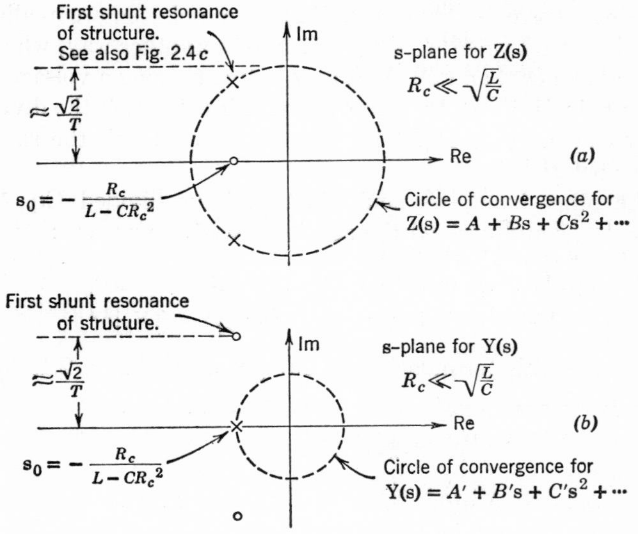

Fig. 2.6. Complex plane (approximate) for the impedance of the structure of Fig. 2.1c.

ing to Fig. 2.4c of our previous analysis of Fig. 2.1b, should be very nearly at $s = \pm j\sqrt{2}/T$, but moved slightly to the left of the imaginary axis because of the loss in the small R_c (see also Chapter 6). These matters are summarized in Fig. 2.6a. Thus, for small R_c, the root s_0 lies deeply inside the circle of convergence and is well approximated by only two series terms.

It follows that $Y(s) = 1/[Z(s)]$ has a *pole* at very nearly $s_0 = -R_c/(L - CR_c^2)$, very close to $s = 0$. Therefore any attempt to make a Taylor series expansion of Y(s) about the point $s = 0$, of the form $Y(s) = A' + B's + C's^2 + \cdots$, will lead to a result having an extremely small radius of convergence when R_c is small, and none at all when $R_c = 0$! This result is illustrated in Fig. 2.6b. Since Eq. 2.23 represents the first two terms of such a Taylor expansion of Y(s) about $s = 0$, we now understand that it will not be accurate over a very significant range of ω when R_c is small. By an argument which is the dual to that above, we would find Eq. 2.22 severely restricted as regards the frequency range of its validity whenever R_c is large. However, the range of validity of Eq. 2.22 is quite broad when R_c is small, and that of Eq. 2.23 is correspondingly broad when R_c is large.

Thus, for either extremely large or extremely small values of R_c, only *one* of Eqs. 2.22 and 2.23 has a reasonable frequency range in which it is valid, the other having its range exceedingly limited by a nearby singularity (pole) in the s-plane. This is why we found that under these conditions the *common* frequency range of validity, for which both expressions would agree numerically, is limited to a small value.

The general lesson we must learn from this example is quite important. Use of Eq. 1.14 to represent a Z(s) approximately in a frequency range near a *pole* of the true Z(s) will lead to rather unsatisfactory results, in the sense that their frequency range of validity will be exceedingly small. If we must consider a physical system at frequencies near one of its impedance poles, we should use Eq. 2.24 above to compute *admittance* instead. We shall then be making Taylor approximations in a region where the actual function and its derivatives are well behaved instead of singular, and we can obtain correspondingly better approximations with the same labor. Of course, $Z(s) = 1/[Y(s)]$; so we get answers for Z by computing Y. An analogous (indeed dual) comment applies to computations near a *zero* of Z(s) [i.e., a *pole* of Y(s)] for which Eq. 1.14 will be far superior to Eq. 2.24 above.

We have spent considerable time on the question of which of the circuits of Fig. 2.5 best represents the structure of Fig. 2.1*c* to the first order in frequency, considering the two limiting situations of very large and very small R_c. One should not lose sight, in all this detail, of the simple fact that when, for example, $R_c > \sqrt{L/C}$, the circuit of Fig. 2.5*a* contains a negative inductance. It is awkward to use a nonphysical circuit to represent a physical structure, even if it is correct for *some* range of ω. The circuit of Fig. 2.5*b* does not suffer from this disadvantage, as long as $R_c > \sqrt{L/C}$; thus it would be the most appealing one to use under these conditions. Similarly, Fig. 2.5*a* would be most appealing when $R_c < \sqrt{L/C}$. Although these aesthetic choices are certainly consistent with our more quantitative discussions for extreme R_c values, we have not shown in similar analytical detail why the "crossover point" in our choice between Figs. 2.5*a* and 2.5*b* should occur just at $R_c = \sqrt{L/C}$. This matter as well as the curious behavior of the foregoing results when $R_c = \sqrt{L/C}$ may be discussed in terms of a more complete analysis using a power series expansion in frequency of the complex fields.

2.2.4 Quasi-Statics Again*

The present solution will be similar to that developed by iteration for the structure of Fig. 2.2. This time, however, we shall take advantage of the simplicity gained by admitting at the outset that we

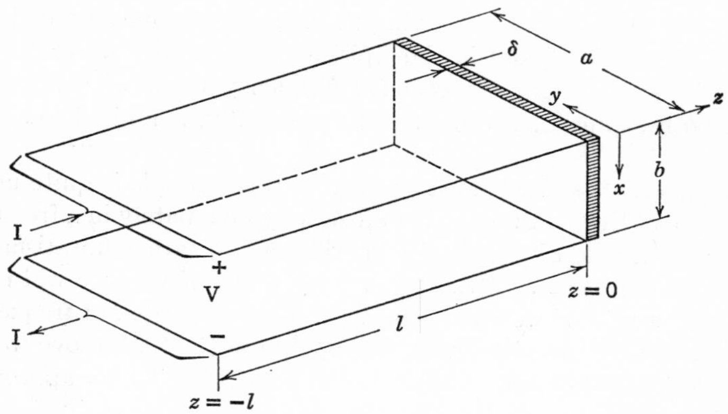

Fig. 2.7. Review of Fig. 2.1c for power series analysis in the frequency domain.

desire a power series expansion of the fields in frequency. This expansion, centered about $\omega = 0$, will be subjected to the same assumptions as the d-c solution, namely, $\mathbf{E} = \boldsymbol{a}_x \mathrm{E}_x$, $\mathbf{H} = \boldsymbol{a}_y \mathrm{H}_y$, $\partial/\partial y \equiv 0$, $\partial/\partial x \equiv 0$, and the neglect of fringing at $z = 0$. The pertinent illustration is Fig. 2.7, and the fields are of interest only in the space $-\infty < y < +\infty$, $0 \leq x \leq b$, and $-\infty < z \leq 0$.

Let

$$\text{(a)} \qquad \mathrm{E}_x = \mathrm{e}_{x0} + \mathrm{e}_{x1}\omega + \mathrm{e}_{x2}\omega^2 + \mathrm{e}_{x3}\omega^3 + \cdots + \mathrm{e}_{x\nu}\omega^\nu + \cdots$$

$$\text{(b)} \qquad \mathrm{H}_y = \mathrm{h}_{y0} + \mathrm{h}_{y1}\omega + \mathrm{h}_{y2}\omega^2 + \mathrm{h}_{y3}\omega^3 + \cdots + \mathrm{h}_{y\nu}\omega^\nu + \cdots$$

where

$$\text{(a)} \qquad \mathrm{e}_{x\nu} = \mathrm{f}_\nu(z, \text{geom.})$$

$$\text{(b)} \qquad \mathrm{h}_{y\nu} = \mathrm{g}_\nu(z, \text{geom.})$$

These expressions are to be substituted in the complex Maxwell equations, which for the present assumptions appear in the following form:

$$\text{(a)} \qquad \nabla \times \mathbf{E} = -j\omega\mu\mathbf{H} \qquad \frac{d\mathrm{E}_x}{dz} = -j\omega\mu\mathrm{H}_y$$

$$\text{(b)} \qquad \nabla \times \mathbf{H} = +j\omega\epsilon\mathbf{E} \qquad \frac{d\mathrm{H}_y}{dz} = -j\omega\epsilon\mathrm{E}_x$$

Performing the substitution and equating like powers of ω on both sides leads to the results tabulated below in Table 1.

TABLE 1. Expansion Coefficients for the Problem of Fig. 2.7

ω^0	ω^1	ω^2	$\cdots$	ω^ν
$\frac{de_{x0}}{dz} = 0$	$\frac{de_{x1}}{dz} = -j\mu h_{y0}$	$\frac{de_{x2}}{dz} = -j\mu h_{y1}$	$\cdots$	$\frac{de_{x\nu}}{dz} = -j\mu h_{y,\nu-1}$
$\frac{dh_{y0}}{dz} = 0$	$\frac{dh_{y1}}{dz} = -j\epsilon e_{x0}$	$\frac{dh_{y2}}{dz} = -j\epsilon e_{x1}$		$\frac{dh_{y\nu}}{dz} = -j\epsilon e_{x,\nu-1}$

It is convenient to define some new parameters before proceeding. In particular, we note that

$$r \equiv \frac{1}{\delta\sigma_c} \quad \text{ohms per square}$$

is the "sheet resistance" of the terminal plane at $z = 0$. This concept is especially useful in the limiting case of $\sigma_c \rightarrow \infty$, $\delta \rightarrow 0$, which we can assume to be the situation in Fig. 2.7. We are relieved by this artifice from considering fields within the sheet, occupying distance δ. From a circuit point of view, however, we still need the total resistance R_c, now expressed as

$$R_c = (br)/a \quad \text{ohms}$$

Finally, instead of frequency, we again introduce

$$\beta_0 = \omega\sqrt{\epsilon\mu} \quad \text{m}^{-1}$$

and as a normalization for impedance

$$\eta = \sqrt{\mu/\epsilon} \quad \text{ohms}$$

Thus, the commonly occurring products $\omega\mu$ and $\omega\epsilon$ become

$$\omega\mu = \beta_0\eta \qquad \omega\epsilon = \beta_0/\eta$$

As regards further boundary conditions, we shall want to prescribe the complex current I_0 at $z = 0$, and intend that this value shall be independent of frequency. Since

$$I_0 = aH_y(z = 0) \qquad \text{for } all \text{ values of } \omega$$

and, therefore,

$$E_x(z = 0) = (I_0R_c)/b \qquad \text{for } all \text{ values of } \omega$$

the power series expressions yield, upon equating like powers of ω in these restrictions,

$$h_{y0}(z = 0) = I_0/a \qquad e_{x0}(z = 0) = (I_0 R_c)/b$$

$$h_{y\nu}(z = 0) = 0 \quad \nu = 1, 2, \cdots \qquad e_{x\nu}(z = 0) = 0 \quad \nu = 1, 2, \cdots$$

The last relations form the boundary conditions to permit the solution of all the equations in Table 1.

Commencing with the ω^0 set in Table 1, we see that h_{y0} and e_{x0} are independent of z and retain their values at $z = 0$ for all points in the structure. From the ω^1 set, in view of the condition $e_{x1}(z = 0) = 0 = h_{y1}(z = 0)$, we have

$$\text{(a)} \qquad \omega e_{x1} = -j\omega\mu \frac{I_0}{a} z = -j\beta_0 z \left(\frac{\eta I_0}{a}\right)$$

$$\text{(b)} \qquad \omega h_{y1} = -j\omega\epsilon \frac{I_0 R_c}{b} z = -j\beta_0 z \left(\frac{I_0 R_c}{\eta b}\right)$$

Similarly from ω^2 terms

$$\text{(a)} \qquad \omega^2 e_{x2} = \frac{I_0 R_c}{b} \frac{(j\beta_0 z)^2}{2!}$$

$$\text{(b)} \qquad \omega^2 h_{y2} = \frac{I_0}{a} \frac{(j\beta_0 z)^2}{2!}$$

and from ω^3

$$\text{(a)} \qquad \omega^3 e_{x3} = \frac{-I_0 \eta}{a} \frac{(j\beta_0 z)^3}{3!}$$

$$\text{(b)} \qquad \omega^3 h_{y3} = -\frac{I_0 R_c}{b\eta} \frac{(j\beta_0 z)^3}{3!}$$

It is clear that the general expressions for E_x and H_y become

$$\text{(a)} \quad E_x = \frac{I_0 R_c}{b}\left[1 - \left(\frac{\eta b}{a R_c}\right)(j\beta_0 z) + \frac{(j\beta_0 z)^2}{2!} - \left(\frac{\eta b}{a R_c}\right)\frac{(j\beta_0 z)^3}{3!} + \cdots\right]$$

$$\text{(b)} \quad H_y = \frac{I_0}{a}\left[1 - \left(\frac{a R_c}{b\eta}\right)(j\beta_0 z) + \frac{(j\beta_0 z)^2}{2!} - \left(\frac{a R_c}{b\eta}\right)\frac{(j\beta_0 z)^3}{3!} + \cdots\right]$$

and for voltage and current

$$\text{(a)} \qquad V = bE_x = I_0 R_c \left\{\left[1 + \frac{(j\beta_0 z)^2}{2!} + \frac{(j\beta_0 z)^4}{4!} + \cdots\right]\right.$$
$$\left. - \frac{\eta}{r}\left[(j\beta_0 z) + \frac{(j\beta_0 z)^3}{3!} + \frac{(j\beta_0 z)^5}{5!} + \cdots\right]\right\}$$
$$= I_0 R_c[\cosh (j\beta_0 z) - \chi \sinh (j\beta_0 z)]$$

(b) $$\mathrm{I} = a\mathrm{H}_y = \mathrm{I}_0[\cosh(j\beta_0 z) - \frac{1}{\chi}\sinh(j\beta_0 z)]$$

with

$$\chi \equiv \eta/r$$

The impedance at $z = -l$ is going to involve $j\beta_0 l = j\omega\sqrt{\epsilon\mu}\, l = j\omega T$, with T again designating the time an electromagnetic wave takes to traverse the distance l. In terms of $\mathrm{s} = j\omega$, then, we have

$$\mathrm{Z}_{-l}(\mathrm{s}) = \frac{\mathrm{V}(-l)}{\mathrm{I}(-l)} = R_c\left(\frac{\cosh \mathrm{s}T + \chi \sinh \mathrm{s}T}{\cosh \mathrm{s}T + (1/\chi)\sinh \mathrm{s}T}\right)$$

At low frequencies, such that $|\mathrm{s}T| \ll 1$ (or $\omega T \ll 1$ in the sinusoidal steady state), this expression becomes

$$\mathrm{Z}_{-l}(\mathrm{s}) \approx R_c\left[\frac{1 + \chi T\mathrm{s}}{1 + (T/\chi)\mathrm{s}}\right] = R_c\chi^2\left[\frac{\mathrm{s} + (1/\chi T)}{\mathrm{s} + (\chi/T)}\right]$$

When χ is large, corresponding to a low-resistance termination, $|(\mathrm{s}T)/\chi| \lll 1$ and

$$Z_{-l}(\mathrm{s}) \approx R_c(1 + \chi T\mathrm{s})\left(1 + \frac{T\mathrm{s}}{\chi}\right)^{-1} \approx R_c\left[1 + T\left(\chi - \frac{1}{\chi}\right)\mathrm{s}\right]$$

Noting that

$$\chi \equiv \frac{\eta}{r} = \frac{(b\eta)/a}{R_c} = \frac{\sqrt{L/C}}{R_c}$$

we see the last form of $\mathrm{Z}_{-l}(\mathrm{s})$ become

$$\mathrm{Z}_{-l}(\mathrm{s}) \approx R_c + L\left(1 - \frac{R_c^2 C}{L}\right)\mathrm{s}$$

in agreement with Fig. 2.5. Similarly, when χ is small, corresponding to a high-resistance termination, $|\chi T\mathrm{s}| \lll 1$, and

$$\mathrm{Y}_{-l}(\mathrm{s}) \approx G_c\left[1 + T\left(\frac{1}{\chi} - \chi\right)\mathrm{s}\right] = G_c + C\left(1 - \frac{G_c^2 L}{C}\right)\mathrm{s}$$

as found also in Fig. 2.5.

2.2.4.1 MORE ABOUT ANALYSIS AND APPROXIMATIONS.* In the event that χ is neither very large nor very small, $\mathrm{Z}_{-l}(\mathrm{s})$ must be left as a fraction. The approximate form of the fraction given above for low frequencies shows a zero and a pole on the negative-real axis of the s-plane:

$$s_{\text{zero}} = -\frac{1}{\chi T} = -\frac{R_c}{L}$$

$$s_{\text{pole}} = -\frac{\chi}{T} = -\chi^2\frac{R_c}{L} = -\frac{1}{R_c C} = \chi^2 s_{\text{zero}}$$

The network equivalents of this approximation to the impedance can avoid negative elements only by using one when $\chi > 1$ (Fig. 2.8*a*) and the other when $0 < \chi < 1$ (Fig. 2.8*b*), although both are valid in either case. The easiest way to check the resistive element values in these networks is by considering $s = 0$ and $s = \infty$ first; then the zero *or* the pole may be matched to determine the reactive elements. The choice between the networks, if they are to have positive elements, depends upon whether the pole or zero is nearest the origin. Alternatively, it is based on whether the resistance increases or decreases with frequency, again easily observed at $s = 0$ and $s = \infty$. The networks of Fig. 2.8 have the advantage of giving correctly, *and in mutual agreement,* the behavior of the actual system at $s = 0$ for $R_c = 0$ or ∞. This was a considerable problem in connection with Fig. 2.5, as will be recalled.

It is important to understand in this respect that Fig. 2.8 comes directly from *field expressions* calculated correctly to first order in s; i.e., from power series expansions of **E** and **H** (or V and I) about $s = 0$, carried to the first two terms only. This is what we mean *precisely* by a "quasi-static approximation"; thus, Fig. 2.8 gives such an approxima-

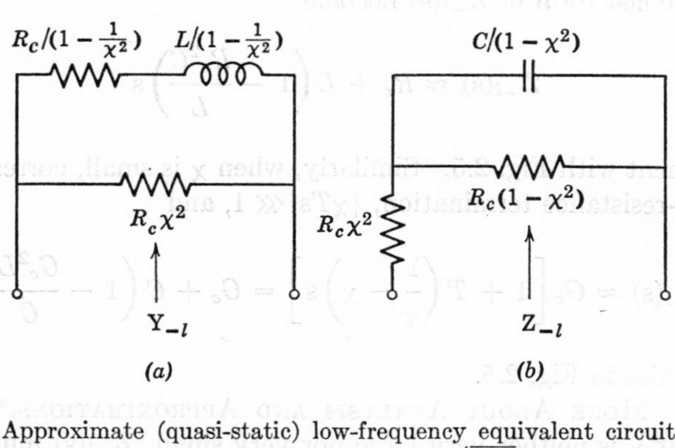

Fig. 2.8. Approximate (quasi-static) low-frequency equivalent circuits for Figs. 2.1*c* or 2.7. (*a*) Appealing when $\chi > 1$ (or $R_c < \sqrt{L/C}$), low-resistance termination; (*b*) appealing when $0 \leq \chi < 1$ (or $R_c > \sqrt{L/C}$), high-resistance termination.

tion to the actual device impedance. Figure 2.5, however, comes essentially from a power series expansion of the *impedance* itself, carried to the first order in s, and we have just seen that this kind of approximation is sometimes more restrictive than the quasi-static one. It turns out that the two are equivalent for nondissipative structures (Figs. 2.1*a* or 2.1*b* in our case) but somewhat different for dissipative ones (Fig. 2.1*c* here), the quasi-static result being the more valid in the presence of losses.

Nevertheless, the quasi-static approximation for Fig. 2.7 still gives a somewhat false picture of the true situation by giving a pole *and* a zero on the negative-real axis of the s-plane. Indeed, it suggests that the reason why $Z_{-l}(s) = R_c =$ constant when $\chi = 1 (R_c = b\eta/a = \sqrt{L/C})$ is that this zero and pole merge and cancel. An examination of the exact expression for $Z_{-l}(s)$ will clear up this matter.

Referring back to the exact $Z_{-l}(s)$, then, we see that it does become constant $(= R_c = \sqrt{L/C})$ when $\chi = 1$, and this occurs because numerator and denominator become identical. In one sense, the idea of zeros and poles merging and canceling is supported by this observation. But let us examine the location of these zeros and poles, starting with the zeros.

$$\cosh s_0T = -\chi \sinh s_0T$$

$$e^{s_0T}(1 + \chi) = e^{-s_0T}(\chi - 1)$$

$$e^{2s_0T} = \frac{\chi - 1}{\chi + 1}\begin{cases} > 0 & \text{for } \chi > 1 \\ < 0 & \text{for } 0 \leq \chi < 1 \end{cases}$$

Let $s_0 = \Omega_0 + j\omega_0$, so that

$$e^{j2\omega_0T}e^{2\Omega_0T} = \frac{\chi - 1}{\chi + 1}\begin{cases} 0 < \dfrac{\chi - 1}{\chi + 1} < 1 & \text{for } \chi > 1 \\ -1 \leqslant \dfrac{\chi - 1}{\chi + 1} < 0 & \text{for } 0 \leq \chi < 1 \end{cases}$$

Thus when $\chi > 1$,

$$j2\omega_0T = \pm j2m\pi \qquad m = 0, 1, 2, \cdots, \qquad \chi > 1$$

or

$$\omega_0 = \pm\frac{m\pi}{T} \qquad m = 0, 1, 2, \cdots, \qquad \chi > 1$$

and

$$\Omega_0 = \frac{1}{2T}\ln\left(\frac{\chi - 1}{\chi + 1}\right) = -\frac{1}{2T}\ln\left(\frac{\chi + 1}{\chi - 1}\right)$$

$$\equiv -\frac{\varphi_e}{T} < 0 \qquad \text{for } \chi > 1$$

But when $0 \leq \chi < 1$,

$$j2\omega_0 T = \pm j(2n+1)\pi \qquad n = 0, 1, 2, \cdots, \qquad 0 \leq \chi < 1$$

or

$$\omega_0 = \pm(2n+1)\frac{\pi}{2T} \qquad n = 0, 1, 2, \cdots, \qquad 0 \leq \chi < 1$$

and

$$\Omega_0 = \frac{1}{2T}\ln\left(\frac{1-\chi}{1+\chi}\right) = -\frac{1}{2T}\ln\left(\frac{1+\chi}{1-\chi}\right)$$

$$\equiv -\frac{\varphi_o}{T} < 0 \qquad \text{for } 0 \leq \chi < 1$$

The groups of zeros for $\chi > 1$ and $0 \leq \chi < 1$ are shown on Figs. 2.9*a* and 2.9*b* respectively. Note that they lie on a line parallel to the imaginary axis, and are spaced by $\Delta\omega_0 = \pi/T$ along this line regardless of the value of χ. Note also that the lossless cases occur for either $\chi \to \infty$ (which reduces Fig. 2.1*c* to Fig. 2.1*b*) or $\chi \rightharpoonup 0$ (which reduces Fig. 2.1*c* to Fig. 2.1*a*). The effect of either $\chi \rightharpoondown 1$ or $\chi \rightharpoonup 1$ (indeed of $\chi \to 1$) is to increase the size of the damping part, Ω_0, by pushing the line of zeros off toward $\Omega = -\infty$.

The pole locations are easy to find now, inasmuch as the exact expression for $Z_{-l}(s)$ shows that the denominator differs from the numerator only by placing $1/\chi$ for χ. Thus poles for $\chi > 1$ in Fig. 2.9*a* lie at the same points as zeros did for $\chi < 1$ in Fig. 2.9*b*, and vice-versa. It is easy to see that for any value of χ, zeros and poles lie on the same

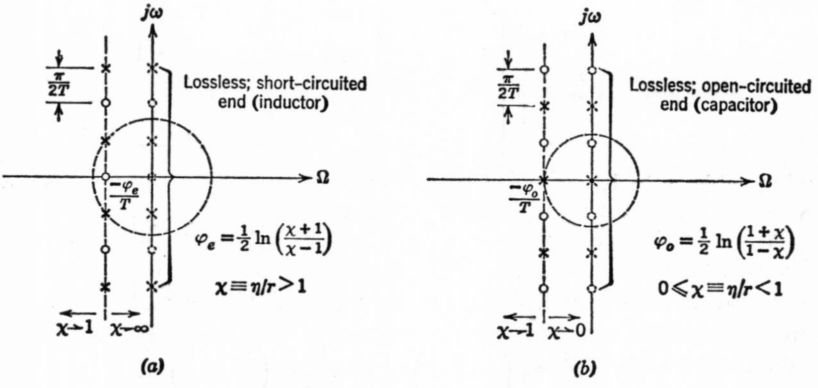

Fig. 2.9. Complex s-plane for $Z_{-l}(s)$ in Fig. 2.7. (*a*) Low-resistance termination; (*b*) high-resistance termination. Dashed circles show radii of convergence for Taylor series expansions of $Z_{-l}(s)$ about $s = 0$.

vertical line since φ_e and φ_o differ by $\chi \rightarrow 1/\chi$, and the same change is required again to pass from zero to pole locations for a given value of χ.

It is clear from Fig. 2.9 that either a zero or a pole lies on the negative real axis, depending upon the value of χ, but never do we have both at the same time. Moreover, zeros and poles are always spaced by $\pi/2T = \Delta\omega$, regardless of the value of resistance R_c (compare this situation in the lossless case $\chi = \infty$ with the uneven spacing predicted by Fig. 2.4*d* and Eq. 2.20, or with the value of the spacing in Fig. 2.6 and Eq. 2.18). The zeros and poles never merge at all, but move off together to $\Omega = -\infty$ in the s-plane as $\chi \rightarrow 1$, maintaining all the while a fixed spacing $\Delta\omega = \pi/2T$ between successive zeros and poles.

The dashed circles in Fig. 2.9 show the problem of trying to make power series expansions of the *impedance* about $s = 0$ in the two cases $0 < \chi < 1$ and $\chi > 1$. These sketches really validate exactly the ideas discussed from the approximate analysis and physical point of view in connection with Fig. 2.6. We see now why the results of Figs. 2.5*a* and 2.5*b* are correct, and mutually agree, when $\chi \rightarrow 1$, as the singularity nearest to $s = 0$ moves off to $-\infty$ in the exact solution.

A glance at the field expressions, and the voltage and current expressions, shows that when $\chi = 1$ (i.e., $R_c = (b\eta)/a$):

$$\chi = 1 \begin{cases} \text{(a)} \quad E_x = \dfrac{I_0 R_c}{b} e^{-j\beta_0 z} = H_y(z = 0)\eta e^{-j\beta_0 z} \\[2ex] \text{(b)} \quad H_y = \dfrac{I_0}{a} e^{-j\beta_0 z} = H_y(z = 0) e^{-j\beta_0 z} \end{cases}$$

and

$$\chi = 1 \begin{cases} \text{(a)} \quad V = I_0 R_c e^{-j\beta_0 z} = I_0 \left(\dfrac{\eta b}{a}\right) e^{-j\beta_0 z} = I_0 \sqrt{\dfrac{L}{C}}\, e^{-j\beta_0 z} \\[2ex] \text{(b)} \quad I = I_0 e^{-j\beta_0 z} \end{cases}$$

It is the traveling-wave nature of these results, and the standing-wave character of Eqs. 2.21, that now drive us to develop further a wave point of view for the behavior of systems like those in Fig. 2.1.

We shall therefore proceed at once to a different kind of exact field analysis of such structures. This analysis will eventually lead us not only to a better understanding of the approximations leading to lumped circuits but also into the general problems of distributed circuits, wave motion, oscillation, and energy transmission.

2.3 Another Exact Field Solution and the Concept of a "Distributed Circuit"

The situations shown in Fig. 2.1 have in common the need for a field solution to Maxwell's equations inside a region bounded on top and bottom by parallel, perfectly conducting plates. They differ from each other in the nature of one end wall. Yet this structural difference alone appears to create violent differences in their a-c electrical behavior. In connection with obtaining a simple a-c field solution that will reflect only this important difference between the structures, and that will not be complicated by extraneous difficulties, we can automatically eliminate the "fringing" problem at the sides of the plates by assuming that they extend to infinity in the $\pm y$ directions. We can then merely focus our attention on a section of width a in each case. Indeed, this is what we really meant by "neglecting fringing" in our previous solutions. More precisely, we imply that our solution will be carried out for $-\infty < y < \infty$; but it must have the property that any section of width a, chosen arbitrarily from the range $-\infty < y < \infty$ in each case, will behave in the same way. In fact, we must therefore ask for a field solution in which the vectors $\boldsymbol{E}$ and $\boldsymbol{H}$ will be *independent of y*. This requirement is, in effect, a *boundary condition* we are imposing arbitrarily upon the solution. Such boundary conditions are always essential in the solution of partial differential equations, like Maxwell's equations. With different boundary conditions, we would, of course, generally find different solutions. The present conditions are expressed mathematically by requiring

$$\frac{\partial \boldsymbol{E}}{\partial y} \equiv \frac{\partial \boldsymbol{H}}{\partial y} \equiv 0 \tag{2.25}$$

for all values of x, y, and z inside the structure. Referring to Fig. 2.2, we shall now consider the "inside" of the structure to be defined by

$$-\infty < y < +\infty$$
$$0 \leq x \leq b$$
$$-l \leq z \leq 0$$

but we shall later restrict our interest to any section of width a in the y direction.

Now we know from Eq. 1.25 that the electric field *at* a perfect conductor must be perpendicular to it. This fact constitutes a required boundary condition for the electric field at the upper and lower plates in our problem. It does not, however, require that the electric field everywhere between the plates also be perpendicular to them. Nevertheless, we would like a solution that resembles as much as possible our d-c solutions, inasmuch as we wish it to reduce properly to these when the time variation vanishes. Hence we shall require arbitrarily that our solution satisfy the more restrictive boundary condition

$$E_y \equiv E_z \equiv 0$$

or

$$\boldsymbol{E} = \boldsymbol{a}_x E_x \tag{2.26}$$

for all values of x, y, and z inside the structure. The necessary boundary conditions at the metal plates will then be satisfied automatically.

Since we have previously found solutions that meet the conditions of Eqs. 2.25 and 2.26 for three very different end-wall conditions at $z = 0$, let us see how far we can go with the present solution on the basis of Eqs. 2.25 and 2.26 before we need to specify precisely the nature of this end wall.

In the charge-free and source-free region of space between the plates, Maxwell's equations (Eqs. 1.15 and 1.20) become

$$\text{(a)} \qquad \nabla \times \boldsymbol{E} = -\mu \frac{\partial \boldsymbol{H}}{\partial t}$$

$$\text{(b)} \qquad \nabla \times \boldsymbol{H} = \epsilon \frac{\partial \boldsymbol{E}}{\partial t} + \sigma \boldsymbol{E} \tag{2.27}$$

$$\text{(c)} \qquad \nabla \cdot \boldsymbol{D} = 0 = \nabla \cdot \boldsymbol{E} = \nabla \cdot \boldsymbol{B} = \nabla \cdot \boldsymbol{H}$$

where, for the sake of generality, we have added the possibility of losses in the dielectric by including the conductivity σ. In rectangular coordinates, using the conditions of Eqs. 2.25 and 2.26, Eqs. 2.27*a* and 2.27*b* become

$$\text{(a)} \begin{cases} \text{(i)} & \dfrac{\partial E_x}{\partial z} = -\mu \dfrac{\partial H_y}{\partial t} \\[2ex] \text{(ii)} & \dfrac{\partial H_x}{\partial t} = 0 \\[2ex] \text{(iii)} & \dfrac{\partial H_z}{\partial t} = 0 \end{cases}$$

$$\text{(b)} \begin{cases} \text{(i)} & -\dfrac{\partial H_y}{\partial z} = \epsilon \dfrac{\partial E_x}{\partial t} + \sigma E_x \\[2ex] \text{(ii)} & \dfrac{\partial H_x}{\partial z} - \dfrac{\partial H_z}{\partial x} = 0 \\[2ex] \text{(iii)} & \dfrac{\partial H_y}{\partial x} = 0 \end{cases} \tag{2.28}$$

But we are particularly interested in a time-varying solution; so Eqs. 2.28*a*(ii) and 2.28*a*(iii) are of no interest. They tell us what we already know: namely, that we could always put a separate magnet near the plates and have a *static* magnetic field between them, as long as the arbitrary condition in Eq. 2.25 is met. For time-varying fields, therefore, we may take $H_x \equiv H_z \equiv 0$. Incidentally, this decision eliminates Eq. 2.28*b*(ii), which also relates to the same possibility of a static magnetic field in which we have no interest here.

From Eq. 2.28*b*(iii), however, we do learn that H_y (which is now the only component of $\boldsymbol{H}$ left) does not depend upon x. As a matter of fact, the same is true of E_x (which is the only component of $\boldsymbol{E}$) because, according to Eq. 2.27*c*,

$$\frac{\partial E_x}{\partial x} = 0 \tag{2.29}$$

Thus we find again that a time-varying field solution is possible in which both $\boldsymbol{E}$ and $\boldsymbol{H}$ have the same directions as they would have for direct current, and like the d-c solutions they do not vary with x. Of course they do not vary with y either, but we imposed this as a boundary condition. The important fact associated with the time variation follows from Eqs. 2.28*a*(i) and 2.28*b*(ii); namely, both E_x and H_y must depend upon z if they vary with time. It is this result which we failed to include in our first "static" discussion of Fig. 2.1, but which showed up in the "successive approximation" and power series solutions. The nature of this variation is contained in Eqs. 2.28*a*(i) and 2.28*b*(ii), repeated below for reference:

$$\text{(a)} \qquad \frac{\partial E_x}{\partial z} = -\mu \frac{\partial H_y}{\partial t}$$

$$\text{(b)} \qquad \frac{\partial H_y}{\partial z} = -\epsilon \frac{\partial E_x}{\partial t} - \sigma E_x \tag{2.30}$$

To solve these equations, we shall have to consider further boundary conditions relating to both z and t. Among other things, the nature of the end wall at $z = 0$ will be important in this connection. Since actual solution of Eqs. 2.30 (or their equivalent) will occupy a great deal of our time later on, we shall not solve them here; but it is worth while to examine them carefully, nevertheless. A glance at Fig. 2.2 and Eq. 2.30*a* shows that E_x must differ at two different values of z, because a time-varying magnetic field links the intervening space. This is just Faraday's law. Similarly, Eq. 2.30*b* shows that H_y must differ at two different values of z, because a conduction current, σE_x, and a displacement current, $\epsilon\, \partial E_x/\partial t$, pass through the intervening space. This is just Ampere's law, as extended by Maxwell to include displacement current. In our original "static" analysis of Fig. 2.1 we simply forgot that a time-varying magnetic field sets up an electric field that cannot be accounted for from electro*statics* alone. We also forgot that a time-varying electric field constitutes a displacement current, and that such a current sets up a magnetic field that cannot be accounted for from magneto*statics* alone. Of course our successive approximation and power series solutions specifically recognized these important facts.

It is now profitable to view Eqs. 2.30 from a different angle. No one likes to be forced to think about electric and magnetic fields in connection with a-c circuit problems. In spite of our difficulties with z variation, is there some way of getting back to our old friends "voltage" and "current"? The *general* answer to this question, in an *arbitrary* electromagnetic field, is NO. Happily, however, there are many cases in which it is possible. A surprisingly large number of practical situations do lend themselves to this simplification, as is clear at least from the plain fact that lumped-circuit theory often does actually work! In many other practical cases, though, lumped-circuit theory does not work; but it is still possible to *define* a "voltage" and a "current" sensibly. Even when this is possible, however, the definitions may look very unfamiliar in some cases. In others, an *almost* conventional definition may be made. For example, in our problem of the parallel planes (Fig. 2.1), let us define voltage V in an almost conventional manner, as the line integral of the electric field from the top plate to the bottom plate, *along any line that remains in a plane of fixed z.* Then V becomes a function of z and t:

$$V(z, t) = \int_0^b E_x \, dx \,\Big|_z = bE_x \tag{2.31}$$

since, by Eq. 2.29, E_x is independent of x. Similarly, let us define current I as the total current in the $+z$ direction along a section of width a of the top plate, *across a line in any plane of fixed z.* Thus I is also a function of z and t. Since the surface current density on a perfect conductor is equal to the component of the magnetic field parallel to the conductor (Eq. 1.26), we have

$$I(z, t) = aH_y \tag{2.32}$$

Equations 2.31 and 2.32 are simply more formal statements of what we assumed in Sec. 2.2.2, Eqs. 2.13 and 2.17.

If we now use Eqs. 2.31 and 2.32 in Eq. 2.30, we find

$$\text{(a)} \qquad \frac{\partial V}{\partial z} = -\left(\frac{\mu b}{a}\right)\frac{\partial I}{\partial t}$$

$$\text{(b)} \qquad \frac{\partial I}{\partial z} = -\left(\frac{\epsilon a}{b}\right)\frac{\partial V}{\partial t} - \left(\frac{\sigma a}{b}\right)V \qquad (2.33)$$

Comparison of Eq. 2.33 with Eq. 2.6 shows that $(\mu b)/a$ is an *inductance per unit length along z*. In other words, at a fixed time, the voltage of Eq. 2.33 *decreases* by dV in a distance dz because of a time rate of *increase* of current through a series inductance $[(\mu b)/a]\,dz$. Moreover, this series inductance per unit length is (very remarkably) equal to the "d-c" inductance per unit length of Eq. 2.4! Similarly, comparison of Eq. 2.33 with Eqs. 2.5 and 2.2 shows that $(\epsilon a)/b$ is a *shunt capacitance per unit length along z*, accounting for the fact that the conduction current I on the plates must decrease with distance because of the displacement current flowing from the top to the bottom plate. In the same way, it is easy to see that $(\sigma a)/b$ is a *shunt conductance per unit length*, accounting for the conduction of current from the top to the bottom plate through the dissipative dielectric. From this point of view, the structures of Fig. 2.1 can be looked upon, for the purpose of a-c analysis, as *distributed circuits*. Instead of separate "lumped" L and C (and G, if the dielectric has losses), we imagine a smooth distribution along z of inductance, capacitance, and conductance simultaneously, which we describe by *per-unit-length values on a differential-circuit basis*. The amazing fact, however, is that these values can apparently be calculated from *static* considerations alone! We shall have much more to say about the definitions of voltage, current, and distributed-circuit parameters as our work progresses.[1]

2.4 Transmission Line as a Distributed Circuit

2.4.1 Parameters of the Line

As we have seen from the preceding developments, no physical structure can be treated as a lumped-circuit element over an arbitrarily wide range of frequencies. There are some, however, which do admit

[1] For more advanced readers, and when the pressure for steady-state transmission-line theory is not raised unbearably by a concurrent laboratory program, it is possible and quite desirable to take up at this point the two-wire-line sections of Chapter 9.

of distributed-circuit interpretations with much greater validity. In these we think of the ordinary lumped-circuit parameters as being "distributed" simultaneously along one space dimension, so much per unit length on a differential basis. In the idealized, parallel, perfectly conducting plane configuration, we found that (subject to certain important boundary conditions) L, C, and G per unit length could be calculated (or measured) on a static basis and would then suffice for an *exact* analysis at all frequencies. There are a great many other situations of similar character to which distributed-circuit notions may be applied either exactly or approximately. Some of these we shall examine in more detail from the field point of view in Chapters 7–9. The simplest general type (to be taken up thoroughly in Chapter 9) is exemplified by two parallel wires embedded in a homogeneous medium, and extending indefinitely along the $\pm z$ directions. The wires need not be of the same diameter; indeed, the coaxial arrangement of one solid wire inside a larger hollow cylinder is a common practical form of the system. The important features are that the structure has (at least) two parallel conductors embedded in a uniform medium, and that every transverse cross section of the system is identical. Used in almost every long-distance power or signal transmission system, such configurations are called *transmission* lines. Short lengths of transmission line are also used as circuit elements at high frequencies, when conventional elements become unmanageable.

From a naive point of view, we might expect transmission lines to be characterized not only by a series L, shunt C, and shunt G per unit length, as we found in connection with our simple parallel-plane example, but also by a series R per unit length representing the resistance of the actual metal conductors (which are never really perfectly conducting). From the field point of view, however, the presence of such a loss in the conductors complicates considerably any *exact* field solution of the transmission-line problem. Nevertheless, we shall indicate in Chapters 8 and 9 how the addition of a series resistance R per unit length to the distributed-circuit model of the line may account *approximately* for the effect of conductor losses. For the sake of generality, we shall include it in our present model without further justification at this point.

Therefore, the basic parameters of our two-wire line become L, R, C, and G, measured respectively in h/m, ohms/m, f/m, and mhos/m. We shall for the time being treat them as constants, independent of frequency, so that, in principle, they may be determined either from calculations or from measurements on a static basis. As a practical matter, this assumption is not always valid—for two reasons: 1. The occurrence of skin effect in imperfect conductors means that the current density in them redistributes itself over their cross sections as the frequency is changed. 2. Some of the basic parameters σ, ϵ, and μ of most solid or liquid dielectrics are functions of frequency. Consequently, in most practical transmission lines, one or more of L, R, C, and G will be frequency-dependent to some extent. Just how this fact is to be reconciled with our simpler assumptions will become clear only when we discuss the relationships between solutions in the time domain and those in the frequency domain.

2.4.2 A Distributed-Circuit Derivation of the Transmission-Line Equations

Viewed as a distributed circuit, the physical two-wire transmission line with constant parameters R, L, G, and C, shown schematically in Fig. 2.10*a*, may be represented in the form of Fig. 2.10*b*. The spatial coordinate is z and the time coordinate is t. The voltage $V(z, t)$ is chosen positive when conductor 1 is at a higher potential than conductor 2. The conductor current $I(z, t)$ is defined as positive when it flows in the positive z direction in conductor 1. Since we assume that there are only two conductors, a current $I(z, t)$ in conductor 1 must be accompanied by an equal and opposite current flowing in conductor 2.

The voltage $V(z, t)$ drops by an incremental amount in the incremental length dz. This voltage drop is caused by the current $I(z, t)$ flowing through the incremental resistance Rdz and the incremental

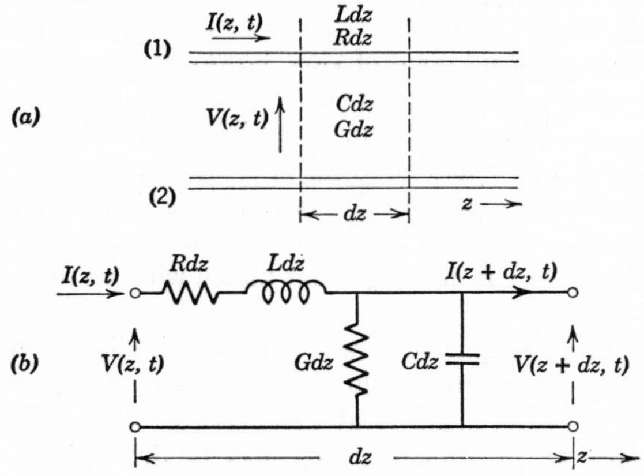

Fig. 2.10. A transmission-line increment.

inductance Ldz presented to the current by the section of transmission line of length dz.

$$dV(z, t) = -RdzI(z, t) - Ldz \frac{\partial I(z, t)}{\partial t}$$

Henceforth the z and t dependence will be assumed tacitly, and the parenthesis (z, t) will be omitted for brevity.

$$dV = -RdzI - Ldz \frac{\partial I}{\partial t}$$

Dividing by dz gives us

$$\frac{\partial V}{\partial z} = -RI - L \frac{\partial I}{\partial t} \tag{2.34a}$$

Equation 2.34a is the result of the application of Kirchhoff's voltage law, Faraday's induction law, and Ohm's law to the distributed-parameter problem. Compare with Eq. 2.33a.

The presence of the incremental conductance Gdz and the incremental capacitance Cdz within the section dz of the transmission line causes a decrease in the conductor current $I(z, t)$, since an incremental amount of current is shunted through the conductance Gdz and capacitance Cdz. Thus

$$dI = -GdzV - Cdz \frac{\partial V}{\partial t}$$

to the first order in dz. In other words:

$$\frac{\partial I}{\partial z} = -GV - C \frac{\partial V}{\partial t} \tag{2.34b}$$

Equation 2.34b expresses the law of continuity of total current. Compare with Eq. 2.33b.

Since the foregoing equations are consistent with both circuit and field theory, they must be consistent with the law of conservation of energy. We can check this as follows. The magnetic energy stored *per unit length* is $\frac{1}{2}LI^2 = u_m$, and the electric energy stored per unit length is $\frac{1}{2}CV^2 = u_e$. The total power dissipated per unit length is $RI^2 + GV^2 = p_d$. Thus

$$p_d + \frac{\partial}{\partial t}(u_m + u_e) = \left(RI + L\frac{\partial I}{\partial t}\right)I + \left(GV + C\frac{\partial V}{\partial t}\right)V$$

$$= -\left(I\frac{\partial V}{\partial z} + V\frac{\partial I}{\partial z}\right) = -\frac{\partial P}{\partial z} \tag{2.35}$$

where we have used Eqs. 2.34a and 2.34b, and where $P = VI$ is the power flowing across any plane of constant z. Equation 2.35 (a form of Poynting's theorem) expressed in words reads:

The spatial rate of decrease of the power flow along the line is equal to the power dissipated per unit length, plus the time rate of increase of the stored linear energy density, $u_m + u_e$.

Compare this result with the corresponding ones obtained for circuits and fields in Chapter 1.

2.4.3 Transmission-Line Equations in the Frequency Domain

The voltage $V(z, t)$ and current $I(z, t)$ are scalar functions of position and time. Therefore, as discussed in Sec. 1.2.2.1, we may express them for the case of sinusoidal time variation in the form

$$V(z, t) = \text{Re}\,[\mathrm{V}(z)e^{j\omega t}]$$

$$I(z, t) = \text{Re}\,[\mathrm{I}(z)e^{j\omega t}]$$

where now $\mathrm{V}(z)$ and $\mathrm{I}(z)$ are the complex amplitudes of the sinusoids as functions of position z on the line. They give the magnitudes and phases of the sinusoids at each value of z. Since Eqs. 2.34a and 2.34b (like all the circuit and field equations discussed in Chapter 1) are linear, and since R, L, C, and G are real constants, the differential equations in z which must be satisfied by $\mathrm{V}(z)$ and $\mathrm{I}(z)$ can be found by substituting $\mathrm{V}(z)e^{j\omega t}$ for $V(z, t)$, and $\mathrm{I}(z)e^{j\omega t}$ for $I(z, t)$ in the differential Eqs. 2.34a and 2.34b. After cancellation of the factor $e^{j\omega t}$, we find

$$\text{(a)} \qquad \frac{d\mathrm{V}(z)}{dz} = -(R + j\omega L)\,\mathrm{I}(z)$$

$$\text{(b)} \qquad \frac{d\mathrm{I}(z)}{dz} = -(G + j\omega C)\,\mathrm{V}(z) \qquad (2.36)$$

It is interesting to note that, although we have derived Eqs. 2.36 from Eqs. 2.34, they could have been written directly from a frequency-domain schematic of the incremental section dz (Fig. 2.11a). The complex voltage drop $-d\mathrm{V}$ through the length dz is $\mathrm{I}(R + j\omega L)\,dz$, leading directly to Eq. 2.36a. The decrease in conductor current $-d\mathrm{I}$ is $\mathrm{V}(G + j\omega C)\,dz$, which gives Eq. 2.36$b$ at once. From this point of view it becomes obvious that Eqs. 2.36 are really more general than Eqs. 2.34; for Eqs. 2.36 would not be altered in form at all if the line parameters were functions of frequency $R(\omega)$, $L(\omega)$, $C(\omega)$, and $G(\omega)$,

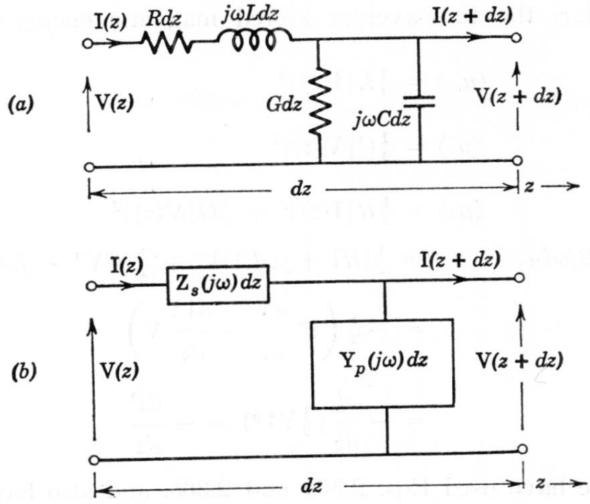

Fig. 2.11. Frequency-domain representations of transmission-line increments.

while Eqs. 2.34 would no longer be valid under these conditions. The added generality can be emphasized by observing that the series combination $R(\omega) + j\omega L(\omega)$ can be written $R(\omega) + jX(\omega)$, which is simply an arbitrary impedance $Z_s(j\omega)$ (s for series), and $G(\omega) + j\omega C(\omega) = G(\omega) + jB(\omega)$ is simply an arbitrary admittance $Y_p(j\omega)$ (p for parallel). The most general system to which the differential equations apply is therefore represented in the frequency domain by Fig. 2.11b. The equations for it read:

$$\text{(a)} \qquad \frac{dV(z)}{dz} = -Z_s(j\omega)\, I(z)$$

$$\text{(b)} \qquad \frac{dI(z)}{dz} = -Y_p(j\omega)\, V(z) \qquad (2.37)$$

where $Z_s(j\omega)$ and $Y_p(j\omega)$ are arbitrary impedance and admittance functions respectively. We shall not immediately pursue these equations in detail beyond the case in which R, L, C, and G are constants (represented by Eqs. 2.36), but we shall have occasion to consider the frequency dependence of the parameters again later.

Returning to Eqs. 2.36, therefore, we observe that they must be consistent with the conservation of complex power (the complex Poynting theorem), since they are consistent with both circuit and field relations.

In particular, the time-average stored magnetic energy[1] per unit length is

$$\langle u_m \rangle = \tfrac{1}{4}L|\mathrm{I}(z)|^2$$

Similarly,

$$\langle u_e \rangle = \tfrac{1}{4}C|\mathrm{V}(z)|^2$$

and

$$\langle p_d \rangle = \tfrac{1}{2}R|\mathrm{I}(z)|^2 + \tfrac{1}{2}G|\mathrm{V}(z)|^2$$

$$\begin{aligned}\langle p_d \rangle + 2j\omega\langle u_m - u_e \rangle &= \tfrac{1}{2}(R\mathrm{I} + j\omega L\mathrm{I})\mathrm{I}^* + \tfrac{1}{2}(G\mathrm{V}^* - j\omega C\mathrm{V}^*)\mathrm{V} \\ &= -\tfrac{1}{2}\left(\mathrm{I}^*\frac{d\mathrm{V}}{dz} + \frac{d\mathrm{I}^*}{dz}\mathrm{V}\right) \\ &= -\frac{d}{dz}(\tfrac{1}{2}\mathrm{VI}^*) = -\frac{d\mathrm{P}}{dz}\end{aligned} \tag{2.38}$$

in which we have used Eqs. 2.36*a* and 2.36*b*, and also have defined $\mathrm{P} \equiv \frac{1}{2}\mathrm{VI}^*$. If we put, as usual,

$$\mathrm{P} = \langle P \rangle + jQ$$

and integrate Eq. 2.38 from some position $z = z_1$ to another $z = z_2$, we find from its real and imaginary parts

$$\begin{aligned}&\text{(a)} \quad \langle P(z_1)\rangle - \langle P(z_2)\rangle = \int_{z_1}^{z_2} \langle p_d \rangle \, dz \\ &\text{(b)} \quad Q(z_1) - Q(z_2) = 2\omega \int_{z_1}^{z_2} \langle u_m - u_e \rangle \, dz = 2\omega\langle W_m - W_e \rangle\end{aligned} \tag{2.39}$$

where

$$\langle W_m \rangle = \int_{z_1}^{z_2} \langle u_m \rangle \, dz$$

$$\langle W_e \rangle = \int_{z_1}^{z_2} \langle u_e \rangle \, dz$$

Equation 2.39*a* shows that the time-average power transmitted down the line decreases with distance on account of the losses in the series and shunt resistance. Equation 2.39*b* shows that the reactive power transmitted down the line changes with distance according to the difference between time-average magnetic and electric energies stored in the intervening length of line. These results should be compared with those found in Chapter 1 for both circuits and fields.

[1] If L and/or C are functions of frequency, the stored-energy interpretations are invalid. See also footnote to Eq. 1.58.

PROBLEMS

Problem 2.1. Consider a lossless transmission line of length l and parameters L and C per unit length. It is open-circuited at the load end ($z = 0$), and we are interested in the sinusoidal steady state for which complex V(z) and I(z) obey:

$$\left.\begin{aligned} \frac{d\mathrm{V}}{dz} &= -j\omega L\mathrm{I} \\ \frac{d\mathrm{I}}{dz} &= -j\omega C\mathrm{V} \end{aligned}\right\} \qquad z < 0$$

(a) Using successive approximation techniques, solve the above equations for V and I correct to terms in $(\beta z)^3$, with $\beta = \omega\sqrt{LC}$. *Do not find the exact solution and develop the power-series expansion from this solution.* Check your results by using the direct method of power series expansions in ω of V and I. (b) Give a dimensioned sketch of the approximate $|\mathrm{V}|$ and $|\mathrm{I}|$ distributions found in (a). (c) From the results of (a) compute the approximate input impedance (at $z = -l$) of the line and give a dimensioned plot of the poles and zeros in the complex s-plane. (d) Now compare the graphical results of (b) and (c) with those obtained from the *exact* solution of the line problem. Then show what analytical approximations must be made in the exact solutions to reduce them to the same analytical results found in (a) and (c). (e) Find the length l (in wave lengths) for which the approximate admittance, correct to first power in ω, differs from the exact value by: (i) 1%; and (ii) 10%. (f) Repeat all the above parts for the case of a short circuit at the load end ($z = 0$).

Problem 2.2. When working on a power line, normal safety procedures call for placing a short circuit across it and grounding the short circuit. If the breaker is closed: (a) Will the man in Fig. 2.12 be electrocuted if he does not ground the

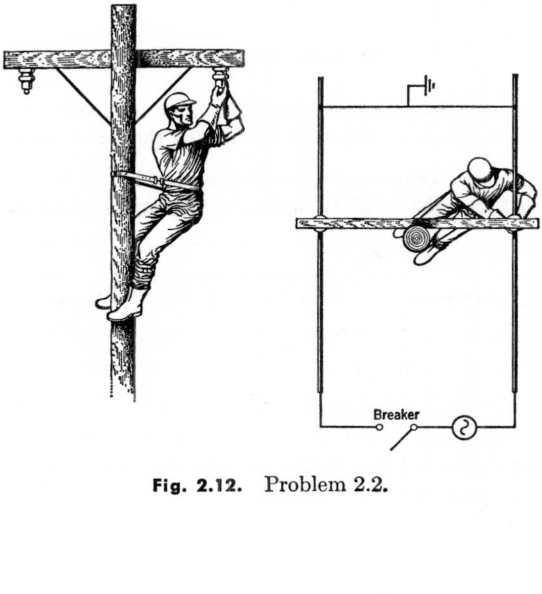

Fig. 2.12. Problem 2.2.

short-circuiting bar? (b) How far can he be away from the short-circuiting bar without being in danger?

Problem 2.3. A "one-turn" inductor (Fig. 2.13) is made from perfectly conducting parallel plates, short-circuited at $z = 0$. It is filled with a linear magnetic medium having parameters σ, ϵ_0, and μ, so that eddy-current losses will occur. Fringing is to be neglected. (a) Regarding the structure as a transmission line, find the per-unit-length constants G, L, and C in terms of the dimensions and medium parameters. (b) Using the transmission-line equations for the system, find the *quasi-static* current and voltage distributions on it. Let I_0 be the current in the short circuit. (c) Determine the quasi-static input admittance from the results of (b) above. Express your answer as a function of complex frequency, s, and the parameters $L' = lL$, $C' = lC$ and $G' = lG$. Show the zeros and poles of this admittance in the s-plane and exhibit an equivalent circuit for the structure. (d) Repeat for the case of an open circuit at $z = 0$.

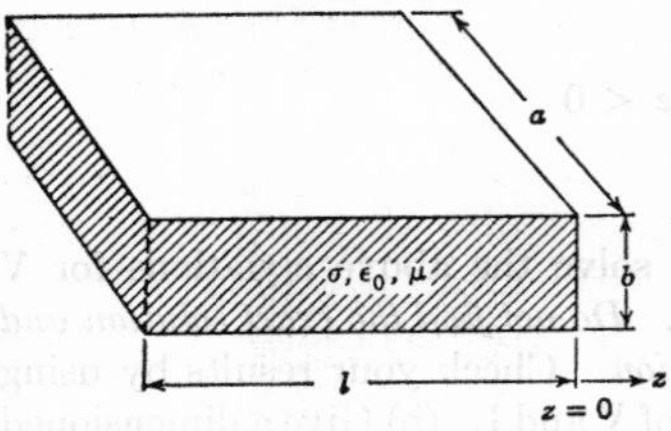

Fig. 2.13. Problem 2.3.

Problem 2.4. A tubular "paper" capacitor (Fig. 2.14) is made by rolling up two strips of aluminum foil with paper in between. Leads are connected to the ends of the foil as shown (only the foil is illustrated). (a) What is the order of magnitude of the first resonant frequency? Is it a zero or a pole of the input impedance?

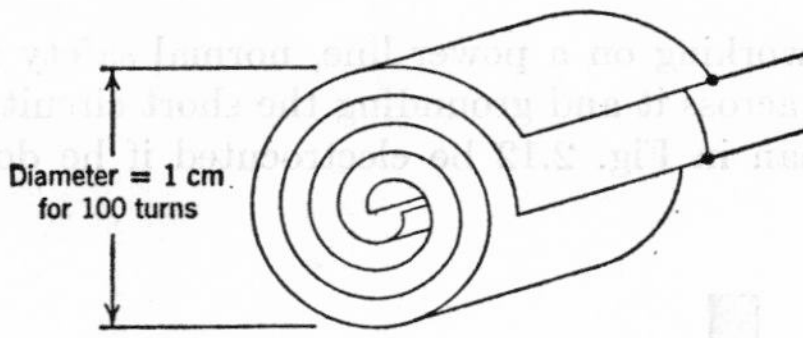

Fig. 2.14. Problem 2.4.

(b) How could you increase the frequency of this first resonance by redesigning the component using the same materials and achieving the same capacitance?

Problem 2.5. A perfectly conducting coaxial cable of length l is terminated at the end $z = 0$ by a thin resistive sheet of R ohms/square (Fig. 2.15). A sinusoidal

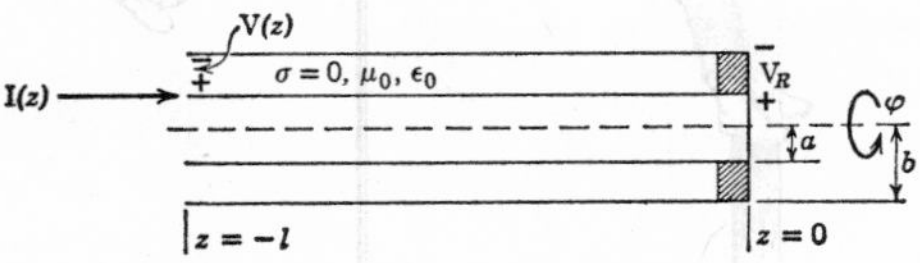

Fig. 2.15. Problem 2.5.

source is applied at $z = -l$ such that the amplitude of the load voltage at $z = 0$ is a constant V_R, independent of frequency. The complex field vectors may be written as power series in the variable ω:

$$\mathbf{E} = \sum_{i=0}^{\infty} \omega^i \mathbf{e}_i, \qquad \mathbf{H} = \sum_{i=0}^{\infty} \omega^i \mathbf{h}_i$$

(a) Find the series for **E** and **H** up to the third power of ω, assuming $E_z = H_z \equiv 0$ and $\partial/\partial\varphi \equiv 0$. (b) Define, and then find series for the complex voltage and current $V(z)$ and $I(z)$ to the same power of ω as in (a). (c) Put the result of (b) in a form which clearly demonstrates that $V(z)$ and $I(z)$ are functions only of the dimensionless variable βz, where $\beta = \omega\sqrt{\epsilon\mu}$. (d) By recognition of familiar series, put the results of (b) in closed form, and verify that they agree essentially with the theory of the parallel-plate case treated in the text. (e) Find an expression for $Z(z)$, the impedance at any point z looking in the positive z-direction. (f) What value of R ohms/square will terminate the line so that the voltage magnitude $|V(z)|$ is independent of z? What are the impedance values at $z = 0$ and $z = -l$ for this particular value of R?

Problem 2.6. A circular parallel-plate air capacitor has perfectly conducting plates. It is fed by terminal wires arranged as shown in Fig. 2.16*a*. (a) Find in terms of

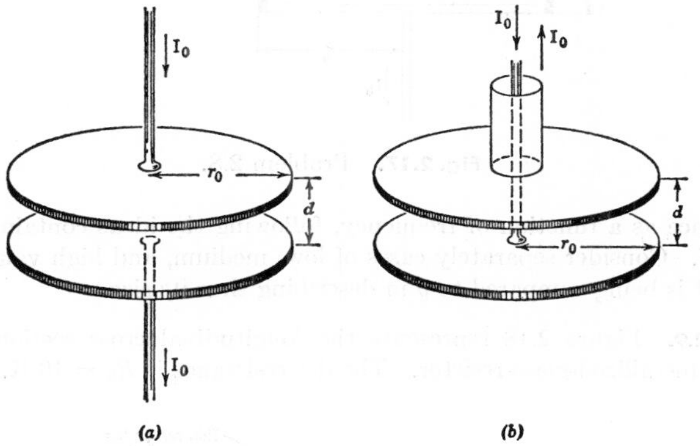

Fig. 2.16. Problems 2.6 and 2.7.

I_0 and the fixed parameters of the structure, the electric and magnetic fields between the plates, correct to terms in ω^3. Neglect fringing. (b) On the basis of your solution in (a), calculate the complex power flowing radially *outward*. Explain the connection between your answer and the fact that the device should be a capacitor at low frequencies. Pay careful attention to algebraic signs. How then should we regard the power as being fed into this device by the source outside? (c) In view of your conclusions, sketch roughly the surface currents on the metal, and the electric and magnetic fields both inside and outside the capacitor at low frequencies. Show the direction of the Poynting vector at important places in the system. (d) Show that your results in (a) and (b) lead to the correct quasi-static capacitance of the device. (e) Compute approximately, on the basis of (a) and (b) alone, the first resonant frequency of an air capacitor with $r_0 = 2.5$ cm and $d = 0.1$ cm. Is this resonance a zero or a pole of the impedance? (f) On the basis of the results of (c), discuss very carefully the problem of finding experimentally the resonance considered in (e).

In particular, would you expect the first resonance measured experimentally for a device like Fig. 2.16*a* to occur at a frequency above, below, or just at the one computed in (e)?

Problem 2.7. Repeat all the pertinent parts of Prob. 2.6 if the capacitor is driven as shown in Fig. 2.16*b*. Assume that the radius of the feeding line is much less than r_0. Note that you cannot expect your solution inside the plates to be valid very close to the center in this case. Discuss carefully the trouble that arises with this "capacitor" if we try to make vanishingly small the radius of the center feeding system.

Problem 2.8. Except for the conductance sheet, the device shown in longitudinal section in Fig. 2.17 is made from perfect conductors and perfect dielectrics. Discuss

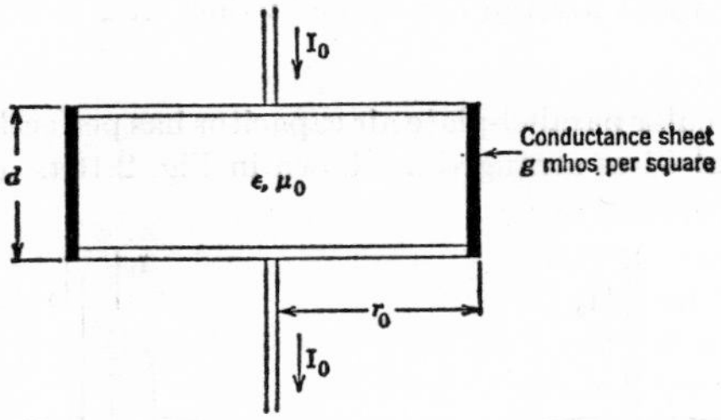

Fig. 2.17. Problem 2.8.

its impedance as a function of frequency, following the ideas contained in Probs. 2.6 and 2.7. Consider separately cases of low, medium, and high values of g, and define what is being compared to g in describing thus its size.

Problem 2.9. Figure 2.18 represents the longitudinal cross section of a small cylindrical metallized-glass resistor. The d-c resistance is $R_0 = 10$ K. The meas-

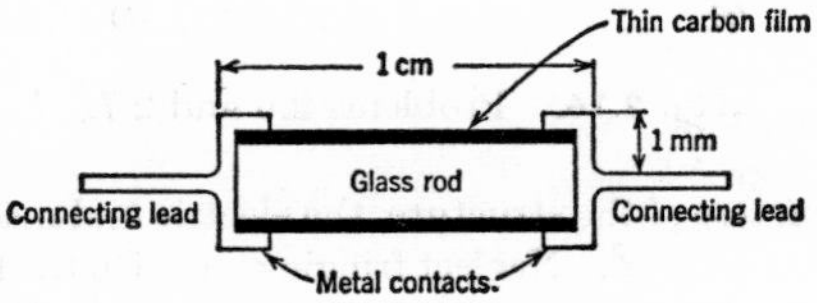

Fig. 2.18. Problem 2.9.

ured value of actual resistance $R(f)$ versus frequency f in *megacycles per second* yields data that fit on the curve

$$\frac{R(f)}{R_0} = \frac{1}{1 + (f/10)^2}$$

up to about $f = 100$, where measurements were stopped. (a) How would you explain these results? In particular, how would you expect the reactance $X(Z = R + jX)$ of this device to vary with frequency? (b) If another resistor of the same physical size and general construction were made, but with $R_0 = 10$ ohms (just the carbon film is altered), would you expect similar behavior over the same frequency range? Why? (c) Check your results of (b) by estimating the inductance and inductive reactance of a square loop of perfectly conducting wire, 1 cm on a side.

Problem 2.10. Consider the flow of air in a smooth-walled rigid hollow pipe whose axis lies along z. (a) What boundary conditions on excess pressure p and/or velocity $\boldsymbol{u}$ are imposed by the pipe wall? (b) Show that a solution is always possible in which p and $\boldsymbol{u}$ depend *only* on z, t (not on x, y). Find the direction of $\boldsymbol{u}(z, t)$ in this solution, and write the differential equations which apply in this case. (c) Show that a solution of the form found in (b) is possible when p and $\boldsymbol{u}$ are constants, independent of z, t. Interpret physically such a solution. What would determine the values of p and $\boldsymbol{u}$? (d) Find the electric transmission-line analogy of the solution in (c). (e) Is any solution at all possible for which p and $\boldsymbol{u}$ have *no* space variation, but do vary with time?

Problem 2.11. Consider the differential equations applicable to Prob. 2.10*b*, and apply them to the complex sinusoidal steady state. Let the tube be driven by a vibrating piston at $z = -l$ and be closed rigidly at $z = 0$. In acoustics, the "volume velocity" or volume of air flowing through the tube cross section per second is analogous to electric current. The average excess pressure at the cross section is like the voltage. The ratio of this pressure to the volume velocity is defined as the "acoustic impedance" Z_A. (a) Convert the above differential equations so they relate pressure and volume velocity directly. Identify the parameters per unit length of the system. (b) Make a quasi-static analysis of the pressure and volume velocity along the tube, and show that, at $z = -l$, Z_A behaves like a lumped compliance. (c) Carry out the analysis to terms in ω^3, showing the appropriate pressure and volume velocity space distributions as well as the zeros and poles of the acoustic impedance seen at $z = -l$. Compare the results with those of Prob. 2.1. (d) How would you arrange the tube so it would behave like a lumped mass? Discuss any serious problems that might arise in trying to achieve this result experimentally.

Problem 2.12. (a) What is the time-domain equivalent of the statement that $L(\omega)$, $C(\omega)$, $R(\omega)$, and $G(\omega)$ are functions of frequency? (b) Show that R, G, C, L must be even functions of ω. (c) What considerations suggest that $Z_s(s) = R + sL$ and $Y_p(s) = G + sC$ must be functions of s which are real for real s? Compare results with those of Prob. 1.8.

C H A P T E R T H R E E

Steady-State Waves on Lossless Transmission Lines

Most of the important properties of wave motion are highlighted when losses are absent from the transmission medium. The simplest examples of wave motion occur in problems involving only one space dimension, in addition to the time. Since the lossless electric transmission line embodies both of these desirable features, and since it is at the same time a reasonable approximation to some practical transmission lines, it will serve well as a vehicle for our discussions in this chapter and the next one. First we shall devote our attention to the behavior of such lines in the sinusoidal steady state, and then take up transients in the next chapter.[1]

3.1 Solution of the Equations

A lossless transmission line operating in the sinusoidal steady state is characterized by Eqs. 2.36 with $R = G = 0$.

$$\text{(a)} \qquad \frac{d\mathrm{V}(z)}{dz} = -j\omega L\, \mathrm{I}(z)$$

$$\text{(b)} \qquad \frac{d\mathrm{I}(z)}{dz} = -j\omega C\, \mathrm{V}(z) \qquad (3.1)$$

About such a line, the complex Poynting theorem (Eqs. 2.38 or 2.39a) tells us in advance that

$$\frac{d\langle P\rangle}{dz} = 0 \qquad (3.2)$$

[1] These chapters may profitably be studied in reverse order. Our choice here is based primarily on the requirements of a concurrent laboratory subject.

which means simply that the time-average flow of power cannot be a function of position. There are no losses to absorb energy.

We may eliminate either $\mathrm{I}(z)$ or $\mathrm{V}(z)$ from Eqs. 3.1. Let us eliminate $\mathrm{I}(z)$ by differentiating Eq. 3.1*a* with respect to z and then substituting for $d\mathrm{I}(z)/dz$ from Eq. 3.1*b*. There results

$$\frac{d^2\mathrm{V}(z)}{dz^2} + \omega^2 LC\, \mathrm{V}(z) = 0 \tag{3.3}$$

with the general solution

$$\mathrm{V}(z) = \mathrm{V}_+ e^{-j\beta z} + \mathrm{V}_- e^{j\beta z} \tag{3.4}$$

where

$$\beta = \omega\sqrt{LC} \tag{3.5}$$

and V_+, V_- are arbitrary complex constants.

According to Eq. 3.1*a*, $\mathrm{I}(z)$ is now determined by the relation

$$\mathrm{I}(z) = -\frac{1}{j\omega L}\frac{d\mathrm{V}(z)}{dz} = \sqrt{\frac{C}{L}}\,(\mathrm{V}_+ e^{-j\beta z} - \mathrm{V}_- e^{j\beta z})$$

$$= \frac{\mathrm{V}_+}{Z_0} e^{-j\beta z} - \frac{\mathrm{V}_-}{Z_0} e^{j\beta z}$$

or

$$\mathrm{I}(z) = Y_0 \mathrm{V}_+ e^{-j\beta z} - Y_0 \mathrm{V}_- e^{j\beta z} \tag{3.6}$$

in which we have defined

$$Z_0 \equiv \frac{1}{Y_0} \equiv \sqrt{\frac{L}{C}} \tag{3.7}$$

Since V_+ and V_- are independent constants, determinable only by boundary conditions and not by the differential equations, the complete solution for the voltage and current consists of two independent sets of relations

$$\begin{cases} \mathrm{V}_+(z) = \mathrm{V}_+ e^{-j\beta z} \\ \mathrm{I}_+(z) = Y_0 \mathrm{V}_+ e^{-j\beta z} \end{cases} + \begin{cases} V_-(z) = \mathrm{V}_- e^{+j\beta z} \\ \mathrm{I}_-(z) = -Y_0 \mathrm{V}_- e^{+j\beta z} \end{cases} \tag{3.8}$$

Our method of study will consist of examining and interpreting each solution separately, then the two in some particularly simple combinations, and finally the most general superposition of them.

3.2 Traveling Waves

The solution denoted by the subscript (+) in Eq. 3.8 comprises a complex voltage and current, each having constant amplitude and a phase that changes linearly with z. At any point along the line, however, the voltage and current of this solution are in phase, their ratio being $[\mathrm{V}_+(z)]/[\mathrm{I}_+(z)] = Z_0$. We call Z_0 the *characteristic impedance* of the line. It depends only on the line parameters (Eq. 3.7).

With $\mathrm{V}_+ = |\mathrm{V}_+|e^{j\varphi_+}$, Fig. 3.1 shows the diagram of the complex voltage and current at two different points, $z = 0$ and $z = \Delta z$. The voltage and current at the point $z = \Delta z$ are delayed in phase by $\beta\,\Delta z$ radians with respect to the voltage and current at $z = 0$. The coefficient β, giving the phase shift per unit distance, is called the *phase constant.* It depends upon the line parameters and the frequency (Eq. 3.5.)

We can understand more about this solution if the voltage and current are studied as real functions of z at different instants of time. We find

$$\begin{aligned} V_+(z, t) &= \mathrm{Re}\,[\mathrm{V}_+(z)e^{j\omega t}] \\ &= \mathrm{Re}\,[\mathrm{V}_+e^{j(\omega t - \beta z)}] \\ &= |\mathrm{V}_+|\cos(\omega t - \beta z + \varphi_+) \end{aligned}$$

and similarly,

$$I_+(z, t) = \mathrm{Re}\,[\mathrm{I}_+(z)e^{j\omega t}] = Y_0|\mathrm{V}_+|\cos(\omega t - \beta z + \varphi_+)$$

Figure 3.2 represents these expressions. The voltage and current are distributed sinusoidally along the line in the form of a wave. They are

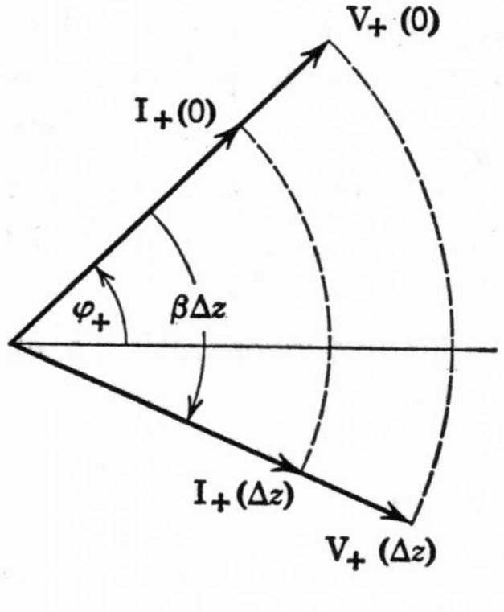

Fig. 3.1. Complex voltage and current amplitudes in a (+) wave.

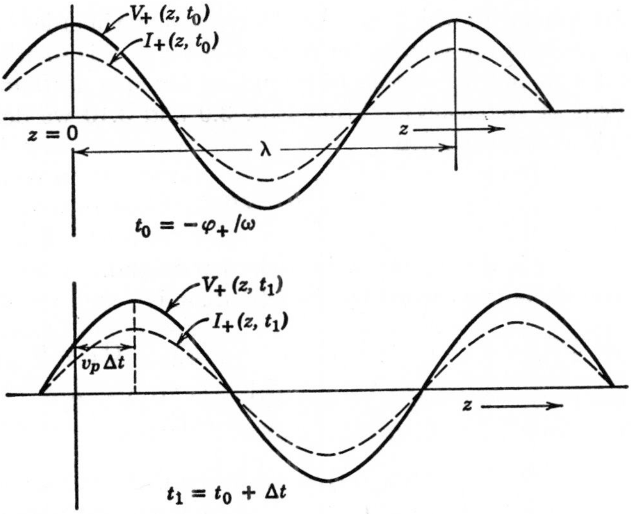

Fig. 3.2. The instantaneous voltage and current of a (+) wave.

periodic functions of z. Indeed, if z is changed at constant t by an amount

$$\lambda = \frac{2\pi}{\beta} \tag{3.9}$$

the voltage or current does not change. Voltages or currents at points spaced an integer multiple of the distance λ apart have equal amplitudes and time phases. λ is called the *wave length.*

Also, if we compare the positions of a crest of the voltage (or current) wave at two different times t_0 and $t_1 = t_0 + \Delta t$, we find that it has moved a distance $\Delta z = (\omega/\beta)\,\Delta t$ to the right (Fig. 3.2). This relation stems from the fact that the voltage, for example, is proportional to $\cos(\omega t - \beta z + \varphi_+)$. If the cos is to remain at its peak (or any other given) value, its argument $(\omega t - \beta z + \varphi_+)$ must remain fixed. Since t increases by Δt, z must increase by Δz such that $\beta\,\Delta z = \omega\,\Delta t$. Indeed, if we wished to move along z at just the right speed to remain on the voltage crest (or any other given voltage value), we should have to move exactly this distance $\Delta z = (\omega/\beta)\,\Delta t$ in time Δt. Our speed v_p would therefore be

$$v_p = \left.\frac{\Delta z}{\Delta t}\right|_{\text{crest}} = \frac{\omega}{\beta} = \frac{1}{\sqrt{LC}} \tag{3.10}$$

which is known as the *phase velocity* of the wave. It is *by definition* simply the speed with which one would have to move in order to remain *at a given phase* of the voltage (or current) wave in this solution. Observe that v_p is not a function of frequency (as long as L and C are independent of frequency). From Eqs. 3.9 and 3.10 we derive the familiar elementary relation

$$v_p = \frac{\omega}{2\pi}\lambda = f\lambda \tag{3.11}$$

where $f \equiv \omega/2\pi$ is the frequency in cycles per second.

The complex power carried by the (+) wave is

$$\mathrm{P}_+ = \tfrac{1}{2}\mathrm{V}_+(z)\mathrm{I}_+^*(z) = \tfrac{1}{2}Y_0|\mathrm{V}_+|^2 = \langle P_+\rangle \qquad Q_+ = 0 \tag{3.12}$$

It does not possess a reactive component, and $\langle P_+\rangle > 0$.

It follows from $Q_+ = 0$ and Eqs. 2.38 or 2.39*b* that

$$\langle u_{m_+}(z)\rangle = \langle u_{e_+}(z)\rangle \tag{3.13}$$

Each section of line stores equal amounts of electric and magnetic energy on a time-average basis. That this is also true on an instantaneous basis can be verified by constructing

$$u_{m_+}(z, t) = \tfrac{1}{2}LI_+^2 = \tfrac{1}{2}LY_0^2|\mathrm{V}_+|^2\cos^2(\omega t - \beta z + \varphi_+) \tag{3.14}$$

$$u_{e_+}(z, t) = \tfrac{1}{2}CV_+^2 = \tfrac{1}{2}C|\mathrm{V}_+|^2\cos^2(\omega t - \beta z + \varphi_+)$$

which, since $LY_0^2 = L(C/L) = C$, proves that

$$u_{m_+}(z, t) = u_{e_+}(z, t) \tag{3.15}$$

At any point, at any instant of time, the electric and magnetic energy densities are equal. Figure 3.3 shows the energy densities at two different instants of time.

A careful examination of Fig. 3.3, along with the following discussion, will show how the energy moves down the line. One might be misled by the equality of the time-average electric and magnetic stored energies per unit length (Eq. 3.13) into believing that each little section dz of the line is in "resonance." We are apt to jump to this conclusion from experience with lossless *lumped* circuits, in which the resonance interpretation would be valid. The instantaneous energies, however, show that the present situation is quite different. In any section of the line, large or small, the electric and magnetic stored energies go up and down *together*. In a resonant, lossless system, the electric energy is large when the magnetic energy is small (and vice versa), with the sum of the two being the *constant* total energy stored in the system. Any

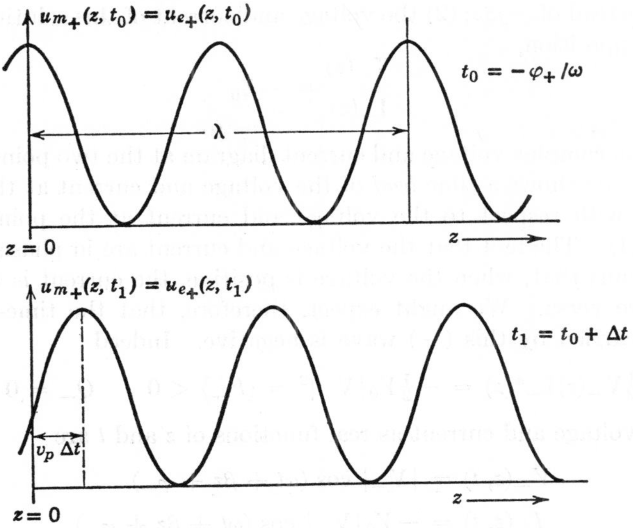

Fig. 3.3. The instantaneous electric and magnetic energy densities in a (+) wave.

small section dz of the line, however, has a total instantaneous energy

$$u_+\,dz = (u_{e_+} + u_{m_+})\,dz = 2u_{e_+}\,dz = 2u_{m_+}\,dz$$
$$= C\,|\mathrm{V}_+|^2\,dz\cos^2(\omega t - \beta z + \varphi_+) \quad (3.16)$$

which goes from zero to $C\,|\mathrm{V}_+|^2\,dz$ and back, twice every fundamental period $2\pi/\omega$. Since there are no losses, this variation can only be ascribed to the motion of the total energy along the line, i.e., to the motion illustrated in Fig. 3.3. More precisely, we may ask at what speed we would have to move an element of length dz along the line in order for the corresponding small section of line dz (which it subtends at each instant) always to have the same total energy stored in it. According to Eq. 3.16, we see again that the argument of the cosine must stay constant, which leads us once more to the phase velocity v_p of Eq. 3.10. Thus we may think of the energy in a (+) wave on our lossless line as moving in the $+z$ direction with the phase velocity $v_p = 1/\sqrt{LC}$. This interpretation of v_p is, however, special for the case at hand, and we shall find situations later in which it is not valid.

The properties of the other independent solution of the transmission-line equations, $\mathrm{V}_-(z)$ and $\mathrm{I}_-(z)$ in Eq. 3.8, can be deduced quite easily from the properties of the (+) wave. The only differences between the two solutions are: (1) The sign of the propagation exponent is

$+j\beta z$ instead of $-j\beta z$; (2) the voltage and current of this solution are in phase opposition,

$$\frac{V_-(z)}{I_-(z)} = -Z_0$$

Thus the complex voltage and current diagram at the two points $z = 0$ and $z = \Delta z$ shows a time *lead* of the voltage and current at the point $z = \Delta z$ with respect to the voltage and current at the point $z = 0$ (Fig. 3.4). The fact that the voltage and current are in phase opposition means that, when the voltage is positive, the current is *negative*, and vice versa. We might expect, therefore, that the time-average power carried by this (−) wave is negative. Indeed

$$P_- = \tfrac{1}{2}V_-(z)I_-^*(z) = -\tfrac{1}{2}Y_0|V_-|^2 = \langle P_- \rangle < 0 \qquad Q_- = 0 \qquad (3.17)$$

The voltage and current as real functions of z and t are

$$V_-(z, t) = |V_-| \cos(\omega t + \beta z + \varphi_-)$$
$$I_-(z, t) = -Y_0|V_-| \cos(\omega t + \beta z + \varphi_-)$$

This wave travels in the $-z$ direction. Its phase velocity is correspondingly

$$v_p = -\frac{1}{\sqrt{LC}}$$

a negative quantity, indicating also the motion of energy along $-z$.

We conclude that the most general solution (Eqs. 3.4 and 3.6) of the transmission-line equations may be viewed as the superposition of two pure traveling waves moving in opposite directions along the line. The amplitudes of these waves, i.e., the complex values of V_+ and V_-, can be determined only by the boundary conditions at the two ends of any finite piece of line. For example, with reference to Fig. 3.5*a*, suppose

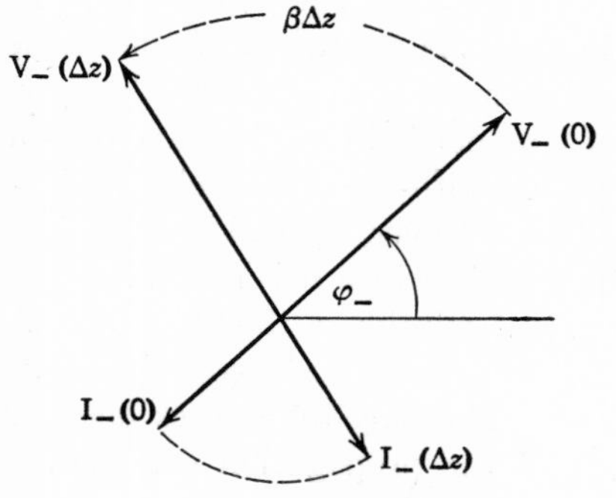

Fig. 3.4. Complex voltage and current amplitudes in a (−) wave with $V_- = |V_-|e^{j\varphi_-}$.

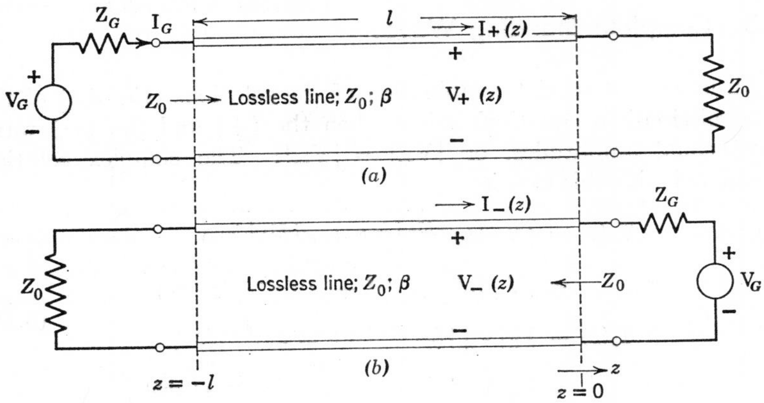

Fig. 3.5. Pure traveling waves on a finite line. (*a*) (+) wave only; (*b*) (−) wave only.

we have a line of finite length l extending from $z = -l$ to $z = 0$, and and we wish to set up on it only a (+) wave. How shall we arrange lumped-circuit sources or loads at the ends?

First, we note that time-average power must flow from left to right (i.e., in the $+z$ direction) for a (+) wave. Hence we shall certainly need a *source* at $z = -l$ and a dissipative *load* at $z = 0$.

Next, the (+) wave solution always has $[V_+(z)]/[I_+(z)] = Z_0$. If a lumped-circuit load is placed at $z = 0$, the voltage across it must be the same as that across the line at $z = 0$, and the current through it must equal that in the line at $z = 0$. These statements are simply Kirchhoff's laws for lumped circuits, it being supposed that the terminal connections themselves have negligible geometric dimensions. It follows that the load must be a device whose impedance is Z_0, which we may most easily consider as a resistor of this value (Fig. 3.5*a*).

Finally, the voltage at the input ($z = -l$) is $V_+e^{+j\beta l}$ and the impedance is again Z_0. This is all that is required by the (+) wave solution. Therefore any type of generator at all will be acceptable. We have represented one in Fig. 3.5*a* by a voltage source V_G in series with any impedance Z_G. Kirchhoff's laws require $V_+e^{+j\beta l} = [Z_0/(Z_0 + Z_G)]V_G$, which determines V_+ (or V_G) and completes the demands of all boundary conditions.

Similarly, it is not difficult to see now that a (−) wave alone can be set up by the arrangement of Fig. 3.5*b*.

It follows that a lossless line terminated in Z_0 has on it a pure traveling wave, regardless of the nature of the driving source.

3.3 Complete Standing Waves

An important special combination of the solutions (Eq. 3.8) of the transmission-line equations arises when the (+) and the (−) wave have equal amplitudes, or $|V_+| = |V_-|$. The complex solution associated with this case is

$$
\begin{aligned}
&\text{(a)} \quad V_s(z) = |V_+|(e^{-j\beta z + j\varphi_+} + e^{+j\beta z + j\varphi_-}) \\
&\qquad\qquad = |V_+|e^{j\varphi_+}(e^{-j\beta z} + e^{j(\varphi_- - \varphi_+)}e^{+j\beta z}) \\
&\text{(b)} \quad I_s(z) = Y_0|V_+|(e^{-j\beta z + j\varphi_+} - e^{+j\beta z + j\varphi_-}) \\
&\qquad\qquad = Y_0|V_+|e^{j\varphi_+}(e^{-j\beta z} - e^{j(\varphi_- - \varphi_+)}e^{+j\beta z})
\end{aligned}
\tag{3.18}
$$

With no loss in generality, it can be assumed that $\varphi_+ = 0$. This corresponds merely to fixing the time reference of the (+) wave voltage. Let us then consider the case $\varphi_- = 0$, which completes the specification of both the time and the space phases of the whole solution. On an *infinite* line this is no real restriction either, inasmuch as the space origin is then just as arbitrary as the time origin. We have

$$
\begin{aligned}
&\text{(a)} \quad V_s(z) = |V_+|(e^{-j\beta z} + e^{+j\beta z}) \\
&\text{(b)} \quad I_s(z) = |V_+|Y_0(e^{-j\beta z} - e^{+j\beta z})
\end{aligned}
\tag{3.19}
$$

Since

$$e^{+j\beta z} + e^{-j\beta z} = 2\cos\beta z$$

and

$$e^{+j\beta z} - e^{-j\beta z} = 2j\sin\beta z$$

we can write Eq. 3.19 in the form

$$
\begin{aligned}
&\text{(a)} \quad V_s(z) = 2|V_+|\cos\beta z \\
&\text{(b)} \quad I_s(z) = -2jY_0|V_+|\sin\beta z
\end{aligned}
\tag{3.20}
$$

Equations 3.20*a* and 3.20*b* show that the current and voltage of this solution are 90° out of time phase all along the line. The complex amplitude of the voltage goes to zero at points z_n spaced half a wave length apart and given by the relation

$$\beta z_n = \frac{2n + 1}{2}\pi$$

where n is any positive or negative integer. Points of zero voltage are called *voltage nodes.* From Eq. 3.20*b* it is apparent that the current distribution has nodes at the points

$$\beta z_n = n\pi$$

where n is any positive or negative integer. The current nodes lie halfway between the voltage nodes. This peculiar behavior of the solution can be stated in words: The voltage and current are 90° out of *space* phase. Hence the voltage and current are 90° out of *both* time and space phase.

The voltage and current as real functions of time and space are

$$\begin{aligned} &(a) \qquad V_s(z, t) = \text{Re}\,[\mathrm{V}_s(z)e^{j\omega t}] = 2|\mathrm{V}_+|\cos \beta z \cos \omega t \\ &(b) \qquad I_s(z, t) = \text{Re}\,[\mathrm{I}_s(z)e^{j\omega t}] = 2Y_0|\mathrm{V}_+|\sin \beta z \sin \omega t \end{aligned} \qquad (3.21)$$

The symmetrical role of the time and space coordinates is noteworthy. Thus, the following two sets of plots would look alike:

1. N plots of the voltage and current distributions as functions of βz at fixed times $t = n(T/N)$, where $T = 2\pi/\omega$ and $n = 0, 1, 2, \cdots, N$.
2. N plots of the voltage and current as functions of ωt at the equally spaced points $z = n(\lambda/N)$, where $\lambda = 2\pi/\beta$ and again $n = 0, 1, 2, \cdots, N$.

Figure 3.6 shows a set of plots drawn as described in (1) above. The set corresponding to (2) can be obtained simply by interchanging βz with ωt and λ with T. From Fig. 3.6 we can see that the voltage and current space distribution plots seem to stand still in space, i.e., they do not convey a feeling of motion down the line. Rather, their amplitudes vary sinusoidally with time, the voltage being a maximum at the time instant when the current in the line is zero, and vice versa. The voltage and current are *complete standing waves.*

The energy densities as functions of z and t are

$$\begin{aligned} &(a) \qquad u_{ms}(z, t) = \frac{L}{2} I_s^2 = \frac{L}{2} Y_0^2 |2\mathrm{V}_+|^2 \sin^2 \beta z \sin^2 \omega t \\ &\qquad\qquad = C|\mathrm{V}_+|^2 \sin^2 \beta z(1 - \cos 2\omega t) \\ &(b) \qquad u_{es}(z, t) = \frac{C}{2} V_s^2 = \frac{C}{2} |2\mathrm{V}_+|^2 \cos^2 \beta z \cos^2 \omega t \\ &\qquad\qquad = C|\mathrm{V}_+|^2 \cos^2 \beta z(1 + \cos 2\omega t) \end{aligned} \qquad (3.22)$$

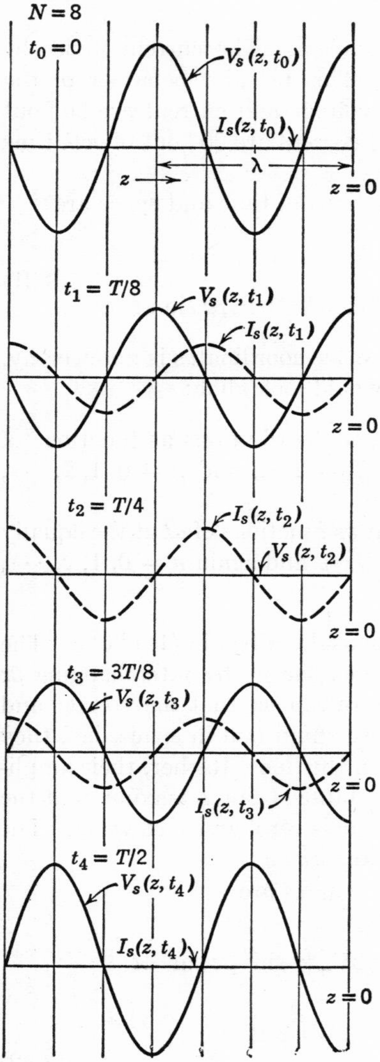

Fig. 3.6. The instantaneous voltage and current in a complete standing wave.

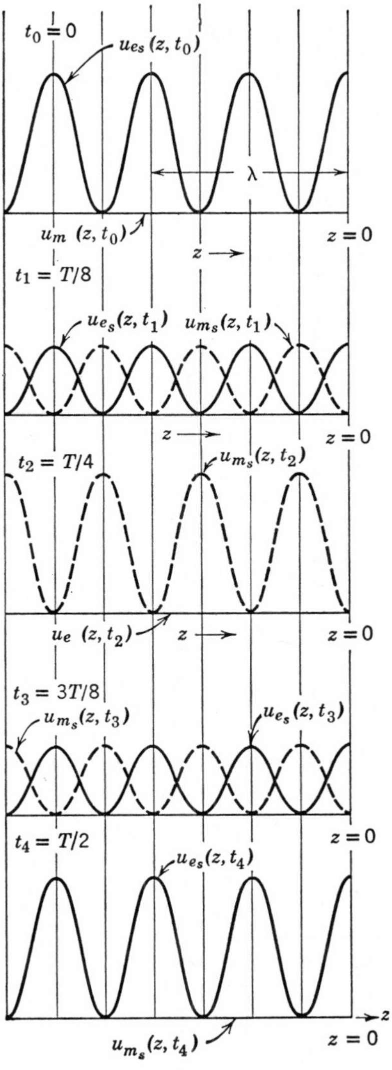

Fig. 3.7. The instantaneous energy densities in a complete standing wave.

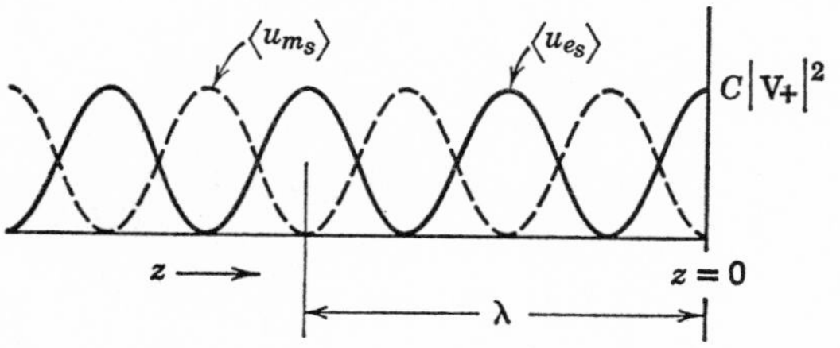

Fig. 3.8. The time average of the energy densities in a complete standing wave.

Figure 3.7 shows these expressions plotted versus z at different instants of time, $t = n(T/N)$. Figure 3.8 gives the time average of the energy densities as functions of z.

The process of energy storage is similar to that in a resonant L-C circuit. The only difference is that in an L-C circuit the electric energy is stored entirely in the capacitor and the magnetic energy in the inductance, whereas the magnetic and electric energy densities of the standing-wave solution overlap in space. Only the predominance (on the time average) of one kind of energy over the other determines the essentially capacitive or inductive character of a section of line.

There is, of course, an instantaneous power flow $P_s(z, t)$ on the line

$$\begin{aligned} P_s(z, t) &= V_s(z, t)I_s(z, t) = 4Y_0|\mathrm{V}_+|^2 \cos \beta z \sin \beta z \cos \omega t \sin \omega t \\ &= Y_0|\mathrm{V}_+|^2 \sin 2\beta z \sin 2\omega t \end{aligned} \tag{3.23}$$

Note that it has a zero time average. The reader, however, can easily check that the energy densities of Eq. 3.22 and the power flow of Eq. 3.23 satisfy the condition of the conservation of energy with $R = G = 0$ (see Eq. 2.35):

$$-\frac{\partial P_s}{\partial z} = \frac{\partial}{\partial t}(u_{ms} + u_{es})$$

The information gained so far by considering the energy and power flow as real functions of time can also be gained from the complex Poynting theorem. The time average of the electric and magnetic energy densities can be found either from Eqs. 3.22 or from Eqs. 3.20, using the relations

$$\begin{aligned} &\text{(a)} \qquad 2\langle u_{ms}\rangle = \frac{1}{2}L|\mathrm{I}_s(z)|^2 = \frac{C}{2}|2\mathrm{V}_+|^2 \sin^2 \beta z \\ &\text{(b)} \qquad 2\langle u_{es}\rangle = \frac{1}{2}C|\mathrm{V}_s(z)|^2 = \frac{C}{2}|2\mathrm{V}_+|^2 \cos^2 \beta z \end{aligned} \tag{3.24}$$

The complex power is

$$\mathrm{P}_s = \frac{1}{2}\mathrm{V}_s(z)\mathrm{I}_s{}^*(z) = \frac{j}{2}|2\mathrm{V}_+|^2 Y_0 \cos \beta z \sin \beta z$$

$$= j|\mathrm{V}_+|^2 Y_0 \sin 2\beta z \tag{3.25}$$

Thus

$$\langle P_s \rangle = 0 \tag{3.26}$$

and

$$Q_s = Y_0 |\mathrm{V}_+|^2 \sin 2\beta z \tag{3.27}$$

The flow of power is purely reactive. It is negative (reactive) throughout the quarter wave of line extending from $z = 0$ to $z > -\lambda/4$, indicating that the electric stored energy predominates within this section of the line. This conclusion follows from Eq. 2.39*b*, taking $z_1 = z$, $z_2 = 0$, and noting from Eq. 3.27 that $Q_s(0) = 0$. In detail:

$$Q_s(z) = 2\omega\langle W_m - W_e\rangle \Big|_z^0 \tag{3.28}$$

Since $Q_s(z) < 0$ for $-\lambda/4 < z < 0$, from Eq. 3.27, it must be true that

$$\langle W_m \rangle \Big|_{z>-\lambda/4}^0 < \langle W_e \rangle \Big|_{z>-\lambda/4}^0 \tag{3.29}$$

When $z = \lambda/4$, the inequality signs in Eq. 3.29 are replaced by equality signs. These energy relations are made quite obvious in Fig. 3.8.

We come now to the question of boundary conditions for this solution on a finite section of line. On account of Eq. 3.26 we know that no source or dissipative load will be necessary to accommodate this complete standing wave since no time-average power flows on the line. Moreover, it is clear from Fig. 3.6 that a short circuit could be placed at any voltage node without disturbing the solution at all. Similarly, the line could be cut apart completely at any current node. In this way we could isolate sections of line, using a soldering iron and scissors, in such a manner that the standing-wave solution on the isolated section would remain undisturbed. For example, we could cut the line in Fig. 3.6 at the current nodes $z = 0$ and $z = -\lambda/2$, obtaining a section $\lambda/2$ long with open circuits at each end; or we could install short circuits at the voltage nodes $z = -\lambda/4$ and $z = -3\lambda/4$, obtaining a section $\lambda/2$ long with short circuits at each end. In both cases, the complete standing wave would persist indefinitely on the isolated line section. The line section becomes a resonator without losses, which simply oscillates forever.

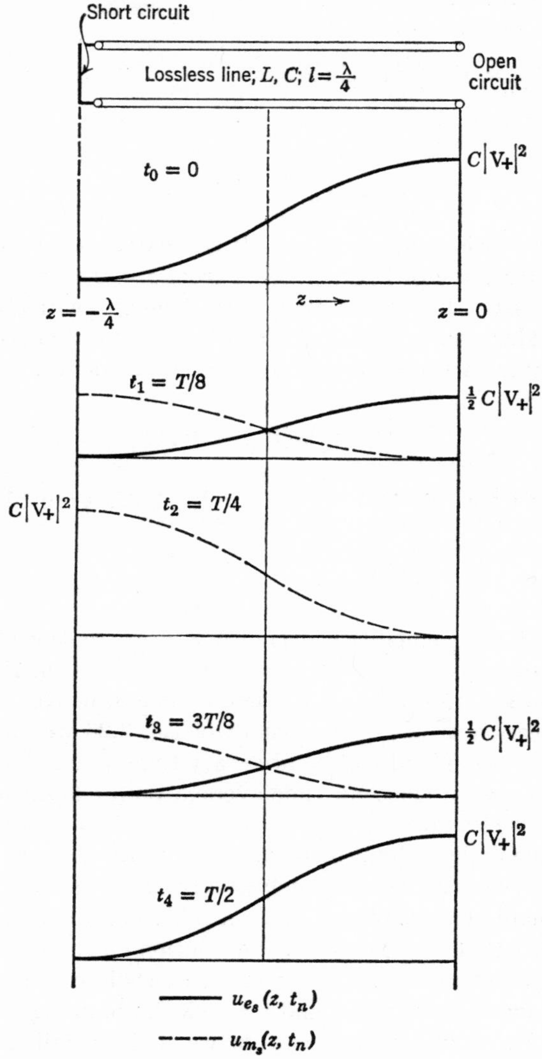

Fig. 3.9. The instantaneous electric and magnetic energy densities in a quarter-wave resonating line.

A little thought will make clear the fact that the shortest piece of line that can be removed, using only short circuits (a soldering iron) and open circuits (a scissors), is a $\lambda/4$ section, one end of which is a voltage node and the other a current node. A case in point is the cutting of Fig. 3.6 at $z = 0$ (current node) and the short-circuiting of it at $z = -\lambda/4$ (voltage node). The energy storage on the $\lambda/4$ section is shown in Fig. 3.9 at five instants of time, ⅛ period apart. We see that the open-circuited end stores predominantly electric energy, whereas the short-circuited end stores primarily magnetic energy. The energy flops back and forth between the two ends. Within one half-cycle (π/ω) the process of energy exchange is completed. The similarity between this situation and that of energy storage in a resonant *L-C* circuit is unmistakable.

Now, open circuits and short circuits are not the only terminations suitable as boundary conditions for the complete standing wave. As indicated in Sec. 3.2, if the infinite line is to be stopped at some point z_1, and the remainder of the line for $z > z_1$ replaced by a lumped impedance, the voltage and current at z_1 required by the line solution must equal the voltage and current taken by the load. In other words, if a particular line solution for $z \leq z_1$ is to remain undisturbed by the replacement, the lumped load must look electrically to the remainder of the line just like the part that was cut off. Thus the impedance of the lumped load must equal the impedance which would have appeared looking in the $+z$ direction at the point z_1 before the line was cut. For the given standing-wave solution $[\mathrm{V}_s(z), \mathrm{I}_s(z)]$, the impedance seen on an infinite line looking toward $+z$ at any point z_1 is

$$\begin{aligned} \mathrm{Z}_s(z_1) &= \frac{\mathrm{V}_s(z_1)}{\mathrm{I}_s(z_1)} = \frac{2|\mathrm{V}_+|\cos\beta z_1}{-2j|\mathrm{V}_+|\,Y_0\sin\beta z_1} \\ &= jZ_0\cot\beta z_1 = jX_s(z_1) \end{aligned} \tag{3.30}$$

We may refer to this impedance as the *generalized impedance* of the complete standing wave. It is pure imaginary and a function of the position z_1. Incidentally, on this basis it should be clear that the generalized impedance of our (+) wave solution in Sec. 3.2 is Z_0, which is pure real and independent of position.

A plot of Eq. 3.30 for the complete standing wave appears in Fig. 3.10. We observe that the reactance necessary to terminate the wave at $z_1 = 0$ is infinite (open circuit), whereas that required at $z_1 = -\lambda/4$ is zero (short circuit). This fact checks our previous discussion regarding open and short circuits placed at current and voltage nodes. Now, however, we find in addition that *any* load reactance $Z_R = jX_R$, with

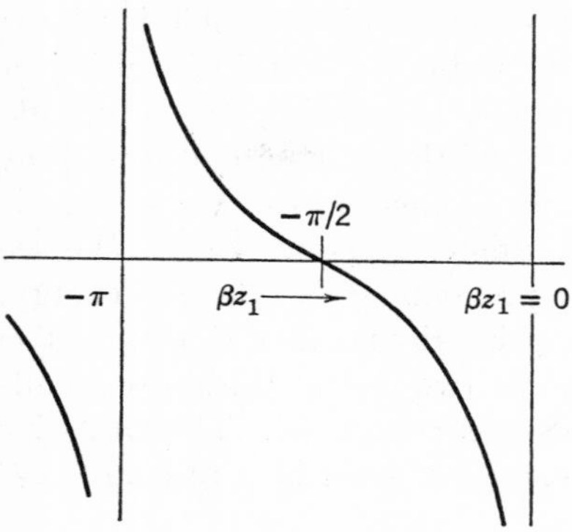

Fig. 3.10. Generalized reactance of complete-standing-wave solution.

$-\infty \leq X_R \leq +\infty$, can be connected at a suitable point $-\lambda/2 \leq z_1 \leq 0$ to terminate the complete standing wave. The appropriate point z_1 is found by requiring that the given load reactance X_R shall equal the generalized complete-standing-wave reactance X_s at z_1:

$$X_R = X_s(z_1) = Z_0 \cot \beta z_1$$

or

$$\beta z_1 = -\pi + \cot^{-1}\left(\frac{X_R}{Z_0}\right)$$

Having located such a reactance at the appropriate z_1, and having removed the line for $z > z_1$, we can now move back to any point $z_2 < z_1$ and install a generator at that point (Fig. 3.11*a*). Any type of generator will do. The requirements imposed by the solution are that

$$V_s(z_2) = 2|V_+| \cos \beta z_2$$

and

$$X_s(z_2) = Z_0 \cot \beta z_2$$

Hence

$$V_G\left[\frac{jX_s(z_2)}{Z_G + jX_s(z_2)}\right] = 2|V_+| \cos \beta z_2$$

is the condition that insures equality of source current and source terminal voltage with line current and line voltage at z_2. Given V_G and Z_G, this relation fixes V_+; or given V_+ and Z_G, it fixes V_G. We conclude that a lossless line terminated in a pure reactance has on it a complete standing-wave solution, regardless of how it is driven at the other end. Also, no time-average power is supplied to the line or load.

As a matter of fact, we indicated previously that, since the complete standing wave carries no time-average power, it should be possible for it to oscillate by itself without a source at all. Having located a load reactance jX_R at z_1 as before, we could move back to any point z_2 (Fig. 3.11*b*) and place there another pure reactance jX_G suitably chosen to meet the boundary conditions at z_2. Equality of I_G with $I_s(z_2)$, and the requirement that $V_s(z_2)$ also be the voltage across jX_G, demand that

$$X_G = -X_s(z_2) = -Z_0 \cot \beta z_2$$

This allows the system to oscillate by itself as a lossless resonator, with a complete standing wave on the line section between z_1 and z_2.

There is another interesting point about the complete standing-wave solution which we should not overlook. We have written the entire solution to the transmission-line differential equations (Eqs. 3.1) in the form

$$\text{(a)} \quad V(z) = V_+ e^{-j\beta z} + V_- e^{j\beta z}$$
$$\text{(b)} \quad I(z) = Y_0 V_+ e^{-j\beta z} - Y_0 V_- e^{j\beta z} \qquad (3.31)$$

which we have thus far interpreted as the superposition of two oppositely directed traveling waves. Another viewpoint, however, results from noting that since

$$e^{\pm j\beta z} = \cos \beta z \pm j \sin \beta z$$

the full solution may be rewritten

$$V(z) = (V_+ + V_-) \cos \beta z - j(V_+ - V_-) \sin \beta z$$
$$I(z) = Y_0(V_+ - V_-) \cos \beta z - j(V_+ + V_-) \sin \beta z$$

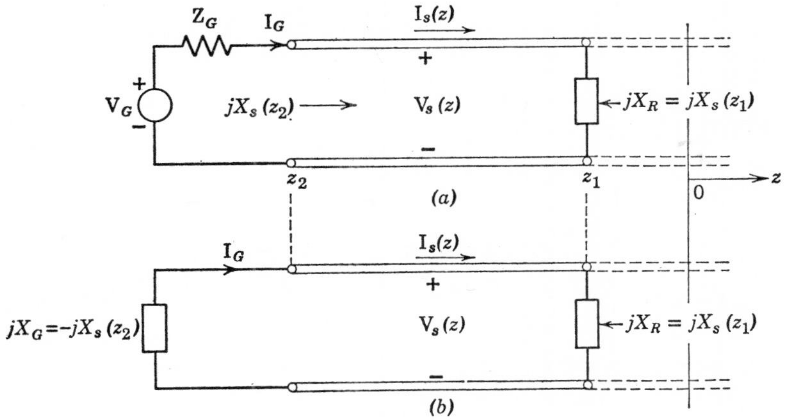

Fig. 3.11. Complete standing waves on a finite line.

Since V_+ and V_- are independent arbitrary complex constants as far as the differential equations are concerned, the factors $(V_+ + V_-) \equiv V_1$ and $-j(V_+ - V_-) \equiv V_2$ are also simply independent arbitrary constants. The solution becomes

$$\begin{aligned} &\text{(a)} \quad V(z) = V_1 \cos \beta z \quad + \quad V_2 \sin \beta z \\ &\text{(b)} \quad I(z) = -jY_0V_1 \sin \beta z \quad + \quad jY_0V_2 \cos \beta z \end{aligned} \tag{3.32}$$

Observe that Eq. 3.32 is divided into two independent solutions for voltage and current. The first is exactly like our complete standing-wave solution in Eq. 3.20 if we set the arbitrary time-phase reference by making V_1 pure real. The second solution is again a complete standing wave whose *space phase* lags that of the first solution by 90°. Its relative amplitude and time phase are arbitrary because of V_2, which will be determined only by boundary conditions. Therefore we may regard the full solution as the superposition of two complete standing waves 90° out of space phase and of arbitrary complex amplitudes. This point of view is in many cases a useful one conceptually, although, analytically, the trigonometric functions $\sin \beta z$ and $\cos \beta z$ are somewhat more difficult to manipulate than the exponential functions $e^{\pm j\beta z}$. For this reason the traveling-wave form of the total solution given in Eq. 3.31 is often the more convenient one for general analysis purposes.

3.4 The Effects of a General Impedance Termination

To study the general solution (Eq. 3.31) of the transmission-line equations, we consider the case of a semi-infinite line with an arbitrary load impedance Z_R at its end. In the following, it is convenient to choose the origin $z = 0$ at the termination, so that the line lies along $z < 0$, as shown in Fig. 3.12. The source is presumably located some-

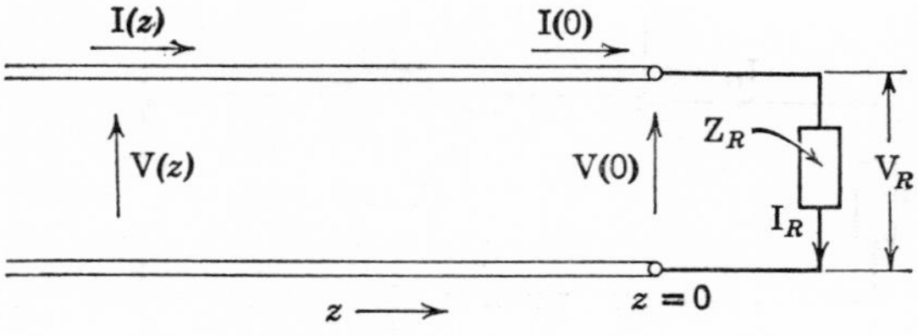

Fig. 3.12. A semi-infinite terminated transmission line.

where at the left, but its precise position need not concern us at the moment.

3.4.1 Load Boundary Condition and Reflection Coefficient

Our interest is first in the nature of the restrictions placed upon the line solution by the boundary conditions at the load. If V_R and I_R are the load voltage and current respectively (Fig. 3.12), then

$$\frac{V_R}{I_R} = Z_R \tag{3.33}$$

Since at the load terminals both the voltage and the current must be continuous, the transmission-line voltage $V(0)$ prescribed by the line solution at $z = 0$ must equal the voltage V_R across the load impedance

$$V(0) = V_R \tag{3.34}$$

and the current $I(0)$ prescribed by the transmission-line solution at $z = 0$ must equal the current I_R flowing in the load impedance

$$I(0) = I_R \tag{3.35}$$

Thus

$$\frac{V(0)}{I(0)} = Z_R \tag{3.36}$$

From Eq. 3.31*a*, however, we have

$$V(0) = V_+ + V_- \tag{3.37}$$

whereas, from Eq. 3.31*b*, we require

$$I(0) = Y_0(V_+ - V_-) \tag{3.38}$$

Equations 3.37 and 3.38, introduced into Eq. 3.36, yield a relation between the constants V_+ and V_-

$$Z_R = Z_0\frac{V_+ + V_-}{V_+ - V_-} = Z_0\frac{1 + (V_-/V_+)}{1 - (V_-/V_+)} \tag{3.39}$$

In particular, the impedance of the termination determines uniquely the ratio of the complex amplitude of the (−) wave voltage to that of the (+) wave at the location of the termination. This ratio is called the *voltage reflection coefficient* (or usually just the *reflection coefficient*) of the termination, and is denoted by

$$\Gamma_R \equiv \frac{V_-}{V_+} \tag{3.40}$$

In terms of the reflection coefficient, Eq. 3.39 can be written

$$Z_R = Z_0 \frac{1 + \bar{\Gamma}_R}{1 - \bar{\Gamma}_R} \qquad \text{or} \qquad \frac{Z_R}{Z_0} = \frac{1 + \bar{\Gamma}_R}{1 - \bar{\Gamma}_R} \tag{3.41}$$

which, solved for $\bar{\Gamma}_R$, yields

$$\bar{\Gamma}_R = \frac{Z_R - Z_0}{Z_R + Z_0} = \frac{(Z_R/Z_0) - 1}{(Z_R/Z_0) + 1} \tag{3.42}$$

The physical significance of Eqs. 3.40 and 3.42 is important. An *incident wave* voltage $V_+(z) = V_+e^{-j\beta z}$ strikes the termination at $z = 0$ with a complex amplitude $V_+(0) = V_+$. In general, the line terminates at $z = 0$ in an impedance Z_R, which does not look electrically to the incident wave like more of the same line. Hence the incident wave alone cannot meet the boundary conditions at the load. The load, therefore, sets up a *reflected wave* voltage $V_-(z) = V_-e^{+j\beta z}$ on the line, its complex amplitude at $z = 0$ being $V_-(0) = V_-$. Given Z_0, the load impedance Z_R does not fix the values of V_+ and V_-, *but it does fix uniquely their ratio* $V_-/V_+ = \bar{\Gamma}_R$. Thus, instead of thinking of a termination as an impedance Z_R, we can just as well think of it as a "reflector," characterized by a complex number $\bar{\Gamma}_R$ which tells us the complex amplitude of the reflected wave voltage it generates at its terminals per unit incident-wave voltage striking it. The numerical value of $\bar{\Gamma}_R$ is every bit as good a description of the load as is its impedance Z_R, since the two are uniquely related by Eqs. 3.41 and 3.42 for any given value of Z_0 (i.e., for any given transmission line). More specifically, it is the ratio (Z_R/Z_0) which governs the value of $\bar{\Gamma}_R$, and conversely. Consequently, it is often advantageous to discuss transmission-line problems in terms of a *normalized impedance* Z_{nR}

$$Z_{nR} \equiv \frac{Z_R}{Z_0} \tag{3.43}$$

the specification of which is precisely equivalent to specifying $\bar{\Gamma}_R$ because of the relations

$$\begin{aligned} &\text{(a)} \qquad \bar{\Gamma}_R = \frac{Z_{nR} - 1}{Z_{nR} + 1} \\ &\text{(b)} \qquad Z_{nR} = \frac{1 + \bar{\Gamma}_R}{1 - \bar{\Gamma}_R} \end{aligned} \tag{3.44}$$

obtained from Eqs. 3.41, 3.42, and 3.43.

3.4.2 Generalized Reflection Coefficient

Knowing now that our load is a "reflector" $\bar{\Gamma}_R$ at $z = 0$, specifying directly the ratio $V_-(0)/V_+(0)$, we wish to investigate the line solution at other points $z < 0$. Since only the ratio of reflected-to-incident-wave voltages is specified at the termination, we are led to ask for this same ratio at other points. Let $\bar{\Gamma}(z)$ represent it, so that

$$\bar{\Gamma}(z) \equiv \frac{V_-(z)}{V_+(z)} \tag{3.45}$$

We call $\bar{\Gamma}(z)$ the *generalized* (voltage) *reflection coefficient* on the line because it gives the ratio of the reflected wave voltage to the incident wave voltage at any point z. Actually the z dependence of $\bar{\Gamma}(z)$ is very simple, because $V_-(z) = V_-e^{j\beta z}$ and $V_+(z) = V_+e^{-j\beta z}$. Therefore,

$$\bar{\Gamma}(z) = \frac{V_-e^{j\beta z}}{V_+e^{-j\beta z}} = \frac{V_-}{V_+}e^{2j\beta z} = \bar{\Gamma}_R e^{2j\beta z} \tag{3.46}$$

As we move away from the load at $z = 0$ toward the generator, located somewhere at $z < 0$, the value of $\bar{\Gamma}(z)$ changes in the complex $\bar{\Gamma}$-plane, as shown in Fig. 3.13. At $z = 0$, it has the value $\bar{\Gamma}(0) = \bar{\Gamma}_R = (|V_-|e^{j\varphi_-}/|V_+|e^{j\varphi_+})$ fixed by the termination, and at values of $z < 0$, it has rotated clockwise by an angle $2\beta|z|$. Note, however, that its magnitude does not change with z, since

$$|\bar{\Gamma}(z)| = |\bar{\Gamma}_R e^{j2\beta z}| = |\bar{\Gamma}_R| \tag{3.47}$$

We can develop a greater familiarity with the significance of the reflection coefficient by writing the general solution (Eqs. 3.31) in terms of $\bar{\Gamma}(z)$.

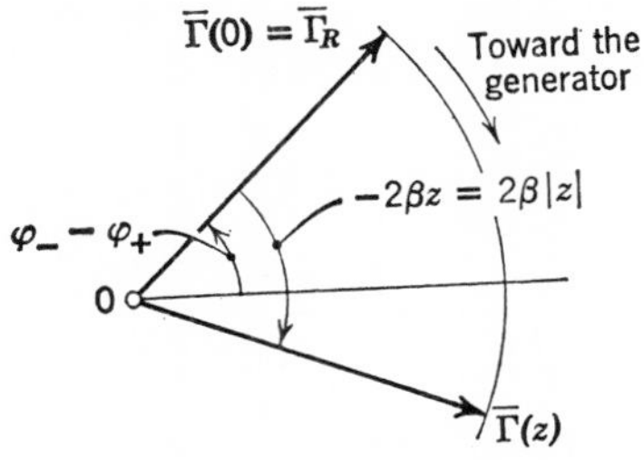

Fig. 3.13. The $\bar{\Gamma}$-plane, showing the z dependence of $\bar{\Gamma}(z)$ for negative values of z when the load is at $z = 0$.

$$\text{(a)} \quad \mathrm{V}(z) = \mathrm{V}_+ e^{-j\beta z}\left(1 + \frac{\mathrm{V}_-}{\mathrm{V}_+} e^{2j\beta z}\right) = \mathrm{V}_+ e^{-j\beta z}[1 + \bar{\Gamma}(z)]$$

$$\text{(b)} \quad \mathrm{I}(z) = Y_0 \mathrm{V}_+ e^{-j\beta z}\left(1 - \frac{\mathrm{V}_-}{\mathrm{V}_+} e^{2j\beta z}\right) = Y_0 \mathrm{V}_+ e^{-j\beta z}[1 - \bar{\Gamma}(z)] \tag{3.48}$$

The complex power flowing in the $+z$ direction at any point z is therefore

$$\mathrm{P}(z) = \tfrac{1}{2}\mathrm{V}(z)\,\mathrm{I}^*(z) = \tfrac{1}{2}Y_0|\mathrm{V}_+|^2[1 - |\bar{\Gamma}(z)|^2 + \bar{\Gamma}(z) - \bar{\Gamma}^*(z)] \tag{3.49}$$

Separating real and imaginary parts, and using Eq. 3.47, we find

$$\text{(a)} \quad \langle P \rangle = \tfrac{1}{2}Y_0|\mathrm{V}_+|^2[1 - |\bar{\Gamma}(z)|^2] = \tfrac{1}{2}Y_0|\mathrm{V}_+|^2[1 - |\bar{\Gamma}_R|^2]$$

$$\text{(b)} \quad jQ = \tfrac{1}{2}Y_0|\mathrm{V}_+|^2[\bar{\Gamma}(z) - \bar{\Gamma}^*(z)] \tag{3.50}$$

Equation 3.50*a* leads to several interesting conclusions:

1. The time-average power $\langle P \rangle$ is independent of z (see also Eq. 3.2).

2. The time-average power $\langle P \rangle$ is exactly the algebraic sum of the power which would be carried by the (+) wave *alone* and that which would be carried by the (−) wave *alone:* i.e., from the definition (Eq. 3.40) of $\bar{\Gamma}_R$ and Eqs. 3.50*a*, 3.12, and 3.17,

$$\langle P \rangle = \langle P_+ \rangle + \langle P_- \rangle \tag{3.51}$$

It follows that $|\bar{\Gamma}_R|^2$ represents physically the fraction of the incident power that is reflected by the load, or the ratio of reflected power to incident power. We must emphasize here that, although Eq. 3.51 is *a posteriori* very reasonable, on the basis of these physical notions of reflection on a lossless transmission line, it is hardly an *a priori* obvious result on general grounds, for it is rare, indeed, that the *powers* produced by two separate causes acting individually upon even a linear system can simply be added algebraically to determine the total power when both causes act together! For example, observe that the reactive power Q in Eq. 3.50*b* does *not* have this property.

3. If the termination or load at $z = 0$ is *passive* (i.e., if it cannot supply power to the line), then $\langle P \rangle \geq 0$. Accordingly, from Eq. 3.50*a*,

$$|\bar{\Gamma}(z)| = |\bar{\Gamma}_R| \leq 1 \qquad \text{for passive terminations} \tag{3.52}$$

Unlike the active (or time-average) power, the reactive power is not, in general, constant along the line. From Eq. 3.50*b* we can construct

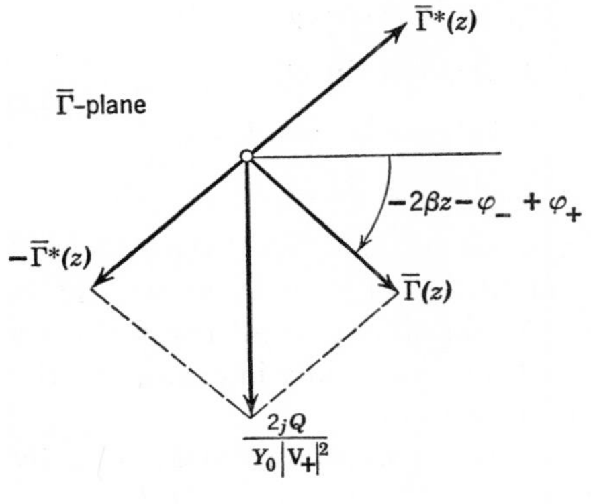

Fig. 3.14. The normalized reactive power for negative values of z, when the load is at $z = 0$.

the diagram of Fig. 3.14 to show $Q(z)$. It is sinusoidal in z, becoming zero when $\bar{\Gamma} - \bar{\Gamma}^* \equiv 2j \operatorname{Im}[\bar{\Gamma}(z)]$ is zero. From the geometry of the diagram, or from Eqs. 3.50*b* and 3.46, we see that

$$\frac{2jQ}{Y_0|\mathrm{V}_+|^2} = 2j|\bar{\Gamma}_R| \sin(2\beta z + \varphi_- - \varphi_+) \tag{3.53}$$

Since, in general, Q changes sign at intervals of $\lambda/4$ in z, the line has some $\lambda/8$ sections that store predominantly electric energy separated by other $\lambda/8$ sections that store predominantly magnetic energy.

3.4.3 Energy and Power

A closer study of the energy conditions on the line, as a function of time and distance, will help us to understand the energy transfer process. We note first that, according to Fig. 3.13, there is always a point z on the line where $\bar{\Gamma}(z)$ is real and positive, no matter what $\bar{\Gamma}_R$ may be. Let us therefore consider only the part of the line to the left of this point by selecting it as $z = 0$ and considering the termination at this point to have a new Γ_R' which is real and greater than zero. By expanding the exponentials in the general solution (Eq. 3.31) in terms of $\cos \beta z \pm j \sin \beta z$, we then have

$$\begin{aligned} &\text{(a)} \qquad \mathrm{V}(z) = \mathrm{V}_+[(1 + \Gamma_R') \cos \beta z - j(1 - \Gamma_R') \sin \beta z] \\ &\text{(b)} \qquad \mathrm{I}(z) = Y_0\mathrm{V}_+[(1 - \Gamma_R') \cos \beta z - j(1 + \Gamma_R') \sin \beta z] \end{aligned} \tag{3.54}$$

If we choose the time reference so that $\mathrm{V}_+ = V_+'$ is positive real, the instantaneous voltage and current become

$$\begin{aligned}
&\text{(a)} & V(z, t) &= V_+'[(1 + \Gamma_R') \cos \beta z \cos \omega t \\
&& &\quad + (1 - \Gamma_R') \sin \beta z \sin \omega t] \\
&\text{(b)} & I(z, t) &= Y_0 V_+'[(1 - \Gamma_R') \cos \beta z \cos \omega t \\
&& &\quad + (1 + \Gamma_R') \sin \beta z \sin \omega t]
\end{aligned} \tag{3.55}$$

The above form is simply the general solution, with the time and space origins chosen specially to make $\varphi_+ = \varphi_- = 0$, as we did in Sec. 3.3 for standing waves. This is again no important restriction on the situation, inasmuch as our present interest is only in examining the energy distribution and its motion on the line.

The instantaneous power may be written in a convenient form by using the trigonometric identities

$$\begin{aligned}
&\text{(a)} & \cos^2 \vartheta &= \tfrac{1}{2}(1 + \cos 2\vartheta) \\
&\text{(b)} & \sin^2 \vartheta &= \tfrac{1}{2}(1 - \cos 2\vartheta)
\end{aligned} \tag{3.56}$$

to eliminate $\sin^2$ and $\cos^2$ wherever they appear. Thus a little algebra shows that

$$\begin{aligned}
P(z, t) &= V(z, t)\, I(z, t) \\
&= \tfrac{1}{2} Y_0 V_+'^2[(1 - \Gamma_R'^2) + (1 - \Gamma_R'^2) \cos 2\beta z \cos 2\omega t \\
&\quad + (1 + \Gamma_R'^2) \sin 2\beta z \sin 2\omega t]
\end{aligned} \tag{3.57}$$

Observe that at points z where the reactive power is zero (Eq. 3.53), the instantaneous power never goes negative. These are the points where $\sin 2\beta z = 0$, or $z = -n(\lambda/4)$, with $n = 0, 1, 2, \cdots$.

Similarly, the instantaneous electric energy density is

$$\begin{aligned}
u_e(z, t) &= \tfrac{1}{2} C V^2(z, t) = \tfrac{1}{4} C V_+'^2\{(1 + \Gamma_R'^2) + 2\Gamma_R' \cos 2\beta z \\
&\quad + [2\Gamma_R' + (1 + \Gamma_R'^2) \cos 2\beta z] \cos 2\omega t \\
&\quad + (1 - \Gamma_R'^2) \sin 2\beta z \sin 2\omega t\}
\end{aligned} \tag{3.58}$$

and the magnetic energy density is

$$\begin{aligned}
u_m(z, t) &= \tfrac{1}{2} L I^2(z, t) = \tfrac{1}{4} C V_+'^2\{(1 + \Gamma_R'^2) - 2\Gamma_R' \cos 2\beta z \\
&\quad + [-2\Gamma_R' + (1 + \Gamma_R'^2) \cos 2\beta z] \cos 2\omega t \\
&\quad + (1 - \Gamma_R'^2) \sin 2\beta z \sin 2\omega t\}
\end{aligned} \tag{3.59}$$

so the total energy density becomes

$$\begin{aligned}
u(z, t) &= u_e(z, t) + u_m(z, t) \\
&= \tfrac{1}{2} C V_+'^2[(1 + \Gamma_R'^2)(1 + \cos 2\beta z \cos 2\omega t) \\
&\quad + (1 - \Gamma_R'^2) \sin 2\beta z \sin 2\omega t]
\end{aligned} \tag{3.60}$$

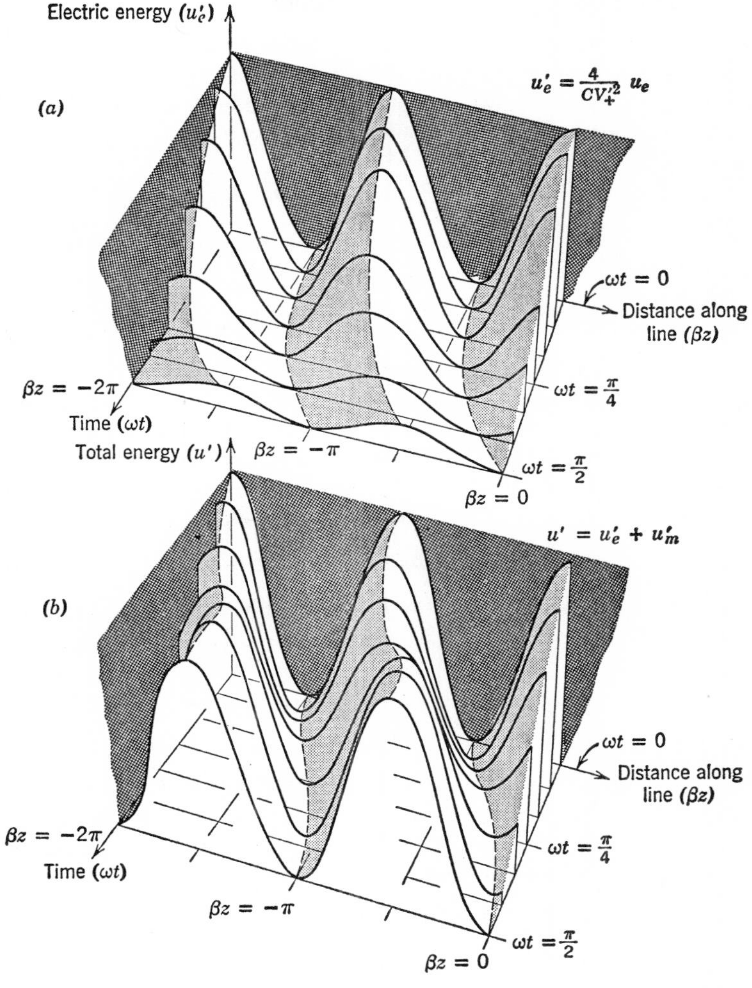

Fig. 3.15. Normalized surfaces of the time-dependent (*a*) electric energy density (u_e') and (*b*) total energy density (u') along a semi-infinite lossless transmission line with partially reflecting termination ($\Gamma_R' = \frac{1}{2}$).

Pictorial models of normalized electric energy density, $u_e' = 4u_e/CV_+'^2$, and normalized total energy density, $u' = 4u/CV_+'^2$, appear in Fig. 3.15. (From Eqs. 3.58 and 3.59 we note that the magnetic energy density may be obtained from the electric one by adding $\pi/2$ to *both* βz and ωt.) Observe that a line $\omega t = \beta z$ drawn in the time-distance plane has a slope which represents phase velocity. None of the energy densities remain constant along such a line; so we cannot view the energy flow in the general (steady-state) case as proceeding with the phase velocity.

From Eq. 3.60, however, it is clear that *the time-average total energy density is not a function of* z

$$\langle u \rangle = \tfrac{1}{2}CV_+'^2(1 + \Gamma_R'^2) \tag{3.61}$$

and of course the time-average power flow is also independent of z (Eq. 3.57)

$$\langle P \rangle = \tfrac{1}{2}Y_0V_+'^2(1 - \Gamma_R'^2) \tag{3.62}$$

Therefore, if we reason that, on the time average, the flow of energy (i.e., the power in joules/sec at a selected point z_1) should be accounted for by a motion of the stored energy from $z < z_1$ to $z > z_1$, then in one average second all the energy stored in a length v_E will cross z_1 if $v_E\langle u \rangle = \langle P \rangle$. Thus we might associate an "average energy" velocity v_E with the flow, such that

$$v_E \equiv \frac{\langle P \rangle}{\langle u \rangle} = \left(\frac{Y_0}{C}\right)\frac{1 - \Gamma_R'^2}{1 + \Gamma_R'^2}$$

$$= v_p\left(\frac{1 - |\bar{\Gamma}_R|^2}{1 + |\bar{\Gamma}_R|^2}\right)$$

This kind of a velocity is of somewhat doubtful physical significance beyond the particular definition we have used to obtain it, since nothing is actually moving with this speed. Still, it reduces to v_p when $\bar{\Gamma}_R = 0$ (i.e., when we have a pure traveling wave), *which provides an additional interpretation of* v_p. Also $v_E < v_p$ when $0 < \bar{\Gamma}_R < 1$, and $v_E = 0$ when $|\bar{\Gamma}_R| = 1$ (i.e., when we have a complete standing wave).

A somewhat clearer picture of the relation between the electric, magnetic, and total energies appears from the sketches in Fig. 3.16. These show the normalized energy densities as functions of position, at various times (Fig. 3.16*a*), as well as the power as a function of position at the same times (Fig. 3.16*b*). Focusing our attention first on the peak marked × of the electric energy and that marked ○ of the magnetic energy, we note that both these distributions always move to the right, but with varying speeds. Indeed, they slide by each

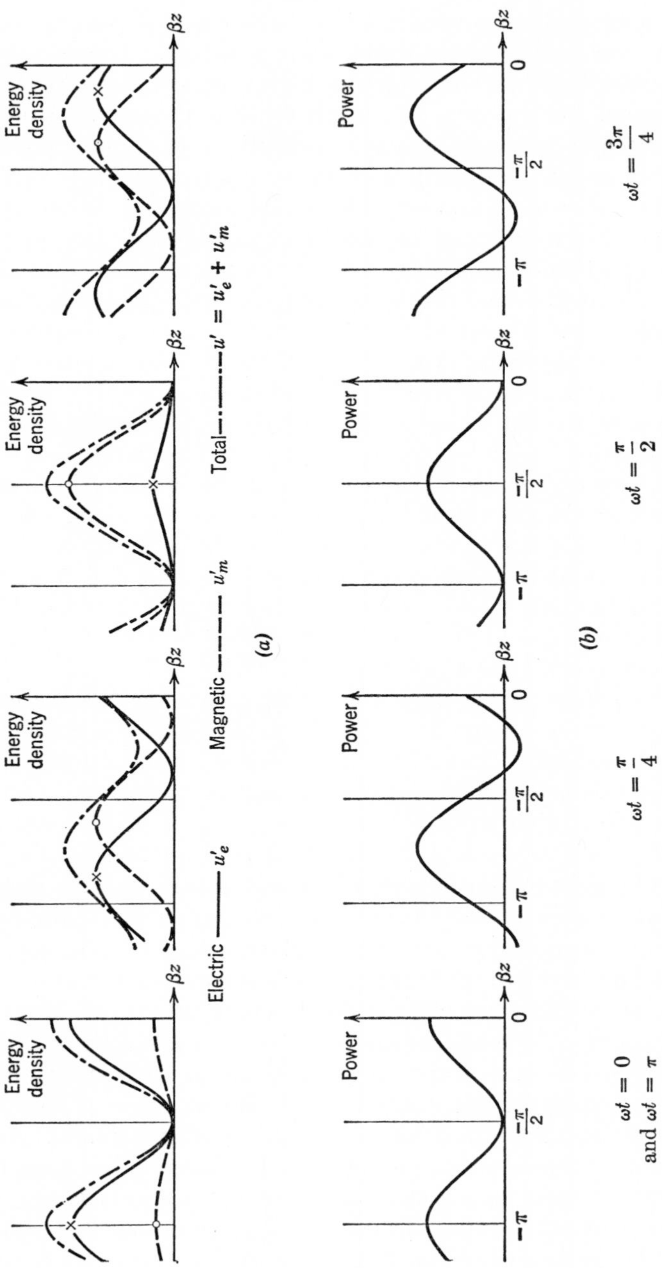

Fig. 3.16. General standing wave; $\Gamma_2' = \frac{1}{2}$ at $z = 0$. (*a*) Stored energy densities (normalized); (*b*) power (normalized).

other with an oscillating relative motion which puts first one ahead of the other, and then reverses their relative positions. The energy motion is a little like a man on a bicycle. His progress may always be in one direction, but his legs pass each other alternately. Moreover the electric and magnetic energy distributions (the "legs") change size as they move, being most *unequal* when they are *in* space phase ($\omega t = 0$ and $\omega t = \pi/2$), and becoming equal when they differ most in space phase ($\omega t = \pi/4$ and $\omega t = 3\pi/4$). Specifically, the one that is smaller at a time when the two are in space phase always runs ahead more rapidly at first, and simultaneously grows in size at the expense of the larger one. Then it slows down somewhat, but it continues to grow in size at the other's expense, until the two are in space phase again. The relative sizes have become interchanged by this time, and the process repeats in a dual way. The complete restoration of the original situation occurs at time intervals of one half-period (i.e., π/ω). However, two differences make the man-on-bike analogy inaccurate. The space variation of the *total* energy distribution (the "man") changes its amplitude as it moves, and it does not proceed at a uniform speed. Figure 3.15 shows this last point more clearly than does Fig. 3.16. Viewing the line as a whole, we see that the energy on it changes from mostly electric ($\omega t = 0$) to mostly magnetic ($\omega t = \pi/2$), and back to mostly electric again ($\omega t = \pi$) in a half-period.

Now, by concentrating upon conditions within the $\lambda/4$ section of line between $\beta z = -\pi$ and $\beta z = -\pi/2$, we can relate the power (Fig. 3.16*b*) to the energies. As we have noted before, at the two boundaries of this section, the instantaneous power never goes negative. It always flows *into* the section at $\beta z = -\pi$ and *out* of it at $\beta z = -\pi/2$. At $\omega t = 0$, there is a large power flowing into the cell and none flowing out. This corresponds to the large total energy hump (mostly electric) moving into it across $\beta z = -\pi$. From $\omega t = 0$ to $\omega t = \pi/4$, the power input to the section decreases and that coming out increases, corresponding primarily to the building up of magnetic energy near $-\pi/2$ in the section at the expense of the electric energy near $-\pi$. However, the total energy in the section does increase because of the *net* power being delivered to it, as shown by motion of the total energy hump into this region of the line. At $\omega t = \pi/4$, the *net* power entering the section is zero, and the total energy stored in it is a maximum. When $\pi/4 < \omega t < \pi/2$, *net* power flows *out* of the section as the large total energy hump (now mostly magnetic) moves toward $\beta z = -\pi/2$. This process actually continues with diminishing force until $\omega t = 3\pi/4$, when the magnetic energy hump has decreased and moved well out of the section, and the electric energy entering it has only partially built up. The *net* power delivered to the line section is again zero, but this time the total energy stored in it is a minimum. From $\omega t = 3\pi/4$ to $\omega t = \pi$, the *net* power input to the section increases, a new hump of total energy (mostly electric) builds up and begins to enter at $\beta z = -\pi$, while the magnetic energy decreases everywhere. At $\omega t = \pi$ the entire process starts over again as it did at $\omega t = 0$. It is instructive to compare Fig. 3.16 with Figs. 3.3, 3.7, and 3.9.

3.4.4 Standing-Wave Parameters

We come now to the general question of making measurements on a transmission line. If we could bridge an ideal a-c voltmeter across the line, and slide it slowly along z, we would measure the *magnitude* of the voltage, $|V(z)|$. Similarly, if we could insert an ideal a-c ammeter in series with either line conductor, and then slide the meter slowly along z, we would measure $|I(z)|$. From Eqs. 3.48 these magnitudes are

$$\begin{aligned} &\text{(a)} \qquad |V(z)| = |V_+|\,|1 + \bar{\Gamma}(z)| \\ &\text{(b)} \qquad |I(z)| = Y_0|V_+|\,|1 - \bar{\Gamma}(z)| \end{aligned} \tag{3.63}$$

Based upon Eqs. 3.63, Fig. 3.17 shows a simple construction in the $\bar{\Gamma}$-plane for determining the variations of $|V(z)|$ and $|I(z)|$. To the complex number $1 + j0$ drawn from the point $(-1, 0)$ to the point $(0, 0)$ are added the complex numbers $\bar{\Gamma}(z)$ and $-\bar{\Gamma}(z)$. The magnitude of the voltage is proportional to $|1 + \bar{\Gamma}(z)|$ and the magnitude of the current to $|1 - \bar{\Gamma}(z)|$. Note that $|V_+|$ is independent of z. We have assumed a passive termination, so that $|\bar{\Gamma}(z)| \leq 1$. The figure shows clearly that the magnitude of the current is smallest at the points where that of the voltage is largest, and vice versa. Moreover,

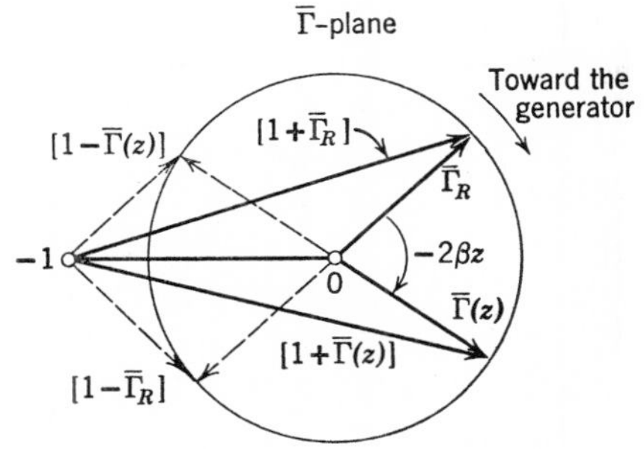

Fig. 3.17. $\bar{\Gamma}$-plane construction of normalized current and voltage as functions of position.

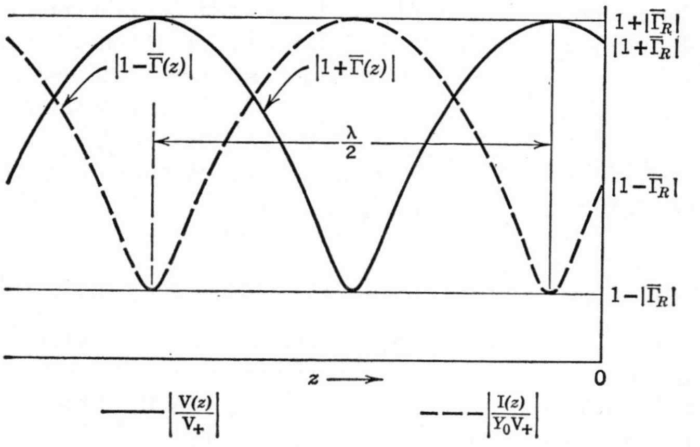

Fig. 3.18. Pattern of the normalized magnitudes of voltage and current along a transmission line.

in conjunction with Eq. 3.48, the figure indicates that the voltage and current are in time phase at the points of maximum or minimum magnitude of either. A rotation of $\bar{\Gamma}(z)$ by an angle π corresponds to a change of $\lambda/4$ in z; thus any voltage maximum is separated by $\lambda/4$ from the two nearest voltage minima.

Another way of showing the variations of $|V(z)|$ and $|I(z)|$ appears in Fig. 3.18, which is simply a plot of the *lengths* of the phasors $[1 \pm \bar{\Gamma}(z)]$ as scaled (or computed) from Fig. 3.17. Such a plot gives what is called the *standing-wave pattern* on the line, although, of course, the solution is not generally a *complete* standing wave at all. Observe carefully, from the geometry of Fig. 3.17, that the patterns of Fig. 3.18 are not generally sinusoidal in shape. The minima are sharper than the maxima; so the minima can usually be located in position more precisely by measurement than the maxima.

In view of Eq. 3.47 and Fig. 3.17, the ratio of the maximum voltage magnitude to the minimum voltage magnitude is

$$\frac{|V(z)|_{\max}}{|V(z)|_{\min}} = \frac{1 + |\bar{\Gamma}_R|}{1 - |\bar{\Gamma}_R|} \equiv \textit{Voltage Standing-Wave Ratio} \text{ (VSWR)} \tag{3.64}$$

It can be measured by recording the largest and smallest readings of our sliding voltmeter. Since only their ratio is involved, however, the instrument need only be guaranteed to read a quantity linearly proportional to the actual voltage; it need not be calibrated absolutely in volts. Thus a measurement of only *relative* voltage yields the standing-wave ratio which, from Eq. 3.64, gives uniquely $|\bar{\Gamma}_R|$.

The angle $-\pi < \varphi \leq \pi$ of $\bar{\Gamma}_R = |\bar{\Gamma}_R| e^{j\varphi}$ can also be determined from these so-called standing-wave measurements. According to Fig. 3.17, $\bar{\Gamma}(z)$ must swing clockwise through an angle $\varphi + \pi$ as we move from the load at $z = 0$ toward the generator to the *first* voltage minimum on the line for $z < 0$. The distance $d_{\min}$ which we must move to reach this first voltage minimum is therefore given by

$$2\beta d_{\min} = \varphi + \pi$$

or

$$\varphi/\pi = 4\left(\frac{d_{\min}}{\lambda}\right) - 1 \tag{3.65}$$

A measurement of $d_{\min}$ from the standing-wave pattern, as well as a determination of the wave length λ from this same pattern (the distance between successive voltage minima is $\lambda/2$), therefore tells us the angle of the complex reflection coefficient of the load. It follows that, by making only *relative* voltage measurements (or current measurements) and some distance measurements on a terminated line, we can determine completely the complex reflection coefficient $\bar{\Gamma}_R$ of any passive load. Indeed, on account of Eqs. 3.42 or 3.44, these data actually determine the complex load impedance *in* Z_0 *units* (i.e., the ratio $Z_R/Z_0 \equiv$ normalized load impedance $\equiv Z_{nR}$). Since we can usually either compute Z_0 from the structure of the line itself or measure it by any one of several simple schemes, we usually know it in advance for any particular line being used. Thus, standing-wave measurements are ordinarily sufficient to determine the terminating impedance Z_R. These measurement properties of the load reflection coefficient, and its relation to the normalized load impedance, are of great importance at high frequencies where the *absolute* measurement of either voltage or current may be almost impossible. Indeed, the possibility of making a direct determination of the load reflection coefficient from standing-wave measurements gives to that coefficient an operational significance that should convince us of its usefulness and importance.

3.4.5 Generalized Impedance

The importance of the generalized reflection coefficient $\bar{\Gamma}(z)$, and its very simple variation with position on the line, should not make us forget about other quantities in which we may be interested. Suppose, for example, that we had a finite length l of line, loaded at $z = 0$ by a termination Z_R (or $\bar{\Gamma}_R$), with input terminals at $z = -l$ (Fig. 3.19).

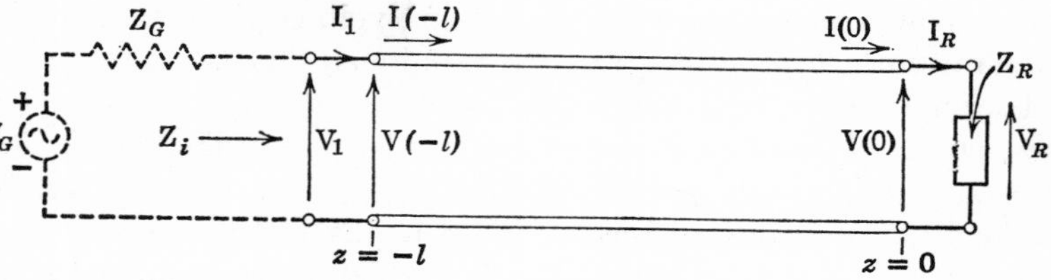

Fig. 3.19. A finite section of terminated transmission line.

In connection with the problem of driving the line from some source at the input, we would often like to know the input impedance Z_i presented to the source. This is, of course, defined as

$$Z_i \equiv \frac{V_1}{I_1} \tag{3.66}$$

But the current I_1 must equal the line current $I(-l)$, and the voltage V_1 must equal the line voltage $V(-l)$; so

$$Z_i = \frac{V(-l)}{I(-l)}$$

$$= Z_0 \frac{1 + \bar{\Gamma}(-l)}{1 - \bar{\Gamma}(-l)} = Z_0 \frac{1 + \bar{\Gamma}_R e^{-2j\beta l}}{1 - \bar{\Gamma}_R e^{-2j\beta l}} \tag{3.67}$$

where we have used Eqs. 3.48 and 3.46. The result (Eq. 3.67) tells us how the input impedance to a line of length l depends upon this length, Z_0, and Z_R (through $\bar{\Gamma}_R$). In fact, along with Eq. 3.42, Eq. 3.67 shows how the line of length l *transforms* a load impedance Z_R into an input impedance Z_i.

Another way of viewing the action of the line stems from the above considerations. Given Z_R and Z_0, or $Z_{nR} = Z_R/Z_0$, we know $\bar{\Gamma}_R$ from Eq. 3.44. We then recall the *generalized impedance* $Z(z)$ from Sec. 3.3, defined as the ratio of voltage $V(z)$ to current $I(z)$ at *any* point z on the line. Thus the use of Eq. 3.48 shows that, for the general solution,

$$Z(z) = \frac{V(z)}{I(z)} = Z_0 \left(\frac{1 + \bar{\Gamma}(z)}{1 - \bar{\Gamma}(z)}\right) \tag{3.68}$$

or

(a) $$\frac{Z(z)}{Z_0} \equiv Z_n(z) = \frac{1 + \bar{\Gamma}(z)}{1 - \bar{\Gamma}(z)}$$

and (3.69)

(b) $$\bar{\Gamma}(z) = \frac{Z_n(z) - 1}{Z_n(z) + 1}$$

The generalized impedance $Z(z)$ is therefore related to the generalized reflection coefficient $\bar{\Gamma}(z)$ exactly as the load impedance Z_R is related to the load reflection coefficient $\bar{\Gamma}_R$; see Eqs. 3.41 to 3.45. The significance of the generalized impedance is seen, in terms of Fig. 3.19 and Eq. 3.67, to be the impedance that would be presented by the terminated line to a generator connected at any point z as input. Since we know from Eqs. 3.46 and 3.44 how $\bar{\Gamma}(z)$ depends upon both z and the normalized load impedance Z_{nR}, Eq. 3.69 tells us how the normalized generalized impedance $Z_n(z)$ varies with position. Equation 3.69, therefore, represents the way in which the load impedance is *transformed* by any length $|z|$ of line.

All features of the variation of the normalized generalized impedance $Z_n(z)$ can be determined from Fig. 3.17, as we can easily verify by glancing quickly at Eq. 3.69. In terms of the figure, and our previous discussion of it, we now learn by inspection that:

1. $Z_n(z)$ is *pure real* and has a *maximum magnitude* at the position $z_{\max}$ of any *voltage maximum* on the line. Indeed,

$$Z_n(z_{\max}) = |Z_n(z)|_{\max} = \frac{1 + \Gamma(z_{\max})}{1 - \Gamma(z_{\max})} = \frac{1 + |\bar{\Gamma}(z)|}{1 - |\bar{\Gamma}(z)|} = \text{VSWR} \tag{3.70}$$

2. $Z_n(z)$ is *pure real* and has a *minimum magnitude* at the position $z_{\min}$ of any *voltage minimum* on the line. Specifically,

$$Z_n(z_{\min}) = |Z_n(z)|_{\min} = \frac{1 + \Gamma(z_{\min})}{1 - \Gamma(z_{\min})} = \frac{1 - |\bar{\Gamma}(z)|}{1 + |\bar{\Gamma}(z)|} = (\text{VSWR})^{-1} \tag{3.71}$$

3. $Z_n(z)$ is periodic in z, with period $\lambda/2$.

$$Z_n\left(z - \frac{\lambda}{2}\right) = Z_n(z) \tag{3.72}$$

4. A section of line $\lambda/4$ in length transforms Z_n into $1/Z_n$. This occurs because $\bar{\Gamma}(z - \lambda/4) = -\bar{\Gamma}(z)$, thereby simply exchanging the positions of the normalized voltage and current phasors in Fig. 3.17. In other words,

$$Z_n\left(z - \frac{\lambda}{4}\right) = \frac{1}{Z_n(z)} \tag{3.73}$$

or

$$Z\left(z - \frac{\lambda}{4}\right) Z(z) = Z_0^2 \tag{3.74}$$

Since voltage maxima and voltage minima are separated by $\lambda/4$, $Z_n(z_{\min})$ and $Z_n(z_{\max})$ in Eqs. 3.71 and 3.70 satisfy Eq. 3.73

3.4.6 Voltage and Current

In some problems, e.g., those involved in the distribution of electric power, it is important to know the actual voltage and current at various points on the line when generators and loads are connected to it. When the distribution system is supposed to be at constant line voltage, for example, it might be necessary to determine how V_R in Fig. 3.19 would vary with load Z_R, if the generator (V_G, Z_G) were fixed. All we have studied so far involves only the *ratio* of incident to reflected waves. The voltage level of either is related to the source (Fig. 3.19). If we know the load impedance Z_R and the characteristic impedance Z_0, however, our previous work serves to determine $\bar{\Gamma}(-l)$ and, therefore, also Z_i. The voltage V_1 is constrained by the generator circuit to be

$$V_1 = \left(\frac{Z_i}{Z_i + Z_G}\right) V_G = \frac{1 + \bar{\Gamma}(-l)}{1 + \bar{\Gamma}(-l) + Z_{nG}[1 - \bar{\Gamma}(-l)]} V_G \tag{3.75}$$

where $Z_{nG} \equiv Z_G/Z_0$. The terminated line, on the other hand, requires

$$V(-l) = V_+ e^{j\beta l}[1 + \bar{\Gamma}(-l)] \tag{3.76}$$

according to Eq. 3.48*a* applied at $z = -l$. Equality of V_1 with $V(-l)$ determines V_+,

$$V_+ = \frac{V_G e^{-j\beta l}}{(1 + Z_{nG}) + (1 - Z_{nG})\bar{\Gamma}(-l)} \tag{3.77}$$

which, in turn, gives V_R as

$$V_R = V_+ + V_- = V_+(1 + \bar{\Gamma}_R) = \frac{V_G(1 + \bar{\Gamma}_R)e^{-j\beta l}}{(1 + Z_{nG}) + (1 - Z_{nG})\bar{\Gamma}(-l)} \tag{3.78}$$

In fact, once we know V_+ from Eq. 3.77, we can determine the line current or line voltage anywhere we wish by using Eqs. 3.48, or the equivalent normalized chart of Fig. 3.17.

The usefulness and simplicity of the variation of $\bar{\Gamma}(z)$ (Eq. 3.46 and Fig. 3.13) are marred somewhat by the need to convert back and forth between $\bar{\Gamma}$ and Z_n, using Eqs. 3.69. Since, in general, $\bar{\Gamma}$ and Z_n are complex numbers, and typical problems may require several applications of these formulas, the numerical work involved becomes tedious. A graphical aid of some sort would be desirable. There are many such aids in existence, but we shall consider only one of them here. It is a graphical representation of the complex transformation (Eq. 3.69), arranged to take maximum advantage of the simplicity of $\bar{\Gamma}(z)$ in Eq. 3.46, and based on the utility of Fig. 3.17.

3.5 The Smith Chart

If we consider the $\bar{\Gamma}$-plane construction of Fig. 3.17, we realize that it would be useful to show on it a network of curves representing the loci of $\bar{\Gamma}$ for various fixed values of Re (Z_n), and loci of $\bar{\Gamma}$ for various fixed values of Im (Z_n). Then any complex value of $\bar{\Gamma}$ would fall on the intersection of two such loci, from which we could read at once the corresponding value of Z_n in rectangular complex form $Z_n = r + jx$. Moreover, since we know that all passive loads $(r \geq 0)$ lead to $|\bar{\Gamma}| \leq 1$, it follows that all the loci mentioned above will be contained within the unit circle $|\bar{\Gamma}| = 1$ of the $\bar{\Gamma}$-plane in Fig. 3.17, as long as we consider only passive terminations of the line. It remains only to determine the details of the loci.

The complex transformation to be represented is

$$\bar{\Gamma} = \frac{Z_n - 1}{Z_n + 1} \tag{3.79}$$

In particular, we wish first to determine the locus of $\bar{\Gamma}$ in the $\bar{\Gamma}$-plane which corresponds to the straight line $Z_n = r_0 + jx$ in the Z_n-plane, where r_0 is fixed ≥ 0 and x runs the range $-\infty \leq x \leq \infty$ (dotted lines in Fig. 3.20*b*). We note that Eq. 3.79 is a linear fractional trans-

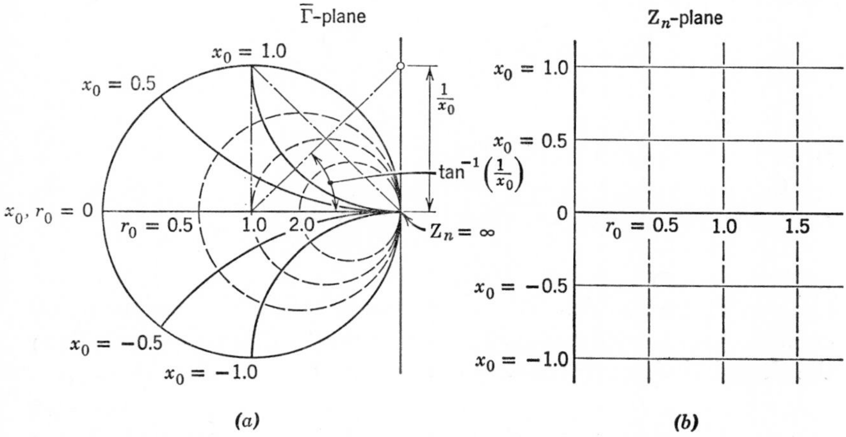

Fig. 3.20. The transformation $\bar{\Gamma} = (Z_n - 1)/(Z_n + 1)$.

formation. It therefore transforms circles (or straight lines) in the Z_n-plane into circles (or possibly straight lines) in the $\bar{\Gamma}$-plane.[1] Moreover, replacement of Z_n by Z_n^* results in changing $\bar{\Gamma}$ to $\bar{\Gamma}^*$. Hence the $\bar{\Gamma}$-map of the line $Z_n = r_0 + jx$ is symmetrical about the real axis of the $\bar{\Gamma}$-plane. Therefore, we look for a circle in the $\bar{\Gamma}$-plane whose diameter lies along the real axis. Such a circle is fixed by any *two* points on it. From Eq. 3.79 we see that for $x = 0$

$$\bar{\Gamma}_{(x=0)} = \frac{r_0 - 1}{r_0 + 1} \tag{3.80}$$

and for $x = \infty$

$$\bar{\Gamma}_{(x=\infty)} = +1 \tag{3.81}$$

The dotted circles in Fig. 3.20*a* have been drawn to conform with the symmetry requirements through points determined by Eqs. 3.80 and 3.81.

Next we wish to find the $\bar{\Gamma}$-map of the straight line $Z_n = r + jx_0$, where x_0 is held fixed at any value and r runs the range $0 \leq r \leq \infty$ (solid lines in Fig. 3.20*b*). In this case, the conjugate symmetry of Eq. 3.79 means that we need consider in detail only $x_0 \geq 0$. Moreover, since the lines $Z_n = r + jx_0$ are orthogonal to the lines $Z_n = r_0 + jx$, the corresponding maps in the $\bar{\Gamma}$-plane must also be orthogonal. This is the "conformal" property of the transformation (Eq. 3.79). Therefore we look for a circle orthogonal to the dotted circles in Fig. 3.20*a*. It passes through the point $\bar{\Gamma} = +1$ because from Eq. 3.79

$$\bar{\Gamma}_{(r=\infty)} = +1 \tag{3.82}$$

When $r = 0$, we have

$$\bar{\Gamma}_{(r=0)} = \frac{jx_0 - 1}{jx_0 + 1} = 1\,\underline{/\pi - 2\tan^{-1} x_0}$$

$$= 1\,\underline{\Big/ 2\left(\frac{\pi}{2} - \tan^{-1} x_0\right)} \tag{3.83}$$

$$= 1\,\underline{\Big/ 2\tan^{-1}\left(\frac{1}{x_0}\right)}$$

lying on the unit circle. The only circles that pass through the points defined by Eqs. 3.82 and 3.83, and can possibly be orthogonal to all the dotted circles, are those shown in Fig. 3.20*a* by solid lines. The

[1] Readers not familiar with this fact are referred to E. A. Guillemin, *The Mathematics of Circuit Analysis*, The Technology Press, Mass. Inst. of Technology, 1949, Ch. VI, Art. 24, 360–378.

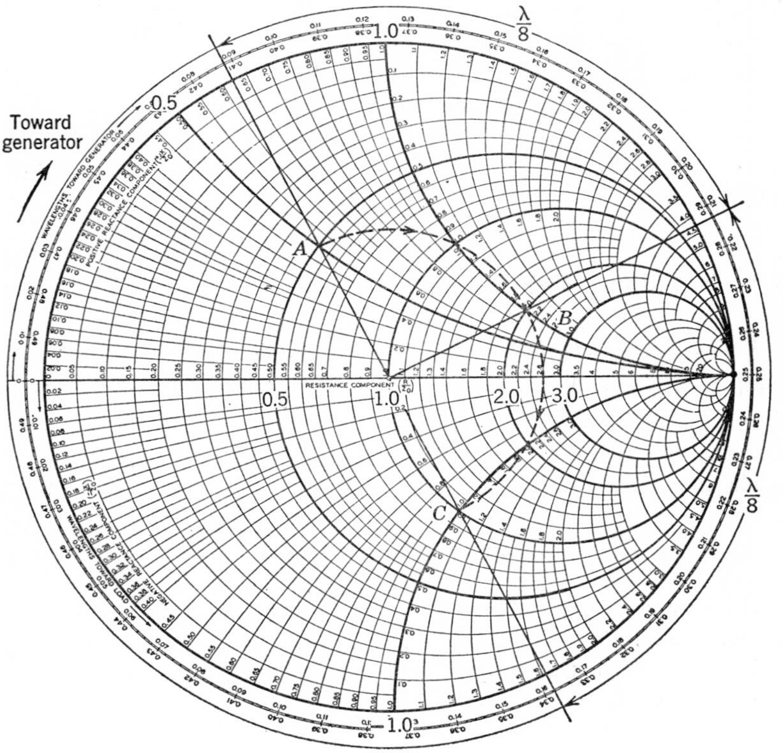

Fig. 3.21. For Example 3.1.

geometry of the figure indicated in dash-dot lines can easily be used, along with Eqs. 3.82 and 3.83, to show that the center of these circles lies at the point $\bar{\Gamma} = 1 + j(1/x_0)$. Their radii are therefore $1/|x_0|$.

The chart of Fig. 3.20*a*, comprised of the dotted and solid circles, and having values of r_0 and x_0 placed on it as shown, is known as the Smith chart. *The construction of Fig. 3.17 should be regarded as superposed upon it to make clear the manner of its use.* Since rotations of $\bar{\Gamma}$ are necessary in solving line problems, a scale of distance in fractions of a wave length is added around the circumference of the unit circle. Inasmuch as motion a distance Δz on the line requires a rotation of $2\beta\,\Delta z = 4\pi\,(\Delta z/\lambda)$, the scale goes *completely around* the unit circle (2π radians) in a range of only *one-half* wave length (i.e., $0 \leq \Delta z/\lambda \leq +\,0.50$). This relieves us of the multiplication by 2 required in the exponent of Eq. 3.46; it is included already in the chart construction.

Also, to ease the burden on memory, the chart carries a designation of the correct rotation direction for motion "toward the generator." It corresponds to the fact shown in Fig. 3.17 that motion along the line *from* the load *toward* the generator (or source) produces a *clockwise* rotation of $\bar{\Gamma}(z)$. A preliminary example worked out on the chart reproduced in Fig. 3.21 will familiarize us with some of these detailed properties.

3.6 Impedance Calculation

EXAMPLE 3.1. LINE IMPEDANCE DETERMINATION

Given the terminating impedance $Z_R = 25 + j25$ ohms, and the characteristic impedance of the line $Z_0 = 50$ ohms, what impedance is seen at the points $\lambda/8$ and $\lambda/4$ away from the termination?

The normalized terminating impedance is

$$Z_{nR} = \frac{Z_R}{Z_0} = 0.5 + j0.5$$

The corresponding value of $\bar{\Gamma}_R$ is found on the Smith chart (point A of Fig. 3.21), but need not be recorded. Indeed, there is no convenient scale provided for this purpose, although a graduated rotating arm is sometimes used.

The value of $\bar{\Gamma}$ a distance $\lambda/8$ away is found by rotating the phasor $\bar{\Gamma}_R$ by an angle $+2\beta z = -(4\pi/\lambda)(\lambda/8) = -\pi/2$ to point B. A protractor is unnecessary since distances in fractions of a wave length are indicated on the Smith chart. Observe that the rotation direction is *toward the generator.*

The normalized impedance at $z = -\lambda/8$ can be read from the Smith chart, and is found to be

$$Z_n = 2.0 + j1.0$$

The actual impedance is therefore

$$Z = Z_n Z_0 = 100 + j50$$

The normalized impedance at a distance $\lambda/4$ away from the termination is obtained by a rotation of $\bar{\Gamma}_R$ by $-(4\pi/\lambda)(\lambda/4) = -\pi$ to point C. The reflection coefficient at this point should be $\bar{\Gamma} = \bar{\Gamma}_R e^{2\beta z} = \bar{\Gamma}_R e^{-j\pi} = -\bar{\Gamma}_R$, which evidently checks the chart result. The normalized impedance read from the chart is

$$Z_n = 1.0 - j1.0$$

It is the inverse of Z_{nR}, as required in general by Eq. 3.73 for $\lambda/4$ lines.

Another important feature of the chart shows up from Eq. 3.70. The standing-wave ratio on the line is numerically equal to the largest normalized impedance magnitude seen anywhere on it. This is always a pure resistance, found at the intersection of the circle $|\bar{\Gamma}| = |\bar{\Gamma}_R| =$ const., with the positive real $\bar{\Gamma}$-axis. In our example above, VSWR $= |Z_n(z)|_{\max} = r_{\max} = 2.6$. This matter is clarified by imagining Fig. 3.17 superposed upon Fig. 3.21.

A significant feature of the chart, as used in Example 3.1 for impedance calculations, is the fact that, as long as we start with an impedance value and end with one, the reflection coefficient itself is never employed *explicitly*. We need never bother to read its complex value at all, even though we are working in the $\bar{\Gamma}$-plane. This peculiar situation results from the fact that we really use the transformation Eq. 3.69*b* or Eq. 3.79 to *enter* the chart with an impedance value, and then we use the *inverse* transformation (Eq. 3.69*a*) to *leave* the chart after rotation. Curiously enough, this means that, for many problems, the chart itself can be used with *normalized admittance* values simply by going blindly ahead *pretending they are normalized impedance values instead!* The underlying theory for this crass-sounding procedure is as follows:

Suppose we were given a normalized load admittance $Y_{nR} = Y_R/Y_0$, and were asked to find the normalized admittance at some other point z on the line. A straightforward procedure would require that we compute first, by hand,

$$Z_{nR} = \frac{Z_R}{Z_0} = \frac{Y_0}{Y_R} = \frac{1}{Y_{nR}} \tag{3.84}$$

Then we would enter the Smith chart at Z_{nR}, thereby determining $\bar{\Gamma}_R$. Next we rotate this reflection coefficient by an angle $2\beta z$, according to the distance z, and read $Z_n(z)$ from the chart. Finally we would have to compute by hand the inversion

$$Y_n(z) = \frac{1}{Z_n(z)} \tag{3.85}$$

to get the desired answer.

The two inversion computations in Eqs. 3.84 and 3.85 are laborious to carry out by hand. Fortunately we can use the Smith chart itself *as a computer* to perform them. We use the fact that normalized impedance is inverted by a $\lambda/4$ section of line, so that a rotation of $\bar{\Gamma}$ by 180° (at constant $|\bar{\Gamma}|$) changes a normalized impedance into its reciprocal.

Hence, to do our problem now, we would enter the Smith chart as a computer with the given value of Y_{nR}, and rotate $\bar{\Gamma}_R$ by 180°, at constant $|\bar{\Gamma}|$. This step computes and locates Z_{nR}. We then rotate $\bar{\Gamma}$ again at constant $|\bar{\Gamma}|$ according to the distance z. This application of the Smith chart to the line action determines $Z_n(z)$. Finally we rotate $\bar{\Gamma}$ by 180° at constant $|\bar{\Gamma}|$ to perform a pure computation of $Y_n(z) = 1/[Z_n(z)]$. Note that we have actually rotated the reflection coefficient at constant $|\bar{\Gamma}|$ by an angle $\pi + 2\beta z + \pi = 2\pi + 2\beta z$ in the whole process. Rotation by 2π on the Smith chart, however, merely brings us back to wherever we started. Only the $2\beta z$ rotation is significant. So why bother with the π rotations at all, as long as we plan to do an even number of them (i.e., *enter* with a Y_n and *leave* again with a Y_n)? The correct answer will obviously be obtained in such cases by simply omitting them.

The final procedure is then simplified:

1. Enter the chart with the complex value Y_{nR}, *exactly as if it were a normalized impedance.*
2. Rotate the reflection coefficient in the regular direction, according only to the distance z.
3. Read $Y_n(z)$ from the chart, *exactly as if it were a normalized impedance.*

The simplified technique of using the Smith chart directly as an admittance chart fails in only one (rare) circumstance. If we actually wished to read the complex reflection coefficient (in magnitude and angle) corresponding to some Y_n, the straightforward *correct* procedure would obviously be to enter with Y_n, rotate $\bar{\Gamma}$ by π to find Z_n, and then measure $\bar{\Gamma}$ at this new point. If we measured $\bar{\Gamma}$ at the original Y_n point, it is clear that we would obtain $-\bar{\Gamma}$ instead of $\bar{\Gamma}$; i.e., we would commit an error of 180° in the angle of the reflection coefficient.

Example 3.2 illustrates the use of the Smith chart on an admittance basis.

EXAMPLE 3.2. SINGLE-STUB MATCHING

A termination $Z_R = 60 - j80$ of a 50-ohm transmission line is to be "matched" to the line by means of a susceptance connected in parallel, at a proper distance from the termination. The susceptance is formed by a short-circuited section of a 50-ohm line of proper length, called a "stub." Figure 3.22 shows the present connections. Two adjustable parameters are necessary, the lengths l_1 and l_2.

The important thing about a lossless stub is that, by varying its length, its susceptance (or reactance) can be made any value desired

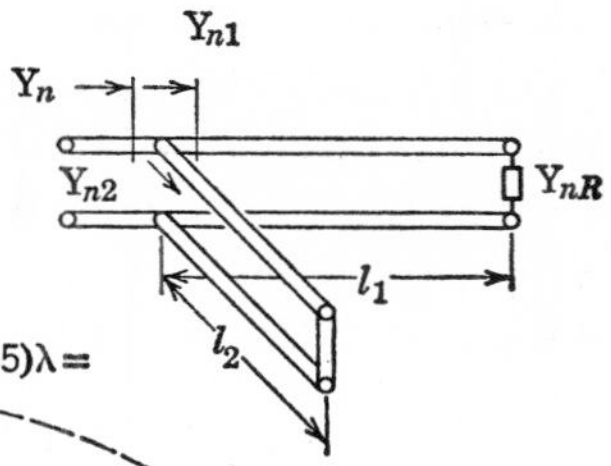

Fig. 3.22. Single-stub matching arrangement for Example 3.2.

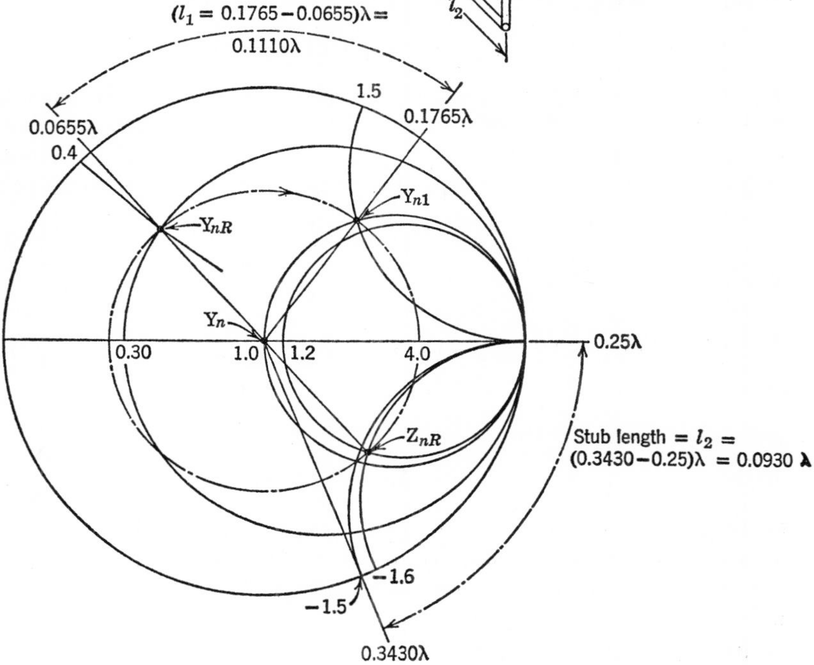

Fig. 3.23. For Example 3.2.

between $-j\infty$ and $+j\infty$. This fact may be verified on either the Smith chart or Fig. 3.10.

"Matching" a transmission line has a special meaning, *not* the one used in circuit theory to indicate equal impedances seen looking both ways from a given terminal pair. In transmission-line work, "matching" means simply *terminating the line in its characteristic impedance.*

To proceed with the adjustment of the stub for the match condition, the normalized impedance of the termination is found first:

$$Z_{nR} = 1.2 - j1.6$$

It is necessary, however, to work with admittances rather than impedances in this problem, because parallel connections are involved at the stub position. The Smith chart itself can be used as a computer to perform the inversion of a complex number, as we outlined previously. Therefore the normalized admittance can be found on the Smith chart by a rotation of $\bar{\Gamma}$ by 180° from the normalized impedance position. In our problem, we see from Fig. 3.23 that this procedure yields

$$Y_{nR} = 0.3 + j0.4$$

In the notation of Fig. 3.22, the admittance Y_{n1} is the admittance of the section of line of length l_1 terminated in Y_{nR}. Y_{n2} is the admittance of the short-circuited line of length l_2. The admittance presented to the line by the stub and the section of line l_1 with Y_{nR} at its end is

$$Y_n = Y_{n1} + Y_{n2} \tag{3.86}$$

For a match, $Y_n = 1$. We shall assume this condition has been met, and work backwards to the load to discover the necessary conditions. The stub being assumed lossless,

$$Y_{n2} = jB_2$$

is pure imaginary. From Eq. 3.86, therefore,

$$Y_{n1} = 1 - jB_2 \tag{3.87}$$

Equation 3.87 gives a condition for the length l_1, as follows. If a match is to be possible, the admittance Y_{n1} *must* lie on the Smith chart circle Re $(Y_n) = 1$. We are now using the Smith chart directly as an *admittance* chart. The amount we must rotate $\bar{\Gamma}$ (at constant $|\bar{\Gamma}|$) from its position at Y_{nR} to reach a point (Y_{n1}) on this circle determines the length l_1. In Fig. 3.23 we find by this procedure

$$l_1 = (0.176 - 0.065)\lambda = 0.111\lambda$$

The admittance of the short-circuited stub must be

$$Y_{n2} = Y_n - Y_{n1} = 1 - (1 + j1.5) = -j1.5$$

to carry us from Y_{n1} to the match position $Y_n = 1$, along the Smith-chart circle Re $(Y_n) = 1$.

The appropriate length of the stub is also found from the Smith chart. It can be seen that a rotation by 0.093λ from a short circuit $(Y_n = \infty)$ gives the appropriate admittance

$$Y_{n2} = -j1.5$$

Again we have used the Smith chart as an admittance chart. Observe that, with the position and length of the stub chosen above, there will be no standing wave on the line to the *left* of the stub. On the section l_1, however, the VSWR is 4.0, according to Fig. 3.23.

A somewhat more ambitious example shows how rather elaborate constraint problems become relatively simple "locus" situations on the Smith chart.

EXAMPLE 3.3. DOUBLE-STUB MATCHING

Constructional requirements are sometimes imposed upon the distance between the generator and the load. The use of a section of line of adjustable length (as in Example 3.2) is then impractical. In such cases, another matching device is used. It consists of two short-circuited stubs connected in parallel with a *fixed* length of the line (Fig. 3.24). The matching is effected by the adjustment of the lengths of the two stubs. The length of the fixed section is usually $\frac{3}{8}$ of a wave length. This matching device has one drawback, however. Not all impedance terminations can be matched by it.

Given the terminating impedance $Z_R = 60 - j80$, and the characteristic impedance of the line and stubs (in this case) $Z_0 = 50$ ohms, what are the lengths of the stubs when a match is achieved? What terminations *cannot* be matched by this double-stub configuration?

The terminating normalized impedance is

$$Z_{nR} = 1.2 - j1.6$$

The normalized load admittance is therefore

$$Y_{nR} = 0.3 + j0.4$$

The admittances indicated in Fig. 3.24 are related by:

$$Y_{n2} = Y_{n1} + Y_{nR}$$

and

$$Y_n = Y_{n3} + Y_{n4}$$

If a match is achieved,

$$Y_n = 1$$

We work the problem backward toward the load, assuming that this condition has been met. In this way we can discover the limitations as we proceed. The normalized admittance of a lossless short-circuited stub is pure imaginary. It follows that, if a match is to be achievable, Y_{n3} must lie on the circle $\text{Re}\,[Y_n] = 1$. Y_{n2} is related to Y_{n3} by the transformation through the $\frac{3}{8}\lambda$ section of line. Y_{n2} must therefore lie on a circle obtained by a rotation of the circle $\text{Re}\,[Y_n] = 1$ through $2\beta\frac{3}{8}\lambda = \frac{3}{2}\pi$ in the *counterclockwise* direction (toward the load). This

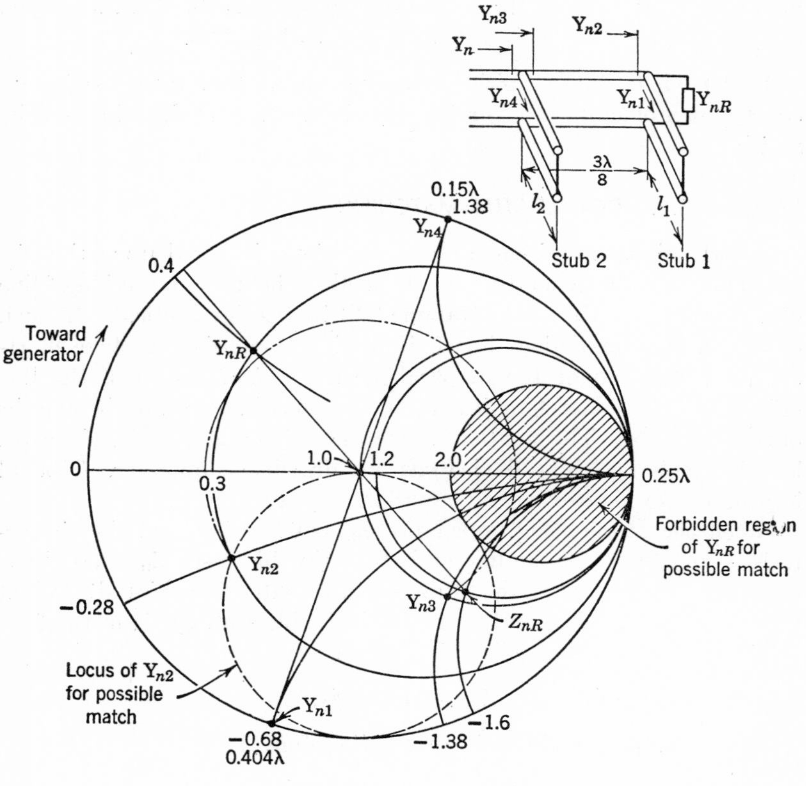

Fig. 3.24. Double-stub matching for Example 3.3.

circle is indicated in Fig. 3.24 as the "locus of Y_{n2} for possible match." From the same figure it can be seen that a pure imaginary normalized admittance—the normalized admittance of stub 1—of value

$$Y_{n1} = -j0.68$$

added to $Y_{nR} = 0.3 + j0.4$ will carry the admittance Y_{nR} along the Smith-chart circle Re $[Y_n] = 0.3$ to

$$Y_{n2} = Y_{nR} + Y_{n1} = 0.3 - j0.28$$

which lies on the "locus of Y_{n2} for possible match" as required.

The length of stub 1 can be found from the Smith chart. A rotation by $l_1 = (0.404 - 0.250)\lambda = 0.154\lambda$ on the unit circle transforms the short-circuited end ($Y_n = \infty$) of stub 1 into the proper normalized admittance $Y_{n1} = -j0.68$, as required above.

The $\frac{3}{8}\lambda$ section of line transforms $Y_{n2} = 0.3 - j0.28$ into

$$Y_{n3} = 1 - j1.38$$

The normalized admittance of stub 2 which cancels the susceptive component of Y_{n3} is

$$Y_{n4} = j1.38$$

The proper length of stub 2 can be found again from the Smith chart. A rotation by $l_2 = (0.25 + 0.15)\lambda = 0.40\lambda$ on the unit circle transforms the short-circuited end of stub 2 into the proper normalized susceptance $j1.38$, reducing the admittance Y_n to 1 as required.

It can be seen from Fig. 3.24 that a normalized admittance Y_{nR} located *inside* the circle Re $[Y_n] = 2$ cannot be brought to lie on the "locus of Y_{n2} for possible match" by the parallel connection of any short-circuited stub, because the two circles are mutually tangent. If the load admittance falls inside the circle Re $[Y] = 2$, it cannot be matched to the line by our device. A different "forbidden region" would be found if the fixed-length section of line were chosen differently. The forbidden region could be reduced by a choice of $l > \frac{3}{8}\lambda$. It turns out, however, that increasing l much beyond $\frac{3}{8}\lambda$ greatly increases the frequency dependence of the match. It makes the match very "narrow-band." Usually we prefer to use the given double stub with $l = \frac{3}{8}\lambda$, but we insert a quarter-wave section of line between stub 1 and the load. The normalized admittance seen to the right of stub 1 is then the reciprocal of its original value, which is outside the forbidden region, and a match may be achieved easily.

A final example relates to the form of data obtained from measurements on a line.

EXAMPLE 3.4. DETERMINATION OF LOAD IMPEDANCE FROM STANDING-WAVE MEASUREMENTS

The VSWR on a terminated line is found to be 1.80. The voltage minimum nearest the load is 10 cm away from it. The distance between successive voltage minima is 50 cm. Find the load impedance Z_R if $Z_0 = 100$ ohms.

At the position of *any* voltage minimum, we know that the normalized impedance is real, minimum, and equal to $(\text{VSWR})^{-1}$. From the Smith chart, the inverse of 1.8 is 0.55 (Fig. 3.25). In particular, this is the impedance at the first voltage minimum.

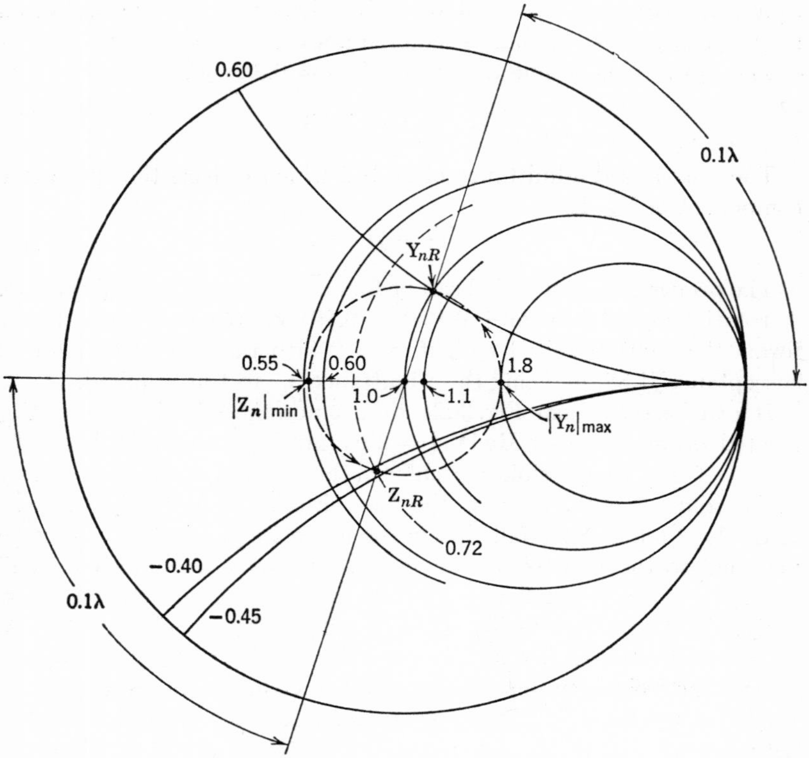

Fig. 3.25. For Example 3.4.

The wave length is *twice* the distance between successive voltage minima. Therefore $\lambda = 100$ cm $= 1$ m. Consequently the first voltage minimum is $\frac{10}{100}\lambda = 0.1\lambda$ away from the load.

Thus we enter the Smith chart at $Z_n = 0.55$, the known normalized impedance at the first voltage minimum, and ask for the normalized impedance at the load end of the line. We must therefore rotate $\bar{\Gamma}$ at constant $|\bar{\Gamma}|$ *counterclockwise, toward the load,* by 0.1λ, and find

$$Z_{nR} = 0.72 - j0.43$$

or

$$Z_R = Z_0 Z_{nR} = 72 - j43 \text{ ohms}$$

We could use the chart instead as an admittance chart. On this basis, the normalized admittance is *largest* at a voltage minimum and equals the VSWR of 1.80. Entering the chart at $Y_n = 1.80$, we rotate $\bar{\Gamma}$ at constant $|\bar{\Gamma}|$ counterclockwise as before. A rotation of 0.1λ brings us to the load terminals, where we read from the chart $Y_{nR} = 1.01 + j0.60$. This is diametrically opposite to our previous value of $Z_{nR} = 0.72 - j0.43$; thus we have found the same answer as before.

PROBLEMS

Problem 3.1. (a) Show that the general solution for the steady-state voltage and current on a lossless line can be considered as the superposition of either: (i) two pure traveling waves; (ii) a forward pure traveling wave and a complete standing wave; (iii) a backward pure traveling wave and a complete standing wave; (iv) two complete standing waves. (b) In each of the above cases, do the traveling-wave components alone account for all the time-average power? Do the standing-wave components alone account for all the reactive power? Prove your answers.

Problem 3.2. (a) Find those complex values of V_{s1} and V_{s2} in Fig. 3.26*b* which will cause exactly the same voltage and current everywhere on the line as occur in Fig. 3.26*a*. (b) Complete and prove the following theorem, based on the results of (a): "The voltage and current at any point on a uniform transmission line may always

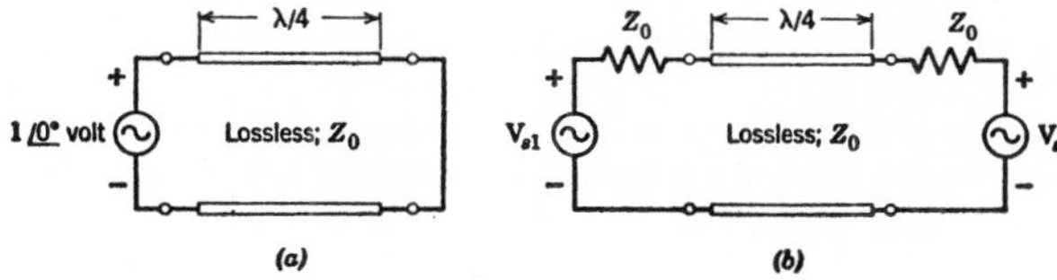

Fig. 3.26. Problem 3.2.

be regarded as the superposition of their values under two auxiliary conditions: the response at the point in question when the line is driven at its input by a ______, and loaded at its output by ______; plus the corresponding response when the line is driven at its output by a ______, and loaded at its input by ______." (c) Show corresponding source arrangements, and develop corresponding theorems, that will represent each of the three views (ii) through (iv) of the line solution given in Prob. 3.1*a*, in the way (*b*) above does for (i).

Problem 3.3. Show that on a lossless line in the steady state the impedance at a voltage minimum is $R_{min} = Z_0/s$ and at a voltage maximum is $R_{max} = Z_0 s$, where Z_0 and s (>1) are respectively the characteristic impedance and VSWR.

Problem 3.4. The peak instantaneous voltage on a certain lossless transmission line must be kept below V_0 volts in order to avoid voltage breakdown. The characteristic impedance is Z_0. If the VSWR is s, how much power can be transmitted by the line without risking breakdown?

Problem 3.5. A lossless transmission line of length l is short-circuited at the load end. The instantaneous current in the short is $\text{Re}\,(I_0 e^{j\omega t})$ and is zero at $t = 0$.

(a) Calculate the time rate of change of the total stored energy on the line, as a function of time. (b) Obtain the result (a) by another method. (c) Express the foregoing result in terms of the input current and the input reactance of the line.

Problem 3.6. At a frequency ω_0, the input admittances B of a certain lumped capacitance C' and a certain $\lambda/8$ section of lossless open-circuited line are equal. (a) If at the frequency in question the same voltage is applied to both devices, find the ratio of their time-average stored electric energies. Explain. (b) Determine the ratio of the frequency derivatives $(\partial B/\partial \omega)_{\omega_0}$ for the two devices considered above, and compare with the ratio of the time-average *total* stored energies in them. See the discussion of Foster's reactance theorem in Chapter 6 for confirmation of your result.

Problem 3.7. At frequency ω_0, a lossless line with $Z_0 = 50$ ohms and $l = 0.63\lambda$ is terminated in an impedance $Z_L = -j50\sqrt{3}$ ohms. If the line is excited by a current source of frequency ω_0, find the time-average power delivered to the line by the source.

Problem 3.8. A lossless line has $Z_0 = 50$ ohms and $l = 0.25\lambda$ at the frequency ω_0. Its output current is $i_2(t) = \cos^3 \omega_0 t$ amp, and its output voltage is $v_2(t) \equiv 0$. Find its input current.

Problem 3.9. The complex power transmitted through a lossless transmission line at a particular point $z = z_0$ is $2 + j1$ w. The complex power carried by the $(+)$ wave is $3 + j0$ w. (a) Find the complex power carried by the $(-)$ wave. (b) Find the VSWR on the line. (c) Find the possible value(s) of (complex) $\bar{\Gamma}(z_0)$.

Problem 3.10. Prove that the normalized input impedance of a lossless line open-circuited at the far end is equal to the normalized input admittance of the same line short-circuited at its far end.

Problem 3.11. (a) Show that the reactive power transmitted through a lossless line has successive maxima and minima at points one-quarter wave length apart. (b) Find the distance between the points found above and the points of maximum voltage on the line. (c) Show that the complex power transmitted through the line at a point of maximum reactive power is the complex conjugate of that transmitted at a point of minimum reactive power. Can we conclude *from this fact alone* that the generalized impedances of the line at these two points will also be just complex conjugates of each other? (d) On a lossless line, *are* the generalized impedances seen at points of maximum and minimum reactive power in fact the complex conjugates of each other?

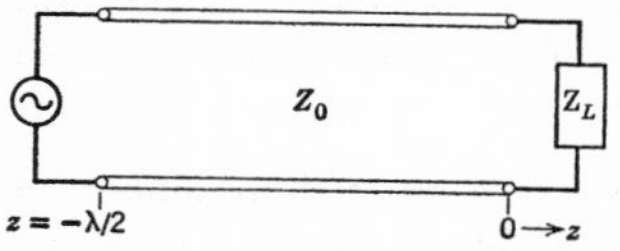

Fig. 3.27. Problem 3.12.

Problem 3.12. With reference to Fig. 3.27, for each case below draw $|V(z)|$ and $|I(z)|$ in the interval $-\lambda/2 \leq z \leq 0$, if: (a) $Z_{nL} = 0;\ \frac{1}{2};\ 1;\ 2;\ \infty$. (b) $Z_{nL} = j;\ -j;\ 1 + j;\ 1 - j;\ 3 + j4$.

Problem 3.13. A quarter-wave-length lossless line is terminated in a load $Z_L = (2 - j50)$ ohms. If $\langle W_m \rangle$, $\langle W_e \rangle$ are the time-average magnetic and electric energies stored in the line (exclusive of the load), is $\langle W_m \rangle \gtreqless \langle W_e \rangle$?

Problem 3.14. The voltage maximum nearest the load of a lossless line terminated in a pure resistance occurs 2 m away from the load. Find the location of the next nearest voltage maximum.

Problem 3.15. In Fig. 3.28, find: (a) The voltage reflection coefficient of the line and termination, as seen at the generator terminals. (b) The line impedance, as seen at the generator terminals. (c) The time-average power to the load. (d) The complex current and voltage at the generator and load ends of the line. (e) The complex power at the generator and load terminals.

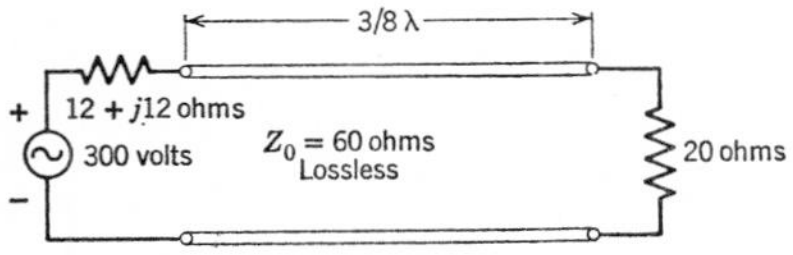

Fig. 3.28. Problem 3.15.

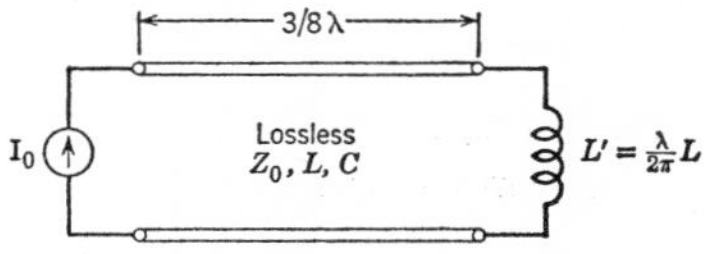

Fig. 3.29. Problem 3.16.

Problem 3.16. With respect to the system shown in Fig. 3.29: (a) Sketch the space variation of the magnitude of voltage and current along the line. (b) Compute the total time-average stored energy in the line and in the load. Check your result by finding the input impedance. (c) Draw a single phasor diagram showing the relative magnitudes and phases of incident, reflected, and total voltages and currents at both ends of the line.

Problem 3.17. A resistance $R = 10Z_0$ terminates a lossless line. The frequency increases by 1%. Find the complex change in input impedance of the line: (a) If the line was $\lambda/2$ long to begin with. (b) If the line was 25λ long to begin with.

Problem 3.18. A lossless transmission line has a load R_L and is driven from a source with open-circuit voltage V_G and internal resistance R_G. Let

$$r_L = R_L/Z_0 \qquad \bar{\Gamma}_L = (r_L - 1)/(r_L + 1)$$

$$r_G = R_G/Z_0 \qquad \bar{\Gamma}_G = (r_G - 1)/(r_G + 1)$$

$$\bar{\vartheta} = j\beta l$$

where l and β are the length and phase constant of the line, respectively. (a) Find the complex amplitudes of the forward and backward voltage waves, as measured at the load. (b) Find the ratio of load voltage to generator voltage V_G. (c) Show that if either r_L or r_G is equal to unity, the load power is independent of l. Interpret the situation for $r_G = 1$ in terms of the Thévènin equivalent of the line system at the load terminals. (d) Plot the result of (b) versus frequency or length (i.e., versus $|\bar{\vartheta}|$) for given values of all other parameters. (e) By a power series expansion of the term $(1 - \bar{\Gamma}_G\bar{\Gamma}_L e^{-2\bar{\vartheta}})^{-1}$, interpret the ratio of (b) in terms of successive reflections from both ends of the line.

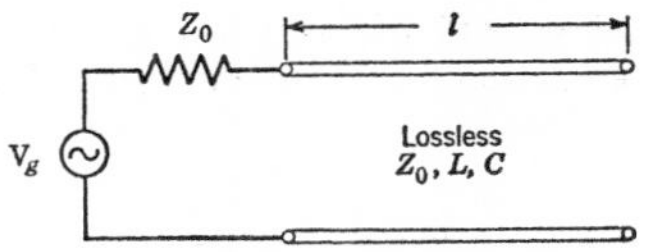

Fig. 3.30. Problem 3.19.

Problem 3.19. (a) For what length l in Fig. 3.30 would the power dissipated in the source resistance be largest? (b) If $l = 3\lambda/8$, find the expression for the instantaneous stored magnetic energy per unit length, $u_m(z, t)$, in terms of the line constants and V_g.

Problem 3.20. Starting with the general solutions for the distribution of voltage and current on lossless lines, obtain the following formulation in terms of the load admittance.

(a) $$V(z) = V_R\left(\cos\beta z - j\frac{Y_R}{Y_0}\sin\beta z\right)$$

(b) $$I(z) = I_R\left(\cos\beta z - j\frac{Y_0}{Y_R}\sin\beta z\right)$$

(c) $$Y(z) = Y_0\left(\frac{Y_R - jY_0\tan\beta z}{Y_0 - jY_R\tan\beta z}\right)$$

Problem 3.21. In an impedance-measuring problem, let $z_{\min}$ be the location of the voltage minimum nearest the load. Then using Prob. 3.20*c*, show that

$$Y_R = Y_0\frac{s + j\tan\beta z_{\min}}{1 + js\tan\beta z_{\min}}$$

where s is the standing wave ratio.

Problem 3.22. It is often convenient to have an equivalent circuit for a length of transmission line. (a) Determine an equivalent T and an equivalent Π circuit for a length l of lossless transmission line. (*Suggestion:* Note the symmetry of the structure and relate the elements of the T or Π to the open- and short-circuit impedance of a length $l/2$ of transmission line, i.e., use Bartlett's bisection theorem.) (b) What do the equivalent circuits become for the particular lengths $l = \lambda/4$, $\lambda/8$, $\lambda/2$, $3\lambda/4$, λ? (c) What are the equivalent circuits for a very short length of the line, i.e., $l \ll \lambda$?

Problem 3.23. In the sinusoidal steady state, the system shown in Fig. 3.31 can be described by the following equations: $V_1 = Z_{11}I_1 + Z_{12}I_2$; $V_2 = Z_{21}I_1 + Z_{22}I_2$. Compute Z_{11}, Z_{12}, Z_{21}, and Z_{22} in terms of l_1, l_2, β, and Z_0.

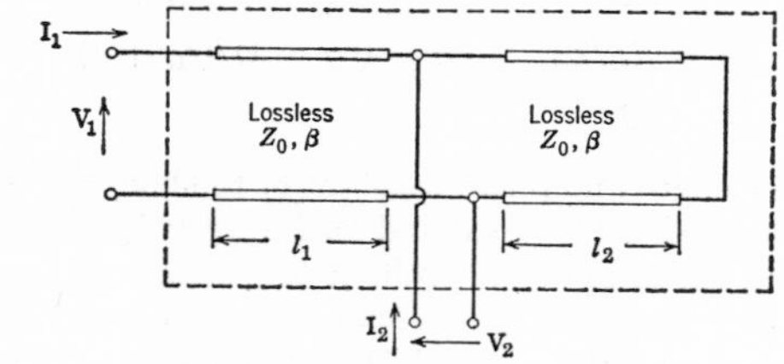

Fig. 3.31. Problem 3.23.

Problem 3.24. In the structure of Fig. 3.32, determine the current in the load and the ratio of the generated-to-transmitted power.

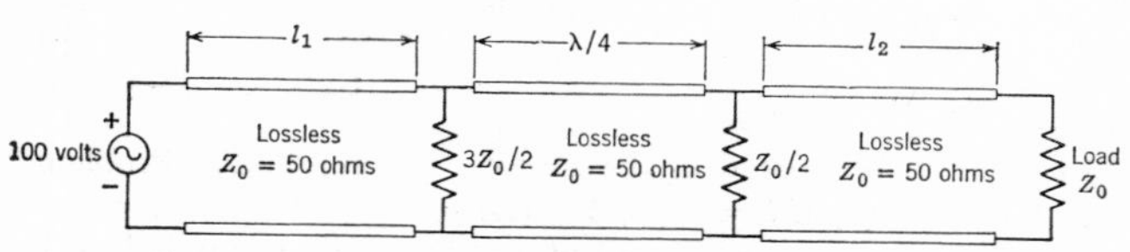

Fig. 3.32. Problem 3.24.

Problem 3.25. In Fig. 3.33, determine the complex reflection and transmission coefficients at the discontinuity C.

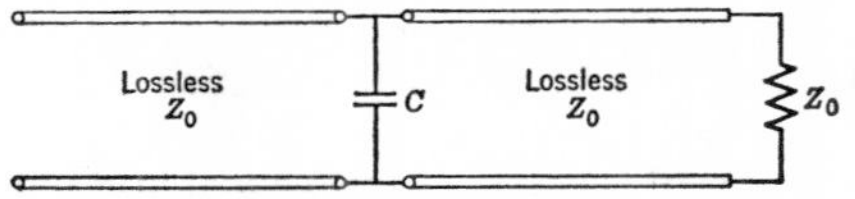

Fig. 3.33. Problem 3.25.

Problem 3.26. For the system in Fig. 3.34, find the magnitudes and phases of the voltages across the terminals a–a', b–b', and c–c'.

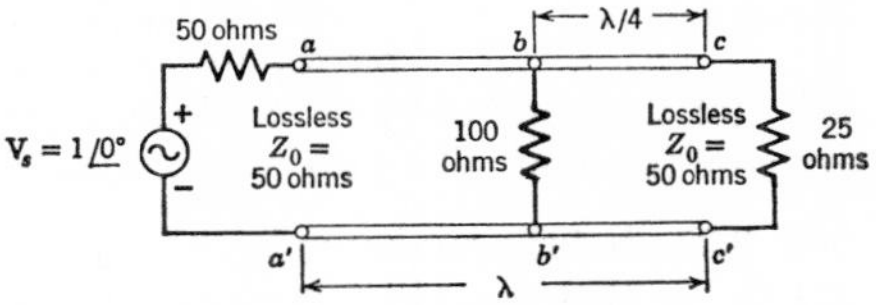

Fig. 3.34. Problem 3.26.

Problem 3.27. In Fig. 3.35, both generators G_1 and G_2 have internal resistances of 120 ohms. The open-circuit voltages of the generators are respectively: $V_1 = 10 \sin \omega t$ (volts); $V_2 = 10 \cos \omega t$ (volts). Find the time-average power furnished by each generator and the power delivered to each load.

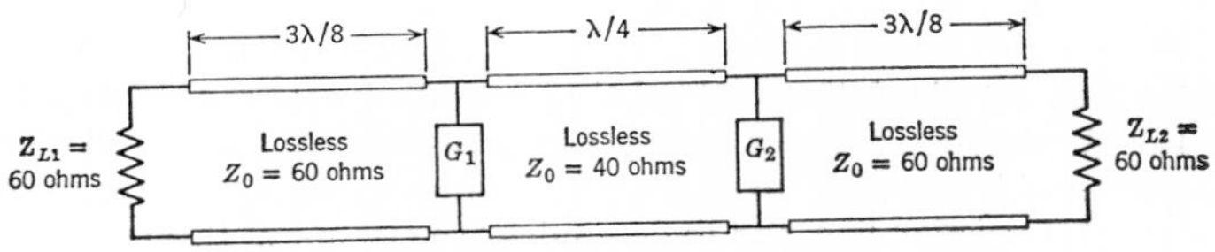

Fig. 3.35. Problem 3.27.

Problem 3.28. A lossless line of length l and characteristic impedance Z_0 has its ends joined together, as shown in Fig. 3.36. Find the impedance Z looking into terminals 1–1′.

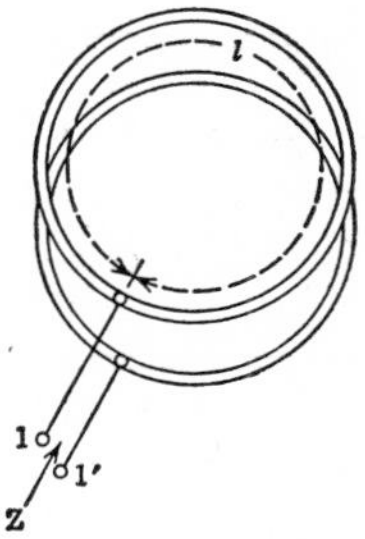

Fig. 3.36. Problem 3.28.

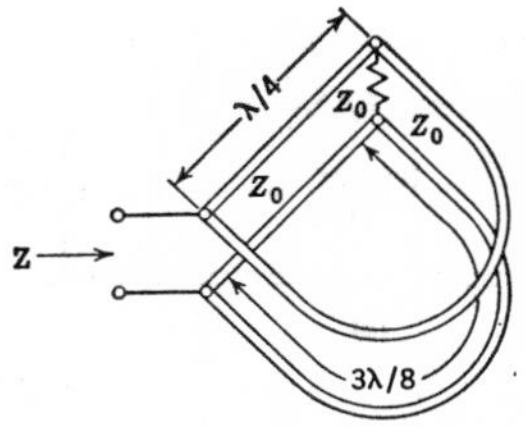

Fig. 3.37. Problem 3.29.

Problem 3.29. Find (Z/Z_0) in Fig. 3.37 if the lines are lossless.

Problem 3.30. Two identical lossless lines of characteristic impedance Z_0 have identical load impedances and differ in length by $\lambda/4$. Their input terminals are connected in parallel to a third line of characteristic impedance $Z_0/2$. (a) Calculate the standing-wave ratio on the third line in terms of the complex reflection coefficients on the other lines. (b) Under what condition or conditions will the input power be divided equally between the two equal load impedances?

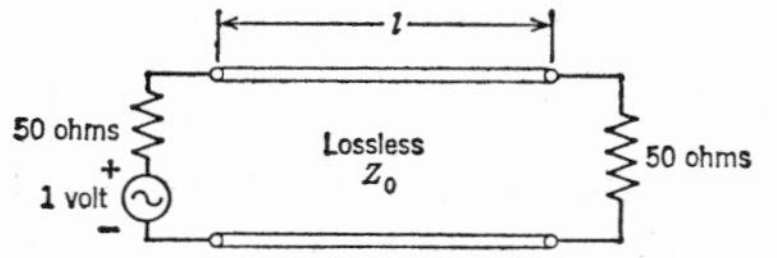

Fig. 3.38. Problem 3.31.

Problem 3.31. In Fig. 3.38, the length l and characteristic impedance Z_0 of the line are at one's disposal. (a) What is the maximum power that can be delivered to the load? (b) Give *all* possible conditions on l and Z_0 for which this maximum is obtained. (c) For each of the conditions above, what is the maximum voltage appearing on the line? The maximum current?

Problem 3.32. In Fig. 3.39, find the minimum length l_2 of line 2, and the relation between Z_{02}, Z_{03}, and Z_{01} so that Z_{in} will remain equal to Z_{01} for *any* l_1 and l_3.

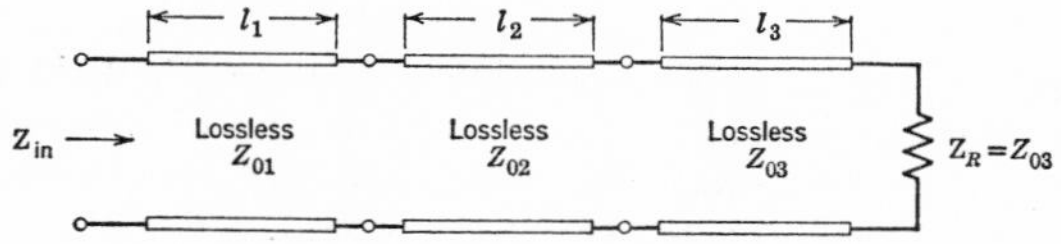

Fig. 3.39. Problem 3.32.

Problem 3.33. In Fig. 3.40*a*: (a) Find the power dissipated in the lumped conductance Y_0. (b) Find the complex voltage across the capacitance C'. In Figs. 3.40*b*, 3.40*c*, and 3.40*d* it is desired to deliver the maximum possible power to terminals a–a' of the transmission line, and to determine the amount of this power. (c) Find

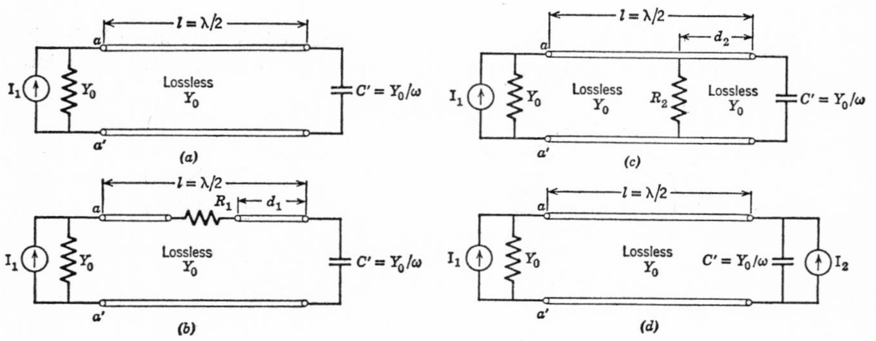

Fig. 3.40. Problem 3.33.

the value of d_1 and R_1 in Fig. 3.40*b*, and the maximum power. (d) Find the value of d_2 and R_2 in Fig. 3.40*c*, and the maximum power. (e) Find the complex value of current source I_2 in Fig. 3.40*d*, and the maximum power.

Problem 3.34. In Fig. 3.41: (a) Find what value of reactance jX shunted across R_L will maximize the power dissipated in R_L. (b) How much average power is delivered to R_L in (a)? (c) What is the VSWR under the conditions above? (d) If the frequency is doubled, but all impedance values and the source voltage are kept at the same values they had above, what is the new power dissipated in R_L?

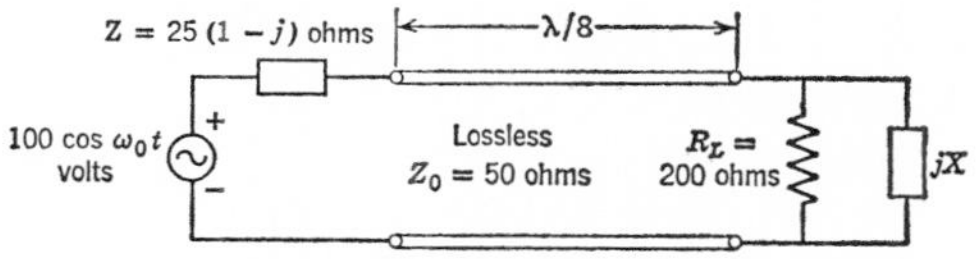

Fig. 3.41. Problem 3.34.

Problem 3.35. For Fig. 3.42, without using any charts: (a) Find Y to deliver the largest possible power to Z_L. (b) Find the VSWR on the line to the right of Y. (c) Find the position of the *current minimum* on the line.

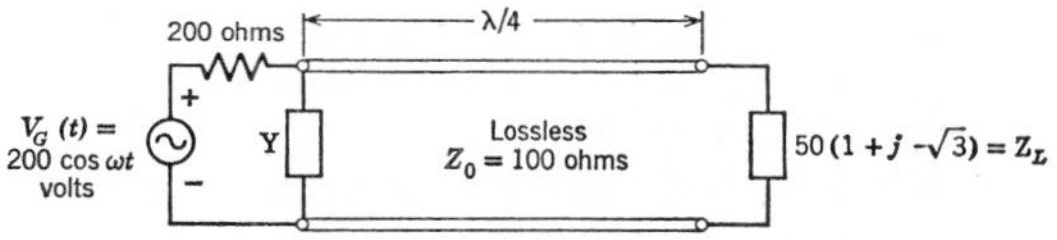

Fig. 3.42. Problem 3.35.

Problem 3.36. In Fig. 3.43, the lines are lossless. Compute the susceptance B, and the characteristic impedance of the right-hand line section, Z_{02}, which will cause the power absorbed by the load R_L to be the maximum possible, for a given V_G. Do not use the Smith chart.

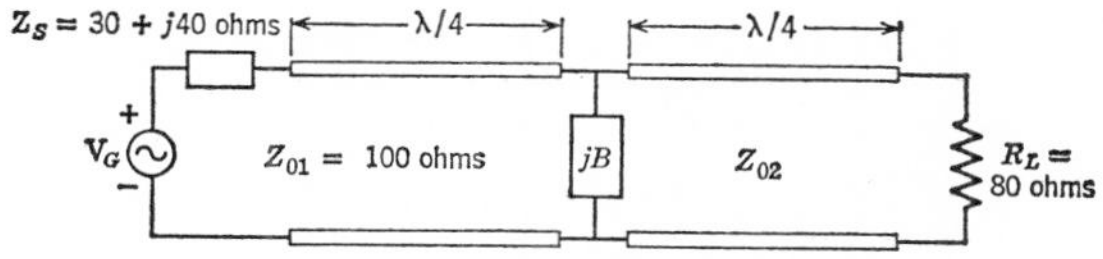

Fig. 3.43. Problem 3.36.

Problem 3.37. The lines in Fig. 3.44 are lossless, and the stub matches the load to the line. $Z_R = 100(1 + j)$ ohms, $I_R = 2 + j0$ amp, and $Z_0 = 100$ ohms for all lines. Find both the magnitude *and the phase* of: (a) The voltage V_1 at the stub location. (b) The current I_S in the short-circuited end of the stub. Show all of the significant voltages and currents used in solving the problem on appropriate phasor diagrams.

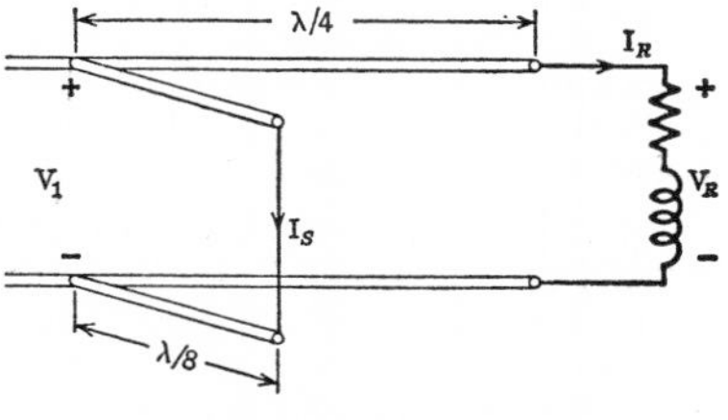

Fig. 3.44. Problem 3.37.

Problem 3.38. A lossless half wavelength line of characteristic impedance 50 ohms is loaded with 150 ohms, and is driven from a source with an internal impedance of 50 ohms. (a) If the breakdown voltage of the line is 100 v, what is the maximum power that can be transmitted to the load? Where is the danger point (or points) on the line as far as breakdown is concerned? (b) Find the line input current under the maximum power conditions of (a). (c) At what points on the line is the magnitude of the impedance equal to Z_0? (d) If a lossless stub is placed at a distance $\lambda/8$ from the load, what is the minimum VSWR that can be obtained to the left of the stub?

Problem 3.39. A lossless transmission line of characteristic impedance $Z_0 = 50$ ohms is to be matched to a load $Z_L = 50/[2 + j(2 + \sqrt{3})]$ ohms by means of a lossless short-circuited stub. Without using a Smith chart, find the stub position and length so that match is obtained. Assume the characteristic impedance of the stub to be: (a) 50 ohms; (b) 100 ohms.

Problem 3.40. For a normalized load admittance $Y_n = 0.40 + jB$, it is found that the reflected wave can be eliminated on a certain line by use of a short-circuited stub of length $\lambda/10$. Find the shortest electrical distance between load and stub, if the line and stub are lossless and have the same characteristic impedance.

Problem 3.41. (a) $Z_0 = 100$ (lossless); VSWR $= 4$; first voltage *maximum*, $\lambda/8$ from load. What is the load impedance? (b) To match this load, a $\lambda/4$ section of different line with characteristic impedance $Z_{01} < Z_0$ is to be inserted somewhere between (in cascade with) the load and the original line. Find: (i) The minimum distance (in terms of λ) between load and matching section; (ii) the characteristic impedance Z_{01} (in terms of Z_0).

Problem 3.42. In a double-stub tuner let the distance from the load to the first stub (No. 1), and also the distance between the stubs, be $\lambda/6$. On a Smith chart: (a) Draw the "locus of Y_{n2} for possible match" as in Fig. 3.24. (b) Locate and cross-hatch the "forbidden region of Y_{nR} for possible match," again as in Fig. 3.24. (c) Demonstrate that the location of a third stub at the load can eliminate the forbidden region.

Problem 3.43. In constant-voltage power-distribution systems, the need to regulate load voltage produces limitations on the amount of power that can be transmitted down a given line. Consider, for example, Fig. 3.45: (a) If $|V_R| = |V_S|$, sketch two possible forms of $|V(z)|$ on the line. (b) Find the locus of complex

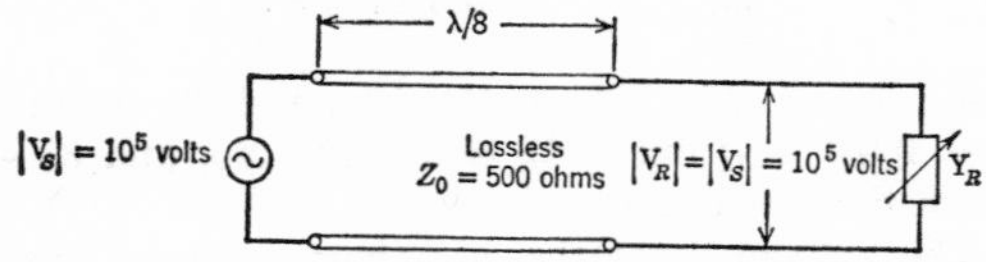

Fig. 3.45. Problem 3.43.

passive load admittances Y_R for which $|V_R| = |V_S|$ is possible. (c) For what passive value of load admittance Y_R consistent with $|V_R| = |V_S|$ is the load time-average power maximum? Find this power, and the corresponding phase of V_R/V_S. (d) Under conditions (c), find the *magnitude* of the voltage at the middle of the line.

Problem 3.44. Solve the single-stub matching problem analytically, for the case where line and stub have the same Z_0.

Problem 3.45. A "double-slug tuner" (Fig. 3.46) consists of two obstacles which, by either partially blocking or changing the dielectric constant of a transmission line, constitute line sections of different characteristic impedance than the original line. The sections are $\lambda/4$ long at their own wave length and can be moved along the line independently. (a) Using a Smith chart, develop a procedure for matching

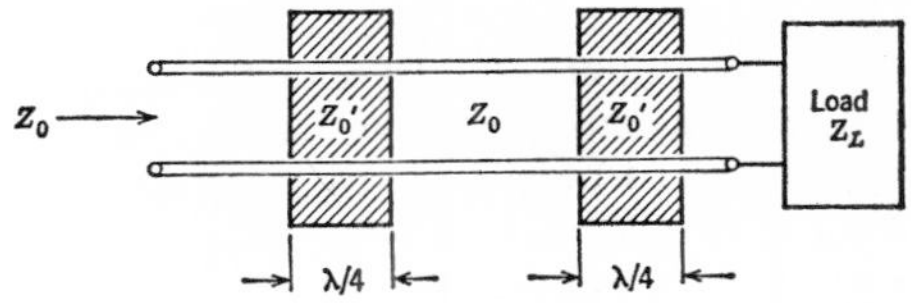

Fig. 3.46. Problem 3.45.

with such a tuner, showing the specific solution for the case $Z_L = Z_0(0.5 + j0.5)$, $Z_0' = Z_0/2$. (b) Are there any values of Z_L which cannot be matched with this choice of Z_0'? If so, indicate the "forbidden region" for Z_L on the Smith chart.

Problem 3.46. Given Z_L and R_0, it is desired to choose Z_{01} and $l < \lambda/4$ for a lossless line to meet the conditions shown in Fig. 3.47. Find the restrictions on Z_L for which this is possible. Find Z_{01} and $l < \lambda/4$ for the case $R_0 = 50$ ohms and $Z_L = 75 - j25$ ohms.

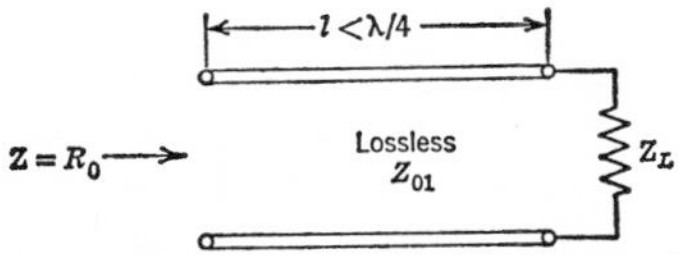

Fig. 3.47. Problem 3.46.

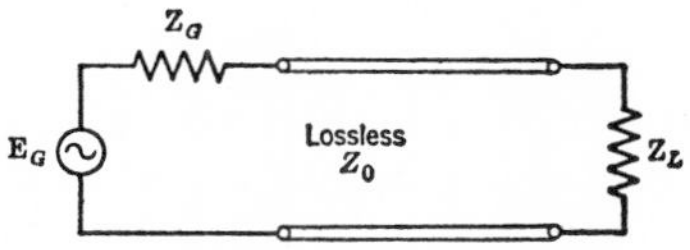

Fig. 3.48. Problem 3.47.

Problem 3.47. (a) For Fig. 3.48 find the location and length of a *single* short-circuited stub having characteristic impedance Z_0 which will cause all of the source available power to be delivered to the load Z_L, if $Z_L/Z_0 = Z_G/Z_0 = 3 + j4$. (b) Determine the VSWR on both sections of line in (a) after the stub is in place. (c) Given Y_{nG} (for example, in the case above $1/Y_{nG} = 3 + j4$), find the region of Y_{nL} values which *cannot* be made to draw all the source available power by using a single stub in the manner prescribed in (a).

Problem 3.48. A transmission line is imagined with *series capacitance C* per unit length, and *shunt inductance L* per unit length. (a) Derive the transmission-line equations. (b) Find the characteristic impedance Z_0. (c) Find the propagation constant and phase velocity. How do you reconcile the results of (b) and (c)? Consider Prob. 6.28 in this connection.

Problem 3.49. The inductance and capacitance per unit length for a particular lossless transmission line are *functions of position*, as follows:

$$\left.\begin{aligned} L &= L_0 e^{-\alpha z} \\ C &= C_0 e^{\alpha z} \end{aligned}\right\} \quad -l \le z \le 0 \qquad \alpha > 0$$

(a) Develop the steady-state solutions for $V(z)$ and $I(z)$. (b) Show that at any given frequency, there is a range of values of α for which waves cannot propagate on the line. Alternatively, express the "cutoff frequency" in terms of given values of L_0, C_0, and α. (c) Find an expression for the termination impedance which, placed at $z = 0$, will prevent reflections. Check your result by considering the case $\alpha \rightarrow 0$, but consider all other cases too. (d) Find an expression for the complex power carried by the line when it is terminated as in (c). Discuss the nature of this complex power, and compare with the case of uniform line. Consider the situations above and below cutoff. (e) Discuss the complex power when the termination at $z = 0$ is an arbitrary passive impedance. Consider specifically the questions of additivity of (+) and (−) real power and complex power when: (i) The line is above cutoff; (ii) The line is below cutoff. (f) Show that an "exponential horn" is the acoustic analog of this problem.

CHAPTER FOUR

Transient Waves on Lossless Transmission Lines

In many ways the most important ideas connected with functions of both time and space show up in transient phenomena. Although, in principle, the steady-state or frequency-domain solution contains all the information about a linear system, a direct study in the time domain, when possible, is often more illuminating. The lossless transmission line, considered from the time-domain point of view, provides at once a clear-cut, relatively simple, and reasonably practical example of one-dimensional wave motion in the transient state.

4.1 Time-Domain Solution of the Differential Equations

The partial-differential equations of a lossless line are given by Eqs. 2.34*a* and 2.34*b* with $R \equiv G \equiv 0$.

$$\text{(a)} \qquad \frac{\partial V}{\partial z} = -L\frac{\partial I}{\partial t}$$

$$\text{(b)} \qquad \frac{\partial I}{\partial z} = -C\frac{\partial V}{\partial t} \qquad (4.1)$$

Either I or V can be eliminated from Eqs. 4.1. We choose first to retain V, eliminating I by partial differentiation of Eq. 4.1*b* with respect to time and of Eq. 4.1*a* with respect to distance.

$$\text{(a)} \qquad \frac{\partial^2 V}{\partial z^2} = -L\frac{\partial^2 I}{\partial z\,\partial t}$$

$$\text{(b)} \qquad \frac{\partial^2 I}{\partial t\,\partial z} = -C\frac{\partial^2 V}{\partial t^2} \qquad (4.2)$$

Since the order of partial differentiation is immaterial for all but the most unusual functions, we can substitute for $\partial^2 I/(\partial z\, \partial t)$ in Eq. 4.2*a* the value of $\partial^2 I/(\partial t\, \partial z)$ given by Eq. 4.2*b*. There results a partial-differential equation for V alone.

$$\frac{\partial^2 V}{\partial z^2} = LC \frac{\partial^2 V}{\partial t^2} \tag{4.3}$$

Equation 4.3 is the *wave equation*, written more compactly by defining

$$v = \frac{1}{\sqrt{LC}} \tag{4.4}$$

in the form

$$\frac{\partial^2 V}{\partial z^2} = \frac{1}{v^2} \frac{\partial^2 V}{\partial t^2} \tag{4.5}$$

By placing $\tau = z/v$, we obtain from Eq. 4.5 an even simpler one

$$\frac{\partial^2 V}{\partial \tau^2} = \frac{\partial^2 V}{\partial t^2} \qquad \text{or} \qquad \frac{\partial^2 V}{\partial \tau^2} - \frac{\partial^2 V}{\partial t^2} = 0 \tag{4.6}$$

To solve Eq. 4.6, we must find the most general function $V(\tau, t)$ whose second derivatives with respect to τ and t are identical. Indeed, we can say more about the requirements on $V(\tau, t)$ by rephrasing Eq. 4.6 as follows:

$$\begin{aligned} \frac{\partial^2 V}{\partial \tau^2} - \frac{\partial^2 V}{\partial t^2} &= \left(\frac{\partial}{\partial \tau} - \frac{\partial}{\partial t}\right)\left(\frac{\partial}{\partial \tau} + \frac{\partial}{\partial t}\right) V \\ &= \left(\frac{\partial}{\partial \tau} + \frac{\partial}{\partial t}\right)\left(\frac{\partial}{\partial \tau} - \frac{\partial}{\partial t}\right) V = 0 \end{aligned} \tag{4.7}$$

from which it is evident that any function V for which either

$$\begin{aligned} &\text{(a)} \qquad \frac{\partial V}{\partial \tau} = \frac{\partial V}{\partial t} \\ &\text{or} \\ &\text{(b)} \qquad \frac{\partial V}{\partial \tau} = -\frac{\partial V}{\partial t} \end{aligned} \tag{4.8}$$

is certainly *one* solution. Quite clearly, Eq. 4.8*a* is satisfied if $V(\tau, t) = f_-(t + \tau)$, where f_- is any functional form at all. Similarly, Eq. 4.8*b* is satisfied by $f_+(t - \tau)$, where f_+ is any functional form at all. We have therefore found two solutions, each containing an arbitrary functional form. Since Eq. 4.6 is linear, however, a sum of solutions is again a solution. Hence,

$$V(\tau, t) = f_+(t - \tau) + f_-(t + \tau) \tag{4.9}$$

is surely a solution. Moreover, it is the most general possible solution, because it contains exactly *two arbitrary functions*. A partial-differential equation of second order (like Eq. 4.6) must, in effect, be solved by integrating twice on one variable, holding the other constant. Hence the two arbitrary "constants" of integration may in fact be arbitrary *functions* of the second variable. Therefore any solution containing the required two "functions of integration" is the general solution.

The fact that Eqs. 4.1 are unchanged by interchanging V with I and L with C (i.e., they are their own duals), means that $I(\tau, t)$ also satisfies Eq. 4.6 with V replaced by I. Therefore

$$\frac{\partial^2 I}{\partial \tau^2} = \frac{\partial^2 I}{\partial t^2} \tag{4.10}$$

and

$$I(\tau, t) = g_+(t - \tau) + g_-(t + \tau) \tag{4.11}$$

where g_+ and g_- are completely arbitrary functions as far as Eq. 4.10 is concerned. However, Eqs. 4.1 establish a relation between the functions g and the functions f. Inserting the solutions given by Eqs. 4.9 and 4.11 into Eqs. 4.1, and recalling that $z = v\tau$, we obtain

$$\begin{aligned} &\text{(a)} \qquad \frac{1}{v}[f_+'(u) - f_-'(w)] = L[g_+'(u) + g_-'(w)] \\ &\text{(b)} \qquad C[f_+'(u) + f_-'(w)] = \frac{1}{v}[g_+'(u) - g_-'(w)] \end{aligned} \tag{4.12}$$

where

$$u \equiv t - z/v \qquad w \equiv t + z/v \tag{4.13}$$

and

$$\begin{aligned} f_+'(u) &= \frac{df_+(u)}{du} \qquad & g_+'(u) &= \frac{dg_+(u)}{du} \\ f_-'(w) &= \frac{df_-(w)}{dw} \qquad & g_-'(w) &= \frac{dg_-(w)}{dw} \end{aligned} \tag{4.14}$$

Observing that

$$\frac{1}{Lv} = \sqrt{\frac{C}{L}} = Y_0 = vC \tag{4.15}$$

where Y_0 is the characteristic admittance defined in Chapter 3, we recast Eqs. 4.12 to read

$$\begin{aligned} &\text{(a)} \quad g_+'(u) - Y_0 f_+'(u) = -[g_-'(w) + Y_0 f_-'(w)] \\ &\text{(b)} \quad g_+'(u) - Y_0 f_+'(u) = +[g_-'(w) + Y_0 f_-'(w)] \end{aligned} \tag{4.16}$$

It follows at once from Eqs. 4.16 that

$$\begin{aligned} &\text{(a)} \quad g_+'(u) = Y_0 f_+'(u) \\ &\text{(b)} \quad g_-'(w) = -Y_0 f_-'(w) \end{aligned} \tag{4.17}$$

Integration of Eqs. 4.17a with respect to u and w shows that

$$\begin{aligned} &\text{(a)} \quad g_+(u) = Y_0 f_+(u) + \text{const.} \\ &\text{(b)} \quad g_-(w) = -Y_0 f_-(w) + \text{const.} \end{aligned} \tag{4.18}$$

The constants in Eq. 4.18 are independent of u and w, and therefore also independent of t and z. They simply express the fact that a direct voltage and current, both independent of z, may exist on a lossless line without being forced by the line to retain any particular ratio. It is clear that this is a case of two perfectly conducting wires feeding direct current to a resistor—any resistor. If we simply agree to remember this physical possibility, we need not carry along explicitly the annoying constants in Eqs. 4.18 when we deal with truly time-varying problems. Thus, dropping them forthwith, we rewrite Eqs. 4.18 in the form

$$\begin{aligned} &\text{(a)} \quad g_+(u) = Y_0 f_+(u) \\ &\text{(b)} \quad g_-(w) = -Y_0 f_-(w) \end{aligned} \tag{4.19}$$

With the new knowledge of Eq. 4.19, we can finally write Eqs. 4.9 and 4.11 in the manner

$$\begin{aligned} &\text{(a)} \quad V(z, t) = f_+(t - z/v) + f_-(t + z/v) \\ &\text{(b)} \quad I(z, t) = Y_0[f_+(t - z/v) - f_-(t + z/v)] \end{aligned} \tag{4.20}$$

as the general solution of Eqs. 4.1. It remains to interpret this solution.

4.2 Traveling Waves

Let us consider Eqs. 4.20 when the arbitrary function $f_-(w)$ is zero. Then the solution becomes simply

$$\text{(a)} \qquad V_+(z, t) = f_+(t - z/v)$$
$$\text{(b)} \qquad I_+(z, t) = Y_0 f_+(t - z/v) \tag{4.21}$$

The behavior of f_+ characterizes that of both V_+ and I_+. In Fig. 4.1*a* we show an arbitrary shape of f_+, plotting it versus z at a time $t = t_0$. At a time Δt later, the voltage will have the form

$$V_+(z, t_0 + \Delta t) = f_+(t_0 + \Delta t - z/v) = f_+\left[t_0 - \left(\frac{z - v\,\Delta t}{v}\right)\right] \tag{4.22}$$

Note that according to Eq. 4.22 a plot of V_+ at time $t_0 + \Delta t$ versus the new variable $z' = z - v\,\Delta t$ must be exactly the same as the one of $V_+(t_0, z)$ plotted versus z in Fig. 4.1*a*. Figure 4.1*b* has been prepared on this basis, but has been arranged to show $V_+(z, t_0 + \Delta t)$ as a function of z also. Evidently, within the time interval Δt, the *shape* of the voltage distribution along the line has not changed, but has simply shifted by the amount

$$\Delta z = v\,\Delta t$$

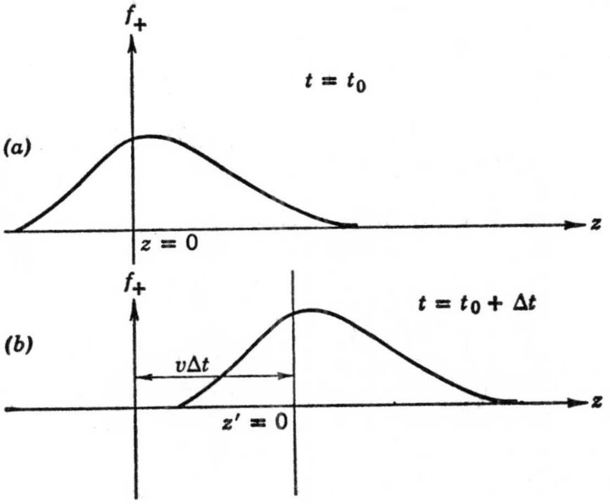

Fig. 4.1. The current or voltage in a (+) wave.

The velocity with which the distribution f_+ moves in the $+z$ direction is therefore given by

$$\frac{\Delta z}{\Delta t} = v = \frac{1}{\sqrt{LC}} \tag{4.23}$$

which is the same as the expression for the phase velocity we found in Chapter 3. The fact that the shape of the voltage (or current) distribution does not change with time is another manifestation of the fact that the phase-velocity expression in the steady state does not contain the frequency, as has been noted previously. The line is absolutely distortionless for the single-traveling-wave solution V_+, I_+.

Another way of looking at the phenomenon of wave motion is to examine the time delay between the start of a disturbance at $z = 0$, $t = 0$ and the moment it first arrives at a point $z = z_1$. Since we have just shown that a disturbance of any shape travels along z unchanged in form, with a constant velocity v, an observer at a distance z_1 from the source ($z = 0$) would have to wait a time

$$t_d = \frac{z_1}{v}$$

before seeing that disturbance. We can therefore consider as a primary feature of wave motion the unavoidable delay in the observation of an event caused by the observer's distance away from the event and by the finite speed of its propagation. The reader should clarify this viewpoint for himself by using Eqs. 4.21 and Fig. 4.1 to prepare sketches of $V_+(0, t)$ and $V_+(z_1, t)$ versus t over the range $-\infty < t < \infty$. The ideas involved in this exercise must be mastered if confusion is to be avoided later.

If we consider next the solution in Eq. 4.20 with the arbitrary function $f_+(u)$ set equal to zero, we have

$$\begin{aligned} &\text{(a)} \qquad V_-(z, t) = f_-(t + z/v) \\ &\text{(b)} \qquad I_-(z, t) = -Y_0 f_-(t + z/v) \end{aligned} \tag{4.24}$$

in which the argument of f_- is $t + z/v$. To maintain a given value of f_- (i.e., to travel with the wave), an observer would have to move with a velocity $v = 1/\sqrt{LC}$ in the *negative* z direction. This is necessary because t is a continuously increasing variable.

The reason for the subscripts + and − on f is now apparent. The first stands for a wave propagating in the positive z direction, and the second stands for one propagating in the negative z direction.

4.3 Boundary Conditions

The general solution (Eq. 4.20) of the transmission-line equations is sufficiently flexible to fit the wide variety of physical boundary conditions we may wish to impose in various particular situations. In fact, since the functions f_+ and f_- are at the moment absolutely unspecified, *all we have really learned from the partial-differential equations themselves is that properly related voltage and current waves of arbitrary shape may travel undistorted in both directions on the line.* The exact shape of these waves will be contained in the specific functional forms of f_+ and f_-, which depend *completely* on the boundary conditions. Under the heading of boundary conditions come three general kinds of statements:

1. Those specifying networks connected to the ends of the line.
2. Those specifying the locations and time variations of voltage or current sources driving the line and its associated networks.
3. Those specifying the complete distributions along z of the voltage and current on the line at an initial instant $t = 0$.

The line-and-load problem is not different in principle from the situation in lumped-circuit theory, where solution of transient problems depends upon both the sources applied to the system and the initial energy stored in it. If the associated networks (like the line itself) are linear systems, we can solve by superposition those problems in which sources and initial conditions occur together.

Although they are similar in principle, the details of the line solution for various types and combinations of boundary conditions exhibit many novelties in comparison with familiar network transients. These novelties require careful study and the working of various examples for their clarification.

4.3.1 Semi-infinite Line with Transient-Voltage Input

One of the simplest kinds of boundary conditions arises in the case of an initially uncharged semi-infinite line driven by a prescribed voltage source $V_1(t)$, as shown in Fig. 4.2. We suppose that the line has no voltage nor current anywhere on it ($z > 0$) at $t = 0$, and we are interested in determining the voltage and current waves along the line for $t \geq 0$, $z \geq 0$.

The general solution (Eq. 4.20) contains two terms: one involving f_+ and the other f_-. These are exhibited separately in Eqs. 4.21 and 4.24. We have already found the meaning of these two terms. One is a wave

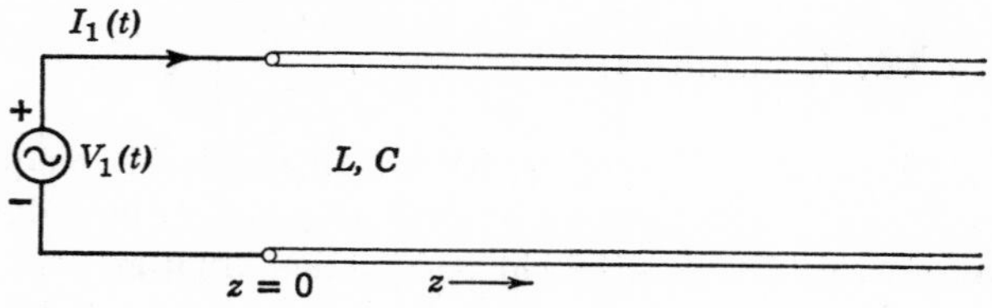

Fig. 4.2. A semi-infinite line driven by a voltage source.

traveling to the right; the other, a wave traveling to the left. It is to be expected that the direction of propagation of each wave is also the direction of the flow of power carried by it. To confirm this assertion, we can use the definition of instantaneous power

$$P(z, t) = V(z, t)\, I(z, t)$$

to find the power carried by a (+) wave alone

$$P_+(z, t) = Y_0 f_+^2(t - z/v) \tag{4.25}$$

and the power carried by a (−) wave alone

$$P_-(z, t) = -Y_0 f_-^2(t + z/v) \tag{4.26}$$

The power in the (+) wave is indeed positive, and that in the (−) wave negative. It is not self-evident, but can be proved quite easily, that the power carried by a combination of a (+) wave and a (−) wave is merely the algebraic sum of the individual powers, as given in Eqs. 4.25 and 4.26.

We conclude that the existence of a (−) wave requires either the presence of a power source at the right-hand end of the line or a termination at that end which would reflect part of the power carried by a (+) wave. Now, what we mean by a semi-infinite line is one with the far end so remote that no waves can reach it or return from it in a finite time. On this basis it seems apparent, on physical grounds, that in the problem at hand the $f_-(t + z/v)$ solution must be identically equal to zero. Let us verify this important physical reasoning by direct analysis.

First, at $t = 0$ we know there is no voltage *nor* current at *any* point $z > 0$. The general solution (Eq. 4.20) tells us, therefore, that

$$\left.\begin{aligned} &\text{(a)} \quad V(z, 0) = 0 = f_+(-z/v) + f_-(+z/v) \\ &\text{(b)} \quad I(z, 0) = 0 = f_+(-z/v) - f_-(+z/v) \end{aligned}\right\} \quad \text{for } z > 0 \tag{4.27}$$

By addition of Eqs. 4.27*a* and 4.27*b* we find

$$f_+(-z/v) = 0 \qquad \text{for } z > 0 \tag{4.28a}$$

and by subtraction

$$f_-(+z/v) = 0 \qquad \text{for } z > 0 \tag{4.28b}$$

Equations 4.28 place restrictions upon the shapes of the functions $f_+(u)$ and $f_-(w)$ as follows:

$$\begin{aligned} &\text{(a)} \qquad f_+(u) = 0 \qquad \text{for } u < 0 \\ &\text{(b)} \qquad f_-(w) = 0 \qquad \text{for } w > 0 \end{aligned} \tag{4.29}$$

The restrictions of Eqs. 4.29 are summarized in Fig. 4.3. The heavy lines show what we now know about the functions $f_+(u)$ and $f_-(w)$, as a result of applying *only* the boundary condition of "initial rest" for the line. The question marks indicate ranges of u and w over which we still have no information about the unknown functions. Actually, however, Eq. 4.29*b* already tells us all we need to know about $f_-(w)$ for the problem at hand. This function enters the general solution in the form $f_-(t + z/v)$, and in our problem only $t > 0$ and $z > 0$ are of interest. In other words, only the behavior of $f_-(w)$ for $w > 0$ is within our view for this problem, and it is then always *zero* according to Eq. 4.29*b*. This means there is never a (−) wave in our example, which checks our initial physical reasoning.

Equation 4.29*a* and Fig. 4.3 now tell us more about the function f_+. This function enters the general line solution in the form $f_+(t - z/v)$,

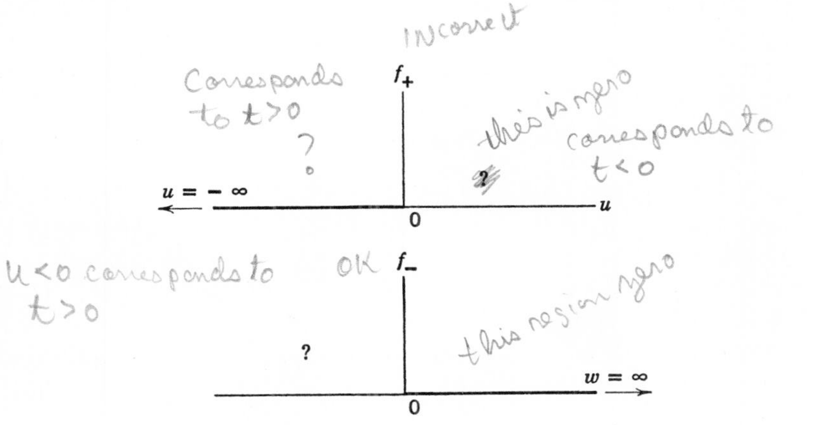

Fig. 4.3. Extent to which $f_+(u)$ and $f_-(w)$ on a semi-infinite line are fixed by the rest conditions $V(z, 0) = 0$ and $I(z, 0) = 0$ for $z > 0$.

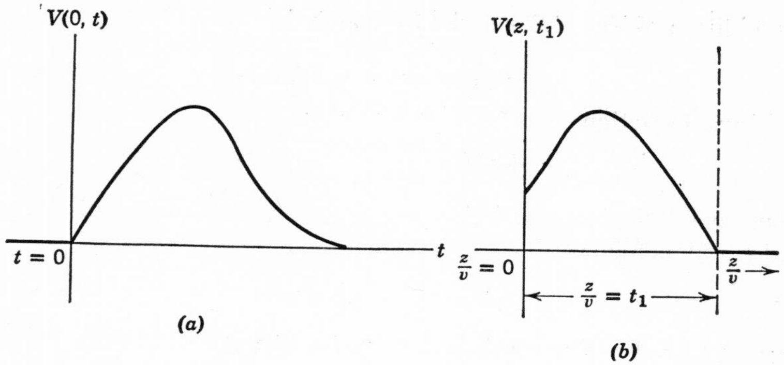

Fig. 4.4. The voltage on a semi-infinite line as a function of z/v and t.

and, although we care only about $t > 0$ and $z > 0$, it is certainly important to consider points and times for which, nevertheless, $z > vt > 0$. Such situations make the argument of f_+ *negative* (i.e., $u < 0$), and we see that f_+ is then zero. Since f_- is always zero, we conclude that there can be no voltage nor current on the line at any point z_1, at least until the time $t_1 = z_1/v$. The interpretation of this result is both simple and important, and we have seen it several times before. There was no energy on the line at $t = 0$, and the only source or termination is at $z = 0$. Whatever this source does at $t = 0^+$ cannot possibly be felt at any point $z_1 > 0$ before the wave (f_+) initiated by the source can reach the point z_1. This requires a time $t_1 = z_1/v$. What happens after that obviously depends upon the source, which must therefore determine for us how $f_+(u)$ behaves when $u \geq 0$.

We are now ready to apply the boundary condition at the source. It is the statement that

$$V(0, t) = f_+(t) = V_1(t) \qquad \text{for } t > 0 \tag{4.30}$$

in which we have already used our knowledge that $f_-(t) = 0$ for $t > 0$. Equation 4.30 determines the shape of $f_+(u)$ for $u > 0$. This shape is exactly that of V_1 as a function of time, and is shown in Fig. 4.4*a* as a plot of $V(0, t)$ versus t. The voltage at any other point, at any time, is consequently given by the solution

$$\begin{aligned} V(z, t) &= f_+(t - z/v) = V_1(t - z/v) && \text{for } t \geq z/v > 0 \\ &= 0 && \text{for } 0 < t < z/v \end{aligned} \tag{4.31}$$

This solution is shown in Fig. 4.4*b* as a plot of $V(z, t_1)$ versus z/v at some fixed time $t_1 > 0$. Observe that the voltage distribution on the line (plotted versus z/v) has the same shape as $V_1(t)$, except that it is turned end for end and, of course, does not extend to the left beyond the input at $z = 0$. The reason for the inversion is made clear by imagining the distribution in Fig. 4.4*b* to be sliding to the right as time goes on. Its rightmost point then reaches a given location z_1 *first* in time, and becomes the start (leftmost point) in a plot of voltage versus time at point z_1.

The final matter of interest is the current wave, given by

$$I(z, t) = Y_0 f_+(t - z/v) \tag{4.32}$$

because $f_- = 0$. This result is exactly like that for the voltage, except for a factor Y_0. Figure 4.4 could therefore also represent the current. One important consequence of these facts is the boundary relation

$$I(0, t) = I_1(t) = Y_0 V(0, t) = Y_0 V_1(t) \tag{4.33}$$

which shows that the semi-infinite line "looks" to the generator like a pure resistance Z_0.

4.3.2 Effect of Source Resistance on the Semi-infinite Line

Only a slight addition to our previous results is required if the generator driving the semi-infinite line has an internal resistance R_s. The problem is shown in Fig. 4.5, where the generator voltage is denoted by $V_s(t)$.

If the line is without energy at $t = 0$, all of our previous calculations are unchanged. Equation 4.30, although still true, does not fix $f_+(t)$ for $t > 0$, because $V_1(t)$ is unknown now. A new boundary condition at $z = 0$ must relate the source voltage V_s to the line input voltage V_1. This relation is simply

$$V_s(t) - I_1(t) R_s = V_1(t) \tag{4.34}$$

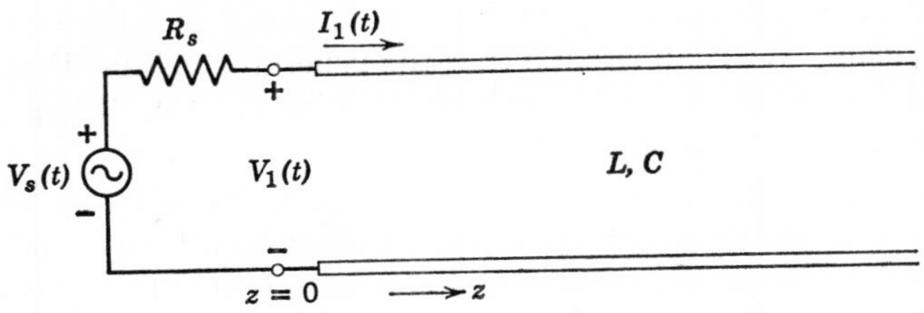

Fig. 4.5. A semi-infinite line driven by a generator with internal resistance.

But, according to Eq. 4.33, which is also still valid, Eq. 4.34 becomes

$$V_s(t) - I_1(t)R_s = I_1(t)Z_0 \tag{4.35}$$

which suggests the equivalent circuit shown in Fig. 4.6. Actually there is little new in it, because we have already seen that the line looks

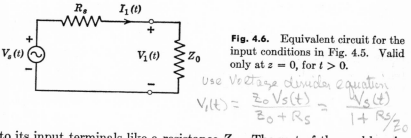

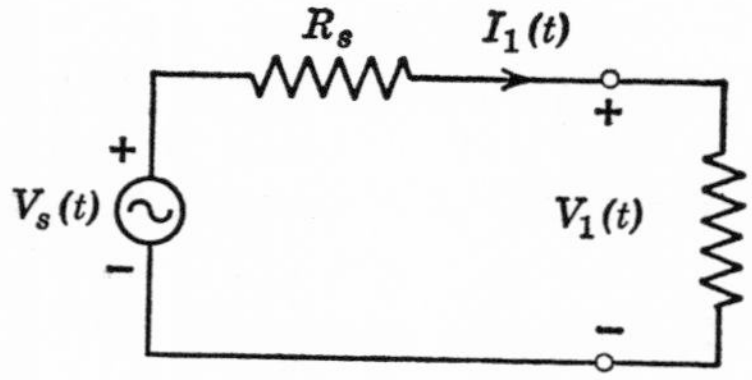

Fig. 4.6. Equivalent circuit for the input conditions in Fig. 4.5. Valid only at $z = 0$, for $t > 0$.

Use Voltage divider equation

$V_1(t) = \frac{Z_0 V_s(t)}{Z_0 + R_s} = \frac{V_s(t)}{1 + R_s/Z_0}$

to its input terminals like a resistance Z_0. The rest of the problem is straightforward, since from Fig. 4.6 or Eqs. 4.34 and 4.35 we have

$$\frac{V_s}{1+\frac{R_s}{Z_0}} = V_1(t) = f_+(t) = \frac{V_s(t)}{1 + Y_0R_s} \qquad \text{for } t \geq 0 \tag{4.36}$$

which fixes $f_+(u)$ for $u > 0$. Thus

$$\begin{aligned} V(z, t) &= 0 && \text{for } 0 < t < z/v \\ &= \frac{V_s(t - z/v)}{1 + Y_0R_s} && \text{for } t \geq z/v \geq 0 \end{aligned} \tag{4.37a}$$

and

$$I(z, t) = Y_0V(z, t) \tag{4.37b}$$

4.3.3 Terminated Transmission Line with Transient Input

We are now prepared to deal with systems like that shown in Fig. 4.7, in which a lossless transmission line is terminated in a linear, passive, lumped load. We shall assume that the generator voltage is zero, and that the line and load are not energized, at times $t < 0$. The generator starts to develop a voltage at $t = 0$.

Physically, if the line is unenergized at $t \leq 0$, the only way for a $(-)$ wave to reach the generator terminals at $z = 0$ is to come from the load end $z = l$. This would certainly take a time l/v. We would expect, therefore, that satisfying the boundary conditions at the input end of the finite line for an excitation starting at $t = 0$ is exactly the same problem as the analogous one for a semi-infinite line, at least during the time interval $0 \leq t < l/v$. Hence the generator at $z = 0$ must simply start a (+) wave at $t = 0$, according to Eq. 4.36 and Fig. 4.6, and this (+) wave must travel toward the load as shown in Eqs. 4.37 and Fig. 4.4. But we can say more. If the load is passive, and has no energy stored in it at $t = 0$, it cannot possibly *originate* a (−) wave until the (+) wave started by the input generator reaches it. This last event requires a time l/v. It follows that the (−) wave at $z = l$ does not even *begin* until $t = l/v$, and it, therefore, cannot affect the generator at $z = 0$ until *another* time interval l/v has elapsed. Consequently, we should expect that the problem of satisfying the boundary conditions at the input end of our finite line for an excitation starting at $t = 0$ is exactly the same as it would be if the line were semi-infinite, for a time interval of at least $0 \leq t < 2l/v$. Briefly, then, we are saying that in our present problem the generator cannot "know" of the existence or nature of the far end of the line until the (+) wave which it originates at $t = 0$ has had a time l/v to reach the load, and there has set up a reflected (−) wave which will take another time interval l/v to get back to the generator again.

Physical reasoning like that employed above is most useful in solving transient problems, and we must develop enough familiarity with it to

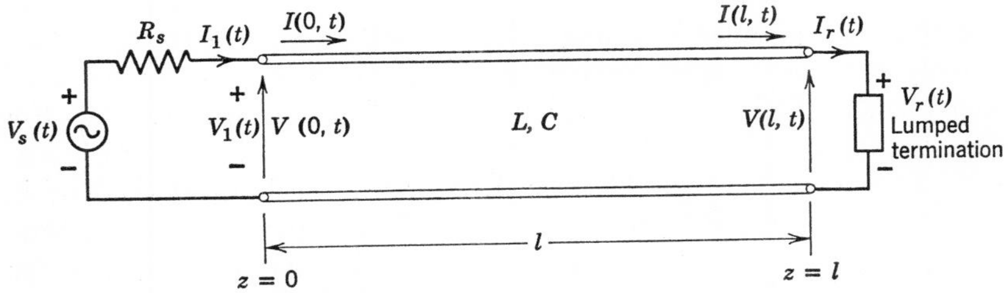

Fig. 4.7. A finite line.

feel confident of its use. Therefore let us once again check very carefully our conclusions by direct mathematical analysis.

The fact that the line in Fig. 4.7 has no voltage nor current anywhere on it at $t = 0$ requires

$$\left.\begin{aligned}&\text{(a)}\quad V(z, 0) = f_+(-z/v) + f_-(z/v) = 0\\&\text{(b)}\quad I(z, 0) = Y_0[f_+(-z/v) - f_-(z/v)] = 0\end{aligned}\right\}\quad \text{for } 0 < z < l \tag{4.38}$$

Therefore

$$f_+(-z/v) = 0 \qquad \text{for } 0 < z < l$$

or

$$f_+(u) = 0 \qquad \text{for } -l/v < u < 0 \tag{4.39}$$

and

$$f_-(z/v) = 0 \qquad \text{for } 0 < z < l$$

or

$$f_-(w) = 0 \qquad \text{for } 0 < w < l/v \tag{4.40}$$

Conditions 4.39 and 4.40 are shown in Figs. 4.8*a* and 4.8*b*. These should be compared with Fig. 4.3 and Eqs. 4.29. The meaning of Eq. 4.40 becomes clear when it is expressed in terms of t and z. The $(-)$ wave voltage is $f_-(t + z/v)$, so that at the input $(z = 0)$ it is just $f_-(t)$. According to Eq. 4.40 this is certainly zero for $0 < t < l/v$. Evidently, even if there were a source (or an initial energy storage) in the termination at $z = l$ (which there is not in our problem), it could not possibly make itself felt at $z = 0$ before the $(-)$ wave transit time l/v. This is part of the physical interpretation of Eq. 4.40, and it certainly checks part of our earlier reasoning.

Now consider the load end of the line, $z = l$. The $(-)$ wave voltage there is $f_-(t + l/v)$. When $t \geq 0$, this function is $f_-(w)$ with $w \geq l/v$. Hence our condition (Eqs. 4.40) of rest for the line tells us nothing about the $(-)$ wave voltage at $z = l$ (see Fig. 4.8*b*). We should not be surprised, since physically a $(-)$ wave at $z = l$ could only be caused by either a $(+)$ wave reaching it from $z = 0$ or by a source (or initial energy storage) in the termination itself. We have not yet put into our formal solution any condition relating to the termination; so the question cannot be resolved completely until we do. However, Eq. 4.39 is of some help here because the $(+)$ wave voltage at the load is $f_+(t - l/v)$. According to Eq. 4.39 this is zero for $0 \leq t < l/v$; thus we see again that the load can receive no information from the generator prior to $t = l/v$.

This exhausts all the useful information contained in the conditions of Eqs. 4.38 that the line was at rest at $t = 0$. We must now consider conditions at the source and load.

At the source $(z = 0)$ we have a generator $V_s(t)$ which is zero for $t < 0$ and some specified function of time for $t \geq 0$, shown in Fig. 4.8*c*. In the interval $0 < t < l/v$ we know from Fig. 4.8*a* that, at $z = 0$, $f_-(t + z/v) = f_-(t) = 0$. Hence the boundary conditions relating the source network to the line input are

$$\left.\begin{aligned} &\text{(a)} \quad V_s(t) - I_1(t)R_s \equiv V(0, t) = f_+(t) \\ &\text{(b)} \quad I_1(t) \equiv I(0, t) = Y_0 f_+(t) \end{aligned}\right\} \qquad \text{for } 0 < t < l/v \tag{4.41}$$

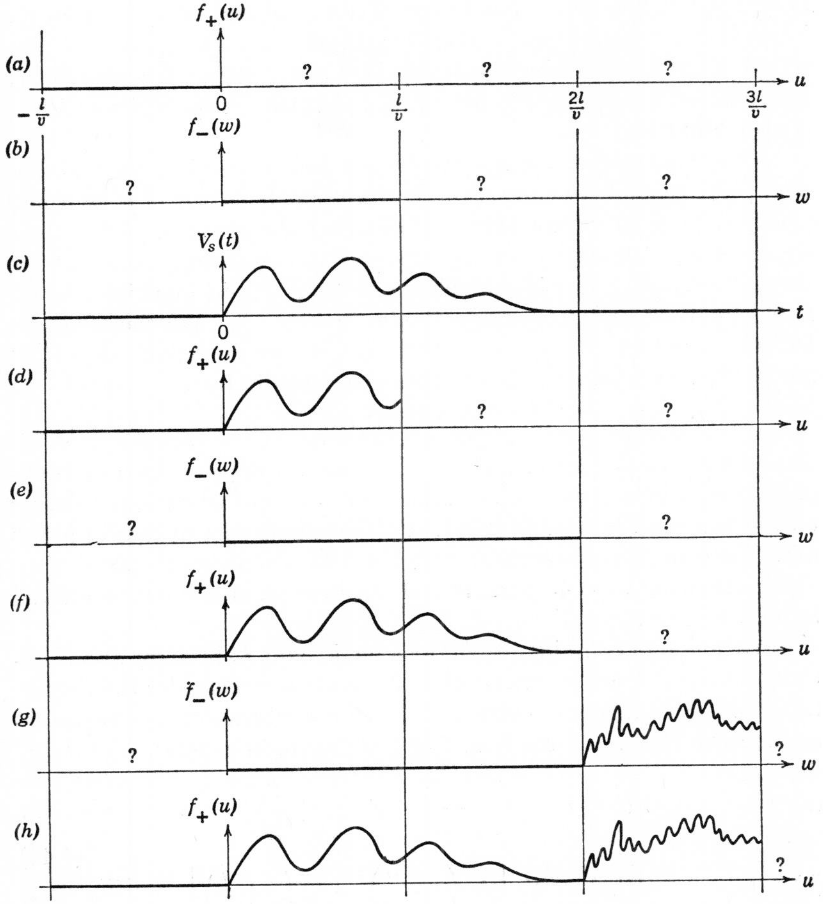

Fig. 4.8. Illustrative stages of the formal transient solution for a finite lossless line. Vertical sizes are not drawn to scale.

These lead to a result just like that of Fig. 4.6 and Eq. 4.36, namely,

$$f_+(t) = \frac{V_s(t)}{1 + Y_0 R_s} \qquad \text{for } 0 < t < l/v \tag{4.42}$$

but more limited in time range of validity, as we predicted on physical grounds earlier. Thus we now know $f_+(u)$ for $0 < u < l/v$, which we show in Fig. 4.8*d*.

We cannot proceed any further with conditions at the input because we do not know $f_-(t)$ beyond $t = l/v$. Physically this is because we have not yet considered conditions at the load end of the line in the interval $0 < t < l/v$; for it is these conditions which would make themselves felt at $z = 0$ during the interval $l/v \leq t < 2l/v$. We must therefore proceed to examine the boundary conditions at $z = l$ if we wish to continue our solution.

At the load end of the line we have a linear, passive, lumped network. In general, to determine the transient behavior of such a network, we have to write a system of differential equations. From these we can always eliminate all currents and voltages except those at the input terminals. The input voltage and current in question are denoted by $V_r(t)$ and $I_r(t)$, respectively, in Fig. 4.7. We therefore are left with a more or less elaborate differential equation connecting $V_r(t)$ and $I_r(t)$, which we may denote here symbolically by

$$I_r(t) = F[V_r(t)] \tag{4.43}$$

The solution of Eq. 4.43 for $t \geq 0$ will generally depend upon both the initial conditions in the network and the voltages or currents arising on the transmission line at $z = l$. In the simple special case of a termination containing only resistance, Eq. 4.43 is of algebraic rather than differential character, but the rest of our discussion will not be altered by this circumstance.

Equation 4.43 contains restrictions on V_r and I_r brought about entirely by the terminating network. We must now consider the restrictions on the voltage and current at $z = l$ brought about by the transmission line itself. At the load $z = l$, we have, in general,

$$\begin{aligned} &\text{(a)} \qquad V(l, t) = f_+\left(t - \frac{l}{v}\right) + f_-\left(t + \frac{l}{v}\right) \\ &\text{(b)} \qquad I(l, t) = Y_0\left[f_+\left(t - \frac{l}{v}\right) - f_-\left(t + \frac{l}{v}\right)\right] \end{aligned} \tag{4.44}$$

The essential boundary conditions at the load are then simply

$$\begin{aligned} &\text{(a)} \qquad V(l, t) = V_r(t) \\ &\text{(b)} \qquad I(l, t) = I_r(t) \end{aligned} \tag{4.45}$$

But solving Eq. 4.44*b* for $f_-(t + l/v)$ yields

$$f_-\left(t + \frac{l}{v}\right) = f_+\left(t - \frac{l}{v}\right) - Z_0 I(l, t) \tag{4.46}$$

which we substitute into Eq. 4.44*a*, using also Eqs. 4.45, to obtain

$$V_r(t) = 2f_+\left(t - \frac{l}{v}\right) - Z_0 I_r(t) \tag{4.47}$$

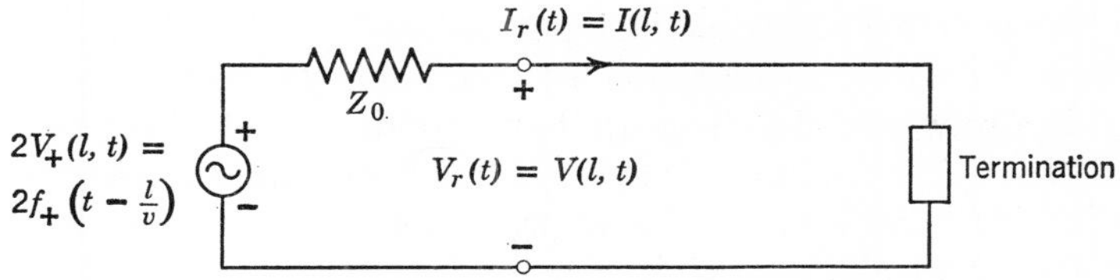

Fig. 4.9. Equivalent circuit at a termination.

Equation 4.47 is extremely important. It says that the transmission line behaves with respect to *any* termination like a voltage source $2f_+(t - l/v)$ in series with an impedance Z_0. In other words, *the effect of the line upon the load can always be calculated from the equivalent circuit shown in Fig. 4.9.* This result allows us to determine both $V_r(t)$ and $I_r(t)$ by simply solving the transient network problem defined by the figure. Such a solution amounts to exactly the same thing as substituting Eq. 4.47 into Eq. 4.43 and solving the resulting differential equation, which then contains either I_r or V_r alone. Note, however, that the whole process can be accomplished only if we know the "incident wave" value at the load, i.e., we must know $f_+(t - l/v)$. This information can come only from our previous considerations at the source end of the line.

Before continuing our discussion of this last point, however, let us observe that Eq. 4.47 is really an extension of the time-dependent form of Thévènin's theorem,[1] in the following particular sense. Imagine for a moment that $f_+(t - l/v)$ is a *specified* incident wave which does not change if we change the termination. Then, if the termination is open-circuited, so that $I_r(t) \equiv 0$, Eq. 4.47 shows that $V_r(t)|_{\text{open circuit}} = 2f_+(t - l/v)$. Now consider artificially forcing $f_+(t - l/v) \equiv 0$, and applying a *source* at the load terminals to measure the "impedance" seen looking back into the line. Inasmuch as we have *forced* $f_+(t - l/v)$ to be zero, the line looks infinitely long to this source. Hence we know from Sec. 4.3.1 that, even though we are working in the time domain, the line may be characterized to the source as an impedance Z_0. From this very special point of view, Eq. 4.47 and Fig. 4.9 may be regarded as the result of replacing the line by its Thévènin equivalent circuit, comprised of its "open-circuit voltage" in series, with the network (in this case simply the resistance Z_0) seen looking back into it with all "internal sources" inactive. The new feature is simply that we regard the load voltage and current as being produced by the wave f_+ rather

[1] See, e.g., Ernst A. Guillemin, *Introductory Circuit Theory*, John Wiley and Sons, New York, 1953, Ch. 5, Art. 3, 235–238.

than by the actual source at the input end of the line. Our right to do so springs from the linear relation between V_r, I_r, f_+, and f_-, a linearity which is always sufficient to permit superposition and, therefore, validates Thévènin's theorem (or other source transformations).

Returning now to our particular problem (Fig. 4.7), we need to consider conditions at the load ($z = l$) in the time interval $0 \leq t < l/v$. In this interval, $f_+(t - l/v)$ is zero, according to Fig. 4.8*d*. If the load network is unenergized at $t = 0$, the transient solution to Fig. 4.9—or to Eqs. 4.43 and 4.47—is $V_r(t) = I_r(t) = 0$, since there is no excitation at all! It follows from Eqs. 4.46 and 4.45*b* that $f_-(t + l/v)$ is also zero during this time interval, or that

$$f_-(w) = 0 \qquad \text{for } l/v \leq w < 2l/v \tag{4.48}$$

as shown in Fig. 4.8*e*. These analytical results indeed check our earlier physical reasoning that a "dead" termination cannot originate a $(-)$ wave during a time interval prior to its reception of the $(+)$ wave from the input generator.

With the foregoing argument completed, we can return to the generator at $z = 0$ to consider the next time interval $l/v \leq t < 2l/v$. The $(-)$ wave voltage at $z = 0$ is $f_-(t)$, which is now known to be zero in this time interval (Fig. 4.8*e*). Hence Eqs. 4.41 and 4.42 and Fig. 4.6 are still valid, but over a time interval $0 \leq t < 2l/v$. This is in accord with our initial physical reasoning that, if the "dead" termination could not start a $(-)$ wave in the interval $0 \leq t < l/v$, the generator could not possibly perceive one in the interval $l/v \leq t < 2l/v$.

Our formal conclusion regarding the extended range of validity of Eq. 4.42 to the time interval $0 \leq t < 2l/v$ is summarized by the equation

$$f_+(u) = \frac{V_s(u)}{1 + Y_0 R_s} \qquad \text{for } 0 \leq u < 2l/v \tag{4.49}$$

shown also in Fig. 4.8*f*.

We have now given a complete analytical justification of all the points covered by the physical reasoning we used at the beginning of this section, and it remains to complete the solution for our specific problem. The following discussion outlines the general steps involved.

The $(+)$ wave started by the generator at $t = 0$, according to Fig. 4.6 and Eqs. 4.41, 4.42, and 4.49, first reaches the load at $t = l/v$. This produces an excitation of the termination according to Fig. 4.9, and we must determine the transient solution for this network. Assuming that we have done this by standard network methods, we shall then know $V_r(t)$ and $I_r(t)$ for as long as we know $f_+(t - l/v)$. Then, according to Eqs. 4.44*a* and 4.45*a*, we find

$$f_-\left(t + \frac{l}{v}\right) = V_r(t) - f_+\left(t - \frac{l}{v}\right) \tag{4.50}$$

which determines the reflected wave at the termination. There will be no reflected wave (i.e., $f_- \equiv 0$) *only if* $V_r(t) = f_+(t - l/v)$. Figure 4.9 or Eqs. 4.44 and 4.45 shows that this can happen only if $V_r(t) = Z_0 I_r(t)$, i.e., if the termination appears from its terminals to be a pure resistance Z_0. Even for transients, then, termination of a lossless line in its characteristic resistance Z_0 is the *only* way to avoid reflections. Otherwise, Eq. 4.50 gives the form of the reflected wave at $z = l$ in the interval $l/v \le t < 2l/v$. This fixes $f_-(w)$ in the range $2l/v \le w < 3l/v$, as indicated in Fig. 4.8*g*.

Once the form of the reflected wave as a function of time is given for $z = l$, we can find its behavior at any other point along the line. Because f_- is a reflected wave, which must travel to the left, its value at the point $z < l$ will be that at the point $z = l$ *delayed* by a time $(l - z)/v > 0$. Thus, from Eqs. 4.50 and 4.47,

$$\begin{aligned} f_-\left(t + \frac{z}{v}\right) &= f_+\left[t - \frac{l}{v} - \left(\frac{l - z}{v}\right)\right] - Z_0 I_r\left[t - \left(\frac{l - z}{v}\right)\right] \\ &= f_+\left(t - \frac{2l}{v} + \frac{z}{v}\right) - Z_0 I_r\left(t - \frac{l}{v} + \frac{z}{v}\right) \end{aligned} \tag{4.51}$$

The right-hand side of Eq. 4.51 is indeed a function of $w = (t + z/v)$, as indicated on the left-hand side and as required of any wave moving to the left on the line.

The reflected wave (Eq. 4.51) first takes on a value other than zero at the generator end of the line at time $t = 2l/v$. The $(-)$ wave voltage $f_-(t)$ at $z = 0$ then begins to show radically new behavior in the interval $2l/v \le t < 3l/v$. It carries information about the load, as indicated in Fig. 4.8*g*. The incident wave $f_+(t)$ can no longer depend only upon what the source $V_s(t)$ is doing because $f_-(t)$ is no longer zero. The basic boundary conditions of current and voltage continuity are

$$\begin{aligned} &\text{(a)} \quad V_s(t) - I_1(t)R_s = V_1(t) = V(0, t) = f_+(t) + f_-(t) \\ &\text{(b)} \quad I_1(t) = I(0, t) = Y_0[f_+(t) - f_-(t)] \end{aligned} \tag{4.52}$$

Eliminating $f_+(t)$ from Eqs. 4.52 yields for us the conclusion that

$$V_s(t) - I_1(t)R_s = 2f_-(t) + Z_0 I_1(t) \tag{4.53}$$

This is really the same in meaning as Eq. 4.47, or Fig. 4.9, except that it is applied at the line input. It says that *as far as the input generator circuit is concerned, the line behaves like a voltage generator* $2f_-(t)$ *in series with a resistance* Z_0. Since, as far as the generator circuit is concerned, the $(-)$ wave is actually impinging upon it—as an "incident" wave—the extended Thévènin interpretation of Eq. 4.53 is just like that we discussed following Eq. 4.47. As a useful exercise, the reader should draw the equivalent circuit suggested by Eq. 4.53, and compare it with Fig. 4.9, noting especially the differences in subscripts and algebraic signs.

From Eq. 4.53, or the corresponding equivalent circuit, we have

$$I_1(t) = \frac{V_s(t)}{R_s + Z_0} - \frac{2f_-(t)}{R_s + Z_0} \tag{4.54}$$

which leads via Eq. 4.52*b* to an expression for the $(+)$ wave at $z = 0$:

$$f_+(t) = Z_0 I_1(t) + f_-(t)$$

or

$$f_+(t) = \frac{V_s(t)}{1 + Y_0 R_s} + \left(\frac{R_s - Z_0}{R_s + Z_0}\right) f_-(t) \tag{4.55}$$

Equation 4.55 is always valid, but is *useful* only as long as we know $f_-(t)$. In our case we know $f_-(t)$ out to $t = 3l/v$ (Fig. 4.8*g*), and, therefore, we can complete $f_+(t)$ in the interval $2l/v \le t < 3l/v$. This we have done in Fig. 4.8*h*, by showing $f_+(u)$ out to $u = 3l/v$.

As should be evident, the general process of formal solution that we have been following must now be repeated over and over again, using *alternately* the boundary conditions and equivalent circuits at the load ($z = l$) and source ($z = 0$) ends of line. In general, we shall thereby "uncover" alternately the functional forms of the general solutions $f_+(u)$ and $f_-(w)$ in intervals of length l/v at a crack, first uncovering this much more of $f_-(w)$ by considerations at the load, and then this much more of $f_+(u)$ by considerations at the generator. This scheme is well illustrated by comparing in sequence the patterns of Figs. 4.8*a*, *b*, *d*, *e*, *f*, *g*, and *h*.

The solution technique used above has the advantage of being quite general, but there is another way of thinking about the whole process which proves exceedingly convenient and useful *whenever the generator and load circuits are linear* (as we may assume they are in Fig. 4.7). The basis of the idea is to use the superposition principle. We can best discuss its use by noting first that $I_1(t)$ and $f_+(t)$ in Eqs. 4.54 and 4.55

are each composed of two terms. The first term is just what would have happened at the generator terminals had the reflected wave $f_-(t)$ been zero (compare Eqs. 4.41 and 4.42). The second term is the contribution made to the terminal behavior by the reflected wave $f_-(t)$ acting alone, with the generator short-circuited $[V_s(t) \equiv 0]$.

Thus we can regard the arrival of the $(-)$ wave at $z = 0$ as a *disturbance* of what would have happened had it not arrived, and compute the effect of the returning wave separately. This we do by setting all other waves on the line equal to zero, and making all sources in the terminations "dead." The currents, voltages, and waves set up by f_- alone we may then designate by using a single prime alongside and preceding the corresponding symbol. Under these conditions, with all sources dead, the generator network can be treated simply as a passive termination, using Fig. 4.9 and Eq. 4.50 with a proper modification of subscripts and algebraic signs (see Eqs. 4.52 and 4.53).

$$\text{(a)} \quad {}'V_1(t) = 2f_-(t) + Z_0{}'I_1(t)$$
$$\text{(b)} \quad {}'f_+(t) = {}'V_1(t) - f_-(t) \tag{4.56}$$

In our particular case, the generator network is simply a pure resistance R_s; thus ${}'V_1(t) = -R_s{}'I_1(t)$ and

$$\text{(a)} \quad {}'V_1(t) = \frac{2R_s f_-(t)}{R_s + Z_0}$$
$$\text{(b)} \quad {}'I_1(t) = -\frac{2f_-(t)}{R_s + Z_0} \tag{4.57}$$

Then

$${}'f_+(t) = \left(\frac{R_s - Z_0}{R_s + Z_0}\right) f_-(t) \tag{4.58}$$

Equations 4.57 and 4.58 should be compared with Eqs. 4.54 and 4.55 to emphasize that the primed solutions must be added to the original unprimed ones to obtain the total solutions.

The ${}'f_+$ wave is regarded as starting from the generator terminals $(z = 0)$ at $t = 2l/v$, when the f_- wave first arrived there. Indeed, it is thought of as the *reflection* produced by the incidence of f_- upon the generator network. The ${}'f_+$ wave then travels toward the load, which it reaches at $t = 3l/v$. A primed reflected wave ${}'f_-$ will then be set up to match the boundary conditions at the load, and this new solution must, again, be added to all the old ones by superposition. The whole

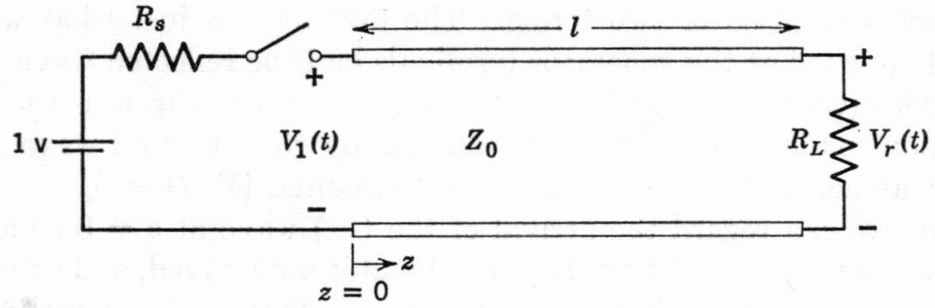

Fig. 4.10. Resistance terminations.

process will, of course, be repeated over and over again, with the complete solution given by the superposition of the effects of all the waves. We would use double primes ($''f_-$ and $''f_+$, etc.) to denote the partial solutions that arise from reflections of $'f_+$ and $'f_-$ at the load and source respectively, and we would continue this multiple-prime notation rule for all successive reflections.

It should be emphasized that equivalent circuits like Fig. 4.9 apply only at one point on the line, i.e., at the load or at the generator for the cases treated thus far. The general line solutions, consisting of positive- and negative-moving traveling waves, must then be used to visualize or find the voltage and current at any other points on the line.

The following two sections (Secs. 4.3.3.1 and 4.3.3.2) present solutions to some specific illustrative examples which fall under the general class defined by Fig. 4.7, and which are covered by the general theory we have discussed so far.

4.3.3.1 Resistance Terminations. Consider the physical arrangement of Fig. 4.10. If the switch is closed at time $t = 0$, a unit step of voltage will appear across the combination of the source resistance and the line. From our previous considerations, an equivalent circuit at the generator may be drawn in the form of Fig. 4.11. We are supposing that there was no current nor voltage on the line prior to $t = 0$. Therefore the requirement that the voltage $V_1(t)$ across the

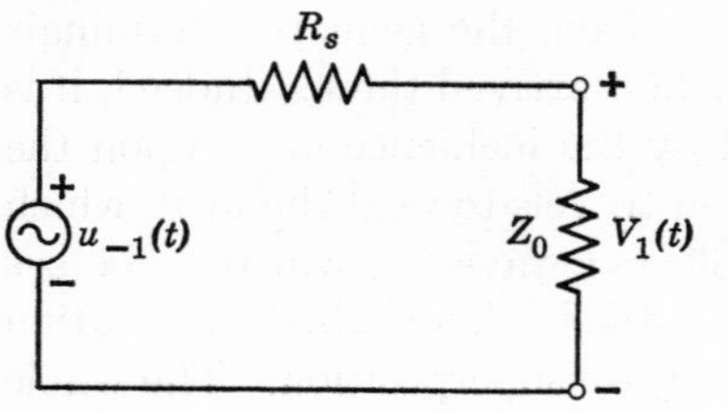

Fig. 4.11. Equivalent circuit for Fig. 4.10 at $z = 0$, valid only for $0 \leq t < 2l/v$.

source terminals equal the line input voltage $V(0, t)$ gives directly the form of the positive traveling wave.

$$V(0, t) = f_+(t) = u_{-1}(t)\left(\frac{Z_0}{Z_0 + R_s}\right) \tag{4.59}$$

At any other point z on the line, the (+) wave must have the same form it had at $z = 0$, but it is delayed by the transit time z/v. Therefore

$$V(z, t) = f_+\left(t - \frac{z}{v}\right) = u_{-1}\left(t - \frac{z}{v}\right)\left(\frac{Z_0}{Z_0 + R_s}\right) \tag{4.60}$$

At $t = l/v$, the wave front of the (+) wave reaches the termination. The load-end equivalent circuit (Fig. 4.12) then becomes useful. From it we find

$$V_r(t) = \frac{2f_+R_L}{R_L + Z_0} = u_{-1}\left(t - \frac{l}{v}\right)\left[\frac{Z_0}{(Z_0 + R_s)}\frac{2R_L}{(R_L + Z_0)}\right] \tag{4.61}$$

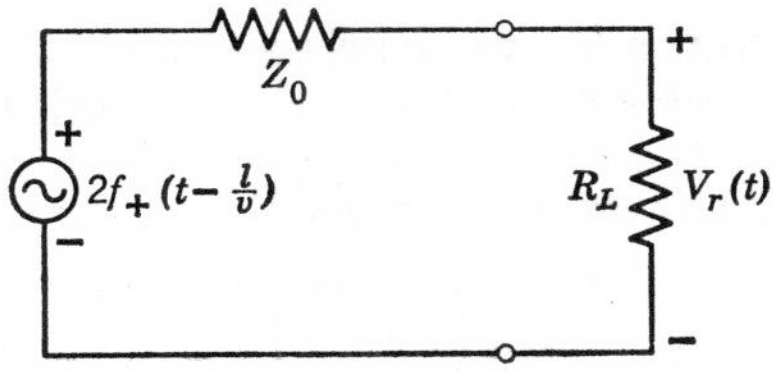

Fig. 4.12. Equivalent circuit for Fig. 4.10 at $z = l$.

Since the terminal voltage at the load must equal that on the line at $z = l$, a reflected wave must be set up such that

$$V_r(t) = V(l, t) = f_+\left(t - \frac{l}{v}\right) + f_-\left(t + \frac{l}{v}\right)$$

or

$$f_-\left(t + \frac{l}{v}\right) = u_{-1}\left(t - \frac{l}{v}\right)\left[\frac{Z_0}{Z_0 + R_s}\left(\frac{R_L - Z_0}{R_L + Z_0}\right)\right] \tag{4.62}$$

At a point $z < l$ elsewhere on the line, the (−) wave must have the same form it has at $z = l$, but it must be *delayed* by the time of transit $(l - z)/v$, since this wave travels to the left. Thus

$$f_-\left(t + \frac{z}{v}\right) = u_{-1}\left(t - \frac{2l}{v} + z\right)\left[\frac{Z_0}{Z_0 + R_s}\left(\frac{R_L - Z_0}{R_L + Z_0}\right)\right] \tag{4.63}$$

gives the (−) wave at any point z on the line.

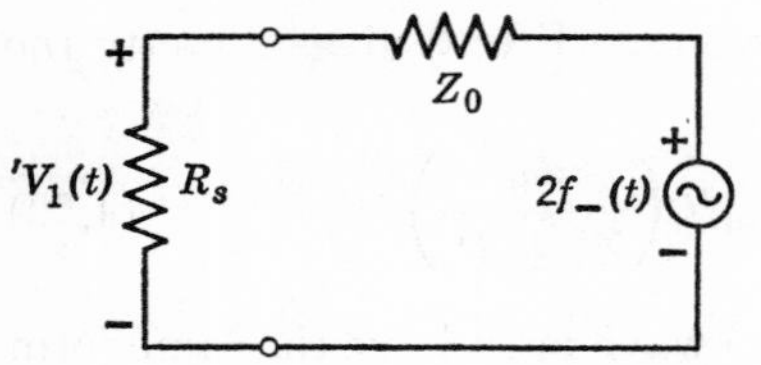

Fig. 4.13. Equivalent circuit for the effect of the reflected wave f_- on the generator in Fig. 4.10.

The generator will not detect the presence of the reflected wave until $t = 2l/v$. At this time, the boundary conditions will again have to be met at the generator. Since the terminations are linear, and the original battery voltage requirements were satisfied by the introduction of the first (f_+) wave, we can satisfy the boundary condition for the f_- wave alone with the source generator short-circuited. An equivalent circuit may be drawn to represent this condition, as shown in Fig. 4.13. From it we find

$$'V_1(t) = \frac{2f_-(t)R_s}{R_s + Z_0} \tag{4.64}$$

Now a $'f_+$ wave arises to meet the boundary condition of voltage continuity from the source network to the line input

$$\begin{aligned} 'f_+(t) &= 'V_1(t) - f_-(t) \\ &= u_{-1}\left(t - \frac{2l}{v}\right)\left\{\frac{Z_0}{Z_0 + R_s}\left[\left(\frac{R_L - Z_0}{R_L + Z_0}\right)\left(\frac{R_s - Z_0}{R_s + Z_0}\right)\right]\right\} \end{aligned} \tag{4.65}$$

where we have used Eqs. 4.63 and 4.64. For another point z on the line, this $'(+)$ wave is delayed by z/v from its form (Eq. 4.65) at $z = 0$.

$$\begin{aligned} 'f_+\left(t - \frac{z}{v}\right) \\ = u_{-1}\left(t - \frac{2l}{v} - \frac{z}{v}\right)\left\{\frac{Z_0}{R_s + Z_0}\left[\left(\frac{R_L - Z_0}{R_L + Z_0}\right)\left(\frac{R_s - Z_0}{R_s + Z_0}\right)\right]\right\} \end{aligned} \tag{4.66}$$

The extension to more reflections is similar. Observe that the superposition technique relieves us of the need to worry about the upper time limits of validity for each expression we write. Instead of "uncovering" the general solutions in sections of time, as we did in the direct analysis leading to Fig. 4.8, we simply *add* new partial (primed) waves to the old solutions. A new partial wave always begins at the instant of time when a wave front of a previous solution reaches a termination and the partial wave originates in space at the termination in question. The voltages and currents anywhere in the system, at any time, are the sums of all partial waves in existence up to that time. This superposition technique for resistance terminations will now be carried through in two specific examples.

EXAMPLE 4.1

In Fig. 4.10, let $R_s = Z_0$ and $R_L = \infty$ (open circuit). Under these conditions we have, at $z = 0$,

$$f_+(t) = \frac{Z_0}{Z_0 + R_s} u_{-1}(t) = \frac{1}{2} u_{-1}(t)$$

and at $z = l$

$$V_r(t) = \frac{2R_L}{R_L + Z_0} f_+\left(t - \frac{l}{v}\right) = 2f_+\left(t - \frac{l}{v}\right) = u_{-1}\left(t - \frac{l}{v}\right)$$

For an open-circuited line, we would expect the final value of the voltage everywhere to be exactly that of the source. In the most general transient situation, this value would only be approached as the long-time limit of the transient solution. In the present example, it has already been reached at $z = l$ when $t = l/v$. The reason for this is that $R_s = Z_0$. From Eq. 4.65 we can see that there will be no reflected wave at the generator end under these conditions. Hence there will be no further transients anywhere on the line after the first reflection from the load end has reached the generator again ($t = 2l/v$). It follows that this first reflection must establish the steady state as it reaches each point on the line.

EXAMPLE 4.2

In Fig. 4.10 let $R_s = \frac{1}{2}Z_0$, $R_L = 2Z_0$. The interesting characteristics of a lossless transmission line depend on the fact that the voltages and currents vary with time. We have seen that in the d-c steady state, such a line is merely a resistanceless connection between source and load (Eq. 4.18 ff.). Figure 4.14 shows the d-c steady-state circuit

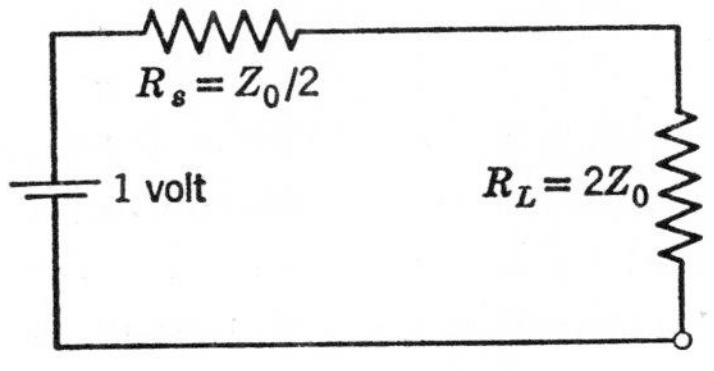

Fig. 4.14. Circuit for d-c steady state, Example 4.2.

for Fig. 4.10 under the present conditions. From it we find the final load voltage ($t = \infty$):

$$V_r(\infty) = \frac{2Z_0}{Z_0/2 + 2Z_0} = \frac{4}{5} \text{ volt}$$

We can now determine how the voltage approaches this final limit. The three important factors needed in connection with Eqs. 4.63 and 4.66 are given below:

$$\frac{Z_0}{R_s + Z_0} = \frac{2}{3}$$

$$\frac{R_L - Z_0}{R_L + Z_0} = \frac{1}{3}$$

$$\frac{R_s - Z_0}{R_s + Z_0} = -\frac{1}{3}$$

At time $t = 0$, the switch is closed and (from Eq. 4.60) an f_+ wave of value $\frac{2}{3}$ v travels down the line. The load does not detect this wave until $t = l/v$.

At this time, an f_- wave is set up in order to satisfy the boundary conditions. From Eq. 4.63 this f_- wave is seen to have a value of of $\frac{2}{3} \times \frac{1}{3} = \frac{2}{9}$ volt.

The total voltage at the load is given by a superposition of the (+) and (−) waves.

$$V_r = \tfrac{2}{3} + \tfrac{2}{9} = \tfrac{8}{9} = 0.889 \text{ volt starting at } t = l/v$$

The negative traveling wave f_- returns to the generator at $t = 2l/v$. Another reflection $'f_+$ occurs, the value of which is given by Eq. 4.65 as $\frac{2}{9}(-\frac{1}{3}) = -\frac{2}{27}$ volt.

At $t = 3l/v$, this new $'f_+$ wave reaches the load again where the boundary conditions must be satisfied independently. The value of the $'f_-$ wave can be seen by applying Eq. 4.63 again, to be given by $(-\frac{2}{27})\frac{1}{3}$, yielding a total V_r as the superposition of all waves

$$\begin{aligned} V_r &= \tfrac{8}{9} - \tfrac{2}{27} - (\tfrac{2}{27})\tfrac{1}{3} \\ &= \tfrac{64}{81} = 0.790 \text{ volt starting at } t = 3l/v \end{aligned}$$

After only two reflections at the load, V_r is already within about 1% of its final steady-state value. In Fig. 4.15, a plot is given of the voltage across the load versus time, out to $t = 5l/v$. Evidently the voltage will keep oscillating about the final value, making smaller and

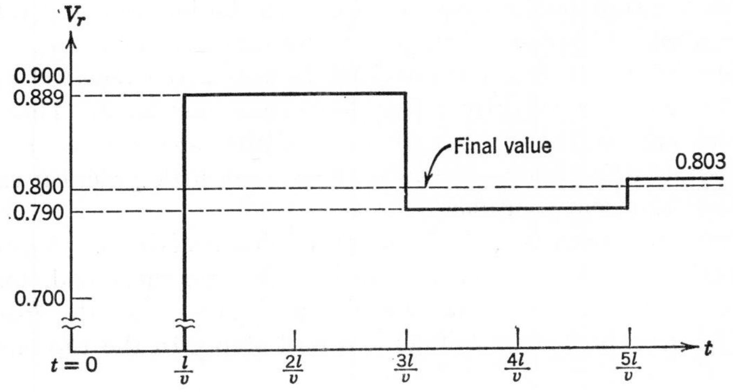

Fig. 4.15. Load voltage for Example 4.2.

smaller excursions on either side of it in successive reflections. It is in this sense that the load voltage approaches the steady-state value 0.8 v.

In the foregoing discussions we notice that the factors in Eqs. 4.62 and 4.65 which relate the reflected waves to the incident waves at the load and generator ends of the line respectively are just the *reflection coefficients* defined for steady-state behavior in Chapter 3. This is, of course, a special situation but a useful one. Ordinarily, we would not expect the steady-state reflection coefficient to have much *direct* significance in transient problems, any more than do the complex impedances involved in its definition. In the special case of purely resistive generator and load impedances on a lossless line, however, Z_0, R_s, and R_L are all real and independent of frequency. Hence the reflection coefficients at each end are also independent of frequency. In terms of a Fourier representation of the time functions, therefore, every sinusoidal component of the incident wave is reflected in precisely the same manner. Consequently, the entire time function representing the incident wave is, in fact, reflected by the very same factor. Thus we can use the ideas of reflection carried over from Chapter 3, *provided we do not forget the special nature of the circumstances in which such use is allowed.* The important distinction in this respect between the resistive and the general cases can be phrased in a different way. If the steady-state complex reflection coefficient of a termination is not the same at all frequencies, the various Fourier components of an incident-wave time function will, in general, be reflected with different relative amplitudes and phases. Hence the reflected wave will be of a different time shape than the incident wave. Therefore, in the time domain we would expect the incident and reflected waves to differ by more than just a multiplicative constant. In the special case of resistive terminations, however, the reflected and incident waves will evidently always have the same shape. These matters will be clarified by the examples of the next section, in which steady-state reflection coefficient notions are of little practical value in obtaining transient solutions.

4.3.3.2 CAPACITIVE AND INDUCTIVE TERMINATIONS. Suppose a lossless transmission line is matched at the generator end, but terminated in an initially uncharged capacitance C at the load end (Fig. 4.16). The source voltage is a unit step. In the time interval

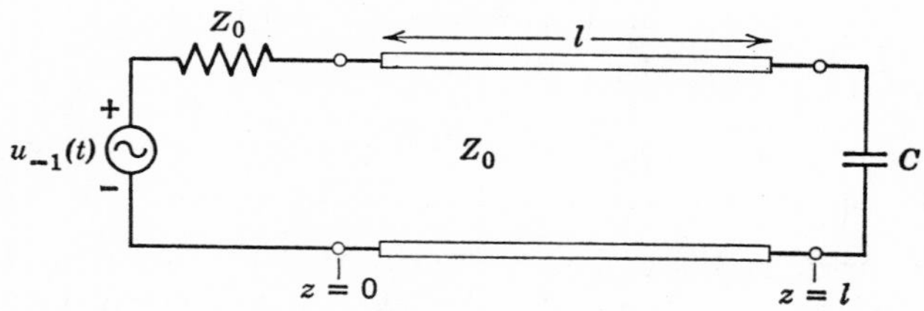

Fig. 4.16. A capacitance termination.

$0 < t < l/v$, the capacitor does not affect the solution, and an incident wave

$$f_+ = \frac{1}{2} u_{-1}\left(t - \frac{z}{v}\right) \tag{4.67}$$

travels down the line. At $t = l/v$ this wave reaches the load, and conditions are represented by Fig. 4.17. We have, therefore,

$$V_r(t) = u_{-1}\left(t - \frac{l}{v}\right) - Z_0 I_r(t) \tag{4.68}$$

and

$$I_r(t) = C\frac{dV_r}{dt} \tag{4.69}$$

so

$$Z_0 C\frac{dV_r}{dt} + V_r = u_{-1}\left(t - \frac{l}{v}\right) \tag{4.70}$$

The particular solution to Eq. 4.70 for $t > l/v$ is 1, and the transient is

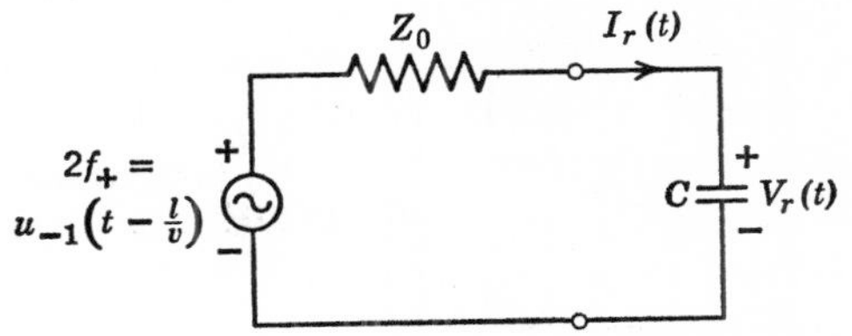

Fig. 4.17. Load-end equivalent circuit for Fig. 4.16.

of the form Ae^{-t/Z_0C}. Since $V_r = 0$ at $t = l/v$, we can evaluate A, and find

$$V_r(t) = u_{-1}\left(t - \frac{l}{v}\right)\left(1 - e^{-(t-l/v)/Z_0C}\right) \tag{4.71}$$

where we use the step-function multiplier to insure $V_r = 0$ for $t < l/v$.

The reflected wave f_- at $z = l$ must be such that $V_r = f_+ + f_-$; thus, from Eqs. 4.67 and 4.71,

$$f_-(l, t) = u_{-1}\left(t - \frac{l}{v}\right)\left(\frac{1}{2} - e^{-(t-l/v)/Z_0C}\right) \tag{4.72}$$

For values of $z < l$, f_- is delayed from its form (Eq. 4.72) by the time $(l - z)/v$. Therefore it becomes

$$f_-(z, t) = u_{-1}\left(t - \frac{2l}{v} + \frac{z}{v}\right)\left(\frac{1}{2} - e^{-(t-2l/v+z/v)/Z_0C}\right) \tag{4.73}$$

At the generator end the line is matched, and the f_- wave satisfies the boundary conditions without further reflection. The entire solution is the sum of Eqs. 4.67 and 4.73, which shows that the line voltage at any point $0 \leq z \leq l$ becomes 1 v when t is sufficiently large. This checks with the open-circuit voltage of the source in Fig. 4.16, as it evidently must in the final d-c steady state. In Fig. 4.18 we show the

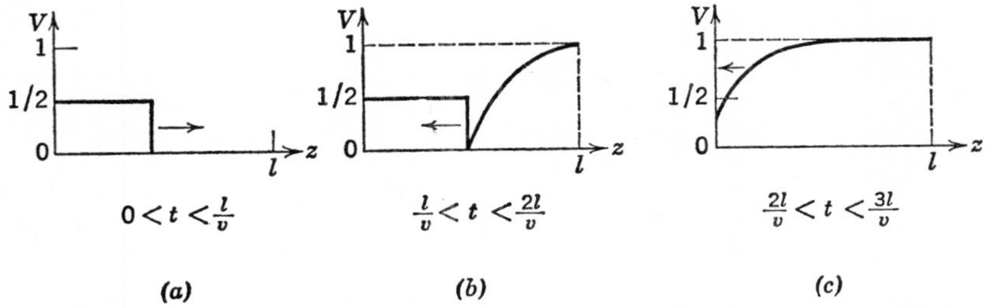

Fig. 4.18. Line voltage versus distance and time for Fig. 4.16.

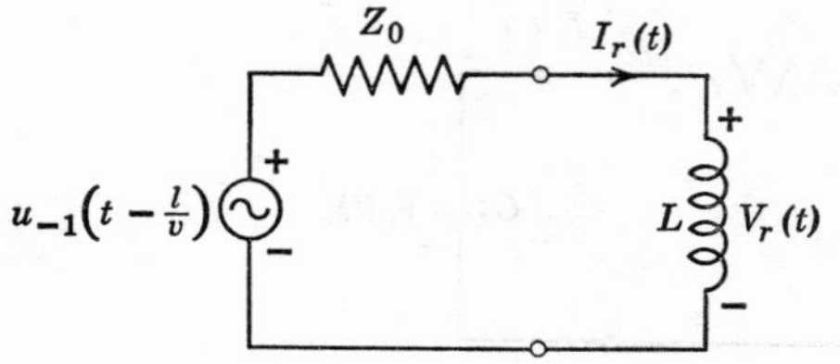

Fig. 4.19. Equivalent circuit for inductive load on the line of Fig. 4.16.

voltage as a function of z for three different times. Notice that the capacitor acts properly like a short-circuit when the incident step voltage first arrives, after which it charges slowly up to the final value of 1 v.

If in Fig. 4.16 we replace the capacitance by an inductance, the first part of the solution (Eq. 4.67) is unchanged. Instead of Fig. 4.17 at the load, however, we have Fig. 4.19. The equation for the circuit is

$$L\frac{dI_r}{dt} + Z_0 I_r = u_{-1}\left(t - \frac{l}{v}\right) \tag{4.74}$$

which, with the boundary condition $I_r = 0$ for $t \le l/v$, has the solution

$$I_r = Y_0 u_{-1}\left(t - \frac{l}{v}\right)\left(1 - e^{-(t-l/v)/Y_0 L}\right) \tag{4.75}$$

But continuity of current in the line and load requires

$$Y_0 f_+ - Y_0 f_- = I_r \qquad \text{at } z = l \tag{4.76}$$

so

$$f_-\left(t + \frac{z}{v}\right) = u_{-1}\left(t - \frac{2l}{v} + \frac{z}{v}\right)\left(e^{-(t-2l/v+z/v)/Y_0 L} - \tfrac{1}{2}\right) \tag{4.77}$$

Using Eqs. 4.67 and 4.77, the total voltage may be plotted versus z for two different times, as shown in Fig. 4.20. The final voltage ($t \to \infty$)

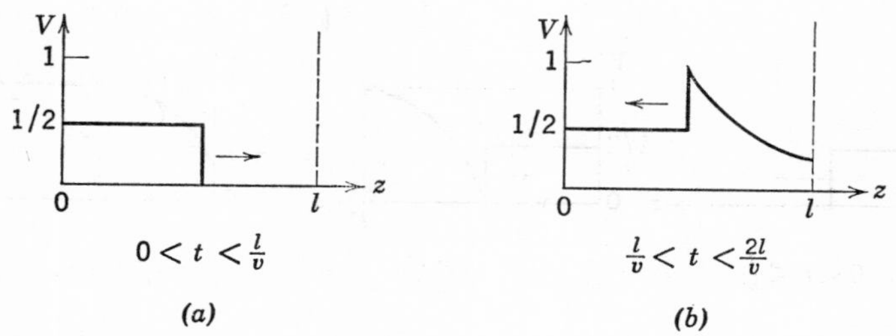

Fig. 4.20. Line voltage versus distance and time for Fig. 4.16 with an inductive load.

is everywhere zero, as it must be for a line short-circuited in the steady state. Also the inductor behaves like an open circuit at its initial excitation.

4.3.4 Initial Conditions

Thus far we have not considered a line with specified voltage and current distributions over it at $t = 0$. The most general problem would include sources (and arbitrary networks) at the ends of the line, in addition to the initial distributions on it; but if the entire system (including the terminations) is linear, we can handle the sources and initial conditions separately and superpose the results. Having discussed previously situations involving sources without initial conditions, we need now study only the problem of initial conditions without sources.

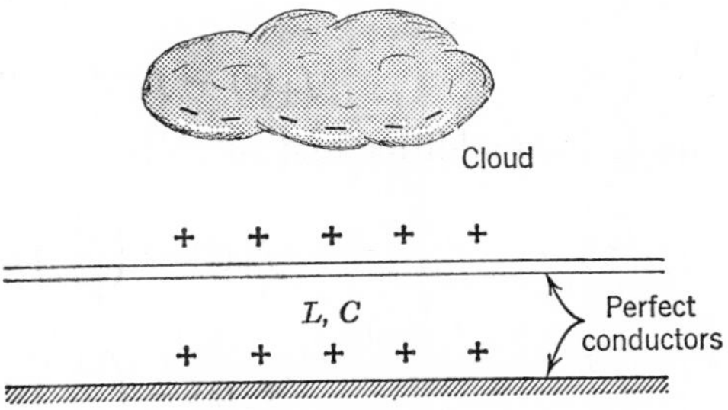

Fig. 4.21. A power line in a lightning storm.

As a first example, consider the infinite power line shown in Fig. 4.21. It consists of a single wire with a ground return, both assumed to be perfect conductors. We suppose that a cloud above the line is charged to a high negative potential. In the period of time during which the charge exists on the cloud, we shall assume that the wire comes to ground potential as a result of small currents flowing over the wet insulators supporting it. Thus the wire will finally have a (known) positive charge distribution $Q(z)$ coulombs per unit length, and no current anywhere along it.

At the moment $t = 0$, when the charge on the cloud is neutralized by lightning, the charge distribution $Q(z)$ on the wire is no longer bound to the cloud. It may be imagined to induce a corresponding negative charge distribution on the ground, causing a voltage across the line at $t = 0$, of

$$V(z, 0) = \frac{Q(z)}{C} \tag{4.78a}$$

and no current anywhere on the line

$$I(z, 0) = 0 \tag{4.78b}$$

Inasmuch as the initial conditions in Eqs. 4.78*a* and 4.78*b* specify functions of z at $t = 0$, rather than functions of t at $z = 0$, it is convenient to rewrite the general solutions (Eqs. 4.20) in the form

$$\text{(a)} \quad V(z, t) = f_+(z - vt) + f_-(z + vt)$$
$$\text{(b)} \quad I(z, t) = Y_0[f_+(z - vt) - f_-(z + vt)] \tag{4.79}$$

Since the functions f_+ and f_- are specified only by the boundary conditions, there is no essential difference between a function of $(t \pm z/v)$ and a function of $(z \pm vt) = \pm v(t \pm z/v)$. Consequently, application of Eqs. 4.79 to Eqs. 4.78 yields

$$\text{(a)} \quad f_+(z) + f_-(z) = \frac{1}{C} Q(z)$$
$$\text{(b)} \quad f_+(z) - f_-(z) = 0 \tag{4.80}$$

which, upon addition and subtraction, become

$$\text{(a)} \quad f_+(z) = \left(\frac{1}{2C}\right) Q(z)$$
$$\text{(b)} \quad f_-(z) = \left(\frac{1}{2C}\right) Q(z) \tag{4.81}$$

According to Eqs. 4.79 and 4.81 we find, for any time t,

$$\text{(a)} \quad V(z, t) = \frac{1}{2C} [Q(z - vt) + Q(z + vt)]$$
$$\text{(b)} \quad I(z, t) = \frac{Y_0}{2C} [Q(z - vt) - Q(z + vt)] \tag{4.82}$$

The original voltage distribution along the line splits up into two waves, each carrying half the original voltage. These waves then move in opposite directions. In Fig. 4.22, we show the voltage and current distributions on the line at $t = 0$ and $t = t_1 > 0$.

The power-line problem just completed shows that the presence of initial distributions of voltage and current on a line means simply that at $t = 0$ *there are already two waves on the line.* If the line is not infinite, these two waves will encounter the terminations and be reflected accordingly.

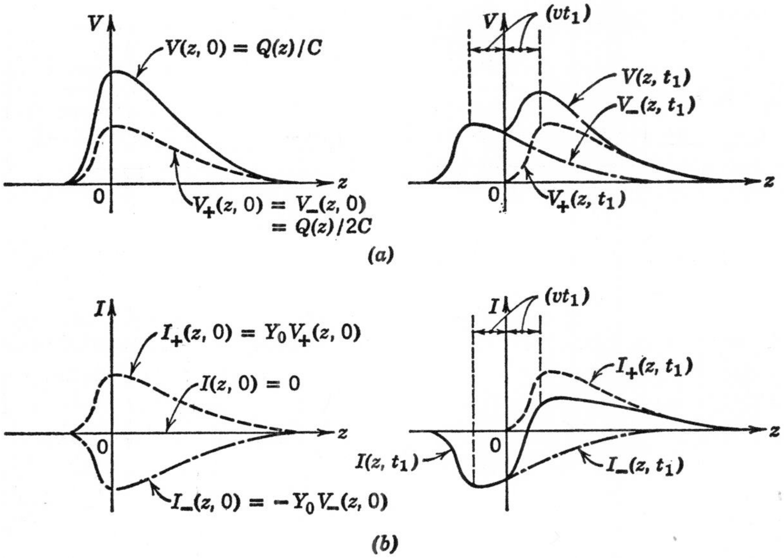

Fig. 4.22. Voltage and current waves on the long power line of Fig. 4.21.

We have mentioned the possibility of using superposition to separate the effects of initial conditions and independent sources. There is a related technique which may be used to simplify many linear problems that involve a sudden change in terminal conditions when, at the same time, complicated initial conditions and/or internal sources are also involved. It will be useful to discuss the scheme involved, which uses essentially superposition ideas to dissociate two aspects of the problem: the effects of the initial conditions and/or internal sources, and the changed terminal conditions. This technique, which amounts to replacing a switch by a source, is applicable not only to transmission lines but also to linear networks in general. We shall therefore discuss the method in appropriately general terms.

In Fig. 4.23 we show the transmission line, or whatever is involved, hidden inside a box, with only a switch S showing outside. This switch is the one used to alter some terminal condition at $t = 0$. Suppose first that the switch is open, as in Fig. 4.23*a*, and that we know (or can easily determine) what *would* be happening in the system and at the terminals from time $t = 0$ on, *if S remained open.* Our problem is to determine what happens in the system and at the terminals, when $t > 0$, if S is suddenly *closed* at $t = 0$.

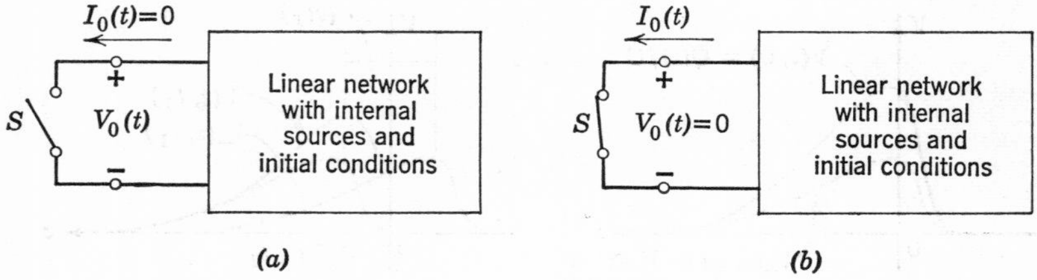

Fig. 4.23. System for a discussion of opening or closing a switch.

After closure of S, as in Fig. 4.23*b*, the response of the network in the box may be viewed as the superposition of two events: (1) its response to the internal sources and/or initial conditions when $I_0(t) \equiv 0$, and (2) its response to the current $I_0(t)$ when the initial conditions and/or internal sources are zero. The first event is the one we know already, inasmuch as it corresponds to the open-switch condition of Fig. 4.23*a*. The second event requires that we establish the current $I_0(t)$ at the network terminals when all the internal sources and/or initial conditions are made zero.

The second situation above could be achieved with a current source equal to $I_0(t)$ applied at the terminals, but, since we do not know the current wave form yet, this idea is of little help. With a little thought, however, we can see that the voltage source $V_0(t)$, arranged as shown in Fig. 4.24*a*, will produce the desired terminal current. As the network is properly de-energized internally in that arrangement, its response during times $t \geq 0$ to the source $V_0(t)$, applied with S closed, is just what must be added to its response in Fig. 4.23*a* to determine the solution for the system of Fig. 4.23*b*. That is, if every current and voltage in the system of Fig. 4.24*a* is added to the corresponding one in Fig. 4.23*a*, there results zero terminal voltage across the network box

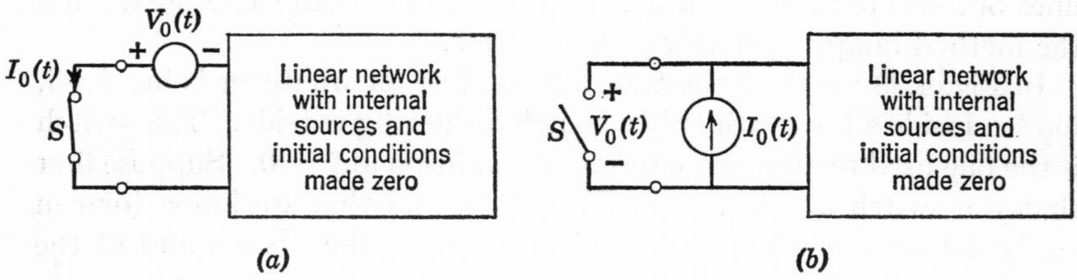

Fig. 4.24. Equivalent networks with respect to switch current or voltage in Fig. 4.23. (*a*) Thévènin equivalent; (*b*) Norton equivalent.

for $t \geq 0$, the full set of initial conditions inside the box at $t = 0$, and activation of all internal sources in the box for $t \geq 0$. Therefore, the sum response, including particularly $I_0(t)$, must be that of Fig. 4.23*b*. But then, as was required, the $I_0(t)$ flowing in Fig. 4.24*a* must indeed be the same as that in Fig. 4.23*b*, inasmuch as $I_0(t) \equiv 0$ in Fig. 4.23*a*. In other words, what happens in Fig. 4.24*a* must represent the *change* produced by the act of switch closure in passing from *a* to *b* in Fig. 4.23.

It is not difficult to show by a dual argument that the arrangement of Fig. 4.24*b* gives rise to the *changes* produced by the opening of an initially closed switch, passing from conditions of *b* to those of *a* in Fig. 4.23.

In either case, the *validity* of the method described depends only upon the linearity of the system, so that superposition applies. The methods will be *useful*, on the other hand, only if the two separate solutions (Figs. 4.23*a* and 4.24*a*, for example) are simpler than the desired one (Fig. 4.23*b*). We shall see how these matters develop in the following two examples.

EXAMPLE 4.3

The line shown in Fig. 4.25 has been in the d-c steady state for a long time, when suddenly at $t = 0$ the switch S is opened. The manner in which the line "discharges" is to be determined.

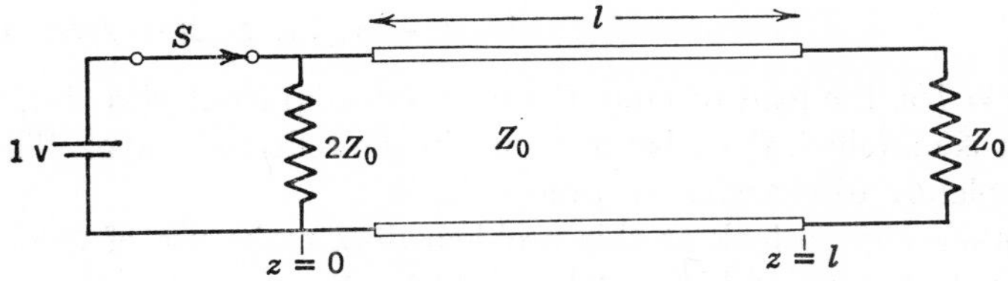

Fig. 4.25. A problem involving initial conditions and terminations, for Example 4.3.

In the d-c steady state, with the switch closed, it is clear that uniform voltage and current

$$\left.\begin{aligned} &\text{(a)} \qquad V(z) = 1 \\ &\text{(b)} \qquad I(z) = Y_0 \end{aligned}\right\} \quad \text{for } t < 0 \qquad (4.83)$$

exist on the line for $0 \leq z \leq l$. Let us first work the problem by the general technique, which does not use superposition.

Evidently, from Eqs. 4.83, we have two waves such that

$$\left.\begin{aligned} &\text{(a)} \qquad f_+ + f_- = 1 \\ &\text{(b)} \qquad Y_0(f_+ - f_-) = Y_0 \end{aligned}\right\} \quad \text{for } 0 \le z \le l \text{ and } t < 0 \tag{4.84}$$

or, by addition and subtraction,

$$\left.\begin{aligned} &\text{(a)} \qquad f_+ = 1 \\ &\text{(b)} \qquad f_- = 0 \end{aligned}\right\} \quad \text{for } 0 \le z \le l \text{ and } t < 0 \tag{4.85}$$

At $t = 0$ when the switch opens, conditions at the load end do not change, and cannot possibly change until $t = l/v$. This fact can be verified by the equivalent circuit of Fig. 4.9. At the source end, however, the corresponding equivalent circuit is that of Fig. 4.13, with $R_s = 2Z_0$ and $f_- = 0$ (by Eq. 4.85*b*). Thus the voltage across the input drops from 1 v (Eq. 4.83) for $t < 0$ to zero volts at $t = 0$. Now, the $(-)$ wave was zero all along the line at $t < 0$; and since it moves to the *left*, its contribution at $z = 0$ is still *zero* when $t > 0$. Consequently, the entire voltage $V(0, t)$ is comprised of the $(+)$ wave $f_+(t)$,

$$V(0, t) = f_+(t) = 1 - u_{-1}(t) \qquad \text{at least for } -\infty \le t < 2l/v \tag{4.86}$$

Then, since $f_- \equiv 0$ at least for $-\infty \le t < (2l - z)/v$,

$$V(z, t) = f_+\left(t - \frac{z}{v}\right) = 1 - u_{-1}\left(t - \frac{z}{v}\right)$$

$$\text{for at least } -\infty \le t < (2l + z)/v \tag{4.87}$$

At $t = l/v$, the load receives the negative step front (Eq. 4.87), but since it is matched, there is no change in the reflected wave. The line is completely discharged by time $t = l/v$.

Another way to look at this problem does make use of the superposition technique of Figs. 4.23 and 4.24. We note that the current through the battery and switch in Fig. 4.25 for $t < 0$ is $+\frac{3}{2}Y_0$ amp and would remain so for $t > 0$ if S stayed closed. Opening the switch at $t = 0$ reduces the battery current to zero. This is equivalent [by Fig. 4.24*b*] to leaving the switch open and initiating a negative current-source step of $-\frac{3}{2}Y_0$ amp in parallel with the switch and, therefore, in series with the battery. The action of this source *alone*, with all previous waves removed from the line and the battery short-circuited, is to send a -1-v step down the line and into the matched load. This result, superposed on the original 1 v existing everywhere on the line, again leads to Eq. 4.86 and shows that the line is discharged completely in time l/v.

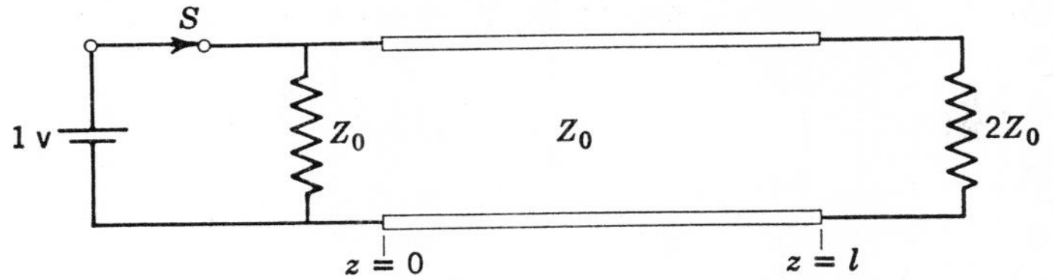

Fig. 4.26. Discharge of a transmission line, for Example 4.4.

EXAMPLE 4.4

The resistors at the input and load in Fig. 4.25 are exchanged, as shown in Fig. 4.26. Again, after the system is in the d-c steady state, switch S is opened at $t = 0$. Therefore, prior to $t = 0$, we have for $0 \leq z \leq l$

(a) $$V(z) = 1$$
(b) $$I(z) = \tfrac{1}{2}Y_0 \qquad (4.88)$$

Again let us solve the problem first without using superposition ideas. Thus, from Eqs. 4.88, we find

(a) $$f_+ = \tfrac{3}{4}$$
(b) $$f_- = \tfrac{1}{4} \qquad \text{for } 0 \leq z \leq l \text{ and } t < 0 \qquad (4.89)$$

At $t = 0$, when S is opened, we see from Fig. 4.27 that the input voltage drops to $\frac{1}{4}$ v. Since f_- moves to the left, its value at $z = 0$, $t \geq 0$ is still $\frac{1}{4}$ v. It follows that the drop in $V_1(t)$ from 1 to $\frac{1}{4}$ v is accounted for by a drop in f_+ at $z = 0$ from $\frac{3}{4}$ v to 0 v.

$$f_+(t) = V_1(t) - f_-(t) = \tfrac{3}{4}[1 - u_{-1}(t)] \qquad \text{at least for } t < 2l/v \qquad (4.90)$$

and

$$f_+\left(t - \frac{z}{v}\right) = \frac{3}{4}\left[1 - u_{-1}\left(t - \frac{z}{v}\right)\right] \qquad (4.91)$$

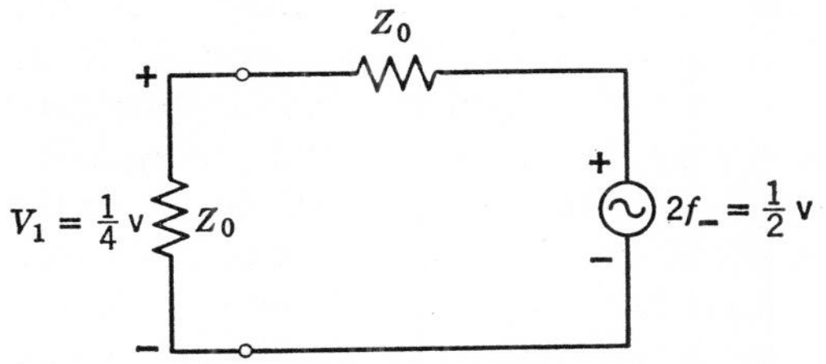

Fig. 4.27. Equivalent circuit for the input of Fig. 4.26 at $t > 0$.

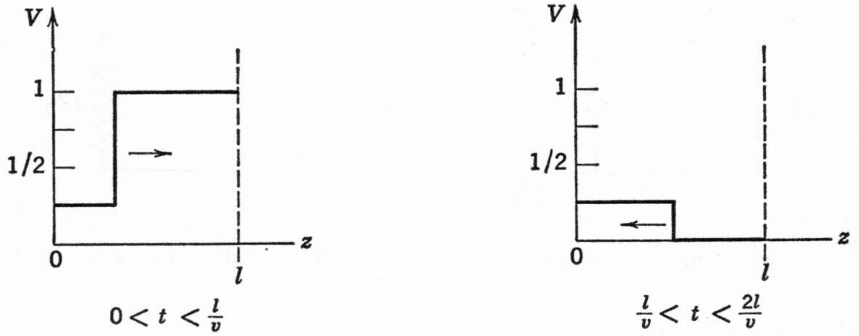

Fig. 4.28. Discharge of the line in Fig. 4.26.

When $t = l/v$, the boundary conditions at the load will require a change of $f_-(t - l/v)$ from its value $\frac{1}{4}$ because the load voltage according to Fig. 4.12 goes from 1 v to zero at that time. Thus $V_r(t) = 1 - u_{-1}$ $(t - l/v)$ and, therefore,

$$f_-\left(t + \frac{l}{v}\right) = \frac{1}{4}\left[1 - u_{-1}\left(t - \frac{l}{v}\right)\right] \tag{4.92}$$

Consequently

$$f_-\left(t + \frac{z}{v}\right) = \frac{1}{4}\left[1 - u_{-1}\left(t - \frac{2l}{v} + \frac{z}{v}\right)\right] \tag{4.93}$$

Equations 4.92 and 4.93 actually hold for all time because the generator end of the line is matched. No further changes in the (+) or (−) waves are required, and the discharge of the line is completed at $t = 2l/v$, as shown in Fig. 4.28.

Now the solution using superposition will be demonstrated. The current through the closed switch in Fig. 4.26 for $t < 0$ is $+\frac{3}{2}Y_0$ amp. Thus, opening the switch at $t = 0$ is like leaving it *closed* and inserting in *series* a current-source step of $-\frac{3}{2}Y_0$ amp. Notice the slight difference between this statement and Fig. 4.24*b* or Ex. 4.3. (Why is this one valid?)

Acting alone, with all other waves removed from the line and the battery short-circuited out, this current source produces an input voltage of $(-\frac{3}{2}Y_0)(\frac{1}{2}Z_0) = -\frac{3}{4}$ v. Observe that the line looks like Z_0 (not $2Z_0$) for this transient, according to the transient-equivalent-circuit principles we have developed. The $-\frac{3}{4}$ v becomes a $'f_+$ wave, superposed upon the original 1 v on the line. It "wipes out" $\frac{3}{4}$ v as it moves toward the load (see Fig. 4.28*a*). At $t = l/v$, the $'f_+$ wave reaches the load where it produces a reflected $'f_-$ wave according to the reflection coefficient

$$'f_- = \left(\frac{2Z_0 - Z_0}{2Z_0 + Z_0}\right) 'f_+ = \frac{1}{3}\left(-\frac{3}{4}\right) = -\frac{1}{4}\text{ v} \qquad (4.94)$$

This wave travels back toward the generator, "wiping out" the ¼ v which was left on the line by the previous waves (see Fig. 4.28*b*). Since the generator is matched, no further ($''f_+$) wave is set up. The line is discharged.

4.4 Another Time-Domain Method *

To treat transient problems most efficiently, it is often convenient to make use of the family of singularity functions: the impulse, step, ramp, etc. These functions have the advantage of being quite easy to handle.

If a unit impulse is designated by $u_0(t)$, each derivative by a subscript one higher, and each integral by a subscript one lower, we have the complete representation of the family. A unit step is therefore $u_{-1}(t)$, as we have denoted it before, and a unit doublet is $u_1(t)$.

The most useful of these functions are the impulse and the step. In the previous discussions, the sources were chosen to be step functions (or batteries connected through switches to the line). We carried out the analyses directly with these sources because they seemed to have a certain physical appeal. It is well to realize, however, that impulses may be easier than steps to trace and superpose in multiple-reflection problems. Since the step or ramp response of a *linear* system is easily found from successive time integrations of the impulse response, it may often be convenient to solve a problem first with an impulse source, and integrate afterward. The reader should follow through this technique on each of the previous examples in the chapter for which it seems appropriate.

There is, however, a much more powerful generalization of the relation between the impulse response of a linear system and its response to any other time-function excitation. This relation is the *superposition integral.*[1] We shall give a cursory presentation of it here for the sake of review and continuity, but it is assumed that the material has been, or will be, studied more carefully elsewhere.

[1] See, for example, Samuel J. Mason and Henry J. Zimmermann, *Electronic Circuits and Signals*, John Wiley and Sons, New York, 1960.

Let $h_0(t)$ be the response of a certain linear system to a unit impulse input $u_0(t)$. The "input" may be, for example, the voltage or current at the sending end of a transmission line. The "response" may refer to the voltage or current at any desired place on the line. To be fairly general, let us denote the input (or "excitation") as a function of time by $E(t)$ and the output by $O(t)$.

Then, for a linear system the superposition principle allows us to write

$$O(t) = \int_0^t h_0(t - \tau)\, E(\tau)\, d\tau \tag{4.95}$$

since an impulse of amount $E(\tau)\, u_0(\tau)\, d\tau$ at time $t = \tau$ would have a response $h_0(t - \tau)\, E(\tau)\, d\tau$ at time t.

An elementary change of variable shows that one can also write

$$O(t) = \int_0^t h_0(\tau)\, E(t - \tau)\, d\tau \tag{4.96}$$

The reader would do well to interpret the two expressions (Eqs. 4.95 and 4.96) graphically and physically to satisfy himself less formally of their equality.

Since we are dealing with transmission lines, where there is a space as well as time dependence, the symbolism may be appropriately modified by writing

$$O(z, t) = \int_0^t h_0(z, \tau)\, E(t - \tau)\, d\tau = \int_0^t h_0(z, t - \tau)\, E(\tau)\, d\tau \tag{4.97}$$

where we have restricted the source to the location $z = 0$.

As an exercise, Eq. 4.97 may be used to show that

$$h_{-1}(z, t) = \int_0^t h_0(z, t - \tau)\, u_{-1}(\tau)\, d\tau = \int_0^t h_0(z, t - \tau)\, d\tau$$

$$= \int_0^t h_0(z, \tau)\, d\tau \tag{4.98}$$

which is a correct conclusion we have already mentioned above.

Since frequently it is the response to a unit step rather than an impulse that is known, it is now convenient to integrate Eq. 4.97 by parts [e.g., let $u = E(t - \tau)$ and $dv = h_0(z, \tau)\, d\tau$]. Noting also that $[dE(t - \tau)]/dt = [-dE(t - \tau)]/d\tau = E'(t - \tau)$, we find the result

$$O(z, t) = E(t)\, h_{-1}(z, 0) + \int_0^t h_{-1}(z, \tau)\, E'(t - \tau)\, d\tau \qquad (4.99)$$

where E' is the derivative of E with respect to its whole argument $(t - \tau)$. The equivalent equation is also valid

$$O(z, t) = E(t)\, h_{-1}(z, 0) + \int_0^t h_{-1}(z, t - \tau)\, E'(\tau)\, d\tau \qquad (4.100)$$

It is worth while to try to arrive at Eqs. 4.99 or 4.100 by direct physical reasoning and then (by again integrating by parts) to derive Eq. 4.97 from there.

*EXAMPLE 4.5 **

Now, for an example of the superposition-integral method, consider an open-circuited transmission line fed by a current source with an internal conductance G, as shown in Fig. 4.29. The response to a cur-

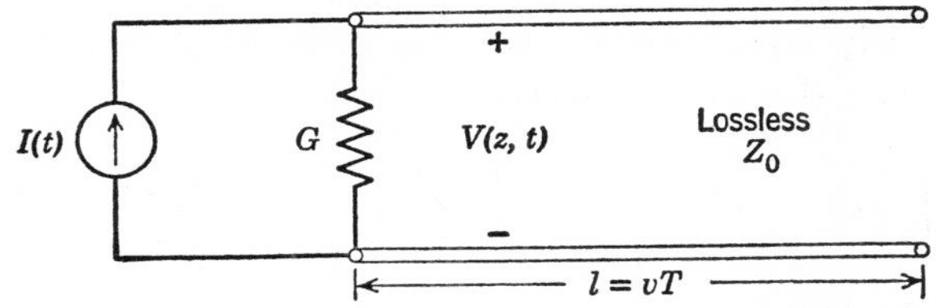

Fig. 4.29. System for study by the superposition integral.

rent impulse $I(t) = u_0(t)$ begins with a voltage impulse $(G + Y_0)^{-1}$ $u_0(t - z/v)$ traveling down the line. At $t = T$ and $z = l$, a reflection takes place, and a wave of voltage $(G + Y_0)^{-1}\, u_0(t - 2T + z/v)$ begins to travel back along the line. At the generator end another reflection takes place, leading to a voltage wave

$$\Gamma(G + Y_0)^{-1}\, u_0(t - 2T - z/v)$$

where Γ is the reflection coefficient at the source end

$$\Gamma = \frac{Y_0 - G}{Y_0 + G}$$

All the information from the similar repeated reflections that occur may be summarized in the statement

$$h_0(z, t) = \sum_{n=0}^{\infty} (G + Y_0)^{-1} \left[\Gamma^n u_0\left(t - 2nT - \frac{z}{v}\right) + \Gamma^n u_0\left(t - 2(n+1)T + \frac{z}{v}\right)\right]$$

$$= \sum_{n=0}^{\infty} (G + Y_0)^{-1}\Gamma^n \left[u_0\left(t - 2nT - \frac{z}{v}\right) + u_0\left(t - (2n+2)T + \frac{z}{v}\right)\right] \qquad (4.101)$$

For the particular case $\Gamma = 0$, i.e., $G = Y_0$, the summation above is to be understood to reduce to

$$h_0(z, t) = (G + Y_0)^{-1} \left[u_0\left(t - \frac{z}{v}\right) + u_0\left(t - 2T + \frac{z}{v}\right)\right]$$

Note that because the line has only resistive terminations, we are still left with impulse functions; but in place of the single one, a series of them has been obtained instead. The integration specified in Eq. 4.97 is then a fairly trivial matter for any excitation since $u_0(\tau - \alpha) = u_0(\alpha - \tau)$ (i.e., the unit impulse is symmetrical) and

$$\int_0^t u_0(\tau - \alpha) f(\tau)\, d\tau = \begin{cases} f(\alpha) & 0 < \alpha < t \\ 0 & \alpha < 0 \text{ or } \alpha > t \end{cases}$$

To be more specific first suppose $I(t) = u_{-1}(t)$, a unit step. Then from Eqs. 4.97

$$V(z, t) = \int_0^t h_0(z, \tau)\, u_{-1}(\tau)\, d\tau$$

and from Eq. 4.101

$$V(z, t) = \sum_{n=0}^{\infty} (G + Y_0)^{-1}\Gamma^n \left[u_{-1}\left(t - 2nT - \frac{z}{v}\right) + u_{-1}\left(t - (2n+2)T + \frac{z}{v}\right)\right]$$

This result should be checked by considering successive reflections directly.

EXAMPLE 4.6 *

Next consider an input to the line of Fig. 4.29 of the form

$$I(t) = u_{-1}(t) \sin \omega_0 t \qquad (4.102a)$$

where

$$\omega_0 = \frac{2\pi}{4T} = \frac{\pi}{2T} \tag{4.102b}$$

This choice of frequency makes the line a quarter-wave length long in the steady state. Then by Eqs. 4.97 and 4.101

$$V(z, t) = \sum_{n=0}^{\infty} (G + Y_0)\Gamma^n \left[u_{-1}\left(t - 2nT - \frac{z}{v}\right) \sin\left(\omega_0 t - n\pi - \omega_0 \frac{z}{v}\right) + u_{-1}\left(t - (2n+2)T + \frac{z}{v}\right) \sin\left(\omega_0 t - (n+1)\pi + \omega_0 \frac{z}{v}\right)\right]$$

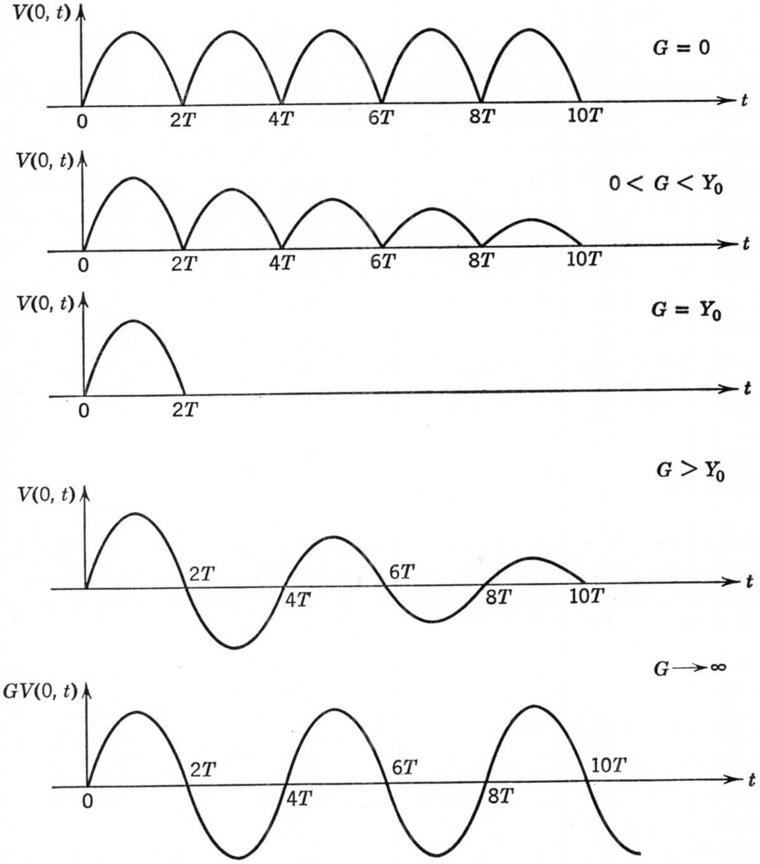

Fig. 4.30. Response of the line system of Fig. 4.29 to a suddenly applied sinusoidal current.

or since $\sin(\vartheta - n\pi) = (-1)^n \sin\vartheta$,

$$V(z, t) = \sum_{n=0}^{\infty} (G + Y_0)^{-1}(-\Gamma)^n \left[u_{-1}\left(t - 2nT - \frac{z}{v}\right) \sin\left(\omega_0 t - \omega_0 \frac{z}{v}\right) - u_{-1}\left(t - (2n+2)T + \frac{z}{v}\right) \sin\left(\omega_0 t + \omega_0 \frac{z}{v}\right)\right] \quad (4.103)$$

A particular value of z for which considerable simplification of Eq. 4.103 occurs is $z = 0$, i.e., at the input terminals

$$V(0, t) = \sum_{n=0}^{\infty} (G + Y_0)^{-1}(-\Gamma)^n [u_{-1}(t - 2nT) - u_{-1}(t - (2n+2)T)] \sin \omega_0 t \quad (4.104)$$

The bracketed function is unity for $2nT < t < (2n + 2)T$ and zero elsewhere. Over this former range of t, the argument $\omega_0 t$ of the sine function goes from $n\pi$ to $(n + 1)\pi$, thereby making $(-1)^n \sin \omega_0 t$ positive in this range. Hence $V(0, t)$ consists of a full-wave rectified sine wave for $\Gamma = +1$. As Γ decreases toward zero, the successive arches become smaller until, for $\Gamma = 0$, only the first remains. For $\Gamma < 0$ the alternate half-waves become negative, and finally, as $\Gamma \to -1$, the wave form approaches just $(1/G) \sin \omega_0 t$. Various cases are illustrated in Fig. 4.30. They should be checked by tracing successive reflections, especially the simple ones for $G = Y_0$, $G = 0$, and $G \to \infty$.

PROBLEMS

Problem 4.1. Show directly, without using Thévènin's or Norton's theorems, that the equivalent circuit of Fig. 4.31 can be used to find the load current and voltage for a lossless transmission line, where $g_+(t - l/v)$ is the positive traveling wave of current at $z = l$.

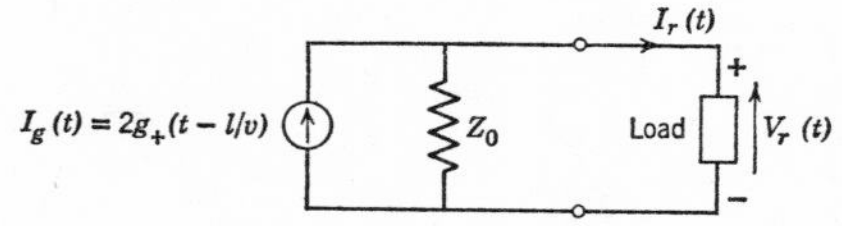

Fig. 4.31. Problem 4.1.

Problem 4.2. Apply Fourier methods to the results of Prob. 3.2 in order to prove, in a manner different from that presented in the text, that the general solution for voltage on a lossless line is $V(t, z) = f_+(t - z/v) + f_-(t + z/v)$, where f_+ and f_- are arbitrary functions. Show also the proper corresponding form for the current.

Problem 4.3. An impulse voltage source drives a lossless line terminated in $2Z_0$. Find the load current as a function of time by three different methods: (a) Direct analysis of successive reflections. (b) Application of Thévènin's theorem at the load end. (c) Application of Norton's theorem at the load end. Describe carefully the great difference between the methods of (b) and (c), and the "equivalent circuit" method used in the text.

Problem 4.4. (a) Find $i_1(t)$ and $i_2(t)$ in Fig. 4.32. (b) Repeat if $R_G = 2Z_0$ and $R_L = Z_0$.

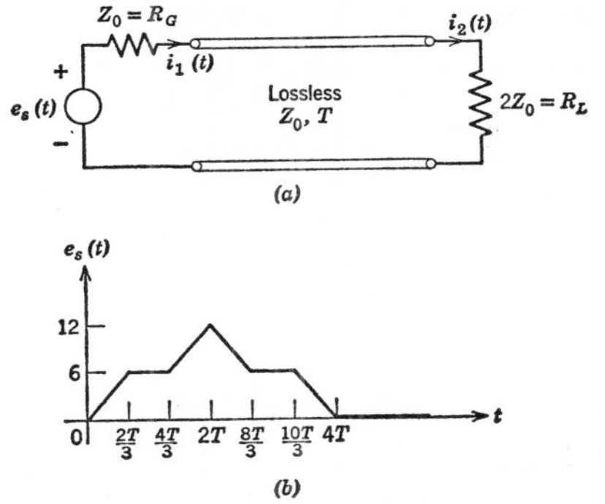

Fig. 4.32. Problem 4.4.

Problem 4.5. If in Fig. 4.33 $e_g(t) = u_{-1}(t) \cos \omega t$, and we write:

$$T = \frac{l}{v} \qquad \Gamma_R = \frac{R_2 - Z_0}{R_2 + Z_0} \qquad \Gamma_S = \frac{R_1 - Z_0}{R_1 + Z_0}$$

(a) Show that

$$V_+(0, t) = \frac{Z_0}{Z_0 + R_1} \operatorname{Re} \left\{ \sum_{n=0}^{\infty} [(\Gamma_R \Gamma_S)^n e^{-j2n\omega T} e^{j\omega t} u_{-1}(t - 2nT)] \right\}$$

where $V_+(0, t)$ is the *incident* (+ wave) voltage on the line at $z = 0$. (b) Write an expression for $V_+(l, t)$, which is the *incident* voltage wave at $z = l$. (c) Show that

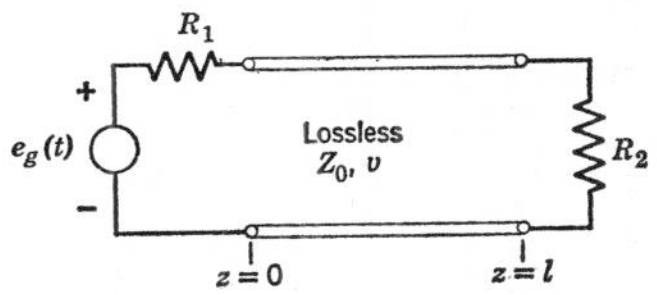

Fig. 4.33. Problem 4.5.

the final steady-state form of $V_+(0, t)$ is $\lim_{t \to \infty} V_+(0, t) = \operatorname{Re}(\mathrm{V}_+ e^{j\omega t})$ and evaluate V_+ in a closed mathematical form (i.e., *not* an infinite series).

Problem 4.6. (a) A d-c source is suddenly connected to the input of a lossless line of length l which is short-circuited at the opposite end. Plot the current through the short circuit for a representative interval of time. (b) Compare the result of

(a) with the response of a suitable parallel L-C circuit to a current step. (c) By making a Fourier-series analysis of the response in (a), construct a lumped network containing many tuned circuits, whose response to a current step approximates the actual one more closely than was achieved in (b).

Problem 4.7. In Figs. 4.34*a* and 4.34*b* assume that steady-state conditions have been reached, i.e., the source has been on for a long time. (a) Sketch the voltage wave form as a function of time that appears on the line at the points $z = 0$, $z = -vT'/4$, $z = -vT'/2$. (b) Discuss the results found in (a) in relation to the steady-state response of the line to *sinusoidal* excitation.

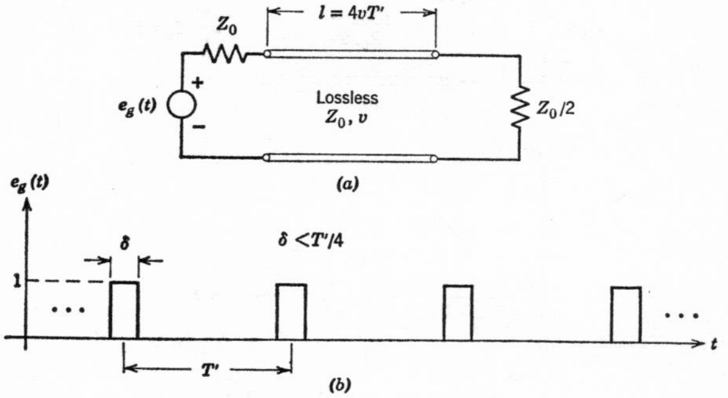

Fig. 4.34. Problem 4.7.

Problem 4.8. (a) In Fig. 4.35*a*, the switch S is closed at $t = 0$. Give dimensioned sketches versus time of the voltage $V_{aa'}$, and the current through R_s for $R_s = Z_0$, $R_s = 9Z_0$, $R_s = Z_0/9$. (b) Compute the energy lost in R_s in each case above. (c) Compare these results with the response of the series R-L-C circuit of Fig. 4.35*b* to the same excitation, indicating how you would choose the values of L and C to establish the closest relationship. (See also Prob. 4.6 for comparison.)

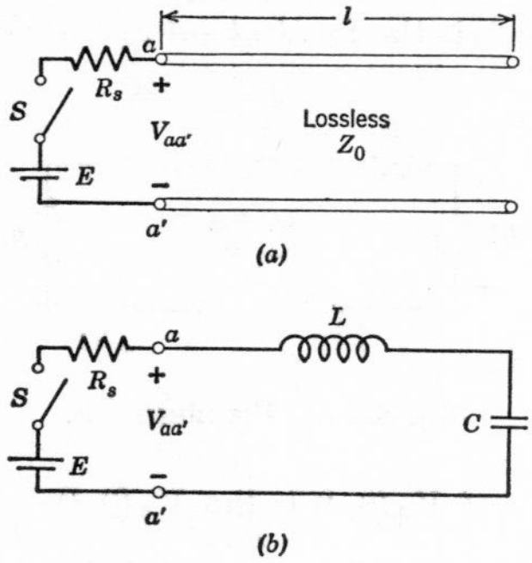

Fig. 4.35. Problem 4.8.

Problem 4.9. The circuit of Fig. 4.36 is an application of the results of Prob. 4.8 to a "surge generator." When the switch S is closed, after having been open a long time, a 1-μsec voltage pulse appears between point A and ground. If the switch is then opened, approximately how long will it take to recharge the line?

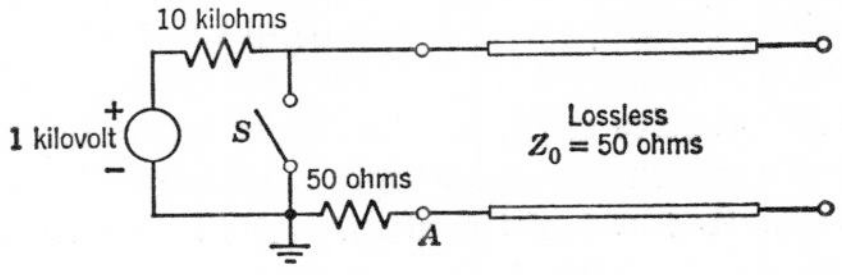

Fig. 4.36. Problem 4.9.

Problem 4.10. In Fig. 4.37: (a) Sketch the voltage distribution on the system at $t = 3T/2$, $5T/2$, and $7T/2$. (b) Plot versus time the voltage at aa'. (c) Repeat (b) and (a) if $e_s(t) = u_{-1}(t)$

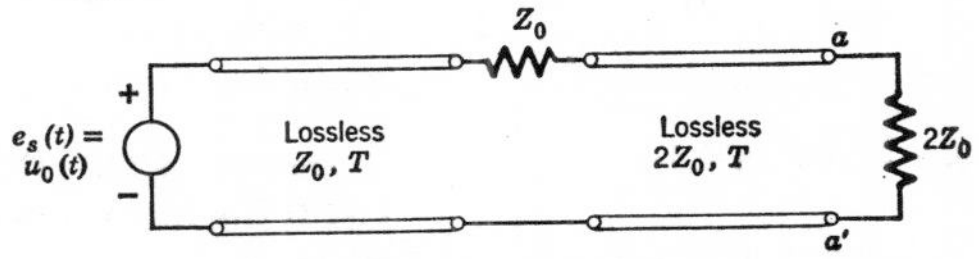

Fig. 4.37. Problem 4.10.

Problem 4.11. A single rectangular pulse $e(t)$ of amplitude 1000v is applied to the system of Fig. 4.38. (a) Determine the maximum voltage that occurs in the cable. (b) Which part of the cable is subjected to the maximum voltage if the pulse width is one-half of the cable transit time (one way)? Two-thirds? Twice?

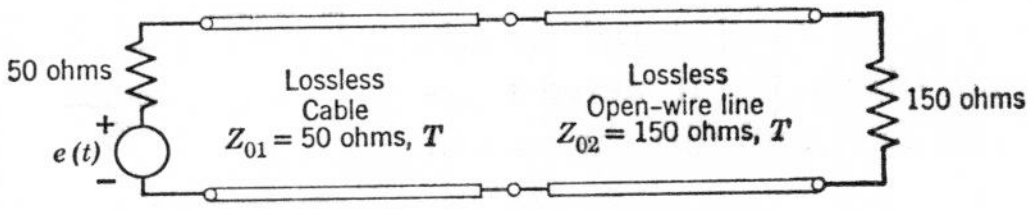

Fig. 4.38. Problem 4.11.

Problem 4.12. Plot $e(t)$ in Fig. 4.39: (a) If S is open during the transient. (b) If S is closed during the transient.

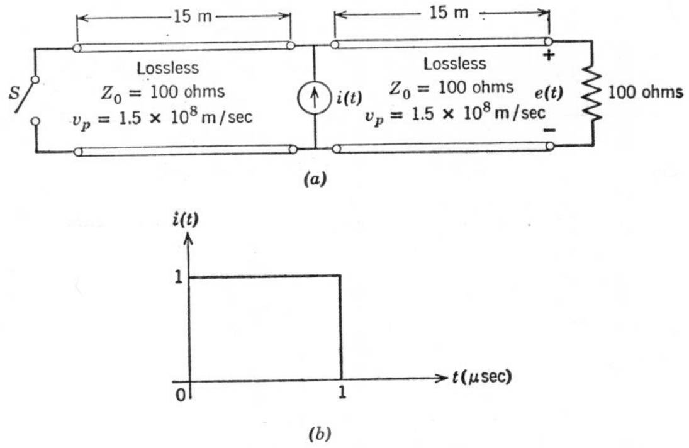

Fig. 4.39. Problem 4.12.

Problem 4.13. A lossless line of characteristic admittance Y_0 and phase velocity 3×10^8 m/sec is terminated in an admittance $Y_0/2$, independent of frequency. (a) At a frequency of 300 mc, determine the position nearest the load, and the shortest length of a short-circuited stub made from the same line material, required to match the line. (b) If the system of (a) is now driven by a source having internal impedance Z_0 at all frequencies and a unit impulse as its open-circuit voltage, determine as functions of time the source terminal voltage, the load voltage, and the current in the short-circuited end of the stub. (c) Plot on a Smith chart the locus of the normalized admittance seen looking toward the load from a point just on the source side of the stub, as the frequency is changed from 150 mc to 600 mc with the system otherwise remaining as set in (a). Over what range of frequency is the VSWR less than 1.2? (d) What do you conclude about the effectiveness of a "single-stub" matching device?

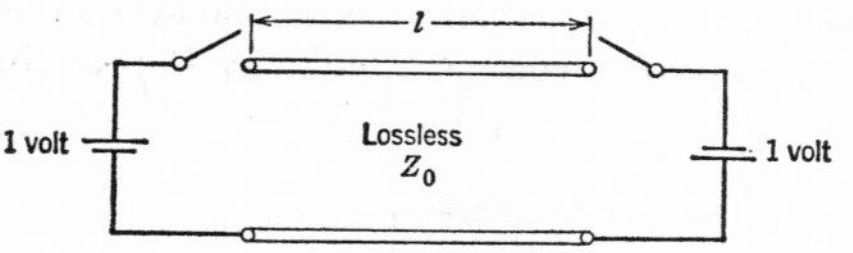

Fig. 4.40. Problem 4.14.

Problem 4.14. In Fig. 4.40: For $t < 0$ the line is uncharged; at $t = 0$ both switches are closed; for $t > 0$ plot: (a) $V(t)$ at center of line. (b) $I(t)$ at center of line. (c) $V(t)$ at the $l/4$ and $3l/4$ points on line.

Problem 4.15. The transmission line in Fig. 4.41 is initially charged to a voltage V with switch S open. Switch S is then closed at time $t = 0$. (a) Sketch the distributions of voltage *and* current on the line at times $t = 0, l/2v, l/v$, and $3l/2v$, where v is the velocity of propagation on the line. (b) On the same sheet, plot the following functions of time: (i) the total stored electric energy on the line; (ii) the total stored magnetic energy on the line; and (iii) the sum of (i) and (ii).

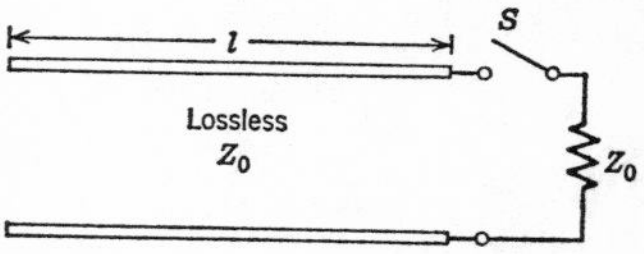

Fig. 4.41. Problem 4.15.

Problem 4.16. As shown in Fig. 4.42, a lossless transmission line of one-way transit time T is placed in the plate circuit of a vacuum tube. The grid voltage is specified in the figure, and the idealization for the tube is also given there. Sketch and dimension $e_0(t)$.

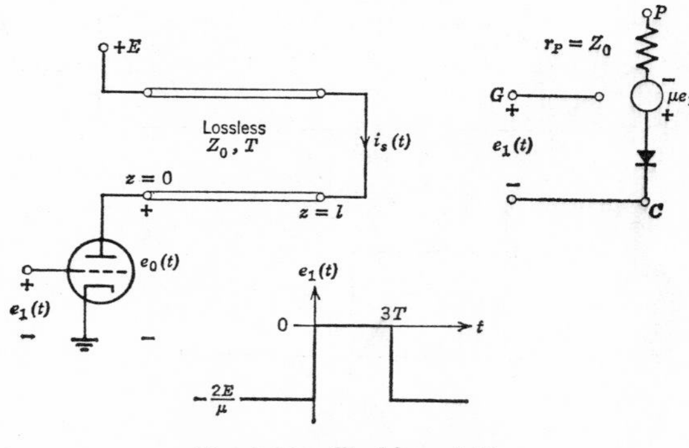

Fig. 4.42. Problem 4.16.

Problem 4.17. Switch S in Fig. 4.43 is closed at $t = 0$, with equilibrium having been established on the lossless line for $t < 0$. Sketch voltages and currents as functions of z for $t < 0$, $t = l/2v$, $t = 3l/2v$, $t = 2l/v$.

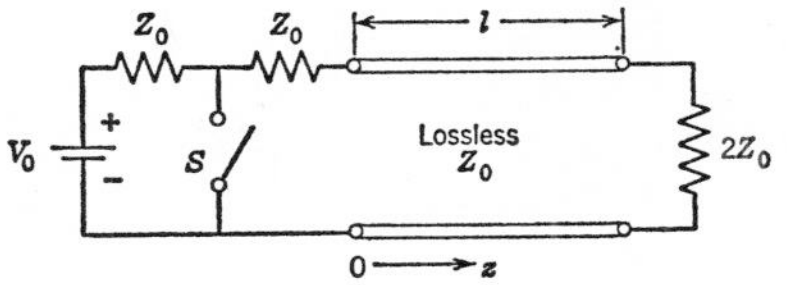

Fig. 4.43. Problem 4.17.

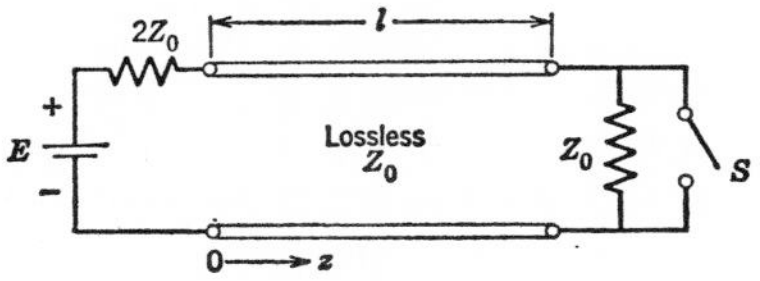

Fig. 4.44. Problem 4.18.

Problem 4.18. In Fig. 4.44 the steady state is established with S closed. At $t = 0$, S is suddenly opened. (a) Sketch the voltage $V(z)$ for t slightly greater than zero. (b) Sketch $V(z)$ just after reflection at the generator end. (c) Write an expression for the voltage at any point on the line and at any instant $t > 0$.

Problem 4.19. The lossless line of Fig. 4.45 is originally charged to 300 v with S open. At $t = 0$, S is suddenly closed. Sketch $V(z)$ at $t = 5l/4v$.

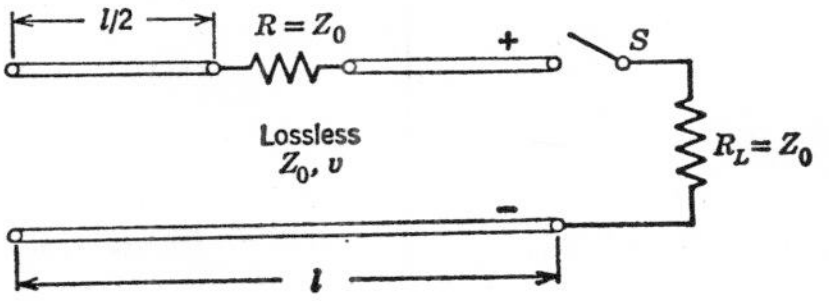

Fig. 4.45. Problem 4.19.

Problem 4.20. The lossless line of Fig. 4.46 is in the steady state with S open for $t < 0$. At $t = 0$, S is closed. (a) Can the superposition technique which replaces the effect of switch closure by that of a voltage source at the switch be used to solve for the system response at $t > 0$ in this problem? Why? (b) Sketch and dimension the voltage across bb' for $t \geq 0$. (c) Sketch the voltage and current distributions on the line at $t = l/2v$, $t = 3l/2v$, and $t = 5l/2v$.

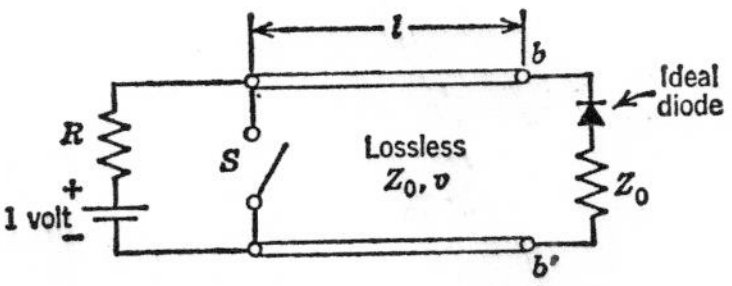

Fig. 4.46. Problem 4.20.

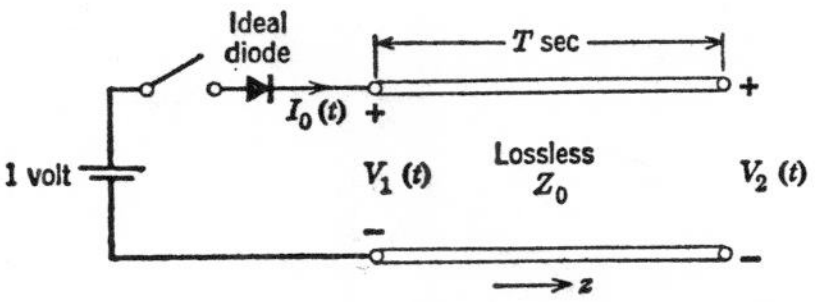

Fig. 4.47. Problem 4.21.

Problem 4.21. The system of Fig. 4.47 is unenergized for $t < 0$. The switch closes at $t = 0$. Find and sketch for all $t \geq 0$: (a) $V_1(t)$. (b) $I_0(t)$. (c) $V_2(t)$. (d) Voltage and current on the line versus z for $t = T/2$, $3T/2$, $2T$, $5T/2$. (e) How much energy is delivered by the battery? (f) Show that energy is conserved in the system.

Problem 4.22. The leftmost line in Fig. 4.48 is originally charged as indicated; the other lines are uncharged. At $t = 0$ the switch is closed. What happens? Illustrate your answer with figures.

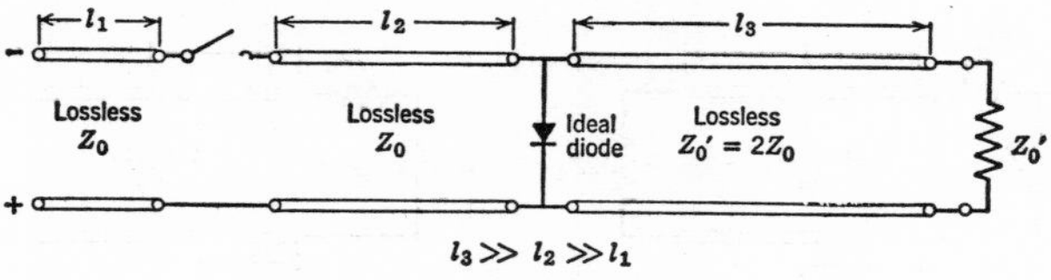

Fig. 4.48. Problem 4.22.

Problem 4.23. Switch S in Fig. 4.49 has been in position 1 for a long time prior to $t = 0$. At $t = 0$, S is thrown to position 2. At $t = 2\pi m/\omega_0$, where m is a large integer, S is thrown back to position 1 again. (a) Sketch to scale $i_1(t)$ and $i_2(t)$ for $t \geq 0$. (b) Sketch to scale the line voltage and current versus z at $t = T/2$; T; $3T/2$; $2T$; $(2\pi m/\omega_0) + T/2$; $(2\pi m/\omega_0) + T$; $(2\pi m/\omega_0) + (3T/2)$; $(2\pi m/\omega_0) + 2T$.

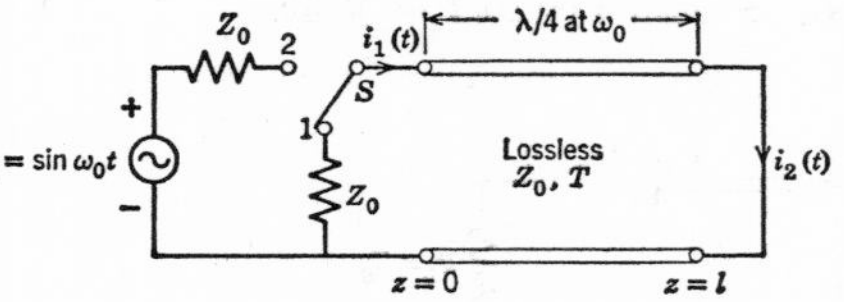

Fig. 4.49. Problem 4.23.

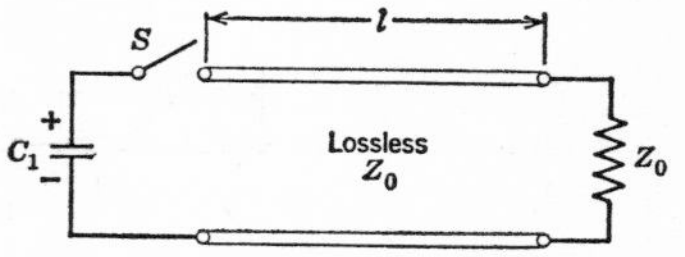

Fig. 4.50. Problem 4.24.

Problem 4.24. Capacitor C_1 in Fig. 4.50 has an initial charge of 1 coulomb and the line is unenergized. Switch S is then closed at $t = 0$. Find the current through the terminating resistance Z_0, and sketch the wave form; label all pertinent amplitudes, times, and time constants.

Problem 4.25. The line *and* capacitor C_1 in Fig. 4.51 are initially charged to $+V$ volts. At $t = 0$ the switch S is closed. Determine, sketch, and dimension: (a) The voltage *and* current distributions versus z at times $t = T/2$, $3T/2$. (b) The voltage and current wave forms versus t at $z = l/2$, for $0 \leq t \leq 5T$. (c) How much energy is dissipated in the load resistance from $t = 0$ to $t = \infty$?

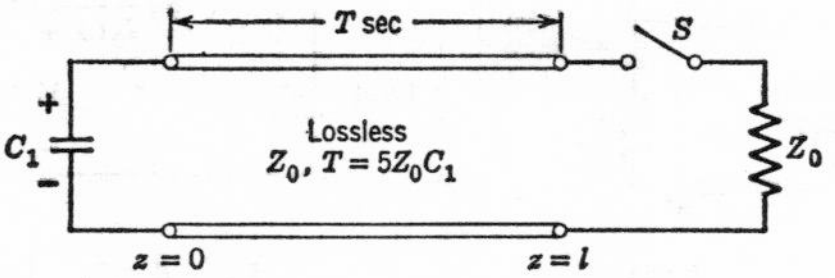

Fig. 4.51. Problem 4.25.

Problem 4.26. The transmission line, closed switch S, and inductor L_1 in Fig. 4.52 have initially a steady current I_0 flowing through them. At $t = 0$, switch S is opened. Determine the currents at $z = 0$ and $z = l$ as functions of time

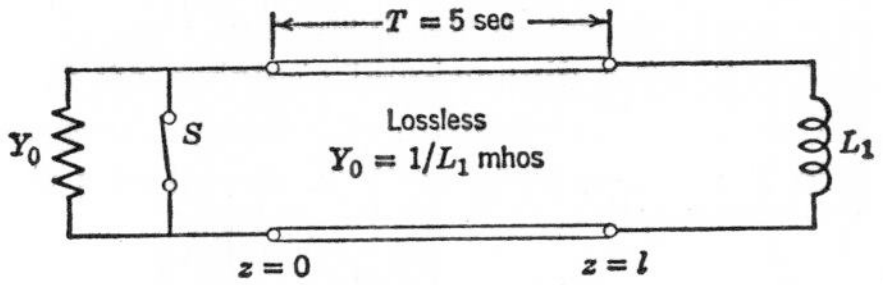

Fig. 4.52. Problem 4.26.

Problem 4.27. Switch S in Fig. 4.53 has been open for a long time, and C_1 is uncharged. S closes at $t = 0$. Determine and sketch: (a) Voltage and current versus time at the load end of the line for $t \geq 0$. (b) Voltage and current versus time at the input end of the line for $t \geq 0$. (c) Voltage and current versus position on the line at $t = T/2$.

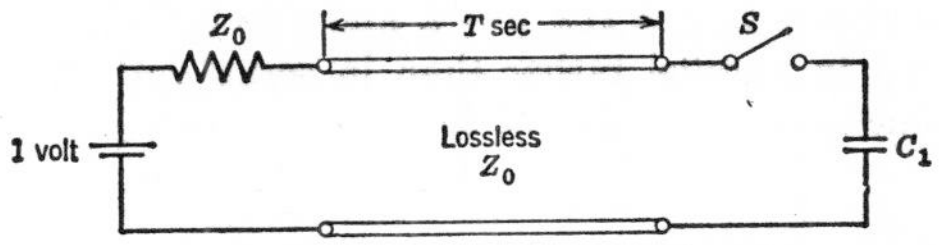

Fig. 4.53. Problem 4.27.

Problem 4.28. Switch S in Fig. 4.54 is *opened* at $t = 0$, after having been closed a long time. (a) Sketch and dimension the voltage distribution along the line at $t = l/2v$ and $t = 3l/2v$. (b) Sketch and dimension the current through the inductor as a function of time. (c) Sketch the voltage across the switch terminals as a function of time.

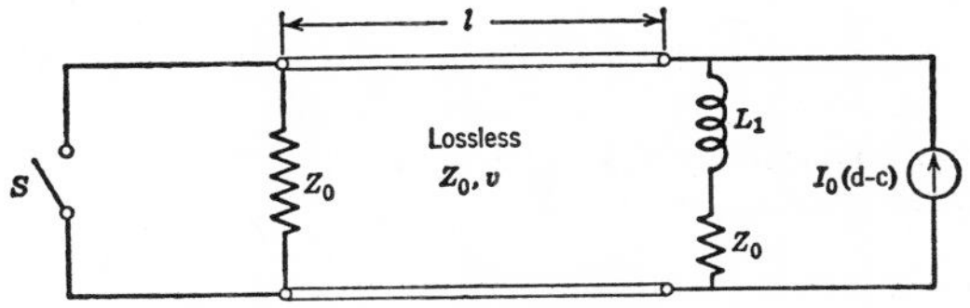

Fig. 4.54. Problem 4.28.

Problem 4.29. In Fig. 4.55: (a) Obtain an expression for the voltage at the load after the arrival of the initial wave. Assume that $Z_0/2L_1 < 1/\sqrt{L_1C_1}$. (b) If $Z_0 = 50$ ohms, $L_1 = 25$ mh, $C_1 = 1.0$ μf, $T = 3333$ μsec, sketch $V(z)$ versus z for $t = T/2$, $3T/2$, and $5T/2$. (c) Calculate the maximum voltage appearing on the line. (d) Repeat if L_1 and C_1 are in parallel instead of series, and $L_1 = 25$ mh, $C_1 = 1.0$ μf as before, but now $Z_0 = 500$ ohms.

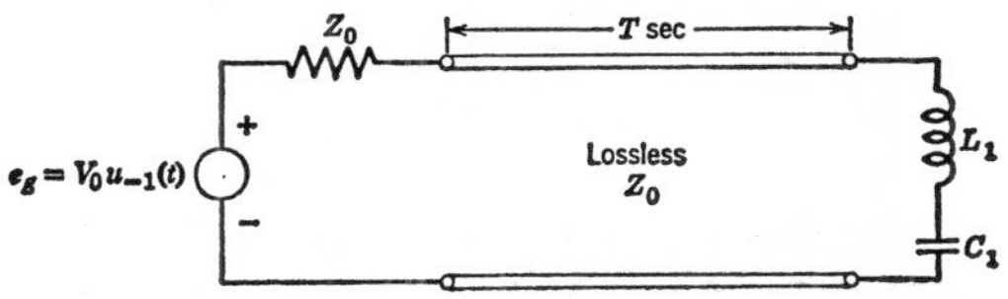

Fig. 4.55. Problem 4.29.

Problem 4.30. No energy is stored in the system of Fig. 4.56 when $t < 0$. If the switch S is closed at $t = 0$, give dimensional sketches of: (a) The voltage across C_1 versus time. (b) The space distribution of voltage along the *entire* line ($0 \leq z < \infty$) at times $t = l/2v$, $t = 3l/2v$, and $t = 2l/v$. (c) Sketch the time functions for

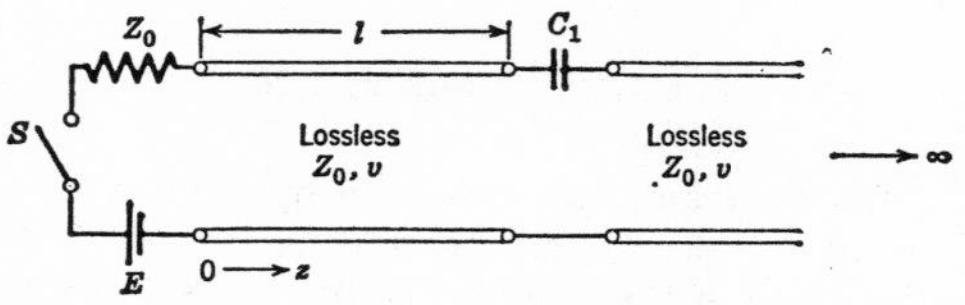

Fig. 4.56. Problem 4.30.

the voltages and currents at $z = l/2$, $z = l$, and $z = 3l/2$. (d) Repeat the above parts if C_1 is placed in parallel rather than in series with the line. (e) Repeat (a)–(d) with an inductance L_1 instead of the capacitance C_1.

C H A P T E R F I V E

Traveling Waves on Dissipative Transmission Lines

From the traveling-wave point of view, the primary effect of dissipation is ordinarily twofold: It results in a diminution of the signal strength with distance, known as *attenuation*, and it causes signals of different frequencies to suffer different time delays per unit distance. The latter feature is called *dispersion*. The dispersion and the frequency dependence of the attenuation produce changes of signal shape with distance in the transient case. These as well as other features of traveling-wave motion in dissipative systems are illustrated by a study of the transmission line with losses. Analysis directly in the time domain is only occasionally possible for such a line, so we shall dwell primarily upon frequency-domain considerations.

5.1 Steady-State Solution

A comparison of Eqs. 2.36 with Eqs. 3.1 shows that the dissipative line is governed by equations which can be obtained from those of a lossless line merely by replacing $j\omega L$ by $(R + j\omega L)$ and $j\omega C$ by $(G + j\omega C)$. It follows that the general solutions for the two cases must have the same relationship. Specifically, instead of Eqs. 3.4 and 3.6, we have for the dissipative line

$$\text{(a)} \qquad \mathrm{V}(z) = \mathrm{V}_+ e^{-\bar{\gamma} z} + \mathrm{V}_- e^{\bar{\gamma} z}$$

$$\text{(b)} \qquad \mathrm{I}(z) = \mathrm{Y}_0(\mathrm{V}_+ e^{-\bar{\gamma} z} - \mathrm{V}_- e^{\bar{\gamma} z}) \qquad (5.1)$$

in which

$$\text{(a)} \qquad \bar{\gamma} = \sqrt{(R + j\omega L)(G + j\omega C)} = \alpha + j\beta$$

$$\text{(b)} \qquad Z_0 = \frac{1}{Y_0} = \sqrt{\frac{R + j\omega L}{G + j\omega C}} = R_0 + jX_0 \tag{5.2}$$

The square root in Eq. 5.2*a* is meant to be that one for which β and ω have the same sign, in which case $\alpha \geq 0$.[1] In Eq. 5.2*b* the root is that one for which $R_0 > 0$. Evidently when the line is lossless, $\bar{\gamma} = j\beta = +j\omega\sqrt{LC}$ and $Z_0 = +\sqrt{L/C}$. In general, we call $\bar{\gamma}$ *the complex propagation constant,* α *the attenuation constant, and* β *the phase constant.*

From Eq. 5.2*a* we see that at $\omega = 0$, $\bar{\gamma} = \sqrt{RG} = \alpha$, and $\beta = 0$. On the other hand, at high frequencies (or with small losses), when $R \ll \omega L$ and $G \ll \omega C$, we have by a binomial expansion

$$\begin{aligned} \bar{\gamma} &= \sqrt{-\omega^2 LC}\,\sqrt{\left(1 + \frac{R}{j\omega L}\right)\left(1 + \frac{G}{j\omega C}\right)} \\ &\approx j\omega\sqrt{LC}\left[1 + \frac{1}{2}\left(\frac{R}{j\omega L} + \frac{G}{j\omega C}\right)\right] \\ &= \frac{1}{2}\left(R\sqrt{\frac{C}{L}} + G\sqrt{\frac{L}{C}}\right) + j\omega\sqrt{LC} \end{aligned} \tag{5.3}$$

Thus, either at high frequencies or when the losses are small,

$$\text{(a)} \qquad \alpha \approx \frac{1}{2}\left(R\sqrt{\frac{C}{L}} + G\sqrt{\frac{L}{C}}\right)$$

$$\text{(b)} \qquad \beta \approx \omega\sqrt{LC} \tag{5.4}$$

In other words, the phase constant is very nearly what it would be for the lossless case. The principal effect of a *small* amount of dissipation upon the propagation constant is to introduce an attenuation constant α. If the line parameters do not vary with frequency, the attenuation produced by these small losses is to a first approximation independent of frequency.

We note that even though we may have $\beta \gg \alpha$ at high frequencies, it would *not* be fair to neglect α entirely by setting $\bar{\gamma} \approx j\omega\sqrt{LC} = j\beta$.

[1] A more detailed treatment of the choice of roots in cases like this one appears in Sec. 6.1.1. The appropriate pages may be read at this point if desired.

The reason is that $\bar{\gamma}$ appears in the solution for V and I in the form $e^{\pm\bar{\gamma}z} = e^{\pm\alpha z}e^{\pm j\beta z}$, and, no matter how small α might be *compared to* β, we may have a long enough line (large enough z) to make the product αz arbitrarily large in magnitude.

Of course it is possible to solve Eq. 5.2*a* explicitly for α and β as functions of R, G, L, C, and ω without introducing any special approximations. The results are obtained in the Problems for the case $R = 0$. They are of the same nature as those of Fig. 8.1. What is more important for us here is the fact that, if the losses are not small, any detailed discussion of $\alpha(\omega)$ and $\beta(\omega)$ is apt to be unrealistic unless some frequency dependence is assigned to R and G (and possibly also to L and C). The appropriate frequency dependence of the line parameters depends greatly upon the actual physical structure of the line under consideration. For our present purposes it is sufficient to recognize that, in general, the attenuation and phase constants of a dissipative line will be more or less complicated functions of frequency, which we may denote simply by $\alpha(\omega)$ and $\beta(\omega)$. In many practical lines these functions are determined experimentally.

The characteristic impedance Z_0 defined in Eq. 5.2*b* is also complex in general. It has the same physical meaning here as it does in the lossless case. If there is no reflection ($\mathrm{V}_- = 0$ in Eqs. 5.1), the impedance of the line is Z_0. In the presence of appreciable loss, Z_0 is evidently a function of frequency. At $\omega = 0$, $\mathrm{Z}_0 = \sqrt{R/G} = R_0$. If, however, the frequency is high or the loss small, we may set $\mathrm{R} \ll \omega L$ and $G \ll \omega \mathrm{C}$ in Eq. 5.2*b*, and find approximately

$$\mathrm{Z}_0 = \sqrt{\frac{L}{C}}\sqrt{\frac{1 + (R/j\omega L)}{1 + (G/j\omega C)}} \approx \sqrt{\frac{L}{C}}\sqrt{\left(1 + \frac{R}{j\omega L}\right)\left(1 - \frac{G}{j\omega C}\right)}$$

$$\approx \sqrt{\frac{L}{C}}\left[1 + \frac{1}{2}\left(\frac{R}{j\omega L} - \frac{G}{j\omega C}\right)\right] \approx \sqrt{\frac{L}{C}} \tag{5.5}$$

Thus the characteristic impedance is, under these special conditions, approximately that of a lossless line.

If we examine only the (+) wave in Eqs. 5.1, we find for the voltage as a function of position and time

$$\begin{aligned} V_+(z, t) &= \mathrm{Re}\,(\mathrm{V}_+ e^{-\alpha z} e^{+j(\omega t - \beta z)}) \\ &= |\mathrm{V}_+| e^{-\alpha z} \cos(\omega t - \beta z + \varphi_+) \end{aligned} \tag{5.6}$$

where $\mathrm{V}_+ = |\mathrm{V}_+|e^{j\varphi_+}$. As shown in Fig. 5.1, this expression represents a wave moving in the $+z$ direction with an amplitude decreasing as it

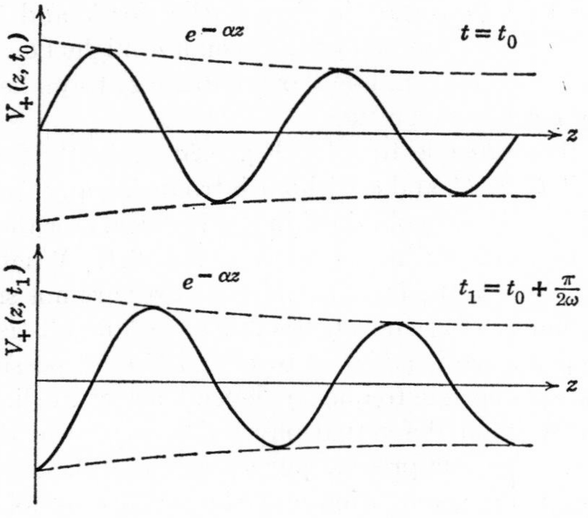

Fig. 5.1. The (+) wave at two instants of time.

progresses. The space rate of decrease of amplitude is given as a fraction by the attenuation constant α.

The complex power carried by this wave is

$$\mathrm{P}_+ = \tfrac{1}{2}|\mathrm{V}_+|^2\mathrm{Y}_0{}^*e^{-2\alpha z} = \langle P_+\rangle + jQ_+ \tag{5.7}$$

showing that the active and reactive parts both decrease with increasing z. In particular, it is clear directly from Eq. 5.7 that

$$-\frac{d\langle P_+\rangle}{dz} = +2\alpha\langle P_+\rangle \qquad \text{or} \qquad 2\alpha = -\frac{(d\langle P_+\rangle)/\langle P_+\rangle}{dz} \tag{5.8}$$

which leads us to interpret 2α as the fractional decrease of transmitted power per unit differential length.

Evidently the attenuation actually arises from the power expended in the series and the shunt loss elements in the line. Thus, per unit differential length at $z = 0$, $\frac{1}{2}|\mathrm{V}_+|^2G + \frac{1}{2}|\mathrm{I}_+|^2R$ watts goes into losses. The power transmitted is $\frac{1}{2}\,\mathrm{Re}\,(\mathrm{V}_+\mathrm{I}_+{}^*) = \frac{1}{2}|\mathrm{V}_+|^2\,\mathrm{Re}\,(\mathrm{Y}_0{}^*)$ watts. Hence by Eq. 5.8, we expect

$$2\alpha = \frac{G + |\mathrm{Y}_0|^2R}{\mathrm{Re}\,(\mathrm{Y}_0)} \approx G\sqrt{\frac{L}{C}} + R\sqrt{\frac{C}{L}} = G\mathrm{Z}_0 + \frac{R}{\mathrm{Z}_0} \tag{5.9}$$

where we have used the small-loss approximation (Eq. 5.5) for Z_0. Note that Eq. 5.9 agrees with Eq. 5.4a, although it was arrived at from quite a different viewpoint.

From Eq. 5.7 we see that the time-average power transmitted by a (+) wave varies as $e^{-2\alpha z}$. Therefore in decibels, the ratio of the power transmitted at z to that at $z + l$ is just

$$\text{Attenuation in db} \equiv 10 \log_{10} e^{2\alpha l} = 20 \log_{10} e^{\alpha l} = 8.68\alpha l \quad (5.10a)$$

Incidentally, the attenuation of a piece of line is sometimes given in units called *nepers,* based on the Napierian logarithm (ln) of the ratio of an input to an output voltage (or current).

$$\text{Attenuation in nepers} \equiv \ln \left|\frac{V_1}{V_2}\right| \quad (5.10b)$$

For the case of a (+) wave on a line, $V_1 = V_+(z)$ and $V_2 = V_+(z + l) = e^{-j\beta l}e^{-\alpha l}V_+(z)$. Therefore,

$$\text{Attenuation in nepers} = \ln e^{\alpha l} = \alpha l \quad (5.10c)$$

It is common to quote the value of α in *nepers per meter,* in view of Eq. 5.10*c*.

It is sometimes convenient to define a "wave length" λ for these waves, even though they do *not* exactly repeat periodically in space. An appropriate definition of wave length is

$$\lambda \equiv \frac{2\pi}{\beta} \quad (5.11)$$

because it does actually give the spacing between points where $V_+(z, t) = 0$, and also reduces to the correct wave length for the lossless line. The attenuation in a distance of one wave length, according to Eqs. 5.10, would then be proportional to $\alpha\lambda$, or, in view of Eq. 5.11, to $(2\pi\alpha)/\beta$. On this basis, Eq. 5.4 shows that the "small loss" condition $R \ll \omega L$ and $G \ll \omega C$ is equivalent to the statement that the attenuation in nepers for one wave length is small.

The second term of Eq. 5.1*a* represents a wave moving to the left, becoming smaller in amplitude as it progresses in that direction. The general solution, then, corresponds to an incident and a reflected wave, each attenuated in the direction of its own progress. As in the lossless case, we define a generalized reflection coefficient

$$\bar{\Gamma}(z) \equiv \frac{V_-e^{+\bar{\gamma}z}}{V_+e^{-\bar{\gamma}z}} = \left(\frac{V_-}{V_+}\right)e^{2\bar{\gamma}z} = \bar{\Gamma}_R e^{2\alpha z}e^{j2\beta z} \quad (5.12)$$

and find

$$\bar{\Gamma}(z) = \frac{Z_n(z) - 1}{Z_n(z) + 1} \quad (5.13)$$

where $Z_n(z)$ is the normalized generalized impedance. There are, how-

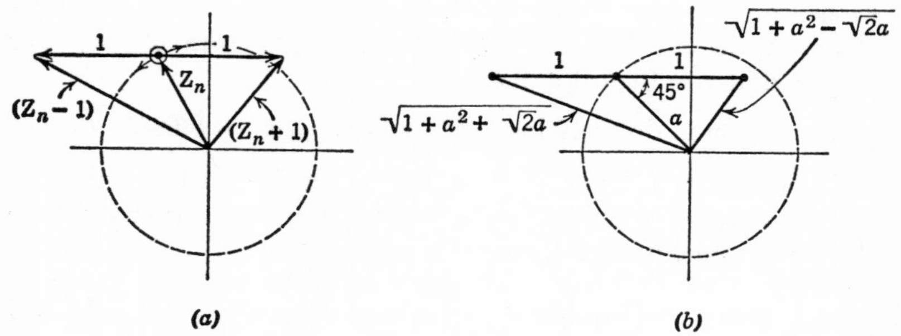

Fig. 5.2. Reflection coefficient on a dissipative line.

ever, some important differences between the detailed properties of both $\bar{\Gamma}(z)$ and $Z_n(z)$ in dissipative lines and lossless lines. These we must now examine.

Suppose that at the load ($z = 0$) there is some passive impedance Z_R, for which, therefore, Re $(Z_R) \geq 0$. The normalized load impedance $Z_{nR} \equiv Z_R/Z_0$ does not have the same angle as Z_R because Z_0 is complex in the general case. From Eq. 5.2*b*, however, we note that for $\omega \geq 0$ the angle of Z_0 cannot be outside the range $\pm 45°$ as long as R, G, L, and C are all non-negative. The angle of a passive Z_R may lie between $\pm 90°$. Thus $Z_{nR} = Z_R/Z_0$ may have an angle anywhere between $\pm 135°$. In Fig. 5.2*a*, we show a phasor diagram exhibiting $Z_n - 1$ and $Z_n + 1$. If we imagine $|Z_n|$ fixed at the arbitrary value a in the figure, and visualize the picture as the angle of Z_n varies from $+135°$ through $0°$ to $-135°$, we see clearly that the ratio $|Z_n - 1|/|Z_n + 1|$ is largest when $\measuredangle Z_n = \pm 135°$. Fixing this angle at $135°$, we ask next what value of $|Z_n| = a$, in Fig. 5.2*b*, leads to the largest $|\bar{\Gamma}|$ (or $|\bar{\Gamma}|^2$). Since

$$|\bar{\Gamma}|^2 = \frac{1 + a^2 + \sqrt{2}\,a}{1 + a^2 - \sqrt{2}\,a} = 1 + \frac{2\sqrt{2}\,a}{1 + a^2 - \sqrt{2}\,a}$$

$$= 1 + \frac{2\sqrt{2}}{(a + 1/a) - \sqrt{2}}$$

it is obvious that $|\bar{\Gamma}|_{\max}$ occurs when $(a + 1/a)$ is minimum. The latter has its minimum value of 2 at $a = 1$; thus

$$|\bar{\Gamma}|_{\max} = \sqrt{\frac{2 + \sqrt{2}}{2 - \sqrt{2}}} = 1 + \sqrt{2} \tag{5.14a}$$

It is worth noting from the geometry of Fig. 5.2*b* that, when $a = 1$, the angle between $Z_n - 1$ and $Z_n + 1$ is 90°. Thus

$$\measuredangle \bar{\Gamma} = \pm\frac{\pi}{2} \qquad \text{when}\,|\bar{\Gamma}| = |\bar{\Gamma}|_{\max} = 1 + \sqrt{2} \tag{5.14b}$$

Contrary to the situation in the lossless case, the magnitude of the reflection coefficient for a dissipative line with a passive load *may* exceed unity, although this is not the most common occurrence. Observe, however, that according to Eq. 5.12 a motion away from the load ($z = 0$) toward the generator ($z < 0$) causes the reflection coefficient not only to rotate in phase but also to become smaller in magnitude by the factor $e^{-2\alpha|z|}$. Accordingly, $|\bar{\Gamma}|$ may not remain above unity for very great distances, even if it is so at the load.

In principle, a Smith chart extended out to $|\bar{\Gamma}| = 1 + \sqrt{2}$, provided with a rotating arm centered at $\bar{\Gamma} = 0$ and divided radially into db, can be used to solve problems on dissipative lines. As shown in Fig. 5.3, one rotates the arm by an angle $2\beta l = 4\pi(l/\lambda)$ exactly as is done on a lossless line, but, at the same time, one must now shrink the radius according to Eqs. 5.12 and 5.10 by

$$8.68\alpha l = 8.68\alpha\lambda\left(\frac{l}{\lambda}\right)\text{ db} \tag{5.15}$$

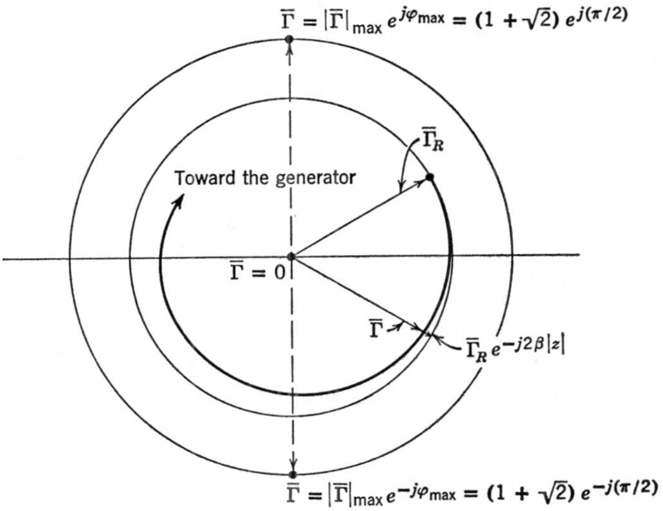

Fig. 5.3. The extended Smith chart for a dissipative line.

The factor $8.68\alpha\lambda$ is the attenuation in "db for one wave length," and must be computed separately for each line (and at each frequency). The factor l/λ, however, then appears in both the rotation and the shrinkage. Evidently the locus traced on such a Smith chart as we move away from the load is a clockwise spiral approaching $\bar{\Gamma} = 0$ at the rate of $4.34\alpha\lambda$ db per revolution ($\frac{1}{2}\lambda$).

The spiral locus means that, regardless of the load impedance, the input impedance of a sufficiently long dissipative line approaches Z_0. Because of the attenuation of both the incident and the reflected waves, the load is effectively "decoupled" from the input. This fact is often used in acoustic and microwave techniques when it is desired to eliminate the effect of load changes upon a generator. Of course, the price of the method is an exceedingly large waste of power.

The fact that $|\bar{\Gamma}|$ may exceed unity on a dissipative line does not violate a condition of power conservation (as it would on a lossless structure). Note that the total complex power transmitted in the presence of both incident and reflected waves is, according to Eqs. 5.1 and 5.12,

$$\begin{aligned} \mathrm{P} &= \tfrac{1}{2}\mathrm{VI}^* \\ &= \tfrac{1}{2}\mathrm{Y}_0^*|\mathrm{V}_+|^2 e^{-2\alpha z} - \tfrac{1}{2}\mathrm{Y}_0^*|\mathrm{V}_-|^2 e^{2\alpha z} \\ &\qquad + \tfrac{1}{2}\mathrm{Y}_0^*(\mathrm{V}_+^*\mathrm{V}_- e^{j2\beta z} - \mathrm{V}_+\mathrm{V}_-^* e^{-j2\beta z}) \\ &= \tfrac{1}{2}\mathrm{Y}_0^*|\mathrm{V}_+|^2 e^{-2\alpha z}[1 - |\bar{\Gamma}|^2 + (\bar{\Gamma} - \bar{\Gamma}^*)] \end{aligned} \tag{5.16}$$

From Eq. 5.16 we find that the time-average power is

$$\langle P \rangle = \mathrm{Re}\,(\mathrm{P}) = \frac{1}{2} G_0 |\mathrm{V}_+|^2 e^{-2\alpha z}\left[1 - |\bar{\Gamma}|^2 + 2\left(\frac{B_0}{G_0}\right)\mathrm{Im}\,(\bar{\Gamma})\right] \tag{5.17}$$

where we have written $\mathrm{Y}_0 = G_0 + jB_0$. In other words, the time-average power transmitted is *not* just the difference between that carried by the incident wave and that carried by the reflected wave, as if they each existed separately. Besides the term $1 - |\bar{\Gamma}|^2$ in Eq. 5.17 there is another one representing interaction between the two waves. When $\bar{\Gamma}$ is restricted, as previously discussed in connection with Fig. 5.2, $\langle P \rangle$ in Eq. 5.17 must remain non-negative, even though $|\bar{\Gamma}|$ may exceed unity at certain special angles. Although this fact can be proved analytically from Eq. 5.17, there is no real need to do so as long as we already know from Eq. 2.39*a* that the transmission line always obeys the law of conservation of energy. The main point of interest lies in the fact that we cannot, in general, superpose the average powers carried by incident and reflected waves on a dissipative line, although we could do so on a lossless line. Note, however, that, if $B_0 = 0$ in Eq. 5.17, it *is* possible to superpose incident and reflected average powers to obtain the total. Therefore, to the extent that Eq. 5.5 represents a valid approximation for small losses, the cross term containing B_0 in Eq. 5.17 may also be dropped from consideration under the same conditions.

5.2 Some Aspects of Transient Response

We have pointed out that, in general, α and β in Eq. 5.2 are more or less complicated functions of frequency. If we imagine a transient signal at one point z on the line to be represented by the superposition of all of its sinusoidal components in the Fourier sense, each of these components will be attenuated and shifted in phase by a different amount as it moves (say) in the $+z$ direction a distance l. Even assuming that there is no reflected wave present, the sinusoidal components of the signal will not, in general, retain either their *relative* amplitudes or *relative* phases as they move. That is to say, the shape of the time function to which they will add up at $z + l$ may be quite different from its shape at z. The line is sometimes said to introduce distortion of a transient signal by this means, although choice of the word "distortion" may be unfortunate here because of possible confusion with nonlinear effects. With the meaning clear, however, we may ask whether the presence of dissipation *must* introduce such distortion, or whether conditions exist under which none will occur.

5.2.1 The Distortionless Line

In considering the latter possibility, we can hardly expect to avoid completely the attenuation of the signal. Inasmuch as there are losses, the signal must get smaller as it moves; but we can demand that this be the *only* change. That is, we can demand that every sinusoidal component be attenuated with distance at the same rate, and that each one suffer exactly the same time delay per unit distance. In this way we preserve the *relative* amplitudes and phases of all the components and so retain the shape of the signal, but we allow the entire time function to shrink in size. With $\bar{\gamma}$ expressed in terms of α and β, we are asking that α be independent of frequency and that β be directly proportional to frequency. Alternatively, we are asking that the complex propagation constant $\bar{\gamma}$ be of the form

$$\bar{\gamma} = K_1 + j\omega K_2 \tag{5.18}$$

in which K_1 and K_2 are independent of ω.

Examination of Eq. 5.2*a* shows that $\bar{\gamma}$ is the square root of the product of two complex numbers, each of which alone is precisely of the form of Eq. 5.18. The condition that the square root yield a result of form similar to either factor is that the angles of the two complex numbers involved be the same at all frequencies:

$$\frac{L}{R} = \frac{C}{G} \tag{5.19}$$

Then

$$\bar{\gamma} = \sqrt{RG}\left(1 + j\omega\frac{L}{R}\right) = \sqrt{RG}\left(1 + j\omega\frac{C}{G}\right)$$

or

$$\text{(a)} \qquad \alpha = \sqrt{RG} = R\sqrt{\frac{C}{L}} = G\sqrt{\frac{L}{C}} = \frac{1}{2}\left(R\sqrt{\frac{C}{L}} + G\sqrt{\frac{L}{C}}\right)$$

$$\text{(b)} \qquad \beta = \omega L\sqrt{\frac{G}{R}} = \omega C\sqrt{\frac{R}{G}} = \omega\sqrt{LC} \tag{5.20}$$

Also

$$Z_0 = \sqrt{\frac{R}{G}} = \sqrt{\frac{L}{C}} \tag{5.21}$$

From Eqs. 5.20*b* and 5.21 it is clear that the line behaves almost exactly as if it were lossless, except for the constant attenuation given by Eq. 5.20*a*. The foregoing relations, exact under the condition of Eq. 5.19, should be compared with the approximate results (Eqs. 5.5 and 5.4).

If our present reasoning is correct, we should find that in the time domain, Eq. 5.19 allows a general solution in the form of two oppositely traveling waves of arbitrary but unchanging shape, each attenuated exponentially with distance in the direction of motion. In short,

$$\text{(a)} \qquad V(z, t) = e^{-\alpha z} f_+\left(t - \frac{z}{v}\right) + e^{\alpha z} f_-\left(t + \frac{z}{v}\right)$$

$$\text{(b)} \qquad I(z, t) = Y_0\left[e^{-\alpha z} f_+\left(t - \frac{z}{v}\right) - e^{\alpha z} f_-\left(t + \frac{z}{v}\right)\right] \tag{5.22}$$

should be a solution to the transmission-line equations (Eqs. 2.34*a* and

2.34*b*) under conditions of Eqs. 5.19, 5.20, and 5.21, and with $v = 1/\sqrt{LC}$. Direct substitution of the (+) wave only, from Eq. 5.22 into Eq. 2.34*a* yields, upon collection of terms,

$$f_+ e^{-\alpha z}(-\alpha + RY_0) + f_+' e^{-\alpha}\left(-\frac{1}{v} + LY_0\right) = 0 \qquad (5.23)$$

where f_+' denotes $(d/du)[f_+(u)]$. In view of Eqs. 5.20*a* and 5.21, the coefficients of f_+ and f_+' in Eq. 5.23 are zero. It is very easy to show a similar result for the (−) wave and to obtain the dual result for Eq. 2.34*b*. Equation 5.22 is therefore the general transient solution under the condition of Eq. 5.19. In these circumstances, the line is said to be *distortionless*. Unless Eq. 5.19 is fulfilled, no *exact* solution is possible in which the waves progress without any change of shape. There will always be some distortion.

Incidentally, it is interesting to observe that Eq. 5.19 is also the condition for minimum attenuation at all frequencies, provided that R and G may be held fixed while L and C are varied by altering the construction of the line in special ways. This fact follows from the exact expression for 2α in Eq. 5.9, using $Y_0 = G_0 + jB_0$:

$$2\alpha = \frac{G + G_0^2 R + B_0^2 R}{G_0} = \frac{G + B_0^2 R}{G_0} + RG_0$$

in which we see that α is a minimum (for given R and G values) if $B_0 = 0$ and $G_0 = Y_0 = \sqrt{G/R}$. These conditions are met by Eq. 5.2*b* only if $R/L = G/C$, which coincides with Eq. 5.19.

The addition of lumped series inductance at periodic space intervals is a scheme often used on telephone lines to achieve (approximately) the condition of Eq. 5.19 for low distortion and low attenuation. If these "loading coils" are spaced less than $\lambda/4$ apart at the highest frequency of importance, they will act very nearly like simple additions to the inductance (L) per unit length.

5.2.2 Group Velocity

Let us now examine the problem of distortion more closely, and ask collaterally what meanings might be attached to the "velocity of propagation" when the signal distorts as it moves. Consider in particular a semi-infinite line along $z > 0$, with a specified voltage source $V_1(t)$ applied to its input $z = 0$ (Fig. 5.4). We wish to study $V_2(t)$, the voltage at a distance l down the line. We suppose there is no re-

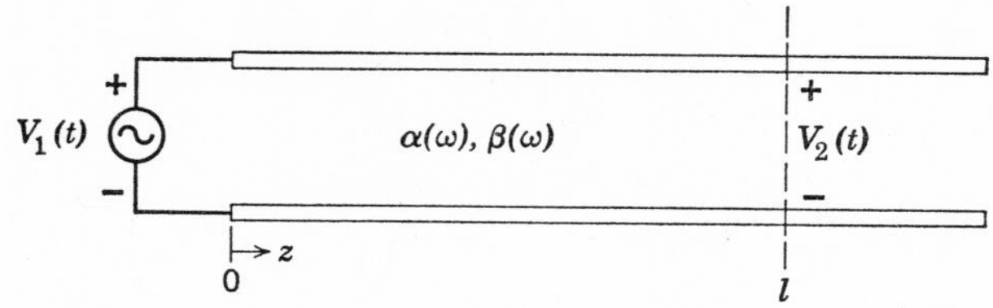

Fig. 5.4. A semi-infinite line, driven by a voltage source.

flected wave. The line has attenuation and phase constants which are functions of frequency $\alpha(\omega)$ and $\beta(\omega)$ respectively.

There is little hope of handling this problem by trying to find general transient solutions to the transmission-line equations in the time domain. In the general case, no such solutions are known. Therefore we approach the matter through superposition ideas, based upon Fourier reasoning. We may consider the transient time function $V_1(t)$ to be broken down into sinusoidal components

$$V_1(t) = \int_0^\infty \mathrm{Re}\,[\mathrm{V}_1(\omega)e^{j\omega t}]\,d\omega = \mathrm{Re}\left[\int_0^\infty \mathrm{V}_1(\omega)e^{j\omega t}\,d\omega\right] \qquad (5.24)$$

where $\mathrm{V}_1(\omega)\,d\omega$ may be regarded as the complex amplitude of the sinusoidal component at frequency ω.

In response to a steady-state signal $\mathrm{V}_1(\omega)$, we know that on the semi-infinite line a voltage $\mathrm{V}_2(\omega)$ will appear at $z = l$, such that

$$\mathrm{V}_2(\omega) = \mathrm{V}_1(\omega)e^{-\bar{\gamma}l} = \mathrm{V}_1(\omega)e^{-\alpha(\omega)l}e^{-j\beta(\omega)l} \qquad (5.25)$$

The entire $V_2(t)$ will then be the sum of all its sinusoidal components

$$V_2(t) = \mathrm{Re}\left[\int_0^\infty \mathrm{V}_2(\omega)e^{j\omega t}\,d\omega\right]$$

$$= \mathrm{Re}\left[\int_0^\infty \mathrm{V}_1(\omega)e^{-\alpha(\omega)l}e^{-j\beta(\omega)l}e^{j\omega t}\,d\omega\right] \qquad (5.26)$$

Now the limits of integration in Eqs. 5.24 and 5.26 suggest that $V_1(t)$ and, consequently, $V_2(t)$ have important frequency components over the entire frequency range 0 to ∞. Actually we are rarely interested in this entire range because the spectrum of $V_1(t)$ itself will undoubtedly be limited if the voltage arises from any real physical generator. Therefore let us assume that the frequency band of interest runs from ω_1 to ω_2, and that ω_0 is some frequency within this band, i.e., $\omega_1 < \omega_0 < \omega_2$. This situation is illustrated in Fig. 5.5*a*.

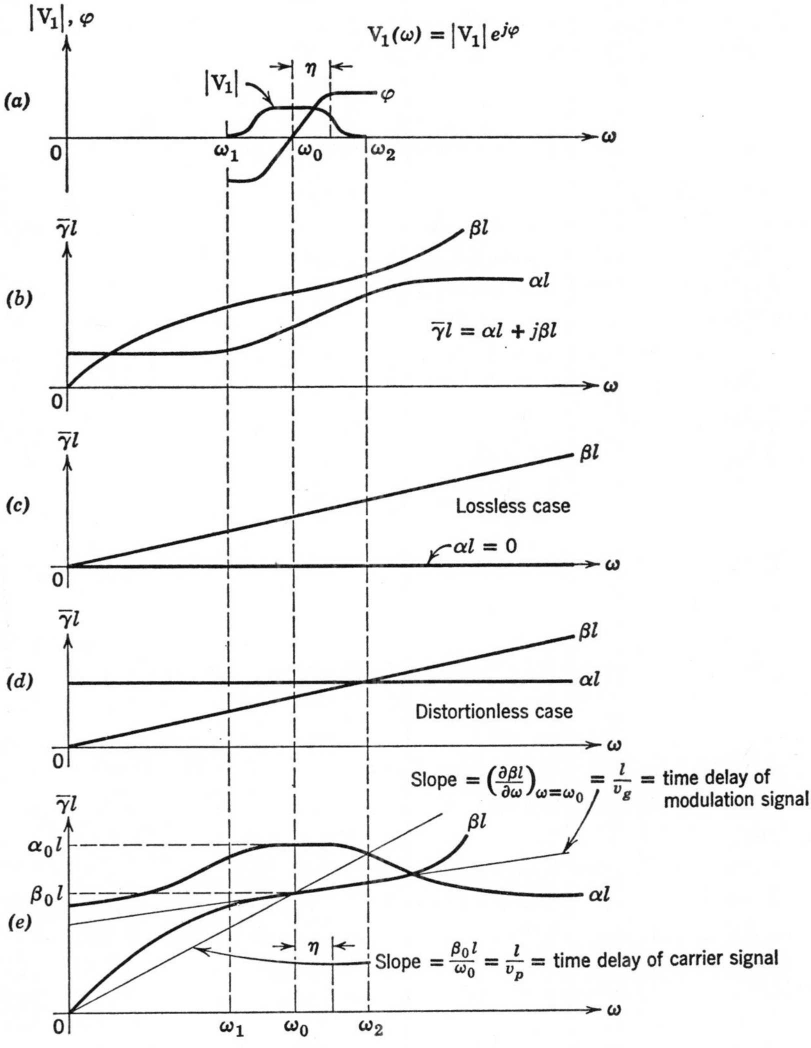

Fig. 5.5. Relationship of distortion to group and phase velocities.

Let us also suppose that the functions $\alpha(\omega)$ and $\beta(\omega)$, an example of the form of which is shown in Fig. 5.5*b*, are analytic in the frequency range $\omega_1 \leq \omega \leq \omega_2$. That is, we assume that they may be represented by power series expansions about the frequency ω_0, which expansions will converge in the range $\omega_1 \leq \omega \leq \omega_2$. Thus

$$\begin{aligned} \alpha(\omega) &= \alpha_0 + \alpha_1\eta + \alpha_2\eta^2 + \cdots \\ \beta(\omega) &= \beta_0 + \beta_1\eta + \beta_2\eta^2 + \cdots \end{aligned} \tag{5.27}$$

In Eq. 5.27, $\eta = \omega - \omega_0$ (the frequency deviation from ω_0), and $\alpha_0 = \alpha(\omega_0)$, $\beta_0 = \beta(\omega_0)$, $\alpha_1 = (d\alpha/d\omega)_{\omega=\omega_0}$, $\beta_1 = (d\beta/d\omega)_{\omega=\omega_0}$, $\alpha_2 = (\frac{1}{2}) \times (d^2\alpha/d\omega^2)_{\omega=\omega_0}$, $\beta_2 = (\frac{1}{2})(d^2\beta/d\omega^2)_{\omega=\omega_0}$, etc.

Introducing into Eq. 5.26 the new limits of integration from ω_1 to ω_2, we have

$$V_2(t) = \text{Re}\left[\int_{\omega_1}^{\omega_2} \mathrm{V}_1(\omega)e^{-\alpha(\omega)l}e^{-j\beta(\omega)l}e^{j\omega t}\,d\omega\right] \tag{5.28}$$

If we then use Eq. 5.27 in Eq. 5.28, and change the variable of integration from ω to η, there results

$$V_2(t) = \text{Re}\left[e^{-\alpha_0 l}e^{j(\omega_0 t - \beta_0 l)}\int_{\omega_1-\omega_0}^{\omega_2-\omega_0} \mathrm{V}_1(\eta)e^{-(\alpha_1\eta l + \alpha_2\eta^2 l + \cdots)}e^{j\eta(t-\beta_1 l - \beta_2\eta l - \cdots)}\,d\eta\right] \tag{5.29}$$

On the same basis, by putting $l = 0$ in Eq. 5.29, we have for $V_1(t)$

$$V_1(t) = \text{Re}\left[e^{j\omega_0 t}\int_{\omega_1-\omega_0}^{\omega_2-\omega_0} \mathrm{V}_1(\eta)e^{j\eta t}\,d\eta\right] \tag{5.30}$$

which is, of course, just Eq. 5.24 modified for the change of the limits of integration to go from ω_1 to ω_2 instead of $-\infty$ to $+\infty$, and the change of variable of integration from ω to η.

In general, there is no simple relation between $V_2(t)$, as given in Eq. 5.29, and $V_1(t)$ in Eq. 5.30. It may occur, however, that the coefficients are so small, the frequency band $\omega_2 - \omega_1$ is so narrow, and the distance l is so short, that for all practical purposes

$$\begin{aligned} \alpha_1\eta l &\approx 0 & & & \beta_2\eta l &\approx 0 \\ & & \text{and} & & & \\ \alpha_2\eta^2 l &\approx 0 & & & \beta_3\eta^2 l &\approx 0 \\ &\vdots & & & &\vdots \\ &\text{etc.} & & & &\text{etc.} \end{aligned} \tag{5.31}$$

when $\omega_1 \leq \omega \leq \omega_2$. Then Eq. 5.29 simplifies to

$$V_2(t) \cong \text{Re}\left[e^{-\alpha_0 l} e^{j(\omega_0 t - \beta_0 l)} \int_{\omega_1-\omega_0}^{\omega_2-\omega_0} V_1(\eta) e^{-j\beta_1 \eta l} e^{j\eta t}\, d\eta\right] \tag{5.32}$$

Comparison of Eqs. 5.32 and 5.30 can now be made on the following basis. We think of Eq. 5.30 as representing $V_1(t)$ in the form of a carrier at frequency ω_0 modulated by a time function represented by the integral. This need not be amplitude modulation alone; it will generally be both phase and amplitude modulation. On this basis, Eq. 5.32 for $V_2(t)$ is seen to represent another modulated signal, differing from $V_1(t)$ in three respects:

1. It is attenuated by the factor $e^{-\alpha_0 l}$.
2. The carrier is shifted backward in phase by $\beta_0 l$ radians, corresponding to a time delay of $(\beta_0 l/\omega_0)$ seconds.
3. The modulation, represented by the integral, has all its frequency components shifted in phase by $\beta_1 l \eta$ radians compared to those in Eq. 5.30. Thus the modulation is simply delayed in time by $\beta_1 l$ seconds—which is a *different* delay from that of the carrier.

Thus, *if the conditions in Eqs. 5.31 are met,* we find that the *modulation* on a carrier will travel down the line essentially unchanged in shape. It will be delayed by a time $\beta_1 l$, which is proportional to distance l. Thus we may define a velocity v_g

$$v_g \equiv \frac{l}{(\beta_1 l)} = \frac{1}{\beta_1} = \left(\frac{d\beta}{d\omega}\right)^{-1}_{\omega=\omega_0} \tag{5.33}$$

which is different from the "phase velocity"

$$v_p = \left(\frac{\beta}{\omega}\right)^{-1}_{\omega=\omega_0} \tag{5.34}$$

A case in which the conditions of Eqs. 5.31 are met approximately is shown in Fig. 5.5*e*. It should be compared with the lossless case and the distortionless case (Figs. 5.5*c* and *d* respectively), in which these conditions are met *exactly*. It is convenient to call v_g the *group velocity* because it is the speed of travel of a modulation signal characterized by a small "group" of sinusoidal components located in a narrow frequency band about ω_0. Since group velocity is simply the inverse slope of the curve β versus ω, it can always be *defined* as such at every frequency ω where $\beta(\omega)$ has a derivative. It will not, however, actually represent the speed of a physical modulation signal unless the approximations in Eqs. 5.31 are valid. In any particular situation, this validity depends upon the shapes of *both* $\alpha(\omega)$ and $\beta(\omega)$ as well as upon the bandwidth $(\omega_2 - \omega_1)$ of the modulation and the maximum length l of line under consideration. In many cases, the group velocity is indeed a physically meaningful concept, as it would tend to be in Fig. 5.5*e*; but a careful study of the restrictions of Eqs. 5.31 also shows that it is not always so, e.g., in Fig. 5.5*b*.

When Eqs. 5.31 cannot be met, even the modulation on a carrier will be distorted in traveling down the line. Just how much distortion occurs in such cases depends upon the magnitude of the terms we have neglected from Eq. 5.29. A detailed analysis of these would be required in order to determine how large they may be allowed to become if the distortion must remain below a tolerable percentage.[1]

5.2.3 An Example [2] of Distortion *

In some coaxial transmission-line applications, it is reasonable to neglect dielectric losses ($G = 0$), but necessary to consider the effect of slightly imperfect conductors. As will be discussed at considerable length in Chapters 8 and 9, the skin effect in these conductors adds both a frequency-dependent series resistance and a so-called internal inductance, also frequency-dependent, to the line. In fact, it adds a skin-effect series impedance per unit length of the form

$$\mathbf{Z}_s = k\sqrt{j\omega} = k\sqrt{\frac{\omega}{2}}\,(1 + j) \tag{5.35}$$

in addition to the normal series inductive reactance $j\omega L$ and shunt-capacitive admittance $j\omega C$.

The conclusion from Eq. 5.35 that $R = k\sqrt{\omega/2}$ is often justifiable experimentally over a rather wide frequency range (from a few hundred kilocycles per second to several thousand megacycles per second). It must, however, fail at very low frequencies, and eventually become unimportant, because of other losses, at very high frequencies.

We shall suppose the losses are small over the entire valid frequency range, which is to say

$$k\sqrt{\frac{\omega}{2}} \ll \omega L \tag{5.36a}$$

[1] Some interesting examples along these lines appear in Max P. Forrer, "Analysis of Millimicrosecond RF Pulse Transmission," *Proc. IRE*, **46**, 1830–1835, Nov. 1958.

[2] This example was done independently by one of the authors, and by R. L. Wigington and N. S. Nahman, in 1956. Priority definitely goes to Wigington and Nahman, however, and their article, *Proc. IRE*, **45**, 166–174, Feb. 1957, contains interesting experimental verification.

This circumstance permits us to use the familiar binomial approximation

$$\bar{\gamma} = \sqrt{(Z_s + j\omega L)j\omega C} = \sqrt{-\omega^2 LC}\sqrt{1 + \frac{Z_s}{j\omega L}}$$

$$\approx j\omega\sqrt{LC}\left(1 + \frac{1}{2}\frac{k\sqrt{j\omega}}{j\omega L}\right) \quad (5.36b)$$

$$= j\omega\sqrt{LC} + \frac{1}{2}\frac{k}{Z_0}\sqrt{j\omega}$$

similar to that in Eq. 5.3, and

$$Z_0 \approx \sqrt{\frac{L}{C}} \quad (5.36c)$$

similar to Eq. 5.5.

Now the response $V_2(t) = h_0(t)$ to a unit impulse $V_1(t) = u_0(t)$ in Fig. 5.4 may be found directly by transform methods in this case. Since the excitation and response will surely be zero before $t = 0$, we may use one-sided Fourier transforms with the complex frequency $\mathrm{s} = j\omega$ (i.e., one-sided Laplace transforms). Thus

$$\mathrm{V}_1(\mathrm{s}) = 1 \quad (5.37)$$

and

$$\mathrm{V}_2(\mathrm{s}) = \mathrm{h}_0(\mathrm{s}) = \mathrm{V}_1(\mathrm{s})e^{-\bar{\gamma}l} \approx e^{-\mathrm{s}T}e^{-(kl/2Z_0)\sqrt{\mathrm{s}}} \quad (5.38)$$

where $T = l\sqrt{LC} = l/v$ as usual. The factor $e^{-\mathrm{s}T}$ represents a straightforward time delay of T seconds, so defining $t' = t - T$ we have

$$V_2(t') = h_0(t') = \mathcal{L}^{-1}(e^{-(kl/2Z_0)\sqrt{\mathrm{s}}}) \quad (5.39)$$

where $\mathcal{L}^{-1}$ represents the inverse Laplace transform operation. This particular transform happens to be tabulated,[1] so

$$V_2(t') = h_0(t') = \frac{K}{\sqrt{\pi t'^3}}e^{-(K^2/t')}u_{-1}(t') \quad (5.40)$$

where

$$K \equiv \frac{kl}{4Z_0} > 0 \quad (5.41)$$

Writing

$$\tau \equiv \frac{t'}{K^2} \quad (5.42)$$

we have

$$K^2V_2(\tau) = \frac{e^{-1/\tau}}{\sqrt{\pi\tau^3}} = K^2h_0(\tau) \quad (5.43)$$

[1] See, for example, R. V. Churchill, *Modern Operational Mathematics in Engineering*, McGraw-Hill Book Co., New York, 1944, Appendix III, pair 82.

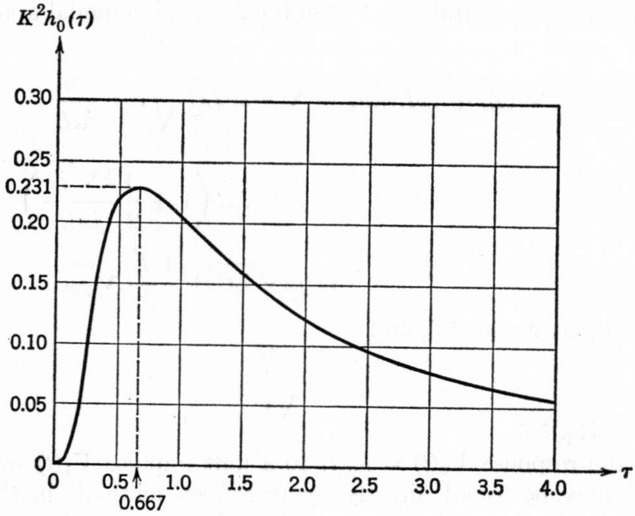

Fig. 5.6. Normalized impulse response of the line of Fig. 5.4, including skin-effect series impedance.

A sketch of the normalized impulse response (Eq. 5.43) appears in Fig. 5.6.

The unit step response $h_{-1}(t')$, rather than the impulse response, shows up some important features of the distortion. A direct integration of Eq. 5.40 with respect to t', from $t' = 0$ to an arbitrary value of t', yields

$$V_2(\tau) = h_{-1}(\tau) = \operatorname{erfc}\left(\frac{1}{\sqrt{\tau}}\right) \tag{5.44}$$

after using Eqs. 5.41 and 5.42. The function

$$\operatorname{erfc} x \equiv \frac{2}{\sqrt{\pi}} \int_x^\infty e^{-y^2}\, dy = 1 - \operatorname{erf} x \tag{5.45}$$

is tabulated (in Peirce's table of integrals, for example), and an illustration of the normalized unit step response (Eq. 5.44) appears in Fig. 5.7.

We observe that the rise is very slow compared to an exponential, and is characterized by an extremely "long-tailed" asymptotic approach to the steady-state final value. Moreover, the time of rise from 10% to 90% is given by the tables as $\Delta\tau \approx 120$, or

$$\Delta t' = \Delta t = K^2\, \Delta\tau \approx \frac{7k^2 l^2}{Z_0{}^2} \tag{5.46}$$

The significant point about this result is that the rise time increases as the *square* of the distance l, and as the *square* of the attenuation level (k/Z_0). These properties turn out to be important limitations to using transmission-line sections for generating pulses.

There is another lesson to be learned from this example before leaving it. One might have been tempted to handle this problem by going back to Eq. 5.4 and putting $\beta \approx \omega\sqrt{LC}$ and $\alpha \approx \frac{1}{2}R/Z_0 \approx k\sqrt{\omega}/2\sqrt{2}\,Z_0$, i.e., *including* the skin-effect *resistance* variation with frequency and *neglecting* the internal skin-effect *inductance* altogether. This in some ways sounds reasonable, because the small internal inductance is competing in the phase velocity with the much larger normal inductance L, whereas the resistance (however small) stands alone as the sole cause of the attenuation. Nevertheless, the suggested procedure is basically unsound. It would leave us with a propagation constant $\bar{\gamma}$ of the form $\bar{\gamma} = a\sqrt{\omega} + j\omega b$ which is *not a real function of* $j\omega = s$. Such a complex function of s could not arise in a real physical system, inasmuch as in such a system the frequency ω is always accompanied by j, as a result of time derivatives, in differential equations with *real* coefficients.

Incidentally, the foregoing problem not only becomes nonphysical but also mathematically much less tractable unless $\sqrt{j\omega}$ is included instead of just $\sqrt{\omega}$. Here is a case where an apparently reasonable approximation actually complicates a situation; thus the danger of making approximations in the frequency domain is suitably empha-

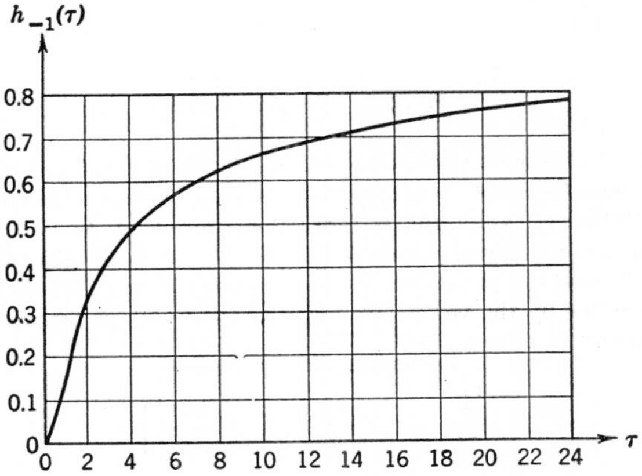

Fig. 5.7. Normalized step response of the line of Fig. 5.4, including skin-effect series impedance.

sized. Of course, approximations still remain in our solution at high and low frequencies, and in the neglecting of dielectric losses $G(\omega)$. But at least what we do retain makes physical sense.

PROBLEMS

Problem 5.1. Because of skin effect, and the availability of rather low-loss dielectrics, some transmission lines behave as if their parameters depended upon frequency in the following way: $R = R_c(1 + k\omega^{1/2})$, $L = L_c[1 + (R_c/L_c)k\omega^{-1/2}]$, $C = C_c$, $G = 0$, where R_c, L_c, C_c, and k are positive constants. Letting

$$\frac{\bar{\gamma}}{(R_c\sqrt{C_c/L_c})} = \alpha'(\eta) + j\beta'(\eta) \qquad \frac{Z_0}{\sqrt{L_c/C_c}} = R_0'(\eta) + jX_0'(\eta)$$

$$\eta \equiv \omega L_c/R_c \qquad \xi \equiv k\sqrt{\frac{R_c}{L_c}}$$

determine, and plot versus η: α', β', R_0', and X_0'. It is helpful to consider the case $k = 0$ ($\xi = 0$) first (see also Chapter 8), and then take up very high and very low frequencies with $\xi > 0$. Intermediate situations may subsequently be examined numerically.

Problem 5.2. Determine admittance Y_1 and impedance Z_2 in Fig. 5.8*a* so the network will behave with respect to terminals *a–a'* and *b–b'* exactly like the dissipative line of Fig. 5.8*b*. Check your results by considering them in the limiting case $|\bar{\gamma}l| \ll 1$.

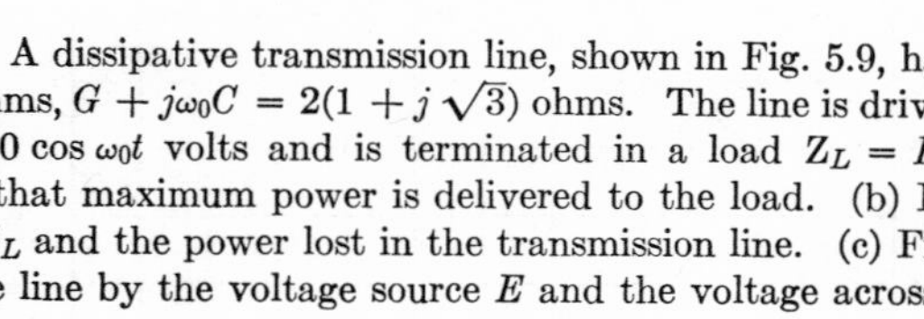

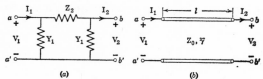

Fig. 5.8. Problem 5.2.

Problem 5.3. A dissipative transmission line, shown in Fig. 5.9, has $R + j\omega_0 L = (\sqrt{3} + j)/2$ ohms, $G + j\omega_0 C = 2(1 + j\sqrt{3})$ ohms. The line is driven by a voltage source $E = 100 \cos \omega_0 t$ volts and is terminated in a load $Z_L = R_L + jX_L$. (a) Choose Z_L so that maximum power is delivered to the load. (b) Find the power dissipated in Z_L and the power lost in the transmission line. (c) Find the current supplied to the line by the voltage source E and the voltage across Z_L, each as a function of time.

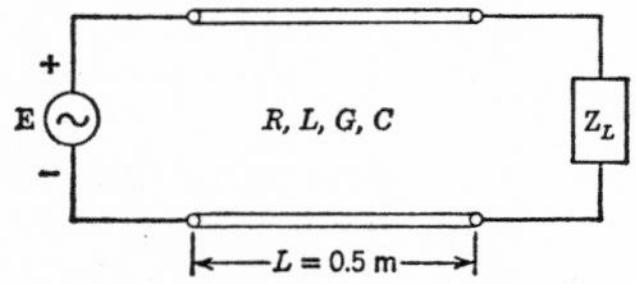

Fig. 5.9. Problem 5.3.

Problem 5.4. The line of Fig. 5.10 is operated in the a-c steady state with applied rms voltage $|E| = 1000$ v. Line parameters are $\omega L = 100$ ohms/m, $\omega C = 2.78 \times 10^{-2}$ mho/m, $R = 1.20$ ohm/m, $G = 0$. (a) Assuming the line to be lossless, find the reading of ammeter A. (b) Including losses, find the reading of A.

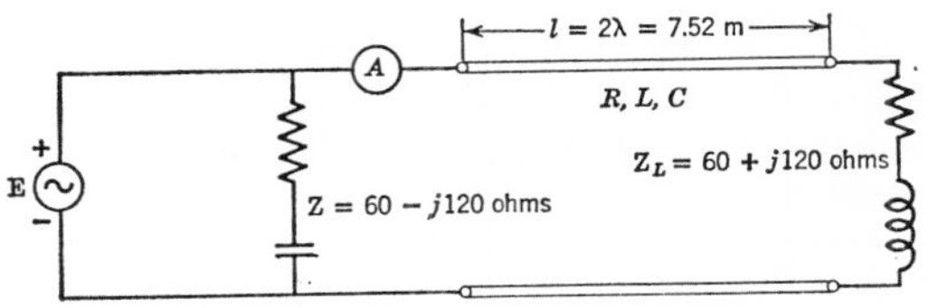

Fig. 5.10. Problem 5.4.

Problem 5.5. The input impedance of a flexible coaxial cable is measured at 3000 mc by connecting it as load for a second line, of characteristic impedance 50 ohms, on which standing-wave measurements can be made. The cable is approximately 100 ft long, and measurements are performed, in succession, with the cable short-circuited and open-circuited at the far end. The following data are obtained: *Far end of cable short-circuited:* Distance between successive minima on second line = 5.0 cm; VSWR = 4.3; distance from the cable input to first voltage minimum on second line = 1.25 cm. *Far end of cable open-circuited:* VSWR = 4.3; distance from the cable input to first voltage minimum on second line = 3.75 cm. (a) What is the characteristic impedance of the cable? (b) What is the attenuation constant (per 100 ft) of the cable?

Problem 5.6. For a *slightly* dissipative line with an unknown load, the VSWR measured around the voltage minimum nearest to the load is 3. Ten wave lengths away, the VSWR is 2. (a) What is the approximate value of the attenuation constant per wave length of the line? (b) If the minimum for the first measurement is 0.2λ from the load, where would a short-circuited stub be placed in order to match the load? (c) Does the fact that the line is slightly dissipative make the matching stub position more or less critical, in terms of degree of mismatch at the input to the line? Why?

Problem 5.7. (a) Extend the Smith chart to fill up the circle $|\bar{\Gamma}| \leq 1 + \sqrt{2}$. (b) Find the loci on the Smith chart corresponding to the condition $\measuredangle Z_n \equiv \vartheta =$ constant, for various values of ϑ in the range $-3\pi/4 \leq \vartheta \leq 3\pi/4$, which are the only allowed values for *passive* loads on a dissipative line. (c) Show that for a given value of ϑ, in the range $3\pi/4 \geq |\vartheta| > \pi/2$, the largest value of $|\bar{\Gamma}|$ always occurs when $\measuredangle\bar{\Gamma} = \pi/2$ and $|Z_n| = 1$. Show also that when $|\vartheta| < \pi/2$, the *smallest* value of $|\bar{\Gamma}|$ occurs when $\measuredangle\bar{\Gamma} = \pi/2$. (d) Considering the limitation on $\measuredangle Z_n$ for passive loading, determine from the above parts the largest value of $|\bar{\Gamma}|$ for each given value of $\measuredangle\bar{\Gamma}$. Check your result by using Eq. 5.17. (e) In the permitted region of the $\bar{\Gamma}$-plane found in (d), plot the loci corresponding to $|Z_n|$ constant.

Problem 5.8. A dissipative transmission line is terminated in an arbitrary passive impedance. (a) What is the minimum total attenuation of the line, in decibels, required to insure that in the worst possible case the reflection coefficient at the input has a magnitude of 0.01 or less? (b) For a line which just meets the minimum conditions specified in (a), what is the largest percentage variation of the magnitude of the input impedance, as the load impedance is varied over all passive values in the worst case? (See Prob. 5.7.)

Problem 5.9. An "*R-C* cable" is a very dissipative line for which the approximation $G \equiv L \equiv 0$ is justified at the frequency under consideration. Suppose the frequency is such that a certain piece of *R-C* cable is $\lambda/8$ long (i.e., $\beta l = \pi/4$), and $Z_0 = (1 - j)/\sqrt{2}$. (a) Find the passive complex load impedance Z_R for which $|\bar{\Gamma}_R|$ is a maximum. (b) Find the complex value of $\bar{\Gamma}_R$ under conditions (a). (c) Find the complex input impedance of the cable under conditions (a). (d) Approximately to scale, draw a phasor diagram of the incident, reflected, and total voltage and current at the load. (e) With the *same* reference phasor as in (d), draw the corresponding phasor diagram at the input end of the line. (f) From the diagram (e), obtain the phase angle between the input voltage and current exactly. Check with (c).

Problem 5.10. A lossless line is connected to a 1-m slightly dissipative line, as shown in Fig. 5.11. The standing-wave ratio on the lossless line is found to be 5/3. What is the attenuation constant of the dissipative line?

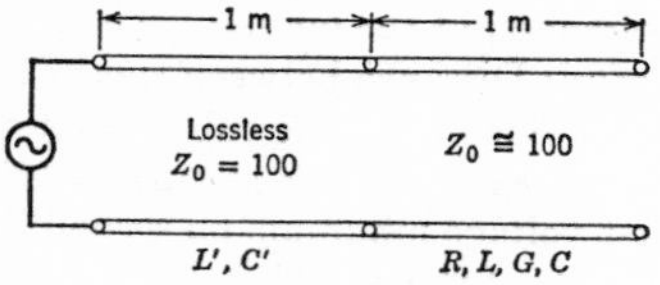

Fig. 5.11. Problem 5.10.

Problem 5.11. A *distortionless* line of characteristic impedance Z_0 and transit time T, open-circuited at the far end $z = l$, is driven by a current impulse at the input $z = 0$. The attenuation constant is α. (a) Find the input and output voltages as functions of time. (b) Repeat (a) for a current-step drive. Check your result by the superposition integral. (c) After the current source in (b) has been on for a long time, it is suddenly removed at $t' = 0$. Find and sketch the terminal voltages as functions of time t', and the voltage as a function of position when $t' = T/2$, $3T/2$. (d) If the system was unenergized for $t < 0$, and the current drive at $z = 0$ is of the form $u_{-1}(t) \sin \omega_0 t$, with $\omega_0 = \pi/2T$, find $V(0, t)$ and $V(l, t)$. Check your results by using those of (a) and the superposition integral. Discuss the solution for large and small values of α.

Problem 5.12. In Fig. 5.12*a*, the attenuation constant is $\alpha = 10^{-2}$/m for both lines; the velocity of propagation is $v = 10^8$ m/sec for both lines. $Z_0C_L = 10^{-8}$ sec. The form of $e_s(t)$ is given in Fig. 5.12*b*. (a) Plot the voltage V_R as a function of time. (b) Plot the line voltages as functions of position for $t = 0.5$ μsec, 1.5 μsec, 2 μsec.

Problem 5.13. A dissipative line of length l, driven by a voltage unit step, is open at the far end. The parameters R, L, G, and C, are assumed to be independent of frequency. (a) By methods of complex Fourier analysis (or Laplace transform), set up an integral expression for the transient voltage at the open end. (b) Evaluate the result (a) as $t \to \infty$. Check the answer against the steady-state response for some simple cases. (c) Show that no signal appears at the end of the line until $t = l\sqrt{LC}$, and that there is a sudden rise in voltage at this time. (The "initial

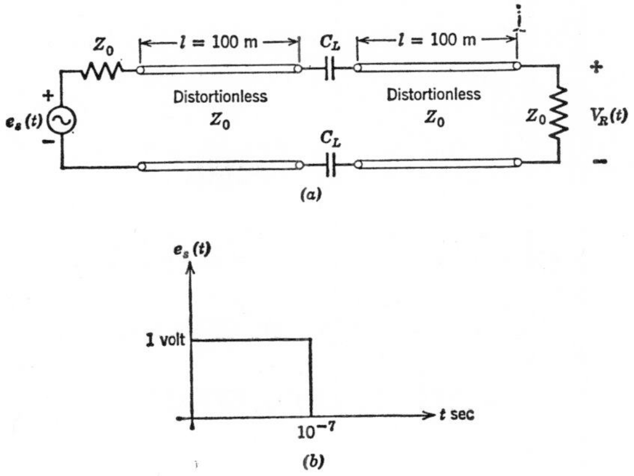

Fig. 5.12. Problem 5.12.

value theorem" from Laplace transform theory may be applied to the output transform multiplied by $e^{s\tau}$, considering values of τ in the range $0 \leq \tau \leq l\sqrt{LC}$. Alternatively, the result may be obtained directly from contour integration of the inverse transform integral.) (d) Repeat (c) if the parameters are functions of frequency such that as $\omega \to \infty$: $L(\omega) \to L_0$, $C(\omega) \to C_0$, $\omega L \gg R$, $\omega C \gg G$. Interpret these results physically.

Problem 5.14. (a) Show that if $\mathrm{f}_c(t)$ is any *complex* function of time, then $f(t) = \mathrm{Re}\,[\mathrm{f}_c(t)e^{j\omega_0 t}]$ represents a general amplitude- and phase-modulated wave with carrier frequency ω_0. Discuss the relationship of this fact with the interpretation of Eqs. 5.30 and 5.32. (b) Carry out the entire discussion of group velocity by starting out with $V_1(t) = \int_{-\infty}^{\infty} \mathrm{V}_1(\omega)e^{j\omega t}\,d\omega$ in place of Eq. 5.24. Be sure to make clear and prove any essential properties of $\mathrm{V}_1(\omega)$, $\alpha(\omega)$, and $\beta(\omega)$ for $\omega < 0$ which you are required to use in the development. See also Probs. 1.7 and 2.12. (c) If $f(t)$ in (a) and $V_1(t)$ in (b) actually represent the *same* time function, discuss carefully the relationship between the Fourier transform $\mathrm{F}_c(\omega)$ of $\mathrm{f}_c(t)$, and the quantity $\mathrm{V}_1(\omega)$. In particular, consider the case in which the bandwidth of $\mathrm{F}_c(\omega)$ is greater than ω_0.

CHAPTER SIX

Natural Oscillations, Standing Waves, and Resonance

There are three important characteristics of an extremely selective linear system. Two of these are commonly emphasized in lumped-network theory. One is the occurrence of very slightly damped complex natural frequencies, representing sustained free oscillations which die out slowly with time. Another is the rapid and radical change with frequency of the steady-state input impedance of the system at some specified terminal pair. These two features generally merge in the functional dependence upon complex frequency of the input impedance or admittance of the system at the selected terminal pair.

The third feature, sometimes overlooked in lumped networks, is of great importance in understanding distributed resonant systems. This feature is the peculiarly characteristic *space pattern* of the fields (or currents and voltages) which accompanies the sharply selective behavior of the system, either at a complex natural frequency or in the neighborhood of a resonance. The violin string that may vibrate at various frequencies, associated directly with the number of nodes occurring along it, illustrates this point. It also shows that, when we study the space patterns of selective behavior in distributed systems, we shall be concerned primarily with standing waves.

Since even small losses become important in sharply selective behavior, we may, on one hand, think of our topic as an examination of standing waves in the presence of dissipation. This description makes clear the connection between the present chapter and the preceding one. On the other hand, the complex natural frequencies and their associated space patterns also play an important part in the transient behavior of the system. Therefore we shall also find it necessary on occasion to refer back to Chapter 4.

6.1 Free Oscillations and Natural Frequencies

A linear, time-invariant system has a number of complex natural frequencies at which it (or sometimes a part of it) may oscillate freely. Such *free oscillations* are characterized by the time dependence e^{st}, and may have been excited by the establishment of appropriate initial conditions (energy storage) in the system at a time in the past. Since no independent sources are present during a free oscillation, no particular entry points or terminals need be singled out for special attention.

The complex natural frequencies are often determined from the search for nontrivial solutions to the homogeneous (source-free) differential equations of the system. These frequencies are the various complex exponents s_i from which may be constructed the corresponding exponential functions $A_i e^{s_i t}$ comprising the general *complementary solution* to the system differential equations.

A finite lumped network has a differential equation of finite order for any current or voltage in it. The corresponding characteristic equation (a polynomial) has a finite number of complex roots, which are the natural frequencies s_i of the network. The number of these (some of which may have equal values) corresponds to the number of arbitrary constants A_i in the complementary solution. If the system model is to be meaningful physically, this number must also represent the number of *independent* initial conditions specifiable in the system, which, in turn, must be the same as the number of *independent* energy-storage points in it.[1] Only with these equalities can we be sure of a unique response of the system to every permissible initial state. We say that a finite lumped network has a finite number of *degrees of freedom.*

The voltage or current in a distributed circuit, like the transmission line, is described by a partial-differential equation involving time and space. From one point of view, such a circuit has an infinite number of (differentially small) independent circuit elements in which energy may be stored initially. From another point of view, one may specify

[1] Each independent pure inductance cut set or pure capacitance tie set reduces by one the number of independent initial energy storages in the system. Each independent pure inductance tie set or pure capacitance cut set also effectively reduces by one the number of interesting degrees of freedom by leading to a trivial isolated "d-c" oscillation.

any initial space distributions of voltage and current on the line, and there is an infinite number of points involved in the specification of any one such distribution. With either view, we might well expect to find an infinite number of complex natural frequencies for any linear, time-invariant system that contains a distributed circuit like a transmission line. The presence of distributed elements gives such a system an infinite number of degrees of freedom.

Whether lumped or distributed, an *isolated* lossless network has pure imaginary complex natural frequencies, $s_i = j\omega_i$. The natural, or free, oscillations are undamped sinusoids; indeed, once started, they are identical with a particular steady-state solution. No energy leaves or enters the system, so the total energy in it remains the same at all times during a free oscillation. Equations 1.5 or 1.38 show that there is a periodic interchange between the electric and magnetic forms of the energy, and Eqs. 1.12 or 1.59 show that the time-average values of the two forms are equal.

If a lumped or distributed system of the kind we are discussing is *passive*, it can dissipate, but not generate, power. If it is dissipative, the energy stored initially in it must be transformed with time into heat. Any free oscillation taking place in the system must die out with time, and the complex natural frequencies s_i *must have negative real parts*. Whereas the natural frequencies of a lossless system lie on the imaginary axis of the s-plane, those of a dissipative system lie in the left half-plane.

6.1.1 An Illustrative Case

To illustrate the ideas described above, along with some more detailed aspects of the situation, let us discuss the behavior of the transmission-line system shown in Fig. 6.1. We look for solutions to the differential equations 2.34*a* and 2.34*b* in which the desired time

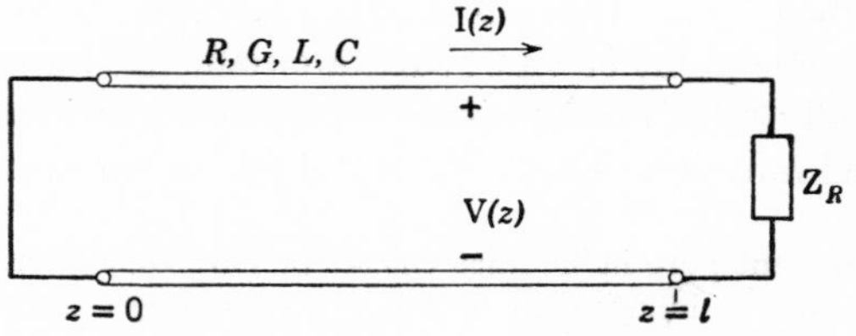

Fig. 6.1. A structure with an infinite number of natural frequencies.

dependence is e^{st}. In other words, we are asking that the current and voltage everywhere on the line be of the form

$$\text{(a)} \qquad V(z, t) = \text{Re}\,[\mathrm{V}(z)e^{st}]$$
$$\text{(b)} \qquad I(z, t) = \text{Re}\,[\mathrm{I}(z)e^{st}] \tag{6.1}$$

Since this form is just like the steady state, except that s replaces $j\omega$, the solution of the transmission-line equations has the same property:

$$\text{(a)} \qquad \mathrm{V}(z) = \mathrm{V}_+e^{-\bar{\gamma}z} + \mathrm{V}_-e^{+\bar{\gamma}z}$$
$$\text{(b)} \qquad \mathrm{I}(z) = \mathrm{Y}_0(\mathrm{V}_+e^{-\bar{\gamma}z} - \mathrm{V}_-e^{+\bar{\gamma}z}) \tag{6.2}$$

where

$$\bar{\gamma} = \sqrt{(R + sL)(G + sC)} \tag{6.3a}$$

$$\mathrm{Y}_0 = \frac{\bar{\gamma}}{R + sL} = \sqrt{\frac{G + sC}{R + sL}} \tag{6.3b}$$

The choice of the square root in Eq. 6.3a (and automatically in 6.3b) is, however, a little more complicated than it has been in the past, because s may now have any complex value. The correct establishment of the root sign depends upon two factors: first, our agreement that $e^{-\bar{\gamma}z}$ shall represent a traveling wave along $+z$ when $s = j\omega$; second, the analytic properties of the root function of s in Eq. 6.3a.

The form of $\bar{\gamma}$ may be rewritten, for convenience in the discussion of the second point above, as follows:

$$\frac{\bar{\gamma}}{+\sqrt{LC}} = \sqrt{(s + \alpha_1)(s + \alpha_2)} \tag{6.3c}$$

where

$$\alpha_1 = \frac{R}{L} \qquad \alpha_2 = \frac{G}{C} \tag{6.3d}$$

The geometry of the s-plane pertinent to Eq. 6.3c appears in Fig. 6.2a, where the various factors are considered in their polar forms

$$\left.\begin{aligned} s &= \rho_0 e^{j\varphi_0} \\ s + \alpha_1 &= \rho_1 e^{j\varphi_1} \\ s + \alpha_2 &= \rho_2 e^{j\varphi_2} \end{aligned}\right\} \qquad -\pi < \varphi_i \leq \pi \qquad i = 0, 1, 2 \tag{6.3e}$$

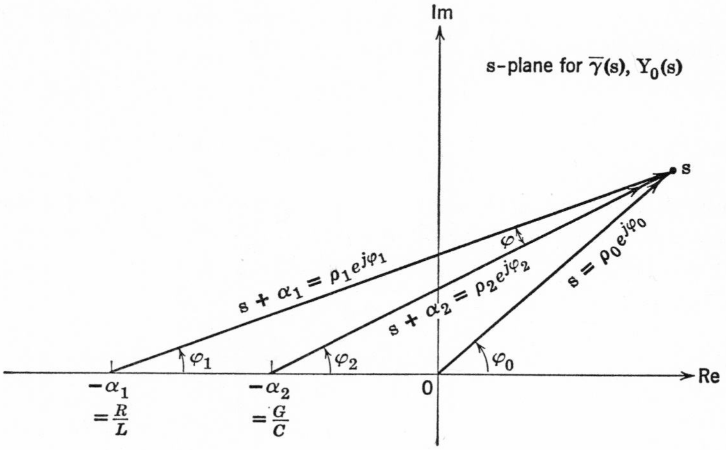

Fig. 6.2a. Geometry for determining the choice of square root in $\bar{\gamma}/+\sqrt{LC} = \sqrt{(s + \alpha_1)(s + \alpha_2)}$ and

$$Y_0/+\sqrt{C/L} = \bar{\gamma}/[+\sqrt{LC}(s + \alpha_1)] = \sqrt{(s + \alpha_2)/(s + \alpha_1)}.$$

Accordingly, we may write for the two choices of the root in Eq. 6.3*c*

$$\left.\begin{aligned} \frac{\bar{\gamma}^{(1)}}{+\sqrt{LC}} &= +\sqrt{\rho_1\rho_2}\, e^{j[(\varphi_1+\varphi_2)/2]} \\ \frac{\bar{\gamma}^{(2)}}{+\sqrt{LC}} &= -\sqrt{\rho_1\rho_2}\, e^{j[(\varphi_1+\varphi_2)/2]} \end{aligned}\right\} \qquad -\pi < \varphi_i \leq \pi \qquad i = 1, 2 \tag{6.3f}$$

Now it is quite clear from the geometry of Fig. 6.2*a* that if $0 \leq \varphi_0 \leq \pi$, then $0 \leq \varphi_1 \leq \pi$ and $0 \leq \varphi_2 \leq \pi$. Correspondingly

$$0 \leq (\varphi_1 + \varphi_2)/2 \leq \pi$$

Similarly, if $-\pi < \varphi_0 < 0$, we have $-\pi < (\varphi_1 + \varphi_2)/2 < 0$. It follows that the choice of the positive (+) root $\bar{\gamma}^{(1)}$ in Eq. 6.3*f* selects that value of $\bar{\gamma}$ having an imaginary part of the *same* sign as that of the imaginary part of s, while choice of $\bar{\gamma}^{(2)}$ leads to Im $(\bar{\gamma}^{(2)})$ and Im (s) having *opposite* algebraic signs.

If we write $s = \Omega + j\omega$, and $\bar{\gamma}(s) = \alpha(\Omega, \omega) + j\beta(\Omega, \omega)$, the time function corresponding to $e^{-\bar{\gamma}z}$ becomes

$$\text{Re}\,(e^{st-\bar{\gamma}z}) = e^{\Omega t-\alpha z}\cos(\omega t - \beta z)$$

which will correspond in phase delay to a traveling wave along $+z$ when $\Omega = 0$ (s $= j\omega$) only if ω and β have the *same* algebraic sign. This requires that the choice $\bar{\gamma}^{(1)}$ be made for the root, rather than $\bar{\gamma}^{(2)}$. As we can also see from Fig. 6.2*a*, $\alpha^{(1)} \geq 0$ for $\Omega = 0$. When s $= j\omega$, the attenuation produced by $\bar{\gamma}^{(1)}$ is along $+z$, consistent with the phase delay. Thus the expression for $\bar{\gamma}^{(1)}$ in Eq. 6.3*f* describes in detail the sense in which $\bar{\gamma}$ is to be computed from Eq. 6.3*a* for any complex value of s.

The corresponding calculation of Y_0 from Eq. 6.3*b* follows directly from the choice of $\bar{\gamma}^{(1)}$:

$$\frac{Y_0}{+\sqrt{C/L}} = \frac{(\bar{\gamma}^{(1)}/+\sqrt{LC})}{s + R/L} = \frac{+\sqrt{\rho_1\rho_2}\, e^{j[(\varphi_1+\varphi_2)/2]}}{\rho_1 e^{j\varphi_1}}$$

$$= +\sqrt{\frac{\rho_2}{\rho_1}}\, e^{j[(\varphi_2-\varphi_1)/2]} \qquad -\pi < \varphi_i \leq \pi, \qquad i = 1, 2$$

Observing that $\varphi_2 - \varphi_1 = \varphi$ in Fig. 6.2*a*, the fact that evidently $-\pi < \varphi \leq \pi$ for all s leads us to conclude that $-\pi/2 < (\varphi_2 - \varphi_1)/2 \leq \pi/2$ when $\alpha_2 < \alpha_1$ (as in Fig. 6.2*a*), or $-\pi/2 \leq (\varphi_2 - \varphi_1)/2 < \pi/2$ when $\alpha_1 < \alpha_2$. In either case, Re $Y_0 \geq 0$ for all values of s.

To Eqs. 6.2 we must now apply the boundary conditions at the ends of the line. First, at $z = 0$, the voltage must be zero for all time. This requires in Eq. 6.2*a* that $V_+ = -V_-$, and, therefore,

$$\text{(a)} \qquad V(z) = -2V_+ \sinh \bar{\gamma}z$$
$$\text{(b)} \qquad I(z) = 2V_+Y_0 \cosh \bar{\gamma}z \qquad (6.4)$$

But next, at $z = l$, we require

$$V(l) = Z_R I(l) \qquad (6.5)$$

which, from Eq. 6.4, yields either

$$Z_R = -Z_0 \tanh \bar{\gamma}l \qquad (6.6)$$

or

$$Y_R = -Y_0 \coth \bar{\gamma}l \qquad (6.7)$$

if we wish any solution besides the trivial one $V_+ = 0$. Equations 6.6 or 6.7 must be obeyed if the source-free line system in Fig. 6.1 is to oscillate at some complex frequency s_i, with other than identically zero voltage and current amplitudes. Of course, the amplitudes do not appear in the results because the level of excitation in the absence of sources may be chosen arbitrarily, perhaps by some initial conditions

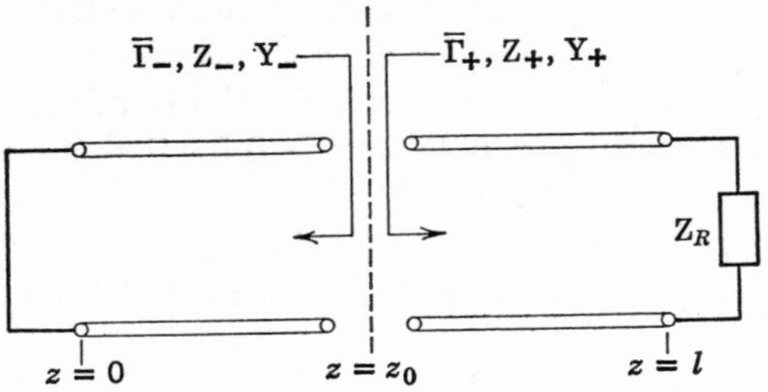

Fig. 6.2b. For the alternate interpretation of the condition for natural oscillation.

in the past. This fact suggests that these results may be obtained in a different fashion.

Because the amplitude of oscillation is not fixed by a linear source-free system, it must be possible to determine directly the oscillation condition in terms of ratios, like impedance or reflection coefficient, which are insensitive to amplitude. We can do this, in fact, by considering conditions at an arbitrary point z_0 during a natural oscillation at complex frequency s_i. The ratio of voltage to current at point z_0, at the frequency s_i, is, on one hand, fixed by the part of the system to the *right* ($z > z_0$ in Fig. 6.2*b*). Specifically it is fixed by the impedance $Z_+(z_0)$ seen "looking to the right" in Fig. 6.2*b*, which depends upon the corresponding section of line and the load Z_R:

$$\frac{V(z_0)}{I(z_0)} = Z_+(z_0) \tag{6.8}$$

On the other hand, this ratio is also fixed by the section of line and load (short circuit) which lies to the *left* of z_0. The effect of these is contained in the impedance $Z_-(z_0)$ seen "looking to the left" in Fig. 6.2*b*; but because of the different sign convention for current employed in defining $Z_-(z_0)$ in that figure and $I(z)$ for the line in Fig. 6.1, we must put

$$\frac{V(z_0)}{-I(z_0)} = Z_-(z_0) \tag{6.9}$$

Accordingly, if the natural oscillation is to take place with a nonzero value of $I(z_0)$, we require

$$Z_-(z_0) = -Z_+(z_0) \qquad \text{for } I(z_0) \neq 0 \tag{6.10a}$$

We must not forget that a satisfactory nontrivial oscillation of the system might also take place with $I(z_0) = 0$ and $V(z_0) \neq 0$. That is, such an oscillation may just happen to have a *current node* at the particular point z_0 we chose for investigation. Then the condition of oscillation with a nonzero voltage at this point is simply

$$Z_-(z_0) = Z_+(z_0) = \infty \qquad \text{for } I(z_0) = 0 \tag{6.10b}$$

The other possible case, when point z_0 happens to be a voltage node $[V(z_0) = 0]$ with nonzero current, is actually covered by Eq. 6.10*a* when $Z_-(z_0) = -Z_+(z_0) = 0$.

The conditions of Eqs. 6.10 may be phrased in terms of admittances, of course, in which case we would find

$$\frac{I(z_0)}{V(z_0)} = Y_+(z_0) = -Y_-(z_0) \qquad \text{for } V(z_0) \neq 0 \tag{6.11a}$$

or

$$Y_+(z_0) = Y_-(z_0) = \infty \qquad \text{for } V(z_0) = 0 \tag{6.11b}$$

Inasmuch as the voltage node condition (Eq. 6.11*b*) is actually covered by Eq. 6.10*a*, and the current node condition (Eq. 6.10*b*) is contained in Eq. 6.11*a*, we may summarize the criterion for natural oscillation in a single statement:

$$\begin{aligned} &\text{either} \qquad Z_+(z_0) = -Z_-(z_0) < \infty \\ &\text{or} \qquad Y_+(z_0) = -Y_-(z_0) < \infty \end{aligned} \tag{6.12}$$

It is noteworthy that Eqs. 6.6 and 6.7 are just Eq. 6.12 written for the system of Fig. 6.1 at the point $z_0 = l$.

One could (and should, as an exercise) verify directly that application of Eq. 6.12 at an *arbitrary* point z_0 of the line of Fig. 6.1 leads again to Eq. 6.6 or 6.7. It is simpler, and even more instructive, however, to relate Eq. 6.12 to a condition upon the reflection coefficients "looking both ways" at a point z_0, and apply that condition directly instead. Thus, with regard to Fig. 6.2*b*, the reflection coefficient "looking to the right" is

$$\bar{\Gamma}_+(z_0) \equiv \frac{V_- e^{\bar{\gamma} z_0}}{V_+ e^{-\bar{\gamma} z_0}} = \frac{Z_+(z_0) - Z_0}{Z_+(z_0) + Z_0} \tag{6.13}$$

and "looking to the left" is

$$\bar{\Gamma}_-(z_0) \equiv \frac{V_+ e^{-\bar{\gamma} z_0}}{V_- e^{\bar{\gamma} z_0}} = \frac{Z_-(z_0) - Z_0}{Z_-(z_0) + Z_0} \tag{6.14}$$

because "incident" and "reflected" waves must always be described with respect to the load at the far end of the line section being considered. Therefore, either directly from the voltage ratios in Eqs. 6.13 and 6.14 or from the impedance expressions there, combined with Eq. 6.12, it is clear that

$$\bar{\Gamma}_-(z_0) = \frac{1}{\bar{\Gamma}_+(z_0)} \tag{6.15}$$

Application of Eq. 6.15 with the specific end conditions of the present line involves finding first that

$$\bar{\Gamma}_+(z_0) = \bar{\Gamma}_R e^{-2\bar{\gamma}(l-z_0)} \tag{6.16a}$$

and then that

$$\bar{\Gamma}_-(z_0) = (-1)e^{-2\bar{\gamma}z_0} \tag{6.16b}$$

Therefore Eq. 6.15 becomes

$$\bar{\Gamma}_R e^{-2\bar{\gamma}l} = -1 \tag{6.17a}$$

which, incidentally, happens to be just what we would have found by writing Eq. 6.15 directly at the point $z_0 = 0$! But since

$$\bar{\Gamma}_R = \frac{Z_R - Z_0}{Z_R + Z_0} \tag{6.17b}$$

Eq. 6.17*a* becomes

$$Z_R = -Z_0 \tanh \bar{\gamma}l$$

as before.

We see from the preceding discussion that Eqs. 6.12 and 6.15 furnish alternative ways of arriving quickly at the essential equation which determines the complex natural frequencies of a transmission-line system. It takes some experience and practice to achieve facility in choosing the easiest formulation, at the most appropriate point, in a given problem.

In connection with our present results, it is now important to emphasize that the conditions of Eqs. 6.6 or 6.7, taken with Eqs. 6.3*a* and *b* and the presumably known function $Z_R(s)$, are transcendental equations in the complex unknown s. The transcendental functions tanh and coth have periodic properties that suggest an infinite number of possible solutions s_i. It is particularly significant that for each such allowed value s_i, there is a definite corresponding value $\bar{\gamma}_i$ for $\bar{\gamma}$, and Z_{0i} for Z_0, determined by Eqs. 6.3*a* and *b*. Accordingly, *with each complex natural frequency* s_i, *there is associated a definite space pattern of current and voltage on the line*, given in this case by Eqs. 6.4. Although the shape of the distributions of voltage and current is fixed at any natural frequency, we have pointed out that the amplitude remains arbitrary; it can be set at any desired value by the manner in which the oscillation is initiated.

The features we have emphasized about the solution for free oscillations of the system in Fig. 6.1 have their counterparts in the natural oscillations of finite lumped networks. Most of them are relatively obvious, except perhaps the correlation between space variation and natural frequency. Even in this respect, however, a little thought shows that the situation is not very different. In the network, the *ratios* of complex currents and voltages at all branches are fixed at a natural frequency, although there remains one undetermined constant to set the level of excitation of the whole network. One may say that the shape of the distribution of current and voltage, regarded as a function of branch number, is fixed, but the over-all amplitude is undetermined. Clearly, the lumped-network analog of our present space-position concept is the idea of branch number. Both notions serve to identify a particular "element" in the system, with relation to the other elements, once the topology of the interconnection is specified.

An examination of particular cases of Fig. 6.1 will be helpful at this point in the discussion.

6.1.1.1 A LOSSLESS SYSTEM. If the line of Fig. 6.1 is lossless, and the impedance Z_R is a short circuit, the problem becomes that illustrated in Fig. 6.3*a*. The natural frequencies are determined by the relation of Eq. 6.6, e.g., in the form

$$Z_0 \tanh \bar{\gamma} l = 0 \qquad Z_0 = \sqrt{L/C} \neq 0 \tag{6.18}$$

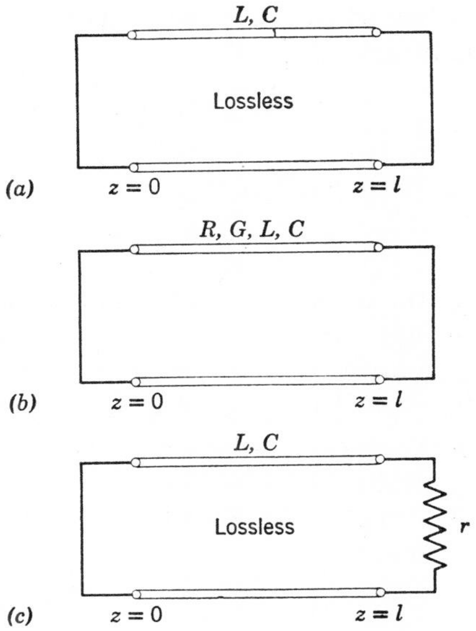

Fig. 6.3. Some simple distributed structures, of the type shown in Fig. 6.1, for which natural oscillations may be studied exactly.

In terms of the exponentials defining the tanh, Eq. 6.18 becomes

$$e^{2\bar{\gamma}l} = 1 \tag{6.19}$$

which leads to

$$2\bar{\gamma}l = \pm j2m\pi \qquad m = 0, 1, 2, \cdots \tag{6.20}$$

The value of $\bar{\gamma}$ is pure imaginary, which we can write as $j\beta$, and find

$$\beta l = \pm m\pi \qquad m = 0, 1, 2, \cdots \tag{6.21}$$

Accordingly, the space distributions of voltage and current are pure sinusoids, as given by Eqs. 6.4 and 6.21:

$$\text{(a)} \qquad V(z) = \mp 2jV_{+} \sin\left(\frac{m\pi z}{l}\right)$$

$$\text{(b)} \qquad I(z) = 2V_{+}Y_0 \cos\left(\frac{m\pi z}{l}\right) \tag{6.22}$$

These distributions are complete standing waves, in which the line length l is an integer multiple of half-periods of the space variation.

Of course, the natural frequencies should be imaginary in the lossless case. This is borne out by the fact that, from Eq. 6.3*a*, the definition of $\bar{\gamma}$ is

$$\bar{\gamma} = +\,s\sqrt{LC} \tag{6.23}$$

It follows from the above, with the allowed values of $\bar{\gamma}$ given by Eqs. 6.20 and 6.21, that

$$s_m = \pm\, j\omega_m = \pm\, j\frac{m\pi}{l\sqrt{LC}} = \pm\, j\frac{m\pi}{T} \qquad m = 0, 1, 2, \cdots \tag{6.24}$$

where we have written

$$T \equiv l\sqrt{LC} = l/v \tag{6.25}$$

for the one-way transit time of waves on the line. The disposition of the allowed values of s_m from Eq. 6.24 may be shown as a pattern of dots in the s-plane, in the manner of Fig. 6.4*a*. The possibility of a "d-c" oscillation, $s = 0$, should not be overlooked. The circulation of direct current, at zero voltage, makes sense here.

The pure imaginary values found for the complex natural frequencies make it possible in the present case to speak of a wave length

$$\lambda_m = \frac{2\pi v}{\omega_m} \tag{6.26}$$

on the basis of which the condition of natural oscillation becomes the "multiple half-wave-length condition"

$$l = m\left(\frac{\lambda_m}{2}\right) \qquad m = 0, 1, 2, \cdots \tag{6.27}$$

Actually, for a simple situation such as that shown in Fig. 6.3a, it is much easier to determine the natural frequencies ω_m from the recogni-

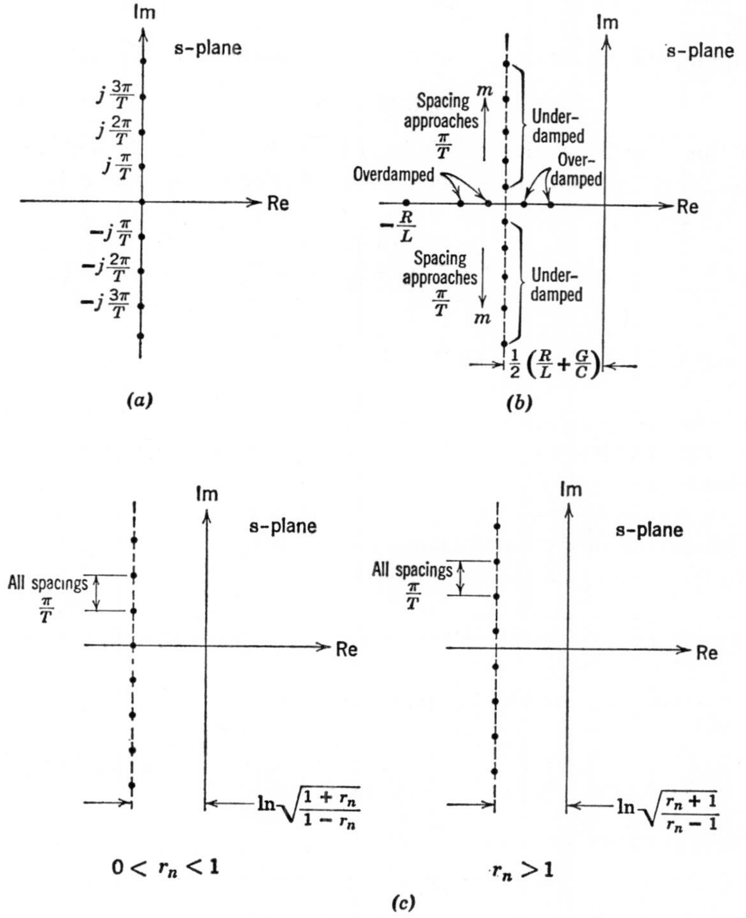

Fig. 6.4. Location of complex natural frequencies for the structures of Fig. 6.3. Parts (a), (b), (c) above apply to (a), (b), (c) respectively in Fig. 6.3.

tion that the network can only oscillate freely in its various "half-wave modes" of space distribution than to proceed in the reverse way as we have done here. There is a good deal to recommend this "mode" technique, which takes advantage of the space patterns of natural oscillation, although in complicated cases it may not be so easy to apply.

6.1.1.2 DISTRIBUTED LOSSES. The effect of adding losses to the system of Fig. 6.3*a* may be studied in a simple case by considering that the line has some (constant) series resistance and shunt conductance per unit length. The end loads remain short circuits, as shown in Fig. 6.3*b*.

Again, the condition of free oscillation is given by Eq. 6.6

$$Z_0 \tanh \bar{\gamma} l = 0 \tag{6.28}$$

but this time the characteristic impedance Z_0 of Eq. 6.3*b* is a function of s. This function can be zero. One solution of Eq. 6.28 is therefore

$$Z_0 = \sqrt{\frac{R + sL}{G + sC}} = 0 \tag{6.29}$$

or

$$s_0 = -\left(\frac{R}{L}\right) \tag{6.30}$$

The remaining solutions of Eq. 6.28 are those given in Eq. 6.20 or Eq. 6.21, so that the distributions of voltage and current are again complete standing waves in space, as given by Eqs. 6.22. The line length l is as before an integer multiple of half-periods of the space variation; but now the frequencies of oscillation are *not* pure imaginary!

In fact, from Eqs. 6.20 and 6.3*a*, we find

$$\bar{\gamma} = \sqrt{(R + sL)(G + sC)} = \pm j\frac{m\pi}{l} \qquad m = 0, 1, 2, \cdots \tag{6.31}$$

or, squaring and rearranging terms, we have

$$s^2 + \left(\frac{R}{L} + \frac{G}{C}\right)s + \left[\frac{RG}{LC} + \left(\frac{m\pi}{l\sqrt{LC}}\right)^2\right] = 0 \qquad m = 0, 1, 2, \cdots$$

The solutions of the quadratic are

$$s_m = -\frac{1}{2}\left(\frac{R}{L} + \frac{G}{C}\right) \pm \sqrt{\left[\frac{1}{2}\left(\frac{R}{L} - \frac{G}{C}\right)\right]^2 - \left(\frac{m\pi}{T}\right)^2} \qquad m = 1, 2, \cdots \tag{6.32}$$

Observe that the choice $m = 0$ has been omitted from Eq. 6.32. This omission has been desirable because the two roots given by that equation for $m = 0$ are

$$s_{01} = -\left(\frac{R}{L}\right) \qquad s_{02} = -\left(\frac{G}{C}\right)$$

The root s_{01} duplicates s_0 in Eq. 6.30. The root s_{02} is extraneous because Z_0 becomes infinite at that complex frequency; actually the value of $Z_0 \tanh \bar{\gamma} l$ in Eq. 6.28 as $s \to s_{02}$ remains finite and approaches $(R + s_{02}L)l = l(R - (LG)/C)$. The latter is not zero unless the line is "distortionless" ($R/L = G/C$); in that case Eq. 6.32 shows that $s_{01} = s_{02} = s_0$ anyway.

Thus we learn that the system of Fig. 6.2*b* has at least one negative-real complex frequency s_0, for which not only $Z_0 = 0$ but also $\bar{\gamma} = 0$ (by Eq. 6.31). The corresponding space distributions, given by Eqs. 6.31 and 6.22, involve a circulating current *independent of position*, which decays exponentially with time. There is no voltage at all. This solution is evidently the damped counterpart of the "d-c" oscillation we found possible in the lossless case of Fig. 6.3*a*!

The remaining complex frequencies represent damped oscillations, again of the *same space patterns* as those of the lossless system. Depending upon the value of $R/L - G/C$ in Eq. 6.32, some of the lower order oscillations may be overdamped. These arise from small values of m, if at all:

$$m < \frac{T}{2\pi}\left(\frac{R}{L} - \frac{G}{C}\right) \qquad \text{overdamped} \qquad (6.33a)$$

The high-order oscillations are underdamped

$$m > \frac{T}{2\pi}\left(\frac{R}{L} - \frac{G}{C}\right) \qquad \text{underdamped} \qquad (6.33b)$$

and, for large enough m, their frequencies approach

$$s_m \to -\frac{1}{2}\left(\frac{R}{L} + \frac{G}{C}\right) \pm j\frac{m\pi}{T} \qquad \text{for large } m \qquad (6.33c)$$

In general, the disposition of the natural frequencies might appear as shown in Fig. 6.4*b*.

Actually the values of m for underdamped oscillations need not be very large in many practical cases. The critical value, from Eqs. 6.33*a* and 6.33*b*, is

$$m_{\text{crit}} = \frac{T}{2\pi}\left(\frac{R}{L} - \frac{G}{C}\right) = \frac{1}{2\pi}\left(lR\sqrt{\frac{C}{L}} - lG\sqrt{\frac{L}{C}}\right)$$

$$= \frac{1}{2\pi}\,(lR/Z_0' - lGZ_0') \qquad (6.34)$$

where Z_0' is the characteristic impedance which the line would have in the absence of the losses. Evidently, if the "total" resistance Rl of the line is much less than Z_0' *and* the total conductance Gl is much less than Y_0', the allowed integer values of m ($\neq 0$) always exceed the bound (Eq. 6.34). Under such conditions, the damping term in Eq. 6.32 is also much smaller than the oscillatory (imaginary) term, and the free oscillations are all very nearly undamped (except for the root s_0, of course). Viewed in the time domain, the apparent oscillation frequency is almost the same as it would be for the lossless case, although the oscillation amplitude decays slowly with time.

We must hasten to add, however, that small losses are not the only condition under which the natural oscillations are all (but one) underdamped. If the line is distortionless, Eq. 6.33*c* becomes exact; the imaginary parts of all the natural frequencies are exactly what they would be without losses at all, even if the value of the damping is extremely large! The point is that the *difference* between R/L and G/C appears under the square root in Eq. 6.32, whereas the *sum* appears in the damping term. This distinction is not important when either R or G (but not both) is negligibly small, so that only one kind of loss dominates; but in many cases, both terms are significant enough to reduce noticeably the effect of loss on the imaginary part of the complex natural frequency.

Because of the importance of slightly damped systems, as resonators or tuned elements, it is significant that the effect of *small* losses upon the imaginary part of the complex natural frequency is much less than its effect on the damping. More precisely, starting from a lossless case, the introduction of small losses shifts the natural frequency further into the left half-plane than it does along the imaginary axis. With reference to Figs. 6.3*a*, 6.3*b*, 6.4*a*, and 6.4*b* let us denote by ω_{m0} the mth radian frequency of undamped natural oscillation in the lossless case, by Ω' the damping factor in the presence of loss, and by Ω'' the correction term under the root in Eq. 6.32:

(a) $$\omega_{m0} \equiv \frac{m\pi}{T}$$

(b) $$\Omega' \equiv \frac{1}{2}\left(\frac{R}{L} + \frac{G}{C}\right) \tag{6.35}$$

(c) $$\Omega'' \equiv \frac{1}{2}\left(\frac{R}{L} - \frac{G}{C}\right)$$

Then Eq. 6.32 may be written

$$s_m = -\Omega' \pm j\sqrt{{\omega_{m0}}^2 - \Omega''^2} \tag{6.36}$$

When the damping is small, we mean

$$\Omega' \ll \sqrt{{\omega_{m0}}^2 - \Omega''^2} \tag{6.37a}$$

which implies also

$$\Omega' \ll \omega_{m0} \tag{6.37b}$$

But it follows from Eqs. 6.35*b* and 6.35*c* that Eq. 6.37*b* requires

$$\Omega'' \leq \Omega' \ll \omega_{m0} \tag{6.37c}$$

Binomial expansion of the root in Eq. 6.36, taking for example the frequencies in the upper half-plane only, then yields

$$s_m \approx -\Omega' + j\left[\omega_{m0} - \frac{1}{2}\left(\frac{\Omega''}{\omega_{m0}}\right)\Omega''\right] \tag{6.38}$$

Compared with the lossless case, in which $s_{m0} = j\omega_{m0}$, the natural frequencies in the presence of small losses are shifted into the left half-plane by the amount Ω', but down the imaginary axis only by the amount $(\Omega''/2\omega_{m0})\Omega''$. In view of Eqs. 6.37*b* and 6.37*c*, the downward shift is much less than that to the left, even in the worst case when $\Omega'' = \Omega'$. Indeed, the downward shift is less than the shift to the left by at least the ratio of the damping constant Ω' itself to the radian frequency ω_{m0} of oscillation in the lossless case. In a situation of small damping, this ratio would be at least one order of magnitude.

6.1.1.3 Lumped Losses. A rather different manner of introducing loss into the system of Fig. 6.3*a* is illustrated in Fig. 6.3*c*; the line remains nondissipative, but a resistance r replaces the short circuit at one end ($z = l$).

To determine the natural frequencies of this system, it is most convenient to employ the reflection coefficient relation of Eq. 6.15. This is most easily accomplished at the point $z = l$. Thus looking to the left at this point, we have

$$\bar{\Gamma}_{-} = -e^{-2\bar{\gamma}l} \qquad (6.39a)$$

because of the short circuit at $z = 0$. Looking to the right, however,

$$\bar{\Gamma}_{+} = \frac{r - Z_0}{r + Z_0} = \frac{r_n - 1}{r_n + 1} \qquad (6.39b)$$

where

$$r_n \equiv \frac{r}{Z_0} \qquad (6.39c)$$

Note that $Z_0 = \sqrt{L/C}$ here, so that r_n is a positive real number. Therefore $|\bar{\Gamma}_{+}| \leq 1$.

According to Eq. 6.15, with the specific forms of Eqs. 6.39*a* and 6.39*b*, we find

$$e^{2\bar{\gamma}l} = \frac{1 - r_n}{1 + r_n} \qquad (6.40)$$

Expressing $\bar{\gamma}$ in the manner of Eq. 6.23, and employing the notation of Eq. 6.25, we have for $s = \Omega + j\omega$

$$\bar{\gamma}l = sl\sqrt{LC} = \Omega T + j\omega T \qquad (6.41)$$

in the light of which Eq. 6.40 becomes

$$e^{2\Omega T}e^{2j\omega T} = \frac{1 - r_n}{1 + r_n} \qquad (6.42)$$

Since the right-hand side of Eq. 6.42 is always real, there are only two cases to consider: $0 \leq r_n \leq 1$ and $r_n > 1$.

In the first case, the right-hand side of Eq. 6.42 has a real value between 0 and +1, which demands

$$\left.\begin{aligned} &\text{(a)} \qquad \omega_m T = \pm m\pi \qquad m = 0, 1, 2, \cdots \\ &\text{(b)} \qquad \Omega_m T = -\frac{1}{2}\ln\left(\frac{1 + r_n}{1 - r_n}\right) \leq 0 \end{aligned}\right\} \quad 0 \leq r_n \leq 1 \qquad (6.43)$$

The particular value $r_n = 0$ makes Fig. 6.3*c* the same as Fig. 6.3*a*. The result (Eq. 6.43*b*) indicates $\Omega_m = 0$ under this condition and Eq. 6.43*a* checks with Eq. 6.24. Interestingly enough, it happens that the spacing of the various complex frequencies in the direction parallel to the imaginary axes is the same for any $0 < r_n < 1$ as it is for the lossless case $r_n = 0$. When $r_n \neq 0$, all the frequencies are displaced into the left half-plane by *equal* amounts $\Omega_m T$, because Ω_m depends upon r_n but not upon m. According to Eq. 6.43*b*, this leftward displacement is small when r_n is small, and becomes infinite as $r_n \to 1$. When $r_n = 1$ the line is matched at $z = l$ for all frequencies, and no free oscillations are possible with finite s.

Finally, the situation $r_n > 1$ renders negative real the right-hand side of Eq. 6.42. It follows that

$$\left.\begin{aligned} &\text{(a)} \qquad \omega_m T = \pm(2m-1)\frac{\pi}{2} \qquad m = 1, 2, \cdots \\ &\text{(b)} \qquad \Omega_m T = -\frac{1}{2}\ln\left(\frac{r_n+1}{r_n-1}\right) < 0 \end{aligned}\right\} \quad r_n > 1 \qquad (6.44)$$

In the event that $r_n = +\infty$, the line of Fig. 6.3*c* is open-circuited at $z = l$. The system is lossless, and we expect undamped oscillations at frequencies for which the line length l is an odd multiple of a quarter wave length. The lack of damping shows up in Eq. 6.44*b*, where $\Omega_m = 0$ when $r_n = \infty$. Equation 6.44*a* is precisely the odd quarter-wave-length condition mentioned above. Again, the imaginary parts of the natural frequencies are independent of the particular value of $r_n > 1$, being always equal to their values in the lossless case $r_n = +\infty$. The natural frequencies are all displaced to the left by a common amount when $1 < r_n < +\infty$, this amount becoming infinite monotonically as $r_n \to 1$. As a matter of fact, the value of the damping constant Ω_m in Eq. 6.44*b* when r_n has some value $a > 1$ happens to be exactly the same as that given by Eq. 6.43*b* when r_n has the reciprocal value $1/a < 1$.

The general arrangement of the natural frequencies for the two cases of Eqs. 6.43 and 6.44 is illustrated in Fig. 6.4*c* for intermediate values of r_n within the respective ranges.

In the present example, it is still true that a definite value $\bar{\gamma}_m$ of $\bar{\gamma}$ accompanies each natural oscillation of complex frequency s_m. However, as shown by Eq. 6.41 (in the light of Eqs. 6.43 and 6.44), the values $\bar{\gamma}_m$ are *complex*. The space pattern of voltage and current is, therefore, not the simple complete standing wave we found in the previous examples of Figs. 6.3*a* and 6.3*b* (Eqs. 6.20 and 6.31), but rather is comprised of the component (+) and (−) waves (Eq. 6.2), each of which shows both attenuation and phase shift in space. There are no perfect nodes of current or voltage. This behavior is required in order that energy stored on the line may travel to the end $z = l$, the only place at which its dissipation is possible, rather than being trapped between perfect nodal points.

6.2 Forced Oscillations, Poles, and Zeros

We may be interested not only in the free, or natural, oscillations of a lumped or distributed network but also in its response to an excitation applied at some terminal pair or point of entry into the system. Although the excitation may have any time dependence in general, the theory of Fourier transforms permits us to restrict our discussion to the case of exponential time dependence e^{st} without real loss of generality. Then various driving point or transfer impedances $Z_{ij}(s)$ [or admittances $Y_{ij}(s)$] are the important items for consideration, and the designation of corresponding terminal pairs or entry points becomes essential.

The relationships between free and forced oscillations are important, and may be discussed first with reference to the driving-point impedance of any linear, time-invariant, source-free network, as shown in Fig. 6.5a. Supposing that either the terminal voltage or current is a

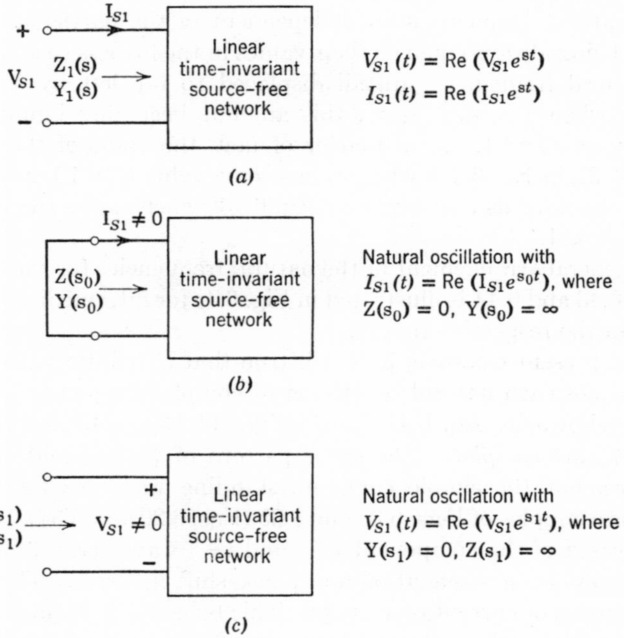

Fig. 6.5. Interpretation in terms of natural frequencies of zeros and poles of impedance or admittance.

specified source, with time dependence e^{st}, we search for that *particular solution* for the other terminal variable that also has the *same* time dependence e^{st}. In other words, leaving quite open for the moment the question of which is the "source," we discuss the relationship between the complex amplitudes V_{S1} and I_{S1} in the expressions

$$\begin{aligned} &\text{(a)} \qquad V_{S1}(t) = \operatorname{Re}(\mathrm{V}_{S1}e^{st}) \\ &\text{(b)} \qquad I_{S1}(t) = \operatorname{Re}(\mathrm{I}_{S1}e^{st}) \end{aligned} \tag{6.45}$$

This relationship is contained as a function of s in the impedance

$$\mathrm{Z}_1(\mathrm{s}) \equiv \frac{\mathrm{V}_{S1}}{\mathrm{I}_{S1}} \tag{6.46a}$$

or the admittance

$$\mathrm{Y}_1(\mathrm{s}) \equiv \frac{\mathrm{I}_{S1}}{\mathrm{V}_{S1}} \tag{6.46b}$$

Regarded as a function of complex frequency s, the impedance $\mathrm{Z}_1(\mathrm{s})$, for example, will be zero at some frequencies and infinite at others. The corresponding complex frequencies are called the zeros and poles respectively of the driving-point impedance. At a zero $\mathrm{s} = \mathrm{s}_0$ of $\mathrm{Z}(\mathrm{s})$, we see from Eq. 6.46*a* that $\mathrm{V}_{S1} = 0$ when $\mathrm{I}_{S1} \neq 0$. Under this condition, one could place a short circuit across the terminals (Fig. 6.5*b*) without disturbing the behavior of the network. Inasmuch as the source has in effect been removed, the behavior of the system is then clearly seen to be that of a *free oscillation* at frequency s_0. In other words, *every zero of the impedance* $\mathrm{Z}(\mathrm{s})$ *at a given pair of network terminals occurs at a complex natural frequency of the system with the terminals in question short-circuited.* Evidently the zeros of $\mathrm{Z}(\mathrm{s})$ are poles of $\mathrm{Y}(\mathrm{s})$, and vice versa, so the above statement remains valid with the words "pole of the admittance $\mathrm{Y}(\mathrm{s})$" substituted for "zero of the impedance $\mathrm{Z}(\mathrm{s})$".

Similarly a zero $\mathrm{s} = \mathrm{s}_1$ of $\mathrm{Y}(\mathrm{s})$ occurs in Eq. 6.46*b* when $\mathrm{I}_{S1} = 0$ and $\mathrm{V}_{S1} \neq 0$. The terminals may now be open-circuited without disturbing the system electrically, and we find that *every zero of the admittance* $\mathrm{Y}(\mathrm{s})$ [*or pole of the impedance* $\mathrm{Z}(\mathrm{s})$] *at a given pair of network terminals occurs at a natural frequency of the system with the terminals in question open-circuited.* This situation is illustrated in Fig. 6.5*c*.

The first suggestion to which we are led by the foregoing remarks is that one may find the zeros and poles of the driving-point impedance [or admittance] at a given pair of terminals by finding instead the natural frequencies of the whole network that arise when the terminals are short-circuited and when they are open-circuited. Alternatively, a second suggestion is that natural frequencies of a given system, as shown in Fig. 6.1*a*, may be found by attaching a pair of terminals to it at a convenient arbitrary point, and computing the impedance or admittance seen at these terminals. Depending upon the manner in which the terminal pair is created, the natural frequencies may be either zeros or poles of the impedance (also poles or zeros of the admittance respectively) seen at the terminals. Referring again to the network in Fig. 6.1*a*, whose natural frequencies might be of interest, the two general ways of creating a temporary terminal pair for determining these frequencies are illustrated in Figs. 6.6*a* and 6.6*b*. In Fig. 6.6*a*, a "soldered" terminal pair is added to Fig. 6.1*a* in such a way that it bridges two nodes or points of the original system. Obviously the

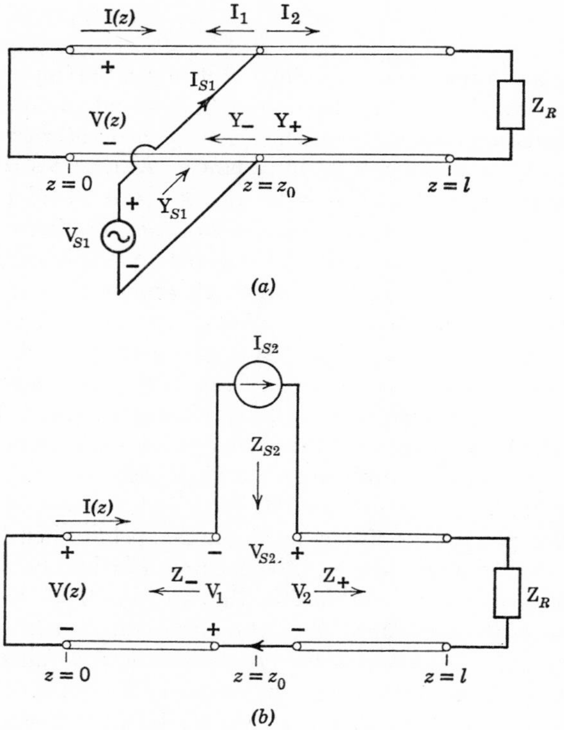

Fig. 6.6. Two ways of attaching a new terminal pair to a given system: (*a*) "Solder" or "bridging" pair, made by attaching leads to two of the nodes of the original system; (*b*) "Scissors" or "cut" pair, made by cutting open a branch of the original system and drawing out the loose ends.

natural frequencies of the original network are those for which the added terminals do not alter the system electrically. Certainly if the terminals are left *open*, they meet this condition. *Therefore, the zeros of the admittance* Y_{S1} *(or the poles of the impedance* $Z_{S1} = Y_{S1}^{-1}$*) seen at a soldered terminal pair are natural frequencies of the system originally defined without the terminals.*

The situation is somewhat different in Fig. 6.6*b*, where a new terminal pair is created with "scissors" by cutting open a branch of the original network and drawing out the loose ends. The added terminal pair does not modify the original system electrically if it is left *short-circuited.* Accordingly, *the zeros of the impedance* Z_{S2} *(or the poles of the admittance* $Y_{S2} = Z_{S2}^{-1}$*) seen at a terminal pair created with scissors are natural frequencies of the system originally defined without the terminals.*

We have said so far that *every* zero of a driving-point impedance corresponds to a natural frequency of oscillation with the terminals short-circuited, and that *every* pole of the driving-point impedance corresponds to a natural frequency of oscillation with the terminals open. Careful reading will show, however, that we have deliberately left unsettled the converse questions of whether every natural frequency of the system with a given pair of terminals short-circuited leads to a zero of the corresponding driving-point impedance, and whether every natural frequency of the system with a given pair of terminals open leads to a pole of the corresponding driving-point impedance. It is now appropriate to deal with these questions.

It must be pointed out first that in most cases all the natural oscillation frequencies with a given pair of terminals short-circuited do lead to zeros of the corresponding driving-point impedance, and all natural frequencies with a given pair of terminals open do lead to poles of the corresponding impedance. But we must hasten to add that there are exceptions to this rule. These exceptions arise *only* when the chosen terminal pair in the system happens to fall at a point where it does not "couple" to one or more of the natural modes of oscillation. We mean by this statement that *the terminals in question happen to be so located that one or more of the system natural frequencies remain unchanged whether the terminal pair in question is open-circuited or short-circuited (or indeed connected to any passive load impedance).* More specifically, with reference to Fig. 6.7, consider a linear system undergoing a natural oscillation at frequency s_ν. Let the current amplitude I_2 in some branch 2–2′ be nonzero, but suppose the voltage amplitude V_1 between some pair of nodes 1–1′ *is* zero at this frequency. Evidently, the natural oscillation at frequency s_ν will not be disturbed by connect-

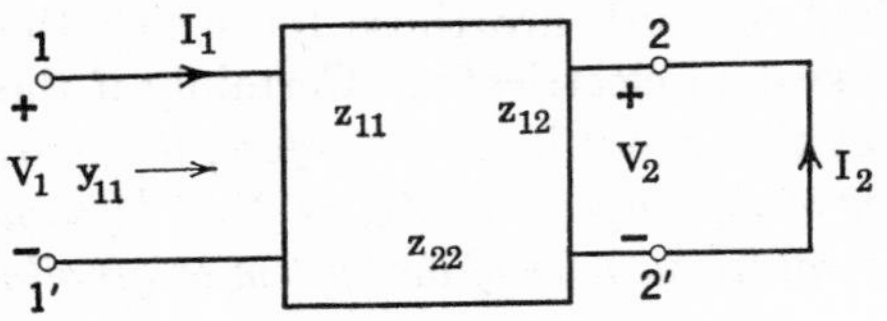

Fig. 6.7. For the discussion of the relationship between free and forced oscillations.

ing any passive impedance to terminals 1–1′, inasmuch as no current will flow because V_1 is zero at this frequency. The question is then whether or not the admittance of the system at terminals 1–1′ must be zero at this frequency.

If we view the system between terminals 1–1′ and 2–2′ as a two-terminal-pair reciprocal network, described in terms of its open-circuit impedance parameters z_{11}, z_{12}, z_{22}, as shown in Fig. 6.7, the admittance at terminals 1–1′, with terminals 2–2′ short-circuited, is

$$y_{11} = \frac{z_{22}}{\Delta(z)} = \frac{z_{22}}{(z_{11}z_{22} - z_{12}^2)} = \frac{1}{z_{11} - z_{12}^2/z_{22}} \tag{6.47}$$

But because the system by hypothesis does undergo a natural oscillation of frequency s_ν with $I_1 = 0$ (terminals 1–1′ open) and $I_2 \neq 0$, we must have, on one hand,

$$z_{22}(s_\nu) = 0 \tag{6.48}$$

This is a consequence of the fact that $I_2 \neq 0$ and $V_2 = 0$, in the manner of Fig. 6.5*b*. On the other hand, with $I_2 \neq 0$ and $V_1 = 0$ (when $I_1 = 0$), it must be true that

$$z_{12}(s_\nu) = 0 \tag{6.49}$$

Application of Eqs. 6.48 and 6.49 to Eq. 6.47 shows that $y_{11}(s_\nu)$ is *indeterminate*. That is, the value of $y_{11}(s_\nu)$ depends upon that of $z_{11}(s_\nu)$ and the manner in which z_{12} and z_{22} approach zero as $s \to s_\nu$. This situation contrasts with the more common one in which $V_1(s_\nu) \neq 0$, so that $z_{12}(s_\nu) \neq 0$. In such normal cases, $y_{11} = 0$ whenever $z_{22} = 0$, according to Eq. 6.47, and the natural frequency of the network with terminals 1–1′ open is definitely a zero of the admittance seen at these terminals.

We conclude that some natural frequencies of a network may not show up at particular "solder" terminal pairs as poles of the impedance there (and similarly not as impedance zeros at particular "cut" terminal pairs). The characteristic of such unusual terminal pairs is that they fall at nodes between which there is no voltage (or in branches through which there is no current) at one or more natural frequencies.

As we have seen, this situation arises, in turn, from a vanishing transfer impedance (or admittance) between two parts of the system at a natural frequency. In lumped networks, these circumstances are usually produced either by a trivial case of completely separate parts (decoupled at all frequencies), by a numerical coincidence or symmetry of of element values, or by a "balanced bridge" arrangement. Simple examples of each type are shown in Figs. 6.8*a* through 6.8*d*.

Note that in each case of Fig. 6.8 the impedance Z seen at the terminals is neither infinite, which might be expected at such a natural frequency of the system with the terminals open, nor zero, which would normally accompany a natural frequency with the terminals short-circuited.

It is also significant that the distribution among branches of the currents and voltages, which accompanies the natural modes of oscillation referred to in Fig. 6.8, is quite different from that which results when the given terminal pair is driven by a source *at the same frequency.* Thus in Fig. 6.8*a* the natural oscillation at $\omega = 1/\sqrt{LC}$ takes place entirely in the left-hand tuned circuit, whereas the driven oscillation occurs only in the right-hand one. In Fig. 6.8*b*, no current flows in resistors $2R$ during the natural "oscillation" but, during a driven one, current is absent from leftmost element R and both elements $C/2$. Figure 6.8*d* similarly exhibits different unexcited branches under free and forced oscillation conditions, G being without current during the former while $L/2$ is without current during the latter. Matters are a little more subtle in Fig. 6.8*c*, however. Here, the natural oscillation at the frequency in question occurs when the two tuned circuits are oscillating with equal amplitudes, but in a bucking relative phase which produces zero voltage across R. On the other hand, the response to a drive from the terminals, at the same frequency, produces aiding voltage drops across the tuned circuits to account for the terminal voltage.

In the network of Fig. 6.8*c*, then, as in all the others, the space pattern characteristic of the natural oscillation is not excited by driving the system from the terminal pair shown, even though a nontrivial driven response does occur at the corresponding frequency. This *lack of excitation of natural-mode space pattern* provides another motivation for the statement that the given terminal pair is *decoupled* from the natural oscillation in question.

For a distributed system, the special decoupled terminal pairs with respect to any natural oscillation are those which fall at *perfect voltage or current nodes* of the natural-mode space pattern. Such perfect nodes always occur in lossless isolated systems; they are the means by which

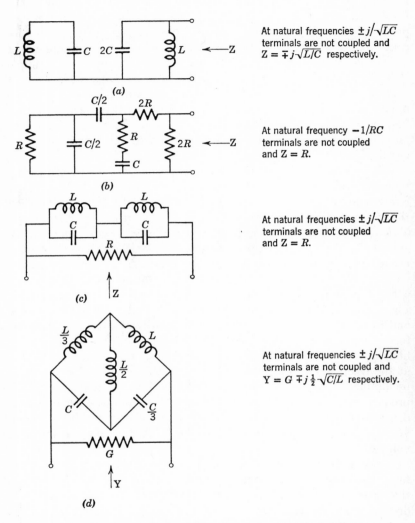

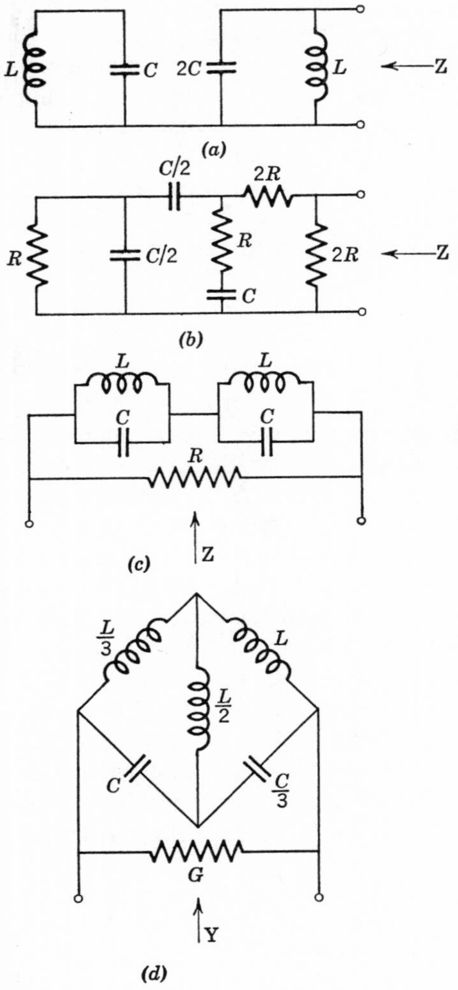

At natural frequencies $\pm j/\sqrt{LC}$ terminals are not coupled and $Z = \mp j\sqrt{L/C}$ respectively.

At natural frequency $-1/RC$ terminals are not coupled and $Z = R$.

At natural frequencies $\pm j/\sqrt{LC}$ terminals are not coupled and $Z = R$.

At natural frequencies $\pm j/\sqrt{LC}$ terminals are not coupled and $Y = G \mp j\frac{1}{2}\sqrt{C/L}$ respectively.

Fig. 6.8. Examples of networks in which a given pair of terminals fails to couple to some of the natural frequencies of the system. (*a*) Trivial case of separate parts; (*b*) case of numerical coincidence of element values, without symmetry; (*c*) case of numerical coincidence of element values with symmetry; (*d*) case of bridge balance.

energy is trapped for all time in various regions of space, namely, those regions between the nodes.

We can see in detail that the nodal points are indeed decoupled by referring again to the example of Fig. 6.3*a*. One of the natural oscillations has a space pattern defined by a wave length $\lambda_2 = l$, and a corresponding frequency $s_2 = j2\pi/T$. The voltage and current standing-wave patterns for this circumstance are shown in Fig. 6.9. The use of

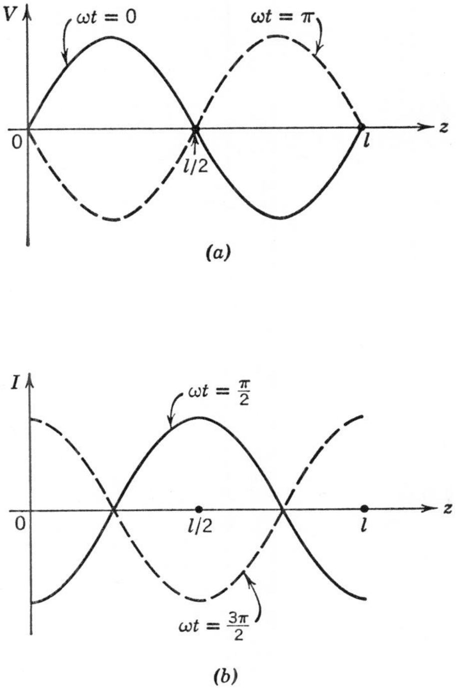

Fig. 6.9. (*a*) Voltage and (*b*) current patterns on the lossless system of Fig. 6.3*a* in the "one-wave-length" mode of oscillation.

solid and dashed lines indicates the 180° phase reversal between the voltage oscillations in successive half-wave intervals, and the in-phase character of the currents in the two line sections. Obviously, connecting a pair of open-circuited terminals across the line at the voltage node position $z = l/2$ cannot disturb the oscillation, which means that s_2 is a natural frequency of the system with such an added terminal pair left open. But any passive impedance, including a short circuit, could be hung across the line at $z = l/2$ without altering the free oscillation because zero current flows in the added element (leaving the *line* current continuous at $z = l/2$), and the resulting zero voltage also remains compatible with the free oscillation condition. In other words, the added terminal pair satisfies the condition of being "decoupled" from the natural oscillation, according to the first criterion developed in connection with Fig. 6.8.

To pursue further the notion that the added terminal pair is decoupled, consider the system behavior with a short circuit at $z = l/2$, as shown in Fig. 6.10*a*. This short circuit allows current to flow through it without voltage drop, thereby permitting an arbitrary discontinuity in line current at the center. Indeed the two halves of the line become completely uncoupled from each other. The sections may individually be excited, and freely oscillate, with arbitrary relative amplitude and phase, at one and the same frequency s_2. An interesting simple possi-

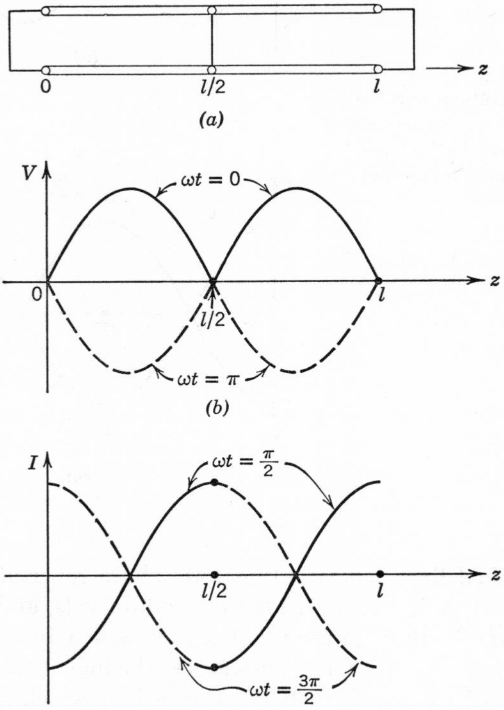

Fig. 6.10. (*b*) Possible voltage and (*c*) current space patterns on the line system of Fig. 6.3*a* with a short circuit at the center (as shown in (*a*)), or with a current source drive at the center.

bility is illustrated in Figs. 6.10*b* and 6.10*c*, in which the even line voltage and odd (discontinuous) line current patterns should be compared with the natural mode patterns of Fig. 6.9. The main point for our immediate discussion, however, is that, with the short circuit in place, an infinite number of different patterns may be excited on the system, all at the same frequency. When more than one space pattern can exist on a system at a single natural frequency, the corresponding natural oscillation is said to be *degenerate.* The degenerate character of the short-circuited system of Fig. 6.10 at frequency s_2 is in some contrast with the nondegenerate nature of the oscillations of the networks in Fig. 6.8 when their given terminals are short-circuited; but one can still say that the original free oscillation *need not* be disturbed when the added terminal pair is short-circuited, and therefore that terminal pair is properly referred to as decoupled from the oscillation in question.

The next comparison between our transmission-line example and the network examples of Fig. 6.8 may be based on the impedance seen at the center terminals of the line at frequency s_2. The impedance for the line system is zero, being the parallel combination of two short-circuited half-wave lines. It is significant that its value is finite rather than infinite, which again supports the idea that the terminal pair is decoupled from the natural oscillation. On this basis, one would not expect to find any evidence of the natural oscillation from the value of the impedance at the natural frequency. Certainly we do not find such evidence in Fig. 6.8, and it is fair to say that the present zero value *itself* is no more significant. But when we inquire specifically how the present zero value arises, and note that it occurs from simultaneous zeros of Z_+ and Z_- (Eq. 6.12 and Fig. 6.2*b*), we know that we have found a natural frequency of the system with the added terminals open-circuited, as well as short-circuited. This peculiar situation for the lossless line may be confusing unless it is well understood. The difficulty is that, in general, a zero of the impedance seen at a pair of terminals indicates a natural frequency of the system with the terminals short-circuited, and usually does *not* indicate a natural frequency with the terminals open-circuited. We expect the latter to show as a pole of the impedance. The coincidence that, for a terminal pair at a voltage node position on a line, both the short- and open-circuit natural frequencies happen to coincide *as well as* lead to a *zero* of the terminal impedance, is somewhat disconcerting. We must be aware of this possible difficulty, arising from decoupled nodes, when we try to find natural frequencies of line systems by examining the impedance or admittance at an added pair of terminals. Careful use of the conditions of Eq. 6.12 or 6.15 essentially circumvents the difficulty.

The center point of the line in Fig. 6.3*a* is, of course, not the only one at which a solder terminal pair is decoupled from some of the system natural frequencies; nor is the mode with $\lambda = l$ the only one involved in such decouplings. Indeed, since the various natural modes are characterized by frequencies for which $\lambda_m/2 = l/m$, $m = 0, 1, 2, \cdots$, the $m + 1$ voltage nodal positions $z = n(\lambda_m/2) = (nl)/m$, $n = 0, 1, 2, \cdots, m$, are places at which a solder pair is decoupled from the mth free oscillation. The places $z = 0$ and $z = l$, which lie directly at the short circuits, are rather trivial locations for solder pairs inasmuch as such pairs are decoupled from *all* the natural oscillations. In quite another way, the "d-c" oscillation $m = 0$ is also a trivial case since it is decoupled from solder terminals at *every* point on the line. Setting aside this d-c oscillation, we can show that a solder terminal pair at an arbitrary point z_0 on the line is either decoupled from an infinite number of the system natural frequencies or it couples to every one of them!

Specifically if $z_0 = (p/q)l$, where p/q is a nonzero rational proper fraction in lowest terms, a solder pair at z_0 is decoupled from every natural oscillation m' for which $\lambda_{m'} = 2l/\nu q$, $\nu = 1, 2, 3, \cdots$; i.e., from the oscillations for which $m' = \nu q$, $\nu = 1, 2, 3, \cdots$. But if $z_0 = \alpha l$, where $0 < \alpha < 1$ and α is irrational, a solder pair at z_0 never falls at a voltage node, and the terminal pair couples to *every* natural oscillation. Since there are more irrational than rational numbers, our earlier statement that most terminal pairs couple to every natural oscillation is still valid! Needless to say, a similar discussion may be applied to cut terminal pairs.

Whereas all isolated lossless distributed systems exhibit perfect nodes in their natural oscillations, most dissipative systems do not. For example, we saw that the line system of Fig. 6.3*c* did not, because of the need to transfer stored energy from all places on the line to the end load, where it is dissipated. One special case occurs, however, in systems with uniform loss, i.e., where the ratio of stored-energy density to dissipated-power density is the same at every point in space. These systems may have perfect nodes, because the local stored energy can all be dissipated right on the spot without having to be transferred to another place for that purpose. The system of Fig. 6.3*b* is a case in point. As we might expect, though, addition of pure resistance in the terminations destroys the perfect character of the nodes. Therefore, in the general case of a dissipative distributed structure, we do not often expect to encounter completely decoupled situations of the special type discussed above for lossless systems. Still, one must realize that real systems are never exactly like the idealized models we treat for the purpose of establishing understanding. Whether or not the idealization that neglects losses, and therefore leads to perfect nodes in the theory, is valid for a particular problem, depends upon the specific question being asked. We shall learn more about this matter later on, in connection with the problem of coupling to a resonant system.

6.3 Transient Response

The meaning and significance of forced and free oscillations as well as the relations between them may be clarified by studying some transient problems with a point of view different from that of Chapter 4.

6.3.1 The Driven System

6.3.1.1 IMPULSE DRIVE. A simple case of a system driven by a source appears in Fig. 6.11. Suppose the input current is specified as an impulse at $t = 0$

$$I_0(t) = u_0(t) \tag{6.50}$$

and we wish to find $V(z, t)$ for $t \geq 0, 0 \leq z \leq l$.

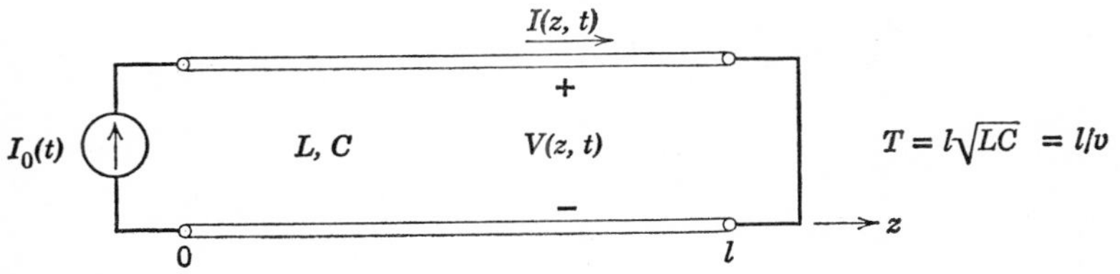

Fig. 6.11. A driven lossless system.

The interesting part of this problem for us now is, as we have said, the point of view we are about to use, not the answer itself. We know the answer, by the methods of Chapter 4. It is shown in Fig. 6.12. It may be thought of conveniently as the positive-time half of an even periodic function $f(t)$, composed of impulse groups, with period $4l/v = 4T$. The rest of such a complete periodic function, lying on the negative-time axis, would be obtained by simply reflecting the part shown about the ordinate axis. Although this negative-time portion of the function has no physical significance in the present problem, it is useful to think about because it makes the whole time function

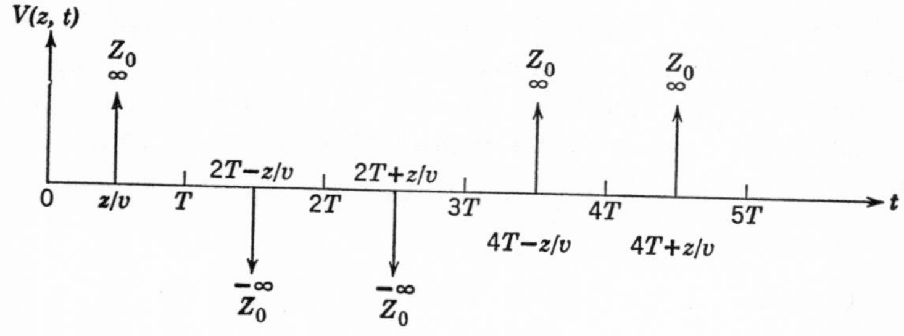

Fig. 6.12. Voltage response as a function of time at a point z of the system of Fig. 6.11 when $I_0(t)$ is a unit impulse at $t = 0$.

periodic in $-\infty < t < \infty$. For this entire time interval we know, therefore, that a Fourier *series* representation of $f(t)$ would be a sum of cosine terms only (because of the even character about $t = 0$), with fundamental period

$$\tau = \frac{4l}{v} = 4T \tag{6.51a}$$

or fundamental radian frequency

$$\omega_1 = \frac{2\pi}{\tau} = \frac{2\pi v}{4l} = \frac{\pi}{2T} \tag{6.51b}$$

This frequency is one for which the wave length on the line would be

$$\lambda_1 \equiv \frac{2\pi v}{\omega_1} = 4l \tag{6.51c}$$

so the line would be a quarter wave length long.

Moreover, we notice that the periodic function $f(t)$ defined above by extension of Fig. 6.12 also has the additional symmetry property $f(t + \tau/2) = -f(t)$, which means that its Fourier series has only the odd harmonics. Thus $f(t)$ is of the form

$$f(t) = \sum_{m=1}^{\infty} A_m \cos\left[(2m - 1)\omega_1 t\right] \tag{6.52}$$

Accordingly, by Fourier series theory and the two symmetry properties mentioned above,

$$\begin{aligned} A_m &= \frac{2}{\tau}\int_{-\tau/2}^{\tau/2} f(t)\cos\left[(2m-1)\omega_1 t\right] dt \\ &= \frac{2}{T}\int_0^T f(t)\cos\left[(2m-1)\omega_1 t\right] dt \end{aligned} \tag{6.53}$$

The integration in Eq. 6.53 may be carried out by inspection of Fig. 6.12:

$$A_m = \frac{2Z_0}{T} \cos\left[(2m-1)\frac{\omega_1 z}{v}\right] = \frac{2Z_0}{T}\cos\left[(2m-1)\frac{\pi z}{2l}\right] \tag{6.54}$$

Since physically $V(z, t)$ is only the part of $f(t)$ lying in the range $t \geq 0$, and is actually zero for $t < 0$, we may put

$$V(z, t) = u_{-1}(t) f(t) \tag{6.55a}$$

whence Eqs. 6.52, 6.54, and 6.55*a* yield

$$V(z, t) = \sum_{m=1}^{\infty} \frac{2Z_0}{T} u_{-1}(t) \cos\left[(2m-1)\frac{\pi z}{2l}\right] \cos\left[(2m-1)\omega_1 t\right] \tag{6.55b}$$

Examination of Eq. 6.55*b* shows that it is a sum of terms, each of which corresponds in both space and time dependence to one of the natural modes of oscillation of the system of Fig. 6.11 with the input terminals *open*. That is, each term represents one of these modes starting suddenly at $t = 0$, and oscillating thereafter at a frequency for which the line length is an odd multiple of a quarter wave length. Since each term has the same amplitude coefficient $2Z_0/T$, we say that the current impulse has excited *equally* all the natural modes of oscillation corresponding to an open-circuit condition at the driving terminals. The impulse current source does not excite any of the natural frequencies appropriate to the system with the driving terminals short-cuited.

We developed Eq. 6.55*b* in terms of the time function, illustrated in Fig. 6.12, which would be observed at some given point z on the line On this basis, the Fourier series (Eq. 6.52) represents a periodic function of time whose coefficients A_m are functions of position z (Eq. 6.54). In another way, however, the result (Eq. 6.55*b*) may also be viewed as a Fourier series in z, whose coefficients are functions of time. Indeed, aside from the factor $u_{-1}(t)$, $V(z, t)$ is exactly the same function of $z/2l$, with $t/2T$ regarded as a parameter, as it is of $t/2T$, with $z/2l$ as a parameter. This means that, for $t_0 \geq 0$, $V(z, t_0)$ is given by Eq. 6.55*b* as the even periodic function made by reflecting Fig. 6.12 about the ordinate axis to continue it along the negative abscissa, with z/v interchanged with t everywhere (including the areas of the impulses). The result is shown in Fig. 6.13 at a time $t_0 < T$. In this case, though, whereas the function (Eq. 6.55*b*) is defined mathematically for all values of z, only the region $0 \leq z \leq l$ has physical significance in the present problem. If we focus our attention upon this region, labeled "actual line" in Fig. 6.13, and observe the movement of all the im-

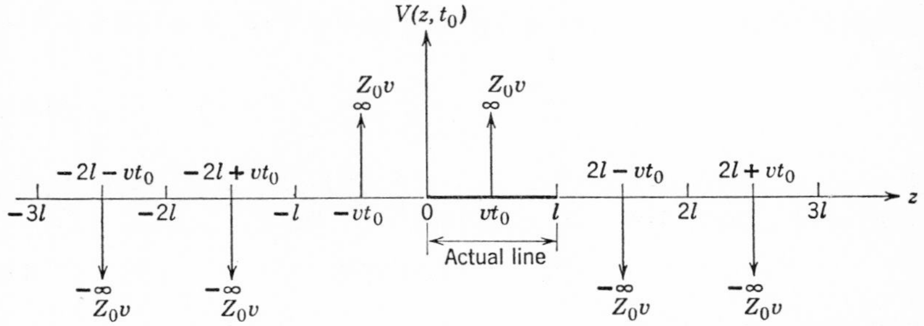

Fig. 6.13. Voltage response as a function of position at a time $0 < t_0 < T$ of the system of Fig. 6.11 when $I_0(t)$ is a unit impulse at $t = 0$.

pulses as t_0 increases, we can see how they enter and cross the physically significant section of the axis alternately from opposite ends, with the correct polarity to account properly for the actual reflections from the short-circuited and open-circuited terminations.

What we have done so far is to solve the problem posed in Fig. 6.11 by using the methods of Chapter 4. We then re-expressed the answer in a form (Eq. 6.55*b*), which could be interpreted, instead, as a sum of certain of the natural modes in time and space with constant coefficients (here all equal in value because we chose impulse excitation). None of this discussion has employed the impedance concept. We now wish actually to solve the problem over again in a manner which does so. Fourier or Laplace transform techniques are appropriate.

Since all the processes in the problem occur for $t \geq 0$, we may describe all time functions by one-sided transforms ($0 \leq t < \infty$). The impulse current (Eq. 6.50) has such a one-sided (Laplace) transform spectrum given by

$$\mathrm{I}_0(\mathrm{s}) \equiv \int_0^\infty u_0(t)e^{-\mathrm{s}t}\,dt = 1 \tag{6.56}$$

which is a uniform spectrum. For each complex frequency s, the complex voltage at a point z on the short-circuited line, the voltage of which results from the contribution of $\mathrm{I}_0(\mathrm{s})$ at this frequency, is given by steady-state analysis in the form

$$\mathrm{V}(z, \mathrm{s}) = 2\mathrm{V}_+(\mathrm{s})e^{-\bar{\gamma}l}\sinh\bar{\gamma}(l - z) \tag{6.57}$$

where $\bar{\gamma} = \mathrm{s}\sqrt{LC} = \mathrm{s}/v$. Similarly,

$$\mathrm{I}(z, \mathrm{s}) = 2\mathrm{V}_+(\mathrm{s})Y_0e^{-\bar{\gamma}l}\cosh\bar{\gamma}(l - z) \tag{6.58}$$

which, in view of Eq. 6.56 gives us

$$\mathrm{I}(0, \mathrm{s}) = \mathrm{I}_0 = 1 = 2\mathrm{V}_+(\mathrm{s})Y_0e^{-\bar{\gamma}l}\cosh\bar{\gamma}l \tag{6.59}$$

Elimination of $V_+(s)$ between Eqs. 6.57 and 6.59 leads to the result

$$V(z, s) = +\frac{Z_0 \sinh \bar{\gamma}(l - z)}{\cosh \bar{\gamma} l} = Z_{0z}(s) \tag{6.60}$$

which is also the *transfer impedance* between drive terminals at $z = 0$ and solder terminals at z. This transfer impedance is a *meromorphic* function of s, because it has only ordinary poles for finite s but an essential singularity at $s = \infty$.

To determine $V(z, t)$, we must invert the transform (Eq. 6.60). The inversion of transforms may be accomplished most easily by rewriting or expanding them as a sum of terms, for each of which we know the inverse by inspection. One way to do this in the present case is to put the hyperbolic functions into exponential form, and carry out the division by longhand.[1] Thus

$$V(z, s) = Z_0\left(\frac{e^{-\bar{\gamma}z} - e^{-2\bar{\gamma}l+\bar{\gamma}z}}{1 + e^{-2\bar{\gamma}l}}\right) \tag{6.61a}$$

and the division is

$$\begin{array}{r|l}
 & e^{-\bar{\gamma}z} - (e^{-2\bar{\gamma}l+\bar{\gamma}z} + e^{-2\bar{\gamma}l-\bar{\gamma}z}) + (e^{-4\bar{\gamma}l+\bar{\gamma}z} + e^{-4\bar{\gamma}l-\bar{\gamma}z}) - \cdots \\ \hline
1 + e^{-2\bar{\gamma}l} & e^{-\bar{\gamma}z} - e^{-2\bar{\gamma}l+\bar{\gamma}z} \\
 & \underline{e^{-\bar{\gamma}z} + e^{-2\bar{\gamma}l-\bar{\gamma}z}} \\
 & (-e^{-2\bar{\gamma}l+\bar{\gamma}z} - e^{-2\bar{\gamma}l-\bar{\gamma}z}) \\
 & \underline{(\qquad \text{``} \qquad) - (e^{-4\bar{\gamma}l+\bar{\gamma}z} + e^{-4\bar{\gamma}l-\bar{\gamma}z})} \\
 & + (e^{-4\bar{\gamma}l+\bar{\gamma}z} + e^{-4\bar{\gamma}l-\bar{\gamma}z}) \\
 & + - - - - - - - - - - - \cdots
\end{array}$$

i.e.,

$$V(z, s) = Z_0[e^{-sz/v} - e^{-2sT+sz/v} - e^{-2sT-sz/v} + e^{-4sT+sz/v} + e^{-4sT-sz/v} - \cdots] \tag{6.61b}$$

Such an expression represents $V(z, s)$ as a sum of *traveling waves* in both directions. But inasmuch as $e^{-s\tau}$ is the Laplace transform of a unit impulse at time τ, it is clear that Eq. 6.61*b* is exactly the transform of the function illustrated in Fig. 6.12! It should not be surprising that the frequency-domain resolution (Eq. 6.61*b*) of the line voltage into traveling waves leads directly to Fig. 6.12, which is the result of apply-

[1] The effect of a factor $(1 + e^{-as})^{-1}$ in producing periodicity is discussed in most texts on the applications of Laplace transforms. The development here essentially duplicates the usual proof.

ing the time-domain traveling-wave analysis of Chapter 4 to the same problem.

Another summation form for the transform (Eq. 6.60) is the more familiar *partial-fraction* expansion. The zeros of the denominator are needed for this expansion, and these are given by the condition

$$\cosh \bar{\gamma} l = 0 \tag{6.62a}$$

or

$$e^{2\bar{\gamma} l} = -1 \tag{6.62b}$$

Therefore the zeros in question occur for

$$2\bar{\gamma} l = \pm j(2m - 1)\pi \qquad m = 1, 2, 3, \cdots \tag{6.63a}$$

or

$$\mathrm{s} = \pm j \frac{(2m - 1)\pi}{2T} \qquad m = 1, 2, 3, \cdots \tag{6.63b}$$

which are again the natural frequencies of oscillation of the system with the drive terminals open, and also the poles of the transfer impedance for most values of z.

The partial-fraction term of Eq. 6.60 which involves the pole at $\mathrm{s} = \mathrm{s}_m$ is of the form $\mathrm{a}_m/(\mathrm{s} - \mathrm{s}_m)$. But

$$\begin{aligned} \mathrm{a}_m &= \lim_{\mathrm{s} \to \mathrm{s}_m} \left\{ \frac{[Z_0 \sinh \bar{\gamma}(l - z)](\mathrm{s} - \mathrm{s}_m)}{\cosh \bar{\gamma} l} \right\} \\ &= \frac{Z_0 \sinh \bar{\gamma}_m (l - z)}{T \sinh \bar{\gamma}_m l} = \frac{Z_0}{T} \cos\left[(2m - 1)\frac{\pi z}{2l}\right] \end{aligned} \tag{6.64}$$

where L'Hôpital's rule has been used to evaluate the otherwise indeterminate ratio $(\mathrm{s} - \mathrm{s}_m)/\cosh \bar{\gamma}_m l = 0/0$, and the values of $\bar{\gamma}$ from Eq. 6.63*a* are included. Therefore the entire expansion reads

$$\mathrm{V}(z, \mathrm{s}) = \sum_{m=1}^{\infty} \frac{Z_0}{T} \cos\left[(2m - 1)\frac{\pi z}{2l}\right] \left(\frac{1}{\mathrm{s} - j\omega_m} + \frac{1}{\mathrm{s} + j\omega_m}\right) \tag{6.65}$$

in which $\omega_m \equiv (2m - 1)\pi/2T$ and the $\pm$ signs of Eq. 6.63*b* are included through the use of two terms for each m value. Equation 6.65 represents $\mathrm{V}(z, \mathrm{s})$ as a sum of *complete standing waves.*

Knowing that a Laplace transform $1/(\mathrm{s} + \mathrm{s}_0)$ has the inverse $u_{-1}(t)e^{-\mathrm{s}_0 t}$, we invert the sum (Eq. 6.65) term by term to get

$$V(z, t) = u_{-1}(t) \sum_{m=1}^{\infty} \frac{2Z_0}{T} \cos\left[(2m - 1)\frac{\pi z}{2l}\right] \cos\left[(2m - 1)\omega_1 t\right]$$

This result agrees precisely with Eq. 6.55*b*. It should not be a surprise that a standing-wave representation (Eq. 6.65) in the frequency domain leads directly to a standing-wave representation (Eq. 6.55*b*) in the time domain.

As we have indicated previously, the natural modes of oscillation represent the force-free, or complementary solutions, for a given system. In case the system is driven by a "pulse" source starting at $t = 0$, but with a value that becomes zero again at some finite time δ and *remains* zero thereafter, only the complementary solutions are needed to describe the entire response for $t > 0$. Of course the pulse shape affects the values of the constants multiplying the complementary solutions. As a matter of fact, this is true for any driving function whose magnitude eventually falls off faster than exponentially with time.[1] One may say that the particular solution in such cases simply has the same form as the complementary one. The system response to these excitations can always be duplicated by suitable initial conditions without the source. The impulse problem treated above is illustrative of this class of drives.

If, on the other hand, the driving function does not ultimately fall off faster than exponentially, the particular solution will be quite different from the complementary ones. In familiar examples, the particular solution is often called the "steady state"; but this nomenclature is rather misleading because the particular function may tend to zero with time. Nevertheless, the additional response of the system produced by the drive in these cases is more closely related in form to the driving function than to the complementary ones. In this sense, we may say that the drive actually "forces" the system into a response which could not be duplicated by suitable initial conditions alone.

6.3.1.2 Sudden Sinusoidal Drive.* As a particular case of forcing, take

$$I_0(t) = \text{Re}\,[u_{-1}(t)e^{\mathrm{s}_0 t}] \tag{6.66}$$

in Fig. 6.11. We allow s_0 to be any complex frequency, and we can drop the Re restriction during the analysis, if we remember to take it in the answer. The corresponding spectrum of $u_{-1}(t)e^{\mathrm{s}_0 t}$ is

$$\mathrm{I}_0(\mathrm{s}) = \frac{1}{\mathrm{s} - \mathrm{s}_0} \tag{6.67}$$

[1] We are using here the property that the complex frequency spectrum $\mathrm{I}_0(\mathrm{s})$ for such functions is analytic for all finite s. This property of Laplace or Fourier transforms is proved in most books on transform theory or in the theory of complex functions defined by integrals.

The voltage spectrum at point z is therefore

$$\mathrm{V}(z, \mathrm{s}) = \mathrm{I}_0(\mathrm{s})\mathrm{Z}_{0z}(\mathrm{s}) = \frac{\mathrm{Z}_0 \sinh \bar{\gamma}(l - z)}{(\mathrm{s} - \mathrm{s}_0) \cosh \bar{\gamma} l} \tag{6.68}$$

We could solve the problem for $V(z, t)$ by using the traveling-wave point of view, either in the time domain by methods of Chapter 4, or in the frequency domain by the exponential expansion of $\mathrm{Z}_{0z}(\mathrm{s})$ in the manner of Eq. 6.61*b*. These methods lead most quickly to a form of $V(z, t)$ which we can visualize and sketch. They should be pursued by the reader for at least those cases discussed below.

We adopt here instead the standing-wave viewpoint to emphasize the roles of the particular solution and the natural oscillations. Therefore we make a partial-fraction expansion of Eq. 6.68, supposing first that s_0 is *not* equal to any of the open-circuit natural frequencies s_m given by Eq. 6.63*b*. Thus, at a pole $\mathrm{s} = \mathrm{s}_m = j\omega_m$, treatment of Eq. 6.68 in the manner of Eq. 6.64 shows that

$$\mathrm{a}_m = \frac{Z_0}{T(j\omega_m - \mathrm{s}_0)} \cos\left(\omega_m \frac{z}{r}\right) \tag{6.69}$$

while at $\mathrm{s} = \mathrm{s}_0$ we have a term $\mathrm{a}_0/(\mathrm{s} - \mathrm{s}_0)$, with

$$\mathrm{a}_0 = \frac{Z_0 \sinh \bar{\gamma}_0(l - z)}{\cosh \bar{\gamma}_0 l} \tag{6.70}$$

It follows that

$$\mathrm{V}(z, \mathrm{s}) = \frac{Z_0 \sinh \bar{\gamma}_0(l - z)}{\cosh \bar{\gamma}_0 l(\mathrm{s} - \mathrm{s}_0)} - \sum_{m=1}^{\infty} \frac{Z_0}{T} \cos\left(\omega_m \frac{z}{v}\right) \times \left[\frac{1}{(\mathrm{s}_0 - j\omega_m)(\mathrm{s} - j\omega_m)} + \frac{1}{(\mathrm{s}_0 + j\omega_m)(\mathrm{s} + j\omega_m)}\right] \tag{6.71}$$

where again for each m we have included the two roots $\mathrm{s} = \pm j\omega_m$ from Eq. 6.63*b*. Term-by-term inversion, remembering to take the real part of the answer, yields the result

$$V(z, t) = u_{-1}(t) \left\{ \mathrm{Re}\left[\frac{Z_0 \sinh \bar{\gamma}_0(l - z)}{\cosh \bar{\gamma}_0 l} e^{\mathrm{s}_0 t}\right] - \frac{Z_0}{T} \sum_{m=1}^{\infty} \cos\left(\omega_m \frac{z}{v}\right) \mathrm{Re}\left(\frac{e^{j\omega_m t}}{\mathrm{s}_0 - j\omega_m} + \frac{e^{-j\omega_m t}}{\mathrm{s}_0 + j\omega_m}\right)\right\} \tag{6.72}$$

Observe that the term involving $e^{s_0 t}$ (under the Re sign) is exactly the transfer impedance evaluated at $s = s_0$ times the driving current. This is the particular solution or steady-state response to a drive $e^{s_0 t}$, which we often interpret as what would happen if current $e^{s_0 t}$ had been applied *forever*, from $t = -\infty$ on. Since, however, the current is actually applied suddenly at $t = 0$, the rest of the natural open-circuit oscillations are excited to take up the difference between the dead condition of the system when $t < 0$ and the behavioral demands of the applied current at $t \geq 0$. The summation term includes these natural oscillations, which are excited this time with *different* amplitudes and phases (compare Eqs. 6.72 and 6.55*b*). The oscillation most strongly excited is that one for which $|s_0 - s_m|$ is smallest, i.e., the one whose complex natural frequency is closest in the s-plane to that of the applied source.

A pleasant feature of the solution in Eq. 6.72 arises if $s_0 = -\Omega$, i.e., the source is a real decreasing exponential which starts suddenly from unit value at $t = 0$. The particular solution is then also a decreasing real exponential, decaying as $e^{-\Omega t}$; but the natural oscillations are of course *undamped*. Here is a case in which the transient part (complementary part) of the solution lasts forever, while the steady-state part (particular part) vanishes quickly. Obviously it is better to think of these parts as simply "complementary" and "particular" rather than in terms of the doubtful and misleading notions of relative duration contained in words like "steady state" or "transient."

A final feature of our problem deserves comment here. Suppose $s_0 = j\omega_1$. The drive frequency (after $t = 0$) then coincides with one of the natural frequencies, and *also* with a pole of the transfer impedance to all points $0 \leq z < l$. One expects *two* results: (1) The natural oscillation in question should be excited infinitely strongly and (2) the particular solution voltage should become infinite with the transfer impedance. These expectations are supported by Eq. 6.72. The first term becomes infinite because $\cosh \bar{\gamma}_1 l = 0$, and the first natural oscillation amplitude ($m = 1$) becomes infinite because $s_0 = j\omega_1$. One does not, of course, expect the total response to jump to infinity right away. After all, the drive is only a 1-amp sinusoid turned on at $t = 0$, with an unenergized line present at the start. The two infinities in Eq. 6.72 *must* cancel as $s_0 \rightarrow s_1 = j\omega_1$. Indeed it is the function of the natural oscillations to mediate gradually between the infinite demands of the source and the zero state of the system.

Let us examine the two offending terms together as $s_0 \rightarrow s_1$, $\bar{\gamma}_0 \rightarrow \bar{\gamma}_1$. That is, consider

$$\lim_{\substack{s_0 \to s_1 \\ \bar{\gamma}_0 \to \bar{\gamma}_1}} \left[\frac{\sinh \bar{\gamma}_0(l - z)}{\cosh \bar{\gamma}_0 l} e^{s_0 t} - \frac{\cosh \bar{\gamma}_1(z)}{T(s_0 - s_1)} e^{s_1 t} \right]$$

$$= \lim_{\substack{s_0 \to s_1 \\ \bar{\gamma}_0 \to \bar{\gamma}_1}} \left\{ \frac{\cosh \bar{\gamma}_1 z}{T(s_0 - s_1)} e^{s_1 t} \left[\frac{T \sinh \bar{\gamma}_0(l - z)}{\cosh \bar{\gamma}_1 z} \left(\frac{s_0 - s_1}{\cosh \bar{\gamma}_0 l} \right) e^{(s_0 - s_1)t} - 1 \right] \right\}$$

But by L'Hôpital's rule

$$\lim_{\substack{s_0 \to s_1 \\ \bar{\gamma}_0 \to \bar{\gamma}_1}} \left(\frac{s_0 - s_1}{\cosh \bar{\gamma}_0 l} \right) = \lim_{\substack{s_0 \to s_1 \\ \bar{\gamma}_0 \to \bar{\gamma}_1}} \left(\frac{1}{T \sinh \bar{\gamma}_0 l} \right) = \frac{1}{T \sinh \bar{\gamma}_1 l}$$

and since

$$\sinh \bar{\gamma}_1(l - z) = \sinh \bar{\gamma}_1 l \cosh \bar{\gamma}_1 z$$

we find that our limit becomes

$$\lim_{\substack{s_0 \to s_1 \\ \bar{\gamma}_0 \to \bar{\gamma}_1}} \left[\frac{e^{s_1 t} \cosh \bar{\gamma}_1 z}{T(s_0 - s_1)} (e^{(s_0 - s_1)t} - 1) \right] = \frac{t e^{s_1 t} \cosh \bar{\gamma}_1 z}{T} = \frac{t e^{j\omega_1 t} \cos (\omega_1 z/v)}{T}$$

Thus Eq. 6.72 may be rewritten as follows:

$$V(z, t) = u_{-1}(t) \frac{Z_0}{T} \operatorname{Re} \left[\left(t e^{j\omega_1 t} + \frac{j e^{-j\omega_1 t}}{2\omega_1} \right) \cos \left(\omega_1 \frac{z}{v} \right) \right.$$
$$\left. + j \sum_{m=2}^{\infty} \cos \left(\omega_m \frac{z}{v} \right) \left(\frac{e^{j\omega_m t}}{\omega_1 - \omega_m} + \frac{e^{-j\omega_m t}}{\omega_1 + \omega_m} \right) \right] \tag{6.73}$$

$V(z, t)$ does become infinite eventually; but, after all, it grows rather slowly, being only linearly proportional to time in the troublesome term. Solution of this problem by traveling-wave methods shows that the voltage oscillation amplitude increases primarily in jumps occurring at intervals $2T$, as the successive reinforcing reflections arrive at a given point z from the end $z = 0$.

6.3.2 Initial Conditions *

We have in the previous section discussed the relationship between the natural modes of oscillation of the system of Fig. 6.11 and its response to time-function current excitation. Voltage excitation instead would raise no new difficulties; but obviously the specification of both voltage and current time functions at the same space position would overspecify the problem.

It is quite clear that superposition techniques allow us to handle as many different current-source time functions at different locations as we wish, and the same applies to voltage sources. Indeed, we may have both types of sources at once, as long as no two of them fall at the same point. In this sense, we have, in principle, studied the role of natural oscillations in the system response to given time-function drives at any points in the system.

A somewhat different view of the natural oscillations arises in connection with the system response to prescribed distributions of voltage and current at the initial instant $t = 0$. As might be expected, this view emphasizes more the spacial than the temporal features of the natural modes, although we have seen that the two features are bound closely together.

For purposes of concrete illustration, suppose we establish our "initial" distributions by looking in at the system of Fig. 6.11 at a time $t_0 < T$ after application of a current impulse at $z = 0$. We would see the voltage distribution in space pictured in the "actual line" section of Fig. 6.13. If the instant in question is our new time origin $t' = 0$, and the impulse position is z_0 $(=vt_0)$, this voltage distribution is given by

$$V(z, 0) = Z_0 v u_0(z_0) \tag{6.74}$$

The corresponding current distribution could be found from our previous work or by inspection, with the use of methods of Chapter 4. It is

$$I(z, 0) = v u_0(z_0) \tag{6.75}$$

We do not suggest that Eq. 6.75 must always go with Eq. 6.74; on the contrary, the two distributions may always be chosen arbitrarily and quite independently. Here we select the particular pair of Eqs. 6.74 and 6.75 only to facilitate comparison of our results with the previous solutions.

Since, when $t' \geq 0$, the line is open-circuited at $z = 0$ and short-circuited at $z = l$, it must be possible to meet the initial energy-storage conditions by a sum of the free solutions which satisfy these end boundary conditions. That is, the voltage and current functions must be sums of the natural modes, with arbitrary relative amplitudes and phases.

$$\text{(a)} \qquad V(z, t') = \text{Re} \sum_{m=1}^{\infty} \mathfrak{a}_m'(\cos \beta_m z) e^{j\omega_m t'} \qquad t' \geq 0$$

$$\text{(b)} \qquad I(z, t') = \text{Re} \sum_{m=1}^{\infty} -jY_0 \mathfrak{a}_m'(\sin \beta_m z) e^{j\omega_m t'} \qquad t' \geq 0 \tag{6.76}$$

Observe that the current and voltage amplitudes in each term are related as in a natural oscillation, so that only *one complex coefficient* a_m' must be determined for each m value. This implies existence of two real numbers per m value, however, and will permit fitting both the initial voltage *and* initial current distributions.

The idea is to evaluate a_m' in Eq. 6.76 so that $V(z, 0)$ and $I(z, 0)$ agree with the demands of Eqs. 6.74 and 6.75 respectively. Setting $t' = 0$, and dividing a_m' into real and imaginary parts,

$$\mathrm{a}_m' = A_m' + jB_m' \tag{6.77}$$

we find from Eqs. 6.76, 6.74, and 6.75 the relations

$$\begin{aligned}
&\text{(a)} \qquad V(z, 0) = Z_0 v u_0(z_0) = \sum_{m=1}^{\infty} A_m' \cos \beta_m z \\
&\text{(b)} \qquad I(z, 0) = v u_0(z_0) = \sum_{m=1}^{\infty} Y_0 B_m' \sin \beta_m z
\end{aligned} \tag{6.78}$$

where it is recalled that $\beta_m = \omega_m/v = [(2m - 1)\pi]/2l$.

To determine A_m' and B_m' we must first observe that the series on the right do *not* have to equal the expressions on the left for *all* values of z. All that is required by the physical problem is that Eqs. 6.74 and 6.75 hold in the range $0 \le z \le l$. The series expansions must therefore also represent the prescribed distributions in this z range, and are free to do what they please elsewhere.

Fortunately, we know from the theory of Fourier series that an odd-harmonic cosine series of period $4l$ like that in Eq. 6.78*a* may represent almost any function over one quarter of its period ($4l$), although the series then repeats the same shape, with or without reversed polarity, as required to create a complete periodic function $f(z)$ with the properties: $f(z + 4l) = f(z)$, $f(-z) = f(z)$, and $f(z + 2l) = -f(z)$. Thus if the series correctly represents $Z_0 v u_0(z_0)$ in the interval $0 \le z \le l$, its behavior outside this interval must be that shown in Fig. 6.13 (with $z_0 = vt_0$). Similarly, the odd-harmonic sine series (Eq. 6.78*b*) of period $4l$ may represent practically any function in the range $0 \le z \le l$, but then repeats it, with or without reversal, to make up a complete periodic function $g(z)$ with the properties: $g(z + 4l) = g(z)$, $g(-z) = -g(z)$, $g(z + 2l) = -g(z)$. Therefore, if such a sine series represents correctly $v u_0(z_0)$ in the range $0 \le z \le l$, it must look like Fig. 6.14 outside this interval.

It also happens that because of their symmetry odd-harmonic cosines are orthogonal not only over whole periods but also over a quarter-

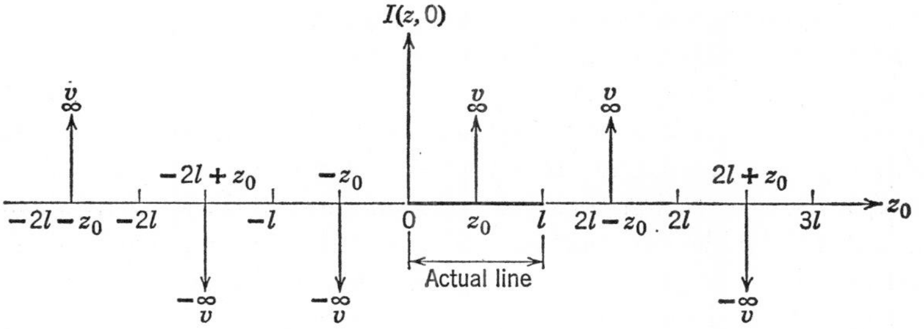

Fig. 6.14. Initial current conditions at $t' = 0$ as rendered by the odd-harmonic sine series of Eq. 6.78*b*.

period interval. The same is true for odd-harmonic sines. Consequently we can determine A_m' and B_m' easily by the usual multiplication and integration (0 to l) method of Fourier series. We find

$$\text{(a)} \qquad A_m' = \frac{2Z_0}{T} \cos \beta_m z_0$$

$$\text{(b)} \qquad B_m' = \frac{2Z_0}{T} \sin \beta_m z_0 \qquad (6.79)$$

$$\text{(c)} \qquad \mathrm{a}_m' = A_m' + jB_m' = \frac{2Z_0}{T} e^{j\beta_m z_0}$$

Therefore

$$V(z, t') = \sum_{m=1}^{\infty} \frac{2Z_0}{T} \cos \beta_m z \cos (\omega_m t' + \beta_m z_0) \qquad t' \geq 0 \quad (6.80)$$

with a similar expression for current. We may compare Eqs. 6.80 and 6.55*b* by noting that they should agree when $t \geq t_0$ in Eq. 6.55*b* and $t' \geq 0$ in Eq. 6.80. Specifically, the substitutions $t' = t - t_0$ and $z_0 = vt_0$ should turn Eq. 6.80 into Eq. 6.55*b*. That they do stems from the relation

$$\omega_m t' + \beta_m z_0 = \omega_m (t - t_0) + \frac{\omega_m}{v} (vt_0) = \omega_m t$$

which does indeed convert Eq. 6.80 into Eq. 6.55*b* for $t > t_0$.

It is unfortunate, in a way, that our familiarity with Fourier series tends to obscure the implications of the fact that Eqs. 6.78 could be solved at all. To appreciate this point, forget for the time being that the functions being summed are the familiar trigonometric ones. Instead, remember only that one set is simply the collection of space dependencies of the voltage in the various natural oscillations of a physical system, and the other set is the collection of corresponding current space patterns. Indeed, members of the two sets are related by the physical properties of the line system through its differential equations. Yet it is presumably possible, from sums of these natural oscillation space patterns, to represent almost any other function of position over the region between the boundaries of the system to which the natural oscillations pertain. That is, the space patterns can be used as a basis for the "expansion" of other arbitrary functions over appropriate ranges of the independent variables.

This important property of natural-mode space patterns is not restricted to transmission lines but arises in nearly every physical linear boundary-value problem involving distributed structures. It *has to* arise thus because, on physical grounds, the set of natural modes or complementary solutions *must* be able to meet every permissible (arbitrary) initial state of the system.

Mathematically, of course, the expansion property may be regarded in another light. If we consider that our present functions, sine and cosine, arose from the solution to a second-order ordinary differential equation (Eq. 3.3) for voltage or current, and that the particular set of permitted periods was determined from boundary conditions, we can understand the nature of the situation. A large class of second-order equations, coupled with a number of kinds of boundary conditions, generates solution families whose members are mutually orthogonal over the interval between boundaries, and may serve as the functions in a generalized Fourier series (or integral) expansion of almost any other function on the same interval. The details of such matters are contained in what is known as the Sturm-Liouville theory.[1]

Because of the importance of resonant systems, one may wish to select for excitation only one particular natural frequency of a distributed system. Our work in Sec. 6.3.1 showed that choosing a steady-state excitation frequency near the desired complex natural frequency is one way to succeed. The development of the present section shows that excitation in a space pattern close to that of the desired natural oscillation is another way. A combination of the two would be most desirable. Cases of degeneracy, in which more than one space pattern can exist at the same natural frequency, must evidently be thought of from both standpoints.

[1] See, for example, Ruel V. Churchill, *Modern Operational Mathematics in Engineering*, McGraw-Hill Book Co., New York, 1944.

6.4 Points of View Involving Energy and Power

Distributed systems often lead us to consider the approximate behavior of an impedance near one of its many resonances, or the description of the resonant behavior of a structure for situations where circuit concepts are only partly applicable. The use of energy and power relations is most helpful in these problems.

6.4.1 Some General Properties of Impedance and Admittance

We may extend to complex values of s the formulation in terms of energy for $Z(j\omega)$ [or $Y(j\omega)$] developed in Chapter 1. We shall do it in terms of fields rather than circuits, employing a purely formal procedure similar to that involved in developing Poynting's theorem.[1]

Maxwell's equations appropriate to a source-free region in which the fields are of the form $\mathbf{E}(x, y, z; \mathrm{s})e^{\mathrm{s}t}$, $\mathbf{H}(x, y, z; \mathrm{s})e^{\mathrm{s}t}$ may be written:

$$\begin{aligned} &\text{(a)} \quad \nabla \times \mathbf{E} = -\mathrm{s}\mu\mathbf{H} \\ &\text{(b)} \quad \nabla \times \mathbf{H} = (\sigma + \mathrm{s}\epsilon)\mathbf{E} \end{aligned} \tag{6.81}$$

The parameters ϵ, μ, σ are permitted to vary with position, $\epsilon(x, y, z)$, etc., but passivity requires ϵ, $\mu > 0$ and $\sigma \geq 0$.

To form something like the Poynting vector, we manipulate Eqs. 6.81 to get $\nabla \cdot (\mathbf{E} \times \mathbf{H}^*)$ on the left side. That is, we form from Eq. 6.81 the products

$$\mathbf{H}^* \cdot \nabla \times \mathbf{E} = -\mathrm{s}\mu \, \| \mathbf{H} \|^2 \tag{6.82a}$$

$$\mathbf{E} \cdot \nabla \times \mathbf{H}^* = \sigma \, \| \mathbf{E} \|^2 + \mathrm{s}^*\epsilon \, \| \mathbf{E} \|^2 \tag{6.82b}$$

in which

$$\| \mathbf{E} \|^2 \equiv \mathbf{E} \cdot \mathbf{E}^* \qquad \| \mathbf{H} \|^2 \equiv \mathbf{H} \cdot \mathbf{H}^* \tag{6.82c}$$

Subtraction of Eq. 6.82*a* from Eq. 6.82*b* and the writing of $\mathrm{s} = \Omega + j\omega$, yield

$$\begin{aligned} -\nabla \cdot (\mathbf{E} \times \mathbf{H}^*) = \sigma \, \| \mathbf{E} \|^2 &+ \Omega(\mu \, \| \mathbf{H} \|^2 + \epsilon \, \| \mathbf{E} \|^2) \\ &+ j\omega(\mu \, \| \mathbf{H} \|^2 - \epsilon \, \| \mathbf{E} \|^2) \end{aligned} \tag{6.83}$$

Integration over simply connected volume V, bounded by closed surface S with outward normal $\boldsymbol{n}$, results in the relation

$$-\int_S \boldsymbol{n} \cdot (\mathbf{E} \times \mathbf{H}^*) \, da = F_0 + \mathrm{s}T_0 + \mathrm{s}^*V_0 \tag{6.84}$$

[1] The development here parallels that given for circuits by Ernst A. Guillemin, *Introductory Circuit Theory*, John Wiley and Sons, New York, 1953, Chapter 10.

where

$$F_0 \equiv \int_V \sigma \parallel \mathbf{E} \parallel^2 dv \geq 0$$

$$T_0 \equiv \int_V \mu \parallel \mathbf{H} \parallel^2 dv \geq 0 \tag{6.85}$$

$$V_0 \equiv \int_V \epsilon \parallel \mathbf{E} \parallel^2 dv \geq 0$$

Observe that when $s \neq j\omega$, neither $\mathbf{E} \times \mathbf{H}^*$ nor any of the various terms like $\epsilon \parallel \mathbf{E} \parallel^2$ appearing above actually has the significance of time-average power or stored energy. Their algebraic expressions are nevertheless the same (except for a factor ¼ or ½) as the time-average quantities would be for $s = j\omega$, and they become interpretable as such energies or powers in that case (only).[1] In any case, however, the quantities F_0, T_0, and V_0 are non-negative because of their *mathematical forms*. Noting that F_0, T_0, and V_0 are all *real* functions of s, we find that the foregoing remarks indicate

$$\left.\begin{aligned} [F_0(s)]_{s=j\omega} &= F_0(j\omega) = 2\langle P_d \rangle \\ [T_0(s)]_{s=j\omega} &= T_0(j\omega) = 4\langle W_m \rangle \\ [V_0(s)]_{s=j\omega} &= V_0(j\omega) = 4\langle W_e \rangle \end{aligned}\right\} \quad \omega \neq 0 \tag{6.86}$$

It will be convenient to make one change in Eq. 6.84, by defining

$$V_0'(s) \equiv ss^*V_0(s) = |s|^2 V_0(s) \geq 0 \tag{6.87}$$

which, it should be noted, involves a change in dimensions. On this basis, Eq. 6.84 becomes

$$-\int_S \boldsymbol{n} \cdot (\mathbf{E} \times \mathbf{H}^*) \, da = F_0 + sT_0 + \frac{V_0'}{s} \tag{6.88}$$

Now suppose the region V of space in question is surrounded nearly everywhere by an opaque wall, such that $\boldsymbol{n} \cdot (\mathbf{E} \times \mathbf{H}^*) = 0$ on it. This condition might, for example, arise from a perfect conductor (short circuit) on which $\boldsymbol{n} \times \mathbf{E} = 0$, or from the somewhat less physical dual "magnetic wall" (open circuit) on which $\boldsymbol{n} \times \mathbf{H} = 0$, or from

[1] The stored energy *interpretations* would fail anyway if ϵ and μ were functions of frequency, as pointed out in connection with Eq. 1.58. This circumstance would *not* detract from our present argument regarding F_0, even if σ were a function of frequency, but it *would* alter the conclusions about T_0 and V_0. See Prob. 1.6–1.9, 2.12, and 6.28.

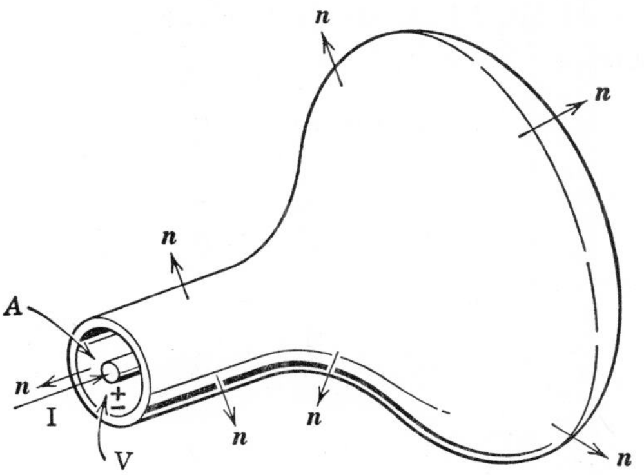

Fig. 6.15. Region of space with a terminal pair for discussion of impedance in terms of fields.

various combinations of these over different portions of the bounding surface. But, as shown in Fig. 6.15, let there be one opening A from which a pair of wires emerges where a voltage and current may be defined accurately. Then with due attention to the meaning of the *outward* normal $\boldsymbol{n}$ in Eq. 6.88, and the directions of voltage and current selected as references in the figure, we find [1]

$$-\int_S \boldsymbol{n}\cdot(\mathbf{E} \times \mathbf{H}^*)\,da = -\int_A \boldsymbol{n}\cdot(\mathbf{E} \times \mathbf{H}^*)\,da$$

$$= \mathrm{VI}^* = |\mathrm{I}|^2\mathrm{Z} = |\mathrm{V}|^2\mathrm{Y}^* \qquad (6.89)$$

That is to say

$$\text{(a)} \qquad \mathrm{Z(s)} = \left(\frac{F_0}{|\mathrm{I}|^2}\right) + \left(\frac{T_0}{|\mathrm{I}|^2}\right)\mathrm{s} + \left(\frac{V_0'}{|\mathrm{I}|^2}\right)\mathrm{s}^{-1}$$

$$(6.90)$$

$$\text{(b)} \qquad \mathrm{Y(s)} = \left(\frac{F_0}{|\mathrm{V}|^2}\right) + \left(\frac{V_0}{|\mathrm{V}|^2}\right)\mathrm{s} + \left(\frac{T_0'}{|\mathrm{V}|^2}\right)\mathrm{s}^{-1}$$

where

$$T_0' \equiv |\mathrm{s}|^2 T_0 \geq 0 \qquad (6.91)$$

again involves a dimensional change.

Because in Z(s) of Eq. 6.90*a* the field integrals F_0, T_0, V_0' are normalized on the square of the input terminal current, they should be

[1] Further discussion of the conditions required to define this voltage and current appears in Chapter 9.

thought of as calculated in terms of this current. The resulting normalized coefficients

$$\begin{aligned} r_0(s) &\equiv \frac{F_0}{|I|^2} \geq 0 \\ l_0(s) &\equiv \frac{T_0}{|I|^2} \geq 0 \\ s_0(s) &\equiv \frac{V_0'}{|I|^2} \geq 0 \end{aligned} \tag{6.92}$$

are then seen to be *independent* of I or V because of the linearity of the system, but they will, in general, be more or less complicated functions of frequency s. The symbols r_0, l_0, and s_0 have been chosen to remind us of the dimensional character of the corresponding quantities: ohms, henrys, darafs. Similarly, in Eq. 6.90*b* for Y(s), F_0, V_0, T_0' should be regarded as calculated in terms of the input *voltage* V. Linearity again makes these field integrals proportional to $|V|^2$, so the normalized coefficients

$$\begin{aligned} g_0(s) &\equiv \frac{F_0}{|V|^2} \geq 0 \\ \mathrm{r}_0(s) &\equiv \frac{T_0'}{|V|^2} \geq 0 \\ c_0(s) &\equiv \frac{V_0}{|V|^2} \geq 0 \end{aligned} \tag{6.93}$$

are functions of s only. Again the symbols are chosen to relate to the corresponding dimensions: mhos, reciprocal henrys, and farads. It follows that we may write

$$\begin{aligned} &\text{(a)} \qquad Z(s) = r_0 + l_0 s + \frac{s_0}{s} \\ &\text{(b)} \qquad Y(s) = g_0 + c_0 s + \frac{\mathrm{r}_0}{s} \end{aligned} \tag{6.94}$$

or, in terms of real and imaginary parts, and $s = \Omega + j\omega$,

$$\left.\begin{aligned} &\text{(a)} \qquad Z(s) = r_0 + \Omega(l_0 + v_0) + j\omega(l_0 - v_0) \\ &\text{(b)} \qquad Y(s) = g_0 + \Omega(c_0 + t_0) + j\omega(c_0 - t_0) \end{aligned}\right\} \tag{6.95}$$

where we have put

$$\text{(a)} \qquad \mathsf{v}_0 \equiv \frac{\mathrm{s}_0}{|\mathrm{s}|^2} = \frac{V_0}{|\mathrm{I}|^2} \geq 0$$

$$\text{(b)} \qquad \mathsf{t}_0 \equiv \frac{\mathrm{r}_0}{|\mathrm{s}|^2} = \frac{T_0}{|\mathrm{V}|^2} \geq 0 \qquad (6.96)$$

In spite of the complicated dependence upon s hidden in the quantities r_0, l_0, s_0, v_0, $\cdots$, etc., their essentially *positive real* character permits us to deduce a great deal from Eqs. 6.94 and 6.95. From Eq. 6.95*a*, for example, it is clear that, regarded as a function of s,

$$\mathrm{Re}\,[Z(\mathrm{s})] = r_0 + \Omega(l_0 + v_0) > 0 \qquad \text{for } \Omega > 0 \qquad (6.97a)$$

and

$$\mathrm{Re}\,[Z(\mathrm{s})] = r_0 \geq 0 \qquad \text{for } \Omega = 0 \qquad (6.97b)$$

with similar conclusions for the admittance from Eq. 6.95*b*. That is to say, *passive driving-point impedance or admittance functions of* s *have positive real parts when* s *lies in the right-half of its complex plane, and non-negative real parts when* s *lies on the imaginary axis.* Such a fact is most important in network synthesis problems because it restricts the class of complex functions amenable to circuit realization. Coupled with the theory of complex variables, for instance, the aforementioned character of Z(s) or Y(s) implies that neither has poles in the right-half of the s-plane. We can arrive at the same conclusion in a simpler way, directly from Eq. 6.94. Thus a zero of Y(s), which is a pole of Z(s), occurs when in Eq. 6.94*b*

$$Y(\mathrm{s}) = 0 = g_0 + c_0\mathrm{s} + \frac{\mathrm{r}_0}{\mathrm{s}}$$

or

$$\mathrm{s}_\nu = -\frac{g_0}{2c_0} \pm j\sqrt{\frac{\mathrm{r}_0}{c_0} - \left(\frac{g_0}{2c_0}\right)^2} = \Omega_\nu \pm j\omega_\nu \qquad (6.98)$$

Evidently Eq. 6.98 shows that s_ν lies either in the *left* half of the s-plane or on the imaginary axis, even though the equation is not really a solution for s_ν *explicitly*. Similarly, zeros of Z(s), i.e., poles of Y(s), are also seen from Eq. 6.94*a* to lie in the left-half s-plane, or on the imaginary axis. Of course, if the system is without losses, $g_0 = r_0 = 0$, and all the zeros or poles of Z(s) or Y(s) lie on the imaginary axis. These analytical results merely confirm the statements made on the basis of physical reasoning in Sec. 6.1.

It is interesting to observe how a zero of Y(s), for instance, shows in

Eq. 6.95*b*. Equating to zero of real and imaginary parts separately leads to

$$\text{(a)} \qquad \Omega_\nu = -\frac{g_0}{c_0 + t_0}$$
$$\text{(b)} \qquad c_0 = t_0 \tag{6.99}$$

or

$$\text{(a)} \qquad \Omega_\nu = \frac{-g_0}{2c_0} = \frac{-g_0}{2t_0} = \frac{-F_0}{T_0 + V_0}$$
$$\text{(b)} \qquad T_0 = V_0 \tag{6.100}$$

which agrees, as far as it goes, with Eq. 6.98. One must, in general, avoid the temptation to interpret Eq. 6.100 in terms of ratios or equalities of energies, because of the comments made previously about the meanings of F_0, T_0, and V_0. In the absence of loss, and with constant ϵ and μ, however, the damping term vanishes and the natural frequency [here with the terminals open, corresponding to $Y(s) = 0$] is pure imaginary. In this case, the equality of T_0 and V_0 *does* indicate equality of time-average stored electric and magnetic energies, as we have so often mentioned before. Later we shall discuss how, in an *approximate* manner, one may retain the energy interpretations when losses are present, but small.

For the moment we turn to an important precise property of lossless systems. Let us first break $Z(s)$ into its real and imaginary parts, for any complex value of $s = \Omega + j\omega$:

$$Z(s) = R(\Omega, \omega) + jX(\Omega, \omega) \tag{6.101}$$

If $Z(s)$ is analytic at the point s in question, the Cauchy-Riemann conditions require

$$\text{(a)} \qquad \frac{\partial X(\Omega, \omega)}{\partial \omega} = \frac{\partial R(\Omega, \omega)}{\partial \Omega}$$
$$\text{(b)} \qquad \frac{\partial X(\Omega, \omega)}{\partial \Omega} = -\frac{\partial R(\Omega, \omega)}{\partial \omega} \tag{6.102}$$

Now for a system without loss, Eq. 6.95*a* shows that

$$R(\Omega, \omega) = \Omega(l_0 + v_0) \tag{6.103}$$

so that by direct calculation

$$\left(\frac{\partial R}{\partial \Omega}\right)_{\Omega=0} = l_0 + v_0 \tag{6.104}$$

Accordingly by Eqs. 6.104 and 6.102*a*

$$\left(\frac{\partial X}{\partial \omega}\right)_{\Omega=0} = \mathfrak{l}_0 + \mathfrak{v}_0 \tag{6.105}$$

If we consider $Z(j\omega)$ only, so that, in the lossless case,

$$Z(j\omega) = jX(\omega) = j\omega(\mathfrak{l}_0 - \mathfrak{v}_0) = \frac{4j\omega\langle W_m - W_e\rangle}{|I|^2} \tag{6.106}$$

we find from Eq. 6.105 that

$$\frac{dX}{d\omega} = \frac{T_0 + V_0}{|I|^2} = \frac{4\langle W_e + W_m\rangle}{|I|^2} = \frac{4\langle W_{\text{total}}\rangle}{|I|^2} > 0 \tag{6.107}$$

as long as $\omega \neq 0$, and indeed from Eqs. 6.104 and 6.106 that

$$\frac{dX}{d\omega} \geq \left|\frac{X}{\omega}\right| > 0 \tag{6.108}$$

A similar development for the admittance shows that, in the lossless case,

$$Y(j\omega) = jB(\omega) = j\omega(\mathfrak{c}_0 - \mathfrak{t}_0) = \frac{4j\omega\langle W_e - W_m\rangle}{|V|^2} \tag{6.109}$$

and

$$\frac{dB}{d\omega} = \mathfrak{c}_0 + \mathfrak{t}_0 = \frac{T_0 + V_0}{|V|^2} > 0 \tag{6.110a}$$

or

$$\frac{dB}{d\omega} = \frac{4\langle W_{\text{total}}\rangle}{|V|^2} \qquad \text{for } \omega \neq 0 \tag{6.110b}$$

so that

$$\frac{dB}{d\omega} \geq \left|\frac{B}{\omega}\right| > 0 \tag{6.111}$$

The content of Eqs. 6.107 through 6.111 for *lossless* systems is known as *Foster's reactance theorem*.[1] It not only shows that the slope versus frequency of a pure reactance or susceptance function is positive but also that it is not less than that of a pure inductor or capacitor having the same reactance value at the frequency in question (Eqs. 6.108 and 6.111). In addition, it relates the reactance and susceptance slopes precisely to the normalized stored energies (Eqs. 6.107 and 6.110).

[1] For a different treatment see Robert M. Fano, Lan Jen Chu, Richard B. Adler, *loc. cit.* It is instructive to consider the case $\epsilon(\omega)$, $\mu(\omega)$, $\sigma \equiv 0$ by both methods in regard to the stored energy for such situations. See also Prob. 6.28 for a similar consideration.

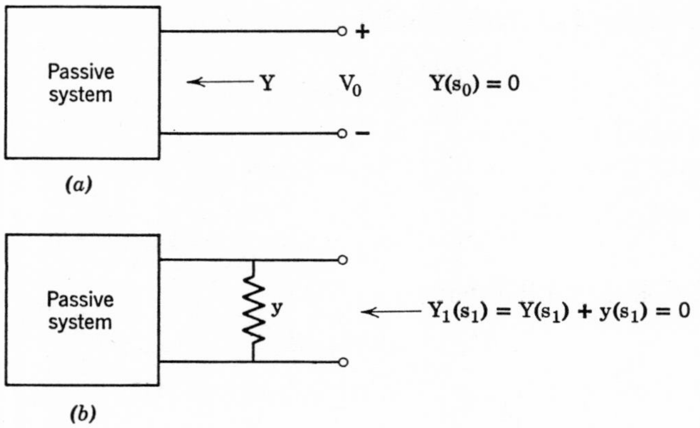

Fig. 6.16. Effect upon natural frequency of adding small admittance to system.

6.4.2 Approximations Using the Energies [1]

In dealing with the effects of added small losses, or with weak coupling between one system and another, it is convenient to have available simplifying approximate techniques to evaluate changes in natural frequencies produced by the alterations. For example, suppose we consider a system as shown in Fig. 6.16*a* having a natural frequency s_0 with a certain pair of terminals open. Let V_0 ($\neq 0$) be the complex open-circuit terminal voltage at this frequency. Then the admittance Y at the terminals obeys the relation

$$Y(s_0) = 0 \tag{6.112}$$

Now suppose y is a *small admittance* added to the terminals, as shown in Fig. 6.16*b*, and let the new natural frequency of the system nearest to the old one s_0 be

$$s_1 = s_0 + \Delta s \tag{6.113}$$

We suppose that y is such that

$$|\Delta s| \ll |s_0| \tag{6.114}$$

In other words, the change in natural frequency produced by adding y where there was previously an open circuit is small compared to the

[1] In a minimum treatment, only the portions of this section dealing with pure resistance perturbations need be treated, and the detailed proof for several simultaneous perturbations may be omitted.

original frequency. Because s_1 is a natural frequency, we know that

$$Y(s_1) + y(s_1) = 0 \tag{6.115}$$

But for small Δs compared with s_0, we have to first order in Δs

$$\text{(a)} \qquad Y(s_1) = Y(s_0) + \left(\frac{dY}{ds}\right)_{s_0} \Delta s + \cdots \cong \left(\frac{dY}{ds}\right)_{s_0} \Delta s$$

$$\text{(b)} \qquad y(s_1) = y(s_0) + \left(\frac{dy}{ds}\right)_{s_0} \Delta s + \cdots \cong y(s_0) + \left(\frac{dy}{ds}\right)_{s_0} \Delta s \tag{6.116}$$

in view of Eq. 6.112. Thus Eq. 6.115 becomes

$$\Delta s \cong \frac{-y(s_0)}{\{[d(Y + y)]/(ds)\}_{s_0}} \tag{6.117}$$

in which everything is evaluated at the *original* natural frequency s_0.

The most interesting applications of Eq. 6.117 arise when the system Y is nondissipative. Then $s_0 = j\omega_0$ and, according to Foster's reactance theorem,

$$\left(\frac{dY}{ds}\right)_{s_0} = \left(\frac{dB}{d\omega}\right)_{\omega_0} = \frac{4\langle W_e + W_m\rangle_{\omega_0}}{|V_0|^2_{\omega_0}} = \frac{4\langle W_{\text{total}}\rangle_{\omega_0}}{|V_0|^2_{\omega_0}} \tag{6.118}$$

for $\omega_0 \neq 0$. We shall wish to discuss two subcases worthy of special mention. First, and probably foremost, the one in which y is just a constant conductance

$$y = g \tag{6.119a}$$

leads to

$$\left(\frac{dy}{ds}\right)_{s_0} = 0 \tag{6.119b}$$

and

$$\Delta s \cong \frac{-g/2|V_0|^2_{\omega_0}}{2\langle W_{\text{total}}\rangle_{\omega_0}} = \frac{-\langle P_d\rangle_{\omega_0}}{2\langle W_{\text{total}}\rangle_{\omega_0}} \qquad \text{for } \omega_0 \neq 0 \tag{6.119c}$$

In other words, the new natural frequency s_1 is

$$s_1 \cong \Omega_1 + s_0 = \Omega_1 + j\omega_0 \qquad \omega_0 \neq 0 \tag{6.120a}$$

with

$$\Omega_1 \cong \frac{\langle P_d\rangle_{\omega_0}}{2\langle W_{\text{total}}\rangle_{\omega_0}} \qquad \omega_0 \neq 0 \tag{6.120b}$$

and

$$\langle P_d\rangle_{\omega_0} = \tfrac{1}{2}g|V_0|^2_{\omega_0} \qquad \omega_0 \neq 0 \tag{6.120c}$$

The effect of adding the loss produces *only damping to the first order*—no change in the imaginary part of the natural frequency occurs to this order. *Moreover, the change in the real part (the damping) can be calculated to first order by using the voltage and current (or field) distributions which existed in the lossless system before any change was made!*

In particular, the "loss" $\langle P_d \rangle_{\omega_0}$ is not in any sense the actual loss at the new damped frequency but is computed as the time-average power which *would be* dissipated in the added conductance if the open-circuit voltage V_0 at frequency $s_0 = j\omega_0$ were impressed upon it in the sinusoidal steady state. This is the significance of Eq. 6.120*c*.

Note also that $\langle W_{\text{total}} \rangle_{\omega_0}$ is the total average stored energy of the system when it is oscillating at the undamped frequency ω_0. This energy will, of course, be proportional to the square of the amplitude of oscillation, and can be expressed in terms of any convenient voltage, current, or field in the system. It is usually advisable, in connection with using Eqs. 6.119*c* or 6.120, to express $\langle W_{\text{total}} \rangle_{\omega_0}$ in terms of $|V_0|^2$, the open-circuit voltage at the place where the loss is to be added. Then the value $|V_0|^2$ will cancel easily (as it must) in the ratios of Eq. 6.119*c* or 6.120*b*.

Regardless of the variable in terms of which we express $\langle P_d \rangle_{\omega_0}$ (or $\langle W_{\text{total}} \rangle_{\omega_0}$), however, the essential point is that the calculation is based *only* on the condition of the *lossless counterpart* of the dissipative system at one of its natural frequencies. *The lossless counterpart of the dissipative system is defined here as the one created by open-circuiting the dissipative element.* Of course, the effect of the loss considered in this manner

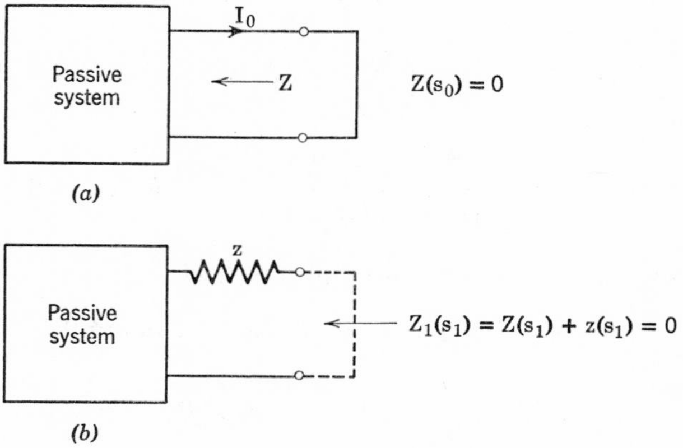

Fig. 6.17. Effect upon natural frequency of adding small impedance to system.

must not produce a large change $|\Delta s|/|s_0|$ when calculated by Eq. 6.119*c* or 6.120, otherwise the whole idea of the first-order theory here presented is invalidated.

On the other hand, there is a perfectly good dual result which may be employed when small impedances rather than small admittances are added to a system. The situation is shown in Fig. 6.17, where frequency s_0 applies to the system with the terminal pairs shown short-circuited. The addition of a small impedance z in (series with) the short circuit changes the natural frequency by Δs, supposed small, such that

$$\Delta s \cong \frac{-z(s_0)}{\{[d(Z + z)/(ds)]\}_{s_0}} \tag{6.121}$$

In case the addition is a small constant resistance

$$z = r \tag{6.122a}$$

added to a lossless system Z for which $s_0 = j\omega_0$, we find

$$\Delta s \cong \frac{-r/2|I_0|^2_{\omega 0}}{2\langle W_{\text{total}}\rangle_{\omega 0}} \qquad \text{for } \omega_0 \neq 0 \tag{6.122b}$$

or

$$s_1 \cong \Omega_1 + j\omega_0 \qquad \omega_0 \neq 0 \tag{6.123a}$$

with

$$\Omega_1 = \frac{-\langle P_d\rangle_{\omega 0}}{2\langle W_{\text{total}}\rangle_{\omega 0}} \qquad \omega_0 \neq 0 \tag{6.123b}$$

and

$$\langle P_d\rangle_{\omega_0} = \tfrac{1}{2}r|I_0|^2_{\omega 0} \qquad \omega_0 \neq 0 \tag{6.123c}$$

In these results, the *lossless counterpart* of the dissipative system, whose undamped natural frequency ω_0 appears in the results and whose current and voltage distribution is implied in the calculation of $\langle W_{\text{total}}\rangle_{\omega_0}$ and $\langle P_d\rangle_{\omega_0}$, is the one obtained by *short-circuiting the dissipative element.*

It becomes clear that one is usually faced with a matter of judgment in deciding whether to use essentially Eq. 6.117 or Eq. 6.121, because we are usually given the dissipative system and we are asked to find some damped complex natural frequency s_1. Granted that we would like to make an approximate calculation, based upon a "lossless counterpart" of the system, how shall we go about removing the loss? Essentially the method is one of trial and error to determine whether short-circuiting or open-circuiting the element in question leads to the smallest value of Δs. Matters are complicated somewhat by the fact that various loss elements in the system may have to be treated differently in this respect and, moreover, that these differences also vary with the particular natural frequency being considered. The only "rule" we can discuss in this connection is the important, but somewhat subtle, implication arising from comparisons of Eqs. 6.120 and 6.123 with Eq. 6.100. The latter, an exact result calculated at the actual damped frequency, will agree most closely with that one of the approximate formulations of Eqs. 6.120 or 6.123 for which the system voltage and current distributions are changed least in passing from the dissipative to the nondissipative system. Thus, if the situation seems to be one in which short-circuiting out a resistor will not much alter the current through its branch, Eqs. 6.123 and their reference lossless system are indicated. On the other hand, if open-circuiting the resistor does not alter greatly the voltage across its nodes, the system pertinent to Eqs. 6.120 is suggested. It is on this basis that the decision is often made very easily by inspection.

A somewhat different picture of our results so far is worth giving at this point to clarify the nature of the approximations. Let us now imagine a system actually to be oscillating freely at one of its complex natural frequencies $s_0 = -\Omega_0 + j\omega_0$. Then all currents and voltages in it will, in the time domain, be of the general form $A_0 e^{-\Omega_0 t} \cos (\omega_0 t + \varphi)$, where A_0 and φ may be functions of space position (z for a transmission-line). The total instantaneous stored energy $W(t)$ in the system must correspond, requiring that

$$W(t) = e^{-2\Omega_0 t}[W_0 + W_2 \cos (2\omega_0 t + 2\varphi)]$$

where W_0 and W_2 are real constants. Consequently, the instantaneous power $P(t)$ dissipated in the system must be given by

$$P(t) = -\frac{dW(t)}{dt} = 2\Omega_0 W(t) + 2\omega_0 W_2 e^{-2\Omega_0 t} \sin (2\omega_0 t + 2\varphi)$$

Our interest often lies particularly in the case of small damping, for which the amount of decay in current or voltage amplitude during the time of one alternating cycle is small, i.e., in time $2\pi/\omega_0$, the current or voltage amplitude decay factor $e^{-\Omega_0 t}$ itself changes by the factor $e^{-2\pi\Omega_0/\omega_0}$, which we wish to assume is nearly 1. Evidently this demands that

$$\Omega_0 \ll \frac{\omega_0}{2\pi}$$

If this inequality does hold, we can define approximate "time-average" energy and power by computing their averages in time over only a *single* alternation. These "averages" will themselves then be functions of time, which we shall denote by $\overline{W}(t)$ and $\overline{P}(t)$. With the small damping approximation, $W(t)$ and $P(t)$ so averaged yield

(a) $$\overline{W}(t) \cong W_0 e^{-2\Omega_0 t}$$

(b) $$\overline{P}(t) \cong 2\Omega_0 \overline{W}(t)$$

leading to

$$\Omega_0 \cong \frac{\overline{P}(t)}{2\overline{W}(t)} = \frac{P_0}{2W_0}$$

where we have written $\overline{P}(t) = P_0 e^{-2\Omega_0 t}$. Observe that P_0 and W_0 are *exactly* F_0 and $(T_0 + V_0)$ of Eqs. 6.85 and 6.100. The approximation here only consists in *interpreting* them as energy and power in an appropriate "average" sense.

We can relate this procedure to that implied by Eqs. 6.120 or 6.123, as follows. When the system is undergoing a natural oscillation with small damping, the oscillation can be thought of as an undamped sinusoid whose amplitude changes very gradually with time. If the system were lossless, there would be no damping, and a certain average stored energy W_0 would reside in it. On the other hand, in the presence of small losses, there would be an average power P_0 dissipated, if we had a true sinusoidal steady state. During a natural oscillation, however, this power must be drawn from the stored energy, and accounts (on the average) for its time rate of decrease. Over *short* times, the system appears to be oscillating steadily; over *long* times, it is seen to be damped. We treat this situation by discussing the long-time changes in the short-time averages. Therefore we may calculate the (short-time) average power dissipated ($\approx P_0$) and the (short-time) average total energy stored ($\approx W_0$), by regarding the system as being *driven* in the sinusoidal steady state at frequency $s = j\omega_0$. The fact that these calculations may, in turn, be performed by using the current and voltage distributions in the (undamped) system natural oscillation at a frequency near ω_0, *which arises with the losses removed in an appropriate manner*, is reasonable, but it is very far from obvious. It is really this point for which the results of Eqs. 6.117 and 6.121 are needed. There are other ways to arrive at the same conclusion, but the one adopted here in connection with those equations seems simplest and most direct for our purposes.

The second important application of Eq. 6.117 or 6.121 arises in determining the effect on a natural frequency of perturbing a system with *reactive* elements. If not only Y is reactive but also y in Eq. 6.117, we have

$$\mathrm{y}(\mathrm{s}_0) = \mathrm{y}(j\omega_0) = jb(\omega_0) \tag{6.124}$$

and for $\omega_0 \neq 0$

$$\left[\frac{d(B+b)}{d\omega}\right]_{\omega_0} = \frac{4\langle W_e + W_m + w_e + w_m\rangle_{\omega_0}}{|\mathrm{V}_0|^2_{\omega_0}} \tag{6.125}$$

where the small w's refer to stored energies in the element y. But when $\omega_0 \neq 0$

$$|\mathrm{V}_0|^2_{\omega_0} b(\omega_0) = 4\omega_0\langle w_e - w_m\rangle_{\omega_0} \tag{6.126}$$

As long as $\omega_0 \neq 0$, therefore

$$\Delta \mathrm{s} \cong \frac{-j\omega_0\langle w_e - w_m\rangle_{\omega_0}}{\langle W_e + W_m + w_e + w_m\rangle_{\omega_0}} \cong \frac{-j\omega_0\langle w_e - w_m\rangle_{\omega_0}}{\langle W_{\text{total}}\rangle_{\omega_0}} \tag{6.127}$$

where the second form is usually (but not always) valid if Δs is indeed small. Equation 6.127 relates to the change in natural frequency resulting from adding a small susceptance across a pair of nodes otherwise open-circuited. A similar result, obtained from Eq. 6.121, applies to adding a small reactance in a branch otherwise short-circuited, provided $\omega_0 \neq 0$.

$$\Delta \mathrm{s} \cong \frac{-j\omega_0\langle w_m - w_e\rangle_{\omega_0}}{\langle W_e + W_m + w_e + w_m\rangle_{\omega_0}} \cong \frac{-j\omega_0\langle w_m - w_e\rangle_{\omega_0}}{\langle W_{\text{total}}\rangle_{\omega_0}} \tag{6.128}$$

Again the decision whether to "remove" the perturbing element by short- or open-circuiting it depends upon which produces less change in the over-all voltage and current distributions in the system.

When the added element y is neither entirely real nor entirely reactive, *but when its frequency dependence is relatively weak compared to that of the nondissipative* Y, we may put

$$\left|\left(\frac{d\mathrm{y}}{d\mathrm{s}}\right)\right|_{j\omega_0} \ll \left(\frac{dB}{d\omega}\right)_{\omega_0} \tag{6.129}$$

and find from Eq. 6.117 for $\omega_0 \neq 0$

$$\Delta \mathrm{s} \cong \frac{-\frac{1}{2}|\mathrm{V}_0|^2_{\omega_0}\mathrm{y}(j\omega_0)}{2\langle W_{\text{total}}\rangle_{\omega_0}} \tag{6.130}$$

with an exactly similar result applying to the case of Eq. 6.121. Evidently, the criterion for smallness of the effect is that *both* the real and reactive powers taken by the perturbing element must be small compared to the peak energy stored in the unperturbed system, the voltage (or current in the dual case) at the branch being perturbed serving as the common amplitude level for both calculations.

The preceding simple development may at first leave us in some doubt about what to do when small dissipative admittances or impedances are added simultaneously at several points in the original lossless system. The "first one," in a manner of speaking, can always be added to a lossless system, but the rest are then added to a slightly dissipative one. We should not, however, be too surprised to find that a first-order calculation can be done, neglecting these mutual interaction effects, because we have already made the interpretation that the success of the previous method depends upon the fact that the added elements do not individually alter very much the current and voltage (or field) distributions throughout the system. Of course, in the same interpretation, the whole first-order method will not be expected to be valid unless all the changes *together* also do not produce large variations from the unperturbed distributions. Accordingly, the appropriate generalization of Eq. 6.130 or its impedance counterpart should be obtained by simply adding up the effects which each element would have produced alone, neglecting all cross effects. Specifically, if the added elements have relatively weak frequency dependence, we expect to find to first order

$$\Delta \mathrm{s} \cong - \left[\frac{\sum_k \frac{1}{2} |\mathrm{I}_k|^2_{\omega_0} \mathrm{z}_k(j\omega_0) + \sum_i \frac{1}{2} |\mathrm{V}_i|^2_{\omega_0} \mathrm{y}_i(j\omega_0)}{2\langle W_{\text{total}} \rangle_{\omega_0}} \right]_{\omega_0 \neq 0} \tag{6.131}$$

in which the sums on k and i extend respectively over previously short-circuited branches in which small impedances z_k have been added, and over previously open-circuited node pairs across which small admittances y_i have been added. As a practical matter, to use Eq. 6.131 we must express $\langle W_{\text{total}} \rangle_{\omega_0}$, $|\mathrm{I}_k|^2_{\omega_0}$ and $|\mathrm{V}_i|^2_{\omega_0}$ all in terms of a *single* current or voltage (or field) somewhere in the lossless system (the one for which $\mathrm{z}_k = \mathrm{y}_i = 0$). This requires knowing in detail the space pattern of the natural undamped oscillation at frequency ω_0.

By way of actually proving Eq. 6.131, which we have so far only made plausible by physical interpretation, let us examine the situation illustrated in Fig. 6.18. In Fig. 6.18*b*, a small impedance z_k has been added to a short-circuited branch k, and a small admittance y_i has been added across an open-circuited node pair i, in a system whose natural oscillation in Fig. 6.18*a* previously occurred at complex frequency s_0. The new natural frequency of oscillation is s_1, with $\Delta \mathrm{s} \equiv \mathrm{s}_1 - \mathrm{s}_0$ being regarded as small.

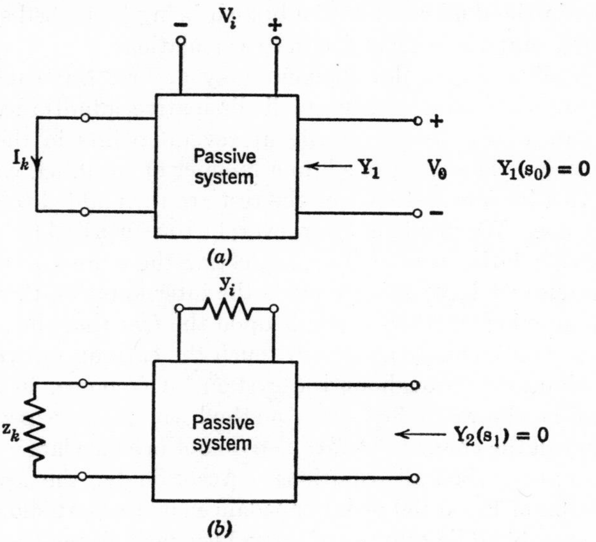

Fig. 6.18. Effect upon natural frequency of adding both a small admittance and a small impedance in (b) to the original system (a).

Now at any frequency s, we have for small z_k and y_i

$$Y_2(s) \cong Y_1(s) + \left(\frac{\partial Y_2}{\partial z_k}\right)_{\substack{z_k, y_i = 0 \\ s = s}} z_k(s) + \left(\frac{\partial Y_2}{\partial y_i}\right)_{\substack{y_i, z_k = 0 \\ s = s}} y_i(s)$$

But with Δs small

$$0 = Y_2(s_1) \cong Y_2(s_0) + \left(\frac{dY_2}{ds}\right)_{s_0} \Delta s$$

From the expression for $Y_2(s)$ above, however, we have directly

$$Y_2(s_0) = \left(\frac{\partial Y_2}{\partial z_k}\right)_{\substack{z_k, y_i = 0 \\ s = s_0}} z_k(s_0) + \left(\frac{\partial Y_2}{\partial y_i}\right)_{\substack{z_k, y_i = 0 \\ s = s_0}} y_i(s_0)$$

because $Y_1(s_0) = 0$. Then, by straightforward differentiation of $Y_2(s)$ with respect to s, we find

$$\left(\frac{dY_2}{ds}\right)_{s_0} = \left(\frac{dY_1}{ds}\right)_{s_0} + \left(\frac{\partial Y_2}{\partial z_k}\right)_{\substack{z_k, y_i = 0 \\ s = s_0}} \left(\frac{dz_k}{ds}\right)_{s_0} + \left(\frac{\partial Y_2}{\partial y_i}\right)_{\substack{y_i, z_k = 0 \\ s = s_0}} \left(\frac{dy_i}{ds}\right)_{s_0}$$
$$+ \left(\frac{\partial^2 Y_2}{\partial z_k \, \partial s}\right)_{\substack{z_k, y_i = 0 \\ s = s_0}} z_k(s_0) + \left(\frac{\partial^2 Y_2}{\partial y_i \, \partial s}\right)_{\substack{y_i, z_k = 0 \\ s = s_0}} y_i(s_0)$$

It is clear from the requirement $Y_2(s_1) = 0$ at the new natural frequency, that

$$\Delta s \cong -\frac{Y_2(s_0)}{(dY_2/ds)_{s_0}}$$

and that Δs is therefore proportional to z_k and y_i. This means that products $(\Delta s)z_k$ and $(\Delta s)y_i$ are of *second order* in magnitude and may be dropped.

But if $y_i = 0$, two-terminal-pair network theory shows that

$$(Y_2)_{y_i=0} = y_{11}' - \frac{(y_{1k}')^2}{(1/z_k) + y_{kk}'}$$

where y_{11}', y_{1k}', and y_{kk}' refer to the special terminal loading situations of Fig. 6.18*a*, i.e., *terminals k short-circuited and terminals i open-circuited.* Accordingly, by differentiation,

$$\left(\frac{\partial Y_2}{\partial z_k}\right)_{\substack{y_i=0\\ z_k=0}} = \left[\frac{-(y_{1k}')^2}{(1 + y_{kk}'z_k)^2}\right]_{z_k=0} = -(y_{1k}')^2$$

Similarly, if $z_k = 0$,

$$(Y_2)_{z_k=0} = y_{11} - \frac{y_{1i}^2}{y_i + y_{ii}}$$

where y_{11}, y_{1i}, and y_{ii} now refer to the more standard-notation conditions in which *both terminal pairs k and i are short-circuited.* Accordingly, by differentiation again,

$$\left(\frac{\partial Y_2}{\partial y_i}\right)_{\substack{z_k=0\\ y_i=0}} = \left[\frac{+y_{1i}^2}{(y_i + y_{ii})^2}\right]_{y_i=0} = +\left(\frac{y_{1i}}{y_{ii}}\right)^2$$

Now, we observe from Fig. 6.18*a* that

$$y_{ik}'(s_0) = \left(\frac{I_k}{V_0}\right)_s$$

and

$$\frac{y_{1i}(s_0)}{y_{ii}(s_0)} = \left(\frac{V_i}{V_0}\right)_{s_0}$$

If the original system is lossless, $s_0 = j\omega_0$, $y_{1k}'(j\omega_0)$ is *pure imaginary* and $(y_{1i}/y_{ii})_{j\omega_0}$ is *pure real.* Hence in the lossless case

$$[y_{1k}'(j\omega_0)]^2 = -\frac{|I_k|^2_{\omega_0}}{|V_0|^2_{\omega_0}} \qquad \frac{y_{1i}^2(j\omega_0)}{y_{ii}^2(j\omega_0)} = \frac{|V_i|^2_{\omega_0}}{|V_0|^2_{\omega_0}}$$

and we know also that

$$\left(\frac{dY_1}{ds}\right)_{j\omega_0} = \frac{4\langle W_e + W_m\rangle_{\omega_0}}{|V_0|^2_{\omega_0}}$$

It follows that for an originally lossless system

$$\Delta s \cong \frac{|I_k|^2_{\omega_0} z_k(j\omega_0) + |V_i|^2_{\omega_0} y_i(j\omega_0)}{4\langle W_e + W_m\rangle_{\omega_0} + |I_k|^2_{\omega_0}(dz_k/ds)_{j\omega_0} + |V_i|^2_{\omega_0}(dy_i/ds)_{j\omega_0}}$$

which for a weak enough frequency dependence of z_k and y_i does indeed give the form of Eq. 6.131. It is quite obvious that the extension to more elements z_k and y_i is straightforward, requiring only the consideration of more partial derivatives

$$(\partial Y_2/\partial z_\nu)_{\text{all } z_\nu, y_\mu = 0}$$

and

$$(\partial Y_2/\partial y_\mu)_{\text{all } y_\mu, z_\nu = 0}$$

but never more than two-terminal-pair network analysis with certain specified combinations of terminal pairs open-circuited, and others short-circuited, in each case.

The most important feature of the proof just completed is that it supports a physical picture of the large "inertia" of the current and voltage (or field) space distributions of natural oscillations in nondissipative systems; for it is on that basis that we could write Eq. 6.131 *completely by inspection.*

6.4.2.1 SMALL LOSSES. A simple example, for the case shown in Fig. 6.19*a*, will make clearer the implications of these approximate calculations regarding the effect of small losses. We wish to locate approximately the lowest oscillatory complex natural frequency of the structure when the losses in R, G, and $r \ll |Z_0|$ are small. For this purpose, we examine first *the lossless counterpart of the system*, shown in

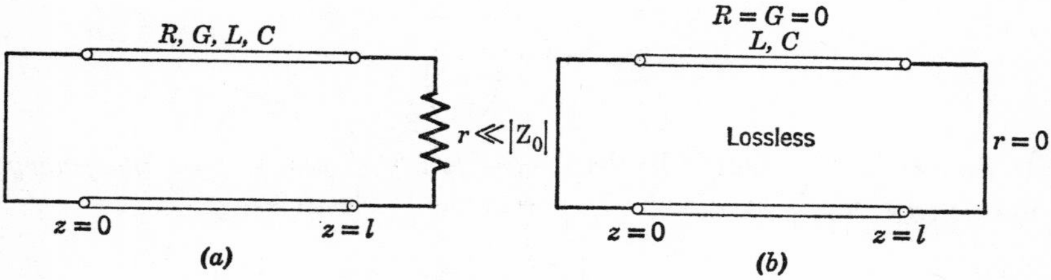

Fig. 6.19. (*a*) A simple low-loss system and (*b*) its lossless counterpart.

Fig. 6.19b. We know from Sec. 6.1.1.1 that the lowest (nonzero) natural frequency of oscillation for the lossless system occurs when

$$l = \frac{1}{2}\lambda = \frac{\pi}{\beta} = \frac{\pi}{\omega_0\sqrt{LC}} = \frac{\pi v}{\omega_0}$$

or

$$\omega_0 = \frac{\pi v}{l} = \frac{\pi}{T} \tag{6.132a}$$

Under this condition, the complex current and voltage on the line have the "half-wave" forms

$$\mathrm{I}(z) = \mathrm{I}_0 \cos\left(\frac{\pi z}{l}\right) \tag{6.132b}$$

$$\mathrm{V}(z) = -jZ_0\mathrm{I}_0 \sin\left(\frac{\pi z}{l}\right) \tag{6.132c}$$

where I_0 is the current in the short-circuit at $z = 0$.

The average stored magnetic energy in the line of Fig. 6.19b at its "half-wave" resonance is therefore

$$\begin{aligned}\langle W_m\rangle_{\omega_0} &= \frac{1}{4}\int_0^l L|\mathrm{I}(z)|^2\,dz = \frac{1}{4}L|\mathrm{I}_0|^2\int_0^l \cos^2\left(\frac{\pi z}{l}\right)dz \\ &= \frac{1}{8}Ll|\mathrm{I}_0|^2 \end{aligned}\tag{6.133}$$

The average stored electric energy need not be calculated separately because the line is in resonance, i.e., $\langle W_e\rangle_{\omega_0} = \langle W_m\rangle_{\omega_0}$. Hence

$$\langle W_{\text{total}}\rangle_{\omega_0} = \langle W_e + W_m\rangle_{\omega_0} = \langle 2W_m\rangle_{\omega_0} = \tfrac{1}{4}Ll|\mathrm{I}_0|^2 \tag{6.134}$$

which will appear in the denominator of Eq. 6.131. Note that in the lossless case the instantaneous magnetic energy oscillates between zero and twice its average value. Hence we are equally free to use for $\langle W_{\text{total}}\rangle$ the *peak* magnetic stored energy (or the equal *peak* electric stored energy).

If the losses are small in Fig. 6.19a, the distributions of current and voltage on the line will differ little from those in Fig. 6.19b (as given in Eq. 6.132). Thus the average power dissipated is nearly

$$\begin{aligned}\langle P_d\rangle &= \tfrac{1}{2}|\mathrm{I}(l)|^2 r + \tfrac{1}{2}\int_0^l G|\mathrm{V}|^2\,dz + \tfrac{1}{2}\int_0^l R|\mathrm{I}|^2\,dz \\ &= \tfrac{1}{2}|\mathrm{I}_0|^2 r + \tfrac{1}{4}lGZ_0^2|\mathrm{I}_0|^2 + \tfrac{1}{4}lR|\mathrm{I}_0|^2 \\ &= \tfrac{1}{2}|\mathrm{I}_0|^2(r + \tfrac{1}{2}GlZ_0^2 + \tfrac{1}{2}Rl)\end{aligned}\tag{6.135}$$

as we are directed to compute it by Eq. 6.131. Note that we have chosen I_0 as the quantity in terms of which to express all energy and power, and that this choice was made automatically by our choice of Eq. 6.132 for V(z) and I(z). Accordingly,

$$-\Delta s_0 \cong -\Omega_0 \cong \frac{r + \frac{1}{2}GlZ_0{}^2 + \frac{1}{2}Rl}{lL} = \left(\frac{r}{l} + \frac{1}{2}R\right)\frac{1}{L} + \frac{1}{2}\frac{G}{C} \tag{6.136}$$

which, together with Eq. 6.132a, defines completely (but approximately) the first oscillatory complex natural frequency s_0 in Fig. 6.19a.

It is interesting to compare the approximate result (Eq. 6.136) with our previous exact results for Figs. 6.3b and 6.3c respectively. In the first case, $r \equiv 0$, and Eq. 6.136 gives for Ω_0 *exactly* the correct result (Eq. 6.35b) for Fig. 6.3b! This occurs because the dissipative system in this case actually does have the *same* voltage and current distribution on it in the damped oscillation as the lossless one does in the undamped one! Thus the exact Eq. 6.100a and the approximate Eq. 6.131 are, in fact, identical. But we do not get the correct imaginary part for s_0 because we have, by definition, missed the second-order correction involving Ω'' from Eqs. 6.35c and 6.36.

The situation of Fig. 6.3c corresponds to $R \equiv G \equiv 0$, and we know that the current and voltage distributions in the exact solution are *not* identical with the lossless ones this time. Therefore we expect Eq. 6.136 to give only an approximation to the exact solution (Eq. 6.43b). The latter yields for the case $r_n \ll 1$ (very small loss, with r_n nevertheless <1) the following damping when $m = 1$ (first oscillatory natural frequency):

$$\begin{aligned} \Omega_1 T &= -\tfrac{1}{2}[\ln(1 + r_n) - \ln(1 - r_n)] \\ &\cong -\tfrac{1}{2}(2r_n) = -r_n \end{aligned}$$

or

$$\Omega_1 \cong -\frac{r}{Z_0 l\sqrt{LC}} = -\frac{r}{lL}$$

which does indeed agree with Eq. 6.136. Again, we cannot say much about the imaginary part of the complex frequency (which happens to be exactly the lossless value $j(\pi/T)$ by Eq. 6.43a for $m = 1$) because changes in it caused by small losses are at worst of second order.

Observe that in the example just above

$$\frac{|\Omega_0|}{\omega_0} \approx \frac{r}{lL(\pi/l\sqrt{LC}\,)} = \frac{r}{\pi Z_0}$$

which is really small only if $r \ll Z_0$. This does not mean that the condition $r \gg Z_0$ is necessarily unmanageable by our approximation methods. Rather it means that such circumstances must be referred to a lossless counterpart with end $z = l$ *open-circuited* rather than short-circuited. The lowest natural frequency of the lossless system is then the "quarter-wave" oscillation $l = \lambda/4$ or $\omega_0' = (\pi v)/(2l) = \pi/2T$. If V_0 is the voltage across the open end, the total energy stored is

$$\langle W_{\text{total}}\rangle_{\omega_0} = \tfrac{1}{4}Cl|\mathrm{V}_0|^2$$

and the power dissipated is

$$\langle P_d\rangle \approx \frac{1}{2}\frac{|\mathrm{V}_0|^2}{r} + \frac{l}{4}G|\mathrm{V}_0|^2 + \frac{l}{4}\frac{R}{Z_0^2}|\mathrm{V}_0|^2$$

Therefore by Eq. 6.131

$$\Delta s_0 \cong \Omega_0' \cong -\left(\frac{1}{rlC} + \frac{G}{2C} + \frac{R}{2L}\right)$$

which is exact again for the case $r = \infty$, and approximate when $G \equiv R \equiv 0$. In the latter situation, the correct result (Eq. 6.44) reads for the approximation $r_n \gg 1$

$$\Omega_1'T = \frac{1}{2}[\ln(r_n - 1) - \ln(r_n + 1)]$$

$$= \frac{1}{2}\left[\ln\left(1 - \frac{1}{r_n}\right) - \ln\left(1 + \frac{1}{r_n}\right)\right] \cong -\frac{1}{r_n}$$

or

$$\Omega_1' \cong -\frac{1}{rlC}$$

in agreement with the conclusion from Eq. 6.131. As regards the status of the approximation this time, we have

$$\frac{|\Omega_1'|}{\omega_0'} \cong \frac{2}{\pi}\frac{Z_0}{r}$$

which is small only when $r \gg Z_0$, the opposite case to our previous one.

6.4.2.2 SMALL REACTIVE EFFECTS.* The effect of reactive perturbations may now be compared with the previous example by considering the problem of Fig. 6.20. First we suppose that C' in Fig. 6.20*a* is small, and look for the lowest natural frequency. The nearest unperturbed arrangement is Fig. 6.20*b*, in which C' has been removed by *opening* it ($C' \to 0$). The lowest corresponding natural frequency corresponds to the quarter-wave oscillation $\omega_0 = \pi/(2T)$. If V_0 is the

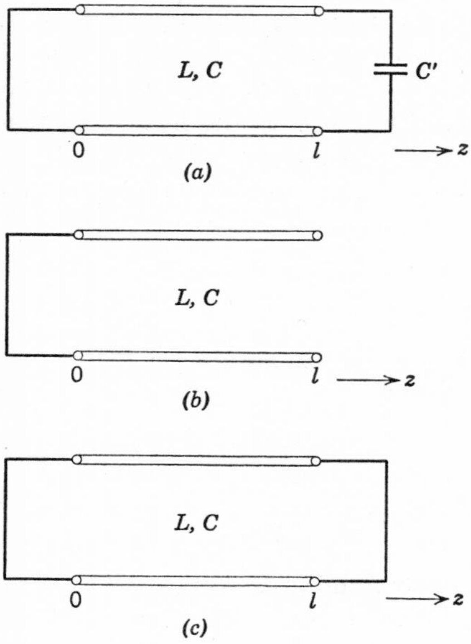

Fig. 6.20. Effect of reactive perturbation C' on a lossless system. The unperturbed system is (b) if C' is small and (c) if C' is large.

voltage at the open end, the average total stored energy is $\frac{1}{4}Cl|\mathrm{V}_0|^2$ and the term $|\mathrm{V}_0|^2\mathrm{y}(j\omega_0)$ is $j\omega_0 C'|\mathrm{V}_0|^2$. Thus the change $\Delta \mathrm{s}_0 = j\,\Delta\omega_0$ is given by

$$\Delta\omega_0 = \frac{-\omega_0 C'}{Cl} \tag{6.137}$$

and the new frequency ω_0' is

$$\omega_0' = \omega_0 + \Delta\omega_0 = \omega_0\left(1 - \frac{C'}{Cl}\right) \tag{6.138}$$

which is lower than the original ω_0 by the fraction C'/Cl. This ratio, of course, compares the added capacitance C' with the total static line capacitance lC.

It is easy to see that a reduction of natural frequency makes sense in the above case. With a small C' in place, the line itself must store a little more magnetic than electric energy if the *whole system* is to have equal amounts of both types of energy. Thus, the line length must be a little less than an odd multiple of a quarter wave length. Therefore, the frequency is lower than that for which the line length is exactly an odd multiple of $\lambda/4$.

The case for large values of C' is also worth examining. The thought is that the impedance of C' is then small at any natural frequency of interest, and the corresponding unperturbed line structure should be the one shown in Fig. 6.20*c*. One must keep in mind, however, the one-to-one correspondence which is presumed to exist between natural frequencies in the perturbed and unperturbed systems when asking, for example, for the lowest natural frequency of Fig. 6.20*a*. Although it is certain that $\omega = 0$ is *not* a natural frequency of that system, it *is* a natural frequency of Fig. 6.20*c*. Presumably, therefore, a large enough value of C' will lead to a lowest natural oscillation of Fig. 6.20*a* which is "close" to zero in frequency. But, of course, "close" cannot mean a small percentage here.

Moreover, since the unperturbed frequency is zero, the added element C' has $z(0) = \infty$, $y(0) = 0$ which makes trouble in Eqs. 6.117, 6.121, or 6.131. A review of the previous derivations show that they break down in the present case, because $Z(s_0) = Z(0) = 0$, while $y(s_0) = y(0) = 0$—a combination of circumstances not contemplated earlier. The derivation is easy enough to correct, on the other hand, if we recognize that Δs_0 now becomes a change from *zero*, so Δs_0 might as well be called s_0' itself. Accordingly

$$\text{(a)}\quad Z(s_0') = Z(0) + \left(\frac{dZ}{ds}\right)_{s=s_0}(s_0' - s_0) + \cdots \cong \left(\frac{dZ}{ds}\right)_{s=0} s_0' \tag{6.139}$$

$$\text{(b)}\quad z(s_0') = \frac{1}{y(s_0')} = \frac{1}{y(s_0) + (dy/ds)_{s=s_0}(s_0' - s_0) + \cdots}$$

$$\cong \frac{1}{(dy/ds)_{s=0}s_0'}$$

and by the method based upon Fig. 6.17*b*, we have

$$s_0'^2 \cong -\frac{1}{(dy/ds)_{s=0}(dZ/ds)_{s=0}} \tag{6.140}$$

Applied to Fig. 6.20, Eq. 6.140 requires that we find

$$y = C's$$

and

$$\frac{dy}{ds} = C' = \left(\frac{dy}{ds}\right)_{s=0} \tag{6.141}$$

Also since

$$Z = Z_0 \tanh sT$$

the derivative is

$$\left(\frac{dZ}{ds}\right)_{s=0} = \left(\frac{Z_0 T}{\cosh^2 sT}\right)_{s=0} = Ll \tag{6.142}$$

Alternatively, these derivatives could be calculated from the energy forms of Foster's reactance theorem, but because the unperturbed frequency is zero, some factors of 2 must be changed from our previous results. Thus, although Eqs. 6.105 and 6.110*a* are correct, Eqs. 6.107 and 6.110*b* must be changed for the case $\omega = 0$ to read

$$\begin{aligned}
&\text{(a)} \quad \left(\frac{dX}{d\omega}\right)_{\omega=0} = \frac{2\langle W_e + W_m\rangle_{\omega=0}}{I^2} \\
&\text{(b)} \quad \left(\frac{dB}{d\omega}\right)_{\omega=0} = \frac{2\langle W_e + W_m\rangle_{\omega=0}}{V^2} \qquad (6.143) \\
&\text{(c)} \quad \langle W_e\rangle_{\omega=0} = \frac{1}{2}\int_{\text{Vol}} \epsilon|\mathbf{E}|^2\,dv \qquad \langle W_m\rangle_{\omega=0} = \frac{1}{2}\int_{\text{Vol}} \mu|\mathbf{H}|^2\,dv
\end{aligned}$$

Then for the element C', we would have by Eq. 6.143*b*

$$\left(\frac{db}{d\omega}\right)_{\omega=0} = \left(\frac{dy}{ds}\right)_{s=0} = \frac{2\frac{1}{2}C'V^2}{V^2} = C' \qquad (6.144a)$$

and for the line section by Eq. 6.143*a*

$$\left(\frac{dX}{d\omega}\right)_{\omega=0} = \left(\frac{dZ}{ds}\right)_{s=0} \quad \frac{2\frac{1}{2}LlI^2}{I^2} = Ll \qquad (6.144b)$$

in agreement with Eqs. 6.141 and 6.142. Note that in this case however, we cannot calculate both energies with a common current or voltage on both elements. At zero frequency, the line contains only magnetic energy and the capacitor contains only electric energy; thus the line has no voltage across it and the capacitor has no current through it. The fact is that, in the present case, the voltage and current distributions for the lowest oscillation frequency of Fig. 6.20*a* are quite *different* than they are at zero frequency in Fig. 6.20*c*. Therefore the approximation implied by Eq. 6.140 is a *quasi-static* one rather than one arising from a perturbation of the space pattern of a natural oscillation. This is clear when we use Eqs. 6.141 and 6.142 in Eq. 6.140 and find

$$-s_0'^2 = \omega_0'^2 = \frac{1}{(Ll)C'} \qquad (6.145)$$

which evidently is the result of the (electrically short) line section acting like an inductor, resonating with the C' as a capacitor.

Equation 6.145 shows that for large C' the lowest frequency of Fig. 6.20*a* is above that of Fig. 6.20*c*. The same is true of the second natural frequency. For Fig. 6.20c, this is the "half-wave" frequency $\omega_1 = \pi/T$, and Eq. 6.121 or Eq. 6.131 gives for large C' in Fig. 6.20a

$$\Delta s_1 = \frac{(-|I_0|^2)/(j\omega_1 C')}{Ll|I_0|^2} = \frac{j}{\omega_1 LlC'} \tag{6.146}$$

or

$$\omega_1' = \omega_1 + \Delta\omega_1 = \omega_1\left(1 + \frac{1}{\omega_1^2 LlC'}\right)$$

$$= \omega_1\left(1 + \frac{lC}{\pi^2 C'}\right) \tag{6.147}$$

The small-change requirement in this case occurs if $C' \gg Cl$, in contrast with the results of Eq. 6.138.

Of course the problem of Fig. 6.20a can be solved exactly, and for comparison purposes here it is worth while to do so. Use of Eq. 6.11a at $z = l$, and acknowledgement that only pure imaginary values of s can be solutions for this lossless system, lead to the relation

$$-jY_0 \cot \beta l = -j\omega C'$$

or

$$Y_0 \cot \beta l = \omega(l\sqrt{LC})\frac{C'}{(l\sqrt{LC})} = \beta l\sqrt{\frac{C}{L}}\left(\frac{C'}{Cl}\right)$$

or

$$\cot \beta l = \left(\frac{C'}{Cl}\right)\beta l \tag{6.148}$$

Observe that $\beta l = \omega l\sqrt{LC} = \omega T$, so variation of βl for a fixed line type and length is essentially variation of frequency. We may plot both sides of Eq. 6.148 versus βl, as shown in Fig. 6.21, and discover the natural frequencies from the values of βl at which the curves intersect. Only the positive half of the βl-axis is shown, because the odd character of both curves means that corresponding to any intersection $\beta_i l > 0$, there is always another one at $-\beta_i l < 0$. The complex natural frequencies always come in conjugate pairs.

Given a particular value of C', it is clear that the high natural frequencies (m large) always approach from above $\omega_m' T = m\pi$, i.e., those of the short-circuited system, Fig. 6.20c; but the smaller is C', the larger m must be before the approach is at all close. In another way, given a particular natural frequency of interest, for instance the lowest, ω_0', it always lies between the corresponding ones (lowest) of Figs. 6.20b and 6.20c. Specifically, it approaches the latter from above when $C' \to \infty$ and the former from below when $C' \to 0$.

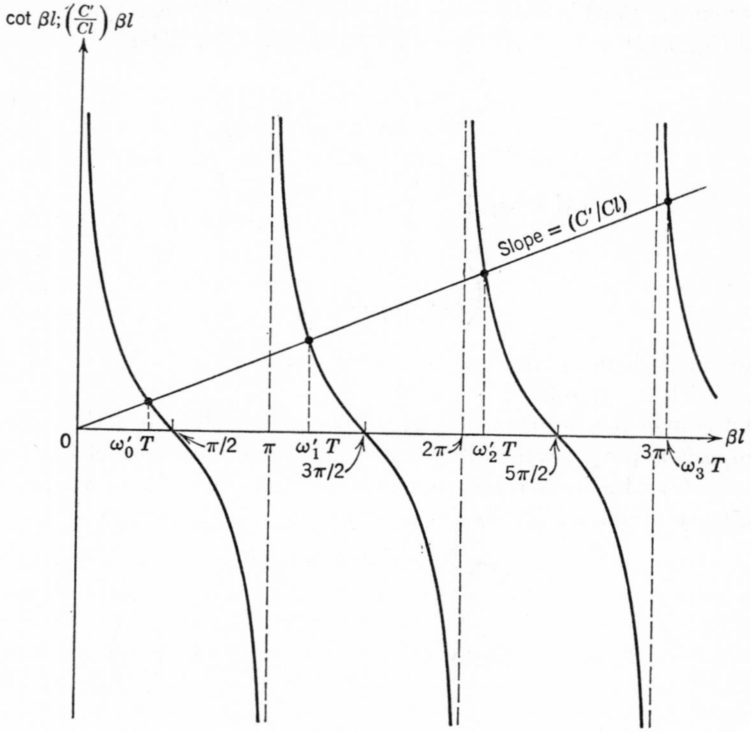

Fig. 6.21. Graphical solution for the natural frequencies of the system of Fig. 6.20*a*.

The exact results confirm completely our previous approximate treatments, whenever the approximations apply, and the exact solution in this case gives us more insight into the general features of the problem. In many other problems, however, we are not so fortunate as to be able to make exact solutions without encountering excessive complexity. The approximations are practically the only methods available, and the need for a thorough understanding of them justifies the extent of our consideration here.

6.5 Resonance

In the previous sections, we have considered the complex natural frequencies of a system and their relationships to various of its properties, including the zeros and poles of the impedance seen at any given terminal pair. We have further developed useful approximate methods for finding such frequencies.

All this work does not, however, tell us directly how the impedance $Z(j\omega)$ at the terminals behaves in the sinusoidal steady state. In connection with resonance, we may be interested in just this question.

We mean by resonance that phenomenon through which, in a certain narrow frequency range, $Z(j\omega)$ or $Y(j\omega)$ undergoes radical changes with frequency. In general, this requires that $j\omega$ pass very near a complex zero s_0 of either $Z(s)$ or $Y(s)$. If the complex frequency s_0 in question is a zero of $Z(s)$, we say that the sudden "low-impedance" behavior of $Z(j\omega)$ describes a "series resonance"; if s_0 is a zero of $Y(s)$, we say that the radical "high-impedance" behavior of $Z(j\omega)$ [or the low values of $Y(j\omega)$] represents a "parallel resonance" or "shunt resonance."

6.5.1 Isolated Simple Resonance

For example, in the case of a series resonance, we are interested in the behavior of $Z(j\omega)$ near an impedance zero (short-circuit natural frequency) at $s_0 = -\Omega_0 + j\omega_0$. Since in the steady state we cannot get closer to s_0 than its distance Ω_0 from the imaginary axis, we are concerned with $Z(j\omega)$ near $\omega = \omega_0$. In Fig. 6.22, we show a part of the s-plane pertinent to this discussion. In general, $Z(s)$ has an infinite number of poles and zeros besides the zero at s_0. In the figure, two of the *nearest* ones (poles, for example) are indicated by s_1 and s_3. It is important to note, however, that the scale may be very poor in regard to their distances from s_0. Reference to the input impedance of a short-circuited lossless line reminds us that to go from the first zero of the input impedance above zero frequency (half-wave condition) to the first pole above it (three-quarter-wave condition) requires that the wave length change from $\lambda = 2l$ to $\lambda = \frac{4}{3}l$. This corresponds to an increase of 33% in frequency. Similarly, the parallel resonance next below the series one at ω_0 would occur for $\lambda = 4l$, i.e., a factor of 2 lower in frequency! To achieve strong resonance, on the other hand, we are interested in low-loss situations where $\Omega_0 \ll 1/(2\pi)\omega_0 \approx \frac{1}{7}\omega_0$. Hence, if Fig. 6.22 were drawn to scale for input impedances at $z = 0$ of lines like those of Figs. 6.3*b* or 6.3*c*, the distances between s_0 and s_1 or s_3 in low-loss cases would be at least $10\Omega_0$, and more likely $100\Omega_0$ or even $1000\Omega_0$ (see also Fig. 6.4).

We shall therefore assume first the simplest situation, namely, that *all* other natural frequencies of our system, with the input either open- or short-circuited, are very far removed from the one of interest (s_0),

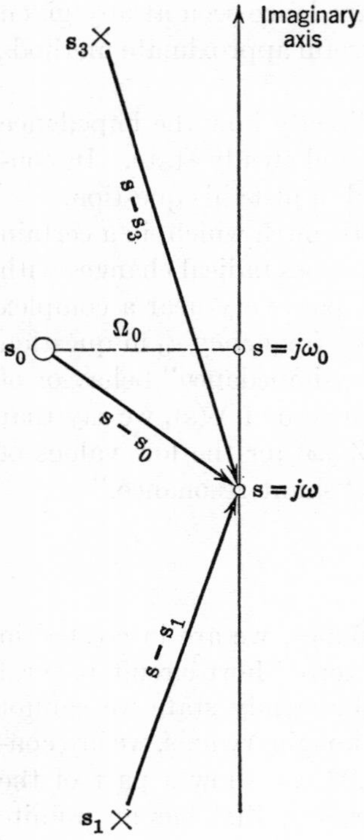

Fig. 6.22. s-plane for Z(s) near an isolated zero.

compared to its distance Ω_0 from the imaginary axis. Moreover, it will be necessary to suppose they are so disposed with respect to s_0 that $Z(j\omega_0)$ is very nearly pure real. This latter condition would, for example, occur quite accurately if all the other natural frequencies were very close to the imaginary axis, although this disposition is not the only satisfactory one. In short, we are not only assuming that the natural frequency of interest is *well isolated* from all the others, compared with its distance from the imaginary axis, but also that the steady-state frequency at which the input reactance to the system becomes zero practically coincides with the imaginary part of the complex natural frequency. Although these conditions are very often met in low-loss systems, it is well to realize that they are not at all guaranteed.

In the system of Fig. 6.1, for example, it is possible for the termination network Z_R itself to have several exceedingly sharp resonances very close together, while the line losses are (by comparison with those of the network) rather large. In such a case, the distances between several natural frequencies of the system with the input ($z = 0$) short-

or open-circuited might well be comparable to or smaller than the distance of any one from the imaginary axis. Fortunately it is usually easy by the methods previously discussed to make an approximate calculation of the other natural frequencies nearest the one of interest to determine whether or not they are too close. We shall proceed now on the basis that they are not. Then we can relate the impedance behavior to things we already know about the natural frequencies.

From Fig. 6.22, when $s = j\omega$ and $\omega \approx \omega_0$, only the phasor $s - s_0$ varies appreciably as ω changes; all other phasors remain substantially constant. Hence

$$Z(j\omega) \cong K(j\omega - s_0) = K\Omega_0 + jK(\omega - \omega_0) \tag{6.149}$$

where K is a *real* constant, on account of our assumption that $Z(j\omega_0)$ is pure real.[1] From either Eq. 6.149 or Fig. 6.22 it is directly clear that the bandwidth w of the resonance between frequencies at which $|Z| = \sqrt{2}\,|Z|_{\min} = \sqrt{2}\,K\Omega_0$ is simply

$$w = 2\Omega_0 \tag{6.150}$$

in radians/sec. Defining the quality or sharpness of this resonance by

$$Q \equiv \frac{\omega_0}{w} \tag{6.151}$$

we find, from Eqs. 6.150, 6.120*b*, or 6.123*b*, and the relation $\omega_0 = 2\pi/T_0$ that

$$\begin{aligned} Q &= \frac{\omega_0}{2\Omega_0} \cong \frac{\omega_0 \langle W_{\text{total}} \rangle_{\omega_0}}{\langle P_d \rangle_{\omega_0}} = \frac{2\pi \langle W_{\text{total}} \rangle_{\omega_0}}{\langle P_d \rangle_{\omega_0} T_0} \\ &\cong \frac{2\pi (W_{m\,\text{peak}})_{\omega_0}}{\langle P_d \rangle_{\omega_0} T_0} = \frac{2\pi (W_{e\,\text{peak}})_{\omega_0}}{\langle P_d \rangle_{\omega_0} T_0} \end{aligned} \tag{6.152}$$

Note that all the stored energies and the power dissipated are calculated *at* frequency ω_0, those voltage and current distributions being used that would exist if the entire system were lossless.

If, now, we imagine driving the actual system at the input with current I_0 at frequency ω_0, Eq. 6.149 requires

$$\langle P_d \rangle_{\omega_0} = \tfrac{1}{2} |I_0|^2 K\Omega_0 \tag{6.153}$$

[1] If K is not pure real but has a small imaginary part, it is easy to show that the reactance zero is shifted slightly from ω_0, and the real part becomes a weak function of frequency. Note that K is real for lossless cases by reason of Foster's reactance theorem.

Thus, from Eqs. 6.153 and 6.120*b* or 6.123*b*,

$$K = \frac{2\langle P_d\rangle_{\omega_0}}{|\mathrm{I}_0|^2\Omega_0} = \frac{4\langle W_{\text{total}}\rangle_{\omega_0}}{|\mathrm{I}_0|^2} \tag{6.154}$$

which determines K in terms of either energy or power.

Observe that if we had discussed directly the reactance $X(\omega)$ of our input impedance for ω near ω_0, a Taylor expansion about ω_0 would have given

$$X(\omega) = \left(\frac{dX}{d\omega}\right)_{\omega_0} (\omega - \omega_0) + \cdots \tag{6.155}$$

since $X(\omega_0) = 0$. Comparison of Eqs. 6.155 and 6.149, and use of Eq. 6.154 show that

$$\left(\frac{dX}{d\omega}\right)_{\omega_0} \cong \frac{4\langle W_{\text{total}}\rangle}{|\mathrm{I}_0|^2} = K = \left(\frac{dZ}{ds}\right)_{s_0} \tag{6.156}$$

The content of our approximations is therefore to apply Foster's reactance theorem to the imaginary part of a slightly dissipative impedance! Such a step is, of course, not strictly correct but becomes more nearly so as the loss vanishes.

For our present purposes, Eq. 6.156 provides an alternate approach to the determination of the impedance of a system near one of its series resonances. When the system is slightly dissipative, we have previously determined the constants K and Ω_0 (or Q) in terms of the approximate energy stored and power dissipated in it at resonance for one input ampere; but, to calculate these, we have used approximately the voltage and current distributions that would exist in the system if it were completely lossless. According to Eq. 6.156, however, we could instead (and to the same approximation) compute the input reactance of the system without losses, and from its slope at ω_0 find the constant K directly. The power dissipated per input ampere can then be calculated as before, yielding at once the value of $K\Omega_0$. The complete (approximate) expression of $Z(j\omega)$ in the form of Eq. 6.149 is therefore determined.

Thus, by the original method, we find for the first series resonance of the input impedance Z ($z = 0$) in Fig. 6.19*a*, from Eqs. 6.135 and 6.153,

$$\frac{2\langle P_d\rangle}{|\mathrm{I}_0|^2} = r + \frac{1}{2}\left[R + G\left(\frac{L}{C}\right)\right] l = K\Omega_0 \tag{6.157}$$

and, from Eqs. 6.134 and 6.154,

$$\frac{4\langle W_{\text{total}}\rangle}{|\mathrm{I}_0|^2} = Ll = K \tag{6.158}$$

By the alternate method we would still use Eq. 6.157, but we would calculate K from Eq. 6.156 in terms of $(dX/d\omega)_{\omega_0}$ as seen at the input ($z = 0$) of the lossless counterpart, Fig. 6.19*b*, of our system. Looking into the line of Fig. 6.19*b* at $z = 0$ (with the short circuit removed from these terminals!), we see that

$$\mathrm{Z}(j\omega) = jX(\omega) = jZ_0 \tan \beta l = jZ_0 \tan (\omega\sqrt{LC}\, l) \qquad (6.159)$$

leading to

$$\frac{dX}{d\omega} = Z_0 \frac{\sqrt{LC}\, l}{\cos^2 \beta l} = \frac{Ll}{\cos^2 \beta l} \qquad (6.160)$$

But at ω_0, $\beta l = (2\pi l/\lambda) = \pi$ for this case, and $\cos^2 \beta l = 1$. Hence

$$\left(\frac{dX}{d\omega}\right)_{\omega_0} = Ll = K \qquad (6.161)$$

which agrees with Eq. 6.158.

In view of the form of Eq. 6.149, it is possible to construct an equivalent circuit which represents approximately the behavior of the line system near any given isolated series resonance. The circuit is shown in Fig. 6.23*a*. If it is to represent the line in the sense that $\mathrm{Z}_e = \mathrm{Z}$ for ω near ω_0, we must have

$$\frac{1}{\sqrt{L_e C_e}} = \omega_0 \qquad (6.162)$$

so that the resonant frequencies will be the same. Then the power input to the circuit at the resonant frequency ω_0 must agree with that to the line at the same frequency and *with the same input current* I_0.

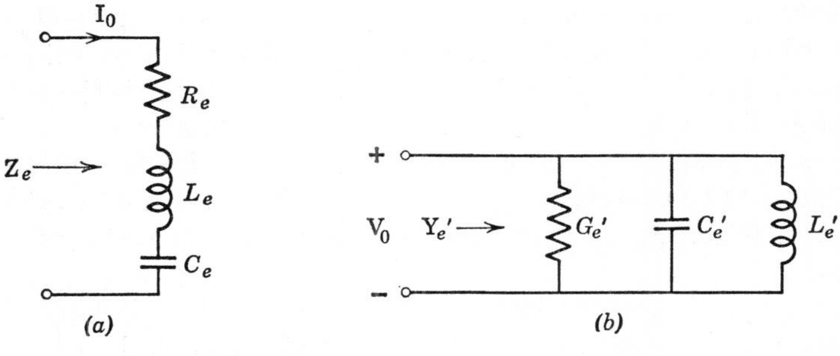

Fig. 6.23. Equivalent circuit for (*a*) an isolated series resonance and (*b*) an isolated parallel resonance.

For the line of Fig. 6.19*a*, the current at $z = 0$ and $z = l$ in Fig. 6.19*b* is I_0. Thus,

$$\frac{1}{2}|I_0|^2 R_e \cong \langle P_d \rangle = \frac{1}{2}|I_0|^2 \left[r + \frac{1}{2}\left(R + G\frac{L}{C}\right) l \right] \tag{6.163}$$

or

$$R_e = r + \frac{1}{2}\left(R + G\frac{L}{C}\right) l \tag{6.164}$$

Finally, the total average energy stored at resonance must be the same in both systems *for the same input current:*

$$\tfrac{1}{2}L_e|I_0|^2 = \tfrac{1}{4}Ll|I_0|^2 \tag{6.165}$$

or

$$L_e = \tfrac{1}{2}Ll \tag{6.166}$$

Equations 6.162, 6.164, and 6.166 determine completely the elements of Fig. 6.23*a* in terms of those in the original system of Fig. 6.19*a*. Observe that, instead of Eq. 6.165 we could make either Q or Ω_0 agree for both systems, since the power dissipated for one input ampere already agrees by Eq. 6.163. It is also possible to make $(dX/d\omega)_{\omega_0}$ the same, instead of using Eq. 6.165. On this basis, for Fig. 6.23*a*,

$$X = j\left(\omega L_e - \frac{1}{\omega C_e}\right) \tag{6.167}$$

and

$$\left(\frac{dX}{d\omega}\right)_{\omega_0} = L_e + \frac{1}{\omega_0{}^2 C_e} = 2L_e \tag{6.168}$$

which, with Eq. 6.161, again yields Eq. 6.166.

An exactly dual procedure may be employed to discuss the behavior of a system near an isolated shunt resonance. Since the input current is zero at such a resonance in the lossless case, all the energy and power calculations must be carried out with the input terminals *open*, preferably on a unit-input-*voltage* basis. Also, inasmuch as the impedance of the lossless reference system is infinite at a parallel resonant frequency ω_0, power series expansions about s_0 or ω_0, like Eqs. 6.149 and 6.155, cannot be made directly on an impedance basis. The *admittance*, however, is zero at s_0; therefore we can proceed in terms of $Y(j\omega)$ instead.

The appropriate equivalent circuit obtained near an isolated shunt resonance is that shown in Fig. 6.23*b*, with which we deal on the basis of the input admittance Y_e and input voltage V_0 rather than the input impedance and current.

It is worth remarking that, from the point of view of the behavior of $Z(j\omega)$ alone, simple series and shunt resonances are distinguished respectively by whether the real part of the impedance $R(\omega)$ or of the admittance $G(\omega)$ remains more nearly constant near the frequency ω_0 where the imaginary parts $X(\omega)$ and $B(\omega)$ are zero. Of course they are also distinguished by the algebraic sign of the imaginary part of either the impedance or admittance at frequencies adjacent to the resonant frequency ω_0.

6.5.2 Coupling to Resonant Systems *

Since in practice a resonant device must be driven from a generator with some internal impedance, it is important to understand how the resonator and the source interact. For example, a resonant circuit may have to be driven from a lossless transmission line of characteristic impedance Z_0, with a source at the input end whose internal impedance is matched, i.e., is equal to Z_0. The relationship between the resonant resistance of the load and the source resistance is then of great significance, as regards the manner in which the power delivered to the resonant device varies with frequency.

Thus, if we drive the series resonant circuit in Fig. 6.23*a* from a matched line system (Fig. 6.24*a*), equivalent to a generator of internal impedance Z_0, as shown in Fig. 6.24*b*, we have a load power given by

$$\langle P_{\text{out}} \rangle = \frac{\frac{1}{2}|\text{E}_0|^2 R_e}{(R_e + Z_0)^2 + X_e^2} \tag{6.169}$$

where

$$X_e = j\left(\omega L_e - \frac{1}{\omega C_e}\right) \cong j2\omega_0 L_e \left(\frac{\omega - \omega_0}{\omega_0}\right) \tag{6.170}$$

The load power at resonance ($X_e = 0$) is a maximum when $R_e = Z_0$, and is small when either $R_e \gg Z_0$ or $R_e \ll Z_0$. On the other hand, the width of the resonance curve $\langle P_{\text{out}} \rangle(\omega)$ is narrowest when $Z_0 \ll R_e$ and becomes very wide when $Z_0 \gg R_e$. We say that the line and load are greatly *undercoupled* when $Z_0 \ll R_e$ because the line does not much affect the load power curve, and the load is so nearly an open circuit at all frequencies that it does not much affect the line voltage or current (i.e., the reflection coefficient). When $Z_0 \gg R_e$, however, the system is said to be greatly *overcoupled;* the line dominates the width of $\langle P_{\text{out}} \rangle(\omega)$, and the load impedance goes from nearly a short circuit (at resonance) to nearly an open circuit (way off resonance), thereby changing radically the voltage and current on the line (i.e., the reflec-

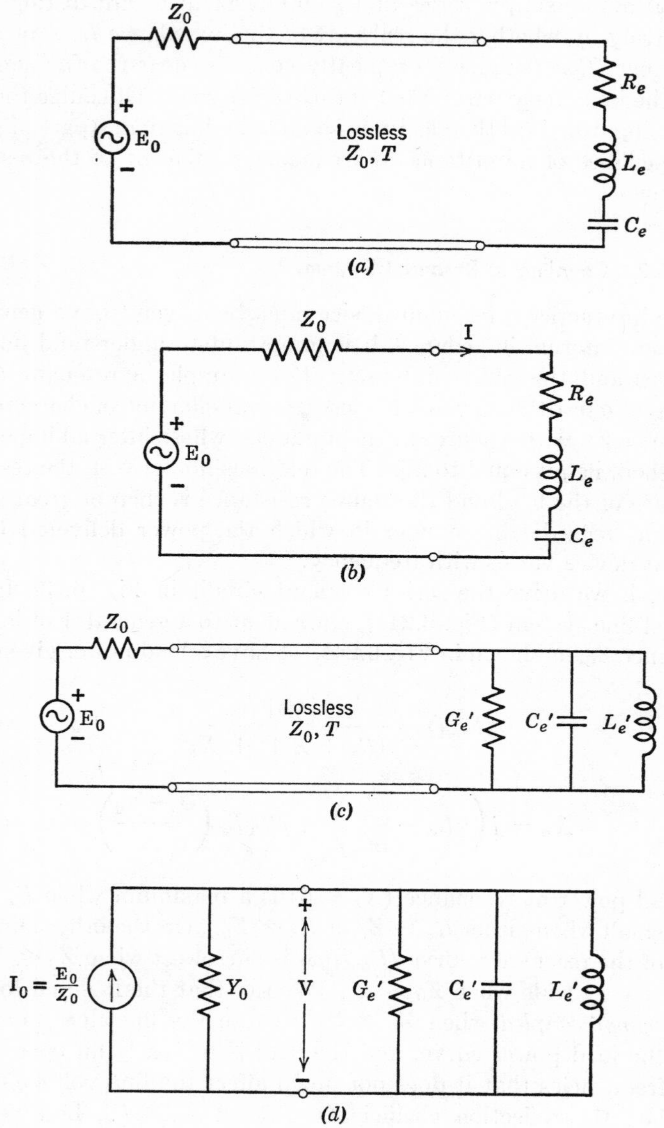

Fig. 6.24. Simple resonant systems driven from resistive sources.

tion coefficient). For a series resonance, therefore, we use the terms "undercoupled," "critically coupled," and "overcoupled" for $R_e > Z_0$, $R_e = Z_0$, $R_e < Z_0$ respectively.

Just the reverse is true for a shunt resonance, illustrated in Figs. 6.24*c* and 6.24*d*. The source is most conveniently represented for this case in the Norton equivalent form, and

$$\langle P_{\text{out}} \rangle = \frac{\frac{1}{2}|\mathrm{I}_0|^2 G_e'}{(G_e' + Y_0)^2 + B_e'^2} \tag{6.171}$$

with

$$B_e' = j\left(\omega C_e' - \frac{1}{\omega L_e'}\right) \approx j2\omega_0' C_e' \left(\frac{\omega - \omega_0'}{\omega_0'}\right) \tag{6.172}$$

Evidently here consideration of increasing mutual influence between source and load shows that undercoupling goes with $G_e' > Y_0$, critical coupling with $G_e' = Y_0$, and overcoupling with $G_e' < Y_0$.

Observe that the variation of $\langle P_{\text{out}} \rangle$ in Eq. 6.169 is produced by the variation of circuit current I in Fig. 6.24*b*,

$$\mathrm{I} = \frac{\mathrm{E}_0}{Z_0 + \mathrm{Z}_e}$$

or by the variation of the total impedance

$$\mathrm{Z}_T = Z_0 + \mathrm{Z}_e \tag{6.173}$$

Similarly, the important variation for Fig. 6.24*c* and Eq. 6.171 is that of the circuit voltage

$$\mathrm{V} = \frac{\mathrm{I}_0}{Y_0 + \mathrm{Y}_e'}$$

or the total admittance

$$\mathrm{Y}_T' = Y_0 + \mathrm{Y}_e' \tag{6.174}$$

This means we are actually interested in the complex natural frequencies of these circuits with the source impedance Z_0 *(or* Y_0*) connected, but with the drives zero* ($\mathrm{E}_0 = \mathrm{I}_0 = 0$). These complex frequencies are associated with the imaginary part $\omega_0 = 1/\sqrt{L_e C_e}$ or $\omega_0' = 1/\sqrt{L_e' C_e'}$, and real part $\Omega_0 = (R_e + Z_0)/2L_e$ or $\Omega_0' = (G_e' + Y_0)/2C_e'$ in the two systems. Accordingly, the important bandwidth for the driven system appears in the so-called *loaded Q*'s

$$\text{(a)} \qquad Q_L = \frac{\omega_0 L_e}{R_e + Z_0}$$

$$\text{(b)} \qquad Q_L' = \frac{\omega_0' C_e'}{G_e' + Y_0} \tag{6.175}$$

rather than the *unloaded* Q's

$$\text{(a)} \qquad Q_u = \frac{\omega_0 L_e}{R_e}$$

$$\text{(b)} \qquad Q_u' = \frac{\omega_0' C_e'}{G_e'} \qquad (6.176)$$

which characterize the original resonant elements. Note that "unloading" a series-resonant system means *short-circuiting* its terminals ($Z_0 = 0$), while "unloading" a shunt-resonant system means open-circuiting its terminals ($Y_0 = 0$).

If we think of the natural oscillation of the circuits of Fig. 6.24 with the impedance Z_0 connected, it not only makes sense to define the Q's of Eqs. 6.175 and 6.176, but also the Q which would result if the original resonant system were lossless and only the load Z_0 (or Y_0) were connected. In other words the Q in question is the one resulting from the transfer of power out of the resonant system into the connected line or source impedance. It is called the *external* Q,

$$\text{(a)} \qquad Q_{\text{ext}} = \frac{\omega_0 L_e}{Z_0}$$

$$\text{(b)} \qquad Q_{\text{ext}}' = \frac{\omega_0' C_e'}{Y_0} \qquad (6.177)$$

There is of course a relationship between the reciprocals of the Q's which we have so far defined. This relationship stems from the additivity of the various losses per unit stored energy in the highly resonant systems under consideration. Specifically,

$$\text{(a)} \qquad \frac{1}{Q_L} = \frac{1}{Q_u} + \frac{1}{Q_{\text{ext}}}$$

$$\text{(b)} \qquad \frac{1}{Q_L'} = \frac{1}{Q_u'} + \frac{1}{Q_{\text{ext}}'} \qquad (6.178)$$

To describe the strength of the coupling about which we have spoken previously it is convenient to introduce a parameter ζ, which is the ratio of the source impedance (or admittance) to the resonant impedance (or admittance) for the series (or parallel) resonant systems respectively. This ratio ζ may also be expressed in terms of the various

Q's which we have introduced in Eqs. 6.175 through 6.177, as indicated in the equations below.

$$\text{(a)} \qquad \zeta = \frac{Z_0}{R_e} = \frac{Q_u}{Q_{\text{ext}}} = \frac{Q_u}{Q_L} - 1$$

$$\text{(b)} \qquad \zeta' = \frac{Y_0}{G_e'} = \frac{Q_u'}{Q_{\text{ext}}'} = \frac{Q_u'}{Q_L'} - 1 \qquad (6.179)$$

In Table 6.1 we have summarized the conditions defining the various coupling strengths in terms not only of the *coupling parameter* ζ but also the various equivalent Q ratios related to it in Eq. 6.179.

TABLE 6.1. Conditions Defining Coupling Strengths

	(ζ, ζ')	$\left(\frac{Q_u}{Q_{\text{ext}}}, \frac{Q_u'}{Q_{\text{ext}}'}\right)$	$\left(\frac{Q_u}{Q_L}, \frac{Q_u'}{Q_L'}\right)$
Undercoupled	< 1	< 1	between 1 and 2
Critically coupled	$= 1$	$= 1$	$= 2$
Overcoupled	> 1	> 1	> 2

An interesting significance may be given to the loaded Q, with reference to Figs. 6.24*a* and 6.24*c*. If either of these systems is excited by a unit impulse of voltage, the resonant system at the end of the line will be excited similarly, a travel time T later. An impulse will return to the source, without further reflection at that point, because of the matched source impedance. Behind the returning impulse will be the damped oscillation, executed by the resonant circuit at the load end of the line in response to its excitation by the original impulse. The decrement of this damped oscillation is, of course, the one characterized by the loaded Q (Q_L, Q_L') of the circuit in Fig. 6.24*b* or 6.24*d*.

The situation just described suggests a transient technique to measure directly the loaded Q of resonant loads on transmission lines. Probably one could not, as a practical matter, use impulses to excite the loads. The reason is that the load circuits shown in Figs. 6.24*a* and 6.24*c* may actually be representations of a physical load system only in the neighborhood of one of its natural frequencies. When a physical system contains other resonances, the use of an impulse would tend to excite these as well as the one of interest represented by the equivalent circuits of Figs. 6.24*a* and 6.24*c*. The resulting transient would contain an admixture of all of the excited natural frequencies. Accordingly, a pulse of radio-frequency signal of frequency ω_0 could be used instead. The duration of the pulse should be sufficient to bring the loads on the transmission lines to a steady state at frequency ω_0. When the pulse is turned on or off, therefore, the build-up or decay transient will contain principally the natural oscillation characterized by the equivalent circuits discussed in the foregoing.

The parameters defined in Eqs. 6.175 through 6.179 are of particular importance in describing the behavior of a resonant system that is not simply made up out of lumped-circuit elements. As we have seen from our study of transmission lines, the normalized impedance that appears in wave-reflection coefficients or standing-wave ratios retains a strong physical significance, even if the impedance value in ohms does not. The complex natural frequencies, and the Q's related to them (or to the bandwidth and mean frequency) also retain a full physical significance whether or not impedance concepts apply in detail. Accordingly, we should write the load impedances Z_e and Y_e' in the normalized form as follows:

$$\text{(a)} \qquad \frac{Z_e}{Z_0} = \frac{1}{\zeta}(1 + j2Q_u\xi) = \frac{Q_{\text{ext}}}{Q_u}(1 + j2Q_u\xi)$$

$$\text{(b)} \qquad \frac{Y_e'}{Y_0} = \frac{1}{\zeta'}(1 + j2Q_u'\xi') = \frac{Q_{\text{ext}}'}{Q_u'}(1 + j2Q_u'\xi') \qquad (6.180)$$

where we have used the notation,

$$\text{(a)} \qquad \xi \cong \frac{\omega - \omega_0}{\omega_0}$$

$$\text{(b)} \qquad \xi' \cong \frac{\omega - \omega_0'}{\omega_0'} \qquad (6.181)$$

It is worth observing that the total impedance or admittance, including the Z_0 or Y_0, which total we call Z_T or Y_T' respectively, may also be written on a normalized basis, as follows:

$$\text{(a)} \qquad \frac{Z_T}{Z_0} = \frac{Q_{\text{ext}}}{Q_L}(1 + j2Q_L\xi)$$

$$\text{(b)} \qquad \frac{Y_T'}{Y_0} = \frac{Q_{\text{ext}}'}{Q_L'}(1 + j2Q_L'\xi') \qquad (6.182)$$

One may say that the coupling between the resonant load and the transmission line or source impedance is weak when the external Q is much higher than the unloaded Q. Needless to say, if ideal transformers were available, one could always adjust the coupling between a line and resonant load to any value desired. As a practical matter, however, one is usually forced to employ some sort of reactive "tapping" procedure, or a voltage-divider arrangement, or a real transformer in order to couple into a resonant system. These techniques, at least, would be the ones used to couple into resonant lumped circuits. The equivalent devices for coupling into the electric or magnetic fields of a resonant transmission-line system could be small wire loops and probes, or, for coupling into resonant boxes or cavities, simply a small hole cut into the side wall at an appropriate point.

6.5.2.1 ILLUSTRATIONS OF COUPLING.* Examples of the techniques of coupling to some very simple resonant systems are shown in Fig. 6.25. The first two parts illustrate the situation for networks, and the last two show elementary coupling to a transmission-line resonator. The important point to notice about all of these figures is that matters have been so arranged in them that, whether the impedance Z_0 is large or small, the natural behavior of the system is not very greatly changed. That is to say, the terminals at Z_0 or Y_0 may be either open-circuited or short-circuited without materially altering the behavior of the rest of the system. This circumstance has been guaranteed for the circuit cases by the fact that the coupling reactances have been chosen to be either very large (L_c' or C_c') or very small (L_c or C_c) compared to other reactances in the system at the frequencies of interest. In *all* the arrangements, in fact, there is either very little voltage across or very little current through element Z_0 or Y_0, regardless of its value. It is this feature which guarantees that the coupling will be loose or weak, because, as we have already seen in Sec. 6.2, the extreme case of no coupling at all arises when the terminal pair of the coupled system occurs at a voltage or a current node.

On one hand, the weak coupling appears to add a complication to the resonant phenomenon. As seen from the terminals of the line element Z_0 or Y_0, the input impedance to the resonator must surely have a closely spaced zero and pole. That is, after all, what we mean by the statement that the natural behavior of the system with these terminals either open- or short-circuited is substantially the same. On the other hand, as viewed from the same terminals, the resonant system looks like a rather simple coupling element in series or parallel with the balance of the network for which the frequency dependence is exceedingly sharp. Indeed, over the range of frequencies of interest, the reactance of the coupling element itself may be regarded as unchanged in value, while the rest of the system behaves as a simple resonant cir-

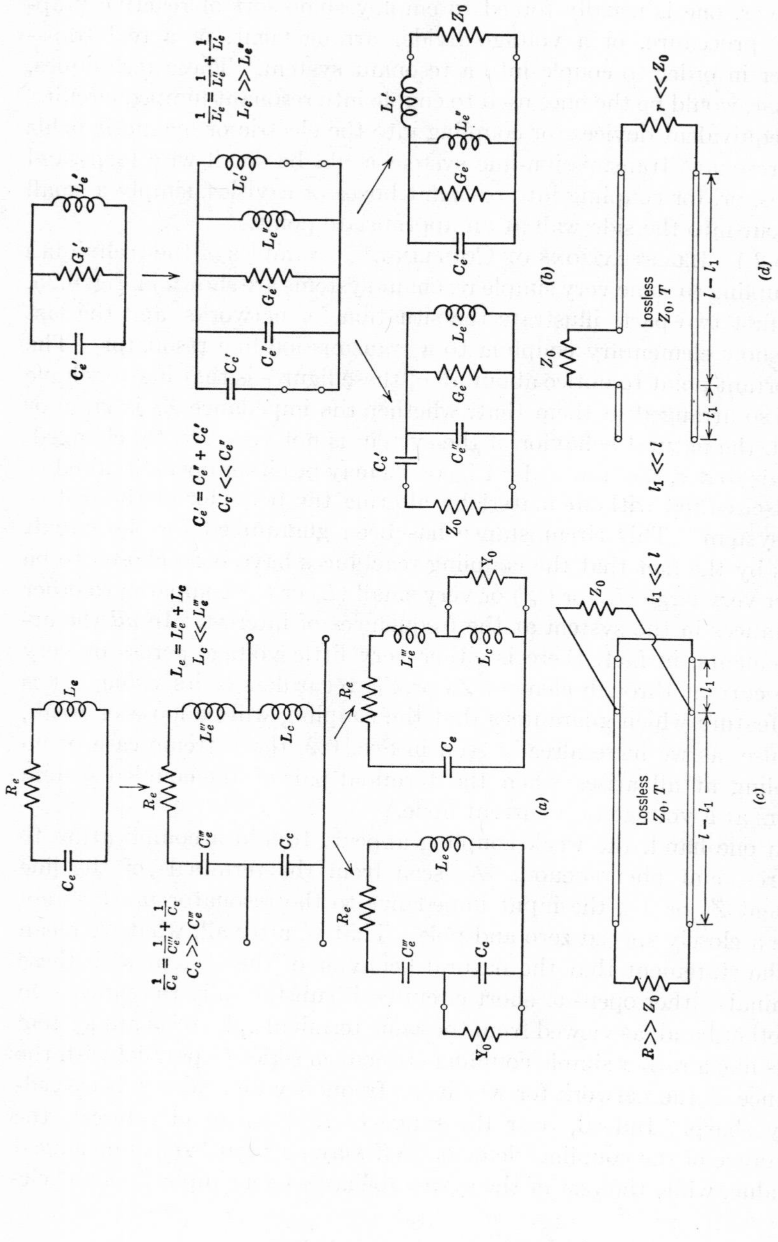

Fig. 6.25. Some methods of coupling to a resonator. Parts (*a*) and (*b*) are dual circuit schemes. Parts (*c*) and (*d*) are dual transmission-line methods.

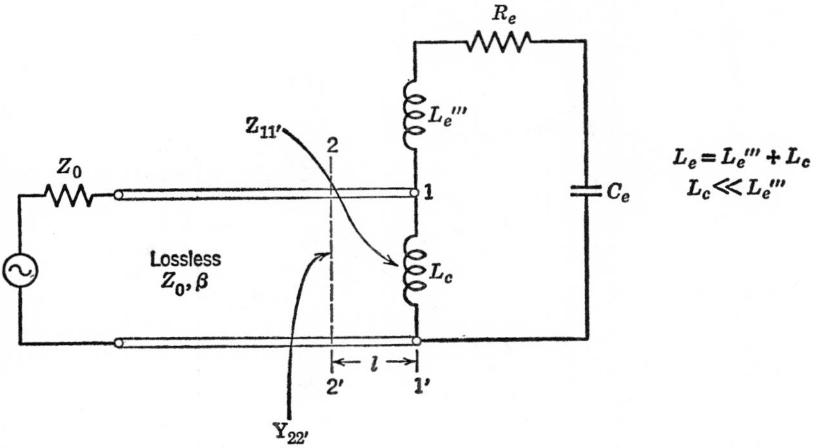

Fig. 6.26. Choice of new reference plane 2–2′ to describe a resonance with coupling at 1–1′.

cuit. It turns out that this feature enhances considerably our ability to analyze the entire system.

To be specific, let us consider the coupling scheme defined by the lower right-hand drawing in Fig. 6.25*a*. The actual transmission-line arrangement is redrawn in Fig. 6.26 for convenience. Prior to making any coupling, the circuit loading the line of Fig. 6.26 could have been described by its impedance Z_e in Fig. 6.23*a* for which the analytical expression appears in Eq. 6.180*a*. Observe that the Q_{ext} in that equation is actually the one which would be obtained by coupling the transmission line and resonant circuit in the manner of Fig. 6.24*a* rather than in the manner of the present arrangement in Fig. 6.26.

Regarding the impedance seen in Fig. 6.26 at terminals 1–1′, denoted by $Z_{11'}$, we observe that it has a pole at the natural frequency of the system with these terminals left open. That is, the pole occurs at the complex frequency s_{pole}, given by

$$\text{(a)} \qquad s_{\text{pole}} = -\Omega_{\text{pole}} + j\omega_{\text{pole}}$$

$$\text{(b)} \qquad \Omega_{\text{pole}} = \frac{R_e}{2L_e} \qquad (6.183)$$

$$\text{(c)} \qquad \omega_{\text{pole}} = \sqrt{\omega_0{}^2 - \Omega_{\text{pole}}{}^2} \cong \omega_0 \equiv \frac{1}{\sqrt{L_e C_e}}$$

where it has been assumed that the damping is small.

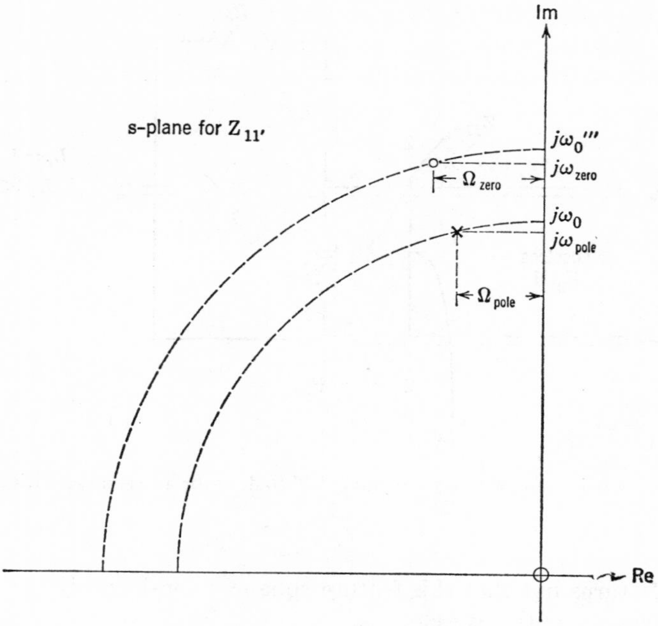

Fig. 6.27. Closely spaced zero and pole characterizing weak coupling to a resonant system (drawn for the case of $Z_{11'}$ in Fig. 6.26, but not to scale).

Similarly, there is a zero of the impedance $Z_{11'}$, occurring at the complex natural frequency of the system with the terminals 1–1′ short-circuited. The pertinent frequency s_{zero} may be written,

$$\text{(a)} \quad s_{\text{zero}} = -\Omega_{\text{zero}} + j\omega_{\text{zero}}$$

$$\text{(b)} \quad \Omega_{\text{zero}} = \frac{R_e}{2L_e'''} = \frac{R_e}{2L_e}\left(\frac{L_e}{L_e'''}\right) = \Omega_{\text{pole}}\left[1 + \left(\frac{L_c}{L_e'''}\right)\right] \qquad (6.184)$$

$$\text{(c)} \quad \omega_{\text{zero}} = \sqrt{\omega_0'''^2 - \Omega_{\text{zero}}^2} \cong \omega_0''' \equiv \frac{1}{\sqrt{L_e'''C_e}}$$

$$= \omega_0\sqrt{1 + \frac{L_c}{L_e'''}} \cong \omega_0\left[1 + \frac{1}{2}\left(\frac{L_c}{L_e'''}\right)\right]$$

In this case the assumption has been made not only that the damping is small under short-circuit conditions but also that the coupling is weak. This means the coupling inductance L_c is much smaller than the total inductance L_e.

The relationship between the various frequencies that occur in the previous equation is illustrated in Fig. 6.27. This figure is not drawn to scale, especially in respect to the high Q which we are really assuming for both the zero and pole natural frequencies, but also in respect to the relative horizontal and vertical distances between the two frequencies. As a result of the small coupling, the relative spacing along the imaginary axis is actually greater than that along the negative real axis by the factor Q_u, the unloaded Q of the system. Specifically

$$\frac{\omega_{\text{zero}} - \omega_{\text{pole}}}{\Omega_{\text{zero}} - \Omega_{\text{pole}}} \cong \frac{\omega_0''' - \omega_0}{\Omega_{\text{zero}} - \Omega_{\text{pole}}} \cong \frac{1}{2}\frac{\omega_0}{\Omega_{\text{pole}}} = Q_u \tag{6.185}$$

It is quite clear that the behavior of the impedance $Z_{11'}(j\omega)$ in the neighborhood of the closely spaced zero and pole is not going to be as simple as that characterizing a well-isolated resonance, illustrated in the s-plane by Fig. 6.22. On the other hand, it is not a great deal more complicated, because in this frequency heighborhood one expects to be able to represent $Z_{11'}(s)$ by the first *two* terms of its Laurent expansion about the pole. This expansion reads

$$Z_{11'}(s) \cong K_1 + \frac{K_2}{s - s_{\text{pole}}} \tag{6.186}$$

or

$$\text{(a)} \qquad Z_{11'}(s) \cong \frac{K_1[s - (s_{\text{pole}} - K_2/K_1)]}{s - s_{\text{pole}}} = \frac{K_1(s - s_{\text{zero}})}{s - s_{\text{pole}}}$$

$$\text{(b)} \qquad Y_{11'}(s) \cong \frac{s - s_{\text{pole}}}{K_1(s - s_{\text{zero}})} = \frac{1}{K_1} + \frac{(s_{\text{zero}} - s_{\text{pole}})}{K_1(s - s_{\text{zero}})} \tag{6.187}$$

in which the ratio $|K_2|$ to $|K_1|$ must be small if the zero and the pole are to be close together. In taking only the first two terms, as represented by Eq. 6.186, we are assuming tacitly that the radius of convergence of the whole expansion is a good deal greater than the actual distance of the critical frequencies from the imaginary axis where we plan to use the expansion results. This assumption, in turn, implies that all the other critical frequencies, both zeros and poles, are widely separated from the ones under consideration. The need to include the zeros in this statement springs from our wish to have Eq. 6.186 represent equally well the admittance $Y_{11'}(s)$ in this frequency range. Inverting Eq. 6.186 to yield Eq. 6.187*b* shows that its validity will be limited by that of the Laurent expansion of $Y_{11'}(s)$ about the complex frequency s_{zero} (which is a pole of the admittance). Bearing in mind that the radius of convergence of the full expansions in question extends out to the nearest pole beyond the one used as expansion center, it is clear that the use of only two terms depends upon the smallness of the region of interest as compared with the area of convergence of the entire expansion. In the present case, of course, the region of interest extends only a few damping constants away from either the zero or the pole.

With reference to Fig. 6.26 again, we may write the admittance $Y_{11'}$ in the following way

$$Y_{11'} = -jB_c + Y' \tag{6.188}$$

where

$$\begin{aligned} &\text{(a)} \quad & B_c &= \frac{1}{\omega L_c} > 0 \\ &\text{(b)} \quad & Y' &= \frac{1/R_e}{1 + j2Q_u'''\xi'''} \\ &\text{(c)} \quad & Q_u''' &= \frac{\omega_0'''L_e'''}{R_e} \\ &\text{(d)} \quad & \xi''' &= \frac{\omega - \omega_0'''}{\omega_0'''} \end{aligned} \tag{6.189}$$

The point is that B_c is a very slowly varying function of ω, in comparison with Y'. It will therefore be valid to treat the small admittance B_c as a constant, so that Eq. 6.188 agrees in form with Eq. 6.187. Also, for the discussion of its effect upon the connected transmission line, it is convenient to normalize the admittance $Y_{11'}$ with respect to the characteristic admittance Y_0 as follows:

$$y_{11'} \equiv \frac{Y_{11'}}{Y_0} = -j\frac{B_c}{Y_0} + \frac{Y'}{Y_0} = -jb_c + y' \tag{6.190}$$

where

$$\begin{aligned} &\text{(a)} \quad & b_c &\cong \frac{Z_0}{\omega_0'''L_c} \\ &\text{(b)} \quad & y' &= \frac{\zeta}{1 + j2Q_u'''\xi'''} \end{aligned} \tag{6.191}$$

Now y' is a linear fractional transformation of the frequency variable $j\xi'''$, so that, as the frequency moves along its j-axis (a straight line), the complex value of y' describes a circle in its plane. Thus $y_{11'}$ describes the locus shown in Fig. 6.28, where it has been assumed that b_c remains constant over the frequency range of significance. Observe in the figure the two points marked A and B at which the admittance

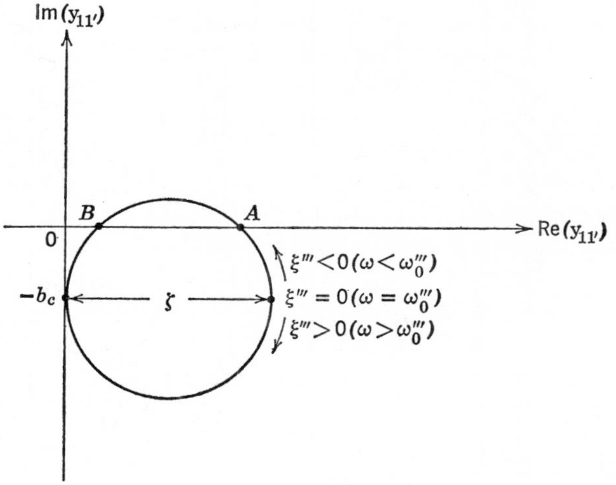

Fig. 6.28. Locus of normalized load admittance $y_{11'} = Z_0 Y_{11'}$ in Fig. 6.26, under the assumption that $\omega L_c \cong \omega_0''' L_c$ at all frequencies of interest

becomes pure real, and notice that A occurs at a positive frequency lower than ω_0''', while B occurs at a negative frequency.

Inasmuch as the reflection-coefficient transformation is also of the linear fractional type in terms of the normalized admittance, it follows that the Smith-chart presentation of the locus of load admittance $y_{11'}$ is also a circle. The Smith-chart locus circle must be tangent to the unit circle of that chart, just as the admittance locus in Fig. 6.28 is tangent to the imaginary axis. This is the conformal property of the reflection-coefficient transformation. From Eqs. 6.190 and 6.191 we see that the admittance locus in the Smith chart must pass through the point $-jb_c$ when $\xi''' = \pm\infty$, and through the point $\zeta - jb_c$ when $\xi''' = 0$. The circle meeting all three requirements is the solid one shown in Fig. 6.29. Observe carefully that the Smith chart, originally laid out on the impedance basis, is here being used directly as an admittance chart. Therefore reflection coefficients *cannot* be read directly from the chart in this application.

6.5.2.2 Choice of Reference Terminals.* Referring once again to Fig. 6.26, we are now in a position to make use of the fact that, with the load circuit operating at frequencies considerably removed from either the zero or pole neighborhoods, the total termination becomes simply the reactance of the small inductance L_c. Under such *detuned* conditions, it is evidently possible to add a section of length l of the lossless

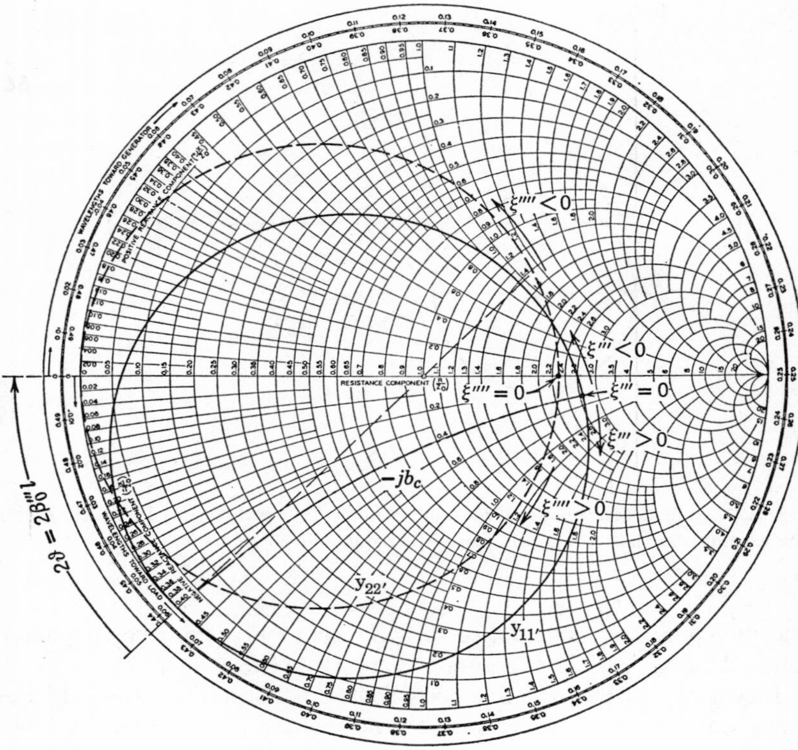

Fig. 6.29. Smith-chart presentation of the admittance $y_{11'}(\xi''')$ (solid circle) and its reduction to $y_{22'}$ (dashed circle) by adding a short section of lossless line. Note admittance use of chart.

transmission line to the system in such a way that, at a new pair of terminals 2–2′, the admittance seen looking to the right becomes zero. That is, the terminals 2–2′ represent the position of a line-current node or line-voltage maximum at a frequency for which the load system is considerably detuned. We say that the terminals 2–2′ have been chosen to lie on the line at the position of *detuned open circuit.*

Inasmuch as the distance l cannot in this case exceed a quarter wave length, and since, usually, a resonant load will have a much narrower band character than the transmission line, the electrical length of the piece of line will not change greatly over the significant frequency range for load variations. It is therefore quite often reasonable to consider that the electrical length of the added line section remains constant, as we have already assumed for the reactance of the coupling element L_c. Under these conditions, the admittance $\mathrm{y}_{22'}$ may be obtained from $\mathrm{y}_{11'}$ at every frequency within the band of interest by a simple rotation of the reflection coefficient through the fixed angle $2\beta_0'''l$. Accordingly, in Fig. 6.29 the dashed circle has been obtained in this manner from the solid circle, the electrical distance $\beta_0'''l$ having been chosen to bring to zero value the normalized admittance $-j0.40$ of the original detuned load system ($b_c = 0.40$ in the figure, for the purposes of presentation). The most interesting feature about the resulting dashed locus is, of course, that it represents the admittance variation of a simple series-resonant circuit of the form shown in Fig. 6.23*a*. In other words, *the choice of the detuned-open position on the line as the terminal pair 2–2′ reduces the coupled resonator in Fig. 6.26 to an equivalent simple series-tuned circuit loading the system in the manner of Fig. 6.24a.*

From the nature of the relationship between the two circles in Fig. 6.29, implied by the rotation of the reflection coefficient at each point, it is clear that the frequency of zero reactance at terminals 2–2′ differs from ω_0'''. Also the conductance at the new resonant frequency is somewhat different from $1/R_e$, the respective values in Fig. 6.29 being 2.3 and 2.7 on a normalized basis.

We can understand in more detail the effect of adding the section l of line to the system, as considered above, by treating the problem analytically. The line section of electrical length $\vartheta = \beta_0'''l$ transforms $\mathrm{y}_{11'}$ to $\mathrm{y}_{22'}$ (both normalized to Y_0) through the relation

$$\mathrm{y}_{22'} = \frac{\mathrm{y}_{11'} + j\tan\vartheta}{1 + j\mathrm{y}_{11'}\tan\vartheta} \tag{6.192}$$

If l is chosen to be the detuned-open position, we must have $\mathrm{y}_{22'} = 0$ when $\mathrm{y}_{11'} = -jb_c$. Thus

$$\tan\vartheta = b_c = \frac{Z_0}{\omega_0'''L_c} \tag{6.193}$$

We shall again assume that ϑ does not vary with frequency, and recall that b_c is assumed to be constant. The validity of these assumptions is rather good in high-frequency distributed systems (transmission-line or wave-guide resonators) where even the loaded Q's are apt to be 300 or more. The interesting frequency range (bandwidth) is then about ⅓% of center frequency, so that the electrical length of l and value of b_c could only change by this same percentage. When l is not large electrically to begin with, these fractional changes are also small numerically (we shall say more about this point shortly).

In view of Eq. 6.193, Eq. 6.192 becomes

$$\mathrm{y}_{22'} = \frac{\mathrm{y}'}{1 + b_c^2 + jb_c\mathrm{y}'}$$

or

$$\mathrm{z}_{22'} = \frac{1 + b_c^2}{\mathrm{y}'} + jb_c$$

With Eq. 6.191*b* we have

$$\mathrm{z}_{22'} = \frac{1 + b_c^2}{\zeta}\,(1 + j2Q_u'''\xi''') + jb_c$$

$$= \frac{R_e(1 + b_c^2)}{Z_0}\left\{1 + j2Q_u'''\left[\xi''' + \frac{b_c\zeta}{2Q_u'''(1 + b_c^2)}\right]\right\} \tag{6.194}$$

But

$$\xi''' + \frac{b_c\zeta}{2Q_u'''(1 + b_c^2)} = \frac{\omega - \omega_0'''\{1 - (b_c\zeta)/[2Q_u'''(1 + b_c^2)]\}}{\omega_0'''}$$

$$= \frac{\omega - \omega_0''''}{\omega_0''''}\left[1 - \frac{b_c\zeta}{2Q_u'''(1 + b_c^2)}\right] \tag{6.195a}$$

where

$$\omega_0'''' = \omega_0'''\left[1 - \frac{b_c\zeta}{2Q_u'''(1 + b_c^2)}\right] < \omega_0''' \tag{6.195b}$$

Therefore we can get Eq. 6.194 into the general form of Eq. 6.180*a* for a simple series-tuned circuit

$$\mathrm{z}_{22'} = \frac{R_e''''}{Z_0}\,(1 + j2Q_u''''\xi'''') \tag{6.196a}$$

if

$$R_e'''' = R_e(1 + b_c^2) > R_e \tag{6.196b}$$

and

$$\xi'''' = \frac{\omega - \omega_0''''}{\omega_0''''} \tag{6.196c}$$

and

$$Q_u'''' = Q_u'''\left[1 - \frac{b_c\zeta}{2Q_u'''(1 + b_c^2)}\right] = \frac{\omega_0''''L_e'''}{R_e}$$

$$= \frac{\omega_0''''[L_e(1 + b_c^2)]}{R_e(1 + b_c^2)} = \frac{\omega_0''''L_e''''}{R_e''''} < Q_u''' \tag{6.196d}$$

where

$$L_e'''' = L_e(1 + b_c^2) > L_e \tag{6.196e}$$

The reciprocal of Eq. 6.196*a* describes the dashed circle of Fig. 6.29. Evidently $R_e'''' > R_e$ by the factor $(1 + b_c^2)$, which checks the values read from Fig. 6.29. Also since $b_c > 0$, $\omega_0'''' < \omega_0'''$, a fact which also agrees with the figure.

The addition of the extra piece of line l in Fig. 6.26 to change the coupled load resonator into a simpler one as viewed from the new terminals may seem to be a somewhat artificial procedure. For a lumped-circuit load, it *is* rather artificial. On the contrary, if we are dealing with distributed systems, the line and tuned load arrangement of Fig. 6.26 may be simply a representation for the behavior of a system in which line and load are *not* clearly divided from each other physically. Then there is no particular reason for treating one place on the line as "the terminals" of the resonator rather than another place, provided the choices are all kept within a fairly short distance of the visible end of the line. By such a short distance, in connection with impedance variations, we mean one whose electrical length does not change *numerically* by a significant fraction of a wave length (say, $<0.05\lambda$) over the frequency range (bandwidth) of interest. This criterion makes sense if we remember that the impedance goes through a whole cycle of variation every time the electrical length of the line increases by $\lambda/2$, regardless of the *percentage* frequency variation this represents. If a resonator's loaded Q is 1000, the fractional change in wave length over the band is $\Delta\lambda/\lambda_0 = -\Delta\omega/\omega_0 = 10^{-3}$. The fractional change in electrical length of l is also 10^{-3}, but the actual numerical change is $\Delta(l/\lambda) = -(l/\lambda_0)(\Delta\lambda/\lambda_0) = 10^{-3}(l/\lambda_0)$. This change is less than 0.05λ if $l/\lambda_0 < 50$! Even if the loaded Q were as low as 100, and we reduced our tolerable change in the electrical length of l to 0.01λ, we still only need hold $l/\lambda_0 < 1$. This gives us a good deal of freedom, considering again that the generalized impedance goes through a complete period of space variation along the line in a distance $\lambda/2$.

The latitude in terminal-pair selection referred to above may be illustrated by exhibiting a physical system for which it arises. Figure 6.30 is a section through a coaxial line and resonator structure. Let us say we are interested in frequencies around that for which the resonator is a wave length long ($l \approx \lambda$), and the distance l_1 and the dimensions of the line cross sections are much less than λ. One might well ask where the appropriate terminal pair a–a' should be located in space for the purpose of discussing *the* impedance of the resonator. That is, where does the "input line" end, and the "resonator" begin?

We might argue first that a–a' should be pushed to the right until it reaches the outside of the cavity wall because the input line ends *geometrically* near that point. Unfortunately, however, as we have pointed out in previous chapters, the problem of defining voltage, current, and impedance (even normalized) depends upon field con-

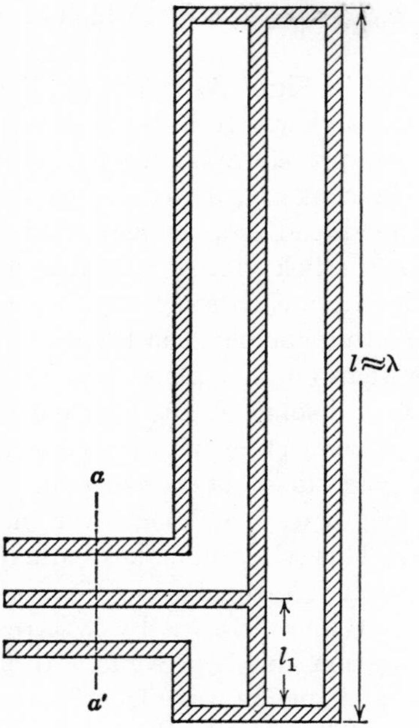

Fig. 6.30. Cross-section of a physical resonator with coupling.

figurations as well as geometry. In Fig. 6.30, fringing fields will exist in the neighborhood of the joints and corners, and these fields are *not* the simple ones which define a transmission-line voltage and current (see Chapter 9). Thus, *electrically*, the input line does not really extend right up to the cavity wall, and we are on this account tempted to push a–a' to the left, at least until it is clear of the fringing fields.

The important point for us now is that a considerable leeway is allowable in selecting the "reference plane" a–a', which we will call the resonator "terminals," and that, within this range, no one can really defend one choice more than another. The only important restriction which we could insist upon is that the principal frequency variation of the impedance, at whatever terminals we elect to call the resonator terminals, should be produced mostly by the large physical resonator l and not primarily by the input line, coupling wires, or corners. Since the latter are assumed to occupy regions much smaller than a wave length in dimensions, the acceptable positions of a–a' on this basis are governed only by our previous considerations of changes in electrical length over the frequency range of interest.

All this still does not tell us how the impedance at a–a' for Fig. 6.30 will behave as a function of frequency, given a reasonable choice of

a–a'. The precise answer to this question depends, of course, on what is usually a rather difficult field solution to the whole problem. Without doing this solution, however, we can see that, if the hole made by the input line in the cavity wall is small, and the distances l_1 and to the plane a–a' are small compared to the wave length, the natural oscillation of the whole system will not be altered much by having a–a' either open- or short-circuited. That is, the total energies stored in the input line to the right of a–a' and in the space about the corners and junctions are small compared to those in the rest of the system. This means we have a system *weakly coupled* to a–a', a corresponding close-spaced zero and pole observed from these terminals, and, consequently, a frequency behavior described by a circle in the Smith chart like the solid one of Fig. 6.29, tipped at some angle 2ϑ with respect to the real axis. Thus a new choice of a–a' can bring the circle into the dashed position (a–a' must be chosen as the detuned-open position for this purpose), and a simple series resonance occurs. The thing we cannot calculate very well is exactly *where* the detuned-open position will be for the geometry of Fig. 6.30, but we can certainly determine it by experiment.

Having seen that any system of untuned weak coupling to a high-Q resonator may always be described by a simple series-resonant circuit from a terminal pair at the detuned-open position of the input line, it is now apparent that an additional quarter-wave change of the terminal pair either toward or away from the resonator will produce *parallel-resonant* behavior! Accordingly, *any system of untuned weak coupling to a high-Q resonator always leads to a system description as a simple series-tuned circuit from a terminal pair or reference plane at the detuned-open position, or as a simple parallel-tuned circuit from the detuned-short position.* We have our free choice; one cannot therefore insist *a priori* that a given physical distributed resonator with untuned coupling is either like a series- or a shunt-tuned circuit.

Note, however, that *the coupling parameter, resonant frequency, and all the Q's are invariant to the two choices of reference plane described above,* provided that the frequency dependence of the quarter-wave section of line between them is negligible over the particular frequency band of the system. It is partly this invariance that makes ζ, ω_0, and the Q's such excellent descriptive parameters for a single resonance.

PROBLEMS

Problem 6.1. (a) Show a graphical construction that will bring out *all* the natural frequencies of the system in Fig. 6.31. (b) If $(T_2/T_1) = 1$, sketch the voltage and current distributions on the system for the three lowest natural frequencies when:

(i) $Z_{02}/Z_{01} = 1$; (ii) $Z_{02}/Z_{01} > 1$; (iii) $Z_{02}/Z_{01} < 1$. Comment on significant similarities and differences. (c) To which of the infinite series of natural modes possible under conditions (b) would a "solder pair" at the middle junction be *coupled*, in each case (i)–(iii)? A "scissors pair"? (d) Consider all the *uncoupled* cases of (c), and, for each one, find the steady-state impedance (or admittance) seen at the terminals. Comment on the relationship between these values and the natural frequencies of the system with the corresponding terminals open-circuited and short-circuited. (e) Repeat each of the above parts for all combinations of the conditions $Z_{02}/Z_{01} \gtrless 1$ and $T_2/T_1 \gtrless 1$ not previously covered. (f) If for an arbitrary choice of Z_{02}/Z_{01} and T_2/T_1 the natural frequencies of the system of Fig. 6.31 are indexed in order of increasing magnitude, ω_{0i} ($i = 0, 1, 2, \cdots$), what quantitative feature of the voltage (or current) space pattern is directly proportional to i? (g) Repeat the whole problem if the short circuit on one end of the system of Fig. 6.31 is replaced by an open circuit.

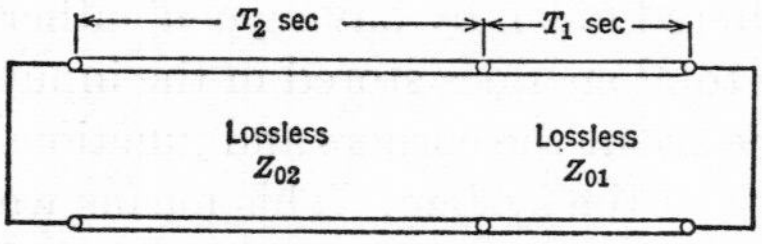

Fig. 6.31. Problem 6.1.

Problem 6.2. (a) For the system of Fig. 6.32, show that the entire set of natural oscillations can be divided into two groups, based upon the symmetry of the voltage space distribution with respect to the center terminal pair a–a′: (i) odd-symmetric; (ii) even-symmetric. (b) Relate the natural frequencies and corresponding space patterns appropriate for (a) to the results of Prob. 6.1.

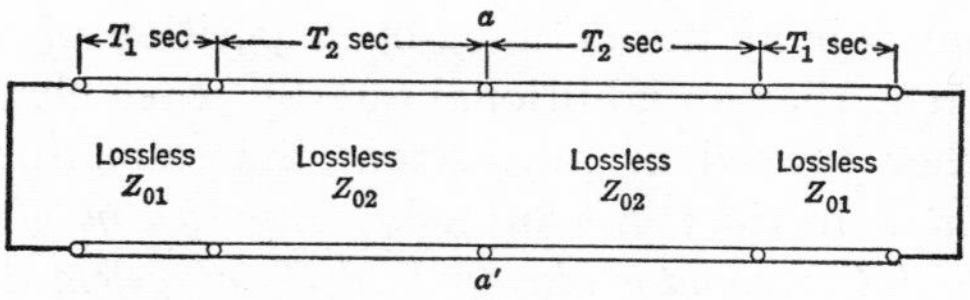

Fig. 6.32. Problem 6.2.

Problem 6.3. Carry out the analyses and interpretations of Sec. 6.3.1, for Fig. 6.11, in the case of a unit-impulse voltage-source excitation. Discuss the significant differences between the responses to current and voltage excitations, particularly in regard to which natural modes are excited by the source.

Problem 6.4. The ends of a transmission line of length l are connected together to form a ring. (a) Determine the three lowest natural frequencies of the system. (b) Sketch two possible current distributions at the lowest nonzero natural frequency. (c) Repeat (a) and (b) for the case where the line is twisted half a turn before the ends are connected together. (d) Relate the standing-wave degeneracy found in (b) and (c) to the possibility of wave solutions which simply travel around the loop. (e) How would you choose sources to excite the loop from a *single* position to start a *single* unit voltage impulse circulating around it? (f) Determine and sketch the voltage time function seen at any point on the ring, under conditions (e), for both the twisted and untwisted cases. (g) Express the solution to (e) and (f), in both cases, as a superposition of all the relevant natural modes, and be sure to determine all but one of the constants.

Problem 6.5. The lossless line of Fig. 6.33 is oscillating freely at its lowest natural frequency with S open. S is closed suddenly when $i(t) = 0$ and is increasing. Sketch the voltage across R as a function of time.

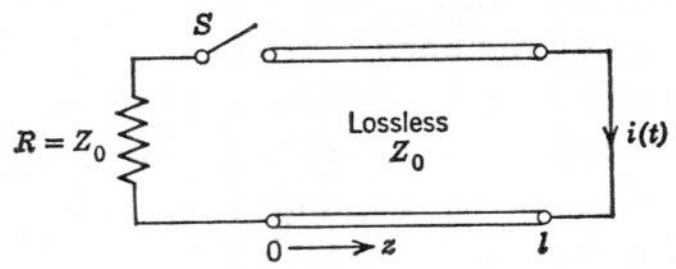

Fig. 6.33. Problem 6.5.

Problem 6.6. A distortionless line is short-circuited at $z = 0$ and open-circuited at $z = -l$. Its characteristic impedance is Z_0, its transit time one way is T, and its attenuation constant is α. At $t = 0$ the initial conditions are: $V(0, z) = -V_0 \sin(\pi z/2l)$, $I(0, z) = 0$. Find the voltage and current everywhere on the line for $t > 0$.

Problem 6.7. (a) Carry out with suitable modifications the analyses and interpretations of Sec. 6.3.1, for Fig. 6.11, but for the case of a *distortionless* line. (b) Repeat the analysis of Sec. 6.3.2 with suitable modifications for the distortionless line. (c) Comment on the difference between the time and space parts of this problem, as compared to their similarity in the case of a lossless line.

Problem 6.8. A transmission line of length l, with constant parameters R, L, G, and C is open-circuited at one end and short-circuited at the other. It is to be driven from a solder-pair tap at distance l_1 from the short circuit. (a) Find the *exact* locations of all the complex zeros and poles of the impedance Z(s) seen at the tap. For various interesting values of the ratios l_1/l, R/L, and G/C, discuss the locations of these frequencies in the s-plane. Be sure to examine all regions of the s-plane, including those near to, at intermediate distances from, and far away from the origin. (b) Assuming that the first pole is underdamped, plot carefully the locus in the s-plane of the nearest zero to it, as a function of l_1 (i.e., l_1/l for fixed l), especially when l_1/l is small. (c) Compare the results of (a) and (b) with answers obtained by approximate energy methods.

Problem 6.9. (a) Write an *exact* expression for the admittance Y(s) seen at the tap in Prob. 6.8. Put the expression into a form which shows clearly that its zeros and poles are respectively the natural frequencies of the system with the tap terminals open-circuited and short-circuited. (b) Assuming that the first zero of Y(s) is only slightly damped, show *analytically* how to approximate the expression of (a), for frequencies on the $j\omega$-axis near this zero, to yield a simple-tuned circuit when (l_1/l) is not too small. Discuss all the approximations very carefully, explaining words like "large" and "small" quantitatively whenever possible. Check by approximate energy methods. (c) Extend the results of (b) to the case when (l_1/l) is small, modifying the equivalent circuit as required. Again, justify analytically every approximation you make. Check by a combination of quasi-statics and approximate energy methods. (d) Examine the equivalent circuits of (b) and (c) in the lossless case $R = G = 0$. Show that they agree with what we would develop directly for the lossless line system. State carefully the approximations in the lossless case. (e) By simply adding to the lossless equivalent circuits of (d) a resistance R_i in series with every inductance L_i, such that $R_i/L_i = R/L$, and a conductance G_j in parallel with every capacitance C_j, such that $G_j/C_j = G/C$, show that the lossless circuits come back to the proper equivalent circuits found in (b) and (c) for the dissipative structure. Are any new approximations necessary over and above those required in the *lossless* case? Explain why, and reconcile with your results of (b) and (c).

Problem 6.10. A lossless line of length l is terminated on one end by a resistance R_1, and on the other end by a resistance R_2. It is to be driven from a solder-pair tap at distance l_1 from R_1. (a) Find the *exact* locations of all complex zeros and poles of the admittance Y(s) seen at the tap. Discuss carefully, and show with sketches, the locations of these complex frequencies in the s-plane for all interesting values of R_1, R_2, and l_1. (b) Show especially how the first zero and pole move as a function of l_1 when l_1/l is near 0 or near 1. (c) Write an *exact* expression for Y(s) which shows clearly that its zeros and poles are respectively the natural frequencies of the system with the tap terminals open-circuited and short-circuited. (e) Make careful approximations *analytically* in Y(s) to reduce its behavior for $s = j\omega$ near its first zero to that of a simple-tuned circuit, whenever such a reduction is possible. State precisely when it is possible, and explain the physical meaning of each approximation. Check by approximate energy methods. (f) Extend the results of (e) to include the modified equivalent circuit required to represent $Y(j\omega)$ near its first zero when l_1/l is near 0 or near 1. Check by a combination of quasi-statics and approximate energy methods.

Problem 6.11. (a) A slightly dissipative line of length l has attenuation constant α and phase velocity v. It is short-circuited on one end, and open-circuited on the other end. Justify the following approximate derivation of Q_n for the nth natural frequency ω_{0n} of the line. Let $\langle P_+\rangle$ and $\langle u_+\rangle$ represent respectively the time-averaged power carried and energy stored per unit length by a single traveling wave on the line. Then: (i) $\langle P_{\text{diss}}\rangle = 2\alpha\langle P_+\rangle$ w/m, one wave alone; (ii) $\langle P_{\text{diss}}\rangle = 4\alpha l\langle P_+\rangle$ w total, both waves; (iii) $\langle u_+\rangle = \langle P_+\rangle/v$ joules/m, one wave alone; (iv) $\langle W_{\text{tot}}\rangle = 2l\langle u_+\rangle = 2l\langle P_+\rangle/v$ joules, both waves; (v) $Q_n \equiv (\omega_{0n}\langle W_{\text{tot}}\rangle)/\langle P_{\text{diss}}\rangle = (\omega_{0n}l/v)/2\alpha l = n\pi/4\alpha l$; $n = 1, 2, 3, \cdots$. (b) If the ends are terminated in pure resistances, one of which has normalized resistance r_n for example, add to $\langle P_{\text{diss}}\rangle$ two more terms, each of the form

$$(1 - |\bar{\Gamma}|^2)\langle P_+\rangle \approx \begin{cases} 4r_n\langle P_+\rangle & \text{if } r_n \ll 1 \\ \dfrac{4\langle P_+\rangle}{r_n} & \text{if } r_n \gg 1 \end{cases}$$

Problem 6.12. (a) A slightly dissipative transmission line of length l and characteristic impedance Z_0 is short-circuited at one end. Measured at the other end, the Q of the system at the first (parallel) resonance is approximately 200. The short-circuit is now replaced by a small resistance R_1. The resonant frequency remains about the same, but the Q is now approximately 50. Find the numerical value of the normalized resistance R_1/Z_0. (b) Using the results of (a), and *assuming* that the line parameters R, L, G, C are independent of frequency, determine and plot versus n the Q of the nth parallel resonance when: (i) the load is a short circuit; (ii) the load is R_1. (c) Repeat (b) for the series resonances. (d) Determine and plot versus n the resonant impedance of the system under conditions (b). (e) Repeat (d) for the series resonances. (f) Repeat (a) if the data given there apply to the first *series* resonance instead of to the first parallel resonance.

Problem 6.13. A transmission line with parameters R, L, and C per unit length is to be used as a high-Q parallel-resonant element at high frequencies. It is terminated in a shorting bar, which nevertheless has resistance R_1. If because of skin effect R and R_1 are proportional to $\omega^{1/2}$, but L and C may still be treated as constants, determine and plot versus frequency the Q and resonant impedance at the input end of the line if: (a) The line length l is adjusted so the first parallel resonance is always used to achieve the desired element. (b) The line length l is left fixed, but successively higher parallel resonance modes are used to achieve the desired element. (c) What establishes limits on the usefulness of each of the above schemes at high frequencies?

Problem 6.14. A lossless line of length l and characteristic impedance Z_0 is a quarter wave length long at radian frequency $\omega_0 = 2\pi \times 10^8$. It is serving to "transform" a load impedance $Z_0/100$ into an input impedance $100Z_0$ at this frequency. If the input is actually driven by a constant voltage source E_1 in series with $100Z_0$: (a) Find the output-to-input power and voltage ratios at center frequency $\omega_0 = 2\pi \times 10^8$ rad/sec. (b) Find the half-power bandwidth at the load, in cps. (c) Determine the general product of maximum impedance ηZ_0 and radian bandwidth w for this quarter-wave transformation from Z_0/η. (d) If the line is slightly dissipative, having a phase velocity 3×10^8 m/sec and a complex propagation constant $\bar{\gamma}$ whose angle in polar form is $(\pi/2) - 1.65 \times 10^{-3}$ radians over the frequency band of interest, determine the new answers to (a) and (b). Determine also the maximum possible input impedance. (e) Check your results by making an exact analysis of a lossless line terminated in R_L and driven from a voltage source in series with R_S. In particular, determine the *smallest* transformation ratio for which the approximation methods are accurate to within 5%. Extend your results to slightly dissipative lines.

Problem 6.15. A lossless transmission line of length l is open-circuited at one end and is terminated at the other end in a resistance R_1. A resistance of the same size R_1 is also connected in *series* with the line at a distance $2l/3$ from the open end. The propagation time for length l of the line is 10^{-7} sec. (a) Find the lowest two natural frequencies of the system, and the Q's at these frequencies, when $R_1 = Z_0/100$. Sketch for each case the approximate voltage and current distributions. (b) Find the lowest natural frequency of the system and the Q at this frequency, when $R_1 = 100Z_0$. Sketch the approximate voltage and current distributions.

Problem 6.16. A slightly dissipative line (R, L, G, C) of length l is terminated by a resistance $Z_0/100$ at one end, shunted by one of value $100Z_0$ at the center, and driven from the other end. Find the frequency of first parallel resonance at the input, and an equivalent circuit which represents the impedance behavior at neighboring frequencies.

Problem 6.17. A lossless line of length l is excited from one end, with the other end open. The first *series* resonance is found. (a) Will this resonant frequency be raised or lowered if a pure inductance is connected across the open end of the line? Explain in terms of energies. (b) Work out the approximate shift if the inductance is "large" (compared to what?). (c) What can you say if the inductance is "small"? (d) Make an *exact* graphical analysis illustrating all significant features mentioned above.

Problem 6.18. Calculate the lowest natural frequency of the system in Fig. 6.34, if L_1, L_2, and C_1 are small and C_2 is large.

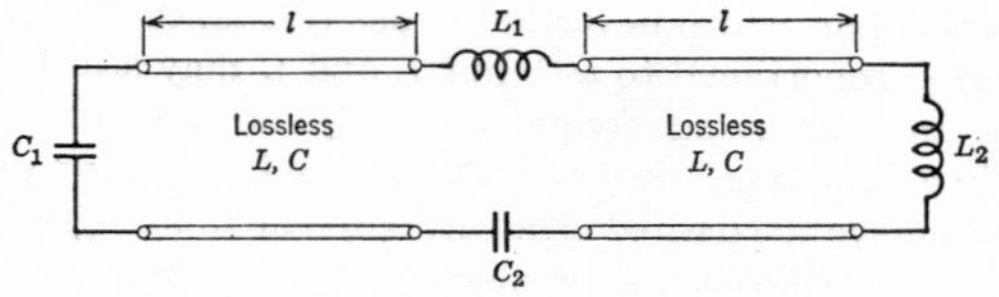

Fig. 6.34. Problem 6.18.

Problem 6.19. In Fig. 6.35: If $C_1 = G_1 = 0$: (a) Find the two lowest *parallel* resonant frequencies at terminals a–a'. (b) Determine an equivalent lumped circuit to represent approximately the admittance at a–a' in the neighborhood of each of the above resonances. If $0 < G_1 \ll Y_0$ and $C_1 = 0$: (c) Determine the approximate Q at each of the above resonances, and explain physically their ratio. If

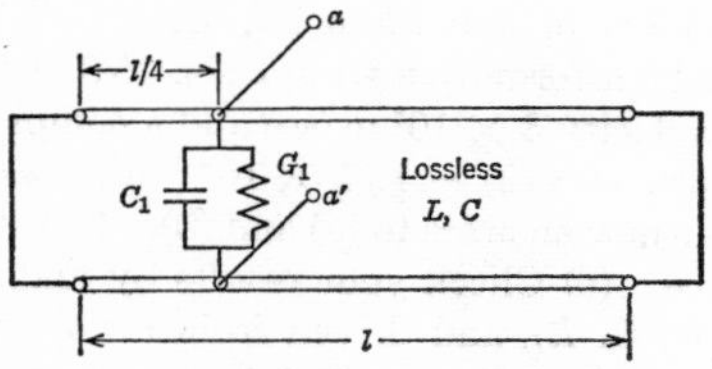

Fig. 6.35. Problem 6.19.

$G_1 = 0$ and $0 < C_1 \ll Cl$: (d) Calculate the approximate *shift* in the two lowest parallel resonant frequencies from their values in (a), and explain physically the ratio of these shifts.

Problem 6.20. A lossless line 10 m long has $C = 60\ \mu\mu\text{f/m}$, $L = 0.6\ \mu\text{h/m}$, and is short-circuited at one end. (a) Find the value of a capacitance C' which, when connected to the other end, makes the system resonate when the line is ⅛ wave length long. Determine also the resonant frequency. (b) Sketch carefully voltage and current distributions on the system at resonance. (c) If dielectric loss becomes significant, so that $G = 10^{-5}$ mhos/m, find the Q of the system at the above resonance. (d) If C' is replaced by a suitable length of lossless open-circuited line having the same L and C as the original line, such that the above resonance frequency is *unchanged*, repeat (b) and (c).

Problem 6.21. A lossless transmission line having parameters L, C per unit length and $v = 1/\sqrt{LC}$ has length $2l$. It is terminated in a short circuit at one end and an open circuit at the other end (terminals a–a'). A capacitance C_1 is connected across the line at the mid-point. (a) Find asymptotic expressions for the lowest two natural frequencies f_1 and f_2 of the system for the cases (i) $C_1 \ll 2Cl$, and (ii) $C_1 \gg 2Cl$. (b) What is the admittance at terminals a–a' at these natural frequencies? (c) From your answers to (a), sketch the form of the two curves relating the frequencies f_1 and f_2 to the capacitance C_1. (d) Find the value of C_1 for which $f_1 = v/16l$, and plot this point on one of the curves in (c). (e) Find the value of C_1 for which $f_2 = 5v/16l$, and plot this point on the other curve in (c). (f) For the values of C_1 and f_2 in (e), sketch the voltage and current space distributions along the line, when a voltage source is connected at terminals a–a'.

Problem 6.22. (a) What is the lowest natural frequency of the lossless system in Fig. 6.36 if $l_1 = 0$? (b) What is it if $l_1 \gg l$? (c) Make an approximate plot of the lowest natural frequency of the system versus l_1, when l_1 is varied between 0 and $5l$. Check your result exactly for the intermediate cases $l_1 = l/2$, $l_1 = l$, $l_1 = 2l$, $l_1 = 5l$.

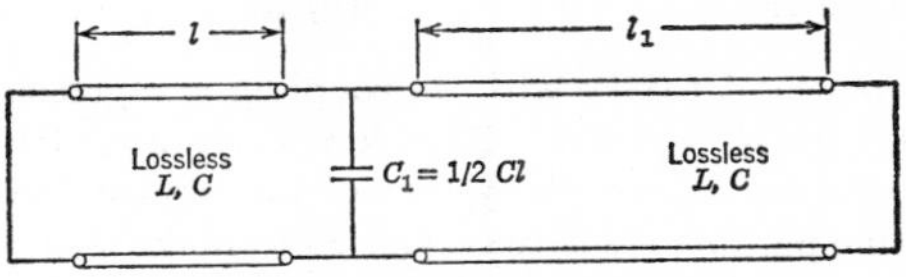

Fig. 6.36. Problem 6.22.

Problem 6.23. A lossless transmission line with $Z_0 = 1$ ohm is terminated in an inductance of 10^{-9} h in series with a 0.1-ohm resistance. The lowest resonant frequency for the combination is $\omega_0 = 10^9$ rad/sec, and the line is ⅛ wave length long at this frequency. If the peak value of the current through the inductance is 1 amp when the system is being driven at resonance: (a) Find the total energy stored in the inductance plus the transmission line by two methods. (b) Find the Q of the system at frequency ω_0. (Should this Q be greater or less than the Q of the inductance alone?) (c) Using the Smith chart, show that the frequency of the next highest resonance above ω_0 is $3.6\omega_0$. Assume that the inductance value is independent of frequency. (d) Repeat (b) if the line has small attenuation, and determine the value of α which reduces the Q to one half its original value.

Problem 6.24. (a) Find L in Fig. 6.37 such that the lowest natural frequency of the system is 100 mc and occurs when $l = \lambda/8$. (b) If L is chosen as in (a), find the Q of the system at $f = 100$ mc. (c) If L is chosen as in (a), find the impedance at a–a' when $f = 200$ mc.

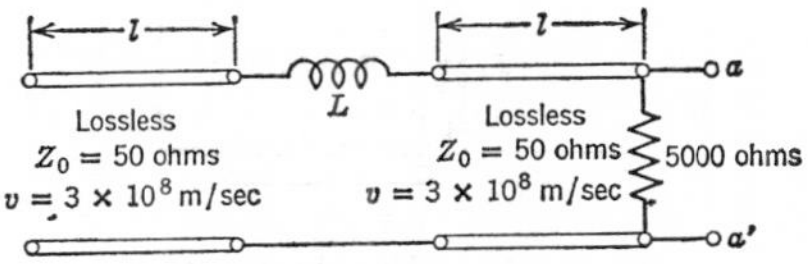

Fig. 6.37. Problem 6.24.

Problem 6.25. Consider Fig. 6.38 when $R_1 = 0$ and $G = 0$. (a) Find the smallest value of l for which $\omega_1 \equiv 1/\sqrt{L_1C_1}$ is a *parallel* resonance of Z. (b) Make a graphical construction to locate all the poles of Z for the value of l in (a). (c) On the same axes as used in (b), locate graphically all the zeros of Z for the same l value.

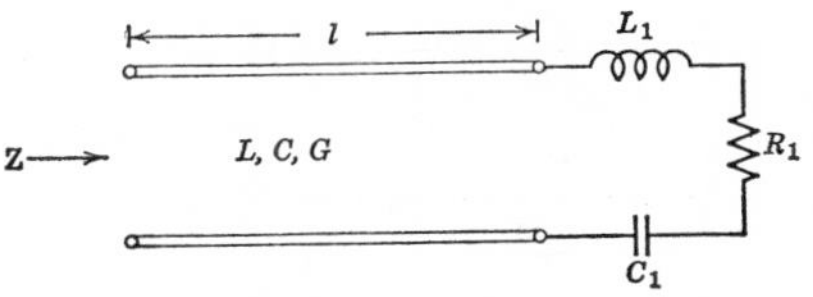

Fig. 6.38. Problem 6.25.

(d) Find all other values of l for which ω_1 is a pole of Z, and repeat (b) and (c) for one of the large ones. (e) Now suppose $0 < R_1 \ll Z_0$, $\omega_1 L_1 \gg R_1$. How do Q and $Z(j\omega_1)$ change for the various lengths discussed in (a) and (d) if: (i) $G = 0$? (ii) $0 < G \ll \omega_1 C$? (f) Find and plot in the s-plane the two *complex* zeros of Z(s) nearest to the pole corresponding to the resonance treated in (e-i) for: (i) the value of l in (a); (ii) the large value of l used in (d). (g) Repeat (f) under conditions (e-ii).

Problem 6.26. A resonant cavity with its coupling is represented by the system shown in Fig. 6.39, with $R_1 = 10Z_0$. At its lowest resonant frequency ω_0, $\omega_0 L_1 \ll Z_0$. (a) When driven from a source system of impedance level Z_0, the value of L_1 in terms of the line parameters is to be chosen to make $Q_{ext} = 100$. What kind of a resonance is involved? Use approximation methods to find this value of L_1.

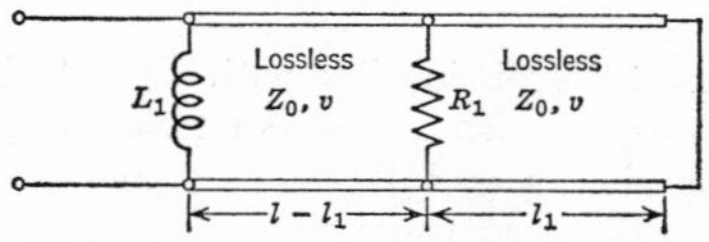

Fig. 6.39. Problem 6.26.

(b) With L_1 adjusted to the value set in (a), find the unloaded Q_u as a function of the position l_1 of R_1. (c) What must be the position of R_1 to receive maximum power from the source? Find the loaded Q_L under this condition.

Problem 6.27. In Fig. 6.40 the slightly dissipative line represents a resonant cavity, and the inductance L_1 its coupling mechanism. When driven from a source system at impedance level $Z_0 = \sqrt{L/C}$ (the same as that of the dissipative line shown), the coupling is to be critical for the lowest high-Q_L resonance. (a) What type of resonance is involved? (b) Find L_1 in terms of the other parameters. (c) For

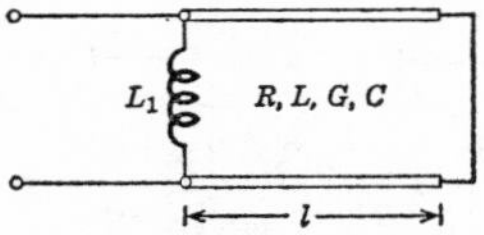

Fig. 6.40. Problem 6.27.

the choice of L_1 made in (b), describe what happens to the coupling at the higher resonances of the same type and the various Q's involved in each case. Assume all the given parameters are independent of frequency.

Problem 6.28. Consider a lossless line which is dispersive (see, for example, Probs. 3.48 and 3.49). It will be described by the frequency-domain equations:

(1) $$\frac{\partial V}{\partial z} = -jX_S(\omega)I$$

(2) $$\frac{\partial I}{\partial z} = -jB_P(\omega)V$$

where $X_S(\omega)$ and $B_P(\omega)$ may be rather complicated series reactance and shunt susceptance functions respectively. These functions arise from physical electromagnetic phenomena which happen to be representable in this case as a distributed-circuit structure. Hence $X_S(\omega)$ and $B_P(\omega)$ must obey Foster's reactance theorem. (a) Find an expression for the time-averaged energy stored per unit length on such a line. Show that your expression for it is guaranteed to be positive, and to reduce to the usual expression in the simple case of a lossless line with constant series L and shunt C. (Refer again to Probs. 1.7 through 1.9, and 2.12.) (b) Show by manipulating Eqs. 1 and 2 above that

$$\frac{\partial}{\partial z}\left(\mathrm{V}\frac{\partial \mathrm{I}^*}{\partial \omega} + \frac{\partial \mathrm{V}^*}{\partial \omega}\mathrm{I}\right) = j\left(|\mathrm{V}|^2\frac{\partial B_P}{\partial \omega} + |\mathrm{I}|^2\frac{\partial X_S}{\partial \omega}\right) = j4\langle u_{\mathrm{tot}}\rangle$$

This result is sometimes called the "energy theorem." Show that if it is applied to the line terminated in some physical reactance $jX_L(\omega)$, it yields Foster's reactance theorem for the input reactance or susceptance of the line. (c) From the propagation constant $\bar{\gamma}$ and the characteristic impedance Z_0 of this line, show that there are, in general, alternate "pass" and "stop" bands. The former are characterized by having $\bar{\gamma} = j\beta(\omega)$ and $\mathrm{Z}_0(\omega)$ real; the latter by having $\bar{\gamma} = \alpha(\omega)$ and $\mathrm{Z}_0(\omega) = jX_0(\omega)$. (d) Apply the energy theorem of (b) to a single traveling wave on the line in a pass band, i.e., to the wave: $\mathrm{V} = \mathrm{V}_+e^{-j\beta z}$, $\mathrm{I} = (\mathrm{V}_+/Z_0)e^{-j\beta z}$, where V_+ is independent of frequency. Show that $(\partial\beta/\partial\omega)^{-1} = \langle P\rangle/\langle u_{\mathrm{tot}}\rangle$. This result proves that "energy velocity" and "group velocity" are the same for all smooth lossless lines. On these, the time-average power and the group velocity must have the same direction along the line, but the phase velocity may be oppositely directed (see Prob. 3.48). (e) Apply the energy theorem to a single wave in a stop band, and discuss the significance of the results. (f) Check the results of this problem by applying them to the special cases treated in Probs. 3.48 and 3.49. See also the lossless wave guide modes treated in Chapter 7, which follow the same theory. (g) What problems arise in identifying stored energy in dissipative lines?

CHAPTER SEVEN

Plane Waves in Lossless Media

Our study of waves has been limited so far to those depending upon only one space dimension. Most of the field aspects of the problem have been suppressed by our use of the voltage and current concepts. It is now our task to consider waves in more than one space dimension, and to pay more attention to the fields.

The *uniform* plane wave provides a simple and important special case of wave motion in space, at the same time retaining, from one point of view, many features characteristic of transmission-line waves. From another point of view, however, uniform plane waves lead to a consideration of the more general *nonuniform* plane waves, and to the whole question of guided waves.

The subject of plane waves is simplest when the medium in which they propagate is lossless. We treat this case in the present chapter, and the effects of loss in the next.

7.1 Uniform Plane Waves in the Time Domain

7.1.1 Introduction

In Chapter 2, we encountered time-varying solutions of Maxwell's equations with electric and magnetic fields, each having only two components. These solutions are in a class we shall subsequently examine further, called TEM waves (transverse electromagnetic waves), so named because the field vectors lie completely in planes *transverse* to the direction of wave propagation. In Chapter 2, Sec. 2.3, moreover, we used the simplest form of such TEM waves to meet the boundary conditions imposed by parallel, perfectly conducting planes. This solution had the second special property that, at any instant of time, *neither* $\boldsymbol{E}$ *nor* $\boldsymbol{H}$ *varied with position in the transverse plane containing them.* This type of solution is known as a *uniform plane wave* (or sometimes an infinite plane wave). It will be compared later with other TEM waves—like those on an open-wire line—in which the strength and direction of $\boldsymbol{E}$ and $\boldsymbol{H}$ do vary with position in the transverse planes containing them (see Chapter 9).

In the solution in Chapter 2, the direction of $\boldsymbol{E}$ in space was chosen normal to the metal planes to meet the boundary conditions they imposed. However, it is clear that, if the planes were drawn apart indefinitely, the entire field in space would remain a solution to Maxwell's equations. Similarly, a uniform plane wave propagating in any direction in free space is also a solution of Maxwell's equations, inasmuch as

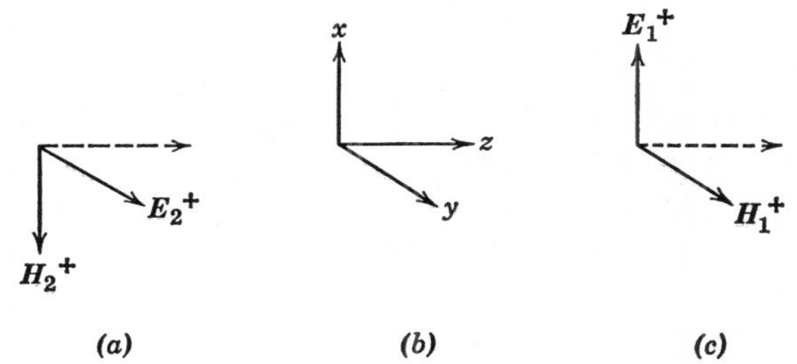

Fig. 7.1. Schematic pictures of two independent uniform plane waves propagating in the $+z$ direction.

the direction of propagation may always be taken as the z-axis to conform with our previous results.

Since Maxwell's equations, as we have agreed to consider them, are linear vector equations, any sum of vector solutions is also a solution. Therefore a superposition of many uniform plane waves traveling in different directions in space is a valid electromagnetic field. Moreover, it is apparent that one uniform plane wave cannot be obtained as a linear combination of others propagating in different directions. Hence there are at least as many linearly independent uniform plane wave solutions to the field equations as there are different directions in space. More than this, even for a given direction of propagation, there are *two* linearly independent uniform plane-wave solutions to Maxwell's equations. Figure 7.1 and the following discussion are addressed to this matter.

7.1.2 Details of the Solution

In terms of the coordinate system of Fig. 7.1*b*, we shall write Maxwell's equations (Eqs. 1.24*b*, 1.25, and 1.31) for a lossless, linear, homogeneous medium with constant parameters ($\sigma \equiv 0,\ \epsilon,\ \mu$) and without sources. We shall consider the case of uniform plane waves, for which, by our definition,

$$\text{(a)}\qquad E_z = H_z \equiv 0 \qquad\qquad (7.1)$$

$$\text{(b)}\qquad \frac{\partial}{\partial x} \equiv \frac{\partial}{\partial y} \equiv 0$$

There remain only the following equations, rearranged for convenience

$$\left.\begin{aligned} &\text{(a)}\quad \frac{\partial E_x}{\partial z} = -\mu \frac{\partial H_y}{\partial t} \\ &\text{(b)}\quad \frac{\partial H_y}{\partial z} = -\epsilon \frac{\partial E_x}{\partial t} \end{aligned}\right\}\text{Set 1} \qquad \left.\begin{aligned} &\text{(c)}\quad \frac{\partial E_y}{\partial z} = \mu \frac{\partial H_x}{\partial t} \\ &\text{(d)}\quad \frac{\partial H_x}{\partial z} = \epsilon \frac{\partial E_y}{\partial t} \end{aligned}\right\}\text{Set 2} \qquad (7.2)$$

There are two *independent* pairs of equations: equations 7.2*a* and 7.2*b* relate *only* E_x and H_y, whereas Eqs. 7.2*c* and 7.2*d* relate *only* E_y and H_x. Each pair is formally identical with the equations of a lossless transmission line. The solutions can be written immediately.

For Eqs. 7.2*a* and 7.2*b*, aside from the obvious d-c solutions, we have

$$\text{(a)}\qquad E_x = f_+\left(t - \frac{z}{v}\right) + f_-\left(t + \frac{z}{v}\right) \qquad\qquad (7.3)$$

$$\text{(b)}\qquad H_y = \frac{1}{\eta}\left[f_+\left(t - \frac{z}{v}\right) - f_-\left(t + \frac{z}{v}\right)\right]$$

where

$$\text{(a)}\qquad \eta = +\sqrt{\frac{\mu}{\epsilon}} \equiv \text{characteristic wave impedance (ohms)} \qquad\qquad (7.4)$$

$$\text{(b)}\qquad v = \frac{1}{+\sqrt{\epsilon\mu}} \equiv \text{velocity of light}$$

Considering the (+) and (−) waves separately, we have

$$\text{(a)}\qquad E_x^+ = \eta H_y^+ = f_+\left(t - \frac{z}{v}\right) \qquad\qquad (7.5)$$

$$\text{(b)}\qquad E_x^- = -\eta H_y^- = f_-\left(t + \frac{z}{v}\right)$$

If subscript 1 is used to denote the solution of Eq. 7.2, Set 1, we have

$$\begin{aligned} &\text{(a)} \qquad \boldsymbol{E}_1 = \boldsymbol{a}_x E_x = \boldsymbol{a}_x E_x^{+} + \boldsymbol{a}_x E_x^{-} = \boldsymbol{E}_1^{+} + \boldsymbol{E}_1^{-} \\ &\text{(b)} \qquad \boldsymbol{H}_1 = \boldsymbol{a}_y H_y = \boldsymbol{a}_y H_y^{+} + \boldsymbol{a}_y H_y^{-} = \boldsymbol{H}_1^{+} + \boldsymbol{H}_1^{-} \end{aligned} \tag{7.6}$$

Equations 7.5 become

$$\begin{aligned} &\text{(a)} \qquad \boldsymbol{E}_1^{+} = \eta \boldsymbol{H}_1^{+} \times \boldsymbol{a}_z \\ &\text{(b)} \qquad \boldsymbol{E}_1^{-} = -\eta \boldsymbol{H}_1^{-} \times \boldsymbol{a}_z \end{aligned} \tag{7.7}$$

Equation 7.5*a* is represented schematically by Fig. 7.1*c*, in which the dotted line shows the direction of propagation ($+z$).

For Eqs. 7.2*c* and 7.2*d*, the obvious modification of algebraic sign yields

$$\begin{aligned} &\text{(a)} \qquad E_y = g_{+}\left(t - \frac{z}{v}\right) + g_{-}\left(t + \frac{z}{v}\right) \\ &\text{(b)} \qquad H_x = -\frac{1}{\eta}\left[g_{+}\left(t - \frac{z}{v}\right) - g_{-}\left(t + \frac{z}{v}\right)\right] \end{aligned} \tag{7.8}$$

The functions $g_{\pm}$ are in no way related to the functions $f_{\pm}$ in Eqs. 7.3 as far as Eqs. 7.2 are concerned, because Eqs. 7.2*c* and 7.2*d* are completely independent of Eqs. 7.2*a* and 7.2*b*. In place of Eqs. 7.5, we now have

$$\begin{aligned} &\text{(a)} \qquad E_y^{+} = -\eta H_x^{+} = g_{+}\left(t - \frac{z}{v}\right) \\ &\text{(b)} \qquad E_y^{-} = \eta H_x^{-} = g_{-}\left(t + \frac{z}{v}\right) \end{aligned} \tag{7.9}$$

If subscript 2 is used to denote the solution of Eq. 7.2, Set 2, we have

$$\begin{aligned} &\text{(a)} \qquad \boldsymbol{E}_2 = \boldsymbol{a}_y E_y = \boldsymbol{a}_y E_y^{+} + \boldsymbol{a}_y E_y^{-} = \boldsymbol{E}_2^{+} + \boldsymbol{E}_2^{-} \\ &\text{(b)} \qquad \boldsymbol{H}_2 = \boldsymbol{a}_x H_x = \boldsymbol{a}_x \boldsymbol{H}_x^{+} + \boldsymbol{a}_x H_x^{-} = \boldsymbol{H}_2^{+} + \boldsymbol{H}_2^{-} \end{aligned} \tag{7.10}$$

Equations 7.9 become

$$\begin{aligned} &\text{(a)} \qquad \boldsymbol{E}_2^{+} = \eta \boldsymbol{H}_2^{+} \times \boldsymbol{a}_z \\ &\text{(b)} \qquad \boldsymbol{E}_2^{-} = -\eta \boldsymbol{H}_2^{-} \times \boldsymbol{a}_z \end{aligned} \tag{7.11}$$

Equation 7.9*a* is represented schematically by Fig. 7.1*a*.

Equations 7.7 and 7.11 prove that $\boldsymbol{E}_{1,2}^{+,-}$, $\boldsymbol{H}_{1,2}^{+,-}$, and the direction of propagation are mutually perpendicular for each of the four separate waves. Their relative orientations are such that the Poynting vector $\boldsymbol{S}_{1,2}{}^{+,-}$ is at every instant in the direction of propagation.

$$\text{(a)}\quad \boldsymbol{S}_{1,2}{}^{+} = \boldsymbol{E}_{1,2}{}^{+} \times \boldsymbol{H}_{1,2}{}^{+} = \eta|\boldsymbol{H}_{1,2}{}^{+}|^2\boldsymbol{a}_z = \frac{1}{\eta}|\boldsymbol{E}_{1,2}{}^{+}|^2\boldsymbol{a}_z \tag{7.12}$$

$$\text{(b)}\quad \boldsymbol{S}_{1,2}{}^{-} = \boldsymbol{E}_{1,2}{}^{-} \times \boldsymbol{H}_{1,2}{}^{-} = -\eta|\boldsymbol{H}_{1,2}{}^{-}|^2\boldsymbol{a}_z = -\frac{1}{\eta}|\boldsymbol{E}_{1,2}{}^{-}|^2\boldsymbol{a}_z$$

We conclude that every uniform plane wave propagating in a given direction (the $+z$ direction, for example) *can always* be decomposed into two waves having the same direction of propagation but with mutually perpendicular electric (and magnetic) fields. Therefore any such solution $\boldsymbol{E}_T{}^{+}$ and $\boldsymbol{H}_T{}^{+}$ is expressible in the form

$$\text{(a)}\quad \boldsymbol{E}_T{}^{+} = \boldsymbol{a}_xE_x{}^{+} + \boldsymbol{a}_yE_y{}^{+} = \boldsymbol{E}_1{}^{+} + \boldsymbol{E}_2{}^{+} \tag{7.13a}$$

$$\text{(b)}\quad \boldsymbol{H}_T{}^{+} = \boldsymbol{a}_yH_y{}^{+} + \boldsymbol{a}_xH_x{}^{+} = \boldsymbol{H}_1{}^{+} + \boldsymbol{H}_2{}^{+} \tag{7.13b}$$

where the subscript T stipulates that $\boldsymbol{E}$ and $\boldsymbol{H}$ lie in planes *transverse* to the direction of propagation. A similar conclusion applies to $(-)$ waves.

7.1.3 Power Considerations and Orthogonality

Having assured ourselves that every uniform plane wave may be expressed as a linear combination of the preceding four solutions, we must now consider some general properties of the energy flow, or Poynting vector, for such linear combinations.

First, the power carried by the general $(+)$ wave (Eq. 7.13) is the algebraic sum of the $(+)$ wave powers carried by each component wave separately. To prove this, note that

$$\begin{aligned}\boldsymbol{S}^{+} &= \boldsymbol{E}_T{}^{+} \times \boldsymbol{H}_T{}^{+} = (\boldsymbol{E}_1{}^{+} + \boldsymbol{E}_2{}^{+}) \times (\boldsymbol{H}_1{}^{+} + \boldsymbol{H}_2{}^{+})\\ &= \boldsymbol{E}_1{}^{+} \times \boldsymbol{H}_1{}^{+} + \boldsymbol{E}_2{}^{+} \times \boldsymbol{H}_2{}^{+} + \boldsymbol{E}_1{}^{+} \times \boldsymbol{H}_2{}^{+}\\ &\quad + \boldsymbol{E}_2{}^{+} \times \boldsymbol{H}_1{}^{+}\end{aligned} \tag{7.13c}$$

But

$$\boldsymbol{E}_1{}^{+} \times \boldsymbol{H}_2{}^{+} = \boldsymbol{E}_2{}^{+} \times \boldsymbol{H}_1{}^{+} \equiv 0 \tag{7.13d}$$

because $\boldsymbol{E}_1{}^{+}$ and $\boldsymbol{E}_2{}^{+}$ are parallel to $\boldsymbol{H}_2{}^{+}$ and $\boldsymbol{H}_1{}^{+}$ respectively. Therefore, as stated above,

$$\boldsymbol{S}^{+} = \boldsymbol{S}_1{}^{+} + \boldsymbol{S}_2{}^{+} \tag{7.13e}$$

Similarly, for the $(-)$ waves:

$$\boldsymbol{S}^{-} = \boldsymbol{S}_1{}^{-} + \boldsymbol{S}_2{}^{-} \tag{7.13f}$$

The fact that the power in a (+) [or (−)] wave of Set 1 adds *without cross terms* to the power in a (+) [or a (−)] wave of Set 2, to give the total (+) [or (−)] wave power when waves of Sets 1 and 2 are present together, is one example of a more general condition known as *orthogonality* of two solutions of a set of differential equations. The usual statement of such a condition involves an integration. For instance, suppose that in a region R of space (or time) ϕ_i and ϕ_j are two different scalar solutions of the same differential equation. Then these functions are said to be orthogonal if

$$\int_R \phi_i \phi_j \, dR = 0 \qquad \text{when } i \neq j$$

If ϕ_i and ϕ_j are vector functions, the multiplication under the integral would be either a dot- or cross-product. Sometimes the integration need not cover the whole region R but may involve only a line or plane in it. In the present case of uniform plane waves, the fields do not vary with position at all over planes normal to the direction of propagation. Accordingly, a result like Eq. 7.13*d* may be integrated over any area in the plane of the fields. Indeed, it may most conveniently be regarded as an orthogonality condition *per unit area.*

As indicated above, the orthogonality of two functions makes sense only if they are different. By this, we mean that one is not simply a (nonzero) constant times the other ($\phi_j \neq a\phi_i$). When they are different in this sense, the functions are said to be *linearly independent.* Evidently, orthogonality is not simply a consequence of the linear independence of the two functions in question. Additional conditions are necessary. For example, in the present field problem, the solutions $\boldsymbol{E}_1$ and $\boldsymbol{E}_2$ are orthogonal specifically because of their perpendicular relation in real space. Such a relationship is by no means implied in the word "orthogonality," as used in the general theory of orthogonal functions, and we shall shortly see examples of orthogonality arising for quite different reasons.

To proceed still further with these ideas, consider the complete combination of (+) and (−) waves of Sets 1 and 2 (Eqs. 7.6 and 7.10). The Poynting vector is

$$\begin{aligned}\boldsymbol{S} &= (\boldsymbol{E}_1 + \boldsymbol{E}_2) \times (\boldsymbol{H}_1 + \boldsymbol{H}_2) \\ &= \boldsymbol{E}_1 \times \boldsymbol{H}_1 + \boldsymbol{E}_2 \times \boldsymbol{H}_2 + (\boldsymbol{E}_2 \times \boldsymbol{H}_1 + \boldsymbol{E}_1 \times \boldsymbol{H}_2) \\ &= \boldsymbol{S}_1 + \boldsymbol{S}_2 + (\boldsymbol{E}_2{}^+ \times \boldsymbol{H}_1{}^- + \boldsymbol{E}_2{}^- \times \boldsymbol{H}_1{}^+ \\ &\qquad + \boldsymbol{E}_1{}^+ \times \boldsymbol{H}_2{}^- + \boldsymbol{E}_1{}^- \times \boldsymbol{H}_2{}^+) \end{aligned} \tag{7.13g}$$

in which we have already used Eq. 7.13*d* and its analog for (−) waves.

Once again, however,

$$\boldsymbol{E}_{2,1}{}^{+} \times \boldsymbol{H}_{1,2}{}^{-} = \boldsymbol{E}_{2,1}{}^{-} \times \boldsymbol{H}_{1,2}{}^{+} \equiv 0 \tag{7.13h}$$

because $\boldsymbol{E}_{2,1}{}^{+}$ and $\boldsymbol{E}_{2,1}{}^{-}$ are parallel respectively to $\boldsymbol{H}_{1,2}{}^{-}$ and $\boldsymbol{H}_{1,2}{}^{+}$. It follows that the (+) waves of one solution type are orthogonal to the (−) waves of the other and that

$$\boldsymbol{S} = \boldsymbol{S}_1 + \boldsymbol{S}_2 \tag{7.13i}$$

Finally, the similarity of Eqs. 7.3 and 7.8 to those of a lossless transmission line reminds us that the power carried by combined (+) and (−) waves of either Set 1 or Set 2 is also simply the algebraic sum of those carried separately by the (+) and (−) waves alone:

$$\boldsymbol{S}_{1,2} = \boldsymbol{S}_{1,2}{}^{+} + \boldsymbol{S}_{1,2}{}^{-} \tag{7.13j}$$

although the reason for this fact is quite different from that underlying Eqs. 7.13*e*, 7.13*f*, or 7.13*i*. Here, indeed, we find an orthogonality relation between two solutions having *parallel* electric fields!

The net result of the power relations, Eqs. 7.13*e*, 7.13*f*, 7.13*i*, and 7.13*j*, is that in a lossless, homogeneous, linear medium with constant parameters the four possible linearly independent uniform plane waves that have a common (undirected) axis of propagation (e.g., the $\pm z$-axis) may always be, and have here been, chosen to be *mutually orthogonal*. That is, the total power (per unit area in this case) carried by the field when all the waves are simultaneously present is simply the vector sum of the powers that each would carry if it existed alone in space. There are no cross terms.

7.1.4 Polarization

We have found that waves of Sets 1 and 2 *can* always be studied separately, not only in the light of normal superposition theory but also in connection with power. It is now appropriate to emphasize another characteristic of these waves which makes them particularly easy to treat separately: the space directions of $\boldsymbol{E}_1$ (or $\boldsymbol{E}_2$) and $\boldsymbol{H}_1$ (or $\boldsymbol{H}_2$) remain along fixed lines in any transverse plane at all times. Thus, as indicated by Eqs. 7.6 and 7.3, $\boldsymbol{E}_1$ always lies along a line parallel to the x-axis. It may reverse direction along this line as time progresses, depending upon the behavior of f_+ and f_-, but no other change of direction occurs. Similar comments apply to $\boldsymbol{H}_1$, $\boldsymbol{E}_2$, and $\boldsymbol{H}_2$. To emphasize this behavior, these waves are referred to as being *linearly polarized*. In most of the recent electrical engineering literature, the direction of polarization is taken to be that of $\boldsymbol{E}$; so we shall say that the wave ($\boldsymbol{E}_1$, $\boldsymbol{H}_1$) is linearly polarized in the x-direction while ($\boldsymbol{E}_2$, $\boldsymbol{H}_2$) is linearly polarized in the y-direction. This situation is illustrated for the (+) waves by the sketches in Figs. 7.1*a* and 7.1*c*. A similar picture would apply for (−) waves. For both together, the linear polarization remains, but propagation direction loses meaning.

The most general uniform plane-wave propagating in a given direction is not linearly polarized. Indeed, Eqs. 7.13*a*, 7.8*a*, and 7.3*a* show that $\boldsymbol{E}_T{}^+$ has, in general, two space components in the transverse plane, and that at any given point in space these components may be quite different functions of the time. At one moment $E_x{}^+$ might be large and $E_y{}^+$ zero, whereas, at a later time, $E_y{}^+$ might be large and $E_x{}^+$ zero. Evidently the tip of the vector $\boldsymbol{E}_T{}^+$ traces some complicated path in the transverse plane, just as the spot on an oscilloscope screen executes an involved pattern when arbitrarily different voltage time functions are impressed upon the vertical and horizontal axes. An extreme case of such complicated behavior in electromagnetic waves is that of "unpolarized" light. Its two electric field components are statistically independent Gaussian random time functions, with equal rms amplitudes and zero average values. Its polarization is therefore "random," i.e., $\boldsymbol{E}_T$ is at any instant as likely to be pointing in one direction as another in the transverse plane.[1] Therefore, in view of the wide latitude of the general case, it is comforting to know that for each direction of propagation we need consider only two independent solutions having mutually perpendicular *linear* polarizations.

7.1.5 Role of Uniform Plane Waves

Since the field of a uniform plane wave extends with constant strength over the infinite area of any transverse plane, the total power carried by such a wave is also infinite. Thus no physical source can possibly be expected to produce exactly such a wave. Nevertheless, we shall find later, from our study of radiation, that at large enough distances from any physical source the field in suitably defined *finite* regions of space approximates closely that of uniform plane waves propagating in appropriate directions. For this reason and others, a thorough understanding of these simple waves in space is fundamental to any study of electromagnetic energy transmission and radiation. There are, however, other simple solutions of Maxwell's equations which are even more important, and it is not our intention to exclude them by this statement.

[1] The reader interested in justifying this statement more carefully than is necessary for the purposes of the present discussion will find support in Prob. 7.6.

We have shown on physical grounds that one or more uniform plane waves propagating in any direction or directions in free space constitutes a solution of Maxwell's equations. If we imagine one such wave propagating in the $+z$ direction, and regard this as an *incident* wave produced by some remote source, we may ask the question: "What happens when an object is placed in the path of this wave?" Unquestionably, the original wave alone will rarely satisfy the boundary conditions imposed by the object. The field in the space about the object will almost surely be modified. From our study of transmission lines, we have learned to interpret this modification as a "reflection" of some sort. In this case, however, the reflected field may or may not be another uniform plane wave. Its nature depends entirely upon the geometry and electrical character of the object.

A perfectly conducting metal sphere placed at $z = 0$ may "scatter" the incident wave in all directions. The reflected field would then be more properly described as a scattered field, and it would not be a simple uniform plane wave.

On the other hand, we should not be surprised to find that a smooth, infinite, plane, perfectly conducting metal sheet placed normal to the direction of propagation will act like an optical mirror. It will simply throw the incident wave back upon itself. The reflected wave is then another uniform plane wave propagating in the opposite direction. If this mirror is not normal to the direction of propagation of the incident wave, optical experience suggests that the reflected field should be another uniform plane wave whose direction of propagation obeys the familar law: angle of reflection equals angle of incidence.

When the object is a dielectric or a real metal rather than a perfect conductor, we expect to be concerned with refraction as well as reflection, although, again, the question of whether or not the reflected and refracted fields are simple uniform plane waves depends upon the shape and electrical character of the object.

To solve the more elaborate among the types of problems described above, uniform plane wave solutions are insufficient. In the sinusoidal steady state or frequency domain, however, it is not difficult to find some more general plane wave solutions to Maxwell's equations. These are called *nonuniform plane waves* in this book, although the nomenclature is not standard and perhaps not sufficiently descriptive. The important theoretical point about these solutions is that, with the uniform plane waves we have already found, *they suffice to solve any steady-state field problem comprising regions of different linear, homogeneous, isotropic, time-invariant materials.* Like the sinusoid (or exponential) in time, these plane waves form a set of building blocks in *space* which may be combined by Fourier-integral methods to meet a wide selection of boundary conditions. Although we will not consider elaborate problems here, it is obviously important to study carefully the character of both uniform and nonuniform plane waves in the sinusoidal steady state, and to apply them to the solution of some simple examples.

Restriction to the sinusoidal steady state, of course, implies no loss of generality, in view of the normal Fourier-integral techniques for time functions.

7.2 Plane Waves in the Sinusoidal Steady State and Frequency Domain

As stated above, the nonuniform plane wave has significance only in the sinusoidal steady state or the frequency domain. The uniform plane wave, however, has significance both in the time domain, as discussed in the preceding section, and in the frequency domain. It is therefore advantageous to begin our study of plane waves in the frequency domain with the uniform plane wave. Our discussion of this subject is to be guided by the fact that we wish to go on into the more general case. The treatment may therefore appear to be less direct than the topic of uniform plane waves itself would require.

7.2.1 Uniform Plane Waves in the Frequency Domain

7.2.1.1 FORM OF THE SOLUTION. The formal similarity of Eqs. 7.2*a*, 7.2*b*, 7.3, and 7.4 to those of a lossless transmission line tells us at once that the complex steady-state solution for x-polarized uniform plane waves traveling in the $\pm z$ directions is

$$\text{(a)} \qquad \mathrm{E}_x = \mathrm{E}_{x0}{}^{+}e^{-j\beta_0 z} + \mathrm{E}_{x0}{}^{-}e^{j\beta_0 z} \tag{7.14}$$

$$\text{(b)} \qquad \mathrm{H}_y = \frac{1}{\eta}\left(\mathrm{E}_{x0}{}^{+}e^{-j\beta_0 z} - \mathrm{E}_{x0}{}^{-}e^{j\beta_0 z}\right)$$

where

$$\beta_0 = +\omega\sqrt{\epsilon\mu} \tag{7.15}$$

and $\mathrm{E}_{x0}{}^{+}$, $\mathrm{E}_{x0}{}^{-}$ are complex constants independent of (x, y, z). For the $(+)$ wave, $E_x{}^{+}(= \mathrm{Re}\,[\mathrm{E}_{x0}{}^{+}e^{j(\omega t-\beta_0 z)}])$ and $H_y{}^{+}(= \mathrm{Re}\,[\mathrm{H}_{y0}{}^{+}e^{j(\omega t-\beta_0 z)}]$ $= \mathrm{Re}\,[(\mathrm{E}_{x0}{}^{+}/\eta)e^{j(\omega t-\beta_0 z)}])$ are in time phase and have the ratio η at every point in space, and at every instant. We may wish to interpret

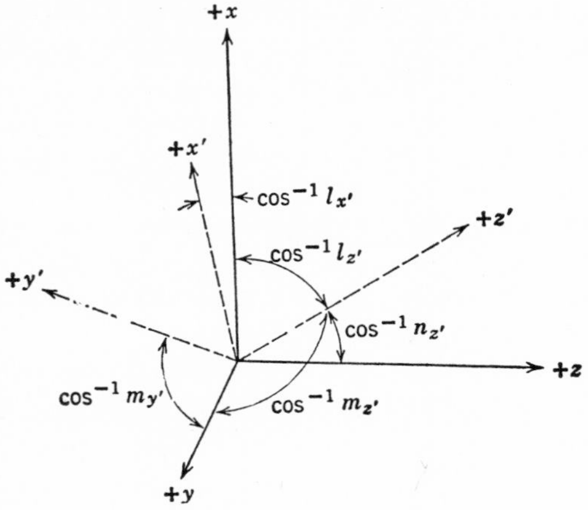

Fig. 7.2. Rotation of coordinate axes.

the (+) solution as an incident wave and the (−) one as a reflected wave. This interpretation certainly is valid under appropriate boundary conditions to which we will return in detail later. Now we prefer to examine the solution itself from another point of view.

Consider just the (+) wave in Eqs. 7.14 and 7.15, and suppose the axes (x, y, z) used to describe it there are not actually the ones we wish to use in our problem. Such a situation arises, for example, in describing simultaneously several uniform plane waves propagating in different space directions. Let us reserve (x, y, z) for the axes we really plan to use, and rewrite the (+) wave part of Eq. 7.14 with new variables (x', y', z') substituted for (x, y, z). From the point of view of the (x, y, z) coordinates, the (x', y', z') axes are then a right-handed set of rotated axes having the same origin as (x, y, z), Fig. 7.2, but with the $+x'$-axis chosen parallel to the (linearly polarized) electric field vector of the (+) wave of interest, and the $+z'$-axis chosen parallel to the direction of propagation. Therefore

$$\text{(a)} \qquad \mathbf{E}^+ = \boldsymbol{a}_{x'} \mathrm{E}_{x'0}{}^+ e^{-j\beta_0 z'}$$

$$\text{(b)} \qquad \mathbf{H}^+ = \frac{1}{\eta} \boldsymbol{a}_{y'} \mathrm{E}_{x'0}{}^+ e^{-j\beta_0 z'} \tag{7.16}$$

With reference to Fig. 7.2 again, let the (real) direction cosines of axis $+z'$ with respect to axes $(+x, +y, +z)$ respectively be $(l_{z'}, m_{z'}, n_{z'})$. Similarly, let $(l_{x'}, m_{x'}, n_{x'})$ and $(l_{y'}, m_{y'}, n_{y'})$ apply correspondingly to axes $+x'$ and $+y'$ respectively. Then because the axes (x, y, z) are a rectangular set, we have for the $+z'$-axis direction cosines

$$l_{z'}^2 + m_{z'}^2 + n_{z'}^2 = 1 \tag{7.17}$$

Similar results for the $+x'$ and $+y'$ axes are obtained by simply replacing z' with x' or y'.

We wish now to describe the field (Eq. 7.16) in terms of the (x, y, z) coordinates and also the unit vectors $\boldsymbol{a}_x$, $\boldsymbol{a}_y$, $\boldsymbol{a}_z$ along the corresponding axes, whereas it is presently described in the coordinates (x', y', z') and the unit vectors $\boldsymbol{a}_{x'}$, $\boldsymbol{a}_{y'}$, $\boldsymbol{a}_{z'}$ along them.

To this end we note first that the position of any point in space is given by its vector displacement $\boldsymbol{r}$ from the origin, and that this *same* vector is expressible in two ways, corresponding to the two sets of axes.

$$\boldsymbol{r} = \boldsymbol{a}_x x + \boldsymbol{a}_y y + \boldsymbol{a}_z z = \boldsymbol{a}_{x'} x' + \boldsymbol{a}_{y'} y' + \boldsymbol{a}_{z'} z' \tag{7.18}$$

Taking the dot product of both sides of Eq. 7.18 with $\boldsymbol{a}_{z'}$, and noting that axes (x', y', z') are a rectangular set, we find

$$z' = l_{z'} x + m_{z'} y + n_{z'} z \tag{7.19}$$

Therefore the phase factor $\beta_0 z'$ in Eq. 7.16 becomes

$$\begin{aligned} \beta_0 z' &= (\beta_0 l_{z'})x + (\beta_0 m_{z'})y + (\beta_0 n_{z'})z \\ &= \boldsymbol{\beta} \cdot \boldsymbol{r} \end{aligned} \tag{7.20}$$

with

$$\begin{aligned} \boldsymbol{\beta} &= \boldsymbol{a}_x(\beta_0 l_{z'}) + \boldsymbol{a}_y(\beta_0 m_{z'}) + \boldsymbol{a}_z(\beta_0 n_{z'}) \\ &= \boldsymbol{a}_x \beta_x + \boldsymbol{a}_y \beta_y + \boldsymbol{a}_z \beta_z \end{aligned} \tag{7.21a}$$

where

$$\beta_x = \beta_0 l_{z'} \qquad \beta_y = \beta_0 m_{z'} \qquad \beta_z = \beta_0 n_{z'} \tag{7.21b}$$

and

$$\boldsymbol{\beta} \cdot \boldsymbol{\beta} \equiv |\boldsymbol{\beta}|^2 = \beta_0^2(l_{z'}^2 + m_{z'}^2 + n_{z'}^2) = \beta_x^2 + \beta_y^2 + \beta_z^2 = \beta_0^2 \tag{7.22}$$

The *propagation vector* $\boldsymbol{\beta}$ is in this case a *real* space vector, whose direction is that of the propagation of the wave and whose magnitude is the phase constant $\beta_0 = \omega\sqrt{\epsilon\mu}$.

Next, from the geometry of Fig. 7.2 we note that

$$\boldsymbol{a}_{x'} = \boldsymbol{a}_x l_{x'} + \boldsymbol{a}_y m_{x'} + \boldsymbol{a}_z n_{x'} \tag{7.23}$$

Consequently

$$\begin{aligned} \boldsymbol{a}_{x'} \mathrm{E}_{x'0}^+ &= \boldsymbol{a}_x l_{x'} \mathrm{E}_{x'0}^+ + \boldsymbol{a}_y m_{x'} \mathrm{E}_{x'0}^+ + \boldsymbol{a}_z n_{x'} \mathrm{E}_{x'0}^+ \\ &= \mathbf{E}_0^+ \end{aligned} \tag{7.24a}$$

where $\mathbf{E}_0^+$ is a complex vector independent of (x, y, z).

From Eqs. 7.24*a* and 7.21*a* we find

$$\boldsymbol{\beta}\cdot\mathbf{E}_0^+ = \beta_0 \mathrm{E}_{x'0}^+(l_{x'}l_{z'} + m_{x'}m_{z'} + n_{x'}n_{z'}) = 0 = \boldsymbol{\beta}\cdot\mathbf{E}^+ \quad (7.24b)$$

since the x'- and z'- axes are perpendicular. The meaning of Eq. 7.24*b* is that the instantaneous electric field $\boldsymbol{E}^+(t)$ is at all times perpendicular to the direction of propagation; but this conclusion is less obvious from Eq. 7.24*b* than it may appear offhand. In general, the vanishing of the dot product of two *complex* vectors does *not* imply any simple perpendicular condition in *real* space. In Eq. 7.24*b*, however, $\boldsymbol{\beta}$ is a *real* vector, so $\boldsymbol{\beta}\cdot\mathbf{E}^+ = 0$ only occurs if $\boldsymbol{\beta}\cdot\mathrm{Re}\,\mathbf{E}^+ = 0$ *and* $\boldsymbol{\beta}\cdot\mathrm{Im}\,\mathbf{E}^+ = 0$. Since $\mathrm{Re}\,\mathbf{E}^+$ and $\mathrm{Im}\,\mathbf{E}^+$ determine the plane in which $\boldsymbol{E}^+(t)$ lies at all times (Sec. 1.2.2.1), $\boldsymbol{\beta}$ is indeed perpendicular to this plane and, therefore, to $\boldsymbol{E}^+(t)$. Of course, this perpendicular relation was true of the original field (Eq. 7.16), and the use of new axes (x, y, z) could hardly be expected to alter it. Nevertheless, the point about interpreting Eq. 7.24*b* is important for our later work.

In connection with the magnetic field (Eq. 7.16*b*), we note that (x', y', z') is a right-handed set of rectangular axes like (x, y, z). Therefore

$$\begin{aligned}\boldsymbol{a}_{y'}\mathrm{E}_{x'0}^+ &= (\boldsymbol{a}_{z'} \times \boldsymbol{a}_{x'})\mathrm{E}_{x'0}^+ = \boldsymbol{a}_{z'} \times (\boldsymbol{a}_{x'}\mathrm{E}_{x'0}^+) \\ &= \boldsymbol{a}_{z'} \times \mathbf{E}_0^+ = (\boldsymbol{a}_x l_{z'} + \boldsymbol{a}_y m_{z'} + \boldsymbol{a}_z n_{z'}) \times \mathbf{E}_0^+ \\ &= \frac{\boldsymbol{\beta}}{\beta_0} \times \mathbf{E}_0^+ \end{aligned} \quad (7.25)$$

where we have also used Eqs. 7.24*a* and 7.21*a*.

In summary, the solution (Eq. 7.16) in the (x, y, z) axis system becomes:

$$\begin{aligned} &\text{(a)} \qquad \mathbf{E}^+ = \mathbf{E}_0^+ e^{-j\boldsymbol{\beta}\cdot\boldsymbol{r}} = \mathbf{E}_0^+ e^{-j\beta_x x} e^{-j\beta_y y} e^{-j\beta_z z} \\ &\text{(b)} \qquad \mathbf{H}^+ = \frac{\boldsymbol{\beta} \times \mathbf{E}^+}{\omega\mu} \end{aligned} \quad (7.26)$$

The fact that $\boldsymbol{\beta}$ is real allows us to interpret Eqs. 7.24*b* and 7.26*b* as showing the mutual perpendicularity of $\boldsymbol{\beta}$, $\boldsymbol{H}^+(t)$, and $\boldsymbol{E}^+(t)$, by the same kind of argument which followed Eq. 7.24*b*.

Observe from Eq. 7.24*a* that $\mathbf{E}_0^+$ in Eq. 7.26 has three space components in the (x, y, z) system. They all have the same time-phase angle in the present case because $l_{x'}$, $m_{x'}$, $n_{x'}$ are real numbers. Hence, as we expect, rotation of the coordinates has not altered the linear polarization of the wave (Eq. 7.16). It is worth remarking here, however, that we may also consider the second type of polarization in the

(x', y', z') system, for which the complex steady-state solution comes from Eq. 7.8 in the form $\mathbf{E}_2{}^+ = \boldsymbol{a}_{y'}\mathrm{E}_{y'0}{}^+ e^{-j\beta_0 z'}$. If we combine the two fields to make $\mathbf{E}_3{}^+ = (\boldsymbol{a}_{x'}\mathrm{E}_{x'0}{}^+ + \boldsymbol{a}_{y'}\mathrm{E}_{y'0})e^{-j\beta_0 z'}$, the result in the (x, y, z) system is again of the form $\mathbf{E}_3{}^+ = \mathbf{E}_{03}{}^+ e^{-j\boldsymbol{\beta}\cdot\boldsymbol{r}}$. This time, though,

$$\mathbf{E}_{03}{}^+ = \boldsymbol{a}_x(l_{x'}\mathrm{E}_{x'0}{}^+ + l_{y'}\mathrm{E}_{y'0}{}^+) + \boldsymbol{a}_y(m_{x'}\mathrm{E}_{x'0}{}^+ + m_{y'}\mathrm{E}_{y'0}{}^+) + \boldsymbol{a}_z(n_{x'}\mathrm{E}_{x'0}{}^+ + n_{y'}\mathrm{E}_{y'0}{}^+)$$

where the three space components do *not*, in general, have the same time phase. The point is: Eq. 7.26 is valid whether or not the uniform plane wave in question is linearly polarized. Consequently, the mutual perpendicularity of $\boldsymbol{\beta}$, $\boldsymbol{H}^+(t)$, and $\boldsymbol{E}^+(t)$ remains true regardless of the type of polarization involved.

7.2.1.2 Phase, Wave Length, and Phase Velocity. From Eq. 7.26, we learn that the time phase of the electric (or magnetic) field varies with position only as $\boldsymbol{\beta}\cdot\boldsymbol{r}$. With reference to Fig. 7.3*a*, a *surface of constant phase* is therefore one for which

$$\text{(a)} \qquad \boldsymbol{\beta}\cdot\boldsymbol{r} = \text{const} = \beta_0 r \cos\psi$$

$$\text{or} \qquad\qquad (7.27)$$

$$\text{(b)} \qquad r\cos\psi = \text{const}$$

where $r = |\boldsymbol{r}|$. Equation 7.27*b* restricts $\boldsymbol{r}$ to those points in space where its projection along the direction of $\boldsymbol{\beta}$ is a constant. The tip of $\boldsymbol{r}$ must therefore lie in a plane perpendicular to $\boldsymbol{\beta}$. Such planes, normal to the propagation direction, are the *planes of constant phase.* For the *uniform* plane wave, they are *also planes of constant magnitude* since $\mathbf{E}_0{}^+$ does not vary with position. The electric and magnetic fields lie in these planes (Eqs. 7.24*b* and 7.26*b*).

The direction normal to the planes of constant phase is the one in

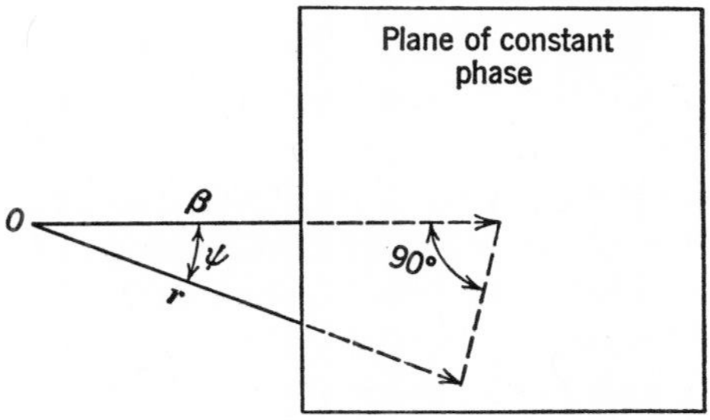

Fig. 7.3a. Equiphase surface for a plane wave.

which the phase φ of the field changes most rapidly with position. To see this, we write with reference to Eqs. 7.26*a*, 7.27, and Fig. 7.3*a*,

$$\varphi = \boldsymbol{\beta}\cdot\boldsymbol{r} + \text{const} \tag{7.28}$$

$$\left(\frac{\partial\varphi}{\partial r}\right)_{\psi=\text{const}} = \beta_0 \cos\psi \tag{7.29}$$

$$\left(\frac{\partial\varphi}{\partial r}\right)_{\substack{\text{max}\\ \text{vs }\psi}} = +\beta_0 \qquad \text{for } \psi = 0, \text{ or } \boldsymbol{r}\,\|\,\boldsymbol{\beta} \tag{7.30}$$

Hence another description of the propagation vector $\boldsymbol{\beta}$ of a uniform plane wave in a lossless medium is a vector whose direction is that of the maximum rate-of-change of phase with position, and whose magnitude β_0 equals that maximum rate of change. In other words, by definition of the gradient operator,

$$\nabla\varphi = \boldsymbol{\beta} = \boldsymbol{a}_x\beta_x + \boldsymbol{a}_y\beta_y + \boldsymbol{a}_x\beta_z \qquad \text{for a uniform plane wave} \tag{7.31}$$

The meaning of the component phase constants β_x, β_y, and β_z is the space-rate-of-change of phase as measured along the respective axes. According to Eqs. 7.21*a* and 7.28,

$$\varphi = \beta_x x + \beta_y y + \beta_z z + \text{const}$$

so

$$\begin{aligned} \beta_x &= \boldsymbol{a}_x\cdot\nabla\varphi = \frac{\partial\varphi}{\partial x} \\ \beta_y &= \boldsymbol{a}_y\cdot\nabla\varphi = \frac{\partial\varphi}{\partial y} \\ \beta_z &= \boldsymbol{a}_z\cdot\nabla\varphi = \frac{\partial\varphi}{\partial z} \end{aligned} \tag{7.32a}$$

Observe that since the direction cosines l, m, and n cannot exceed 1 in magnitude (also in view of Eq. 7.22),

$$|\beta_{x,y,z}| \leq \beta_0 \tag{7.32b}$$

Instead of considering the space-rate-of-change of phase along various directions, we can consider the distances between successive equiphase surfaces on which the field phase differs by exactly 2π. These distances will be *apparent wave lengths* because the phase shift is linearly proportional to distance in any direction. Measured along the propagation direction (Fig. 7.3*b*), the distance λ, in which the phase changes by 2π, is the conventional wave length. Measured along any

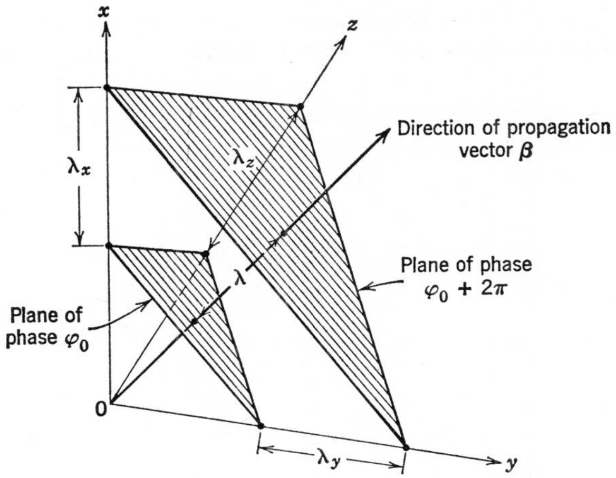

Fig. 7.3b. Various wave-length concepts for a plane wave.

other direction, the distances between the same two planes will be *greater* than λ, as shown in Fig. 7.3*b*, because the space-rate-of-change of phase along such a direction is *less* than it is along the propagation direction. Thus

$$\begin{aligned} \lambda &\equiv \frac{2\pi}{\beta} \\ \lambda_x &\equiv \frac{2\pi}{|\beta_x|} \geq \lambda \\ \lambda_y &\equiv \frac{2\pi}{|\beta_y|} \geq \lambda \\ \lambda_z &\equiv \frac{2\pi}{|\beta_z|} \geq \lambda \end{aligned} \tag{7.33}$$

and, in view of Eqs. 7.22 and 7.33,

$$\frac{1}{\lambda_x^2} + \frac{1}{\lambda_y^2} + \frac{1}{\lambda_z^2} = \frac{1}{\lambda^2} \tag{7.34}$$

Similarly, since in the time domain (for the sinusoidal steady state) a particular phase front (plane) moves uniformly with time at speed $v = 1/\sqrt{\epsilon\mu} = \omega/\beta_0$ along the propagation direction, its intercept along any other direction also moves uniformly with time. As is shown in

Fig. 7.3*b*, however, the apparent speed measured along any of the coordinate axes will be *greater* than that along $\boldsymbol{\beta}$. In the time of one cycle $(1/f)$, a phase front moves a distance along x of $2\pi/|\beta_x| = \lambda_x$, because the total phase of the field involves $\omega t - \beta_x x$. Analogous comments apply for the other directions y, and z. Thus the various *apparent phase velocities* are $v_x = f\lambda_x$, etc.; or

$$
\begin{aligned}
v_x &\equiv \frac{\omega}{\beta_x} = \left(\frac{1}{l_{z'}\sqrt{\epsilon\mu}}\right) \geq v \\
v_y &\equiv \frac{\omega}{\beta_y} = \left(\frac{1}{m_{z'}\sqrt{\epsilon\mu}}\right) \geq v \qquad (7.35) \\
v_z &\equiv \frac{\omega}{\beta_z} = \left(\frac{1}{n_{z'}\sqrt{\epsilon\mu}}\right) \geq v
\end{aligned}
$$

and in view of Eqs. 7.22, again

$$
\frac{1}{v_x^2} + \frac{1}{v_y^2} + \frac{1}{v_z^2} = \frac{1}{v^2} = \epsilon\mu \qquad (7.36)
$$

There is a familiar analogy to the fact that the apparent wave length and phase velocity along any direction but that of the wave propagation *exceed* the same quantities along the propagation direction. Visualize water waves striking a beach obliquely. The distance between successive crests, for example, is large along the beach, especially if the waves are only slightly off normal incidence. Also to keep up with a given crest, one has to run much faster along the beach than the waves progress in their own normal direction.

In view of the manner of obtaining the field expressions (Eqs. 7.26) and the auxiliary relations (Eqs. 7.22 and 7.24*b*) by coordinate rotation from Eqs. 7.16, there is no doubt that in the (x, y, z) axes they constitute a solution to Maxwell's (vector) equations. The entire space dependence of this solution is a *simple exponential*, i.e., the exponent is *linear* in the variables (x, y, z). There remains the question of whether or not we have found the *only* such simple exponential solution to Maxwell's equations in the sinusoidal steady state.

7.2.2 Nonuniform Plane Waves *

7.2.2.1 Character of the Solution.* The most general exponential expressions for the complex fields, with exponent linear in the variables (x, y, z), are of the form

$$\text{(a)} \qquad \mathbf{E} = \mathbf{E}_0 e^{-\bar{\gamma}_x x} e^{-\bar{\gamma}_y y} e^{-\bar{\gamma}_z z} = \mathbf{E}_0 e^{-\bar{\boldsymbol{\gamma}}\cdot\boldsymbol{r}}$$

$$\text{(b)} \qquad \mathbf{H} = \mathbf{H}_0 e^{-\bar{\gamma}_x x} e^{-\bar{\gamma}_y y} e^{-\bar{\gamma}_z z} = \mathbf{H}_0 e^{-\bar{\boldsymbol{\gamma}}\cdot\boldsymbol{r}} \qquad (7.37)$$

where $\mathbf{E}_0$ and $\mathbf{H}_0$ are constant complex vectors, independent of (x, y, z), and the various $\bar{\gamma}$'s are *complex* numbers

$$\begin{aligned} \bar{\gamma}_x &= \alpha_x + j\beta_x \\ \bar{\gamma}_y &= \alpha_y + j\beta_y \\ \bar{\gamma}_z &= \alpha_z + j\beta_z \end{aligned} \qquad (7.38a)$$

The *propagation vector* $\bar{\boldsymbol{\gamma}}$ is then a *complex vector*

$$\begin{aligned} \bar{\boldsymbol{\gamma}} = \boldsymbol{a}_x\bar{\gamma}_x + \boldsymbol{a}_y\bar{\gamma}_y + \boldsymbol{a}_z\bar{\gamma}_z &= (\boldsymbol{a}_x\alpha_x + \boldsymbol{a}_y\alpha_y + \boldsymbol{a}_z\alpha_z) + j(\boldsymbol{a}_x\beta_x \\ &\qquad + \boldsymbol{a}_y\beta_y + \boldsymbol{a}_z\beta_z) \\ &= \boldsymbol{\alpha} + j\boldsymbol{\beta} \end{aligned} \qquad (7.38b)$$

$\boldsymbol{\alpha}$ and $\boldsymbol{\beta}$ being *real* space vectors. We shall call $\boldsymbol{\alpha}$ the *attenuation vector* and $\boldsymbol{\beta}$ the *phase vector*. *With $\bar{\boldsymbol{\gamma}}$ complex, the field (Eq. 7.37) is by definition a nonuniform plane wave.*

We know already that if $\boldsymbol{\alpha} = 0$ the field of Eq. 7.37 can be a solution to Maxwell's equations, provided $\mathbf{E}_0$, $\mathbf{H}_0$, and $\boldsymbol{\beta}$ satisfy some auxiliary conditions. If they do, the resulting solution is interpretable as a *uniform* plane wave traveling in some arbitrary direction in space. If $\boldsymbol{\alpha} \neq 0$, however, we must determine whether or not Eq. 7.37 can be a solution to the equations

$$\text{(a)} \qquad \nabla \times \mathbf{E} = -j\omega\mu\mathbf{H}$$

$$\text{(b)} \qquad \nabla \times \mathbf{H} = j\omega\epsilon\mathbf{E} \qquad (7.39)$$

and, if so, how to interpret this solution.

The first question above can be answered by substituting Eqs. 7.37 into Eqs. 7.39. Expansion of the curl term in Eq. 7.39*a* yields

$$\begin{aligned} \nabla \times \mathbf{E} = \nabla \times (\mathbf{E}_0 e^{-\bar{\boldsymbol{\gamma}}\cdot\boldsymbol{r}}) &= -\mathbf{E}_0 \times \nabla e^{-\bar{\boldsymbol{\gamma}}\cdot\boldsymbol{r}} + e^{-\bar{\boldsymbol{\gamma}}\cdot\boldsymbol{r}}\,\nabla \times \mathbf{E}_0 \\ &= \mathbf{E}_0 \times [e^{-\bar{\boldsymbol{\gamma}}\cdot\boldsymbol{r}}\,\nabla(\bar{\boldsymbol{\gamma}}\cdot\boldsymbol{r})] = (\mathbf{E}_0 \times \bar{\boldsymbol{\gamma}})e^{-\bar{\boldsymbol{\gamma}}\cdot\boldsymbol{r}} \end{aligned}$$

because, since $\mathbf{E}_0$ is independent of (x, y, z), $\nabla \times \mathbf{E}_0 \equiv 0$. A similar calculation is made of $\nabla \times (\mathbf{H}_0 e^{-\bar{\boldsymbol{\gamma}}\cdot\boldsymbol{r}})$. Substitution of these results, with Eqs. 7.37, into Eq. 7.39 yields the final conditions:

$$\text{(a)} \qquad \bar{\boldsymbol{\gamma}} \times \mathbf{E}_0 = j\omega\mu\mathbf{H}_0$$

$$\text{(b)} \qquad \bar{\boldsymbol{\gamma}} \times \mathbf{H}_0 = -j\omega\epsilon\mathbf{E}_0 \qquad (7.40)$$

It is clear that, if the exponential factors in **E** and **H** had been chosen unequal in Eq. 7.37, they would have been forced to equality at this point for any solution to be possible. Equations 7.40 result in any case.

Now dot-premultiplying Eqs. 7.40*a* and 7.40*b* by $\bar{\boldsymbol{\gamma}}$, and noting that

$$\bar{\boldsymbol{\gamma}}\cdot(\bar{\boldsymbol{\gamma}} \times \mathbf{E}_0) = (\bar{\boldsymbol{\gamma}} \times \bar{\boldsymbol{\gamma}})\cdot\mathbf{E}_0 = 0 = \bar{\boldsymbol{\gamma}}\cdot(\bar{\boldsymbol{\gamma}} \times \mathbf{H}_0) = (\bar{\boldsymbol{\gamma}} \times \bar{\boldsymbol{\gamma}})\cdot\mathbf{H}_0$$

because

$$\bar{\boldsymbol{\gamma}} \times \bar{\boldsymbol{\gamma}} \equiv 0$$

we find

$$\text{(a)} \qquad \bar{\boldsymbol{\gamma}}\cdot\mathbf{E}_0 = 0$$

$$\text{(b)} \qquad \bar{\boldsymbol{\gamma}}\cdot\mathbf{H}_0 = 0 \qquad (7.41)$$

Since $\bar{\boldsymbol{\gamma}}$, $\mathbf{E}_0$, and $\mathbf{H}_0$ are complex vectors in general, Eqs. 7.40 and 7.41 do *not* imply any obvious conditions of perpendicularity in space.

To eliminate $\mathbf{H}_0$ from Eqs. 7.40, cross-premultiply Eq. 7.40*a* by $(\bar{\boldsymbol{\gamma}}/j\omega\mu)$ and add Eqs. 7.40*a* and 7.40*b*. There results

$$\bar{\boldsymbol{\gamma}} \times (\bar{\boldsymbol{\gamma}} \times \mathbf{E}_0) = \omega^2\epsilon\mu\mathbf{E}_0 = \beta_0{}^2\mathbf{E}_0$$

or

$$\bar{\boldsymbol{\gamma}}(\bar{\boldsymbol{\gamma}}\cdot\mathbf{E}_0) - (\bar{\boldsymbol{\gamma}}\cdot\bar{\boldsymbol{\gamma}})\mathbf{E}_0 = \beta_0{}^2\mathbf{E}_0$$

In view of Eq. 7.41*a* and our wish to have $\mathbf{E}_0 \neq 0$ (i.e., to find a non trivial solution), we must require

$$\bar{\boldsymbol{\gamma}}\cdot\bar{\boldsymbol{\gamma}} = -\beta_0{}^2 \qquad (7.42)$$

Because $\bar{\boldsymbol{\gamma}}$ may be complex, Eq. 7.42 is really two scalar equations. Using Eq. 7.38, we find

$$(\boldsymbol{\alpha}\cdot\boldsymbol{\alpha} - \boldsymbol{\beta}\cdot\boldsymbol{\beta}) + j2\boldsymbol{\alpha}\cdot\boldsymbol{\beta} = -\beta_0{}^2$$

or since $\boldsymbol{\alpha}$, $\boldsymbol{\beta}$, and $\beta_0{}^2$ are *real*,

$$\text{(a)} \qquad \boldsymbol{\beta}\cdot\boldsymbol{\beta} - \boldsymbol{\alpha}\cdot\boldsymbol{\alpha} = \beta_0{}^2$$

$$\text{(b)} \qquad \boldsymbol{\alpha}\cdot\boldsymbol{\beta} = 0 \qquad (7.43)$$

7.2.2.2 Phase Delay and Attenuation.* To understand Eq. 7.43, we must recall that the space dependence of **E** and **H** is contained completely in the exponential factor

$$e^{-\bar{\boldsymbol{\gamma}}\cdot\boldsymbol{r}} = e^{-\boldsymbol{\alpha}\cdot\boldsymbol{r}}e^{-j\boldsymbol{\beta}\cdot\boldsymbol{r}} \qquad (7.44)$$

The first factor $e^{-\boldsymbol{\alpha}\cdot\boldsymbol{r}}$ is always real, whereas the second factor $e^{-j\boldsymbol{\beta}\cdot\boldsymbol{r}}$ always has unit magnitude. The phase of the second factor varies with position as $\boldsymbol{\beta}\cdot\boldsymbol{r}$. Therefore a *surface of constant phase* is defined by the relation

$$\boldsymbol{\beta}\cdot\boldsymbol{r} = \beta r \cos\psi = \text{const} \tag{7.45}$$

with $\beta = |\boldsymbol{\beta}|$. This equation pertains to a *plane normal to* $\boldsymbol{\beta}$, as illustrated in Fig. 7.3*a*.

The amplitude of the field, on the other hand, varies as $e^{-\boldsymbol{\alpha}\cdot\boldsymbol{r}}$. Accordingly, a *surface of constant amplitude* is defined by the relation

$$\boldsymbol{\alpha}\cdot\boldsymbol{r} = \alpha r \cos\psi' = \text{const} \tag{7.46}$$

in which, by analogy with Eq. 7.45 and Fig. 7.3*a*, ψ' is the angle between $\boldsymbol{r}$ and $\boldsymbol{\alpha}$, and $\alpha = |\boldsymbol{\alpha}|$. Again, a plane is described by Eq. 7.46. *This plane is normal to* $\boldsymbol{\alpha}$.

The point of Eq. 7.43*b* is that the *real* vectors $\boldsymbol{\alpha}$ and $\boldsymbol{\beta}$ are perpendicular in space (if $\alpha, \beta \neq 0$). Moreover, as long as the frequency is not zero ($\omega \neq 0$), $\beta_0 > 0$, and Eq. 7.43*a* guarantees $\beta \neq 0$. Specifically

$$\beta^2 = \beta_0{}^2 + \alpha^2 \geq \beta_0{}^2 \qquad >0 \text{ for } \omega \neq 0 \tag{7.47}$$

Therefore, *if* $\boldsymbol{\alpha} \neq 0$, *the field of a nonuniform plane wave (Eq. 7.37) has planes of constant amplitude (normal to* $\boldsymbol{\alpha}$*) and planes of constant phase (normal to* $\boldsymbol{\beta}$*) that are mutually perpendicular.*

The most rapid space-rate-of-change of amplitude occurs in the direction of $\boldsymbol{\alpha}$ (Eq. 7.46), which is perpendicular to $\boldsymbol{\beta}$ and therefore parallel to the planes of constant phase. This field is still a *plane wave,*

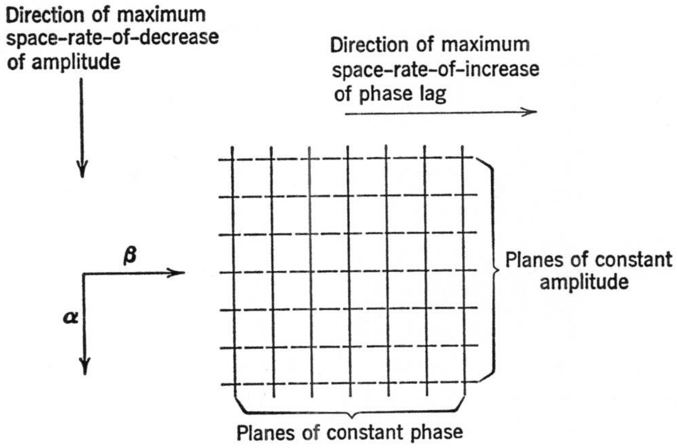

Fig. 7.4. Significant features of the space variation of $e^{-\bar{\boldsymbol{\gamma}}\cdot\boldsymbol{r}} = e^{-\boldsymbol{\alpha}\cdot\boldsymbol{r}}e^{-\boldsymbol{\beta}\cdot\boldsymbol{r}}$ for nonuniform plane waves without losses. Note that $\boldsymbol{\alpha}\cdot\boldsymbol{\beta} = 0$, and that there is no variation at all in the direction normal to the page.

because its equiphase surfaces are planes; but it is *not* a uniform plane wave, because the field strength varies with position over the equiphase planes. Similarly, the phase of the field varies most rapidly over the planes of constant amplitude.

There is evidently no variation of the field in the direction normal to both $\boldsymbol{\alpha}$ and $\boldsymbol{\beta}$ because this direction is characterized by the lines of intersection of planes of constant amplitude with planes of constant phase.

All the foregoing important features of space variation are illustrated in Fig. 7.4.

7.2.2.3 TE AND TM PLANE WAVES.* We have found that at any nonzero frequency, $\boldsymbol{\alpha} \neq 0$ is acceptable to Maxwell's equations as long as $\boldsymbol{\beta}$ obeys Eqs. 7.43. The somewhat puzzling question of how an "attenuation" can take place in a lossless medium will be answered only after we study the details of the field vectors themselves. To do this conveniently, choose a $+z$-axis along $\boldsymbol{\beta}$ and a $+y$-axis along $\boldsymbol{\alpha}$ (always possible, since $\boldsymbol{\alpha} \cdot \boldsymbol{\beta} = 0$). The $+x$-axis must then be along $\boldsymbol{\alpha} \times \boldsymbol{\beta}$. Consequently, as illustrated in Fig. 7.5,

$$\bar{\boldsymbol{\gamma}} = \boldsymbol{\alpha} + j\boldsymbol{\beta} = \boldsymbol{a}_y\alpha + \boldsymbol{a}_z j\beta \qquad \alpha, \beta > 0 \tag{7.48}$$

Then Eq. 7.41*a* becomes

$$\alpha E_{y0} + j\beta E_{z0} = 0 \tag{7.49}$$

Surprisingly enough, Eq. 7.49 does not restrict E_{x0} in any way. It is therefore certainly possible to choose as the electric field one with only

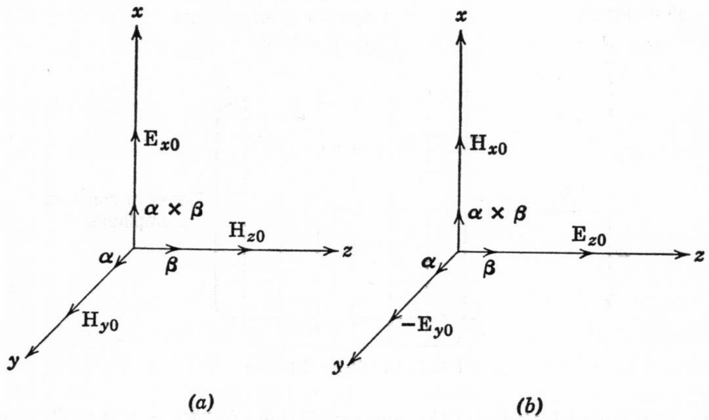

Fig. 7.5. Choice of axes and illustration of field components for (*a*) TE and (*b*) TM plane waves in a lossless medium.

a single component, in the x-direction, and this electric field automatically satisfies Eq. 7.49. Such a field is at all times parallel to $\boldsymbol{\alpha} \times \boldsymbol{\beta}$. It is linearly polarized along the direction in which there is no space variation of the solution. One solution of Eq. 7.49 or Eq. 7.41*a* is accordingly

$$\mathbf{E}_0 = \boldsymbol{a}_x \mathrm{E}_{x0} \tag{7.50}$$

The magnetic field corresponding to the choice (Eq. 7.50) for the electric field is given by Eqs. 7.40*a* and 7.48:

$$\mathbf{H}_0 = \frac{\bar{\boldsymbol{\gamma}} \times \mathbf{E}_0}{j\omega\mu} = \boldsymbol{a}_y \left(\frac{\beta}{\omega\mu}\right) \mathrm{E}_{x0} - \boldsymbol{a}_z \left(\frac{\alpha}{j\omega\mu}\right) \mathrm{E}_{x0} \tag{7.51}$$

This magnetic field automatically satisfies Eq. 7.41*b*, as follows.

$$\bar{\boldsymbol{\gamma}} \cdot \mathbf{H}_0 = \frac{\bar{\boldsymbol{\gamma}} \cdot (\bar{\boldsymbol{\gamma}} \times \mathbf{E}_0)}{j\omega\mu} = \frac{(\bar{\boldsymbol{\gamma}} \times \bar{\boldsymbol{\gamma}}) \cdot \mathbf{E}_0}{j\omega\mu} \equiv 0$$

We have in Eqs. 7.50 and 7.51 a solution to Eqs. 7.39, because restrictions of Eqs. 7.41, 7.40*a*, and 7.43*b* are already met. Exercise of care to choose α and β in accordance with Eq. 7.43*a* (or Eq. 7.47) will then guarantee consistency of Eq. 7.40*b*. The magnetic field defined by Eq. 7.51 is elliptically polarized in the plane (y, z) of $\boldsymbol{\alpha}$ and $\boldsymbol{\beta}$, thereby remaining at all times perpendicular to the electric field described by Eq. 7.50. Since the electric field has no components in the plane of $\boldsymbol{\alpha}$ and $\boldsymbol{\beta}$, in which all space variation takes place, this entire wave solution may be called a *Transverse Electric* (*TE*) plane wave. The important space relations are shown in Fig. 7.5*a*.

A similar (in fact dual) argument may now be applied, starting from Eq. 7.41*b* instead of Eq. 7.41*a*. One solution to the former is

$$\mathbf{H}_0 = \boldsymbol{a}_x \mathrm{H}_{x0} \tag{7.52}$$

with corresponding electric field given by Eqs. 7.40*b* and 7.48.

$$\mathbf{E}_0 = \frac{\mathbf{H}_0 \times \bar{\boldsymbol{\gamma}}}{j\omega\epsilon} = -\boldsymbol{a}_y \left(\frac{\beta}{\omega\epsilon}\right) \mathrm{H}_{x0} + \boldsymbol{a}_z \left(\frac{\alpha}{j\omega\epsilon}\right) \mathrm{H}_{x0} \tag{7.53}$$

This electric field automatically satisfies Eq. 7.41*a*.

Characterized by a magnetic field which is linearly polarized in the direction of no space variation, $\boldsymbol{\alpha} \times \boldsymbol{\beta}$, and an electric field elliptically polarized in the plane of $\boldsymbol{\alpha}$ and $\boldsymbol{\beta}$, the solutions in Eqs. 7.52 and 7.53 may be called a *Transverse Magnetic* (*TM*) plane wave. The relevant illustration is Fig. 7.5*b*.

It is convenient to describe the foregoing TE and TM solutions in a form which does not depend upon the particular set of (x, y, z) axes chosen in Eqs. 7.48 through 7.53. As Fig. 7.5 shows, the fields can be described completely by axes along $\boldsymbol{\alpha}$, $\boldsymbol{\beta}$, and $\boldsymbol{\alpha} \times \boldsymbol{\beta}$, which happen to form a right-handed rectangular system. Using unit vectors $\boldsymbol{a}_\alpha$, $\boldsymbol{a}_\beta$, and $\boldsymbol{a}_{\alpha\times\beta}$ along these directions respectively, and corresponding subscripts E_α, $H_{\alpha\times\beta}$, etc., for the field components, we may summarize our results in the following way:

TE WAVE ($\bar{\boldsymbol{\gamma}} = \boldsymbol{\alpha} + j\boldsymbol{\beta}$; $\boldsymbol{\alpha}\cdot\boldsymbol{\beta} = 0$; $\beta^2 - \alpha^2 = \beta_0{}^2 = \omega^2\epsilon\mu$)

$$\text{(a)} \qquad \mathbf{E} = \boldsymbol{a}_{\alpha\times\beta}E_{\alpha\times\beta,0}e^{-\bar{\boldsymbol{\gamma}}\cdot\boldsymbol{r}} = \boldsymbol{a}_{\alpha\times\beta}E_{\alpha\times\beta,0}e^{-\boldsymbol{\alpha}\cdot\boldsymbol{r}}e^{-j\boldsymbol{\beta}\cdot\boldsymbol{r}}$$

$$\text{(b)} \qquad \mathbf{H} = \frac{\bar{\boldsymbol{\gamma}} \times \mathbf{E}}{j\omega\mu} = \left[\boldsymbol{a}_\alpha\left(\frac{\beta}{\omega\mu}\right) - \boldsymbol{a}_\beta\left(\frac{\alpha}{j\omega\mu}\right)\right]E_{\alpha\times\beta,0}e^{-\boldsymbol{\alpha}\cdot\boldsymbol{r}}e^{-j\boldsymbol{\beta}\cdot\boldsymbol{r}} \tag{7.54}$$

TM WAVE ($\bar{\boldsymbol{\gamma}} = \boldsymbol{\alpha} + j\boldsymbol{\beta}$; $\boldsymbol{\alpha}\cdot\boldsymbol{\beta} = 0$; $\beta^2 - \alpha^2 = \beta_0{}^2 = \omega^2\epsilon\mu$)

$$\text{(a)} \qquad \mathbf{H} = \boldsymbol{a}_{\alpha\times\beta}H_{\alpha\times\beta,0}e^{-\bar{\boldsymbol{\gamma}}\cdot\boldsymbol{r}} = \boldsymbol{a}_{\alpha\times\beta}H_{\alpha\times\beta,0}e^{-\boldsymbol{\alpha}\cdot\boldsymbol{r}}e^{-j\boldsymbol{\beta}\cdot\boldsymbol{r}}$$

$$\text{(b)} \qquad \mathbf{E} = \frac{\mathbf{H} \times \bar{\boldsymbol{\gamma}}}{j\omega\epsilon} = \left[-\boldsymbol{a}_\alpha\left(\frac{\beta}{\omega\epsilon}\right) + \boldsymbol{a}_\beta\left(\frac{\alpha}{j\omega\epsilon}\right)\right]H_{\alpha\times\beta,0}e^{-\boldsymbol{\alpha}\cdot\boldsymbol{r}}e^{-j\boldsymbol{\beta}\cdot\boldsymbol{r}} \tag{7.55}$$

We have unquestionably determined two linearly independent plane-wave solutions to Eqs. 7.39 of the form of Eq. 7.37. Are there any more solutions of this form? In what sense there are not may be seen by considering an arbitrary linear combination of a TE and a TM solution having the same $\bar{\boldsymbol{\gamma}}$. The combined field meets all the conditions imposed by Eqs. 7.41 and 7.40, provided that $\bar{\boldsymbol{\gamma}}$ obeys Eq. 7.42. Moreover, Fig. 7.5 shows that the combined field has three space components of both **E** and **H**, neither of which is necessarily linearly polarized. *A linear combination of a TE and a TM plane wave is the most general simple exponential solution possible for a given value of $\bar{\boldsymbol{\gamma}}$ consistent with Eq. 7.42.*

Obviously, however, even at a fixed frequency (fixed value of β_0), one may choose many different values of $\bar{\boldsymbol{\gamma}}$ consistent with Eqs. 7.42 or 7.43. For example, note that conditions of Eqs. 7.43*a* (or 7.47) and 7.43*b* allow independent choice of the algebraic sign of α and β, given a value of $|\alpha|$ or $|\beta|$ and given a frequency ω (i.e., given a value of β_0). Of course, the relative phases of the field components will vary with the various possible choices, as shown by Eqs. 7.51 and 7.53 (or 7.54 and 7.55). These facts simply express the idea that, as with uniform plane waves, a traveling TE or TM plane wave oriented in any direction in space is a valid solution of Maxwell's equations as long as $\mathbf{E}_0$, $\mathbf{H}_0$, and $\bar{\boldsymbol{\gamma}}$ bear the correct mutual relationships. Needless to say, sums of TE and/or TM plane waves in various space orientations are also solutions to the field equations.

7.2.2.4 Power Considerations and Wave Impedances.* Examination of the complex power for the nonuniform plane waves disclosed above will help to explain the "attenuation."

For example, we have noted that the electric field of the TE wave (Eq. 7.54) is linearly polarized in the direction normal to both $\boldsymbol{\alpha}$ and $\boldsymbol{\beta}$, but the magnetic field is elliptically polarized in the plane of $\boldsymbol{\alpha}$ and $\boldsymbol{\beta}$. Since $E_{\alpha\times\beta,0}$ and $H_{\alpha 0}$ are in time phase, the component of the complex Poynting vector along $\boldsymbol{\beta}$ is *entirely real.* However $E_{\alpha\times\beta,0}$ and $H_{\beta 0}$ are 90° out of time phase, so the component of the complex Poynting vector along $\boldsymbol{\alpha}$ is *entirely reactive.* In other words: *only real (time-average) power flows in the direction* $\boldsymbol{\beta}$ *(normal to the constant-phase planes and parallel to the constant-amplitude planes), whereas only reactive power flows in the direction* $\boldsymbol{\alpha}$ *(normal to the constant-amplitude planes and parallel to the constant-phase planes).* The "attenuation" therefore does *not* act to decrease with distance any space component of the time-average power and, consequently, causes no difficulty with conservation of energy in the lossless medium. Similar comments apply to the TM plane wave (Eq. 7.55). These power relationships are illustrated in Fig. 7.6 for both wave types.

Actually the idea of a decreasing field strength in a lossless situation is quite familiar. Imagine a ladder network of inductors only, driven from one end. Both the voltage and the current will decrease with distance away from the drive terminals; but since the whole network is reactive, no real power enters it from the source, and no power loss is implied by the decreasing voltage and current levels. We have simply a lossless "voltage (or current) divider."

The only new idea in the nonuniform plane wave field is the possibility of having real unattenuated power flow in one direction, analogous to a traveling wave on a lossless transmission line in that direction, and, at the same time, having a lossless voltage divider action in another (perpendicular) direction. This however is just the kind of difference we might expect to be brought about by considering problems in more than one space dimension.

Indeed, it is often convenient to define in terms of field components impedances that describe the nature of the complex-power components in various space directions. Thus, in general, field components $+E_x$ and $+H_y$ (or $+E_y$ and $-H_x$) enter into the $+z$-component of the complex Poynting vector S_z, while $+E_x$ and $-H_z$ (or $+E_z$ and $+H_x$)

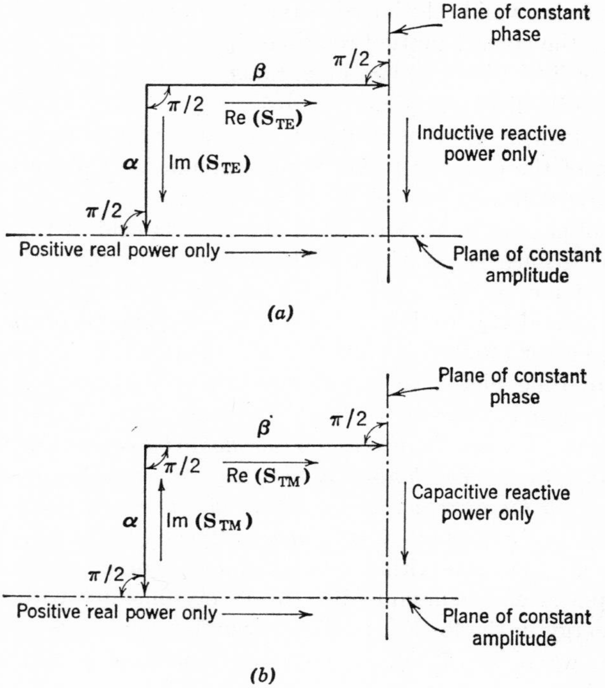

Fig. 7.6. Analysis of the complex Poynting vector for nonuniform plane waves in a lossless medium. (*a*) TE plane wave; (*b*) TM plane wave.

are the pertinent components for S_y. The relevant *wave impedances* may be introduced as follows, using S_z as example:

$$S_z = \frac{1}{2} E_x H_y^* - \frac{1}{2} E_y H_x^* = \frac{1}{2}\left(\frac{E_x}{H_y}\right)|H_y|^2 + \frac{1}{2}\left(\frac{-E_y}{H_x}\right)|H_x|^2 \tag{7.56a}$$

Letting

$$Z_z^{(x,y)} \equiv \frac{E_x}{H_y} \tag{7.56b}$$

and

$$Z_z^{(y,x)} \equiv -\frac{E_y}{H_x} \tag{7.56c}$$

we have

$$S_z = \tfrac{1}{2}|H_y|^2 Z_z^{(x,y)} + \tfrac{1}{2}|H_x|^2 Z_z^{(y,x)} \tag{7.56d}$$

Evidently the real and imaginary parts of the wave impedances give the correct algebraic sign of the real and reactive power associated with the pertinent field components. In many cases, one of the field components responsible for power in a given direction is absent so that the distinction between the two wave impedances [$Z_z^{(x,y)}$ and $Z_z^{(y,x)}$, for example] for that direction need not be made. In such cases the superscript will simply be omitted, it being understood from the context which field components must be involved.

Applying the foregoing ideas to the TE and TM plane waves of Eqs. 7.54 and 7.55, we may define the following wave impedances:

$$\begin{array}{llll} & \text{TE} & & \text{TM} \\ \text{(a)} & Z_\beta \equiv \dfrac{E_{\alpha\times\beta,0}}{H_{\alpha 0}} = \dfrac{\omega\mu}{\beta} & \text{(c)} & Z_\beta \equiv \dfrac{-E_{\alpha 0}}{H_{\alpha\times\beta,0}} = \dfrac{\beta}{\omega\epsilon} \\ \text{(b)} & Z_\alpha \equiv \dfrac{E_{\alpha\times\beta,0}}{-H_{\beta 0}} = \dfrac{j\omega\mu}{\alpha} & \text{(d)} & Z_\alpha \equiv \dfrac{E_{\beta 0}}{H_{\alpha\times\beta,0}} = \dfrac{\alpha}{j\omega\epsilon} \end{array} \tag{7.57}$$

Observe that, if $\beta > 0$, Z_β is positive real, indicating again that only time-average power flows in the $\boldsymbol{\beta}$ direction. If $\alpha > 0$, Z_α is *inductive* for the TE wave, indicating that reactive power only flows in the $\boldsymbol{\alpha}$ direction and that *the TE field stores more time-average magnetic than electric energy per unit volume.* On the other hand, Z_α is *capacitive* for the TM wave, which means that *the TM field stores more time-average electric than magnetic energy per unit volume.* These facts are also included in Fig. 7.6.

The most general lossless plane wave with a given $\bar{\boldsymbol{\gamma}}$ is, as we have already mentioned, a linear combination of a TE and a TM plane wave with respect to the direction of the vectors $\boldsymbol{\beta}$ and $\boldsymbol{\alpha}$. It is not difficult to see in Fig. 7.5 that *the field components missing from these waves make them orthogonal with respect to complex power in the* $\boldsymbol{\beta}$ *and* $\boldsymbol{\alpha}$ *directions.* That is, the complex power in these directions when both waves are present together is the sum of the complex powers which would be carried in these directions by each one separately. Unfortunately, however, this orthogonality does *not* hold for complex power in the direction normal to $\boldsymbol{\alpha}$ *and* $\boldsymbol{\beta}$ (i.e., the direction of $\boldsymbol{\alpha} \times \boldsymbol{\beta}$). For either wave alone $S_{\alpha\times\beta} \equiv 0$, but, when both are present together, Eqs. 7.54 and 7.55 show that

$$S_{\alpha\times\beta} = \frac{1}{2}(E_\alpha H_\beta^* - E_\beta H_\alpha^*) = \frac{j\alpha\beta}{\beta_0^2} H_{\alpha\times\beta,0}\, E_{\alpha\times\beta,0} e^{-2\alpha\cdot r}$$

This is not in general zero, but depends upon $E_{\alpha\times\beta,0}$ and $H_{\alpha\times\beta,0}$, which we have seen are *independent* as far as Maxwell's equations are concerned. Their presence and relative magnitude and phase will therefore be determined completely by boundary conditions.

7.2.3 Relationships between Uniform and Nonuniform Plane Waves *

The results of our work in Secs. 7.2.1 and 7.2.2 may be summarized as follows: The field

$$\text{(a)} \qquad \mathbf{E} = \mathbf{E}_0 e^{-\bar{\boldsymbol{\gamma}}\cdot\boldsymbol{r}} = \mathbf{E}_0 e^{-\bar{\gamma}_x x} e^{-\bar{\gamma}_y y} e^{-\bar{\gamma}_z z}$$

$$\text{(b)} \qquad \mathbf{H} = \mathbf{H}_0 e^{-\bar{\boldsymbol{\gamma}}\cdot\boldsymbol{r}} = \frac{\bar{\boldsymbol{\gamma}} \times \mathbf{E}_0}{j\omega\mu} e^{-\bar{\boldsymbol{\gamma}}\cdot\boldsymbol{r}} = \frac{\bar{\boldsymbol{\gamma}} \times \mathbf{E}}{j\omega\mu} \qquad (7.58)$$

is a sinusoidal steady-state solution to Maxwell's equations without loss if and only if the conditions

$$\text{(a)} \qquad \bar{\gamma}_x^2 + \bar{\gamma}_y^2 + \bar{\gamma}_z^2 = -\beta_0^2 (= -\omega^2 \epsilon\mu)$$

and

$$\text{(b)} \qquad \begin{cases} \bar{\gamma}_x \mathrm{E}_{x0} + \bar{\gamma}_y \mathrm{E}_{y0} + \bar{\gamma}_z \mathrm{E}_{z0} = 0 \\ \text{or} \\ \bar{\gamma}_x \mathrm{H}_{x0} + \bar{\gamma}_y \mathrm{H}_{y0} + \bar{\gamma}_z \mathrm{H}_{z0} = 0 \end{cases} \qquad (7.59)$$

are satisfied. Note that any two of $\bar{\gamma}_x$, $\bar{\gamma}_y$, $\bar{\gamma}_z$ are completely arbitrary complex numbers, with the third fixed (except for algebraic sign) by Eq. 7.59*a*. Similarly, any two of E_{x0}, E_{y0}, E_{z0} (or H_{x0}, H_{y0}, H_{z0}) are also completely arbitrary complex numbers, with the third fixed by Eq. 7.59*b*. Equation 7.58*b* then determines the remaining complex vector $\mathbf{H}_0$ (or $\mathbf{E}_0$).

The following special cases have arisen:

1. If $\bar{\boldsymbol{\gamma}} = j\boldsymbol{\beta}$ (i.e., $\boldsymbol{\alpha} = 0$), the solution is a uniform plane wave, which is one form of the TEM wave with respect to $\boldsymbol{\beta}$.
2. If $\boldsymbol{\alpha} \neq 0$ and $\mathbf{E}_0 = \mathrm{A}\boldsymbol{E}_0$, where A is any complex number and $\boldsymbol{E}_0$ any real vector, the solution has a linearly polarized electric field and is a TE wave with respect to $\boldsymbol{\beta}$ and $\boldsymbol{\alpha}$.
3. If $\boldsymbol{\alpha} \neq 0$ and $\mathbf{H}_0 = \mathrm{B}\boldsymbol{H}_0$, where B is any complex number and $\boldsymbol{H}_0$ any real vector, the solution has a linearly polarized magnetic field and is a TM wave with respect to $\boldsymbol{\beta}$ and $\boldsymbol{\alpha}$.
4. If $\boldsymbol{\alpha} = 0$, together with either $\mathbf{E}_0 = \mathrm{A}\boldsymbol{E}_0$ or $\mathbf{H}_0 = \mathrm{B}\boldsymbol{H}_0$, then the solution is a uniform plane wave with linearly polarized electric and magnetic fields.

We see from the summary that the condition of Eq. 7.59*a* is of great importance. It was met automatically in the case of the uniform plane wave ($\boldsymbol{\alpha} = 0$) by the well-known geometric condition (Eq. 7.17) on the direction cosines ($l_{z'}$, $m_{z'}$, $n_{z'}$) of the propagation direction with respect to the (x, y, z) axes. What, if any, are the geometric consequences of Eq. 7.59*a* when $\boldsymbol{\alpha} \neq 0$?

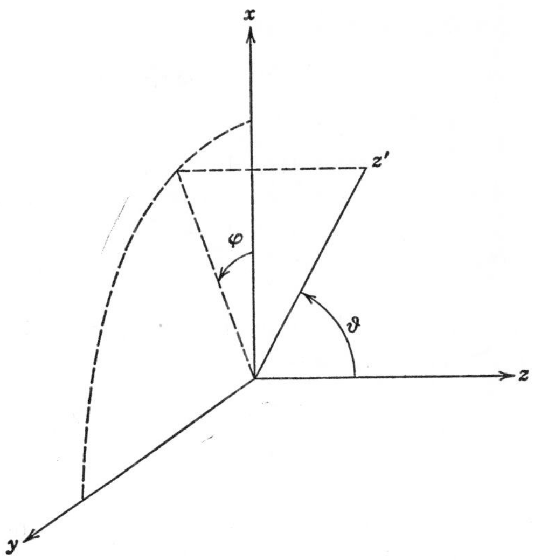

Fig. 7.7. Polar angles for location of a new axis.

With reference to Fig. 7.7, let $+z'$ be an axis along $\boldsymbol{\beta}$ for a *uniform* plane wave. Instead of locating $+z'$ by its direction cosines, use the polar angles ϑ and φ shown. Then the direction cosines of $+z'$ are:

$$\left.\begin{aligned} l_{z'} &= \sin\vartheta\cos\varphi \\ m_{z'} &= \sin\vartheta\sin\varphi \\ n_{z'} &= \cos\vartheta \end{aligned}\right\} \quad 0 \le \vartheta \le \pi \qquad 0 \le \varphi < 2\pi \tag{7.60}$$

It is obvious that $l_{z'}^2 + m_{z'}^2 + n_{z'}^2 \equiv 1$ automatically when expressed thus in terms of ϑ and φ.

The exponential factor in the fields therefore becomes

$$e^{-\bar{\boldsymbol{\gamma}}\cdot\boldsymbol{r}} = e^{-j\boldsymbol{\beta}\cdot\boldsymbol{r}} = e^{-(j\beta_0 \sin\vartheta\cos\varphi)x-(j\beta_0\sin\vartheta\sin\varphi)y-(j\beta_0\cos\vartheta)z} \tag{7.61}$$

so that we may write

$$\begin{aligned} &\text{(a)} & \bar{\gamma}_x &= j\beta_0 \sin\vartheta\cos\varphi \\ &\text{(b)} & \bar{\gamma}_y &= j\beta_0 \sin\vartheta\sin\varphi \\ &\text{(c)} & \bar{\gamma}_z &= j\beta_0 \cos\vartheta \end{aligned} \tag{7.62}$$

with the consequence that, identically,

$$\bar{\gamma}_x^2 + \bar{\gamma}_y^2 + \bar{\gamma}_z^2 \equiv -\beta_0^2 \tag{7.59a}$$

The important point now is the fact that

$$\sin^2 \bar{\vartheta} \cos^2 \bar{\varphi} + \sin^2 \bar{\vartheta} \sin^2 \bar{\varphi} + \cos^2 \bar{\vartheta} \equiv 1 \qquad (7.63)$$

when $\bar{\vartheta}$ and $\bar{\varphi}$ are any arbitrary complex numbers!

In other words, if we allow the notion of *complex polar angles* $\bar{\vartheta}$ *and* $\bar{\varphi}$ for the direction of propagation of a *uniform plane wave,* the corresponding solution becomes a *nonuniform plane wave* with complex $\bar{\gamma}_x$, $\bar{\gamma}_y$, $\bar{\gamma}_z$ defined by Eq. 7.62 and satisfying the condition of Eq. 7.59*a* identically.

An illustrative example of the complex angles is furnished by the case treated in Sec. 7.2.2.3, where

$$\bar{\boldsymbol{\gamma}} = \boldsymbol{a}_y \alpha + \boldsymbol{a}_z j\beta \qquad (\alpha, \beta > 0) \qquad (7.48)$$

and

$$\beta^2 = \beta_0{}^2 + \alpha^2 > \beta_0{}^2 \qquad (7.47)$$

In this case it is easiest to start with Eq. 7.62*a* because $\bar{\gamma}_x = 0$. In other cases, Eq. 7.62*c* might be simpler because it contains only one angle.

Thus in the present example

$$\bar{\gamma}_x = 0 = \sin \bar{\vartheta} \cos \bar{\varphi}$$

so either $\sin \bar{\vartheta} = 0$ or $\cos \bar{\varphi} = 0$, or both. But $\bar{\gamma}_y \neq 0$, so Eq. 7.62*b* shows that $\sin \bar{\vartheta} \neq 0$. Hence we must choose $\cos \bar{\varphi} = 0$. Since, however $\bar{\varphi}$ is a complex angle $\varphi_R + j\varphi_I$, we must set

$$\cos \bar{\varphi} = \cos (\varphi_R + j\varphi_I) = \cos \varphi_R \cosh \varphi_I - j \sin \varphi_R \sinh \varphi_I = 0$$

Because $\cosh \varphi_I \geq 1$, this implies in sequence

$$\begin{aligned}
&\text{(a)} \qquad \varphi_R = \frac{\pi}{2}\left(\text{or } \frac{3\pi}{2}\right) \qquad 0 \leq \varphi_R < 2\pi \\
&\text{(b)} \qquad \sin \varphi_R = +1 \text{ (or } -1) \\
&\text{(c)} \qquad \varphi_I = 0 \\
&\text{(d)} \qquad \sin \bar{\varphi} = \sin \varphi_R = +1 \text{ (or } -1)
\end{aligned} \qquad (7.64)$$

Now with reference to $\bar{\gamma}_y$ in Eq. 7.62*b*, we need to know that

$$\sin \bar{\vartheta} = \sin (\vartheta_R + j\vartheta_I) = \sin \vartheta_R \cosh \vartheta_I + j \cos \vartheta_R \sinh \vartheta_I$$

It follows from this result, with the use of Eqs. 7.48, 7.62*b*, and the condition of Eq. 7.64*d* on $\sin \bar{\varphi}$, that

$$\alpha = {}_{(-)}^{+}\beta_0(j \sin \vartheta_R \cosh \vartheta_I - \cos \vartheta_R \sinh \vartheta_I)$$

Therefore, since α is positive real and $\cosh \vartheta_I \geq 1$, we have in sequence that

$$\begin{aligned}
&\text{(a)} && \sin \vartheta_R \cosh \vartheta_I = 0 \\
&\text{(b)} && \vartheta_R = 0 \,[\text{or } \pi] \qquad 0 \leq \vartheta_R < 2\pi \\
&\text{(c)} && \cos \vartheta_R = 1 \,[\text{or } -1] \qquad (7.65) \\
&\text{(d)} && \alpha = {}^{+}_{(-)}\beta_0 \left({}^{-}_{[+]} \sinh \vartheta_I\right) > 0
\end{aligned}$$

where we have used the curved $(-)$ and square $[+]$ marks in Eq. 7.65*d* to identify the algebraic sign alternatives from Eqs. 7.64*d* and 7.65*c* respectively.

To eliminate some of the various choices of algebraic sign and quadrant of angle remaining, we must look at Eq. 7.62*c*, when $\bar{\gamma}_z = j\beta$, as required by Eq. 7.48. We find

$$\beta = \beta_0(\cos \vartheta_R \cosh \vartheta_I - j \sin \vartheta_R \sinh \vartheta_I) = \beta_0 \cos \vartheta_R \cosh \vartheta_I > 0$$

in view of Eq. 7.65*b*.

Again, since $\cosh \vartheta_I \geq 1$, $\cos \vartheta_R > 0$. This means

$$\vartheta_R = 0 \text{ (not } \pi)$$

in Eq. 7.65*b*, which removes completely the [ ] alternatives in Eqs. 7.65*d*. Accordingly, using the first choice in Eq. 7.65*c*, we have

$$\cosh \vartheta_I = \frac{\beta}{\beta_0} > 1$$

which leaves the algebraic sign of ϑ_I still in doubt.

We are left with a choice of two solutions. Either

$$\varphi_R = \frac{\pi}{2} \qquad \text{from Eq. 7.64}a$$

and

$$\sinh \vartheta_I = -\frac{\alpha}{\beta_0} \qquad \text{from Eq. 7.65}d$$

or

$$\varphi_R = \frac{3\pi}{2} \qquad \text{from Eq. 7.64}a$$

and

$$\sinh \vartheta_I = +\frac{\alpha}{\beta_0} \qquad \text{from Eq. 7.65}d$$

The existence of these two possibilities has no particular physical significance. They are simply consequences of the multiple-valued properties of the inverse trigonometric functions. A *uniform* plane wave at either set of complex angles

$$\bar{\vartheta} = -j \sinh^{-1}\left(\frac{\alpha}{\beta_0}\right)$$

and

$$\bar{\varphi} = \frac{\pi}{2}$$

or

$$\bar{\vartheta} = +j \sinh^{-1}\left(\frac{\alpha}{\beta_0}\right)$$

and

$$\bar{\varphi} = \frac{3\pi}{2}$$

represents physically the *same nonuniform* plane wave. We can choose either solution arbitrarily. Needless to say, the polarization of the uniform plane wave sets the TE, TM, or combined TE–TM character of the resulting wave.

Our understanding of the plane-wave solutions we have found will be weak unless we apply them to some situations involving boundaries. The principal purpose of the following sections of this chapter is therefore to discuss some problems in which it only takes a small number—one, two, three, etc.—of plane waves to meet the boundary conditions. We shall find many useful similarities between these problems and those encountered in transmission lines, but there are also important differences. The differences arise from both the three-dimensional variation of the fields, and their three-dimensional vector character. The examples chosen do illustrate significant physical phenomena in optics and radio transmission, but they are equally important as part of a background of understood cases upon which to develop a physical intuition suitable for dealing with harder field problems.

7.3 Normal Incidence of a Uniform Plane Wave

We suggested in connection with Eqs. 7.14 and 7.15, that we could interpret the (+) solution for a uniform plane wave as an incident wave and the (−) one as a reflected wave. We also suggested in Sec. 7.1.5, however, that such an interpretation would be meaningful only if the boundary conditions in *all space* could be met by Eq. 7.14. Specifically, we must have a physical situation in three-dimensional space for which the boundary conditions require only x-polarized uniform plane waves propagating in the $\pm z$ directions.

Now the boundary conditions for the complex fields involve only components of **E** and **H** tangential to interfaces between different media (Eqs. 1.22 and 1.25). If we take the incident wave (specified by some remote source) as an x-polarized uniform plane wave propagating in the $+z$ direction, its electric and magnetic fields lie in planes normal to the z-axis and extend uniformly throughout these planes. Thus a plane boundary or interface between different media which also extends uniformly in some plane normal to the z-axis should require only similar field components in the reflected wave.

7.3.1 Normal Incidence on a Perfect Conductor

For a first example, consider the situation in Fig. 7.8. A perfectly conducting metal wall occupies the entire x, y plane, and the incident

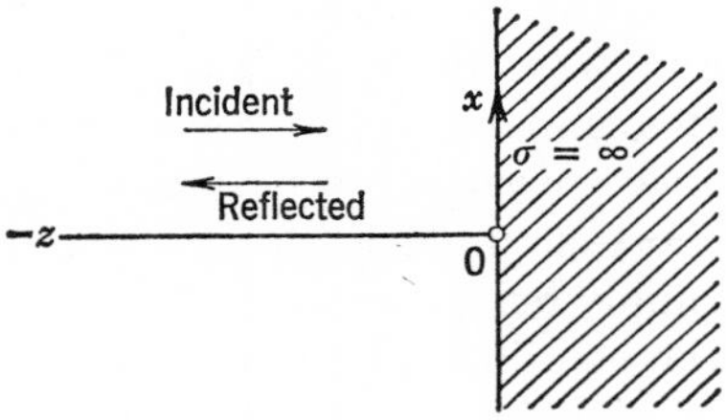

Fig. 7.8. Normal incidence of uniform plane wave on a perfect conductor. The $+y$-axis points out of the paper.

uniform plane wave (from a source at $z = -\infty$) has the specified form

$$\mathbf{E}_i = \boldsymbol{a}_x \mathrm{E}_{x0}{}^{+} e^{-j\beta_0 z} \tag{7.66}$$

in which $\mathrm{E}_{x0}{}^{+}$ is a given constant. In accordance with Eqs. 7.14, the corresponding magnetic field is

$$\mathbf{H}_i = \boldsymbol{a}_y \frac{\mathrm{E}_{x0}{}^{+}}{\eta} e^{-j\beta_0 z} \tag{7.67}$$

and the reflected field will be denoted by

$$\text{(a)} \qquad \mathbf{E}_r = \boldsymbol{a}_x \mathrm{E}_{x0}{}^{-} e^{j\beta_0 z}$$

$$\text{(b)} \qquad \mathbf{H}_r = -\boldsymbol{a}_y \frac{\mathrm{E}_{x0}{}^{-}}{\eta} e^{j\beta_0 z} \tag{7.68}$$

The boundary condition on the metal wall requires the total tangential electric field to vanish when $z = 0$ for *all x and y*. Therefore

$$\mathrm{E}_{x0}{}^{+} + \mathrm{E}_{x0}{}^{-} = 0 \tag{7.69}$$

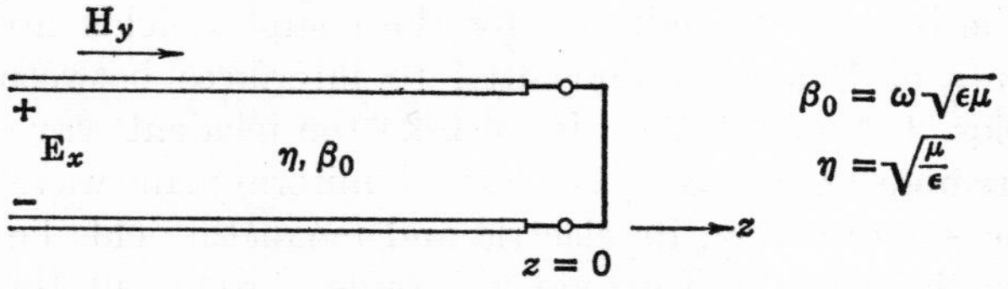

Fig. 7.9. Lossless transmission line analog for Fig. 7.8 with x-polarized uniform plane waves.

and the total field becomes

$$\text{(a)} \qquad \mathbf{E} = \mathbf{E}_i + \mathbf{E}_r = -\boldsymbol{a}_x 2j\mathrm{E}_{x0}^{+} \sin \beta_0 z \tag{7.70}$$

$$\text{(b)} \qquad \mathbf{H} = \mathbf{H}_i + \mathbf{H}_r = \boldsymbol{a}_y \frac{2\mathrm{E}_{x0}^{+}}{\eta} \cos \beta_0 z$$

We see that, besides being perpendicular in space, $\boldsymbol{E}\,[= \text{Re}\,(\mathbf{E}e^{j\omega t})]$ and $\boldsymbol{H}\,[= \text{Re}\,(\mathbf{H}e^{j\omega t})]$ are 90° out of phase with respect to both their time and space variations. Indeed, **E** and **H** have relative phases and space *variations* exactly like the voltage and current respectively on a short-circuited lossless transmission line. As a matter of fact, if we simply remember that **E** lies along the x-axis and **H** lies along the y-axis, the solution (Eq. 7.70) is otherwise identical with that of the transmission line in Fig. 7.9. Note that the "characteristic wave impedance"

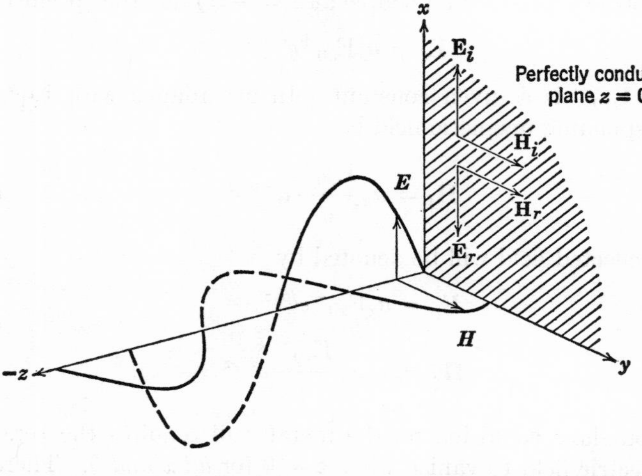

Fig. 7.10. Standing-wave pattern in front of a perfect conductor illuminated by a normally incident, $+x$-polarized uniform plane wave.

η serves as characteristic impedance of the line in this case. In Fig. 7.10 we show space plots of $\boldsymbol{E}$ and $\boldsymbol{H}$ at a moment when neither one has its maximum magnitude. The relation between this figure and Fig. 3.6 should be studied carefully.

A little thought will convince us that, had we started with a y-polarized incident wave, the reflected wave would also have been y-polarized, and the standing-wave pattern of Fig. 7.10 would merely have been rotated about the $+z$-axis clockwise by 90° in space. Moreover, the equivalent transmission line of Fig. 7.9 would have remained the same, except that E_y would have replaced E_x, and $-\mathrm{H}_x$ would have replaced H_y (see Eqs. 7.2 and compare Eqs. 7.3 and 7.8).

If the incident uniform plane wave had electric field components along both the x and the y directions, we would simply treat each component (with its associated magnetic field according to Eqs. 7.5a and 7.9a) separately, and superpose the separate solutions after completing them independently.

7.3.2 Normal Incidence on a Lossless Dielectric

Our second example of a plane boundary normal to the direction of propagation of the incident wave involves two lossless dielectrics, as appear in Fig. 7.11. The incident wave is given in medium 1 as an x-polarized uniform plane wave:

$$\text{(a)} \qquad \mathbf{E}_i = \boldsymbol{a}_x \mathrm{E}_{x1}{}^{+} e^{-j\beta_{01}z}$$
$$\text{(b)} \qquad \mathbf{H}_i = \frac{1}{\eta_1} \boldsymbol{a}_z \times \mathbf{E}_i \qquad (7.71)$$

We must, in this case, allow for a wave transmitted into medium 2 as well as for a reflected wave in medium 1. We shall assume, however, that no $(-)$ wave occurs in medium 2. Some discussion is in order to indicate that the absence of a $(-)$ wave is in fact an assumption. There is a temptation to try to summarize it simply by pointing out that medium 2 extends infinitely far to the right, and by suggesting that no wave has time to come back from such a distance. These statements, however, do *not* explain the absence of a $(-)$ wave *in a lossless medium in the sinusoidal steady state.* The steady-state idea itself supposes that we have, in fact, waited long enough for all initial transients to disappear, even if that takes infinite time! For example, if there were a perfectly conducting plane at $z = z_0 > 0$, our previous work shows that there would certainly be a $(-)$ wave in medium 2, *no matter how large z_0 might be.* If there is to be no $(-)$ wave, the im-

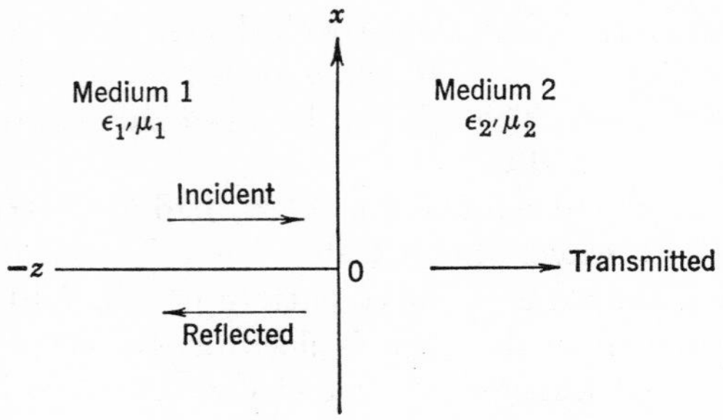

Fig. 7.11. Normal incidence of uniform plane wave on a lossless dielectric.

portant point is not how far medium 2 extends but rather that there cannot be (for example) a metal plane at the *end* of it, even if the end is at $z = +\infty$. The honest way to state our present assumption is *as a boundary condition on the solution when* $z \to +\infty$. Specifically, we are arbitrarily looking for a solution which behaves like a right-going wave $e^{-j\beta_{02}z}$ as $z \to +\infty$.

The immediate question is then whether such a solution can be found. If it can, the ensuing question of what real physical situations might correspond to it is quite another matter. Dealing with this one requires physical judgements, the successful making of which involves more than we can profitably discuss at this point in our argument.

Returning, then, to the solution itself, we take the form of the reflected wave in medium 1 to be

$$\text{(a)} \qquad \mathbf{E}_r = \boldsymbol{a}_x \mathrm{E}_{x1}{}^{-} e^{j\beta_{01}z} \qquad (7.72)$$

$$\text{(b)} \qquad \mathbf{H}_r = -\frac{1}{\eta_1} \boldsymbol{a}_z \times \mathbf{E}_r$$

and the form of the transmitted wave in medium 2 to be

$$\text{(a)} \qquad \mathbf{E}_t = \boldsymbol{a}_x \mathrm{E}_{x2}{}^{+} e^{-j\beta_{02}z} \qquad (7.73)$$

$$\text{(b)} \qquad \mathbf{H}_t = \frac{1}{\eta_2} \boldsymbol{a}_z \times \mathbf{E}_t$$

The boundary conditions at the interface $z = 0$ require both $\mathbf{E}_{\text{tang}}$ and $\mathbf{H}_{\text{tang}}$ to be continuous:

$$\text{(a)} \qquad \mathbf{E}_i + \mathbf{E}_r = \mathbf{E}_t \qquad \text{at } z = 0 \qquad (7.74)$$

$$\text{(b)} \qquad \mathbf{H}_i + \mathbf{H}_r = \mathbf{H}_t \qquad \text{at } z = 0$$

Using Eqs. 7.71, 7.72, and 7.73 in Eq. 7.74, we have

$$\text{(a)} \quad E_{x1}^{+} + E_{x1}^{-} = E_{x2}^{+}$$

$$\text{(b)} \quad \frac{1}{\eta_1}(E_{x1}^{+} - E_{x1}^{-}) = \frac{1}{\eta_2}E_{x2}^{+} \tag{7.75}$$

Multiplication of Eq. 7.75*b* by η_2, and subtraction from Eq. 7.75*a* yields

$$\frac{E_{x1}^{-}}{E_{x1}^{+}} \equiv \bar{\Gamma}_R = \frac{(\eta_2/\eta_1) - 1}{(\eta_2/\eta_1) + 1} \tag{7.76a}$$

which, in turn, upon substitution back into Eq. 7.75*a* also gives

$$\frac{E_{x2}^{+}}{E_{x1}^{+}} \equiv T = \frac{2(\eta_2/\eta_1)}{(\eta_2/\eta_1) + 1} \tag{7.76b}$$

The use of the symbol $\bar{\Gamma}_R$ in Eq. 7.76*a* stems from its clear analogy with the reflection coefficient of the load on a transmission line. In this case, the *reflection coefficient* $\bar{\Gamma}_R$ is *the ratio at the interface of the complex reflected electric field to the complex incident electric field.* Indeed, both Eqs. 7.76*a* and 7.76*b* (or Eqs. 7.75) follow from the equivalent transmission-line system, shown in Fig. 7.12, in which the incident "voltage" E_{x1}^{+} is regarded as given. We can see this clearly by first replacing the second transmission line in Fig. 7.12 according to the modified form of Thévènin's theorem proved for the time domain in Chapter 4. This step yields Fig. 7.13, from which Eq. 7.76*a* becomes apparent.

Next, we also replace line 1 in Fig. 7.13 by its Thévènin equivalent on the same basis, remembering that E_{x1}^{+} (the incident wave) is given. There results Fig. 7.14, exhibiting clearly Eq. 7.76*b*. The symbol T in this equation is called the *transmission coefficient,* defined as *the*

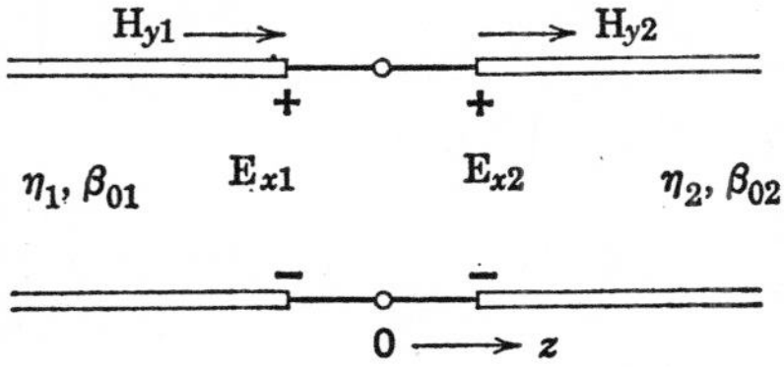

Fig. 7.12. Lossless transmission line analogy for Fig. 7.11.

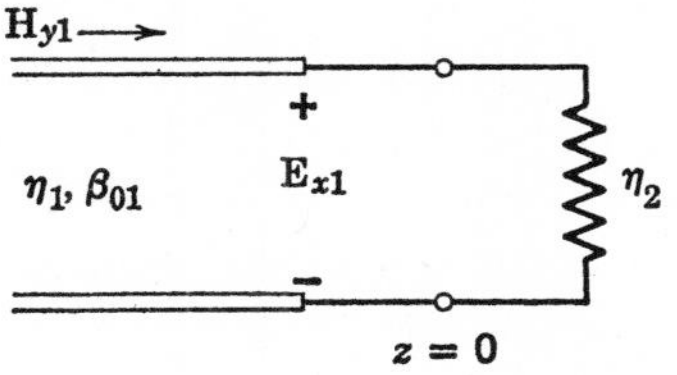

Fig. 7.13. An equivalent circuit for Fig. 7.12 from the point of view of line 1.

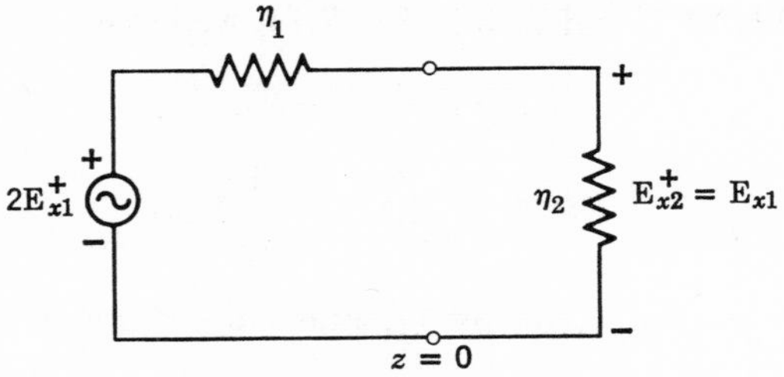

Fig. 7.14. Complete equivalent circuit of Fig. 7.12 at $z = 0$.

ratio at the interface of the complex transmitted electric field to the complex incident electric field.

It follows from Eqs. 7.75*a*, and the definitions of $\bar{\Gamma}_R$ and T in Eqs. 7.76*a* and 7.76*b*, that

$$1 + \bar{\Gamma}_R = \mathrm{T} \tag{7.77}$$

Also we note that the time-average power per unit area carried in the $+z$ direction in medium 1 is

$$\langle S_{z1} \rangle = \frac{|\mathrm{E}_{x1}{}^+|^2 - |\mathrm{E}_{x1}{}^-|^2}{2\eta_1} = \frac{|\mathrm{E}_{x1}{}^+|^2}{2}\,\eta_1(1 - |\bar{\Gamma}_R|^2)$$

and that carried in the $+z$ direction in medium 2 is

$$\langle S_{z2} \rangle = \frac{|\mathrm{E}_{x2}{}^+|^2}{2\eta_2}$$

Since the interface is lossless, these should be equal

$$\left(\frac{\eta_1}{\eta_2}\right)|\mathrm{T}|^2 = 1 - |\bar{\Gamma}_R|^2 \tag{7.78}$$

and direct substitution of Eqs. 7.76*a* and 7.76*b* into Eq. 7.78 shows that the latter is indeed correct. *The transmitted power is equal to the incident power minus the reflected power.*

In view of Eqs. 7.14 and 7.76*a*, it is clear that a generalized reflection coefficient can be defined for situations involving uniform plane waves at normal incidence upon plane boundaries. By analogy with transmission lines, we write

$$\bar{\Gamma}(z) \equiv \frac{\mathrm{E}_x{}^-(z)}{\mathrm{E}_x{}^+(z)} = \frac{\mathrm{E}_{x0}{}^- e^{j\beta_0 z}}{\mathrm{E}_{x0}{}^+ e^{-j\beta_0 z}}$$

$$= \bar{\Gamma}(0)e^{j2\beta_0 z} \tag{7.79}$$

We then *define* a normalized impedance $Z_n(z)$ by the relation

$$Z_n(z) \equiv \frac{1 + \bar{\Gamma}(z)}{1 - \bar{\Gamma}(z)} \tag{7.80}$$

But by Eqs. 7.79 and 7.14 we can re-express $Z_n(z)$ of Eq. 7.80 in the form

$$Z_n(z) = \frac{E_{x0}{}^{+}e^{-j\beta_0 z} + E_{x0}{}^{-}e^{j\beta_0 z}}{E_{x0}{}^{+}e^{-j\beta_0 z} - E_{x0}{}^{-}e^{j\beta_0 z}} = \frac{E_x}{\eta H_y} \tag{7.81}$$

It is often convenient, but by no means necessary, to regard η as a "characteristic impedance," in which case the unnormalized impedance $Z(z)$ becomes equal to what we have previously called the wave impedance looking in the $+z$ direction:

$$Z(z) = Z_z \equiv \frac{E_x}{H_y} \text{ ohms} \tag{7.82}$$

It is on this basis, in fact, that we have made the transmission-line representations in Figs. 7.9 and 7.12.

In any case, simply on the strength of Eqs. 7.79, 7.80, and 7.81, the entire set of procedures for transmission lines becomes applicable to these uniform plane-wave problems. In particular, the Smith chart (and others) can be used to great advantage instead of dealing directly with the boundary conditions at interfaces. The next section illustrates the techniques.

7.3.3 Normal Incidence on Multiple Dielectrics

Consider the problem of a sheet of dielectric of thickness l, upon one side ($z = -l$) of which is incident normally from $z = -\infty$ an x-polarized uniform plane wave (Fig. 7.15). Our interest might be in the percentage of the incident power which is reflected for various values of l, given the frequency and the dielectric constant $\epsilon_2 > \epsilon_0$.

A straightforward attack on the problem would require consideration of the five waves shown, and then a matching of tangential electric and magnetic fields at the two interfaces $z = 0$ and $z = -l$ to determine the amplitudes of the four unknown waves. Since, however, the wave propagation *within* each medium obeys transmission-line equations, and since the boundary conditions of continuity of tangential electric and magnetic fields are exactly like the continuity conditions of voltage and current at a line termination, the problem may be reduced to that shown in Fig. 7.16.

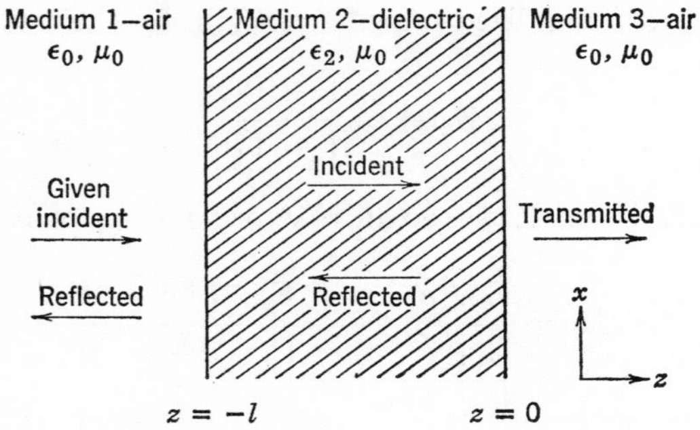

Fig. 7.15. A multiple-interface problem.

We wish to find the magnitude of the reflection coefficient on line 1, which requires that we determine the impedance Z normalized on the characteristic impedance η_0. For definiteness, let $\epsilon_2 = 4\epsilon_0$, and suppose that the frequency is such that the wave length λ_0 *in air* is 3 cm. The wave length in the dielectric medium (line 2) is then $\lambda_2 = \frac{1}{2} \times 3 = 1.5$ cm.

We see at once that, if $l = \frac{1}{2}\lambda_2 = 0.75$ cm (or any integer multiple thereof), line 2 becomes a half-wave line and $Z = \eta_0$. Then line 1 is matched, and there is no reflection at all. There will, however, be reflections for other values of l.

From the point of view of line 2, the normalized load impedance at $z = 0$ is $\eta_0/\eta_2 = \sqrt{\epsilon_2/\epsilon_0} = 2$, shown at point A in Fig. 7.17. As l is increased from zero, the Smith chart shows that Z/η_2 becomes smallest in magnitude and real at point B, when $l = \frac{1}{4}\lambda_2 = 0.375$ cm. Its smallest value is 0.5. From the point of view of line 1, the normalized

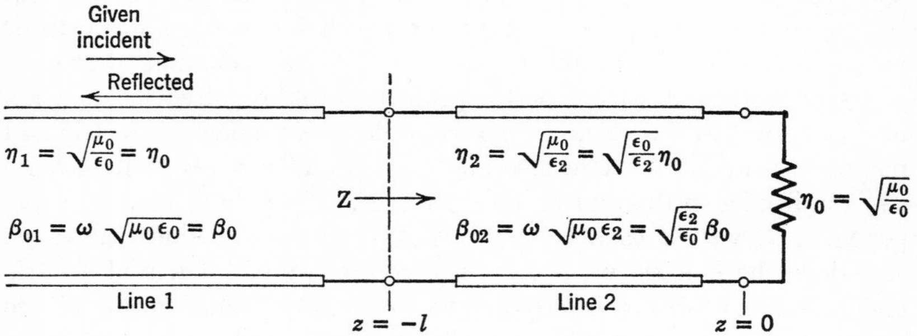

Fig. 7.16. Transmission-line equivalent of Fig. 7.15.

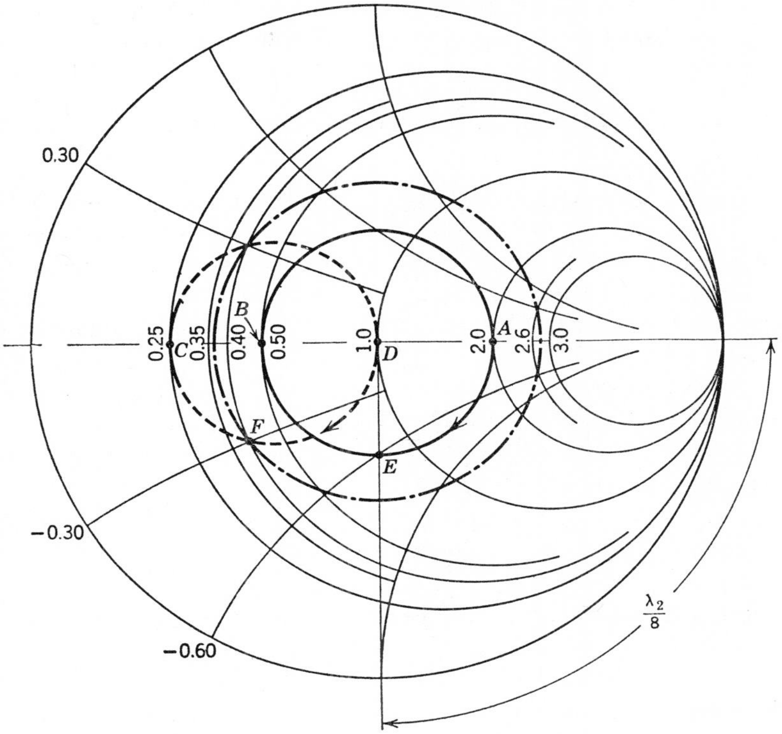

Fig. 7.17. Smith chart for the example of Figs. 7.15 and 7.16.

load impedance is $Z/\eta_0 = (Z/\eta_2)(\eta_2/\eta_0) = \frac{1}{2}(Z/\eta_2)$, which traces the dashed circle on the chart passing through the real axis at normalized resistance values 0.25 and 1.0 when $l = \frac{1}{4}\lambda_2$ and $l = 0$ (points C and D) respectively.[1] The largest reflection coefficient on line 1 thus occurs when $l = \frac{1}{4}\lambda_2$, and has the value

$$|\bar{\Gamma}|_{\max} = \frac{1 - 0.25}{1 + 0.25} = \frac{3}{5} = 0.6$$

Thus the maximum percentage reflected power is $|\bar{\Gamma}|^2_{\max} = 9/25 = 0.36$, or 36%, when $l = 0.375$ cm (or an odd multiple thereof).

For the case $l = \frac{1}{8}\lambda_2 = 0.1875$ cm, the Smith chart shows $Z/\eta_2 = 0.8 - j0.6$ at point E. Normalized on η_0, to refer to line 1, this be-

[1] That the dashed locus *must* be a circle follows from Sec. 3.5, Eq. 3.79 ff., and the reference cited there.

comes $0.4 - j0.3$ (point F), which lies on the dash-dot circle of constant $|\bar{\Gamma}|$ passing through normalized resistance values 0.36 and 2.78. Thus the standing-wave ratio on line 1 is $s = 2.78$, and $|\bar{\Gamma}| = \dfrac{s-1}{s+1} = 1.78/3.78 = 0.471$. This corresponds to $(0.471)^2$ or 22.2% reflected power.

Needless to say, since the dielectric is lossless, the power transmitted into medium 3 (Fig. 7.15) is equal to the difference between the incident and reflected powers in medium 1. Expressed as a percentage of the incident power, that transmitted in our numerical example is 100% for $l = 0$ (or $l = \lambda_2/2$), 77.8% for $l = \lambda_2/8$, and a minimum of 64% for $l = \lambda_2/4$.

Perhaps it is worth while to see the analytical treatment of the problem of Fig. 7.15 (or the equivalent in Fig. 7.16), if only to appreciate the simplification available from the previous Smith-chart solution. Thus, for medium 2 at $z = 0$, we have

$$\Gamma_{R2} = \frac{\eta_0 - \eta_2}{\eta_0 + \eta_2}$$

since Γ_{R2} is real here. At $z = -l$, however,

$$\bar{\Gamma}_2(-l) = \Gamma_{R2}e^{-j2\beta_{02}l}$$

The corresponding normalized impedance *in medium* 2 at $z = -l$ is

$$\frac{Z}{\eta_2} = \frac{1 + \bar{\Gamma}_2(-l)}{1 - \bar{\Gamma}_2(-l)} = \frac{1 + \Gamma_{R2}e^{-j2\beta_{02}l}}{1 - \Gamma_{R2}e^{-j2\beta_{02}l}}$$

But the wave impedance Z *in ohms* must look the same at $z = -l$, whether we consider ourselves to be just in medium 1 or just in medium 2. This is, in fact, a direct result of the boundary condition requiring tangential **E** and tangential **H** to be separately continuous across the interface. In the transmission-line analogy (Fig. 7.16), the continuity of Z follows likewise from the separate continuity of voltage and current. Thus the normalized load impedance seen by medium 1 at $z = -l$ is

$$\frac{Z}{\eta_0} = \frac{\eta_2}{\eta_0}\left(\frac{Z}{\eta_2}\right) = \frac{\eta_2}{\eta_0}\left(\frac{1 + \Gamma_{R2}e^{-j2\beta_{02}l}}{1 - \Gamma_{R2}e^{-j2\beta_{02}l}}\right)$$

leading to a load reflection coefficient in medium 1 at $z = -l$ of

$$\bar{\Gamma}_{R1} = \frac{(Z/\eta_0) - 1}{(Z/\eta_0) + 1} = \Gamma_{R2}\left(\frac{e^{j2\beta_{02}l} - 1}{\Gamma_{R2}{}^2 - e^{j2\beta_{02}l}}\right)$$

where a considerable amount of algebra has been omitted which the reader is urged to reproduce on a separate sheet.

Now we are interested in $|\bar{\Gamma}_{R1}|^2$, which can be found most easily from sketches showing the geometry of the complex numbers in the numerator and denominator of $\bar{\Gamma}_{R1}$. We find

$$|\bar{\Gamma}_{R1}|^2 = \frac{4\Gamma_{R2}{}^2 \sin^2 \beta_{02}l}{(1 - \Gamma_{R2}{}^2)^2 + 4\Gamma_{R2}{}^2 \sin^2 \beta_{02}l}$$

Evidently $|\bar{\Gamma}_{R1}| = 0$ if $\beta_{02}l = 0, \pi, 2\pi, \cdots$, i.e., the medium thickness l is an integer multiple of $\lambda_2/2$. Also it is clear that the whole fraction is largest (as a function of l) when $\sin^2 \beta_{02}l$ is largest, i.e., when $2\beta_{02}l = \pi, 3\pi, 5\pi, \cdots$, or $l = \lambda_2/4, 3\lambda_2/4, 5\lambda_2/4$, etc. Under this condition the (maximum) value of $|\bar{\Gamma}_{R1}|$ is

$$|\bar{\Gamma}_{R1}|_{\text{max}} = \frac{2|\Gamma_{R2}|}{1 + |\Gamma_{R2}|^2}$$

In our numerical example, where $\eta_0/\eta_2 = 2$, we have $\Gamma_{R2} = (2 - 1)/(2 + 1) = \frac{1}{3}$. Hence $|\bar{\Gamma}_{R1}|_{\text{max}} = \frac{2}{3}/(1 + \frac{1}{9}) = \frac{3}{5}$, as found previously from the Smith-chart solution.

If the polarization of the incident wave were along the y-axis, our previous work would remain essentially unchanged. In fact, the simplest procedure would be to relabel the axes so that y became x and $-x$ became y. Since the boundary conditions are not affected physically by the direction of the electric field in the x, y plane, the formal solution of the problem then remains completely unchanged.

In the sinusoidal steady state at frequency ω, the comments about polarization made in Sec. 7.1.4 and Sec. 1.3.2 mean that the most general incident wave might simply have an $E_{x0}{}^+$ *and* an $E_{y0}{}^+$ of different magnitudes and phases. Thus we would make (and later superpose) two solutions to our problem, the difference between these solutions being only a $+90°$ rotation about the $+z$-axis and a magnitude and phase difference corresponding to those of $E_{x0}{}^+$ and $E_{y0}{}^+$.

An interesting practical example of transmission-line thinking applied to plane-wave problems is that of "coated optics." If, in Fig. 7.15, medium 3 were glass rather than air—representing a lens or show window—and medium 1 were still air, one might ask for a coating (medium 2) which would match η_3 to η_0 and thereby eliminate objectionable reflections from the front surface. One solution would be a quarter-wave coating acting as a matching transformer whose characteristic wave impedance η_2 would have to satisfy the relation $\eta_2 = \sqrt{\eta_0\eta_3}$. Such materials exist for visible-light wave lengths, and are used on some optical-instrument lenses. It is important to note that the thickness must be a quarter of a wave length as measured *in the coating material itself*, not of the wave length in glass or air. Moreover, the matching behavior of the coating is exact only at the single frequency for which the film is a quarter wave thick, and when the light waves strike it at normal incidence.

Before leaving the topic of uniform plane waves at normal incidence in lossless media, we should mention that the rather complete analogy of these problems to those of lossless transmission lines applies in the time domain, for transients, as well as in the frequency domain for sinusoidal steady state. This is evident either directly from Eqs. 7.3, 7.4, and 7.8 or from a Fourier point of view applied to our results for the steady state. As long as the characteristic impedance $\eta = \sqrt{\mu/\epsilon}$ and phase velocity $v_p = 1/\sqrt{\epsilon\mu}$ are independent of frequency, the steady-state idea of independent (+) and (−) waves propagating with corresponding time delays proportional to distance, and with voltage-to-current ratios η and $-\eta$ respectively, carries over directly into the time domain. Only the question of rotating polarizations is slightly new; but since such situations can always be resolved into two separate linear polarizations, the difficulty amounts merely to solving twice as many of the same old problems.

Transmission-line analogies are present, but more subtle, in our next topics, which begin with oblique incidence upon plane boundaries and conclude with some examples of guided waves.

7.4 Oblique Incidence of a Uniform Plane Wave

7.4.1 Geometry of Oblique Incidence

We already know that a uniform plane wave propagating in any space direction is a solution to Maxwell's equations, and that it can always be decomposed into two linearly polarized components with mutually perpendicular electric (and magnetic) fields. These facts have not really helped us greatly in our consideration of normal incidence, for two reasons:

1. The direction of propagation of the incident wave could always be chosen as the z-axis, and the normal to the plane boundary would coincide with it (as would the direction of propagation of the reflected and transmitted waves).
2. The polarization of the incident wave could not only be considered linear but also its orientation needed little attention because of the

uniform electrical structure and symmetrical orientation of the boundary with respect to the z-axis.

When we come to oblique incidence, however, the questions of direction of propagation and polarization are by no means trivial physically. Therefore we must acknowledge them in our analysis.

The geometry for a uniform plane wave at oblique incidence upon a uniform plane boundary is shown in Fig. 7.18. The unit vector $\boldsymbol{n}_1$, normal to the boundary, and the propagation vector $\boldsymbol{\beta}_{0i}$ of the incident wave define what is called the *plane of incidence*. The *angle of incidence* ϑ_i is measured in this plane as the acute angle between $-\boldsymbol{\beta}_{0i}$ and $\boldsymbol{n}_1$. A unit vector $\boldsymbol{n}_2$, normal to the plane of incidence, is also parallel to the boundary plane.

Now the electric and magnetic field vectors of the incident wave must lie in planes perpendicular to its propagation vector $\boldsymbol{\beta}_{0i}$. There is only one direction in space which is parallel to these planes and also parallel to the boundary plane. This direction is defined by the unit vector $\boldsymbol{n}_2$, normal to both $\boldsymbol{n}_1$ and $\boldsymbol{\beta}_{0i}$. We shall make the $+x$-axis parallel to $\boldsymbol{n}_2$. For the $+z$-axis we shall choose the line normal to the boundary plane, i.e., that defined by $-\boldsymbol{n}_1$. The y-axis is finally determined by the line perpendicular to $\boldsymbol{n}_2$ and $\boldsymbol{n}_1$ in such a sense that it forms a right-handed coordinate system with the $+z$- and $+x$-axes. The $+y$-axis is therefore in the direction $(-\boldsymbol{n}_1) \times \boldsymbol{n}_2$, parallel to both the boundary plane and the plane of incidence.

It should now be clear that the electric and magnetic fields of the incident wave, which are mutually perpendicular and lie in a plane normal to its direction of propagation, cannot *both* also be parallel (tangential) to the boundary. Thus, the simplest polarizations we can

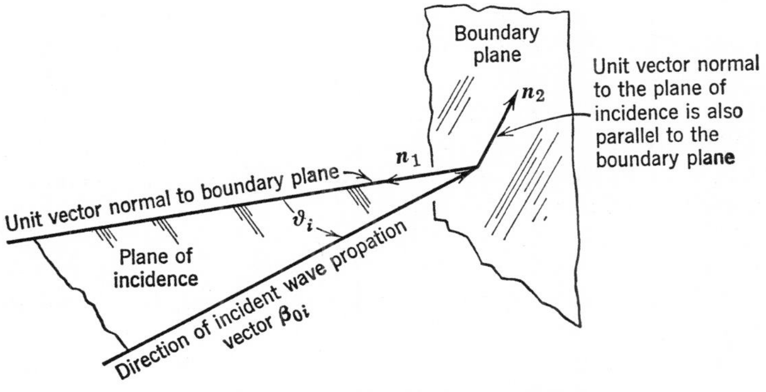

Fig. 7.18. Geometry of oblique incidence.

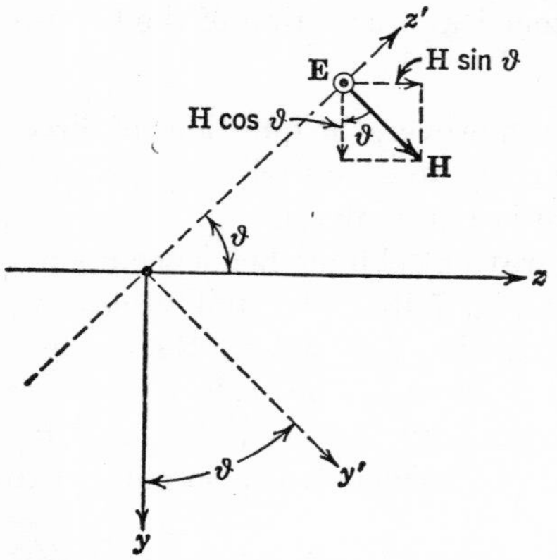

Fig. 7.19. Rotation of coordinates for a uniform plane wave.

have, from the point of view of meeting boundary conditions on tangential field components, are those for which either $\boldsymbol{E}_i$ or $\boldsymbol{H}_i$ is linearly polarized along the x-axis ($\boldsymbol{n}_2$). Then *either* $\boldsymbol{E}_i$ or $\boldsymbol{H}_i$ (but not both) becomes parallel to the boundary plane. Since $\boldsymbol{E}_i$ and $\boldsymbol{H}_i$ are always mutually perpendicular in a linearly polarized uniform plane wave, these two cases, in fact, also represent two mutually perpendicular electric-field polarizations of the incident wave. Therefore, if for every ϑ_i we treat these two cases, we shall be able to handle all possibilities through subsequent use of superposition.

Our choice of the z-axis normal to the boundary surface means that it neither coincides with the direction of propagation of the incident wave nor, in general, with that of the reflected or transmitted waves. This leaves us with the task of expressing analytically the field of a uniform plane wave in a rotated coordinate system, a problem which we have discussed previously in connection with Figs. 7.2 and 7.7.

In the present case, consider the wave propagating along the $+z'$ direction in Fig. 7.19, and take the polarization (direction of $\boldsymbol{E}$) to be linear and along the $+x$-axis, which points out of the paper. The complex vector **E** will then also be along the $+x$-axis, and the magnetic field will be along the $+y'$ axis, as shown. In the (x, y', z') coordinates, which are the familiar ones, we have

$$\text{(a)} \qquad \mathbf{E} = \boldsymbol{a}_x \mathrm{E} e^{-j\beta_0 z'}$$

$$\text{(b)} \qquad \mathbf{H} = \boldsymbol{a}_{y'} \mathrm{H} e^{-j\beta_0 z'} \qquad (7.83)$$

$$\text{(c)} \qquad \mathrm{H} = \frac{\mathrm{E}}{\eta}$$

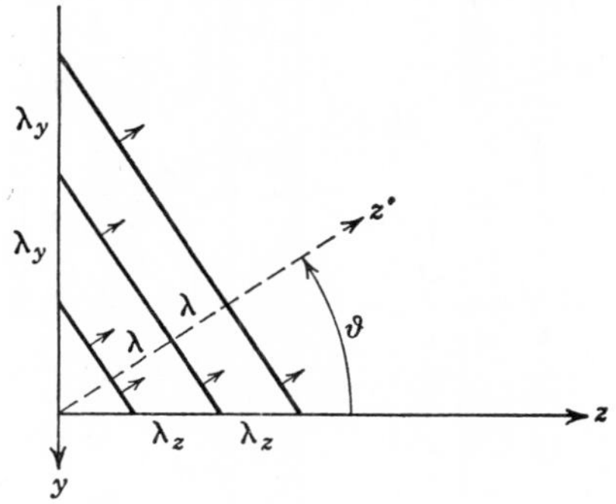

Fig. 7.20. A uniform plane wave moving along the $+z'$ direction.

For the new coordinates (x, y, z) of Fig. 7.19, the x-axis has remained unchanged, but the others have been rotated by the angle ϑ defined on the sketch. Therefore, to express unit vector $\boldsymbol{a}_{y'}$ in terms of unit vectors $\boldsymbol{a}_y$ and $\boldsymbol{a}_z$, we have from the geometry

$$\boldsymbol{a}_{y'} = \boldsymbol{a}_y \cos \vartheta + \boldsymbol{a}_z \sin \vartheta \tag{7.84}$$

Comparing Fig. 7.19 with Fig. 7.2 in regard to z', we note that in Fig. 7.19

$$l_{z'} = 0 \qquad m_{z'} = -\sin \vartheta \qquad n_{z'} = \cos \vartheta \tag{7.85a}$$

so that from Eq. 7.19 there follows the relation

$$z' = z \cos \vartheta - y \sin \vartheta \tag{7.85b}$$

Consequently Eqs. 7.83 become

$$\text{(a)} \qquad \mathbf{E} = \boldsymbol{a}_x \mathrm{E} e^{j\beta_0 y \sin \vartheta} e^{-j\beta_0 z \cos \vartheta}$$

$$\text{(b)} \qquad \mathbf{H} = (\boldsymbol{a}_y \mathrm{H} \cos \vartheta + \boldsymbol{a}_z \mathrm{H} \sin \vartheta) e^{j\beta_0 y \sin \vartheta} e^{-j\beta_0 z \cos \vartheta} \tag{7.86}$$

$$\text{(c)} \qquad \mathrm{H} = \frac{\mathrm{E}}{\eta}$$

which should be compared with Eqs. 7.26 and 7.32a. Observe particularly the relationship between the vector components of **H** in Eq. 7.86b and the corresponding geometric representation of them in Fig. 7.19. Obviously this part of **H** could be written *by inspection* from the figure. The same is true of the exponential factors in Eqs. 7.86a and 7.86b. A special case of Fig. 7.3b for the present situation is shown in Fig. 7.20. The corresponding form of Eqs. 7.33 through 7.36 may be written in view of either Eqs. 7.85a or, preferably, of the geometry of Fig. 7.20 directly:

$$\text{(a)} \qquad \lambda_y \equiv \frac{\lambda}{\sin \vartheta} = \frac{2\pi}{\beta_0 \sin \vartheta} \equiv \frac{2\pi}{|\beta_y|}$$
$$\text{(b)} \qquad \lambda_z \equiv \frac{\lambda}{\cos \vartheta} = \frac{2\pi}{\beta_0 \cos \vartheta} \equiv \frac{2\pi}{|\beta_z|} \qquad (7.87)$$

$$\text{(a)} \qquad |\beta_y| = \beta_0 \sin \vartheta$$
$$\text{(b)} \qquad |\beta_z| = \beta_0 \cos \vartheta \qquad (7.88)$$

$$\text{(a)} \qquad v_y \equiv \frac{\omega}{|\beta_y|} = \frac{\omega}{\beta_0 \sin \vartheta} = \frac{v}{\sin \vartheta}$$
$$\text{(b)} \qquad v_z \equiv \frac{\omega}{|\beta_z|} = \frac{\omega}{\beta_0 \sin \vartheta} = \frac{v}{\cos \vartheta} \qquad (7.89)$$

and because $\sin^2\vartheta + \cos^2\vartheta = 1$,

$$\text{(a)} \qquad \beta_z^{\,2} + \beta_y^{\,2} = \beta_0^{\,2}$$
$$\text{(b)} \qquad \frac{1}{\lambda_z^{\,2}} + \frac{1}{\lambda_y^{\,2}} = \frac{1}{\lambda^2} \qquad (7.90)$$
$$\text{(c)} \qquad \frac{1}{v_z^{\,2}} + \frac{1}{v_y^{\,2}} = \frac{1}{v^2}$$

It should be emphasized in connection with Fig. 7.20 that when the actual wave moves in the arrow directions as time goes on, the phase fronts advance along $+z'$, $+z$, and $-y$. This accounts for the different signs of the exponents containing z and y in Eqs. 7.86.

If the plane wave had the alternate linear polarization (i.e., **H** parallel to the boundary) but still propagated in the $+z'$ direction, it could have **E** along $+y'$ and **H** along $-x$:

$$\text{(a)} \qquad \mathbf{E} = \boldsymbol{a}_{y'}\mathrm{E}e^{-j\beta_0 z'}$$
$$\text{(b)} \qquad \mathbf{H} = -\boldsymbol{a}_x\mathrm{H}e^{-j\beta_0 z'} \qquad (7.91)$$
$$\text{(c)} \qquad \mathrm{H} = \frac{\mathrm{E}}{\eta}$$

In the (x, y, z) coordinates, Eqs. 7.91 become

$$\text{(a)} \qquad \mathbf{E} = (\boldsymbol{a}_y\mathrm{E}\cos\vartheta + \boldsymbol{a}_z\mathrm{E}\sin\vartheta)e^{j|\beta_y|y}e^{-j\beta_z z}$$
$$\text{(b)} \qquad \mathbf{H} = -\boldsymbol{a}_x\mathrm{H}e^{j|\beta_y|y}e^{-j\beta_z z} \qquad (7.92)$$
$$\text{(c)} \qquad \mathrm{H} = \frac{\mathrm{E}}{\eta}$$

It is extremely desirable to learn to write equations like Eqs. 7.86 and 7.92 directly from pictures of the **E** and **H** vectors (in the manner of Fig. 7.19) and of the phase fronts (in the manner of Fig. 7.20). Such pictures are surprisingly helpful in resolving ambiguities of algebraic sign connected with the field components and the propagation exponents along the coordinate axes. *A thorough mastery of these matters now will remove the major difficulties from the forthcoming discussions.*

We are now in a position to examine reflection and refraction of uniform plane waves at oblique incidence upon the interface between lossless media.

7.4.2 Oblique Incidence upon a Perfect Conductor

First we may consider the perfectly conducting metal mirror at $z = 0$ in Fig. 7.21, upon which is incident an x-polarized wave from a direc-

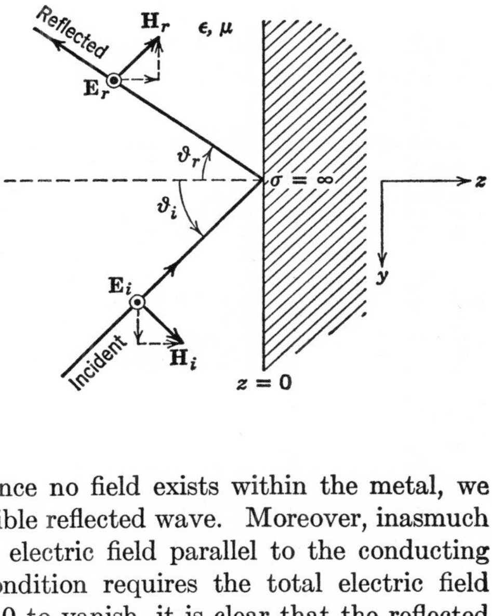

Fig. 7.21. Oblique incidence upon a perfect conductor with polarization parallel to the boundary.

tion ϑ_i with the normal. Since no field exists within the metal, we have only to deal with a possible reflected wave. Moreover, inasmuch as the incident wave has its electric field parallel to the conducting plane, and the boundary condition requires the total electric field parallel to that plane at $z = 0$ to vanish, it is clear that the reflected wave will have the same polarization as the incident wave. Hence the direction of propagation of the reflected wave will lie in the plane of incidence, as shown in Fig. 7.21, at some angle ϑ_r with respect to the boundary normal.

The incident wave is given as

$$\text{(a)} \qquad \mathbf{E}_i = \boldsymbol{a}_x \mathrm{E}_i e^{j\beta_0 y \sin \vartheta_i} e^{-j\beta_0 z \cos \vartheta_i}$$

$$\text{(b)} \qquad \mathbf{H}_i = \left(\boldsymbol{a}_y \frac{\mathrm{E}_i}{\eta} \cos \vartheta_i + \boldsymbol{a}_z \frac{\mathrm{E}_i}{\eta} \sin \vartheta_i\right) e^{j\beta_0 y \sin \vartheta_i} e^{-j\beta_0 z \cos \vartheta_i} \tag{7.93}$$

and the reflected wave is of the form

$$\text{(a)} \qquad \mathbf{E}_r = \boldsymbol{a}_x \mathrm{E}_r e^{j\beta_0 y \sin \vartheta_r} e^{j\beta_0 z \cos \vartheta_r}$$

$$\text{(b)} \qquad \mathbf{H}_r = \left(-\boldsymbol{a}_y \frac{\mathrm{E}_r}{\eta} \cos \vartheta_r + \boldsymbol{a}_z \frac{\mathrm{E}_r}{\eta} \sin \vartheta_r\right) e^{j\beta_0 y \sin \vartheta_r} e^{j\beta_0 z \cos \vartheta_r} \tag{7.94}$$

At $z = 0$, for *all* y (and x), we require $\mathbf{E}_i + \mathbf{E}_r = 0$; i.e.,

$$\mathrm{E}_i e^{j\beta_0 y \sin \vartheta_i} = -\mathrm{E}_r e^{j\beta_0 y \sin \vartheta_r} \tag{7.95}$$

so

$$\text{(a)} \qquad \vartheta_r = \vartheta_i (\equiv \vartheta)$$

$$\text{(b)} \qquad -\mathrm{E}_r = \mathrm{E}_i (\equiv \mathrm{E}) \tag{7.96}$$

Equation 7.96*a* is the familiar optical law that the angles of incidence and reflection are equal. From Eq. 7.96*b* we see that all the incident power is reflected.

The total field for $z \leq 0$ is given by the sum of the incident and reflected waves under conditions of Eq. 7.96.

$$\text{(a)} \qquad \mathbf{E} = -\boldsymbol{a}_x 2j\mathrm{E} e^{j\beta_0 y \sin \vartheta} \sin (\beta_0 z \cos \vartheta)$$

$$\text{(b)} \qquad \mathbf{H} = \frac{2\mathrm{E}}{\eta} [\boldsymbol{a}_y \cos \vartheta \cos (\beta_0 z \cos \vartheta) - \boldsymbol{a}_z j \sin \vartheta \sin (\beta_0 z \cos \vartheta)] e^{j\beta_0 y \sin \vartheta} \tag{7.97}$$

We observe that E_x and H_y are 90° out of time phase. This is to be expected on the basis that the z component of the complex Poynting vector is $\frac{1}{2}\mathrm{E}_x \mathrm{H}_y^*$, which cannot have a real part because no average power is absorbed by the mirror. It is also noteworthy that $\mathrm{E}_x = 0$ whenever $\beta_0 z \cos \vartheta = -m\pi$, $m = 0, 1, 2, \cdots$, or

$$z = \frac{-m\lambda}{2 \cos \vartheta} = -\frac{m\lambda_z}{2} \qquad m = 0, 1, 2, \cdots$$

The electric field has nodal planes parallel to the mirror at multiples of $\frac{1}{2}\lambda_z$ from it. Observed along z, all the field components have standing-wave character; but E_x and H_z have nodes at the same planes, while

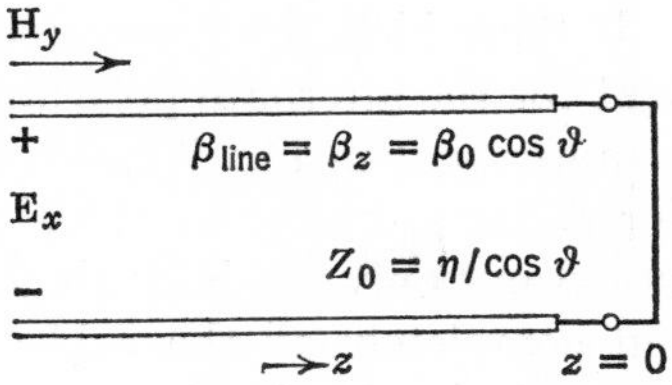

Fig. 7.22. Transmission line analogy for the z dependence and the x and y field components in Fig. 7.21.

the nodes of H_y are displaced $\frac{1}{4}\lambda_z$ from those of E_x and H_z. The y variation of all the field components, on the other hand, has traveling-wave character with effective wave length λ_y and propagation in the $-y$ direction. This fact checks with the y component of the complex Poynting vector $(-\frac{1}{2}E_xH_z^*)$, which is real and negative, indicating average power flowing in the $-y$ direction.

It is therefore convenient to think of the whole field as a pure standing wave extending along $-z$, which, however, also "slides" bodily along $-y$. Indeed, we note that the field components parallel to the mirror are

$$\text{(a)} \qquad E_x = \left[-2jE \sin\left(\frac{2\pi z}{\lambda_z}\right)\right] e^{j(2\pi y/\lambda_y)}$$

$$\text{(b)} \qquad H_y = \left[\frac{2E}{(\eta/\cos\vartheta)} \cos\left(\frac{2\pi z}{\lambda_z}\right)\right] e^{j(2\pi y/\lambda_y)} \tag{7.98}$$

which, examined along z at fixed y, look exactly like the voltage and current standing waves on the transmission line of Fig. 7.22. Of course, we see from Eqs. 7.93 and 7.94 that this line shows only *some* of the features of the actual problem inasmuch as it omits all information relating to the traveling-wave nature of the y dependence, and to the corresponding magnetic field component H_z. Nevertheless, such partial equivalence is often useful. It is however most important to remember that both the characteristic impedance of the line and its phase constant involve the angle of incidence. It is not hard to see why this is so, if we look back at just the incident wave in Fig. 7.21. Whereas $(\| \mathbf{E}_i \| / \| \mathbf{H}_i \|) = \eta$,[1] our transmission line deals only with $|E_{xi}|/|H_{yi}|$, i.e., only with the $+z$-directed wave impedance Z_z. For the case at hand, $|E_{xi}| = \| \mathbf{E}_i \|$, but $|H_{yi}| = \| \mathbf{H}_i \| \cos \vartheta$. Hence the appropriate line impedance is $\eta/\cos\vartheta$. Similarly, the appearance of $\beta_z = \beta_0 \cos\vartheta$ is evident from Fig. 7.20, where the wave length observed along z is seen to be greater than λ by the factor $1/\cos\vartheta$.

[1] $\| \mathbf{A} \| \equiv \sqrt{\mathbf{A}\cdot\mathbf{A}^*} = \sqrt{|A_x|^2 + |A_y|^2 + |A_z|^2} = \sqrt{|\mathcal{A}_r|^2 + |\mathcal{A}_i|^2}$. Hence for a *linearly polarized* vector $\mathcal{A} = \text{Re}\,(\mathbf{A}e^{j\omega t})$, $\| \mathbf{A} \| = |\mathcal{A}|_{\max}$.

If these ideas are understood, it will be clear that a revision of Fig. 7.21 for the alternate polarization in which $\mathbf{H}_i$ is parallel to the mirror requires that in Fig. 7.22 we exchange E_y for E_x; $-H_x$ for H_y; $\eta \cos \vartheta$ for Z_0 instead of $\eta/\cos \vartheta$; and $\beta_{\text{line}} = \beta_0 \cos \vartheta$, as before. The reader is urged not only to check these relationships carefully but also to write out completely all the field components and boundary conditions for this polarization in the manner of Eqs. 7.93 through 7.98.

It is significant to realize in this connection that the familiar law of reflection 7.96*a* comes directly from the form of the exponential factors in Eqs. 7.93 and 7.94 and, therefore, is not influenced by the polarization. Indeed, it is really quite obvious from Fig. 7.20 that, since the boundary condition on the fields has to be satisfied for *all* values of y, a *necessary* (but not sufficient) condition is that all the waves concerned in the problem have the same phase constant (or wave length or phase velocity) *when measured along the boundary* (in the manner of Eqs. 7.87, 7.88, and 7.89). Only in this way can the various waves "keep in step" along the interface, so that the remainder of the boundary conditions can be met at more than just a single point on it.

7.4.3 Oblique Incidence upon an Interface between Lossless Dielectrics

7.4.3.1 Polarization Parallel to the Boundary. The next example for consideration involves incidence of a uniform plane wave upon the interface between lossless dielectrics (Fig. 7.23). The incident wave is x-polarized (parallel to the boundary) as shown, and is given analytically by Eq. 7.93. Again, it is clear that the reflected and transmitted (refracted) waves are also x-polarized, so that all propagation directions lie in the plane of incidence, as indicated by the figure.

If we are to have any hope of establishing the necessary continuity of tangential **E** and **H** across the boundary plane $z = 0$ (for *all* y), the phase velocities of the three waves *as measured along the y axis* must be the same in magnitude *and* sense. Thus, from the figure [which, for convenience, shows λ rather than $v = (\omega\lambda)/(2\pi)$],

$$\text{(a)} \qquad \frac{v_1}{\sin \vartheta_i} = \frac{v_1}{\sin \vartheta_r}$$

$$\text{(b)} \qquad \frac{v_1}{\sin \vartheta_i} = \frac{v_2}{\sin \vartheta_2} \tag{7.99}$$

or

$$\text{(a)} \qquad \vartheta_r = \vartheta_i \equiv \vartheta_1 \tag{7.100}$$

$$\text{(b)} \qquad \sin \vartheta_2 = \left(\frac{v_2}{v_1}\right) \sin \vartheta_1 = \sqrt{\frac{\epsilon_1 \mu_1}{\epsilon_2 \mu_2}} \sin \vartheta_1$$

Equation 7.100*a* is again the law of reflection and Eq. 7.100*b* is the equally well-known optical result called Snell's law of refraction. They arise here from that part of the boundary conditions which merely de-

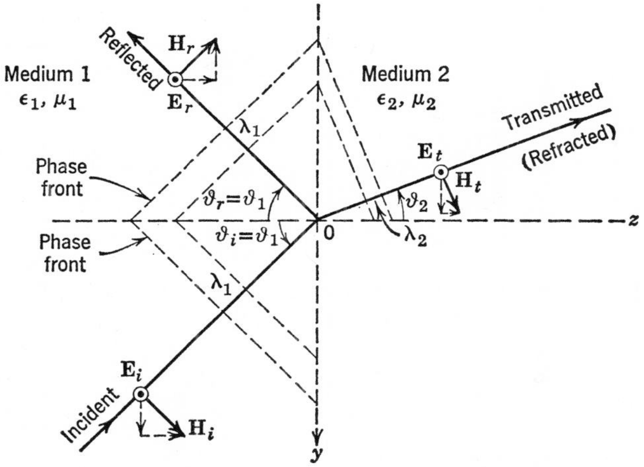

Fig. 7.23. Oblique incidence on an interface between lossless dielectrics with polarization parallel to interface.

mands that the three waves be "in step" along the interface; thus these laws do not depend upon the polarization.

The remainder of the required conditions, however, will involve the specific continuity of E_x and H_y across the boundary $z = 0$; but, since Eq. 7.100 already guarantees the common phase velocity of the three waves along y, it will be sufficient to apply the field continuity condition at any single point. We choose $y = 0$; so

$$\text{(a)} \qquad E_{xi} + E_{xr} = E_{xt} \tag{7.101}$$

$$\text{(b)} \qquad H_{yi} + H_{yr} = H_{yt}$$

But from Fig. 7.23, and the fact that the complete electric and magnetic fields of a uniform traveling plane wave have the ratio η, Eqs. 7.101 may be recast to read

$$\left.\begin{aligned} &\text{(a)} \quad E_{xi} + E_{xr} = E_{xt} \\ &\text{(b)} \quad \frac{E_{xi} - E_{xr}}{(\eta_1/\cos\vartheta_1)} = \frac{E_{xt}}{(\eta_2/\cos\vartheta_2)} \end{aligned}\right\} \quad \text{E parallel to boundary} \tag{7.102}$$

Eliminating first E_{xt} and then E_{xr}, we find from Eq. 7.102

$$\begin{aligned} &\text{(a)} \quad \frac{E_{xr}}{E_{xi}} \equiv \bar{\Gamma}_R = \frac{Z_2 - Z_1}{Z_2 + Z_1} \\ &\text{(b)} \quad \frac{E_{xt}}{E_{xi}} \equiv T = \frac{2Z_2}{Z_2 + Z_1} \end{aligned} \tag{7.103}$$

where

$$\left.\begin{aligned} &\text{(a)} \quad Z_1 = \frac{\eta_1}{\cos\vartheta_1} = \frac{E_{xi}}{H_{yi}} = -\frac{E_{xr}}{H_{yr}} \\ &\text{(b)} \quad Z_2 = \frac{\eta_2}{\cos\vartheta_2} = \frac{E_{xt}}{H_{yt}} \end{aligned}\right\} \quad \begin{gathered}\text{E parallel} \\ \text{to boundary}\end{gathered} \tag{7.104}$$

The similarity of Eqs. 7.103 to those of a transmission-line junction is evident. The fact that the characteristic impedances Z_1 and Z_2 are wave impedances and, therefore, involve the angles of incidence and refraction (Eq. 7.104) is very important. The reason for their appearance has been discussed in connection with Fig. 7.22. If we wish to draw an analogous transmission-line system to represent the z dependence of all the waves, as we did in Fig. 7.22, the propagation constants also must contain these angles, because they will refer to phases only along the z-axis. In Fig. 7.24, we show the line system for this case. As in Fig. 7.22, it is vital to understand that Fig. 7.24 contains no information about either H_z or the y dependence of any of the fields in the actual problem.

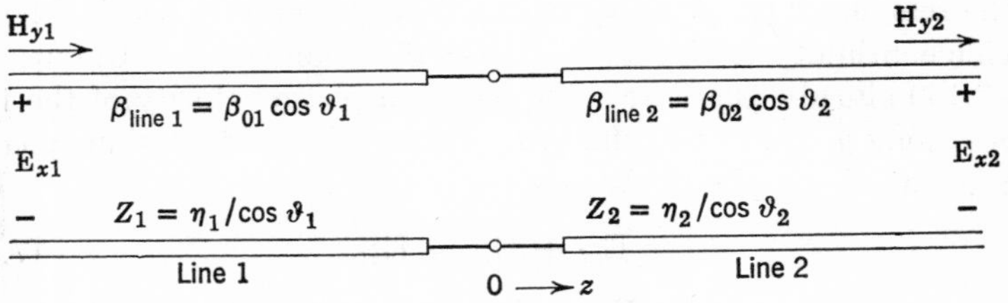

Fig. 7.24. Transmission line analogy for the z dependence and the x and y field components in Fig. 7.23.

It is equally significant that Snell's law (Eq. 7.100) relates ϑ_1 and ϑ_2; thus a knowledge of *either angle* is sufficient to determine the parameters of *both* lines in Fig. 7.24, assuming, of course, that we know ϵ and μ in both media of the physical problem.

In spite of the fact that some information about the actual physical problem (Fig. 7.23) is absent from the transmission-line system of Fig. 7.24, the latter contains the most involved aspects of the situation, and it does so in a form for which the Smith chart is applicable. With just a single interface, the advantages of the transmission-line representation are hardly evident. Equations 7.103 are really all we need in that case. If there are several interfaces, however, the problem of determining $\bar{\Gamma}$ at the first one involves considerations like those we encountered in Fig. 7.15, for which the Smith chart is a great help. In fact, the only added complication in the oblique case is the need to apply Snell's law at each interface to find the directions of the refracted waves in each medium. If this is done at the start of the problem, the parameters of the transmission line for each medium are determined in the manner of Fig. 7.24. From here on, the rest of the work follows exactly the usual transmission-line pattern.

There is one point about the flow of power in the case of oblique incidence which deserves special comment. In Fig. 7.23, all the waves have components of the Poynting vector along $-y$. There is no reason why these components of the power flow should obey the relation that the "incident" power minus that "reflected" equals that "transmitted," for they are all in the same direction—*parallel* to the boundary. On the other hand, all the waves also have components of the Poynting vector along $+z$. It is clear in this case that none of the power flowing normal to the boundary in medium 2 can have originated anywhere but as $+z$-directed power in medium 1. Hence, for these normal components, we *would* expect the aforementioned power relation. In particular, the $+z$-directed incident power is

$$\frac{1}{2}\,\mathrm{E}_{xi}\mathrm{H}_{yi}{}^{*} = \frac{|\mathrm{E}_{xi}|^2 \cos\vartheta_1}{2\eta_1} = \frac{|\mathrm{E}_{xi}|^2}{2Z_1}$$

Similarly, the $+z$-directed reflected power is

$$-\frac{1}{2}\,\mathrm{E}_{xr}\mathrm{H}_{yr}{}^{*} = \frac{-|\mathrm{E}_{xr}|^2}{2Z_1}$$

and that transmitted is

$$\frac{1}{2}\,\mathrm{E}_{xt}\mathrm{H}_{yt}{}^{*} = \frac{|\mathrm{E}_{xt}|^2}{2Z_1}$$

The "conservation" relation for $+z$-directed power therefore reads

$$|E_{xi}|^2 - |E_{xr}|^2 = \frac{Z_1}{Z_2}|E_{xt}|^2$$

or, by definition of $\bar{\Gamma}_R$ and T,

$$1 - |\bar{\Gamma}_R|^2 = \frac{Z_1}{Z_2}|T|^2 \qquad \textbf{E} \text{ parallel to boundary} \qquad (7.105)$$

This is also the conclusion we would reach from Fig. 7.24. Substitution of Eqs. 7.103 into 7.105 shows that the latter is indeed true.

It is also true that the ratio of *total* power per unit area in the reflected wave $[(|E_{xr}|^2)/(2\eta_1)]$ to that in the incident wave $[(|E_{xi}|^2)/(2\eta_1)]$ is just $|\bar{\Gamma}_R|^2$; but the ratio of the *total* power per unit area in the transmitted wave $[(|E_{xt}|^2)/(2\eta_2)]$ to that in the incident wave is $(\eta_2/\eta_1)|T|^2$, which by Eq. 7.105 is obviously *not* equal to $1 - |\bar{\Gamma}_R|^2$. The explanation for this result is the one suggested above: While the *normal* components of the power flow must obey the familiar relation between incident, reflected, and transmitted powers, the *parallel* components need not—and, in general, do not—behave similarly. Hence the *total* power-flow-per-unit-area vectors need not—and, in general, do not—obey the familiar "conservation" relation!

7.4.3.2 Polarization in the Plane of Incidence. The alternate polarization for the problem of Fig. 7.23 places **H** parallel to the boundary and **E** in the plane of incidence. As mentioned previously, Eqs.

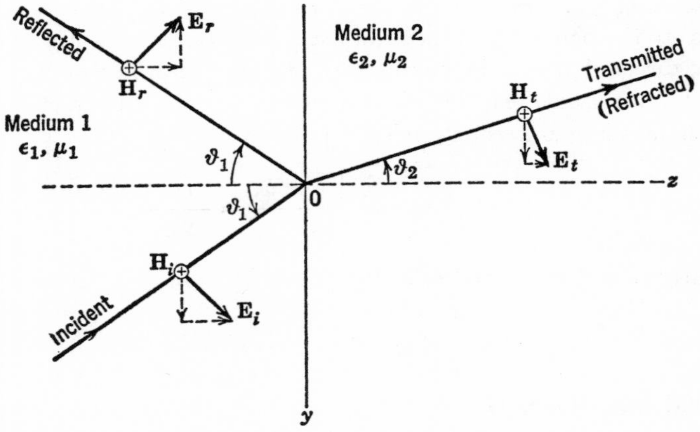

Fig. 7.25a. Oblique incidence on interface between lossless dielectrics with polarization in the plane of incidence.

7.99 and 7.100, relating the propagation constants along y and z, are unchanged. Therefore the directions of propagation of all the waves remain unaltered. On this basis the situation is as shown in Fig. 7.25a.

At the point ($z = 0, y = 0$), the remaining boundary conditions require

$$\left.\begin{array}{ll}\text{(a)} & E_{yi} + E_{yr} = E_{yt} \\ \text{(b)} & H_{xi} + H_{xr} = H_{xt}\end{array}\right\} \quad \textbf{H} \text{ parallel to boundary} \qquad (7.106)$$

In this case, however,

$$\left.\begin{array}{l} H_{xi} = -\left(\dfrac{E_{yi}}{\cos\vartheta_1}\right)\dfrac{1}{\eta_1} \\ H_{xr} = +\left(\dfrac{E_{yr}}{\cos\vartheta_1}\right)\dfrac{1}{\eta_1} \\ H_{xt} = -\left(\dfrac{E_{yt}}{\cos\vartheta_2}\right)\dfrac{1}{\eta_2}\end{array}\right\} \quad \textbf{H} \text{ parallel to boundary}$$

So Eqs. 7.106 become

$$\left.\begin{array}{ll}\text{(a)} & E_{yi} + E_{yr} = E_{yt} \\ \text{(b)} & \dfrac{E_{yi} - E_{yr}}{\eta_1 \cos\vartheta_1} = \dfrac{E_{yt}}{\eta_2 \cos\vartheta_2}\end{array}\right\} \quad \textbf{H} \text{ parallel to boundary} \qquad (7.107)$$

or

$$\begin{array}{ll}\text{(a)} & \dfrac{E_{yr}}{E_{yi}} \equiv \bar{\Gamma}_R' = \dfrac{Z_2' - Z_1'}{Z_2' + Z_1'} \\ \text{(b)} & \dfrac{E_{yt}}{E_{yi}} \equiv T' = \dfrac{2Z_2'}{Z_2' + Z_1'}\end{array} \qquad (7.108)$$

where

$$\left.\begin{array}{ll}\text{(a)} & Z_1' \equiv \eta_1 \cos\vartheta_1 = \dfrac{-E_{yi}}{H_{xi}} = \dfrac{E_{yr}}{H_{xr}} \\ \text{(b)} & Z_2' \equiv \eta_2 \cos\vartheta_1 = \dfrac{-E_{yt}}{H_{xt}}\end{array}\right\} \quad \textbf{H} \text{ parallel to boundary} \qquad (7.109)$$

Consequently, this time the propagation angles enter the impedances Z' differently from the way they entered the previous impedances Z. Compare Eqs. 7.108 and 7.103, on the one hand, and Eqs. 7.109 and 7.104 on the other. In other words: For the same media and angle of incidence, the angles of reflection and refraction are the same for both polarizations; so also are the propagation constants along y and z; *but the percentage power reflected (and transmitted) may be radically different in the two cases.*

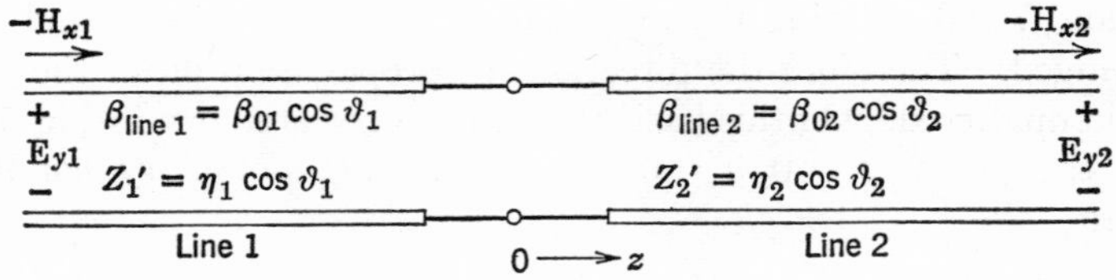

Fig. 7.25b. Transmission line analogy for the z dependence and the x and y field components in Fig. 7.25a.

Unlike any of the cases of normal incidence we have treated, or even of oblique incidence on a perfect conductor, a change of the polarization of the wave incident obliquely on an interface between two dielectrics does much more than merely produce a trivial corresponding change in the polarization of the reflected and transmitted waves. These matters are summarized by comparing the transmission-line analogy of Fig. 7.25a (which we show in Fig. 7.25b) with that of Fig. 7.23 (shown in Fig. 7.24).

An interesting example of the different effects produced by the two polarizations arises if we ask for what angle of incidence the reflected wave will vanish: i.e., all the normally directed incident power will be transmitted.

Taking first Fig. 7.24 (**E** parallel to the boundary), we note that the required condition is a "match" at $z = 0$:

$$Z_1 = Z_2$$

or

$$\eta_2 \cos \vartheta_{1p} = \eta_1 \cos \vartheta_{2p} \qquad \text{E parallel to boundary} \qquad (7.110)$$

where ϑ_{1p} and ϑ_{2p} are the particular angles being sought. Squaring Eq. 7.110 yields

$$\eta_2^2(1 - \sin^2 \vartheta_{1p}) = \eta_1^2(1 - \sin^2 \vartheta_{2p})$$

which, in the light of Snell's law (Eq. 7.100), becomes

$$\sin^2 \vartheta_{1p} = \frac{\eta_1^2 - \eta_2^2}{\eta_1^2[(v_2/v_1)]^2 - \eta_2^2}$$

$$= \frac{(\eta_1/\eta_2)^2 - 1}{[(\eta_1 v_2)/(\eta_2 v_1)]^2 - 1}$$

or

$$\sin^2 \vartheta_{1p} = \frac{(\mu_1 \epsilon_2/\mu_2 \epsilon_1) - 1}{(\mu_1/\mu_2)^2 - 1} \qquad \text{E parallel to boundary} \qquad (7.111a)$$

The angle ϑ_{1p} is sometimes called *Brewster's angle.* As shown by Eq. 7.111*a*, a real solution for ϑ_{1p} does not always exist for this polarization. One important situation for which there is *no* real solution to Eq. 7.111*a* is that in which $\mu_1 = \mu_2$ but $\epsilon_1 \neq \epsilon_2$. This is the very common case of two nonmagnetic dielectrics, say, air and glass. On the other hand, for the very uncommon situation $\epsilon_1 = \epsilon_2$, $\mu_1 \neq \mu_2$, Eq. 7.111*a* becomes

$$\sin^2 \vartheta_{1p} = \frac{1}{(\mu_1/\mu_2) + 1}$$

or

$$\tan \vartheta_{1p} = \sqrt{\frac{\mu_2}{\mu_1}} \qquad \textbf{E} \text{ parallel to boundary and } \epsilon_1 = \epsilon_2 \qquad (7.111b)$$

which *always* has a real solution $\vartheta_{1p} < \pi/2$!

In the second polarization, when **H** is parallel to the boundary, Fig. 7.25*b* shows that a "match" occurs for

$$Z_1' = Z_2'$$

or

$$\eta_1 \cos \vartheta_{1p}' = \eta_2 \cos \vartheta_{2p}' \qquad \textbf{H} \text{ parallel to boundary} \qquad (7.112)$$

Squaring Eq. 7.112 and using Snell's law as before yield

$$\sin^2 \vartheta_{1p}' = \frac{(\eta_2/\eta_1)^2 - 1}{(\eta_2 v_2/\eta_1 v_1)^2 - 1}$$

or

$$\sin^2 \vartheta_{1p}' = \frac{(\mu_2 \epsilon_1/\mu_1 \epsilon_2) - 1}{(\epsilon_1/\epsilon_2)^2 - 1} \qquad \textbf{H} \text{ parallel to boundary} \qquad (7.113a)$$

Again, a real Brewster's angle ϑ_{1p}' for this polarization does not always exist; but, as duality ideas suggest, there is now *no* solution in the uncommon case when $\epsilon_1 = \epsilon_2$, $\mu_1 \neq \mu_2$, while there is *always* a solution in the very common case $\mu_1 = \mu_2$, $\epsilon_1 \neq \epsilon_2$. In fact, under this latter condition, Eq. 7.113*a* becomes

$$\sin^2 \vartheta_{1p}' = \frac{1}{(\epsilon_1/\epsilon_2) + 1}$$

or

$$\tan \vartheta_{1p}' = \sqrt{\frac{\epsilon_2}{\epsilon_1}} \qquad \textbf{H} \text{ parallel to boundary and } \mu_1 = \mu_2 \qquad (7.113b)$$

which is the dual of Eq. 7.111*b*.

Advantage is taken of the circumstances described above to produce polarized light. As we have previously pointed out, ordinary light is not linearly polarized, but has instead (on a statistical basis) about equal amounts of each fundamental linear polarization. A slab of glass, oriented with respect to an incident beam of ordinary light so that the angle of incidence is ϑ_{1p}', will not reflect any of the light polarized in the plane of incidence. The reflected wave will then be polarized entirely parallel to the boundary; but its intensity may be rather low because the relative dielectric constant of glass is not very large ($\approx$4–5). Use of many such slabs stacked together overcomes this difficulty, because at each interface a fixed small fraction of the parallel-polarized light is reflected, thereby removing it from the transmitted beam. Finally, the *transmitted* beam contains the light originally polarized in the plane of incidence, with very little contamination by the parallel-polarized part. The latter appears almost entirely in the net reflected wave; thus the pile of glass slabs effectively separates the unpolarized incident beam into two separate polarized ones. It is on account of this application that Brewster's angle is more commonly known as the *polarizing angle.*

7.4.3.3 Critical Reflection.* We have seen that Snell's law (Eq. 7.100*b*) does not depend upon the polarization. According to this law, whenever $v_2 > v_1$, there exist large angles of incidence ϑ_1 for which $\sin \vartheta_2$ would have to exceed unity. In other words, if the incident wave is in a medium of slower light velocity (larger index of refraction) than that in which the transmitted wave propagates, Snell's law demands

$$\sin \vartheta_2 \geq 1$$

whenever

$$\sin \vartheta_1 \geq \frac{v_1}{v_2} = \sqrt{\frac{\epsilon_2\mu_2}{\epsilon_1\mu_1}} \tag{7.114}$$

or

$$\vartheta_1 \geq \vartheta_c \equiv \sin^{-1}\left(\frac{v_1}{v_2}\right) \tag{7.115}$$

The angle ϑ_c is called the *critical angle.*

Let us examine first the situation of Fig. 7.23 (**E** parallel to the boundary) as $\vartheta_1 \rightarrow \vartheta_c$ when $v_2 > v_1$. We note from the condition $v_2 > v_1$ that in Snell's law $\vartheta_2 > \vartheta_1$ for $\vartheta_1 < \vartheta_c$; hence the direction of propagation of the transmitted wave in this case actually makes a larger angle with the $+z$-axis than does that of the incident wave. The refraction bends the wave *away* from the normal. When $\vartheta_1 = \vartheta_c$, $\sin \vartheta_2 = 1$ and $\vartheta_2 = 90°$. The transmitted wave is then propagating entirely parallel to the boundary, so we would expect that $H_{yt} = 0$. Since $\cos 90° = 0$, Fig. 7.24 shows that $Z_2 = \infty$ and that the incident and reflected electric fields must be equal (see also Eq. 7.103*a*). We have a condition of "total reflection" in which the power reflected back into medium 1 equals that incident in medium 1. The real power transmitted in a normal direction into medium 2 (i.e., along $+z$) is zero, although there is a field in medium 2 representing a uniform plane wave traveling along the $-y$-axis. As such, this field does not vary with z at all.

What must happen physically if we now make $\vartheta_1 > \vartheta_c$?

First, there must be *some* field in medium 2; for, if not, medium 2 would suddenly be acting just like a perfect conductor, merely because we happened to choose some special angles of incidence in medium 1. This is unreasonable. Besides, we should be reminded of the experimental fact in optics that, if a prism is causing an internal critical reflection, and another similar one is pressed closely against one of the sides at which the critical reflection is occurring, most of the light goes into the second prism instead of being critically reflected! There must have been some field just *outside* the original prism surface; otherwise, how could bringing up the second prism cause any change?

Granted that there is *some* field in medium 2, then, it must be such that no real (time-average) power flows in the $+z$ direction as long as $\vartheta_1 \geq \vartheta_c$. This requirement stems from the well-known experimental fact in optics that "total reflection" occurs for *all* angles of incidence greater than the critical angle.

Thirdly, with reference to Fig. 7.23, there is still the problem of insuring that the incident, reflected, and transmitted waves stay "in step" along the y direction. For the reflected wave, this condition is easily met if $\vartheta_r = \vartheta_i$ as usual. For the refracted wave, however, the present situation $\vartheta_1 > \vartheta_c$ makes Snell's law demand $(v_1/\sin\vartheta_1) < v_2$, or $\beta_{01} \sin\vartheta_1 > \beta_{02}$. But the wave in medium 2 must have an effective phase constant *along* y which equals $\beta_{01}\sin\vartheta_1$, and we see that this is *greater* than that (β_{02}) of any possible *uniform plane wave* in medium 2!

Finally, since the incident wave is x-polarized, it should be possible to meet the boundary conditions with reflected and transmitted waves, both also x-polarized.

The conclusion is almost inescapable: The wave in medium 2 must be a *nonuniform plane wave*, arranged to *attenuate* in the $+z$ direction (boundary condition at infinity), have a phase delay along $-y$ with a phase constant $\beta_2 = +\sqrt{\beta_{02}^2 + \alpha^2} > \beta_{02}$, and to be TE with respect to the y (or z) directions. A sketch of the conditions appears in Fig. 7.26.

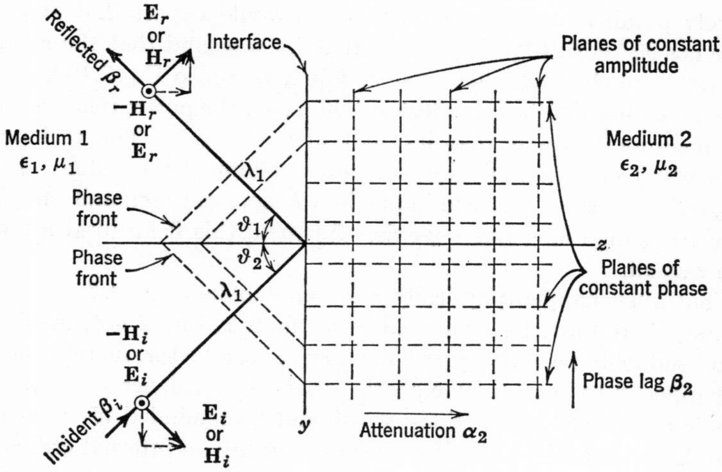

Fig. 7.26. Oblique incidence beyond the critical angle on an interface between lossless dielectrics ($\epsilon_1\mu_1 > \epsilon_2\mu_2$). Drawn either for polarization in the plane of incidence, or parallel to the interface.

Analytically, referring to the foregoing discussion and to Eqs. 7.54, this field must be of the TE form:

$$\text{(a)} \quad \bar{\boldsymbol{\gamma}}_2 = -\boldsymbol{a}_y j\beta_2 + \boldsymbol{a}_z\alpha_2 \qquad \alpha_2, \beta_2 > 0 \qquad \beta_2{}^2 - \alpha_2{}^2 = \beta_{02}{}^2$$

$$\text{(b)} \quad \mathbf{E}_t = \boldsymbol{a}_x \mathrm{E}_{xt} e^{j\beta_2 y} e^{-\alpha_2 z} \tag{7.116}$$

$$\text{(c)} \quad \mathbf{H}_t = \frac{\bar{\boldsymbol{\gamma}} \times \mathbf{E}_t}{j\omega\mu_2} = \left(\frac{\boldsymbol{a}_y\alpha_2 + \boldsymbol{a}_z j\beta_2}{j\omega\mu_2}\right) \mathrm{E}_{xt} e^{j\beta_2 y} e^{-\alpha_2 z}$$

The requirement of phase match along the boundary becomes

$$\beta_2 = \beta_{01} \sin \vartheta_1 \tag{7.117a}$$

and therefore on account of Eqs. 7.116*a* and 7.115

$$\alpha_2 = +\sqrt{\beta_{01}{}^2 \sin^2 \vartheta_1 - \beta_{02}{}^2} = +\beta_{01}\sqrt{\sin^2 \vartheta_1 - \sin^2 \vartheta_c} \tag{7.117b}$$

With regard to the remaining boundary conditions on the tangential fields, we may still apply Eqs. 7.101 and 7.102*a*. For Eq. 7.102*b*, however, matters are altered slightly because, whereas

$$\frac{\mathrm{E}_{xi}}{\mathrm{H}_{yi}} = -\frac{\mathrm{E}_{xr}}{\mathrm{H}_{yr}} = \frac{\eta_1}{\cos \vartheta_1} = Z_1 = \text{real}$$

as before, we now find from Eqs. 7.116b, 7.116c, 7.117b, and 7.115

$$Z_2 \equiv \frac{E_{xt}}{H_{yt}} = \frac{j\omega\mu_2}{\alpha_2} = \frac{j\omega\mu_2}{\beta_{01}\sqrt{\sin^2 \vartheta_1 - \sin^2 \vartheta_c}}$$

$$= \frac{j\eta_2}{\sqrt{(\sin \vartheta_1/\sin \vartheta_c)^2 - 1}} = jX_2 \tag{7.118}$$

which is positive imaginary (inductive) and independent of frequency. Consequently the *form* of Eqs. 7.103 remains unchanged, but the values in it are rather different:

$$\text{(a)} \qquad \frac{E_{xr}}{E_{xi}} \equiv \bar{\Gamma}_R = \frac{Z_2 - Z_1}{Z_2 + Z_1} = \frac{jX_2 - Z_1}{jX_2 + Z_1} = e^{j\psi}$$

$$\text{(b)} \qquad \frac{E_{xt}}{E_{xi}} \equiv T = \frac{2Z_2}{Z_2 + Z_1} = \frac{2jX_2}{Z_1 + jX_2} \tag{7.119}$$

where

$$\psi = 2\tan^{-1}\left(\frac{Z_1}{X_2}\right) \qquad 0 \leq \tan^{-1}\left(\frac{Z_1}{X_2}\right) \leq \frac{\pi}{2} \tag{7.120}$$

The boundary condition results just found are partially summarized in Fig. 7.27a, which represents them in the form of the equivalent transmission line appropriate to the z-axis features of the problem. Note that medium 2 presents a lossless (inductive) load to the line of medium

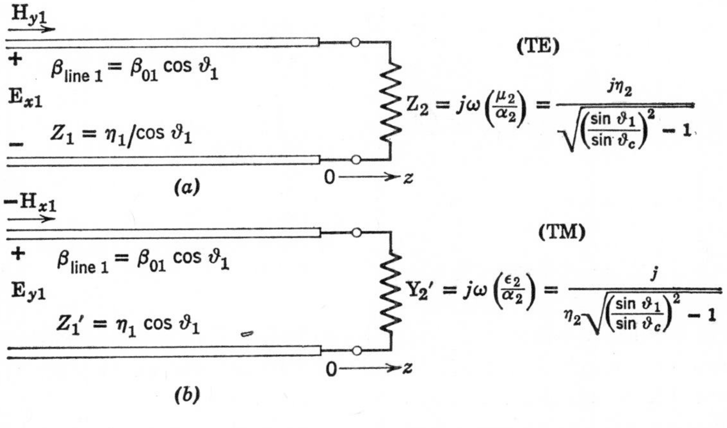

Fig. 7.27. Transmission-line analogies for z dependence and x and y field components in Fig. 7.26; (a) with polarization parallel to interface, and (b) with polarization in the plane of incidence.

1 in this case, which accounts properly for the absence of real power entering medium 2 in the z direction. This is, of course, checked by the fact that $|\bar{\Gamma}_R| = 1$ in Eq. 7.119a. Again, one must not lose sight of the fact that traveling-wave propagation is taking place along the $-y$ direction on *both* sides of the boundary, and that time-average power is flowing accordingly along $-y$ in both media.

In the case of polarization in the plane of incidence, beyond critical angle (Figs. 7.25a and 7.26), the magnetic field will have only an x component. The transmitted nonuniform plane wave in medium 2 will be of the TM type (Eqs. 7.55),

$$\text{(a)} \quad \bar{\boldsymbol{\gamma}}_2 = -\boldsymbol{a}_y j\beta_2 + \boldsymbol{a}_z\alpha_2 \qquad \alpha_2, \beta_2 > 0 \qquad \beta_2^{\,2} - \alpha_2^{\,2} = \beta_{02}^{\,2}$$

$$\text{(b)} \quad \mathbf{H}_t = \boldsymbol{a}_x \mathrm{H}_{xt} e^{j\beta_2 y} e^{-\alpha_2 z} \tag{7.121}$$

$$\text{(c)} \quad \mathbf{E}_t = -\frac{\bar{\boldsymbol{\gamma}} \times \mathbf{H}_t}{j\omega\epsilon_2} = \frac{-(\boldsymbol{a}_y\alpha_2 + \boldsymbol{a}_z j\beta_2)}{j\omega\epsilon_2} \mathrm{H}_{xt} e^{j\beta_2 y} e^{-\alpha_2 z}$$

with Eqs. 7.117a and 7.117b remaining unchanged. Also as before

$$\frac{-\mathrm{E}_{yi}}{\mathrm{H}_{xi}} = \frac{\mathrm{E}_{yr}}{\mathrm{H}_{xr}} \equiv Z_1' = \eta_1 \cos \vartheta_1 = \text{real} \tag{7.109a}$$

but now

$$\frac{-\mathrm{E}_{yt}}{\mathrm{H}_{xt}} \equiv \mathrm{Z}_2' = \frac{\alpha_2}{j\omega\epsilon_2} = -j\eta_2 \sqrt{\left(\frac{\sin \vartheta_1}{\sin \vartheta_c}\right)^2 - 1} = -jX_2' \tag{7.122}$$

which is negative imaginary (capacitive) and independent of frequency. Thus

$$\text{(a)} \quad \frac{\mathrm{E}_{yr}}{\mathrm{E}_{yi}} \equiv \bar{\Gamma}_R' = \frac{\mathrm{Z}_2' - Z_1'}{\mathrm{Z}_2' + Z_1'} = \frac{-Z_1' - jX_2'}{Z_1' - jX_2'} = e^{j\psi'}$$

$$\text{(b)} \quad \frac{\mathrm{E}_{yt}}{\mathrm{E}_{yi}} \equiv \mathrm{T}' = \frac{2\mathrm{Z}_2'}{\mathrm{Z}_2' + Z_1'} = \frac{-2jX_2'}{Z_1' - jX_2'} \tag{7.123}$$

In this case, however, we must put

$$\psi' = -2 \tan^{-1}\left(\frac{Z_1'}{X_2'}\right) \qquad 0 \le \tan^{-1}\left(\frac{Z_1'}{X_2'}\right) \le \frac{\pi}{2} \tag{7.124}$$

which should be compared carefully with Eq. 7.120, recognizing that by our definitions Z_1, X_2, Z_1', and X_2' are *all positive real numbers.* Figure 7.27b summarizes the conditions leading to Eqs. 7.123 and 7.124.

We have now seen for the first time a physical situation giving rise to nonuniform plane waves. They occur here as the refracted wave "on the other side of a reflection beyond critical angle," when the incident and reflected waves are both *uniform* plane waves! This being the case, however, one may wonder whether or not the burden of shifting the refracted wave from a uniform to a nonuniform plane wave, as the angle of incidence passes smoothly from zero to values beyond the critical angle ϑ_c, must always be handled as a brand-new problem—completely independent of the solution for $\vartheta_1 \leq \vartheta_c$.

The answer is that the idea of a *complex angle of refraction*, $\bar{\vartheta}_2 = \vartheta_{2R} + j\vartheta_{2I}$, can be used to make the transition in an almost completely automatic manner. As an example, take the case of polarization parallel to the boundary (Fig. 7.23).

When $\vartheta_1 < \vartheta_c$, or, indeed, before we had any particular reason to expect a "critical angle" problem at all, we would have said that the transmitted wave must be given by a *uniform* plane wave expression similar to Eq. 7.86. Specifically,

$$\text{(a)} \qquad \mathbf{E}_t = \boldsymbol{a}_x \mathrm{E}_{xt} e^{-j\beta_{02} z \cos \vartheta_2} e^{+j\beta_{02} y \sin \vartheta_2}$$

$$\text{(b)} \qquad \mathbf{H}_t = \frac{\mathrm{E}_{xt}}{\eta_2} (\boldsymbol{a}_y \cos \vartheta_2 + \boldsymbol{a}_z \sin \vartheta_2) e^{-j\beta_{02} z \cos \vartheta_2} e^{+j\beta_{02} y \sin \vartheta_2}$$

with Eqs. 7.100, 7.103, and 7.104 used to meet the boundary conditions.

When it came to evaluation of ϑ_2, however, we might discover that in Snell's law

$$\sin \vartheta_2 = \left(\frac{v_2}{v_1}\right) \sin \vartheta_1 = \text{real and} > 1$$

because $v_2 > v_1$, and ϑ_1 was rather large. In that case, let ϑ_2 be complex, and find out what it must be to satisfy Snell's law anyway! Accordingly,

$$\sin \bar{\vartheta}_2 = \sin \vartheta_{2R} \cosh \vartheta_{2I} + j \cos \vartheta_{2R} \sinh \vartheta_{2I}$$
$$= \left(\frac{v_2}{v_1}\right) \sin \vartheta_1 = \text{real and} > 1$$

Therefore the only possibility is

$$\cos \vartheta_{2R} = 0 \qquad \vartheta_{2R} = \frac{\pi}{2}$$

to make

$$\sin \bar{\vartheta}_2 = +\cosh \vartheta_{2I} = \left(\frac{v_2}{v_1}\right) \sin \vartheta_1 = \text{real and} > 1$$

This defines ϑ_{2I}, except for algebraic sign; $\vartheta_{2I} = \pm\cosh^{-1}[(v_2/v_1) \sin \vartheta_1]$. Consequently,

$$\cos \bar{\vartheta}_2 = \cos \vartheta_{2R} \cosh \vartheta_{2I} - j \sin \vartheta_{2R} \sinh \vartheta_{2I} = -j \sinh \vartheta_{2I}$$

Employing the above values for $\sin \bar{\vartheta}_2$ and $\cos \bar{\vartheta}_2$ in the transmitted-field expressions, and using Snell's law, $\beta_{02} \sin \bar{\vartheta}_2 = \beta_{01} \sin \vartheta_1$, in the exponents, we find

$$\text{(a)} \quad \mathbf{E}_t = \mathbf{a}_x \mathrm{E}_{xt} e^{j\beta_{01} y \sin \vartheta_1} e^{-\beta_{02} z \sinh \vartheta_{2I}}$$

$$\text{(b)} \quad \mathbf{H}_t = \frac{\mathrm{E}_{xt}}{\eta_2} (-\boldsymbol{a}_y j \sinh \vartheta_{2I} + \boldsymbol{a}_z \cosh \vartheta_{2I}) e^{j\beta_{01} y \sin \vartheta_1} e^{-\beta_{02} z \sinh \vartheta_{2I}}$$

Now it is reasonable to assume that the solution we want must die out rather than become infinite as $z \to +\infty$; at least this is the only boundary condition at infinity which is consistent with our previous assumptions about the incident wave being the only source in the problem. Thus we may write

$$\text{(c)} \qquad \alpha_2 = \beta_{02} \sinh \vartheta_{2I} > 0 \qquad \vartheta_{2I} > 0$$

$$\text{(d)} \qquad \beta_2 = \beta_{01} \sin \vartheta_1 = \beta_{02} \cosh \vartheta_{2I}$$

so that

$$\frac{\cosh \vartheta_{2I}}{\eta_2} = \frac{\beta_2}{\beta_{02}\eta_2} = \frac{\beta_2}{\omega\mu_2}$$

and

$$\frac{-j \sinh \vartheta_{2I}}{\eta_2} = \frac{-j\alpha_2}{\beta_{02}\eta_2} = \frac{\alpha_2}{j\omega\mu_2}$$

We can therefore convert the last field expressions to

$$\text{(a)} \qquad \mathbf{E}_t = \boldsymbol{a}_x \mathrm{E}_{xt} e^{j\beta_2 y} e^{-\alpha_2 z}$$

$$\text{(b)} \qquad \mathbf{H}_t = \left(\frac{\boldsymbol{a}_y \alpha_2 + \boldsymbol{a}_z j\beta_2}{j\omega\mu_2} \right) \mathrm{E}_{xt} e^{j\beta_2 y} e^{-\alpha_2 z}$$

$$\text{(c)} \qquad \beta_2^2 - \alpha_2^2 = \beta_{02}^2$$

$$\text{(d)} \qquad \beta_2 = \beta_{01} \sin \vartheta_1$$

which agrees precisely with Eqs. 7.116 and 7.117.

The straightforward use of a complex angle of refraction in Snell's law, *plus the election of a choice regarding the behavior of the refracted solution at infinity,* allows us to carry the refracted field solution continuously through from uniform to nonuniform plane waves as the incident angle ϑ_1 passes through the critical value ϑ_c.

One of the roles of the nonuniform plane wave in a lossless medium may be summarized from the examples just treated. It arises when boundary conditions require an effective phase velocity (or wave length) in some space direction which is *less* than that provided at the given frequency by the conventional "free space" phase velocity $1/\sqrt{\epsilon\mu}$ [or wave length $2\pi/(\omega\sqrt{\epsilon\mu})$] in the surrounding medium. Larger phase velocities (or wave lengths) in a given direction can be achieved with uniform plane waves at real oblique angles of propagation; smaller ones, however, require "uniform" plane waves at *complex* angles of propagation—which is to say, actually, *nonuniform* plane waves oriented to provide large space-rates-of-change of phase along the direction in question. The price of such small wave lengths (or rapid phase changes) along one direction, however, is complete attenuation in another (perpendicular) direction. Compressing phase velocity along one axis loses amplitude at right angles!

These matters are illustrated even more forcefully in the following examples of guided waves.

7.5 Guided Waves *

In studying the oblique incidence of uniform plane waves upon plane boundaries, we have so far stressed primarily the behavior of the fields as regards the direction *normal* to the boundary. Examination of these problems from the point of view of the direction *parallel* to the boundary instead leads to the concept of guided waves. This approach is by no means the only one for introducing guided waves, but it does furnish one illuminating view of the problem.

7.5.1 Metallic (Rectangular) Wave Guides *

Consider the problem of an x-polarized uniform plane wave at an oblique angle of incidence ϑ upon a perfect conductor (Fig. 7.21). The total field is given in Eq. 7.97.

This field has the property that $E_x = 0$, not only on the plane $z = 0$ but also on the planes

$$z_m = \frac{-m\lambda}{2\cos\vartheta} = \frac{-m\lambda_z}{2} \qquad m = 1, 2, \cdots \tag{7.125}$$

Thus it will still be a solution both to Maxwell's equations and the boundary conditions if a second perfect conductor is inserted at the position $z_m = -a$, with

$$a = \frac{m\lambda}{2\cos\vartheta} = \frac{m\lambda_z}{2} \qquad m = 1, 2, \cdots \tag{7.126}$$

Under these conditions, the tangential electric field (E_x) vanishes

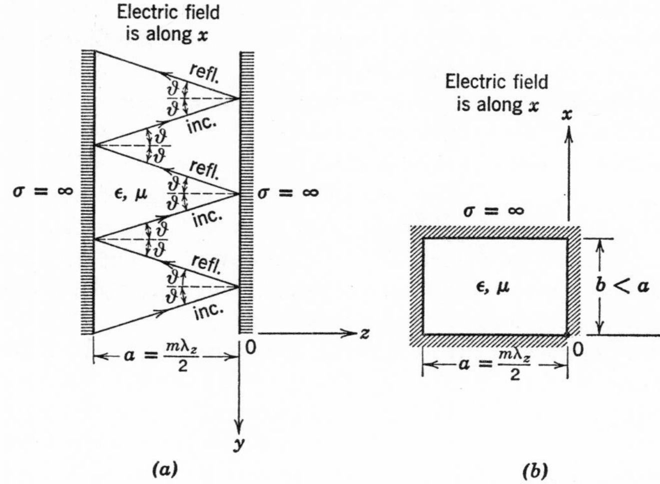

Fig. 7.28. Construction of a rectangular wave-guide solution from the obliquely incident and reflected waves for a metal plane $z = 0$. (*a*) Addition of side wall, $z = -a = -(m\lambda_z/2)$; (*b*) addition of top and bottom walls, $x = 0$ and $x = b < a$.

automatically on the surface of both conductors. Consequently, as shown schematically in Fig. 7.28*a*, the field

$$\text{(a)} \qquad \mathbf{E} = [-\boldsymbol{a}_x 2j\text{E} \sin(\beta_0 z \cos\vartheta)]e^{j\beta_0 y \sin\vartheta}$$

$$\text{(b)} \qquad \mathbf{H} = \frac{2E}{\eta}[\boldsymbol{a}_y \cos\vartheta \cos(\beta_0 z \cos\vartheta) - \boldsymbol{a}_z j \sin\vartheta \sin(\beta_0 z \cos\vartheta)]e^{j\beta_0 y \sin\vartheta} \qquad (7.127)$$

is a solution in the space region $(-a \leq z \leq 0;\ -\infty < y < +\infty)$, *provided* Eq. 7.126 is met. Moreover, this solution constitutes a traveling wave in the $-y$ direction, completely confined in the z direction. The "source" for it must now be regarded as located at $y = +\infty$.

Evidently, two more perfectly conducting planes might just as well be added to the system, one at $x = 0$ and the other at $x = b < a$, as shown in Fig. 7.28*b*. Since these planes are perpendicular to the only component of the electric field (E_x), they do not require any modification of the field solution to meet their boundary conditions.[1]

We now have a traveling wave along $-y$, completely confined in a hollow rectangular pipe defined by $0 \leq x \leq b$ and $-a \leq z \leq 0$ (Fig. 7.28*b*). The structure is serving as a *wave guide*. It is of interest, then,

[1] It is worth while to pause here and consider whether this whole scheme would work if one were to start with the magnetic instead of the electric field along x

to ask how the propagation of this wave in the wave guide varies with frequency, assuming dimensions a and b are fixed. In view of Eq. 7.126, ϑ now becomes merely a parameter, fixed by the value of a and the *free-space* (not vacuum) wave length λ. The latter is really just a measure of the frequency, given the values ϵ and μ of the medium which fills the guide, because by definition

$$\lambda \equiv \frac{2\pi}{\beta_0} \equiv \frac{2\pi}{\omega\sqrt{\epsilon\mu}} \tag{7.128}$$

Thus we shall wish to eliminate ϑ in some of our immediate discussion.

The important questions about this wave are: (1) How does the y variation depend on ω (or λ)? (2) How does the flow of complex power along $-y$ depend upon frequency? (3) How does the field distribution across the transverse dimensions (x, z) vary with frequency?

First, we define the *guide wave length* λ_g as the distance between equiphase surfaces measured along the guide propagation axis ($-y$ in this case). Previously we called this λ_y, so

$$\lambda_g \equiv \lambda_y = \frac{\lambda}{\sin\vartheta} \tag{7.129}$$

But in view of Eq. 7.126

$$\sin\vartheta = \sqrt{1-\cos^2\vartheta} = \sqrt{1-\left(\frac{\lambda}{2a/m}\right)^2} \tag{7.130}$$

and Eq. 7.129 becomes

$$\lambda_g = \frac{\lambda}{\sqrt{1-(\lambda/\lambda_{m,0})^2}} > \lambda \qquad \text{for } \lambda \leq \lambda_{m,0} \tag{7.131}$$

where

$$\lambda_{m,0} \equiv \frac{2a}{m} \tag{7.132}$$

Then using Eqs. 7.126 and 7.129 to eliminate ϑ from the field expressions (Eqs. 7.127), and defining

$$\beta_g \equiv |\beta_y| \equiv \frac{2\pi}{\lambda_g} = \beta_0 \sin\vartheta \tag{7.133}$$

we find

$$\text{(a)} \qquad \mathbf{E} = -\left[\boldsymbol{a}_x 2j\mathrm{E}\sin\left(\frac{m\pi z}{a}\right)\right]e^{j\beta_g y} \qquad m = 1, 2, 3, \cdots$$

$$\text{(b)} \qquad \mathbf{H} = \frac{2E}{\eta}\left[\boldsymbol{a}_y\left(\frac{\lambda}{\lambda_{m,0}}\right)\cos\left(\frac{m\pi z}{a}\right) - \boldsymbol{a}_z j\left(\frac{\lambda}{\lambda_g}\right)\sin\left(\frac{m\pi z}{a}\right)\right]e^{j\beta_g y} \qquad m = 1, 2, 3, \cdots \tag{7.134}$$

It is clear from Eq. 7.134 that the behavior of the fields as a function of the transverse coordinates (x,z) does *not* depend upon frequency. There is actually no x dependence, and the z dependence is sinusoidal with dimension a, an integer number (m) of half-periods. The wave is TE with respect to the guide (y) axis, having an electric field linearly polarized in the transverse plane (along x) and both a transverse (z) component and a longitudinal (y) component of magnetic field.

On account of their m half-period variations along the wide dimension a, and their zero variation along the narrow dimension b, these rectangular wave-guide field solutions are called $TE_{m,0}$ *waves* or *modes.* They are different for different (arbitrary) integer choices of m, so we actually have an infinite set of solutions for a given guide size and a specified frequency. Nevertheless, the transverse variation of the fields depends *only* on m, and not on frequency, so we can easily identify one solution by its m value and then consider its behavior as a function of frequency. Each of the "modes" $TE_{1,0}$, $TE_{2,0}$, $\cdots$, $TE_{m,0}$ may be regarded as a separate rectangular wave-guide solution which varies in its own way with frequency.

To study the frequency variation of a $TE_{m,0}$ mode, refer first to Eqs. 7.131 and 7.132. At very high frequencies, $\lambda \to 0$ and $\lambda_g \to \lambda \to 0$. As the frequency is lowered, however, $\lambda \to \lambda_{m,0}$ and $\lambda_g \to \infty$. When $\lambda > \lambda_{m,0}$, on the other hand, λ_g must become imaginary

$$\lambda_g = \frac{\binom{+}{-} j\lambda}{\sqrt{(\lambda/\lambda_{m,0})^2 - 1}} \qquad \text{for } \lambda > \lambda_{m,0} \tag{7.135}$$

The meaning of Eq. 7.135 stems from Eq. 7.134, where

$$\begin{aligned} e^{j\beta_g y} &= e^{j(2\pi y/\lambda_g)} = e^{\binom{+}{-}\{[(2\pi/\lambda)(\sqrt{(\lambda/\lambda_{m,0})^2-1})]y\}} \\ &= e^{\binom{+}{-}\alpha_g y} \qquad \text{for } \lambda > \lambda_{m,0} \end{aligned} \tag{7.136}$$

shows pure attenuation along $\binom{-}{+}$ y! Actually we would choose the + sign in Eqs. 7.135 and 7.136, since we are discussing the problem for a source at $y = +\infty$. The solution should therefore vanish (rather than grow exponentially) as $y \to -\infty$, and grow (rather than vanish) at $y = +\infty$.

At long wave lengths (low frequencies), the *uniform* plane waves from which we originally constructed the wave guide solution have changed into *nonuniform* plane waves. This assertion is confirmed from Eqs. 7.126 and 7.132 by the fact that $\cos\vartheta > 1$ when $\lambda > \lambda_{m,0}$, and $\sin\vartheta$ is pure imaginary in Eq. 7.130. The angle ϑ becomes complex, $\bar{\vartheta} = \pi + j\vartheta_I$ (or $\bar{\vartheta} = -j\vartheta_I$). We expect no *real* (time-average) power flow down the guide under these conditions and, of course, we never expected any in directions normal to each of the four perfectly conducting walls. Let us check these features of the power flow by examining the wave impedances.

Observe from Eq. 7.134 that

$$\text{(a)} \qquad Z_{-y} \equiv \frac{E_x}{H_z} = \eta \left(\frac{\lambda_g}{\lambda}\right)$$

$$\text{(b)} \qquad Z_z \equiv \frac{E_x}{H_y} = -j\eta \left(\frac{\lambda_{m,0}}{\lambda}\right) \tan\left(\frac{m\pi z}{a}\right) \tag{7.137}$$

Accordingly, since Z_z is always pure imaginary, there is only reactive power flowing in the $\pm z$ directions at all frequencies. Z_{-y} however is positive real when $\lambda < \lambda_{m,0}$ and becomes pure imaginary (inductive) when $\lambda > \lambda_{m,0}$ (λ_g becomes positive imaginary). Real power does flow down the guide at high frequencies ($\lambda < \lambda_{m,0}$), but only (inductive) reactive power flows at low frequencies ($\lambda > \lambda_{m,0}$). Thus the wave guide acts like a *high-pass filter*, with a *cutoff frequency* $\omega_{m,0}$ defined by

$$\omega_{m,0} = \frac{2\pi}{\lambda_{m,0}\sqrt{\epsilon\mu}} = \frac{\pi m}{a\sqrt{\epsilon\mu}} \tag{7.138}$$

The mechanism of propagation ($\lambda < \lambda_{m,0}$) down the guide can be represented in terms of uniform plane waves by the familiar optical type of ray-tracing picture of a single unit-field-strength ray alternately reflected from each side wall, as in Fig. 7.28*a*, in which we focus attention only on the space between $z = 0$ and $z = -a$; or, more symmetrically, by considering two sets of multiply reflected rays, each of half field strength, shown in the similar picture Fig. 7.29*a*. Alternatively, if we mentally remove the side walls at $z = 0$ and $z = -a$, the interference pattern in *all space* of just *two* rays, each of unit field strength, as shown in Fig. 7.29*b*, will produce tangential electric field nodal planes (dotted lines) where the walls used to be (and also at other such planes spaced $\frac{1}{2}\lambda_z$ from these).

In either view, regarding the dimension a and mode index m as fixed, we may use Eqs. 7.126 and 7.132 to determine ϑ as a function of frequency for the mth mode field.

$$\cos \vartheta = \frac{\lambda}{\lambda_{m,0}} \tag{7.139}$$

As long as $\lambda < \lambda_{m,0}$, $a > m\lambda/2$ from Eq. 7.126, and, according to Eq. 7.129 or 7.131, $\lambda_z > \lambda$. This condition can always be met with a real angle ϑ. At high frequencies ($\lambda \to 0$), $\vartheta \to \pi/2$, and the two component uniform plane waves are propagating in directions nearly parallel to the guide axis (y). The wave-guide fields approach something very much like a uniform plane wave (though the side-wall boundary conditions prevent it from actually becoming one; see Eqs. 7.131, 7.133, 7.134, and 7.137 as $\lambda \to 0$). When the frequency is lowered to $\omega_{m,0}$ ($\lambda = \lambda_{m,0}$), $\vartheta = 0$, and the two-component uniform plane waves are propagating directly across the transverse face of the guide, at normal incidence to the side walls. There is no y variation of the fields ($\lambda_g = \infty$) and $H_z = 0$; energy simply rattles back and forth between the side walls, making no progress down the guide at all. Under this condition, $a = (m\lambda_{m,0})/2 = m(\lambda/2)$, so the boundary conditions of Eq. 7.126 demand that λ_z *equal* the free-space wave length λ at this frequency ($\omega_{m,0}$). Further lowering of the frequency will result in the requirement $a < m\lambda/2$, or $\lambda_z < \lambda$, for which the side walls force a compressed wave length condition along z. We expect, therefore, that the two component plane waves will become nonuniform and will produce attenuation along the guide axis (y). This does in fact occur.

Among the $TE_{m,0}$ waves that we have found, it is clear from Eq. 7.138 that, for a given guide dimension a, the $TE_{1,0}$ ($m = 1$) mode has the lowest cutoff frequency $\omega_{1,0}$:

$$\omega_{1,0} = \frac{\pi}{a\sqrt{\epsilon\mu}} \qquad \lambda_{1,0} = 2a \qquad (b < a) \qquad (7.140)$$

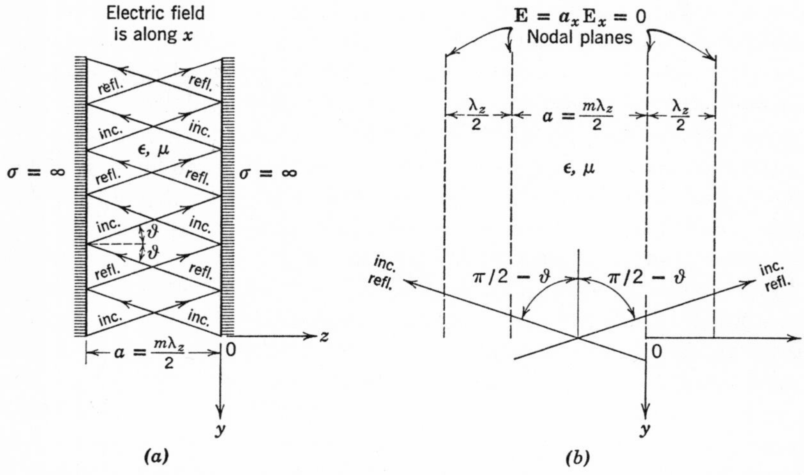

Fig. 7.29. Ray representations of $TE_{m,0}$ propagation in a rectangular wave guide. Note that each ray in (a) represents half the field strength of each one in (b).

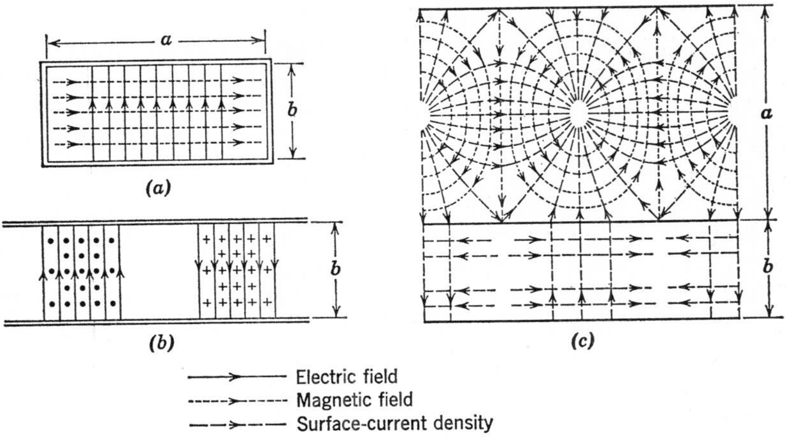

Fig. 7.30. Field lines for $TE_{1,0}$ mode in rectangular wave guide. (*a*) Cross section; (*b*) longitudinal section, along center line; (*c*) views of top and side plates as seen from inside, showing surface-current densities and magnetic fields on inner surfaces.

The fields for the $TE_{1,0}$ mode undergo only a *single* half-period variation across the guide from $z = 0$ to $z = -a$. Some sketches of the field lines from Eq. 7.134 for this important mode (when $\omega > \omega_{1,0}$) appear in Fig. 7.30. These field lines are actually contours to which the instantaneous fields $\boldsymbol{E}$ and $\boldsymbol{H}$ of a single traveling wave are tangent, at some arbitrary moment of time. The contours are spaced closely where the instantaneous field strength is great. In Fig. 7.30*c* we have also shown the wall currents, which clarify the connection between the strong magnetic field at the walls, and the strong electric field (displacement current) along the center regions of the interior.

It turns out, from an extended analysis of all possible rectangular wave-guide modes which behave exponentially ($e^{\bar{\gamma}y}$) along the guide axis $(-y)$[1] that TE solutions exist with standing-wavelike field variations along *either or both* transverse axes (x and z). These are called $TE_{m,n}$ modes ($m = 0, 1, 2, \cdots$; $n = 0, 1, 2, \cdots$; but *not* $m = n = 0$). This nomenclature for the rectangular guide is based on the convention that the first subscript (m) refers to the number of half-period field variations along a path *parallel to the wide dimension* of the cross section, and the second subscript (n) to the number of half-period field variations along a path *parallel to the narrow dimension* of the cross section. By convention also, the wide dimension is denoted by a, the narrow by

[1] The analysis involves the field produced by four plane waves instead of two. It is taken up in the Problems.

b (Fig. 7.28b). The character of some of these modes, namely, the $TE_{0,n}$, is obvious from our solutions here. These are identical in form to the $TE_{m,0}$ modes, except that the whole field solution is turned 90° inside the guide so that the electric field is parallel to the long dimension (a or z). Their cutoff frequencies will be given by Eq. 7.138 with n for m and b for a, but their numerical values will be different, even when $n = m$, because dimension b is less than a. There are also $TM_{m,n}$ modes ($m = 1, 2, \cdots; n = 1, 2, \cdots$) which necessarily have standing-wavelike field variations along both transverse axes. Among all the TE and TM solutions, the $TE_{1,0}$ has the lowest cutoff frequency. This frequency $\omega_{1,0}$, given in Eq. 7.140, defines, in fact, the lowest frequency for which an infinitely long rectangular wave guide will propagate real power. For this reason, the $TE_{1,0}$ is called the *dominant mode* of the wave guide.[1]

It develops that the mode with the next higher cutoff frequency is either the $TE_{0,1}$ or the $TE_{2,0}$, depending upon the particular value of $b/a < 1$. It is then clear from Eq. 7.138 that if $b \le a/2$, $\omega_{2,0} \le \omega_{0,1}$ and the $TE_{2,0}$ is the mode with the next higher cutoff frequency. Under these particular conditions, in the frequency range $\omega_{1,0} \le \omega \le \omega_{2,0}$ ($= 2\omega_{1,0}$) *only* the $TE_{1,0}$ mode can propagate without attenuation along the guide axis. In fact, rectangular wave guides are normally designed with $b \le a/2$ in order to have this full-octave frequency range for single-mode transmission. At frequencies above $2\omega_{1,0}$, in a lossless structure, more than one mode may propagate without attenuation at the same time. This condition complicates considerably the use of the guide for transmission, and is avoided except in rather special circumstances.

A graphical representation of the multimode situation described above is shown in Fig. 7.31 in terms of the variation of β_g (or α_g) with frequency ω for the various possible modes. Analytically, for the $TE_{m,0}$ waves, we have from Eqs. 7.133, 7.131, and 7.138 (or 7.136 and 7.138),

$$\text{(a)} \qquad \beta_g = \frac{2\pi}{\lambda_g} = \frac{2\pi}{\lambda}\sqrt{1 - \left(\frac{\lambda}{\lambda_{m,0}}\right)^2}$$

$$= \sqrt{\epsilon\mu}\sqrt{\omega^2 - \omega_{m,0}{}^2} \qquad \omega \ge \omega_{m,0} \qquad (7.141)$$

$$\text{(b)} \qquad \alpha_g = \sqrt{\epsilon\mu}\sqrt{\omega_{m,0}{}^2 - \omega^2} \qquad \omega \le \omega_{m,0}$$

[1] Sometimes, the nomenclature of wave-guide modes is chosen to emphasize which field *has* a longitudinal component rather than which one *does not* have such a component. In that case, $TE_{m,n}$ modes are called $H_{m,n}$, and $TM_{m,n}$ modes are called $E_{m,n}$. In rectangular guide, then, the $H_{1,0}$ mode is dominant.

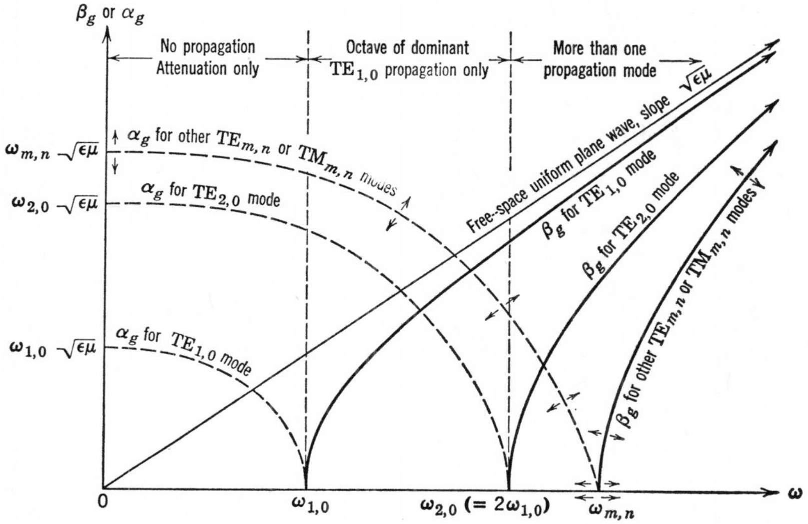

Fig. 7.31. Attenuation (α_g) and phase (β_g) constants versus frequency ω for modes in a lossless rectangular wave guide with narrow dimension not exceeding half the wide one. Note $\beta_{g_{m,n}} < \omega\sqrt{\epsilon\mu}$ and $\alpha_{g_{m,n}} < \omega_{m,n}\sqrt{\epsilon\mu}$.

These relations contain some additional interesting information, also discernible in Fig. 7.31.

Above cutoff frequency, the phase velocity along the guide axis is greater than that of light in the medium filling the guide:

$$v_{\text{phase}} \equiv \frac{\omega}{\beta_g} = \frac{1}{\sqrt{\epsilon\mu}}\left[\frac{1}{\sqrt{1-(\omega_{m,0}/\omega)^2}}\right]$$

$$> \frac{1}{\sqrt{\epsilon\mu}} \qquad \text{for } \omega \geq \omega_{m,0} \tag{7.142}$$

The fact that the phase velocity depends upon frequency makes the wave guide a *dispersive* (though lossless) transmission system. Because there is no attenuation under these conditions, however, the idea of group velocity has a valid meaning (see Chapter 5). For the case at hand,

$$v_{\text{group}} \equiv \left(\frac{\partial\beta_g}{\partial\omega}\right)^{-1} = \left(\frac{\omega\sqrt{\epsilon\mu}}{\sqrt{\omega^2-\omega_{m,0}{}^2}}\right)^{-1} = \frac{1}{\epsilon\mu}\left(\frac{1}{v_{\text{phase}}}\right) = \frac{\sin\vartheta}{\sqrt{\epsilon\mu}}$$

or

$$v_{\text{group}}v_{\text{phase}} = \frac{1}{\epsilon\mu} = v_{\text{light}}{}^2 \tag{7.143}$$

This is to say, the free-space light velocity $1/\sqrt{\epsilon\mu}$ is the geometric mean between the wave-guide group and phase velocities. Whereas the guide phase velocity always exceeds that of light in the medium filling the guide, the group velocity never does so. Indeed the group velocity is just the speed of light in the medium filling the guide, projected along the guide axis from the oblique angle of bounce.

We have just learned that electromagnetic waves can be guided along one axis by confining them from spreading in other directions with the aid of entirely opaque (metal) walls. The now relatively common phenomenon of extraordinary light transmission through solid lucite rods suggests that much less stringent transverse constraints will suffice to guide electromagnetic energy along a single solid. The analysis of some such situations follows.

7.5.2 Nonmetallic Wave Guides *

The possibility of total reflection arises at the interface between different lossless dielectrics when incidence occurs on the side with the greater index of refraction. This suggests that a slab (or perhaps even a rod) of such material with a high index of refraction may be able to confine and guide electromagnetic waves by successive internal reflections at angles ϑ_1 beyond the critical angle. The idea is illustrated for a dielectric slab in Fig. 7.32*a*, to which the nonuniform plane-wave pattern of Fig. 7.26 presumably applies not only for $z > 0$ but also for $z < -a$ (with an appropriate exchange of "left" for "right").

Analytical determination of the conditions for guidance along the $-y$ direction requires application of boundary conditions at both $z = 0$ and $z = -a$. The TM case (magnetic field parallel to the boundaries) is to be treated here, Eqs. 7.121 being the relevant field expressions pertinent to that part of medium 2 in the region $z > 0$. In the part of medium 2 occupying $z < -a$, however, $\boldsymbol{\alpha}_2$ must be directed *left*, to make the solution vanish at $z = -\infty$. Otherwise the entire field could hardly be considered "confined" to the slab in the $\pm z$-directions. Naturally Snell's law (Eq. 7.117*a*) applies at both boundaries.

The equivalent transmission line in Fig. 7.27*b* accounts for the z-axis features of the problem in regard to medium 1 and the boundary conditions at $z = 0$. The tangential field continuity conditions at $z = -a$ are represented in the equivalent transmission line by terminating the line on the left ($z = -a$) in the same impedance as on the right ($z = 0$). This step appears in Fig. 7.32*b*, justified by the symmetry of the entire problem about the plane $z = -\frac{1}{2}a$.

According to Eqs. 7.123*a* and 7.124, we know the reflection coefficient

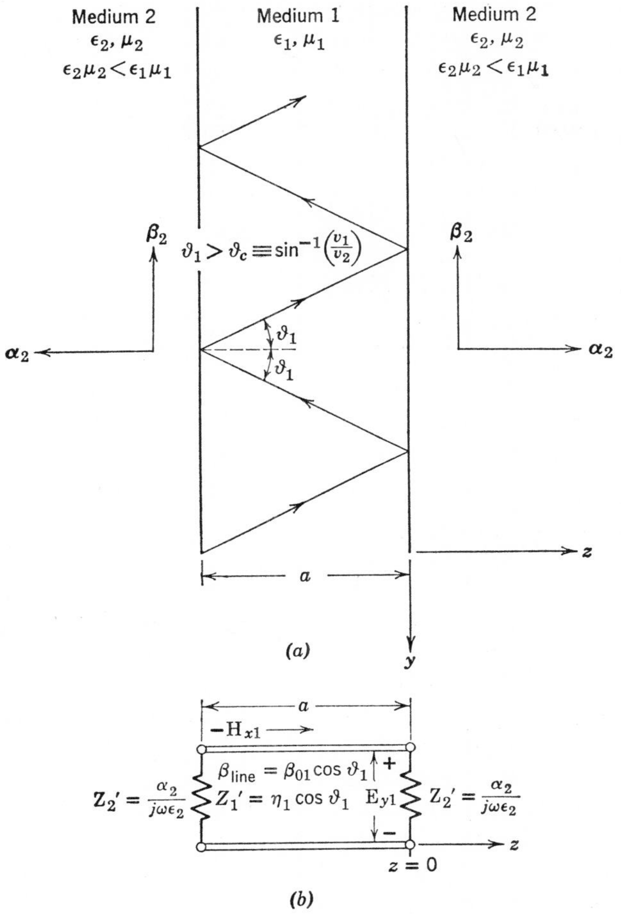

Fig. 7.32. Dielectric slab (medium 1) as a wave guide. (*a*) Mechanism of successive total internal reflection; (*b*) transmission-line representation with respect to *z*-axis; TM case, **H** parallel to boundaries.

$\bar{\Gamma}_R'$ looking to the right at $z = 0$. The reflection coefficient *looking to the right* at $z = -a$ is accordingly required by the transmission line for medium 1 to be

$$\bar{\Gamma}'(-a) = \bar{\Gamma}_R' e^{-j2\beta_z a} = e^{j(\psi' - 2\beta_z a)} \tag{7.144}$$

where we have written $\beta_z \equiv \beta_{\text{line 1}} = \beta_{01} \cos \vartheta_1$ for convenience.

On the other hand, *looking to the left* at $z = -a$, we must see the *same* load impedance Z_2' as appears in the calculation 7.123*a* for $\bar{\Gamma}_R'$.

Inasmuch as "looking to the right" and "looking to the left" at a given point on the line merely exchanges the roles of "incident" and "reflected" waves, the corresponding reflection coefficients are simply reciprocals of each other, as pointed out in Chapter 6 (Eq. 6.15). The presence of Z_2' at $z = -a$ therefore requires

$$\bar{\Gamma}'(-a) = \frac{1}{\bar{\Gamma}_R'} \tag{7.145}$$

In view of Eqs. 7.144 and 7.145 then,

$$e^{j(\psi' - 2\beta_z a)} = e^{-j\psi'}$$

or

$$e^{j2(\psi' - \beta_z a)} = +1 \tag{7.146a}$$

It follows that

$$\psi' - \beta_z a = k\pi \qquad k = 0, \pm 1, \pm 2, \cdots \tag{7.146b}$$

which, by Eq. 7.124, becomes

$$\frac{n\pi}{2} - \tan^{-1}\left(\frac{Z_1'}{X_2'}\right) = \frac{\beta_z a}{2} \qquad n = +1, +2, \cdots \tag{7.147a}$$

and

$$\frac{p\pi}{2} + \tan^{-1}\left(\frac{Z_1'}{X_2'}\right) = -\frac{\beta_z a}{2} \qquad p = 0, +1, +2, \cdots \tag{7.147b}$$

Use of n and p in Eqs. 7.147a and 7.147b is for the purpose of separating explicitly the cases for negative and positive values of k respectively in Eq. 7.146b. Inasmuch as β_z must be positive for Eq. 7.144 to be correct, however, only Eq. 7.147a is appropriate here (bear in mind that $0 \leq \tan^{-1}(Z_1'/X_2') \leq \pi/2$ according to Eq. 7.124). Thus considering separately odd and even values of n in Eq. 7.147a, we find

$$\text{(a)} \qquad \frac{X_2'}{Z_1'} = +\tan\frac{\beta_z a}{2} \qquad n = 1, 3, 5, \cdots$$

$$\tag{7.148}$$

$$\text{(b)} \qquad \frac{X_2'}{Z_1'} = -\cot\frac{\beta_z a}{2} \qquad n = 2, 4, \cdots$$

In the interests of examining the "guide wave length" $\lambda_g(\equiv \lambda_y)$ or the guide propagation constant $\beta_g(\equiv \beta_2 \equiv 2\pi/\lambda_g)$ as functions of frequency ω, we shall write the function X_2'/Z_1' as follows from Eqs.

7.109*a*, 7.122, 7.121*a*, and the definition $\beta_z = \beta_{01} \cos \vartheta_1$

$$\frac{X_2'}{Z_1'} = \frac{\alpha_2}{\omega \epsilon_2 \eta_1 \left(\dfrac{\beta_z}{\beta_{01}}\right)} = \frac{\sqrt{\beta_g^2 - \beta_{02}^2}}{\beta_z} \left(\frac{\epsilon_1}{\epsilon_2}\right) \tag{7.149}$$

But in medium 1

$$\beta_g^2 + \beta_z^2 = \beta_{01}^2 = \omega^2 \epsilon_1 \mu_1 \tag{7.150}$$

so Eq. 7.149 becomes, upon eliminating β_g^2,

$$\frac{X_2'}{Z_1'} = \frac{\epsilon_1}{\epsilon_2} \sqrt{\left(\frac{\beta_{01}^2 - \beta_{02}^2}{\beta_z^2}\right) - 1} \tag{7.151}$$

Equation 7.151 expresses X_2'/Z_1' as a function of β_z [or $(\beta_z a)/2$] for any given frequency ω (which fixes $\beta_{01} = \omega\sqrt{\epsilon_1\mu_1}$ and $\beta_{02} = \omega\sqrt{\epsilon_2\mu_2}$). This allows a graphical solution of Eqs. 7.148*a* or 7.148*b* for the quantity $(\beta_z a)/2$ at the frequency ω. The graphical solution entails plotting both sides of Eqs. 7.148*a* and 7.148*b* against $(\beta_z a)/2$ on the same abscissa to determine intersection points. Figures 7.33*a* and 7.33*c* respectively present these solutions (○) for three different frequencies $\omega_A < \omega_B < \omega_C$, pertaining to the curves marked *A*, *B*, and *C* respectively. Figure 7.33*b* shows the calculation of β_g from Eq. 7.150. This equation plots as a circle, relating β_g to β_z, having a radius directly proportional to ω.

The important restriction from Eq. 7.151 that

$$0 \leq \beta_z \leq \sqrt{\beta_{01}^2 - \beta_{02}^2} = \omega\sqrt{\epsilon_1\mu_1 - \epsilon_2\mu_2} \tag{7.152}$$

should not be overlooked. It defines the points at which curves *A*, *B*, and *C* strike the abscissa in Figs. 7.33*a* and 7.33*c*. Nor should its consequences from Eq. 7.150 be neglected:

$$\beta_{02} < \beta_g < \beta_{01} \tag{7.153}$$

The guide wave length always must lie *between* that of free-space uniform plane waves (or light) in medium 1 and in medium 2. Similarly, the guide phase velocity exceeds that of light in medium 1 (with the higher refractive index), but is less than the (greater) light velocity in medium 2.

Although no generally valid analytical relation for $\beta_g(\omega)$ may be written in the present circumstances, a study of Eqs. 7.151, 7.148, 7.150, and Fig. 7.33 provides satisfying insight into the effect of frequency variation when the dimension *a* is fixed. At a high frequency ω_C, the curve (Eq. 7.151) extends over a large range of β_z, because the

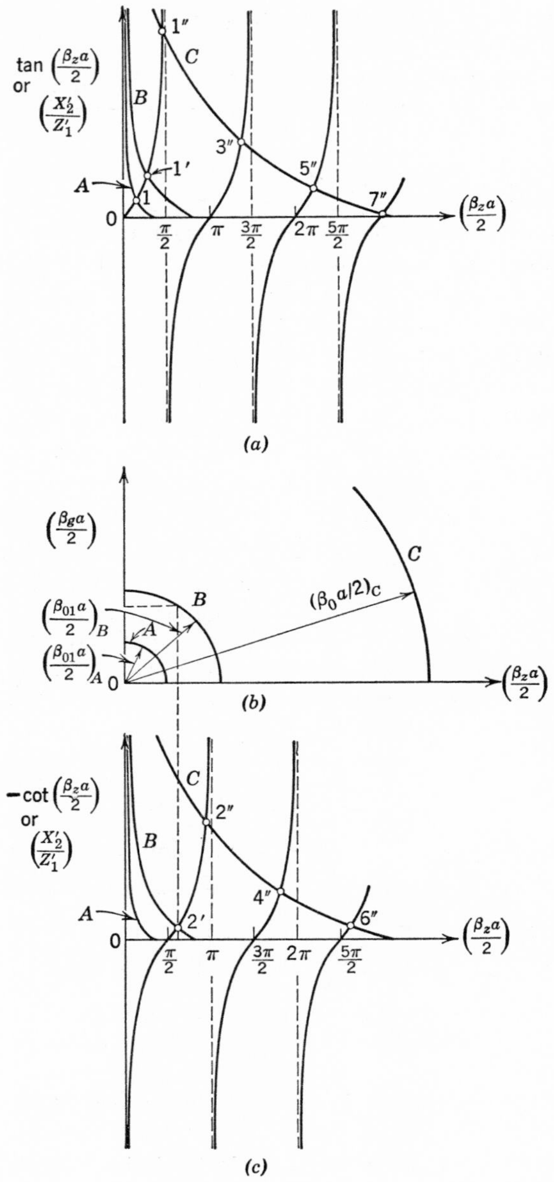

Fig. 7.33. Solution for TM-wave propagation constants for Fig. 7.32. (*a*), (*b*), and (*c*) represent Eqs. 7.148*a* and 7.151, Eq. 7.150, and Eqs. 7.148*b* and 7.151 respectively. Points marked (0) are solutions for β_z. Parts (*a*) and (*b*) also apply to TM waves, and (*c*) and (*b*) to TE waves, for Fig. 7.35.

upper bound (Eq. 7.152) increases directly with ω. There are then many intersections ($1''$ through $7''$) of the curves in Figs. 7.33*a* and 7.33*c*, which represent solutions to Eqs. 7.148. Evidently many of these TM waves can propagate together along the slab at high frequencies.

As frequency is lowered continuously, the various intersection points in Figs. 7.33*a* and 7.33*c* slide downward (see points $1''$, $1'$, and 1, or $2''$ and $2'$), each on its own (positive) branch of the tan or $-$cot function. Accordingly we can identify each such (positive) branch with a field solution or "mode" whose properties change continuously with frequency. At least this appears to be the case for any such branch until a frequency is reached for which the intersection arrives at the abscissa. (Refer to points $2''$ and $2'$ on curves C and B in Fig. 7.33*c*, and then especially to curve A there.) Any additional lowering of the frequency suddenly leaves us with *no solution corresponding to this mode! The wave under consideration simply ceases to exist.*

In surprising contrast with our metal wave-guide solutions, the present ones do *not* go into an attenuating condition *below* cutoff, but rather reach a cutoff (or *critical*) condition and vanish forthwith from the scene.

The explanation for this behavior is found by computing the angle ϑ_1 in Fig. 7.32 at the *critical* frequency ω_{crit} of any mode. At such frequency, the end point of validity of Eq. 7.151 falls at the base of a positive branch of the tan or $-$cot function, as would occur for example at a frequency between ω_A and ω_B on branch $2''$–$2'$ in Fig. 7.33*c*. Then from Eq. 7.151

$$\beta_{z\text{crit}}^2 = \beta_{01\text{crit}}^2 - \beta_{02\text{crit}}^2 \tag{7.154}$$

and from Eq. 7.150

$$\beta_{g\text{crit}}^2 = \beta_{01\text{crit}}^2 - \beta_{z\text{crit}}^2 = \beta_{02\text{crit}}^2 \tag{7.155}$$

But in general

$$\beta_g = \beta_{01} \sin \vartheta_1$$

so

$$\sin (\vartheta_{1\text{crit}}) = \frac{\beta_{g\text{crit}}}{\beta_{01\text{crit}}} = \frac{\beta_{02\text{crit}}}{\beta_{01\text{crit}}} = \frac{v_1}{v_2} \tag{7.156}$$

This result shows that at the critical frequency, the internal reflections have just reached the critical angle. Any lower frequency will bring the incident angle below critical, and there will be no basis for confining the waves to the slab.

A calculation of the critical frequencies in terms of the geometry stems from recognizing that the intersection points on Figs. 7.33*a* and 7.33*c* fall at zeros of the tan or $-$cot functions, in addition to being end points of curves like A, B, or C. These zeros, in both figures, can be summarized by the statement

$$\beta_{zm_{\text{crit}}} = \frac{m\pi}{a} \qquad m = 0, 1, 2, \cdots$$

or (7.157)

$$\lambda_{zm_{\text{crit}}} = \frac{2a}{m} \qquad m = 0, 1, 2, \cdots$$

even values of m arising from Fig. 7.33a and odd values of m from Fig. 7.33c. With Eq. 7.154, Eq. 7.157 becomes

$$\omega_{m_{\text{crit}}} = \frac{m\pi}{a\sqrt{\epsilon_1\mu_1 - \epsilon_2\mu_2}} \qquad m = 0, 1, 2, \cdots \tag{7.158}$$

The existence of *one* TM solution which persist down to zero frequency is a remarkable feature of these results. It occurs in the case $m = 0$ in Eq. 7.158, leading to $\omega_{0_{\text{crit}}} = 0$. This solution corresponds to branch 1″, 1′, 1, in Fig. 7.33a. Thus very low frequency waves may, in principle, be guided by a dielectric sheet.

Before extrapolating this conclusion any further, however, attention is directed to the fact that for any TM mode at high frequencies

$$\beta_{gm} \underset{\omega\to\infty}{\longrightarrow} \beta_{01} = \omega\sqrt{\epsilon_1\mu_1} \tag{7.159}$$

This follows from the feature of Figs. 7.33a and 7.33c that

$$\beta_{zm} \underset{\omega\to\infty}{\longrightarrow} \frac{(m+1)\pi}{a} \qquad m = 0, 1, 2, \cdots \tag{7.160}$$

which remains finite, while in Eq. 7.150 β_{01} (and hence β_g) approaches infinity with ω. At high frequencies, then, one expects most of the real longitudinal power flow to occur within the slab (medium 1), inasmuch as its phase velocity dominates the longitudinal propagation. A check upon this conclusion is supplied by the relation

$$\alpha_2 = \sqrt{\beta_g^{\,2} - \beta_{02}^{\,2}} \underset{\omega\to\infty}{\longrightarrow} \omega\sqrt{\epsilon_1\mu_1 - \epsilon_2\mu_2} \underset{\omega\to\infty}{\longrightarrow} \infty \tag{7.161}$$

The penetration of the field into medium 2, outside the slab, becomes arbitrarily small at high frequencies as the transverse attenuation becomes indefinitely large.

On the other hand, Eqs. 7.161 and 7.155 show that

$$\left.\begin{aligned} &\text{(a)} \quad \beta_g = \beta_{02} \\ &\text{(b)} \quad \alpha_2 = 0 \end{aligned}\right\} \quad \text{at } \omega = \omega_{\text{crit}} \tag{7.162}$$

The field extends uniformly to infinity outside the slab, in medium 2, at the critical frequency. An infinite amount of real longitudinal power is carried *outside* medium 1! Medium 2 dominates the longitudinal propagation constant [Eq. 7.162*a*]. The spread of the field outward from the slab increases as frequency drops toward the critical one, even if the critical frequency is zero. While the lowest TM mode can therefore *in principle* be excited at very low frequencies, the extension of the field far beyond the slab requires that any source supply large amounts of total power to do so. Just as is the case in any critical reflection (*at* the critical angle), the field in medium 2 at frequency ω_{crit} becomes a *uniform plane wave* traveling along $-y$. When $\omega_{\text{crit}} = 0$, this behavior is approached, but never actually reached, as $\omega \rightharpoonup \omega_{\text{crit}} = 0$.

An examination of Fig. 7.32*b* shows the standing-wave character of the field distribution across the z dimension of the slab (in medium 1). The impedance Z_2' is a function of frequency (see Eqs. 7.161 and 7.162):

$$\text{(a)} \qquad Z_2' \underset{\omega \to \infty}{\longrightarrow} -j\eta_2 \sqrt{\left(\frac{v_2}{v_1}\right)^2 - 1}$$

$$\text{(b)} \qquad Z_2' \xrightarrow[\omega \to \omega_{\text{crit}} \neq 0]{} 0 \qquad (7.163)$$

Equation 7.163*b* makes clear the reason for the "half-wave" conditions (Eq. 7.157) at the critical frequency. It also clarifies the fact that $E_{y2} = 0$, which supports the previous contention that the field in medium 2 is a uniform plane wave traveling along $-y$ at this frequency. Characteristically different from the modes in a metal wave guide, these modes *do* change their field distributions over the guide cross section as frequency is altered. Concentrated strongly within the slab at high frequencies, the field energy of a given mode spreads laterally into the surrounding space as the frequency is lowered until, finally, at the critical frequency, it has spread to infinity and no further guided wave of that mode can be supported at reduced frequencies.

Sketches of $\beta_g(\omega)$ for the various TM modes discussed can be made essentially by visualizing carefully Fig. 7.33 and noting Eqs. 7.153, 7.159, 7.160, and 7.162*a*. In this connection it helps, however, to consider analytically the slope $(d\beta_g/d\omega) = v_{\text{group}}^{-1}$, implicitly defined by Eqs. 7.148, 7.151, and 7.150. From the derivative with respect to ω of Eq. 7.150, we have

$$\frac{d\beta_g}{d\omega} = \left(\frac{\beta_{01}}{\beta_g}\right)\left(\frac{d\beta_{01}}{d\omega}\right) - \left(\frac{\beta_z}{\beta_g}\right)\left(\frac{d\beta_z}{d\omega}\right) = \frac{\beta_{01}^2}{\omega\beta_g} - \frac{\beta_z}{\beta_g}\left(\frac{d\beta_z}{d\omega}\right) \qquad (7.164)$$

Because Fig. 7.33 or Eq. 7.160 shows that $(d\beta_z/d\omega) \underset{\omega\to\infty}{\longrightarrow} 0$ for any given mode, and in view of Eqs. 7.159 and 7.160, Eq. 7.164 tells us that

$$\left(\frac{d\beta_g}{d\omega}\right)_{\omega\to\infty} \to \frac{\beta_{01}}{\omega} = \sqrt{\epsilon_1\mu_1} = \frac{1}{v_1} \tag{7.165}$$

The group velocity at high frequencies is dominated by the medium (medium 1) in which most of the energy is concentrated at these frequencies.

At the critical frequency, however, Eqs. 7.162 and 7.154 convert Eq. 7.164 to read

$$\left(\frac{d\beta_g}{d\omega}\right)_{\omega_{\rm crit}} = \frac{\beta_{01}}{\beta_{02}}\left(\frac{d\beta_{01}}{d\omega}\right)_{\omega_{\rm crit}} - \left(\frac{\sqrt{\beta_{01}^2 - \beta_{02}^2}}{\beta_{02}}\right)_{\omega_{\rm crit}}\left(\frac{d\beta_z}{d\omega}\right)_{\omega_{\rm crit}}$$

or

$$\left(\frac{d\beta_g}{d\omega}\right)_{\omega_{\rm crit}} = \frac{v_2}{v_1}\left[\frac{1}{v_1} - \sqrt{1 - \left(\frac{v_1}{v_2}\right)^2}\left(\frac{d\beta_z}{d\omega}\right)_{\omega_{\rm crit}}\right] \tag{7.166}$$

where it is understood that the derivatives are taken as limits for $\omega \rightharpoonup \omega_{\rm crit}$ only.

To compute $d\beta_z/d\omega$, note that from the right side of Eq. 7.148*a* we have on one hand

$$\frac{d}{d\omega}\left(\tan\frac{\beta_z a}{2}\right) = \frac{1}{\cos^2[(\beta_z a)/2]}\frac{a}{2}\frac{d\beta_z}{d\omega} \tag{7.167a}$$

with a similar result for Eq. 7.148*b*:

$$\frac{d}{d\omega}\left(-\cot\frac{\beta_z a}{2}\right) = \frac{1}{\sin^2[(\beta_z a)/2]}\frac{a}{2}\frac{d\beta_z}{d\omega} \tag{7.167b}$$

On the other hand, from the left side of Eqs. 7.148, making use of Eq. 7.151, we find

$$\frac{d(X_2'/Z_1')}{d\omega} = \frac{\epsilon_1(\beta_{01}^2 - \beta_{02}^2)}{\epsilon_2 r\beta_z^3}\left(\frac{\beta_z}{\omega} - \frac{d\beta_z}{d\omega}\right) \tag{7.168a}$$

with

$$r \equiv \sqrt{\left(\frac{\beta_{01}^2 - \beta_{02}^2}{\beta_z^2}\right) - 1} \tag{7.168b}$$

That is, the value of $d\beta_z/d\omega$ is computed by equating Eqs. 7.167 and 7.168*a*, which yields

$$\frac{d\beta_z}{d\omega} = \frac{\beta_z}{\omega}\left(\frac{1}{1 + \{(a\epsilon_2\beta_z^3)/[2\nu^2\epsilon_1(\beta_{01}^2 - \beta_{02}^2)]\}}\right) \leq \frac{\beta_z}{\omega} \tag{7.169a}$$

where

$$\nu = \cos(\beta_z a/2) \text{ or } \sin(\beta_z a/2) \tag{7.169b}$$

according to whether we are dealing with Eq. 7.148a or 7.148b respectively.

In the particular case $\omega \rightarrow \omega_{\text{crit}}$, we observe that $\nu^2 \rightarrow 1$ and $\beta_{z_{\text{crit}}}$ is given by Eqs. 7.154 and 7.157. So Eq. 7.169a becomes

$$\left(\frac{d\beta_z}{d\omega}\right)_{\omega \rightarrow \omega_{\text{crit}}} = \sqrt{\epsilon_1\mu_1 - \epsilon_2\mu_2} = \frac{1}{v_1}\sqrt{1 - \left(\frac{v_1}{v_2}\right)^2} \qquad (7.170)$$

which gives us in Eq. 7.166

$$\left(\frac{d\beta_g}{d\omega}\right)_{\omega_{\text{crit}}} = \frac{1}{v_2} \qquad (7.171)$$

Again, the group velocity is dominated by the medium in which most of the field energy is concentrated.

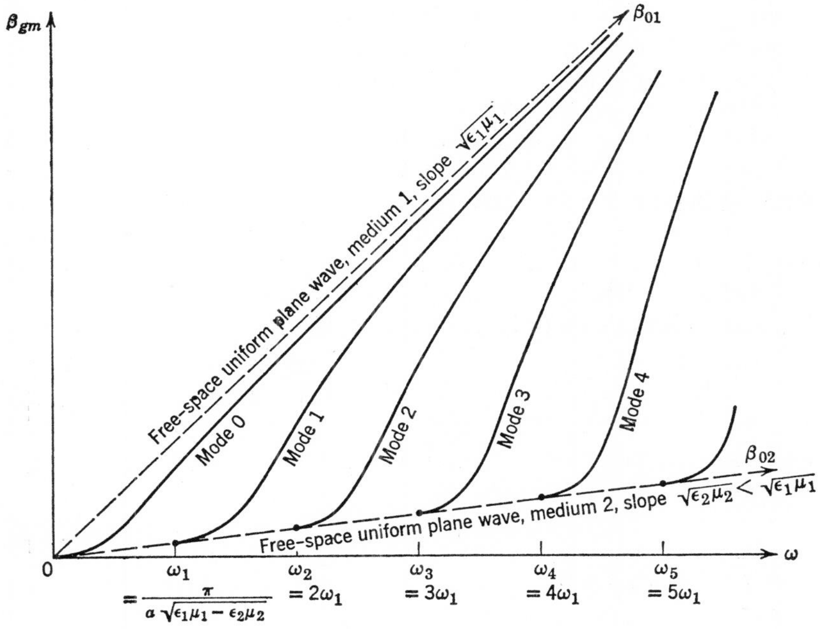

Fig. 7.34. Phase constants β_{gm} versus frequency ω for TM modes for Figs. 7.32 and 7.33. Same general form, with same critical frequencies ω_m, applies to TE waves. *Even-order* modes also apply to TM waves for Fig. 7.35. *Odd-order* modes also apply to TE waves for Fig. 7.35.

At intermediate frequencies $\omega_{crit} < \omega < \infty$, Eqs. 7.164, 7.169*a*, and 7.150 show that

$$\frac{d\beta_g}{d\omega} \geq \frac{{\beta_{01}}^2}{\omega\beta_g} - \frac{\beta_z}{\beta_g}\left(\frac{\beta_z}{\omega}\right) = \frac{\beta_g}{\omega} \tag{7.172}$$

which means that the group velocity cannot exceed the phase velocity in these modes.

Figure 7.34 shows the general form of $\beta_{gm}(\omega)$ based upon all the preceding considerations. It should be compared with Fig. 7.31 for the rectangular metal wave guide. An analysis for TE modes (electric field parallel to the boundaries) yields similar results, with the *same* set of critical frequencies. This case is taken up in the Problems.

7.5.3 Combination Metal and Dielectric Wave Guide *

As our final example, we ask whether a perfectly conducting metal surface, upon which is coated a layer of lossless dielectric, may serve to guide electromagnetic waves. The configuration is shown in Fig. 7.35*a*.

The TM case is most interesting, and we start again with the idea of successive total internal reflection from the boundary at $z = 0$. In this case, however, we shall have metallic reflection at $z = -a/2$, and the equivalent transmission-line picture relevant to the z-axis directions is shown in Fig. 7.35*b*. Compare Figs. 7.35 and 7.32.

Equation 7.144 with $a/2$ in place of a expresses line conditions at $z = -a/2$, *looking to the right*. This time, though, we wish $E_{y1} = 0$ at $z = -a/2$ on account of the metal surface. Looking to the left therefore, the metal requires $\bar{\Gamma}'_{left}(-a/2) = -1$, or looking to the right the metal demands the reciprocal $\bar{\Gamma}'_{right}(-a/2) = (1/-1) = -1$. Using Eq. 7.144 with this condition, and $a/2$ for a yields

$$\bar{\Gamma}'\left(-\frac{a}{2}\right) = e^{j(\psi' - \beta_z a)} = -1 \tag{7.173}$$

which says simply that the metal surface must fall at a node of E_{y1}. Therefore

$$\psi' - \beta_z a = \pm(2n + 1)\pi \qquad n = 0, 1, 2, \cdots \tag{7.174}$$

or in view of Eq. 7.124

$$\frac{X_2'}{Z_1'} = \tan\left(\frac{\beta_z a}{2}\right) \tag{7.175}$$

where only positive values of β_z are of interest. The result (Eq. 7.175) is identical with Eq. 7.148*a*. All other considerations pertinent to the

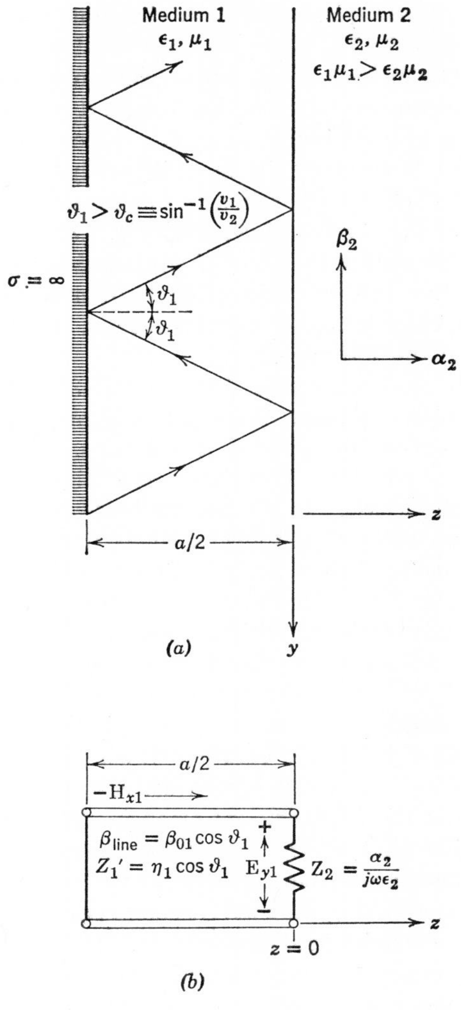

Fig. 7.35. Metal surface with dielectric coating (medium 1) as a wave guide. (*a*) Mechanism of successive reflections, alternately total internal ($z = 0$) and metallic ($z = -a/2$); (*b*) transmission-line representation with respect to z-axis (TM case, **H** parallel to boundaries).

details of solution, *excluding* Eq. 7.148*b* and Fig. 7.33*c*, but *including* Figs. 7.33*a* and 7.33*b*, apply to this case.

In the present problem, then, there are only half as many TM solutions as we had for the dielectric slab alone. In particular, only the cases of *m even* in Eq. 7.157 apply here. This is reasonable because, if we consider the solutions to the dielectric slab problem, symmetry requires that half of them will have nodes of E_{y1} at the middle of the slab and the rest will have maximum values of E_{y1} at the middle. The clearest picture of this situation occurs at the critical frequencies (Eq. 7.157) where the slab width a is seen to be an *even* multiple of $(\lambda_z/2)$ for m even, but an *odd* multiple of $(\lambda_z/2)$ for m odd. With the comments following Eq. 7.163*b*, it is clear that only those solutions with m even do in fact have $E_{y1} = 0$ on the plane $z = -a/2$, and for these cases we could slip in an infinitely thin conducting sheet on this plane to achieve a solution to our present problem. The *even-order* curves of Fig. 7.34 (derived from Fig. 7.33*a*) apply to these TM waves. It is interesting that one of the solutions preserved is that with $\omega_{\text{crit}} = 0$; thus, subject to the general power limitations discussed previously, the dielectric-coated metal sheet can support low-frequency guided waves of the TM type.

Since only Figs. 7.33*a* and 7.33*b* apply to the metal-and-dielectric-guide problem, there are evidently "gaps" in the values of $(\beta_z a)/2$ covered by these TM solutions. For example $\pi/2 < \beta_z a/2 < \pi$ is not possible. These gaps were "filled in" by Fig. 7.33*c* for the dielectric-sheet case. In the present configuration, however, a study of the TE waves (in the Problems) shows that they just fill in the gaps by leading to Fig. 7.33*c*. They lead correspondingly to the odd-order curves of Fig. 7.34. This means, of course, that the TE and TM waves do *not* have the same critical frequencies in this case, and that *there is no TE mode which propagates at very low frequencies (i.e., for which* $\omega_{\text{crit}} = 0$).

An interesting application of the lowest TM mode for "one-wire transmission" was first made by Goubau[1] (Fig. 7.36*a*). He actually uses the similar solution which applies to a round metal wire coated with a dielectric layer (just rust or corrosion will suffice, though it is not the most efficient arrangement). The nature of the field is quite like that of Fig. 7.35*a* if we visualize the metal and dielectric sheet bent around so that the x-axis becomes the perimeter of a circle (φ-coordinate), the z-axis becomes the radial coordinate (ρ), and y becomes the longitudinal direction (z). The coating is thin, and the field usually spreads outward transversely by many wire diameters. Most of the power is carried *outside* the wire and coating. The "launching" arrangement (Fig. 7.36*b*) takes this spreading into account by employing a coaxial-line feed with the outer conductor flared into a horn. A similar "receiving" horn collects the power at the other end. The arrangement is especially good for straight unimpeded runs, as suggested by Fig. 7.36*b*, but it may be used under other circumstances where sharp bends or serious discontinuities do not occur too fre-

[1] George Goubau, "Single-Conductor Surface-Wave Transmission Lines," *Proc. IRE*, **39**, 6, June 1951, pp. 619–624.

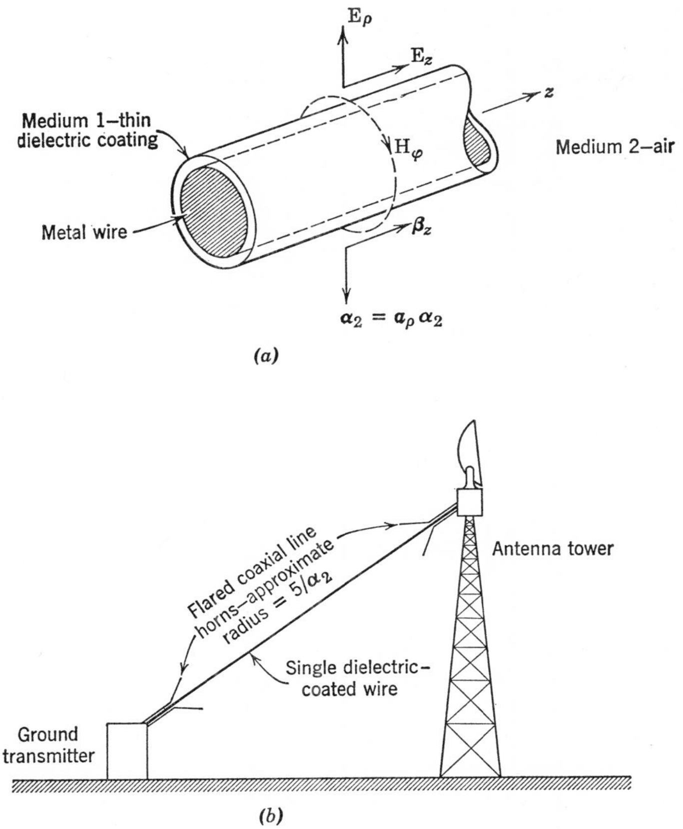

Fig. 7.36. Application of lowest TM mode on dielectric-coated metal surface to "one-wire" transmission. (*a*) Circular form of Fig. 7.35*a*; (*b*) use of single wire, showing flared coaxial horns to launch and receive TM mode.

quently. At these, the critical-reflection guidance feature of the system would fail, and some loss by radiation would result.

We have considered some of the simplest examples of waves guided by lossless media. In more complicated problems, the coordinate systems may become more elaborate, and mixtures of TE and TM waves together may be required to meet the boundary conditions. We have attempted to go only far enough with guided-wave concepts to introduce the major points of physical significance, and to place on firm ground our understanding of uniform and nonuniform plane waves. Some further elaboration of guided waves is taken up in the Problems. To treat additional details here would quickly carry us into the realm of specialized methods and studies.

PROBLEMS

Problem 7.1. (a) Write a complete set of Maxwell's equations for a source-free region with constant parameters ϵ, μ, σ. Specialize the equations to the case in which the electromagnetic fields are independent of x and y. (b) Find the most general solutions to these equations for the z components of the fields and for the charge density. Under what conditions would $\rho = \rho_0$, a constant, be a suitable solution? (c) Find *all* the degenerate solutions for E_x and H_y which have the special forms: (i) $E_x = E_x(z)$; (ii) $E_x = E_x(t)$; (iii) $H_y = H_y(z)$; (iv) $H_y = H_y(t)$. Repeat also for x and y interchanged in (i)–(iv) above. (d) Define the necessary and sufficient conditions to be placed upon Maxwell's equations so that *only* uniform plane waves will be solutions. (e) It has been suggested that because of the possibility of solutions (b) and (c), "light" is not necessarily a transverse wave phenomenon. Do you agree? Support your position with examples.

Problem 7.2. Make perspective sketches, corresponding to those of Figs. 7.1*a* and 7.1*c*, showing the orientations of the field vectors of two orthogonal plane waves traveling in the $-z$ direction.

Problem 7.3. With reference to Sec. 7.1 of the text, consider in the time domain the pair of plane waves, $\boldsymbol{E}_{TA}{}^+ = a\boldsymbol{E}_1{}^+ + b\boldsymbol{E}_2{}^+$, $\boldsymbol{H}_{TA}{}^+ = \frac{1}{\eta}\boldsymbol{a}_z \times \boldsymbol{E}_{TA}{}^+$ and $\boldsymbol{E}_{TB}{}^+ = c\boldsymbol{E}_1{}^+ + d\boldsymbol{E}_2{}^+$, $\boldsymbol{H}_{TB}{}^+ = \frac{1}{\eta}\boldsymbol{a}_z \times \boldsymbol{E}_{TB}{}^+$, in which a, b, c, and d are real constants and $\boldsymbol{E}_1{}^+$ and $\boldsymbol{E}_2{}^+$ are oriented respectively in the $\boldsymbol{a}_x$ and $\boldsymbol{a}_y$ directions. (a) $\boldsymbol{E}_1{}^+$ and $\boldsymbol{E}_2{}^+$ are linearly independent *vectors* because they lie in different (in fact, perpendicular) directions. Are $|\boldsymbol{E}_1{}^+|$ and $|\boldsymbol{E}_2{}^+|$ always linearly independent? Give examples. (b) Determine necessary and sufficient conditions for $\boldsymbol{E}_{TA}{}^+$ and $\boldsymbol{E}_{TB}{}^+$ to be linearly independent. Describe your results in words. (c) Waves A and B (described respectively by $\boldsymbol{E}_{TA}{}^+$ and $\boldsymbol{E}_{TB}{}^+$) are said to be orthogonal in the time domain if the instantaneous power density carried by waves A and B together equals the sum of the power densities of the individual waves. What restrictions must be imposed on $\boldsymbol{E}_{TA}{}^+$ and $\boldsymbol{E}_{TB}{}^+$ if the two waves are to be orthogonal in the time domain? State your result in words. (d) Express the restrictions found in (c) in terms of the components $\boldsymbol{E}_1{}^+$ and $\boldsymbol{E}_2{}^+$ of $\boldsymbol{E}_{TA}{}^+$ and $\boldsymbol{E}_{TB}{}^+$, and interpret the results for all possible circumstances regarding $\boldsymbol{E}_1{}^+$ and $\boldsymbol{E}_2{}^+$.

Problem 7.4. The components of the complex electric field of a uniform plane wave propagating in the $+z$ direction are $\mathrm{E}_{x0}{}^+ = |\mathrm{E}_{x0}{}^+|e^{j\vartheta_x}$, $\mathrm{E}_{y0}{}^+ = |\mathrm{E}_{y0}{}^+|e^{j\vartheta_y}$. (a) Show that the locus in time of the tip of the electric field vector satisfies the equation

$$\left(\frac{E_x{}^+}{|\mathrm{E}_{x0}{}^+|}\right)^2 - 2\frac{E_x{}^+E_y{}^+}{|\mathrm{E}_{x0}{}^+|\,|\mathrm{E}_{y0}{}^+|}\cos\delta + \left(\frac{E_y{}^+}{|\mathrm{E}_{y0}{}^+|}\right)^2 = \sin^2\delta$$

where $\delta = \vartheta_y - \vartheta_x$. (b) Show that the coordinate axes, which in general do not coincide with the principal axes of the ellipse obtained above, can be brought into coincidence with them by a rotation about the z-axes through an angle φ where

$$\tan 2\varphi = \frac{2|E_{x0}^+||E_{y0}^+|}{|E_{x0}^+|^2 - |E_{y0}^+|^2} \cos \delta$$

(c) Under what conditions does the ellipse degenerate into a straight line? a circle? (d) A uniform method of describing the sense in which the $\boldsymbol{E}$-vector traces out the polarization ellipse has not been adopted in the scientific literature. The most common practice is to view the fields looking in the direction *opposite* to the direction of propagation, i.e., in the negative z direction here, and then describe the polarization as being "right-handed" or "left-handed," depending on whether it is clockwise or counterclockwise respectively in this view. Under what analytical conditions is the polarization right-handed?

Problem 7.5. The components in the x, y-plane of the complex electric field of a plane wave are E_x and E_y. The complex ratio $R_E = E_y/E_x$, which can be represented as a point in a complex plane, determines the polarization in the x, y-plane. In other words, each point of what may be called the R-plane has associated with it a different elliptic polarization. (a) Prepare a sketch of the R-plane in which at representative points (e.g., points on the unit, ½ and 2 circles which intersect the axes and the 45° lines) a small ellipse is drawn showing (roughly) the inclination and relative ratio of the axes of the polarization ellipse as well as the direction of rotation about the ellipse. See Prob. 7.4. (b) If the E_x and E_y are components of a uniform plane wave propagating in the $+z$ direction, what is the ratio R_H of the complex magnetic field components in terms of R_E? Describe how the diagram constructed in (a) can be used to obtain the polarization ellipse of the magnetic field corresponding to a given polarization of the electric field. (c) A certain medium has the characteristic that it propagates plane waves in pairs, such that to each wave with polarization in the x, y-plane described by the ratio R_{E1} corresponds another wave with ratio $R_{E2} = 1/R_{E1}$. Describe how the sketch prepared in (a) can be used to relate the polarizations of corresponding waves.

Problem 7.6.[1] Two statistically independent noise voltages, each of zero mean value, are applied respectively to the horizontal and vertical plates of an oscilloscope. (a) Sketch one of the patterns that might be traced on the scope face in a short observation interval. (b) Suppose that the two noise voltages have equal rms values and that each is characterized by a Gaussian amplitude distribution. Show that, if these voltages are regarded as the x and y components of the electric field of a uniform plane wave, this wave will be "randomly" polarized, as defined in the text. (c) Is the polarization still "random" if the two noise voltages are Gaussian, but have different rms values? Explain. (d) Let the probability densities of the two noise voltages be $p(x)$ and $q(y)$ respectively. Find all pairs of densities $[p, q]$ such that the polarization of the uniform plane wave discussed in (b) is "random."

Problem 7.7. The complex electric field vector of an electromagnetic wave in free space (vacuum) is given by the expression $\mathbf{E} = 10^{-4}(\boldsymbol{a}_x - j\boldsymbol{a}_y)e^{-j20\pi z}$ v/m. (a) Find the frequency f. (b) Sketch the instantaneous electric field vector $\boldsymbol{E}(t, z)$ at $z = 0$,

[1] This problem requires a little familiarity with the principles of probability and statistics.

showing on a single diagram its magnitude and orientation at times $t = 0$, $t = 1/4f$, $t = 1/2f$, and $t = 3/4f$. (c) Repeat (b) at $z = 0.025$ m. (d) What is the *type* of polarization of the wave? (e) Find the complex magnetic field **H**. (f) Repeat (b) and (c) for the instantaneous magnetic field vector $\boldsymbol{H}(t, z)$. (g) Find the complex Poynting vector **S** and the instantaneous Poynting vector $\boldsymbol{S}$ for the wave.

Problem 7.8. Given the following complex amplitude for a sinusoidal electric field in a vacuum $\mathbf{E} = 10(\boldsymbol{a}_x + j0.4\boldsymbol{a}_y + j0.3\boldsymbol{a}_z)e^{+j0.6y}e^{-j0.8z}$. (a) What kind of disturbance is represented by this field and what is its frequency? (b) What is its direction of propagation? (c) What is its state of polarization? Explain. (d) Find the associated magnetic field. (e) Find the average power flow per square meter normal to the direction of propagation.

Problem 7.9. A traveling uniform plane wave propagates in air in a direction making equal acute angles with the $+x$-, $+y$-, and $+z$-axes. The electric field vector lies at all times in a plane parallel to the x, y-plane, and at $x = y = z = 0$ has a magnitude, $|\boldsymbol{E}_1(0, 0, 0, t)| = f(t)$. (a) Express analytically the electric and magnetic fields, $\boldsymbol{E}_1(x, y, z, t)$ and $\boldsymbol{H}_1(x, y, z, t)$, of the wave. (b) Express analytically the electric and magnetic fields, $\boldsymbol{E}_2(x, y, z, t)$ and $\boldsymbol{H}_2(x, y, z, t)$, of a second traveling uniform plane wave that is propagating in the same direction as wave 1, has $|\boldsymbol{E}_2(0, 0, 0, t)| = f(t)$, but is orthogonal to wave 1.

Problem 7.10. A uniform plane wave is moving in the z direction with $\boldsymbol{E} = \boldsymbol{a}_x 100 \sin(\omega t - \beta z) + \boldsymbol{a}_y 200 \cos(\omega t - \beta z)$. (a) Express $\boldsymbol{H}$ by use of Maxwell's equations. (b) If the wave encounters a perfectly conducting x, y-plane at $z = 0$, express the resulting $\boldsymbol{E}$ and $\boldsymbol{H}$ for $z < 0$. (c) Find the magnitude and direction of the surface current density on the perfect conductor.

Problem 7.11. A uniform plane wave in free space strikes normally a semi-infinite slab of lossless material. In the free space, the standing wave ratio is 3. In the material, the wave length is shorter by a factor of 6 than it is in free space. Find the relative permeability μ/μ_0 and relative permittivity ϵ/ϵ_0 of the material.

Problem 7.12. A uniform plane wave, $f = 3.75 \times 10^7$ cps, x-polarized, strikes normally a slab of lossless dielectric backed by a perfectly conducting layer. The dielectric has $\epsilon/\epsilon_0 = 4$ and is 1.0 m thick. (a) What is $\bar{\Gamma}$ at the air-dielectric interface? (b) Sketch to scale the amplitudes $|E_x|$ and $|H_y|$ as functions of z outside and inside the dielectric. (c) If the thickness of the dielectric is 2.0 m, what change occurs in (b)?

Problem 7.13. (a) Find the three lowest frequencies at which all the incident power in Fig. 7.37 will be transmitted. The permeability of all media is μ_0. (b) If complete

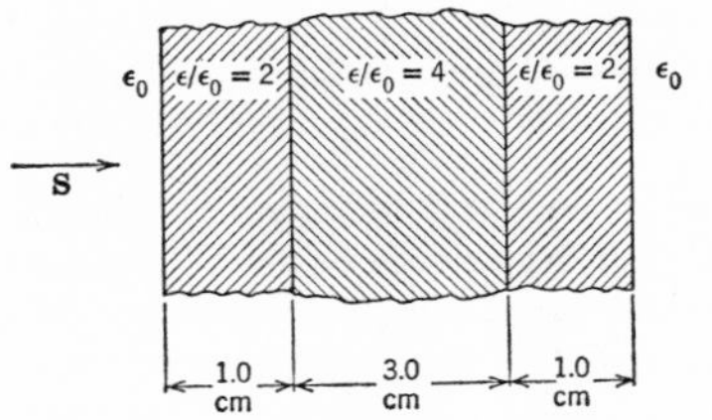

Fig. 7.37. Problem 7.13.

transmission is required for *any* thickness of the center medium, what is the lowest usable frequency? (c) For the situation shown in the figure, find the bandwidth of the transmission between its two lowest percentage values adjacent to and on either side of the frequency of (b). Find also these lowest percentage values of the transmission. (d) Why does the reflection from the modern coated optical lenses tend to be purple in color?

Problem 7.14. A slab of lossless dielectric has constant parameters ϵ_1, μ_1, and thickness l. It is interposed normal to the direction of propagation of a uniform plane wave in free space. (a) Sketch the transmission-line analog of the system. (b) Sketch the transmission efficiency of the sheet versus its thickness measured in *free space* wave lengths. Take $\mu_1 = \mu_0$ and $\epsilon_1 = 2.25\epsilon_0$ as an example. Sketch on the same axes the reflection efficiency. (c) Under what conditions (on ϵ_1, μ_1, l, and the frequency) does the slab behave with respect to points outside it like a lumped capacitor shunted across a free-space transmission line? (d) The results of (a) suggest that a pane of window glass (for which $\epsilon \cong 2.25\epsilon_0$ at optical frequencies) can distort the color of a scene viewed through it, and, particularly, of a scene reflected in it! Why is this suggestion false? Calculate the number of maxima of the transmission or reflection efficiency of a ¼-in. sheet of glass for normal incidence in the wave-length range of visible light (4×10^{-7} m $< \lambda_0 < 7 \times 10^{-7}$ m). (e) Make a rough estimate of the thickness of a film of oil floating on water if parts of the film appear blue and parts appear red when viewed with reflected light. The optical properties of the oil are roughly the same as those of the window glass.

Problem 7.15. (a) A parallel-faced slab of dielectric (medium b) of thickness l separates two different dielectric regions (media a and c). Calculate the squared magnitude of the reflection coefficient, $|\bar{\Gamma}_{12}|^2$, for a monochromatic uniform plane wave at normal incidence from medium a. Assume there is no reflected wave in medium c. Express your result in terms of Γ_1, the reflection coefficient that would apply at boundary 1 between media a and b if this were the only boundary in existence, and Γ_2, the corresponding reflection coefficient at boundary 2 between media b and c. Interpret $1 - |\bar{\Gamma}_{12}|^2$ physically. (b) A source of broadband visible light, whose continuous power density spectrum is fairly flat over many periods of the function $|\bar{\Gamma}_{12}|^2$ found above, illuminates the slab. Assuming that the eye or other optical instrument responds to the mean square value of the reflected field strength, deduce the optical reflection efficiency $\langle|\bar{\Gamma}_{12}|^2\rangle$ of the slab thus measured. Express the answer in terms of parameters $g_1 = \Gamma_1^2$ and $g_2 = \Gamma_2^2$. (c) Derive the result of (b) by adding up the *powers* of the various multiply reflected components of the composite reflected wave. Calculate the optical transmission efficiency of the double boundary in the same way. (d) Optically, in situations like that discussed above, n parallel sheets of glass separated by air constitute $2n$ equally reflecting boundaries in cascade. The sheet thicknesses and spacings are irrelevant. From your result of (b), calculate the optical transmission and reflection efficiencies at normal incidence of such a multiple boundary.

Problem 7.16. A dielectric slab (medium 2 of Fig. 7.38a) extends over the entire x, y-plane. The system between the "input" and "output" planes is to be regarded as a filter whose instantaneous input and output functions are E_{xi} and E_{xo} respectively. The filter input is a normally incident uniform plane wave. (a) For $\epsilon_2 = 36\epsilon_0$, find the impulse response of the system. (b) It is desired to modify the filter so that it has the new impulse response shown in Fig. 7.38b, in which A can be any

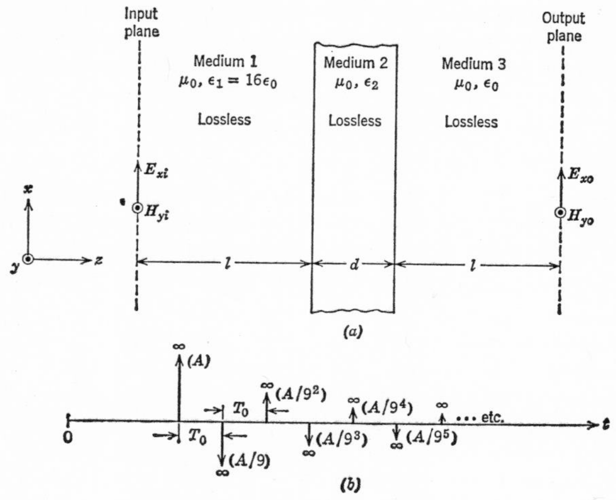

Fig. 7.38. Problem 7.16.

positive constant and $T_0 = \frac{4}{3} \times 10^{-8}$ sec. The time interval between $t = 0$ and the first output impulse is of no concern. Find the thickness d and the permittivity ϵ_2 of the dielectric slab required to obtain the desired impulse response.

Problem 7.17. In the sinusoidal steady state at frequency ω, a linearly polarized uniform plane wave is incident at angle ϑ upon a perfect conductor whose surface defines the x, y-plane. The incident magnetic field is parallel to the conductor, and has an amplitude H_0 which it attains in the $+x$ direction at $x = y = z = t = 0$. (a) Write expressions for the complex-vector electric and magnetic fields $\mathbf{E}_i(x, y, z)$ and $\mathbf{H}_i(x, y, z)$ of the *incident* wave. (b) Write expressions for the complex-vector electric and magnetic fields $\mathbf{E}(x, y, z)$ and $\mathbf{H}(x, y, z)$ of the complete field solution to the problem. (c) Write the expression for the real instantaneous surface current density vector $\boldsymbol{K}(x, y, t)$ amp/m on the metal. (d) Write the expression for the real instantaneous surface-charge density $Q_s(x, y, t)$ coulombs/m^2 on the metal. (e) Repeat for a circularly polarized incident wave for which $\boldsymbol{H}(0, 0, 0, 0) = \boldsymbol{a}_x H_0$. (f) For the case (e), when $\vartheta = 0$, determine the instantaneous and complex Poynting vectors, $\boldsymbol{S}(x, y, z, t)$ and $\mathbf{S}(x, y, z)$ respectively, and interpret them.

Problem 7.18. A uniform plane wave is incident at angle ϑ_1 upon the interface between two lossless dielectrics. The (real) angle of refraction is ϑ_2. When the polarization is parallel to the interface, the reflection coefficient is $\frac{1}{2}$. (a) Find the numerical value of the transmission coefficient. (b) If the propagation direction of the transmitted wave is reversed, thus making this wave become a new incident wave with the same polarization as before, what will be the numerical values of the reflection and transmission coefficients? (c) If $\vartheta_1 = 60°$ and $\vartheta_2 = 30°$ above, repeat (a) and (b) for these same values of $\vartheta_{1,2}$, but with the polarization rotated 90°.

Problem 7.19. A uniform plane wave of frequency ω polarized in the plane of incidence is obliquely incident at angle ϑ from air onto a lossless dielectric with $\mu = \mu_0$ and $\epsilon > \epsilon_0$. Measurement of standing wave ratio s is made with a probe sensitive only to *tangential* components of electric field, and moving normal to the interfaces, for the following two cases: *Case 1*—$\vartheta = 30°$, $s = s_1$; *Case 2*—$\vartheta = 45°$, $s = s_2$. It is found that $s_2/s_1 = 7/3$. Find: (a) The specific permittivity ϵ/ϵ_0. (b) The value of s_1.

Problem 7.20. (a) A flat plastic plate ($\epsilon/\epsilon_0 = 2.25$) is 2 in. thick. A uniform-plane electromagnetic wave of frequency 10 kmc is incident on the plate at an angle ϑ_1 (from the normal). The polarization is parallel to the surface. Find *all* values of the angle ϑ_1 for which reflections are eliminated. (b) Repeat part (a) with the wave polarized in the plane of incidence.

Problem 7.21. An experiment is being designed to illustrate clearly Brewster's angle for a given plastic [$(\epsilon/\epsilon_0) = 2.25$, $\mu = \mu_0$, $\sigma = 0$]. The method consists of measuring directly the reflected uniform plane wave field from a sheet of the material, as a function of incident angle, at a microwave frequency $f = 10$ kmc. The plastic is very expensive, so minimum thickness is desirable. On the other hand, if the reflections are never very strong at any angle, the sensitivity of the null is poor. There is also the problem of avoiding "false" nulls (see Prob. 7.20) within the angle range considered practical (say, 0°–85°). (a) Determine reasonable limits upon the thickness of the plastic sheet. (b) Discuss the effect upon the experiment if the incident plane wave source actually emits a pencil of waves with directions lying in a cone of small half-angle α rad, and if the receiver for the reflected wave can be twisted about its own axis by $\pm\alpha$ without changing its indication of a given uniform plane wave directed at it (twisting by more than $\pm\alpha$ rad gives zero indication). Consider the relationships between the transverse sheet dimensions and the distances between source, sheet, and receiver. Comment also on what factors fix the angle range usable in the experiment.

Problem 7.22. A plane wave which in air has a 10 cm wave length is incident on an air-dielectric interface from the *dielectric* side. The permeability of the dielectric is equal to that of air, and $\epsilon_1 = 4\epsilon_0$. (a) What is the critical angle for no transmitted power into the air? (b) A nonmagnetic coating is to be added to the dielectric to make the interface nonreflecting at normal incidence, again from the dielectric side. What thickness d and dielectric constant ϵ_2 should this coating have? (c) Repeat (a) with the coating determined in (b) in place.

Problem 7.23. Prove or disprove: There is always one and only one angle of incidence of a randomly polarized uniform plane wave on a boundary between two lossless media, characterized by constant parameters (ϵ_1, μ_1) and (ϵ_2, μ_2), at which the reflected wave is linearly polarized. (*Caution:* Bear in mind that if the angle of incidence exceeds the critical angle, the incident wave is totally reflected.) Relate the polarization of the reflected wave to the parameters of the two media for those cases in which the reflected wave is linearly polarized.

Problem 7.24. The boundary between two lossless dielectrics is the plane $z = 0$. The wave impedance and speed of light in the left-hand ($z < 0$) dielectric are η_1 and v_1; in the right-hand dielectric these quantities are η_2 and v_2 respectively. A traveling wave having the electric field

$$\boldsymbol{E}_i = \boldsymbol{a}_x f\left(t + \frac{y \sin \vartheta_1 - z \cos \vartheta_1}{v_1}\right) \qquad z \leq 0$$

is incident on the boundary from the left. (a) Express both electric and magnetic fields of the incident, reflected, and transmitted waves if ϑ_1 is less than the critical angle. (b) Let $f(\tau)$ be the rectangular pulse

$$f(\tau) = \begin{cases} 1, & |\tau| \leq \frac{1}{2} \\ 0, & |\tau| > \frac{1}{2} \end{cases}$$

and let $\vartheta_1 = 60°$, $v_2 = \frac{1}{2}v_1$, and $\eta_2 = \eta_1$. Sketch and dimension the contours in the y, z-plane: amplitude of total electric field = constant. Make your contour "map" at the instant $t = 0$, and indicate with arrows how these contours change as time advances. (c) Now suppose $v_2 > v_1$, and the angle of incidence exceeds the critical angle. The electric field of the incident traveling wave is expressed in terms of the Fourier transform $\mathrm{F}(\omega)$ of the function $f(\tau)$:

$$\boldsymbol{E}_i = \frac{1}{2\pi} \int_{-\infty}^{+\infty} \boldsymbol{a}_x \mathrm{F}(\omega) e^{j\omega\{t+[(y \sin \vartheta_1 - z \cos \vartheta_1)/v_1]\}}\, d\omega$$

or

$$\boldsymbol{E}_i = \frac{1}{2\pi} \int_{0}^{+\infty} 2 \operatorname{Re} [\boldsymbol{a}_x \mathrm{F}(\omega) e^{j\omega\{t+[(y \sin \vartheta_1 - z \cos \vartheta_1)/v_1]\}}]\, d\omega$$

since $\boldsymbol{E}_i$ is a real vector. Express similarly both the electric and magnetic fields of the reflected and transmitted waves. (d) If you had used the first given form in answering (c), would your transmission and reflection coefficients depend on frequency? How? What if you had used the second given form? What about the coefficient of z, corresponding to $-\cos \vartheta_1/v_1$ above, in the expressions for the transmitted fields? Find two *different* arguments to support your answers to these questions. (e) Let $f(\tau)$ be a unit impulse at $\tau = 0$. From your answer to (c), calculate, by actually performing the integrations, the electric field in the x, z-plane on the right of the boundary. Does your calculation show this field to be zero before the arrival of the incident wave at the origin? Explain.

Problem 7.25. Consider Maxwell's equations for a lossless source-free region with constant permittivity ϵ and permeability μ. (a) Show that the rectangular vector components of the steady-state complex electric field $\mathbf{E}$ obey the equation $\nabla^2\mathbf{E} + k^2\mathbf{E} = 0$ ($k = \omega\sqrt{\epsilon\mu}$) and that $\mathbf{H}$ obeys the same equation. (b) Under what conditions on the complex constants $\bar{\gamma}_x$, $\bar{\gamma}_y$, and $\bar{\gamma}_z$ is $\mathbf{E}(x, y, z) = \mathbf{E}_0 e^{-(\bar{\gamma}_x x + \bar{\gamma}_y y + \bar{\gamma}_z z)}$ a solution to the equation in (a), with $\mathbf{E}_0$ a constant complex vector?

Problem 7.26. The complex fields produced in a certain isotropic, homogeneous, linear, time-invariant, source-free medium by a 100-mc sinusoidal source are

$$\mathbf{E} = \mathrm{E}_{0x}\boldsymbol{a}_x - 20\pi[(13 - j45)\boldsymbol{a}_y + (26 + j60)\boldsymbol{a}_z]e^{-\pi[(1+j5)x-(3+j)y+(4-j2)z]}$$
$$\mathbf{H} = (\mathrm{H}_{0x}\boldsymbol{a}_x + \mathrm{H}_{0y}\boldsymbol{a}_y + \mathrm{H}_{0z}\boldsymbol{a}_z)e^{-\pi[(1+j5)x-(3+j)y+(4-j2)z]}$$

and it is known that $|\mathrm{H}_{0z}| = 7$. (a) What can be said about the medium from the *space variation* of the fields? (b) Where is the source? Give a unit vector normal to a plane which separates a region in space where the fields are bounded from one where the fields are arbitrarily large. Let the unit vector point toward the strong-field region. (c) Determine the unspecified field components, E_{0x}, H_{0x}, H_{0y}, and H_{0z}. (d) What are the conductivity, relative permeability, and dielectric constant of the medium? (e) Describe the fields in words. Exactly what kind of wave do they represent? (f) [1] Determine the (complex) polar angles $\bar{\vartheta}$ and $\bar{\varphi}$ of the direction of propagation of the equivalent uniform plane wave.

[1] The considerable algebra required for this part should be done carefully or not at all.

Problem 7.27. A 900-mc nonuniform TE plane wave propagates in air in such a manner that its phase *increases* most rapidly with position in a direction parallel to, and lying in the first quadrant of, the x, y-plane. The amplitude of the x-component of the electric field is 3 v/m throughout the plane $x - 2y + 2z = 0$; in the plane $x - 2y + 2z = 1/\pi$ m, it is $3e$ v/m. (a) Determine the (complex) propagation vector of the wave. (b) Express the most general fields that fit the problem statement. (c) Show that the complex polar angles for the direction of propagation of the equivalent uniform plane wave are $\bar{\vartheta} = \pi/2 - j \sinh^{-1} \frac{1}{3}$ and $\bar{\varphi} = \tan^{-1} \frac{1}{2} + j \tanh^{-1} \frac{1}{3}$, with Re $(\bar{\varphi})$ in the third quadrant. Do not use a slide rule or tables.

Problem 7.28. Consider fields of the form $\boldsymbol{E} = \text{Re}\,\{\mathbf{E}e^{-\Omega t}\}$, $\mathbf{E} = \mathbf{E}_0 e^{-\bar{\boldsymbol{\gamma}}\cdot\boldsymbol{r}}$, $\boldsymbol{H} = \text{Re}\,\{\mathbf{H}e^{-\Omega t}\}$, $\mathbf{H} = \mathbf{H}_0 e^{-\bar{\boldsymbol{\gamma}}\cdot\boldsymbol{r}}$ in a lossless medium with Ω a pure real number and $\mathbf{E}_0$, $\mathbf{H}_0$ complex vectors. (a) If these fields are to satisfy Maxwell's equations, what constraints must be imposed on $\mathbf{E}_0$, $\mathbf{H}_0$, and $\bar{\boldsymbol{\gamma}}$? On $\mathbf{E}_0$ and $\bar{\boldsymbol{\gamma}}$? On $\mathbf{H}_0$ and $\bar{\boldsymbol{\gamma}}$? On $\bar{\boldsymbol{\gamma}}$ alone? (b) Express $\boldsymbol{E}$ and $\boldsymbol{H}$ in the case $\bar{\gamma}_x = \bar{\gamma}_y = \mathrm{E}_{0y} = \mathrm{E}_{0z} = 0$. Describe these fields in words. How could they be excited? Could a "uniform plane wave source" be used? (c) Examine $\boldsymbol{E}$ and $\boldsymbol{H}$ in the more general case, $\bar{\boldsymbol{\gamma}} = \boldsymbol{\alpha} + j\boldsymbol{\beta}$. Interrelate $\boldsymbol{\alpha}$ and $\boldsymbol{\beta}$. Describe these fields in words. How could they be excited? (*Hint:* Study Sec. 7.4.3.)

Problem 7.29. As illustrated in Fig. 7.39, two perfectly conducting and infinitesimally thin sheets in air form a *semi*-infinite parallel-plate wave guide, with mouth in the plane $z = 0$ and sides parallel to the y, z-plane. Two y-polarized, steady-state uniform plane waves (1 and 2) of equal strength are incident upon the mouth of the guide at angles ϑ shown, and their separate equiphase surfaces of the same phase intersect in lines lying in the y, z-plane. (a) If the wave length λ of the incident waves is such that $\sin \vartheta = \lambda/2a = \sqrt{3}/2$ and the peak field strength in each is 1 v/m, find the time-average power flowing down the *inside* of the guide, per meter of the y dimension. (b) Repeat (a) if wave 2 is turned off. (c) Repeat (a) if wave 2 is present but reversed in time phase. Check your result by verifying the relationships between the answers to (a), (b), (c) by a different method than the one used to get them. (d) Suppose ϑ has a given arbitrary real value, not specially related to λ, and we determine by measurement that power $\langle P_A \rangle$ goes down the inside of the guide (per meter of y dimension) when wave 1 *alone* is incident at power density level $\langle S_1 \rangle$. In terms of $\langle P_A \rangle$ find: (i) The power down the guide if wave 2 is turned on with $\langle S_2 \rangle = \langle S_1 \rangle$ and with identical phase at the origin; (ii) The result of (i) if wave 2 has reversed time phase.

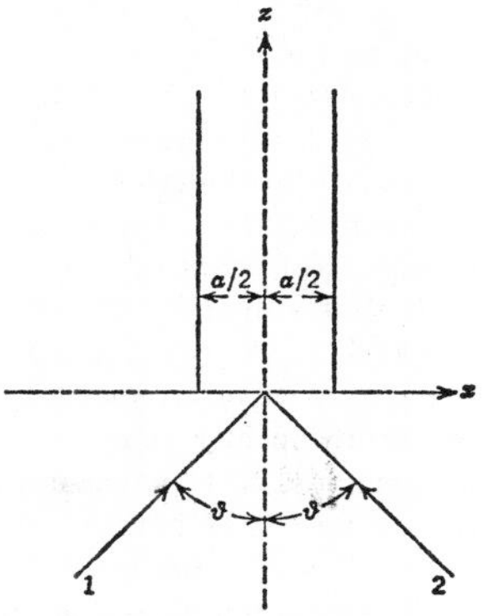

Fig. 7.39. Problem 7.29.

Problem 7.30. A long piece of 3-cm rectangular wave guide (inside dimensions 0.4 in. × 0.9 in.) is filled with polystyrene blocks ($\epsilon = 2.5\epsilon_0$) so that the $\mathrm{TE}_{1,0}$ mode will now propagate in the guide at 5 kmc. One end of the guide is connected to a matched load and the other to a 5-kmc source. In assembling the equipment, two of the blocks are separated so that there is a rectangular air-filled section of guide, having a length l and located halfway between source and load. You are to determine fields and power flow in the vicinity of the air gap. (a) Show that only the $TE_{1,0}$ mode transverse pattern is needed to meet boundary conditions in each of the three sections, and that the longitudinal parts of the problem may be solved by transmission-line analogy. (b) The fields in the air-filled section can be decomposed into two waves—one that decays exponentially with distance along the guide as it approaches the load end of the section, and one that decays exponentially as it approaches the source end. How much mean power is carried toward the load by each of these waves separately? Can any power reach the load when both are present together? (See Prob. 3.49.) (c) Calculate the ratio of the total electric field in the center of the guide at the boundary of the air-filled section nearest the load to the corresponding quantity at the boundary nearest the source. Show that the magnitude of this ratio does not exceed 1. (d) Calculate and sketch, as a function of l, the ratio of the load power to the power of the source wave that is *incident* on the air-filled section from the dielectric-filled one. (e) Calculate the VSWR, for $l = 1$ in., in those sections of the guide in which the VSWR concept applies. From these results, and those of (d), discuss the electrical limitations to making out of this device an attenuator for which loss in decibels will be directly proportional to length l.

Problem 7.31. Consider a $TE_{m,0}$ (+)-wave in a lossless rectangular wave guide of wide dimension a and narrow dimension b. Define "voltage" V_+ as the line integral of the electric field up the center of the cross section, along a line parallel to the narrow side. Define current I_+ so that $\frac{1}{2}V_+I_+^*$ gives correctly the total complex power carried longitudinally through the cross section (as determined from the complex Poynting vector). (a) From the values of $\bar{\gamma}$ and $Z_0 \equiv V_+/I_+$, determine the series impedance per unit length Z_s and the shunt admittance per unit length Y_p of an equivalent transmission line for this mode. (b) Make a circuit diagram representing a length dz of the line in (a). In this diagram *all* element values must be *independent* of frequency. How does the cutoff frequency show in the equivalent circuit? (c) Identify the energy stored in each element of the equivalent circuit with that stored in the actual mode by one space component of the electric or magnetic field. (d) If a different line integral is taken to define V_+', so that $V_+' = KV_+$, but we still choose I_+' such that $\frac{1}{2}V_+'I_+'^*$ represents the correct complex power, reconsider your results in (a)–(c). (e) Could we choose to define "voltage" proportional to transverse *magnetic* field and "current" proportional to transverse *electric* field? Illustrate. (f) If we chose "current" proportional to *longitudinal* magnetic field, what choices are open for "voltage"? Repeat if "voltage" is chosen proportional to *longitudinal* field. Illustrate.

Problem 7.32. In the rarefied upper atmosphere several hundred kilometers above the earth's surface there exists a region of dense, horizontally stratified ionization caused by solar radiation. The several ionized layers are known collectively as the ionosphere. The parameters of an ionospheric layer may be taken to be $\sigma = 0$, $\mu = \mu_0$, $\epsilon = \epsilon_0[1 - (\omega_p/\omega)^2]$, where ω_p, the plasma frequency, is proportional to the square root of the electron density. (a) Obtain the propagation constant $\bar{\gamma}$ and the wave impedance $\bar{\eta}$ of a uniform plane wave in such an ionospheric layer. (b) A uniform plane wave in air of electric field $\boldsymbol{E}_1 = \boldsymbol{a}_x \operatorname{Re}[\mathbf{E}_1 e^{j(\omega t - \beta_0 z)}]$ strikes a hypothetical abrupt boundary $z = 0$ of an ionospheric layer from below. Neglecting reflections from the upper boundary of the layer, express the transmitted and reflected fields in the cases $\omega > \omega_p$ and $\omega < \omega_p$. Sketch the squared magnitude of the reflection coefficient as a function of frequency for $0 < \omega < 2\omega_p$. (The reflection coefficient of an actual ionized layer which does not have abrupt boundaries diminishes much more rapidly with frequency when $\omega > \omega_p$.) (c) On a particular date and time of day, and above a particular geographical location, the densest ionization occurs at an altitude of 400 km, and the plasma frequency of this dense ionization is $\omega_p/2\pi = 5$ mc. Neglecting ionization below this densest layer, determine the greatest possible angle of incidence ϑ (Fig. 7.40) at which radiation from a ground-based transmitter could possibly strike the layer. (d) How far away from the transmitter will the reflected radiation return to earth? (e) What is the highest frequency at which this obliquely incident radiation will be totally

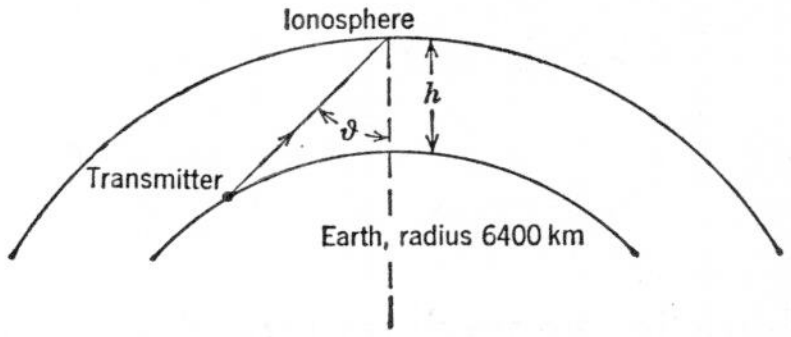

Fig. 7.40. Problem 7.32.

reflected? (f) A radar set is used to measure the distance to the moon at a time when the moon is at the zenith. The bandwidth of the radar signal is small compared with the difference between the center frequency of the radar signal and the plasma frequency of the ionosphere. Calculate and sketch, as a function of the radar signal center frequency, the discrepancy between the measured and actual distances to the moon caused by passage of the radar signal through a uniformly dense layer of ionization 100 km thick. What range of center frequencies may be used if this discrepancy is to be less than 1 km? Note that the model of the ionosphere implied by the problem statement has abrupt boundaries and thus would support multiple reflections of the transmitted signal and of the moon echo, both inside the ionized layer and between its lower boundary and the earth. For the purposes of this problem, it may be supposed that the radar signal is a pulsed sinusoid of short enough duration so that the first-arriving moon echo may be selected from the array of other echoes caused by unwanted or multiple reflections.

C H A P T E R   E I G H T

Plane Waves in Dissipative Media

The presence of loss in the medium modifies somewhat the nature of electromagnetic plane waves. The dispersion introduced by the conductivity makes general solution in the time domain impossible except by Fourier expansion methods. Therefore, as we did for the transmission line with losses, we will explore primarily the solutions for the sinusoidal (or complex-frequency) steady state. Both uniform and nonuniform plane waves arise again, altered from their forms in lossless media to provide for the power dissipation associated with the conductivity. There emerges the relation between these plane waves and the phenomenon of skin effect when we consider waves bounded or guided by real metal conductors. We shall try to emphasize primarily the new features produced by the dissipation, so Chapter 7 is a prerequisite to this one.

8.1 Plane Waves (Frequency Domain)

To provide some contrast with our procedure in the last chapter, we shall commence the present discussion by obtaining the most general simple exponential solution to Maxwell's equations in the sinusoidal steady state. We will then consider various special cases of that solution, which lead to TEM uniform plane waves, and TE or TM (nonuniform) plane waves.

8.1.1 Exponential Solution

8.1.1.1 Form of Solution. In a homogeneous, isotropic, time-invariant, linear, source-free medium with conductivity $\sigma \neq 0$, the steady-state form of the Maxwell equations is given by Eqs. 1.50 and 1.51:

$$\begin{aligned} &\text{(a)} \qquad \nabla \times \mathbf{E} = -j\omega\mu\mathbf{H} \\ &\text{(b)} \qquad \nabla \times \mathbf{H} = \sigma\mathbf{E} + j\omega\epsilon\mathbf{E} = (\sigma + j\omega\epsilon)\mathbf{E} \end{aligned} \tag{8.1}$$

These become identical to Eqs. 7.39 if in the latter $(\sigma + j\omega\epsilon)$ takes the place of $j\omega\epsilon$ or, alternatively, if $[\epsilon - j(\sigma/\omega)]$ replaces ϵ.

We search for a solution to Eqs. 8.1 of the form

$$\begin{aligned} &\text{(a)} \qquad \mathbf{E} = \mathbf{E}_0 e^{-\bar{\boldsymbol{\gamma}}\cdot r} \\ &\text{(b)} \qquad \mathbf{H} = \mathbf{H}_0 e^{-\bar{\boldsymbol{\gamma}}\cdot r} \end{aligned} \tag{8.2}$$

identical with Eqs. 7.37. The notation is as described in connection with Eqs. 7.38.

Replacement of $j\omega\epsilon$ by $(\sigma + j\omega\epsilon)$ in Eqs. 7.40, 7.41, and 7.42 makes them applicable immediately to the case at hand. Taking them in the above order, we find

$$\begin{aligned} &\text{(a)} \qquad \bar{\boldsymbol{\gamma}} \times \mathbf{E}_0 = j\omega\mu\mathbf{H}_0 \\ &\text{(b)} \qquad \bar{\boldsymbol{\gamma}} \times \mathbf{H}_0 = -(\sigma + j\omega\epsilon)\mathbf{E} \end{aligned} \tag{8.3}$$

$$\begin{aligned} &\text{(a)} \qquad \bar{\boldsymbol{\gamma}}\cdot\mathbf{E}_0 = 0 \\ &\text{(b)} \qquad \bar{\boldsymbol{\gamma}}\cdot\mathbf{H}_0 = 0 \end{aligned} \tag{8.4}$$

and

$$\bar{\boldsymbol{\gamma}}\cdot\bar{\boldsymbol{\gamma}} = j\omega\mu(\sigma + j\omega\epsilon) \equiv \bar{\gamma}_0^{\,2} \tag{8.5}$$

as conditions for which nontrivial solutions of the form of Eq. 8.2 satisfy Eqs. 8.1. Special attention to the complex nature of $\mathbf{E}_0$, $\mathbf{H}_0$, $\bar{\boldsymbol{\gamma}}$, and $\bar{\gamma}_0$ is again mandatory to avoid oversimplified geometric interpretations of the cross and dot products in Eqs. 8.3, 8.4, and 8.5.

8.1.1.2 Propagation Constants. The complex number $\bar{\gamma}_0$ in Eq. 8.5 has the dimensions of a propagation constant. It is characterized completely by the medium parameters and the frequency:

$$\bar{\gamma}_0 \equiv \sqrt{j\omega\mu(\sigma + j\omega\epsilon)} = \alpha_0 + j\beta_0 \qquad \beta_0 \geq 0, \alpha_0 \geq 0 \quad \text{when } \omega \geq 0 \tag{8.6}$$

The square-root sign in Eq. 8.6 is chosen by definition to make the imaginary part (β_0) have the same sign as ω. The non-negative character of σ, ϵ, μ then guarantees that α_0 and β_0 have the same sign. A similar situation arose in connection with dissipative lines, Eq. 5.2*a*, and has been discussed carefully in Eq. 6.3 as regards $\bar{\gamma}_0(s)$. From Maxwell's equations and Eqs. 8.2 in the present case, either $\bar{\gamma}_0$ or $-\bar{\gamma}_0$ is equally acceptable in the field solution because only $\bar{\gamma}_0^2$ appears in Eq. 8.5. This fact is reminiscent of the possibility of (+) and (−) wave solutions in Chapter 5.

We remark that β_0 in Eq. 8.6 is *not* the same as β_0 in Chapter 7, unless $\sigma = 0$. Like α_0, β_0 *is here simply one part of a complex number defined by Eqs. 8.6.* It is important to examine these quantities more closely.

Equating real and imaginary parts of the square of Eq. 8.6 yields

$$\text{(a)} \qquad \beta_0^2 - \alpha_0^2 = \omega^2 \epsilon\mu$$

$$\text{(b)} \qquad \alpha_0\beta_0 = \frac{\omega\mu\sigma}{2} \tag{8.7}$$

Elimination of α_0 by substitution from Eq. 8.7*b* into Eq. 8.7*a* leads to the biquadratic in β_0

$$\beta_0^4 - (\omega^2\epsilon\mu)\beta_0^2 - \left(\frac{\omega\mu\sigma}{2}\right)^2 = 0 \tag{8.8a}$$

Inasmuch as β_0 (or α_0) must be real, β_0^2 must be positive. This fact allows rejection of one solution to Eq. 8.8*a*. The remaining solution is

$$\beta_0^2 = \frac{\omega^2\epsilon\mu}{2}\left[\sqrt{1 + \left(\frac{\sigma}{\omega\epsilon}\right)^2} + 1\right] \tag{8.8b}$$

and the corresponding value of α_0^2 from Eq. 8.7*a* is

$$\alpha_0^2 = \frac{\omega^2\epsilon\mu}{2}\left[\sqrt{1 + \left(\frac{\sigma}{\omega\epsilon}\right)^2} - 1\right] \tag{8.8c}$$

Observing the sign convention $\alpha_0, \beta_0 \geq 0$ when $\omega \geq 0$, discussed in connection with Eq. 8.6 ff., we find

$$\text{(a)} \qquad \beta_0 = \omega\sqrt{\epsilon\mu}\left\{\frac{1}{2}\left[\sqrt{1 + \left(\frac{\sigma}{\omega\epsilon}\right)^2} + 1\right]\right\}^{1/2} \geq \omega\sqrt{\epsilon\mu}$$

$$\text{(b)} \qquad \alpha_0 = \omega\sqrt{\epsilon\mu}\left\{\frac{1}{2}\left[\sqrt{1 + \left(\frac{\sigma}{\omega\epsilon}\right)^2} - 1\right]\right\}^{1/2} \tag{8.9}$$

$$= \frac{(\sigma/2)\sqrt{\mu/\epsilon}}{\{[\frac{1}{2}\sqrt{1 + (\sigma/\omega\epsilon)^2} + 1]\}^{1/2}} \leq \beta_0$$

where the second form of α_0 in Eq. 8.9*b* comes most directly from Eqs. 8.9*a* and 8.7*b*.

Evidently $\bar{\gamma}_0$ is actually a function of only two parameters: (a) $\omega\sqrt{\epsilon\mu}$, which is what the propagation constant of light or uniform plane waves would be if σ were zero; and (b) $\sigma/(\omega\epsilon)$, sometimes called the *loss tangent* of the medium, which is the ratio of conduction current $\sigma\mathbf{E}$ to displacement current $j\omega\epsilon\mathbf{E}$ in the dissipative medium. The behavior of $\bar{\gamma}_0$ in Eqs. 8.6 or 8.9 is particularly interesting in two limiting cases.

1. $\sigma/(\omega\epsilon) \ll 1$. The loss tangent is small or the displacement current at the frequency in question is large compared to the conduction current. This is the case of a medium with small loss—a slightly imperfect dielectric or magnetic medium. Binomial expansion of either Eq. 8.6 or 8.9 yields

$$\left.\begin{aligned} &\text{(a)} \quad \beta_0 \approx \omega\sqrt{\epsilon\mu}\left[1 + \frac{1}{8}\left(\frac{\sigma}{\omega\epsilon}\right)^2 + \cdots\right] \approx \omega\sqrt{\epsilon\mu} \\ &\text{(b)} \quad \alpha_0 \approx \frac{1}{2}\sigma\sqrt{\frac{\mu}{\epsilon}}\left[1 - \frac{1}{8}\left(\frac{\sigma}{\omega\epsilon}\right)^2 + \cdots\right] \approx \frac{1}{2}\sigma\sqrt{\frac{\mu}{\epsilon}} \end{aligned}\right\} \quad \frac{\sigma}{\omega\epsilon} \ll 1 \qquad (8.10)$$

a result similar to Eqs. 5.4 for a transmission line with only small shunt losses. If the parameters σ, ϵ, μ are independent of frequency (often not a valid assumption), the condition of small loss tangent always occurs at high enough frequencies in a given medium. The high-frequency value of β_0 approaches that of light in a lossless medium with the same ϵ and μ, and the high-frequency value of α_0 approaches a constant. In any case, as long as $\sigma/(\omega\epsilon) \ll 1$, the result of Eq. 8.10 links the frequency dependence of α_0 and β_0 to that of σ, ϵ, μ.

2. $\sigma/(\omega\epsilon) \gg 1$. The loss tangent is very large or the conduction current exceeds greatly the displacement current. This is the case of a medium with large loss—a slightly imperfect conductor. In a metal like copper, for example, σ is of the order of 10^7–10^8 mhos/m. The dielectric constant appears to be of the order of that of free space ($\epsilon_0 = 8.85 \times 10^{-12} \approx (1/36\pi) \times 10^{-9}$ f/m), as is the magnetic permeability ($\mu_0 \approx 4\pi \times 10^{-7}$ h/m). Thus $\omega\epsilon \ll \sigma$ for ω up to perhaps 10^{17} radians/sec, or frequencies up to 10^{10} mc. Evidently, in the highly conducting metals $\sigma \gg \omega\epsilon$ up to frequencies in the range of visible light ($\approx 6 \times 10^8$ mc). In fact, because of microscopic physical processes, the value of σ changes from its familiar one, employed above, at frequencies somewhat below those we have just mentioned.

With large-loss tangent, then, binomial expansion of Eq. 8.6 or 8.9 leads to the results:

$$\left.\begin{aligned} &\text{(a)} \qquad \beta_0 \approx \sqrt{\frac{\omega\mu\sigma}{2}}\left(1 + \frac{\omega\epsilon}{2\sigma} + \cdots\right) \approx \sqrt{\frac{\omega\mu\sigma}{2}} \\ &\text{(b)} \qquad \alpha_0 \approx \sqrt{\frac{\omega\mu\sigma}{2}}\left(1 - \frac{\omega\epsilon}{2\sigma} + \cdots\right) \approx \sqrt{\frac{\omega\mu\sigma}{2}} \\ &\text{(c)} \qquad \bar{\gamma}_0 \approx \sqrt{j\omega\mu\sigma}\left(1 + j\frac{\omega\epsilon}{2\sigma} + \cdots\right) \approx \sqrt{j\omega\mu\sigma} \end{aligned}\right\} \frac{\sigma}{\omega\epsilon} \gg 1 \qquad (8.11)$$

Again, if σ, ϵ, μ were independent of frequency, a condition of large loss tangent would always occur at low enough frequencies in a given dissipative medium ($0 < \sigma < \infty$). Both β_0 and α_0 would vary directly as $\sqrt{\omega}$. In any event, as long as $(\sigma/\omega\epsilon) \gg 1$, $\beta_0 \approx \alpha_0$, and $\bar{\gamma}_0 \approx \sqrt{j\omega\mu\sigma}$. The angle of $\bar{\gamma}_0$ is very nearly 45°, and Eq. 8.11 relates the frequency dependence of $\bar{\gamma}_0$ to that of σ and μ. Note that the entire propagation constant is then fixed by the attenuation alone.

The conclusions about $\bar{\gamma}_0$, obtained from Eqs. 8.6 through 8.11, are summarized in Fig. 8.1. Figure 8.1*a* shows normalized values, $\alpha_0/[(\sigma/2)(\sqrt{\mu/\epsilon})]$ and $\beta_0/(\omega\sqrt{\epsilon\mu})$, as functions of the loss angle $\xi = \tan^{-1}(\sigma/\omega\epsilon)$, $0 \le \xi \le \pi/2$. This figure remains unaltered whether or not σ, ϵ, μ vary with frequency. On the other hand, Fig. 8.1*b* illustrates $\alpha_0(\omega)$ and $\beta_0(\omega)$ as they would be if σ, ϵ, μ were independent of frequency.

We are now prepared to study further the relations between $\bar{\boldsymbol{\gamma}}$ and $\bar{\gamma}_0$ imposed by Eq. 8.5. We know already from Chapter 7 that, since $\bar{\boldsymbol{\gamma}} = \boldsymbol{\alpha} + j\boldsymbol{\beta}$, the planes of constant amplitude of the solution (Eq. 8.2) are normal to $\boldsymbol{\alpha}$ while the planes of constant phase are normal to $\boldsymbol{\beta}$. Writing $\beta = |\boldsymbol{\beta}|$ and $\alpha = |\boldsymbol{\alpha}|$ once again, we have, equating real and imaginary parts of Eq. 8.5,

$$\begin{aligned} &\text{(a)} \qquad \beta^2 - \alpha^2 = \omega^2\epsilon\mu \\ &\text{(b)} \qquad \boldsymbol{\alpha}\cdot\boldsymbol{\beta} = \alpha\beta\cos\zeta = \frac{\omega\mu\sigma}{2} \end{aligned} \qquad (8.12)$$

in which ζ is the *space angle* between $\boldsymbol{\alpha}$ and $\boldsymbol{\beta}$.

It is immediately apparent from Eqs. 8.12 that as long as neither ω nor σ is zero, neither α nor β can be zero. This is hardly surprising. If there is *loss*, the fields *must* suffer *attenuation* (as well as phase shift) in the sinusoidal steady state.

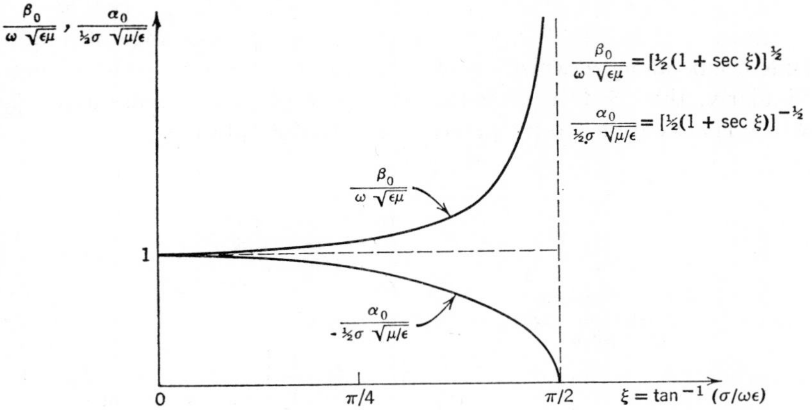

(a)

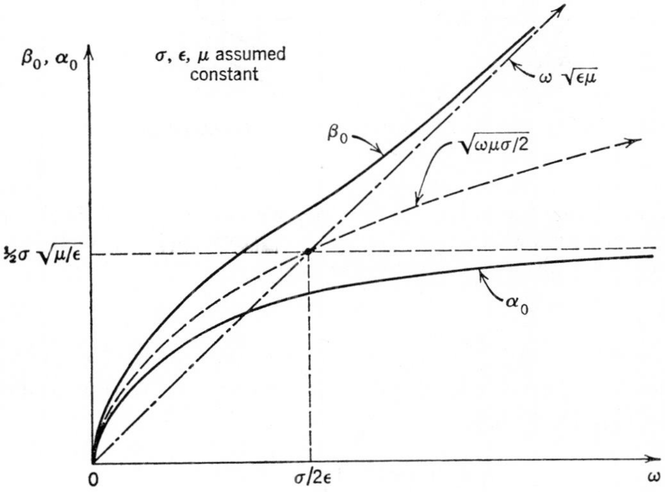

(b)

Fig. 8.1. Real and imaginary parts of $\bar{\gamma}_0 \equiv \sqrt{j\omega\mu(\sigma + j\omega\epsilon)} = \alpha_0 + j\beta_0$ for a dissipative medium. (a) Normalized values versus loss angle ξ, where $\tan \xi$ is the loss tangent of the medium; (b) α_0 and β_0 versus frequency ω if σ, ϵ, μ are independent of frequency.

More surprising, however, is the additional fact that under the same conditions $\cos \zeta \neq 0$! *The planes of constant amplitude and the planes of constant phase cannot be mutually perpendicular in the presence of loss.* As illustrated in Fig. 8.2, the greatest space-rate-of-change of amplitude occurs along $\boldsymbol{\alpha}$, which is *not* parallel to the constant-phase planes. Similarly, the greatest space-rate-of-change of phase occurs along $\boldsymbol{\beta}$, which is *not* parallel to the planes of constant amplitude.

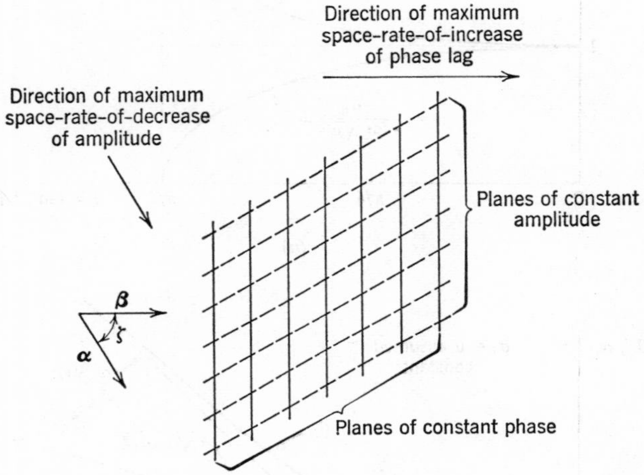

Fig. 8.2. Significant features of the space variation of $e^{-\bar{\gamma}\cdot r} = e^{-\alpha\cdot r}e^{-j\beta\cdot r}$ for non-uniform plane waves with losses. Note that $\boldsymbol{\alpha}\cdot\boldsymbol{\beta} > 0$, and that there is no space variation at all in the direction normal to the page.

Furthermore α and β, being by definition magnitudes of real vectors, must not be negative. It follows from the non-negative character of μ and σ in Eq. 8.12*b* that

$$\left.\begin{aligned} 0 \le \cos\zeta \le 1 \\ \text{or} \qquad 0 \le \zeta \le \frac{\pi}{2} \end{aligned}\right\} \quad \text{if } \omega > 0 \tag{8.13}$$

as is indicated in Fig. 8.2.

Solution of Eqs. 8.12 can only express α and β in terms of the known parameters of the medium (ω, σ, ϵ, μ) *along with the unknown parameter* ζ. Comparison of Eqs. 8.12 with Eqs. 8.7 shows that they would be identical if σ in the latter were replaced by ($\sigma/\cos \zeta$). Accordingly, the solution of Eqs. 8.12 can be written immediately from Eqs. 8.9 with this replacement. Thus

$$\text{(a)}\qquad \beta = \omega\sqrt{\epsilon\mu}\left[\frac{1}{2}\left(\sqrt{1+\left(\frac{\sigma}{\omega\epsilon\cos\zeta}\right)^2}+1\right)\right]^{1/2} \geq \beta_0$$

$$\text{(b)}\qquad \alpha = \omega\sqrt{\epsilon\mu}\left[\frac{1}{2}\left(\sqrt{1+\left(\frac{\sigma}{\omega\epsilon\cos\zeta}\right)^2}-1\right)\right]^{1/2} \qquad \beta \geq \alpha \geq \alpha_0 \tag{8.14}$$

In general, since Maxwell's equations have not completely fixed ζ, its value must be determined finally by boundary conditions. Keep in mind, however, that Eq. 8.13 provides some limitations from the field equations.

For a material with large losses, it is evident from Eq. 8.13 that the corresponding condition $\sigma \gg \omega\epsilon$, also implies $\sigma \gg \omega\epsilon \cos\zeta$. Therefore, by replacing σ by $\sigma/\cos\zeta$ in Eq. 8.11, we find

$$\left.\begin{aligned} &\text{(a)}\qquad \beta \approx \sqrt{\frac{\omega\mu\sigma}{2\cos\zeta}} \geq \beta_0 \\ &\text{(b)}\qquad \alpha \approx \sqrt{\frac{\omega\mu\sigma}{2\cos\zeta}} \geq \alpha_0 \\ &\text{(c)}\qquad \bar{\gamma} \approx \sqrt{\frac{j\omega\mu\sigma}{\cos\zeta}} \end{aligned}\right\} \quad \begin{gathered}\frac{\sigma}{\omega\epsilon} \gg 1 \\ \left(\text{implies also } \frac{\sigma}{\omega\epsilon\cos\zeta} \gg 1\right)\end{gathered} \tag{8.15}$$

On the contrary, with reference to Eq. 8.14, the "small-loss" condition $\sigma \ll \omega\epsilon$ does *not* guarantee $\sigma \ll \omega\epsilon \cos\zeta$. We might have $\cos\zeta$ very small on account of boundary conditions. Hence the analogy of Eq. 8.10 must contain a condition involving $\cos\zeta$:

$$\left.\begin{aligned} &\text{(a)}\qquad \beta \approx \omega\sqrt{\epsilon\mu} \approx \beta_0 \\ &\text{(b)}\qquad \alpha \approx \frac{1}{2}\frac{\sigma}{\cos\zeta}\sqrt{\frac{\mu}{\epsilon}} > \alpha_0 \end{aligned}\right] \quad \begin{gathered}\frac{\sigma}{\omega\epsilon\cos\zeta} \ll 1 \\ \left(\text{implies also } \frac{\sigma}{\omega\epsilon} \ll 1\right)\end{gathered} \tag{8.16}$$

Observe however that, if the condition $\sigma \ll \omega\epsilon\cos\zeta$ is met, then surely $\sigma \ll \omega\epsilon$ (as indicated in Eq. 8.16) and Eqs. 8.10 also apply.

The considerable freedom left in $\cos\zeta$ by the field equations allows us to study the details of our exponential solution under two broad headings (when $\sigma \neq 0$): (1) $\cos\zeta = 1$ ($\zeta = 0$); and (2) $0 < \cos\zeta < 1$ ($0 < \zeta < \pi/2$). The first situation leads to uniform plane waves, whereas the second leads to nonuniform ones.

8.1.2 Uniform Plane Waves

8.1.2.1 FORM OF SOLUTION. If in Eq. 8.12 $\zeta = 0$, $\boldsymbol{\alpha}$ and $\boldsymbol{\beta}$ become parallel. Choose a $+z$-axis along them, and x- and y-axes to make a right-handed rectangular system. Then we may write

$$\bar{\boldsymbol{\gamma}} = \boldsymbol{a}_z(\alpha + j\beta) \equiv \boldsymbol{a}_z\bar{\gamma} \tag{8.17}$$

and Eq. 8.5 becomes

$$\bar{\gamma}^2 = \bar{\gamma}_0{}^2 \tag{8.18}$$

or

$$\bar{\gamma} = \pm\bar{\gamma}_0 \tag{8.19}$$

Because both the real and imaginary parts of $\bar{\boldsymbol{\gamma}}$ are parallel to the z-axis, the planes of constant amplitude and planes of constant phase are parallel to the x, y-plane. There is no variation of either the amplitude or phase of $\mathbf{E}$ (or $\mathbf{H}$) along planes normal to both $\boldsymbol{\alpha}$ and $\boldsymbol{\beta}$ (planes parallel to the x, y-plane). The solution (2) has become a *uniform plane* wave in which $\partial/\partial x \equiv \partial/\partial y \equiv 0$; and there is possible a wave along $+z$ ($\bar{\gamma} = \bar{\gamma}_0$) or one along $-z$ ($\bar{\gamma} = -\bar{\gamma}_0$).

Under the present circumstances, Eqs. 8.4 become

$$\begin{aligned} &\text{(a)} \qquad \bar{\gamma}\boldsymbol{a}_z \cdot \mathbf{E}_0 = 0 = \boldsymbol{a}_z \cdot \mathbf{E}_0 \\ &\text{(b)} \qquad \bar{\gamma}\boldsymbol{a}_z \cdot \mathbf{H}_0 = 0 = \boldsymbol{a}_z \cdot \mathbf{H}_0 \end{aligned} \tag{8.20}$$

Inasmuch as $\boldsymbol{a}_z$ is a *real* vector, Eq. 8.20 *does* imply that the instantaneous electric and magnetic fields $\boldsymbol{E}$ and $\boldsymbol{H}$ have no z components. The fields are at every instant confined to the x, y-plane perpendicular to the direction of propagation ($\pm z$). The wave is therefore TEM, besides being uniform and plane. *The significance of $\bar{\gamma}_0$ is now clear. It is the complex propagation constant of uniform plane waves in the dissipative medium.*

Finally, we come to the last conditions upon the solution, contained in Eqs. 8.3

$$\mathbf{H}_0 = \frac{\bar{\gamma}}{j\omega\mu}\boldsymbol{a}_z \times \mathbf{E}_0 \tag{8.21}$$

This relation will differ, depending upon the choice of sign in Eq. 8.19. Indeed $\mathbf{E}_0$ (or $\mathbf{H}_0$) will have a different sign for $\bar{\gamma} = +\bar{\gamma}_0$ than for $\bar{\gamma} = -\bar{\gamma}_0$. For the moment, let us consider only the (+) wave $\bar{\gamma} = \bar{\gamma}_0$.

We have from Eqs. 8.2, 8.17, 8.19, 8.20, and 8.21:

$$\text{(a)} \qquad \mathbf{E}^+ = \mathbf{E}_0{}^+ e^{-\bar{\gamma}_0 \cdot r} = \mathbf{E}_0{}^+ e^{-\alpha_0 z} e^{-j\beta_0 z}$$

$$\text{(b)} \qquad \mathbf{H}^+ = \frac{\bar{\gamma}_0}{j\omega\mu} \boldsymbol{a}_z \times \mathbf{E}^+ \equiv \frac{1}{\bar{\eta}} \boldsymbol{a}_z \times \mathbf{E}^+ \qquad (8.22)$$

$$\text{(c)} \qquad \boldsymbol{a}_z \cdot \mathbf{E}^+ = \boldsymbol{a}_z \cdot \mathbf{H}^+ = 0$$

in which we have written

$$\bar{\eta} \equiv \frac{j\omega\mu}{\bar{\gamma}_0} = \sqrt{\frac{j\omega\mu}{\sigma + j\omega\epsilon}} \qquad \text{Re}\,(\bar{\eta}) > 0,\ \text{Im}\,(\bar{\eta}) \geq 0 \quad \text{when } \omega \geq 0 \qquad (8.23)$$

The fact that Re $(\bar{\eta}) > 0$ is *forced* by our choice $\beta_0 > 0$ in Eq. 8.6 defining $\bar{\gamma}_0$. The non-negative character of ω, σ, ϵ, μ then guarantee that Im $(\bar{\eta}) \geq 0$. Evidently, $\bar{\eta}$ *is the complex characteristic wave impedance for uniform plane waves in the dissipative medium.* Like $\bar{\gamma}_0$, $\bar{\eta}$ is similar to the analogous quantity (Z_0) in a transmission line with "shunt" losses only ($R = 0$, $G \neq 0$ in Eq. 5.2).

8.1.2.2 POLARIZATION. So far, the uniform plane wave in a dissipative medium seems little different from that in a lossless medium, except for the rather obvious replacement of $j\omega\epsilon$ by $\sigma + j\omega\epsilon$ in both the propagation constant $\bar{\gamma}_0$ and the characteristic wave impedance $\bar{\eta}$. Certainly this replacement is sufficient to convert from one solution to the other; but the effect of the replacement upon the field components themselves is greater than we might guess from a cursory inspection of Eqs. 8.22. Let us examine them more closely.

First, we may separate the complex vectors $\mathbf{E}_0{}^+$ and $\mathbf{H}_0{}^+$ into their real and imaginary parts:

$$\text{(a)} \qquad \mathbf{E}_0{}^+ = \boldsymbol{E}_{0r}{}^+ + j\boldsymbol{E}_{0i}{}^+ \qquad \boldsymbol{a}_z \cdot \boldsymbol{E}_{0r}{}^+ = \boldsymbol{a}_z \cdot \boldsymbol{E}_{0i}{}^+ = 0$$

$$\text{(b)} \qquad \mathbf{H}_0{}^+ = \boldsymbol{H}_{0r}{}^+ + j\boldsymbol{H}_{0i}{}^+ \qquad \boldsymbol{a}_z \cdot \boldsymbol{H}_{0r}{}^+ = \boldsymbol{a}_z \cdot \boldsymbol{H}_{0i}{}^+ = 0 \qquad (8.24)$$

in which $\boldsymbol{E}_{0r}{}^+$, $\boldsymbol{E}_{0i}{}^+$, $\boldsymbol{H}_{0r}{}^+$, and $\boldsymbol{H}_{0i}{}^+$ are *real* space vectors. On account of Eqs. 8.20 or 8.22*c*, they all lie in the x, y-plane. The actual (+)-wave fields, as functions of time and position, are:

$$\text{(a)} \qquad \boldsymbol{E}^+ = \text{Re}\,(\mathbf{E}^+ e^{j\omega t}) = e^{-\alpha_0 z}\,\text{Re}\,(\mathbf{E}_0{}^+ e^{j(\omega t - \beta_0 z)})$$

$$\text{(b)} \qquad \boldsymbol{H}^+ = \text{Re}\,(\mathbf{H}^+ e^{j\omega t}) = e^{-\alpha_0 z}\,\text{Re}\left(\frac{\boldsymbol{a}_z \times \mathbf{E}_0{}^+}{\bar{\eta}} e^{j(\omega t - \beta_0 z)}\right) \qquad (8.25)$$

$$= \frac{e^{-\alpha_0 z}}{|\bar{\eta}|}\,\text{Re}\,(\boldsymbol{a}_z \times \mathbf{E}_0{}^+ e^{j(\omega t - \beta_0 z - \varphi)})$$

where we have written the complex number $\bar{\eta}$ (Eq. 8.23) in polar form

$$\bar{\eta} = |\bar{\eta}|e^{j\varphi} \qquad 0 \leq \varphi \leq \frac{\pi}{4} \tag{8.26}$$

In the light of Eqs. 8.24, Eqs. 8.25 become:

$$\begin{aligned} &\text{(a)} \qquad \boldsymbol{E}^+ = e^{-\alpha_0 z}[\boldsymbol{E}_{0r}{}^+ \cos(\omega t - \beta_0 z) - \boldsymbol{E}_{0i}{}^+ \sin(\omega t - \beta_0 z)] \\ &\text{(b)} \qquad \boldsymbol{H}^+ = \frac{e^{-\alpha_0 z}}{|\bar{\eta}|}[(\boldsymbol{a}_z \times \boldsymbol{E}_{0r}{}^+) \cos(\omega t - \beta_0 z - \varphi) \\ &\qquad\qquad - (\boldsymbol{a}_z \times \boldsymbol{E}_{0i}{}^+) \sin(\omega t - \beta_0 z - \varphi)] \end{aligned} \tag{8.27}$$

In general, vectors $\boldsymbol{E}_{0r}{}^+$ and $\boldsymbol{E}_{0i}{}^+$ in the x, y-plane may have any magnitudes and directions we wish, as far as Maxwell's equations are concerned. An arbitrary choice leads to an elliptically polarized electric field $\boldsymbol{E}^+$ like that presented in Fig. 8.3. This figure is a picture of Eqs. 8.27 drawn in the plane $z = 0$, for the two cases $\varphi = \pi/4$ (large losses) in Fig. 8.3*a* and $\varphi = 0$ (lossless) in Fig. 8.3*b*. One complete period of time variation ($0 \leq t \leq 2\pi/\omega$) is shown.

The vector $\boldsymbol{H}^+$ in Eq. 8.27*b* is of the same over-all form as $\boldsymbol{E}^+$, i.e., it is generally the resultant of two nonperpendicular vectors which vary sinusoidally in magnitude, but are 90° out of time phase. Moreover, the essential vectors defining $\boldsymbol{H}^+$ are $\boldsymbol{a}_z \times \boldsymbol{E}_{0r}{}^+$ and $\boldsymbol{a}_z \times \boldsymbol{E}_{0i}{}^+$, which are, respectively, equal in magnitude and perpendicular to $\boldsymbol{E}_{0r}{}^+$ and $\boldsymbol{E}_{0i}{}^+$. Hence the ellipse traced by $\boldsymbol{H}^+$ has the same shape (more precisely, eccentricity) as that traced by $\boldsymbol{E}^+$, *and its major (or minor) axis is perpendicular to that of the electric-field ellipse.* In Fig. 8.3 the scale of $\boldsymbol{H}^+$ has been chosen arbitrarily as compared with that of $\boldsymbol{E}^+$, so that the sizes of the two ellipses should not be compared.

More interesting, however, is the effect of the phase angle φ of the characteristic admittance $\bar{\eta}$. It appears in $\boldsymbol{H}^+$ as a time phase (Eq. 8.27*b*). It does not affect the direction of rotation of $\boldsymbol{H}^+$ because the *relative* phase of the two sinusoidal vectors defining $\boldsymbol{H}^+$ does not depend upon φ. Thus $\boldsymbol{H}^+$ and $\boldsymbol{E}^+$ always trace geometrically similar ellipses, oriented perpendicularly to each other, and they trace them in the same direction once every period $2\pi/\omega$. Nevertheless, on account of φ, the tips of $\boldsymbol{E}^+$ and $\boldsymbol{H}^+$ are not necessarily at corresponding points of their orbits at the same time. In Fig. 8.3, for the case $\varphi = \pi/4$, the actual space angle between $\boldsymbol{E}^+$ and $\boldsymbol{H}^+$ is *not* 90°; in fact, it varies with time. This occurs because these vectors generally do not rotate by equal angular increments in equal times, as is quite clear from a consideration of Fig. 8.3. Therefore unless the tips of two such vectors tracing similar orbits in the same direction are also *similarly placed*

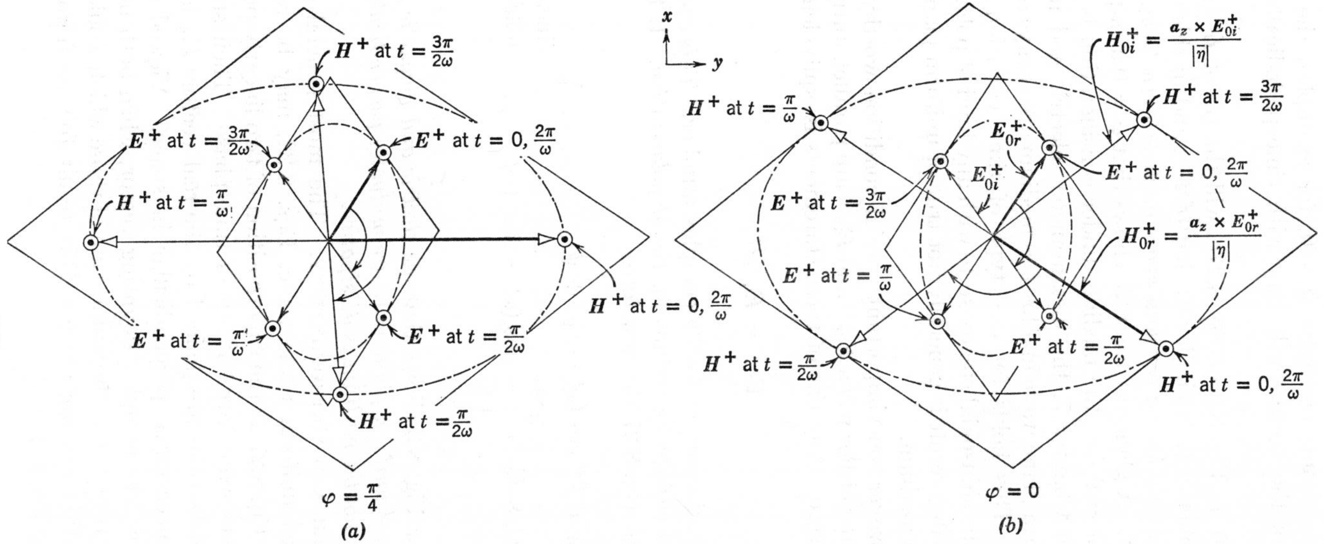

Fig. 8.3. Loci of instantaneous electric and magnetic fields, $\boldsymbol{E}^+(0,t)$ and $\boldsymbol{H}^+(0,t)$, for a uniform plane wave with sinusoidal (ωt) time variation. The characteristic wave impedance is $|\bar{\eta}|e^{j\varphi}$. Fields are shown in (a) for a highly dissipative medium ($\varphi = \pi/4$), and in (b) for a lossless medium ($\varphi = 0$).

in these orbits at some one time ($t = 0$, for example), the relative angular speed of the vectors will vary with time periodically over a cycle.

Only if $\varphi = 0$, in Fig. 8.3 or Eq. 8.27*b*, will the tip of $\boldsymbol{H}^+$ trace its ellipse in a manner exactly similar to that of $\boldsymbol{E}^+$; and in this case, $\boldsymbol{E}^+$ *and* $\boldsymbol{H}^+$ *will not only have a constant space angle between them, but this angle will be 90°.* A glance at Eq. 8.23 shows that $\varphi = 0$ as a practical matter only if $\sigma = 0$, i.e., the medium under consideration is lossless. As we found in Chapter 7, the instantaneous electric and magnetic fields of a uniform plane wave in a lossless medium are always mutually perpendicular. This holds for *any* type of time variation and, in particular, for the sinusoidal steady state with any kind of polarization. Evidently the same result is generally not true of uniform plane waves in a dissipative medium.

There is, however, one important special condition worthy of discussion. It occurs either when $\boldsymbol{E}_{0r}{}^+$ and $\boldsymbol{E}_{0i}{}^+$ are parallel, antiparallel, or when one is zero. Any one of these statements is equivalent to the equation

$$\mathbf{E}_0{}^+ = \boldsymbol{E}_0{}^+ e^{j\psi} \tag{8.28}$$

where $\boldsymbol{E}_0{}^+$ is a *real vector,* and $e^{j\psi}$ a complex number. *This equation is the necessary and sufficient condition for linear polarization.* If Eq. 8.28 holds, we find for Eq. 8.27 the new form

$$\left.\begin{aligned}
&\text{(a)} \quad \boldsymbol{E}^+ = \boldsymbol{E}_0{}^+ e^{-\alpha_0 z} \cos(\omega t - \beta_0 z + \psi) \\
&\text{(b)} \quad \boldsymbol{H}^+ = (\boldsymbol{a}_z \times \boldsymbol{E}_0{}^+) \frac{e^{-\alpha_0 z}}{|\bar{\eta}|} \cos(\omega t - \beta_0 z + \psi - \varphi)
\end{aligned}\right\} \begin{array}{l}\text{linear}\\ \text{polarization}\end{array} \tag{8.29}$$

Equation 8.29 or Fig. 8.3 shows that *the electric and magnetic fields of a linearly polarized uniform plane wave are mutually perpendicular at all times, whether or not the medium is dissipative.*

8.1.2.3 Power and Orthogonality. The most general steady-state $+z$-directed uniform plane wave, Eqs. 8.22, may be regarded now in terms of Eq. 8.27 as the superposition of two linearly polarized $+z$-directed waves which are 90° out of time phase. Their strengths and (linear) polarizations are defined by the real vectors $\boldsymbol{E}_{0r}$ and $\boldsymbol{E}_{0i}$, which are not necessarily perpendicular in space. Viewed in this particular manner, the complex Poynting vector, when both polarizations are present together, is the simple sum of those which would apply to each polarization existing alone. This result follows from direct calculation, using Eq. 8.22:

$$\begin{aligned} \mathbf{S}^+ &= \tfrac{1}{2}\mathbf{E}^+ \times \mathbf{H}^{+*} = \tfrac{1}{2}\bar{\eta}(\mathbf{H}^+ \times \boldsymbol{a}_z) \times \mathbf{H}^{+*} \\ &= \boldsymbol{a}_z \tfrac{1}{2}\bar{\eta}\mathbf{H}^+ \cdot \mathbf{H}^{+*} = \boldsymbol{a}_z \tfrac{1}{2}\bar{\eta}(\boldsymbol{H}_{0r}{}^+ + j\boldsymbol{H}_{0i}{}^+) \cdot (\boldsymbol{H}_{0r}{}^+ - j\boldsymbol{H}_{0i}{}^+)e^{-2\alpha_0 z} \\ &= \boldsymbol{a}_z \tfrac{1}{2}\bar{\eta}e^{-2\alpha_0 z}|\boldsymbol{H}_{0r}{}^+|^2 + \boldsymbol{a}_z \tfrac{1}{2}\bar{\eta}e^{-2\alpha_0 z}|\boldsymbol{H}_{0i}{}^+|^2 \end{aligned} \tag{8.30}$$

The reactive power comes from the imaginary part of $\bar{\eta}$, which is not zero unless the medium is lossless. Evidently, this part of the complex power is directly additive for these two polarizations, in the same way that the real (time-average) part, arising from the real part of $\bar{\eta}$, is additive.

With the choice of the two linear polarizations described above, however, the instantaneous Poynting vector is not necessarily the simple sum of those pertinent to each polarization alone. From Eq. 8.27, we find specifically:

$$\begin{aligned} \boldsymbol{S}^+ = \boldsymbol{E}^+ \times \boldsymbol{H}^+ = \boldsymbol{a}_z \frac{e^{-2\alpha_0 z}}{|\bar{\eta}|} [&|\boldsymbol{E}_{0r}|^2 \cos(\omega t - \beta_0 z)\cos(\omega t - \beta_0 z - \varphi) \\ &+ |\boldsymbol{E}_{0i}|^2 \sin(\omega t - \beta_0 z)\sin(\omega t - \beta_0 z - \varphi) \\ &- \boldsymbol{E}_{0r}\cdot\boldsymbol{E}_{0i}\sin(2\omega t - 2\beta_0 z - \varphi)] \end{aligned} \tag{8.31}$$

in which the cross term containing $\boldsymbol{E}_{0r}\cdot\boldsymbol{E}_{0i}$ will not be identically zero unless $\boldsymbol{E}_{0r}$ and $\boldsymbol{E}_{0i}$ are perpendicular in space.

An alternative view of the general (+) uniform plane wave (Eq. 8.22) comes from this last result, in particular from expressing the complex vectors $\mathbf{E}_0{}^+$ and $\mathbf{H}_0{}^+$ in terms of their complex components along an arbitrary pair of rectangular x, y-axes.

$$\begin{aligned} &\text{(a)} \qquad \mathbf{E}_0{}^+ = \boldsymbol{a}_x\mathrm{E}_{x0} + \boldsymbol{a}_y\mathrm{E}_{y0} \\ &\text{(b)} \qquad \mathbf{H}_0{}^+ = \boldsymbol{a}_x\mathrm{H}_{x0} + \boldsymbol{a}_y\mathrm{H}_{y0} = -\boldsymbol{a}_x\left(\frac{\mathrm{E}_{y0}}{\bar{\eta}}\right) + \boldsymbol{a}_y\left(\frac{\mathrm{E}_{x0}}{\bar{\eta}}\right) \end{aligned} \tag{8.32}$$

in which E_{x0}, E_{y0}, H_{x0}, H_{y0} are *complex* numbers. This is the way we examined the uniform plane wave in Chapter 7. In this form, the complex Poynting vector is

$$\mathbf{S}^+ = \boldsymbol{a}_z \tfrac{1}{2}\bar{\eta}\mathbf{H}^+ \cdot \mathbf{H}^{+*} = \boldsymbol{a}_z \tfrac{1}{2}\bar{\eta}e^{-2\alpha_0 z}(|\mathrm{H}_{x0}|^2 + |\mathrm{H}_{y0}|^2) \tag{8.33}$$

which again contains no cross terms. This resolution of general polarization into two linear polarizations along perpendicular space directions, but with arbitrary time phase, still leads to direct additivity of complex power.

As regards instantaneous power in this case, we would write

$$\boldsymbol{E}^+ = \boldsymbol{a}_x E_x{}^+ + \boldsymbol{a}_y E_y{}^+ = \boldsymbol{E}_1{}^+ + \boldsymbol{E}_2{}^+ \tag{8.34}$$

With reference to Fig. 8.4, we remark that $\boldsymbol{E}_1{}^+$ and $\boldsymbol{E}_2{}^+$ are perpendicular to each other ($\psi = \pi/2$). Inasmuch as the linear polarization feature guarantees that $\boldsymbol{H}_1{}^+$ and $\boldsymbol{H}_2{}^+$ are always perpendicular respectively to $\boldsymbol{E}_1{}^+$ and $\boldsymbol{E}_2{}^+$, as shown in Fig. 8.4, it is clear that when $\psi = \pi/2$

$$\boldsymbol{E}_1{}^+ \times \boldsymbol{H}_2{}^+ = \boldsymbol{E}_2{}^+ \times \boldsymbol{H}_1{}^+ = 0 \tag{8.35}$$

Accordingly

$$\boldsymbol{S}^+ = \boldsymbol{E}^+ \times \boldsymbol{H}^+ = \boldsymbol{E}_1{}^+ \times \boldsymbol{H}_1{}^+ + \boldsymbol{E}_2{}^+ \times \boldsymbol{H}_2{}^+ = \boldsymbol{S}_1{}^+ + \boldsymbol{S}_2{}^+ \tag{8.36}$$

which means direct additivity of instantaneous power. This same general geometry applies to the result of Eq. 8.31, showing failure of orthogonality, if we think of $\boldsymbol{H}_{0r}$ for $\boldsymbol{H}_1{}^+$ and $\boldsymbol{H}_{0i}$ for $\boldsymbol{H}_2{}^+$, etc.

In other words, resolution of a single, traveling, uniform plane wave of arbitrary polarization type into two similar waves with mutually perpendicular linear polarizations guarantees orthogonality of the

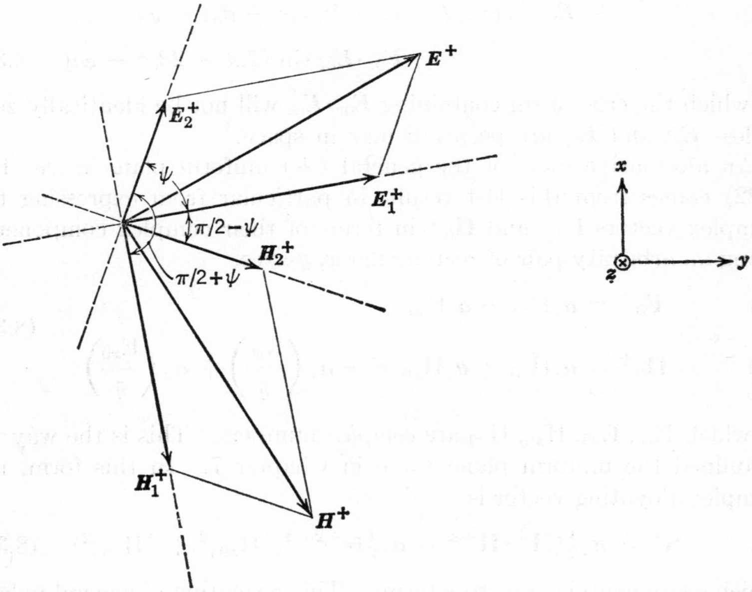

Fig. 8.4. Resolution at one instant of time of a $+z$-directed steady-state uniform plane wave $\boldsymbol{E}^+$, $\boldsymbol{H}^+$ of arbitrary polarization type into two component linearly polarized $+z$-directed waves: $\boldsymbol{E}^+ = \boldsymbol{E}_1{}^+ + \boldsymbol{E}_2{}^+$, $\boldsymbol{H}^+ = \boldsymbol{H}_1{}^+ + \boldsymbol{H}_2{}^+$. Two cases are considered: $\boldsymbol{E}_1{}^+$ and $\boldsymbol{E}_2{}^+$ 90° out of time phase, with ψ arbitrary; $\boldsymbol{E}_1{}^+$ and $\boldsymbol{E}_2{}^+$ perpendicular in space ($\psi = 90°$). Broken lines indicate axis along which corresponding vector tip oscillates harmonically with time.

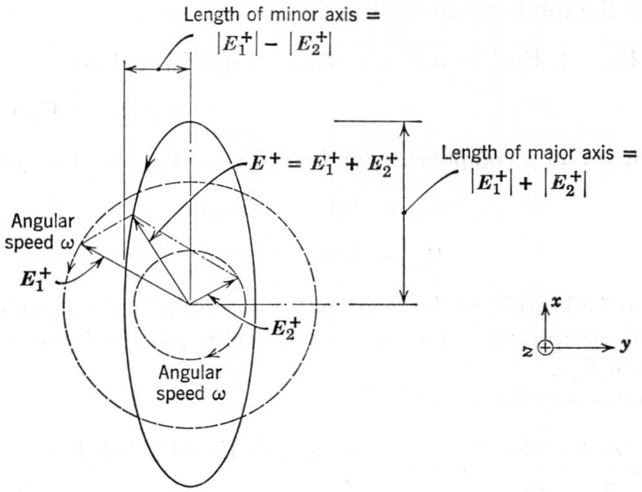

Fig. 8.5. Decomposition of elliptical polarization ($\boldsymbol{E}^+$) into two counterrotating circular polarizations, $\boldsymbol{E}_1{}^+$ and $\boldsymbol{E}_2{}^+$. In the limiting case when $\boldsymbol{E}^+$ is linearly polarized, $|\boldsymbol{E}_1{}^+| = |\boldsymbol{E}_2{}^+|$. Otherwise the sense of rotation of $\boldsymbol{E}^+$ always agrees with that of the larger of $\boldsymbol{E}_1{}^+$ and $\boldsymbol{E}_2{}^+$.

component waves with respect to both instantaneous and complex power.

Occasionally it is helpful to express an elliptically polarized wave as the sum of two oppositely rotating circularly polarized waves, as illustrated in Fig. 8.5. To do so analytically, note first that for a $+z$-directed uniform plane wave with circular polarization in a left-handed screw sense about $+z$

$$\mathbf{E}_{01}{}^+ = \boldsymbol{a}_x \mathrm{E}_{01} + \boldsymbol{a}_y j \mathrm{E}_{01} \tag{8.37a}$$

$$\mathbf{H}_{01}{}^+ = \frac{\boldsymbol{a}_z \times \mathbf{E}_{01}{}^+}{\bar{\eta}} = -\boldsymbol{a}_x j \frac{\mathrm{E}_{01}}{\bar{\eta}} + \boldsymbol{a}_y \frac{\mathrm{E}_{01}}{\bar{\eta}} \tag{8.37b}$$

where E_{01} is any complex number

$$\mathrm{E}_{01} = |\mathrm{E}_{01}| e^{j\psi_{01}} \tag{8.37c}$$

Then, for the opposite sense of rotation

$$\begin{aligned}
&(a) \qquad \mathbf{E}_{02}{}^+ = \boldsymbol{a}_x \mathrm{E}_{02} - \boldsymbol{a}_y j \mathrm{E}_{02} \\
&(b) \qquad \mathbf{H}_{02}{}^+ = \boldsymbol{a}_x j \frac{\mathrm{E}_{02}}{\bar{\eta}} + \boldsymbol{a}_y \frac{\mathrm{E}_{02}}{\bar{\eta}} \\
&(c) \qquad \mathrm{E}_{02} = |\mathrm{E}_{02}| e^{j\psi_{02}}
\end{aligned} \tag{8.38}$$

Thus for the total electric field we have

$$\mathbf{E}_0{}^+ \equiv \mathbf{E}_{01}{}^+ + \mathbf{E}_{02}{}^+ = \boldsymbol{a}_x \mathrm{E}_{x0} + \boldsymbol{a}_y \mathrm{E}_{y0} = \boldsymbol{a}_x(\mathrm{E}_{01} + \mathrm{E}_{02}) + \boldsymbol{a}_y j(\mathrm{E}_{01} - \mathrm{E}_{02}) \quad (8.39)$$

Equating x and y components, and solving for E_{01} and E_{02}, yield

$$\text{(a)} \qquad \mathrm{E}_{01} = \tfrac{1}{2}(\mathrm{E}_{x0} - j\mathrm{E}_{y0})$$
$$\text{(b)} \qquad \mathrm{E}_{02} = \tfrac{1}{2}(\mathrm{E}_{x0} + j\mathrm{E}_{y0}) \quad (8.40)$$

Equation 8.40 furnishes the relations for determining the two circularly polarized components of a given elliptically polarized wave $\mathbf{E}_0{}^+ = \boldsymbol{a}_x \mathrm{E}_{x0} + \boldsymbol{a}_y \mathrm{E}_{y0}$.

The total complex power is

$$\mathbf{S}^+ = \frac{\boldsymbol{a}_z}{2} \frac{\mathbf{E}_0{}^+ \cdot \mathbf{E}_0{}^{+*}}{\bar{\eta}^*} e^{-2\alpha_0 z} = \frac{\boldsymbol{a}_z}{2} \frac{(\mathrm{E}_{01} + \mathrm{E}_{02})(\mathrm{E}_{01}{}^* + \mathrm{E}_{02}{}^*)}{\bar{\eta}^*} e^{-2\alpha_0 z}$$
$$+ \frac{\boldsymbol{a}_z}{2} \frac{(\mathrm{E}_{01} - \mathrm{E}_{02})(\mathrm{E}_{01}{}^* - \mathrm{E}_{02}{}^*)}{\bar{\eta}^*} e^{-2\alpha_0 z}$$
$$= \boldsymbol{a}_z \frac{|\mathrm{E}_{01}|^2}{\bar{\eta}^*} e^{-2\alpha_0 z} + \boldsymbol{a}_z \frac{|\mathrm{E}_{02}|^2}{\bar{\eta}^*} e^{-2\alpha_0 z} = \mathbf{S}_1{}^+ + \mathbf{S}_2{}^+ \quad (8.41)$$

where Eq. 8.39 has been used. The result (Eq. 8.41) shows that the two circularly polarized components are orthogonal in regard to complex power.

Since $\boldsymbol{H}_1{}^+$ is circularly polarized in the same sense (and with the same rotation speed ω) as $\boldsymbol{E}_1{}^+$ (see Eq. 8.37), the instantaneous power $\boldsymbol{E}_1{}^+ \times \boldsymbol{H}_1{}^+$ is not a function of time. The same is true of wave 2. Therefore if the total instantaneous power carried by both waves together were the simple sum of the individual contributions, that total power would be independent of time! In general, this is not true of the instantaneous power (see Eq. 8.31). We conclude that the circularly polarized components of an elliptically polarized wave are not orthogonal as regards instantaneous power.

A review of our results (Eqs. 8.30, 8.31, 8.33, 8.36, 8.41) regarding the orthogonality of the various linearly or circularly polarized components into which a *single*, traveling, uniform plane wave may always be decomposed will show that they apply to a lossless medium ($\sigma = 0$) as well as to a dissipative one ($\sigma \neq 0$). Indeed, for the lossless case, those orthogonality properties relating to instantaneous power were actually covered by the discussion in Sec. 7.1.3.

The orthogonality situation for a (+) and a (−) uniform plane wave, propagating along a common axis in a dissipative medium, is a little more involved than the analogous problem on a dissipative transmission line (Chapter 5). Let the z-axis again be the axis of propagation. Then

$$\text{(a)} \qquad \mathbf{E} = \mathbf{E}^+ + \mathbf{E}^- = \mathbf{E}_0{}^+ e^{-\bar{\gamma}_0 z} + \mathbf{E}_0{}^- e^{\bar{\gamma}_0 z}$$

$$\text{(b)} \qquad \mathbf{H} = \mathbf{H}^+ + \mathbf{H}^- = \boldsymbol{a}_z \times \left(\frac{\mathbf{E}_0{}^+}{\bar{\eta}} e^{-\bar{\gamma}_0 z} - \frac{\mathbf{E}_0{}^-}{\bar{\eta}} e^{\bar{\gamma}_0 z} \right) \qquad (8.42)$$

and the complex power is

$$\mathbf{S} = \frac{1}{2} \mathbf{E} \times \mathbf{H}^*$$

$$= \boldsymbol{a}_z \left[\frac{\mathbf{E}_0{}^+ \cdot \mathbf{E}_0{}^{+*} e^{-2\alpha_0 z} - \mathbf{E}_0{}^- \cdot \mathbf{E}_0{}^{-*} e^{2\alpha_0 z} + j2 \,\mathrm{Im}\, (\mathbf{E}_0{}^- \cdot \mathbf{E}_0{}^{+*} e^{j2\beta_0 z})}{2\bar{\eta}^*} \right] \qquad (8.43)$$

Even if $\bar{\eta}$ were real (lossless medium), the last numerator term (the cross term) would be the entire reactive power unless it vanished identically for all z. When $\bar{\eta}$ is complex ($\sigma \neq 0$), the cross term also contributes to the real power unless it vanishes identically for all z.

The only condition under which the cross term in Eq. 8.43 does vanish identically is

$$\mathbf{E}_0{}^- \cdot \mathbf{E}_0{}^{+*} = 0 \qquad (8.44)$$

Then the waves are orthogonal with regard to the entire complex power. To interpret this condition, let the x- and y-axes be chosen as the principal axes of the polarization ellipse defined by $\mathbf{E}_0{}^-$ for the (−) wave. Also normalize the field strength and choose the time-phase reference so that

$$\mathbf{E}_0{}^- \equiv \boldsymbol{a}_x \mathrm{E}_{x0}{}^- + \boldsymbol{a}_y \mathrm{E}_{y0}{}^- = \boldsymbol{a}_x + \boldsymbol{a}_y jb \qquad (8.45)$$

where b is a positive or negative *real* number. Figure 8.6 shows $\boldsymbol{E}^-(t)$ at $z = 0$ corresponding to Eq. 8.45, for the case $0 < b < 1$. Then if

$$\mathbf{E}_0{}^+ = \boldsymbol{a}_x \mathrm{E}_{x0}{}^+ + \boldsymbol{a}_y \mathrm{E}_{y0}{}^+ \qquad (8.46)$$

Eq. 8.44 demands

$$\mathrm{E}_{x0}{}^{+*} + jb\mathrm{E}_{y0}{}^{+*} = 0$$

or

$$\frac{\mathrm{E}_{y0}{}^+}{\mathrm{E}_{x0}{}^+} = -j \left(\frac{1}{b} \right) \qquad (8.47)$$

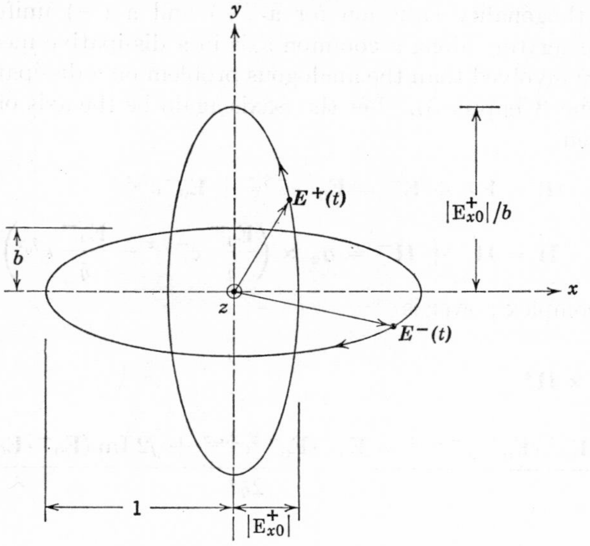

Fig. 8.6. Interpretation of the condition $\mathbf{E_0^- \cdot E_0^{+*}} = 0$ in terms of the instantaneous fields. Note that the ellipses are not necessarily the same size, nor is there any restriction on the relative positions of $\boldsymbol{E}^+(t)$ and $\boldsymbol{E}^-(t)$ on their respective ellipses at any one time. Notation corresponds to Eqs. 8.45 and 8.46, drawn for the case $0 < b < 1$.

The 90° phase relation in Eq. 8.47 shows that the principal axes of the (+) wave polarization ellipse coincide with the x- and y-axes and, therefore, also with the principal axes of the (−) wave polarization ellipse. The fact that, when $b > 0$, for example, E_{y0}^+ *lags* E_{x0}^+ by 90° and E_{y0}^- *leads* E_{x0}^- by 90° means that $\boldsymbol{E}^+$ rotates about the $+z$ direction in the sense *opposite* to that of $\boldsymbol{E}^-$. Finally, the ratio of minor-to-major axis of the (−) wave ellipse is $|b|$ (if, for example, $|b| < 1$), and the same is true of the (+) wave ellipse in Eq. 8.45; but, whereas the major axis is along x in Eq. 8.45, it is along y in Eq. 8.47. In summary, as illustrated in Fig. 8.6, Eq. 8.44 requires that the (+) and (−) wave polarization ellipses have the same shape (eccentricity), mutually perpendicular major (and minor) axes, and *opposite* senses of traversal when viewed from the same end of the propagation axis (or the same sense of traversal when viewed along their own individual propagation directions). Then, and only then, will the (+) and (−) waves be orthogonal as regards complex power (although in a lossless medium the real power is always additive anyway).

One may wonder whether Eq. 8.44 also renders the (+) and (−) waves orthogonal with respect to instantaneous power. If $\boldsymbol{E}$ and $\boldsymbol{H}$ represent the instantaneous fields of a (+) and a (−) uniform plane wave together, then

$$\boldsymbol{S} = \boldsymbol{E} \times \boldsymbol{H} = \left(\frac{\mathbf{E}e^{j\omega t} + \mathbf{E}^*e^{-j\omega t}}{2}\right) \times \left(\frac{\mathbf{H}e^{j\omega t} + \mathbf{H}^*e^{-j\omega t}}{2}\right)$$

$$= \text{Re}\,\mathbf{S} + \tfrac{1}{2}\,\text{Re}\,[(\mathbf{E} \times \mathbf{H})e^{j2\omega t}] \tag{8.48}$$

Clearly, the conditions which eliminate cross terms from the time-average power (Re $\mathbf{S}$) must at least be met to eliminate them from $\boldsymbol{S}$. These conditions are automatically satisfied in a lossless medium, but require Eq. 8.44 in a dissipative one. In addition, the second term in Eq. 8.48 must have no cross terms for any value of t. Such cross terms will vanish for all t if, and only if,

$$\mathbf{E}^+ \times \mathbf{H}^- + \mathbf{E}^- \times \mathbf{H}^+ = 0 \tag{8.49}$$

Referring to Eq. 8.42, Eq. 8.49 becomes

$$\frac{-\mathbf{E}_0{}^+ \cdot \mathbf{E}_0{}^-}{\bar{\eta}} + \frac{\mathbf{E}_0{}^- \cdot \mathbf{E}_0{}^+}{\bar{\eta}} \equiv 0 \tag{8.50}$$

an identity. Consequently, in a dissipative medium the condition of Eq. 8.44 (or Fig. 8.6) is indeed necessary and sufficient to make (+) and (−) waves orthogonal as regards *both* complex and instantaneous power. In a lossless medium, however, the condition of Eq. 8.44 is necessary for the reactive power, the instantaneous and time-average powers for (+) and (−) waves being automatically additive without cross terms (see also Sec. 7.1.3).

It is evident from Fig. 8.6 that Eq. 8.44 is a very special condition, so that, in general, neither the real nor the reactive powers of a (+) and a (−) uniform plane wave in a dissipative medium ($\sigma \neq 0$) can be added algebraically to obtain the net real or reactive power. Occasionally, as in a transmission line, when losses are small and $\bar{\eta}$ is very nearly real, approximately correct results for the *real* power may be obtained by simply adding the real powers algebraically; but this is *an approximation*, however accurate it may sometimes be numerically.

8.1.3 Nonuniform Plane Waves *

8.1.3.1 CHARACTER OF THE SOLUTION.* We pass now to the exponential solution when $\boldsymbol{\alpha}$ and $\boldsymbol{\beta}$ are not parallel ($0 < \zeta < \pi/2$), as illustrated in Fig. 8.2. The field components may be studied most ef-

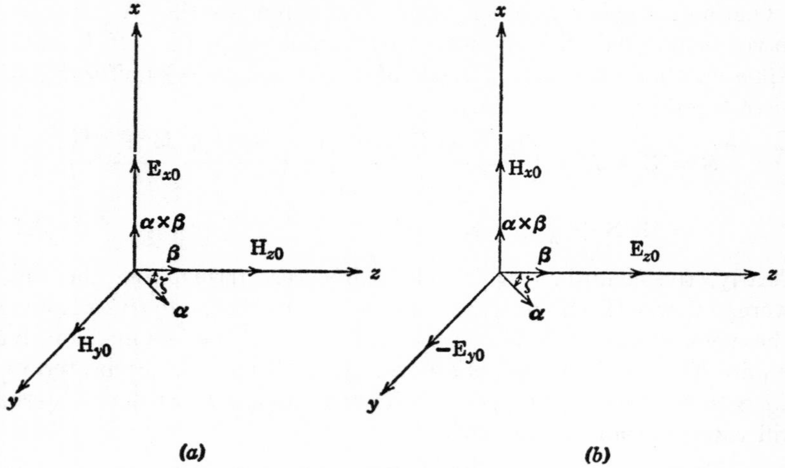

Fig. 8.7. Choice of axes and illustration of field components for (*a*) TE and (*b*) TM plane waves in a dissipative medium.

fectively if we choose a convenient set of rectangular axes. Let $\boldsymbol{\beta}$ be along $+z$, and let the x-axis be parallel to $\boldsymbol{\alpha} \times \boldsymbol{\beta}$. The y-axis will then be in the plane of $\boldsymbol{\alpha}$ and $\boldsymbol{\beta}$, directed to make a right-handed set of coordinates, as illustrated in Fig. 8.7.

Then

$$\bar{\boldsymbol{\gamma}} = \boldsymbol{a}_y\alpha_y + \boldsymbol{a}_z(\alpha_z + j\beta_z) = \boldsymbol{a}_y\alpha_y + \boldsymbol{a}_z\bar{\gamma}_z \tag{8.51}$$

and

$$\bar{\boldsymbol{\gamma}}\cdot\bar{\boldsymbol{\gamma}} = \alpha_y^2 + \bar{\gamma}_z^2 = \bar{\gamma}_0^2 \tag{8.52}$$

The condition of Eq. 8.4*a* requires

$$\bar{\boldsymbol{\gamma}}\cdot\mathbf{E}_0 = (\boldsymbol{a}_y\alpha_y + \boldsymbol{a}_z\bar{\gamma}_z)\cdot(\boldsymbol{a}_x\mathrm{E}_{x0} + \boldsymbol{a}_y\mathrm{E}_{y0} + \boldsymbol{a}_z\mathrm{E}_{z0}) = 0 \tag{8.53a}$$

or

$$\alpha_y\mathrm{E}_{y0} + \bar{\gamma}_z\mathrm{E}_{z0} = 0 \tag{8.53b}$$

As in the lossless case, the conditions of Eq. 8.53 do not restrict that component of the electric field which points in the direction of no space variation. This direction is that of $\boldsymbol{\alpha} \times \boldsymbol{\beta}$, or x in our choice of coordinates. Accordingly one possible electric field solution is certainly

$$\mathbf{E}_{01} = \boldsymbol{a}_x\mathrm{E}_{x0} \tag{8.54a}$$

because this automatically satisfies Eqs. 8.53. The corresponding magnetic field comes from Eq. 8.3a:

$$\mathbf{H}_{01} = \frac{\bar{\boldsymbol{\gamma}} \times \mathbf{E}_{01}}{j\omega\mu} = \frac{\boldsymbol{a}_y\bar{\gamma}_z \mathrm{E}_{x0} - \boldsymbol{a}_z\alpha_y \mathrm{E}_{x0}}{j\omega\mu} \tag{8.54b}$$

which automatically satisfies Eq. 8.4b. From Eqs. 8.54 we then have the TE solution (Fig. 8.7a)

$$\begin{aligned}
&\text{(a)} \quad \bar{\boldsymbol{\gamma}} = \boldsymbol{a}_y\alpha_y + \boldsymbol{a}_z\bar{\gamma}_z \qquad \alpha_y^2 + \bar{\gamma}_z^2 = \bar{\gamma}_0^2 = j\omega\mu(\sigma + j\omega\epsilon)\\
&\text{(b)} \quad \mathbf{E}_0 = \boldsymbol{a}_x \mathrm{E}_{x0}\\
&\text{(c)} \quad \mathbf{H}_0 = \boldsymbol{a}_y\left(\frac{\bar{\gamma}_z}{j\omega\mu}\right)\mathrm{E}_{x0} - \boldsymbol{a}_z\left(\frac{\alpha_y}{j\omega\mu}\right)\mathrm{E}_{x0}
\end{aligned} \tag{8.55}$$

Similarly, the condition of Eq. 8.4b does not restrict the component of $\mathbf{H}_0$ in the $\boldsymbol{\alpha} \times \boldsymbol{\beta}$ direction, which is the direction of no space variation of the fields. Thus a possible solution is

$$\mathbf{H}_{02} = \boldsymbol{a}_x \mathrm{H}_{x0} \tag{8.56a}$$

The corresponding electric field must be given by Eq. 8.3b

$$\mathbf{E}_{02} = \frac{\mathbf{H}_{02} \times \bar{\boldsymbol{\gamma}}}{\sigma + j\omega\epsilon} = -\boldsymbol{a}_y\left(\frac{\bar{\gamma}_z}{\sigma + j\omega\epsilon}\right)\mathrm{H}_{x0} + \boldsymbol{a}_z\left(\frac{\alpha_y}{\sigma + j\omega\epsilon}\right)\mathrm{H}_{x0} \tag{8.56b}$$

which automatically satisfies Eqs. 8.53. This solution in Eq. 8.56 is the TM field (Fig. 8.7b)

$$\begin{aligned}
&\text{(a)} \quad \bar{\boldsymbol{\gamma}} = \boldsymbol{a}_y\alpha_y + \boldsymbol{a}_z\bar{\gamma}_z \qquad \alpha_y^2 + \bar{\gamma}_z^2 = \bar{\gamma}_0^2 = j\omega\mu(\sigma + j\omega\epsilon)\\
&\text{(b)} \quad \mathbf{H}_0 = \boldsymbol{a}_x \mathrm{H}_{x0}\\
&\text{(c)} \quad \mathbf{E}_0 = -\boldsymbol{a}_y\left(\frac{\bar{\gamma}_z}{\sigma + j\omega\epsilon}\right)\mathrm{H}_{x0} + \boldsymbol{a}_z\left(\frac{\alpha_y}{\sigma + j\omega\epsilon}\right)\mathrm{H}_{x0}
\end{aligned} \tag{8.57}$$

Equations 8.55 and 8.57 should be compared with Eqs. 7.55 and 7.57. Once a particular *allowed* value of $\bar{\boldsymbol{\gamma}}$ has been chosen, the sum of a TE and a TM wave having this $\bar{\boldsymbol{\gamma}}$, and with arbitrary relative amplitudes and phases of their fields, represents the most general solution to the complete set of conditions of Eqs. 8.3a, 8.4a, 8.3b, 7.40a, 7.40b, and 7.41a. Such a sum not only meets all the necessary conditions but also both the electric and magnetic fields in it have all three space components. No greater generality is possible within the exponential form (Eq. 8.2) from which we started, once the *directions* of $\boldsymbol{\alpha}$ and $\boldsymbol{\beta}$ are fixed. Of course these directions may be chosen arbitrarily as far as Maxwell's equations are concerned, provided only that condition of Eq. 8.5 is observed.

8.1.3.2 Power.* The complex Poynting vector for nonuniform plane waves exhibits interesting features. For the traveling-wave solution (Eq. 8.2), one may always write

$$\mathbf{S} = \frac{1}{2}\mathbf{E} \times \mathbf{H}^* = \frac{1}{2}\frac{\mathbf{E} \times (\bar{\boldsymbol{\gamma}}^* \times \mathbf{E}^*)}{-j\omega\mu} = \frac{j\bar{\boldsymbol{\gamma}}^*}{2\omega\mu}\mathbf{E}\cdot\mathbf{E}^* + \frac{(\mathbf{E}\cdot\bar{\boldsymbol{\gamma}}^*)\mathbf{E}^*}{2j\omega\mu} \tag{8.58}$$

or alternatively in terms of **H** instead

$$\mathbf{S} = \frac{1}{2}\frac{(\mathbf{H} \times \bar{\boldsymbol{\gamma}}) \times \mathbf{H}^*}{\sigma + j\omega\epsilon} = \frac{\bar{\boldsymbol{\gamma}}\mathbf{H}\cdot\mathbf{H}^*}{2(\sigma + j\omega\epsilon)} - \frac{(\bar{\boldsymbol{\gamma}}\cdot\mathbf{H}^*)\mathbf{H}}{2(\sigma + j\omega\epsilon)} \tag{8.59}$$

Equation 8.58 is convenient for the TE wave, because **E** has only a component $E_{\alpha\times\beta}$ parallel to $\boldsymbol{\alpha} \times \boldsymbol{\beta}$. This makes

$$\mathbf{E}\cdot\bar{\boldsymbol{\gamma}}^* = 0 \qquad \text{for the TE case} \tag{8.60}$$

and, therefore,

$$\mathbf{S}_{\mathrm{TE}} = \frac{|E_{\alpha\times\beta,0}|^2}{2\omega\mu}e^{-2\boldsymbol{\alpha}\cdot\mathbf{r}}(\boldsymbol{\beta} + j\boldsymbol{\alpha}) \tag{8.61}$$

As indicated by this result, and illustrated in Fig. 8.8, the component of the complex Poynting vector along any direction perpendicular to $\boldsymbol{\beta}$ is entirely reactive. It is inductive in directions $\boldsymbol{a}$ for which $\boldsymbol{a}\cdot\boldsymbol{\alpha} > 0$ and $\boldsymbol{a}\cdot\boldsymbol{\beta} = 0$. Only reactive power flows along the planes of constant phase of the TE wave.

In contrast, the component of $\mathbf{S}_{\mathrm{TE}}$ along any direction perpendicular to $\boldsymbol{\alpha}$ is entirely real. Only real power flows along the planes of constant amplitude of the TE wave.

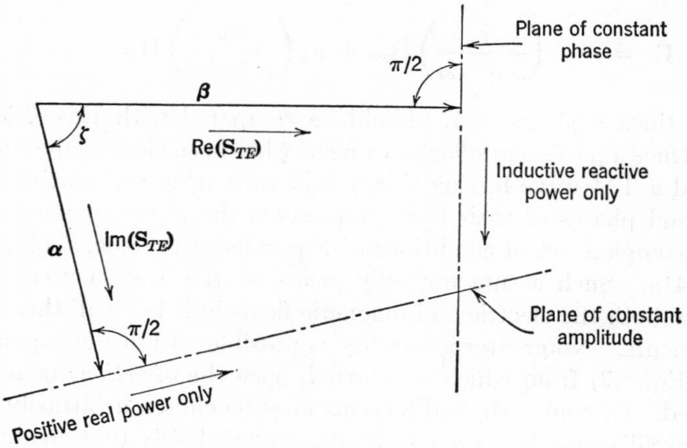

Fig. 8.8. Analysis of the complex Poynting vector for a TE plane wave in a dissipative medium. Compare with Fig. 7.6*a* and Fig. 8.9.

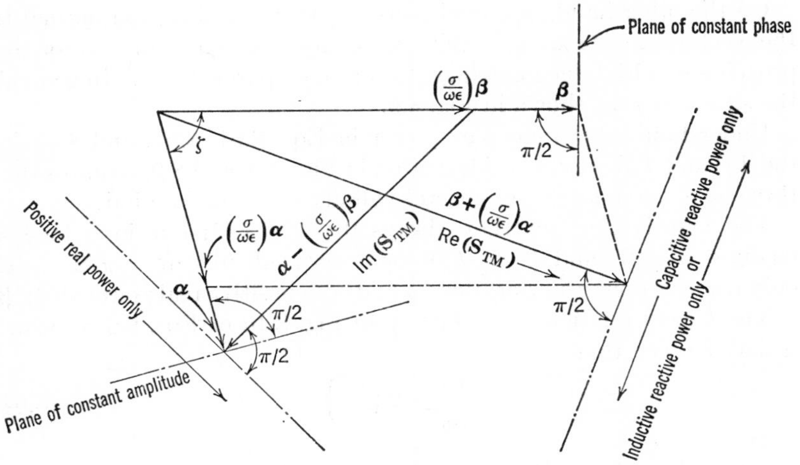

Fig. 8.9. Analysis of the complex Poynting vector for a TM plane wave in a dissipative medium. Compare with Fig. 7.6*b* and Fig. 8.8.

To discuss complex power for the TM wave, Eq. 8.59 is convenient. This results from the fact that **H** has only a component $H_{\alpha\times\beta}$ parallel to $\boldsymbol{\alpha} \times \boldsymbol{\beta}$, which makes $\bar{\boldsymbol{\gamma}}\cdot\mathbf{H}^* = 0$ in this case. Accordingly

$$\mathbf{S}_{TM} = \left(\frac{\boldsymbol{\alpha} + j\boldsymbol{\beta}}{\sigma + j\omega\epsilon}\right)\frac{1}{2}\,|H_{\alpha\times\beta,0}|^2 e^{-2\boldsymbol{\alpha}\cdot\mathbf{r}} \tag{8.62}$$

It follows that for the TM wave both real and reactive power flow parallel to the planes of constant amplitude and both flow parallel to the planes of constant phase.

To find the directions in which there is only real or only reactive power for the TM wave, we divide $\mathbf{S}_{TM}$ into real and imaginary parts by noting that

$$\frac{\boldsymbol{\alpha} + j\boldsymbol{\beta}}{\sigma + j\omega\epsilon} = \frac{(\sigma - j\omega\epsilon)(\boldsymbol{\alpha} + j\boldsymbol{\beta})}{\sigma^2 + (\omega\epsilon)^2}$$

$$= \frac{\omega\epsilon}{\sigma^2 + (\omega\epsilon)^2}\left\{\left[\boldsymbol{\beta} + \left(\frac{\sigma}{\omega\epsilon}\right)\boldsymbol{\alpha}\right] - j\left[\boldsymbol{\alpha} - \left(\frac{\sigma}{\omega\epsilon}\right)\boldsymbol{\beta}\right]\right\} \tag{8.63}$$

The introduction of Eq. 8.63 into Eq. 8.62 shows that only reactive power flows in the planes perpendicular to the vector $[\boldsymbol{\beta} + (\sigma/\omega\epsilon)\boldsymbol{\alpha}]$. If the loss tangent is small, this vector is very nearly $\boldsymbol{\beta}$; but if the loss tangent is large, it becomes very nearly parallel to $\boldsymbol{\alpha}$. An intermediate case is illustrated in Fig. 8.9.

On the other hand, only real power flows in the direction normal to the vector $[\boldsymbol{\alpha} - (\sigma/\omega\epsilon)\boldsymbol{\beta}]$. With small loss tangent, this vector approaches $\boldsymbol{\alpha}$; with large loss tangent it becomes parallel to $\boldsymbol{\beta}$. In general, the situation is as shown in Fig. 8.9.

Understanding the complex power in Eqs. 8.61, 8.62, and 8.63 for the TE and TM waves, as illustrated in Figs. 8.8 and 8.9, requires that three auxiliary relations of general validity be borne in mind:

The first is the fact that always $0 \leq \zeta \leq \pi/2$; but in a lossless medium $\boldsymbol{\alpha}$ ($\neq 0$) and $\boldsymbol{\beta}$ *must* become perpendicular ($\zeta = \pi/2$), with only reactive power flowing along $\boldsymbol{\alpha}$ and only positive real power along $\boldsymbol{\beta}$.

The second relation stems first from Fig. 8.1*a* or Eqs. 8.9, in which it may be seen that

$$\frac{\beta_0}{\alpha_0} \geq 2\left(\frac{\omega\epsilon}{\sigma}\right) \tag{8.64}$$

and then from Eqs. 8.14, in which it is noted that β and α are related just as are β_0 and α_0, except for a replacement of σ by $\sigma/\cos\zeta$. This means the analogous relation to Eq. 8.64 is

$$\alpha \cos\zeta \leq \frac{1}{2}\left(\frac{\sigma}{\omega\epsilon}\right)\beta \tag{8.65}$$

In other words, the projection of $\boldsymbol{\alpha}$ onto $\boldsymbol{\beta}$ is always smaller than $\frac{1}{2}(\sigma/\omega\epsilon)\beta$.

The third relation is again Eq. 8.14, where it is clear that, for specified values of σ, ϵ, μ, and ω, both α and β increase in size (from their minimum values α_0 and β_0) as ζ increases from zero to $\pi/2$.

The importance of the above three conditions lies not only in visualizing the vectors $[\boldsymbol{\alpha} - (\sigma/\omega\epsilon)\boldsymbol{\beta}]$ and $[\boldsymbol{\beta} + (\sigma/\omega\epsilon)\boldsymbol{\alpha}]$ but also in resolving questions about the algebraic sign of real and reactive power components in various space directions. The designations shown in Figs. 8.8 and 8.9 for these components of the complex power should be studied carefully, with reference to Eqs. 8.61, 8.62, and 8.63, to master the space-direction and algebraic-sign problems involved.

It is worth emphasizing the considerable lack of symmetry (or rather duality) we have found between the TE and TM plane waves in a dissipative medium. This situation is brought about by the fact that the loss is associated with the electric field only, through the conductivity σ, with no provision for an analogous loss associated with the magnetic field. It therefore makes considerable difference in the loss whether there is a single (linearly polarized) component of electric field, or two such components comprising an elliptically polarized electric field. Exactly this difference occurs between the electric fields of the TE and TM plane waves; the corresponding difference which also occurs for the magnetic fields simply has no dissipative effect for the kind of medium with which we are dealing.

With the general family of uniform and nonuniform plane waves now at our disposal in both lossless and dissipative media, we can proceed to treat some rather intriguing examples. Once again, in the interests of simplicity, we shall limit the scope to problems in which the boundary conditions can be met with only a few plane waves. Nevertheless, the surprising flexibility available through linear combination of these elementary waves will emerge more clearly than it has in our previous considerations.

8.2 Normal Incidence of Uniform Plane Waves

8.2.1 Some General Remarks

The simplest plane wave problems to consider involve normal incidence of uniform plane waves upon a boundary between two different media. If both media are isotropic and homogeneous, the polarization of the incident wave makes little difference in the physics of this problem (see Sec. 7.4.1). We may therefore limit our consideration arbitrarily to only x-polarized uniform plane waves traveling in the $\pm z$ directions (Sec. 7.3).

If either medium concerned is dissipative, the appropriate field solutions in it must be of the form dictated by Eqs. 8.6, 8.19, 8.22, and 8.23, namely:

$$\text{(a)} \qquad \mathrm{E}_x = \mathrm{E}_x{}^+ e^{-\bar{\gamma}_0 z} + \mathrm{E}_x{}^- e^{\bar{\gamma}_0 z} \tag{8.66}$$

$$\text{(b)} \qquad \mathrm{H}_y = \frac{1}{\bar{\eta}} (\mathrm{E}_x{}^+ e^{-\bar{\gamma}_0 z} - \mathrm{E}_x{}^- e^{\bar{\gamma}_0 z})$$

where when $\omega \geq 0$

$$\text{(a)} \qquad \bar{\eta} = \sqrt{\frac{j\omega\mu}{\sigma + j\omega\epsilon}} = \eta' + j\eta'' \qquad \eta' > 0 \qquad \eta'' \geq 0 \tag{8.67}$$

$$\text{(b)} \qquad \bar{\gamma}_0 = \sqrt{j\omega\mu(\sigma + j\omega\epsilon)} = \alpha_0 + j\beta_0 \qquad \beta_0 \geq 0 \qquad \alpha_0 \geq 0$$

The similarity of Eqs. 8.66 and 8.67 to those of a dissipative transmission line with $R = 0$ (Eqs. 5.1 and 5.2) means that the general features of the wave motion are the same as those of such a line. The effect of the loss is primarily to introduce attenuation and to make the "characteristic impedance" (here the "wave impedance" $\bar{\eta}$) complex. We can see that for small losses, i.e., when $\sigma \ll \omega\epsilon$, the approximate forms of the solutions are essentially those valid for lossless media, modified only by the addition of the attenuation constant α_0. Specifically, from Eqs. 8.10 or Fig. 8.1

$$\left.\begin{aligned} &\text{(a)} \qquad \alpha_0 \approx \frac{1}{2}\sigma\sqrt{\frac{\mu}{\epsilon}} \\ &\text{(b)} \qquad \beta_0 \approx \omega\sqrt{\epsilon\mu} \\ &\text{(c)} \qquad \bar{\eta} \approx \sqrt{\frac{\mu}{\epsilon}} \end{aligned}\right\} \quad \text{for small loss, i.e., } \sigma \ll \omega\epsilon \qquad (8.68)$$

which should be compared with Eqs. 5.4 and 5.5. All the techniques involving the reflection coefficient and the Smith chart for dissipative lines are available in this case. Uniform plane wave propagation in slightly imperfect dielectrics is entirely similar to that of waves on a slightly dissipative transmission line in which $G \ll \omega C$ and $R = 0$.

From the point of view of plane waves, however, there is another very important case, the counterpart of which we have not emphasized in connection with our transmission-line discussions. This is the problem of wave propagation in a highly (but not perfectly) conducting medium like copper or silver. Obviously this problem involves a medium with large losses. The transmission-line counterpart of it would be a very dissipative line, which is, of course, not generally the kind of line used in practice either for energy transmission or for resonant systems. This is why we did not stress such situations in Chapters 5 and 6. It is true that our general transmission-line equations and solutions cover these cases just as well as they do those of small loss. The same is true of Eqs. 8.66 and 8.67 for uniform plane waves. Nevertheless, the specific features arising from the large loss are rather different from those discussed thus far.

In a metal like copper, we have pointed out that the approximation $\sigma \gg \omega\epsilon$ applies quite safely for other than optical problems. Equations 8.6 and 8.11, or Fig. 8.1, under these conditions tell us that:

$$\left.\begin{aligned}
&\text{(a)} && \bar{\eta} \approx \sqrt{\frac{j\omega\mu}{\sigma}} = \sqrt{\frac{\omega\mu}{2\sigma}}\,(1+j) \\
&\text{(b)} && \bar{\gamma}_0 \approx \sqrt{j\omega\mu} = \sqrt{\frac{\omega\mu\sigma}{2}}\,(1+j) \approx \sigma\bar{\eta} \\
&\text{(c)} && \eta' \approx \eta'' \approx \sqrt{\frac{\omega\mu}{2\sigma}} \\
&\text{(d)} && \alpha_0 \approx \beta_0 \approx \sqrt{\frac{\omega\mu\sigma}{2}} \approx \sigma\eta' \approx \sigma\eta''
\end{aligned}\right\}\quad \sigma \gg \omega\epsilon \tag{8.69}$$

We note that the characteristic wave impedance $\bar{\eta}$ has an angle of 45° at all frequencies for which the approximation is valid, as does the propagation constant $\bar{\gamma}_0$. The attenuation constant α_0 is just as much a function of frequency as the phase constant β_0. In particular, α_0 and β_0 are proportional to $\sqrt{\omega}$ if σ and μ are constant.

Obviously there is no point in trying to think of the transient behavior of such a dissipative medium in terms at all similar to those appropriate for a lossless one; the dispersion is staggering, so wave shapes change radically with position and time. Moreover, the attenuation (in decibels) in one wave length (on a steady-state basis) is $8.68\ \alpha_0\lambda = 8.68 \times (2\pi\alpha_0)/\beta_0 = 2\pi \times 8.68 = 54.5$ db! Thus it is more appropriate to think of these "waves" in space as attenuating exponentially, with some associated phase shift, rather than as undergoing progressive phase shift, with slight attenuation, as customary for a small-loss medium. There is perhaps some question about whether one ought to bother about a "wave length" concept at all, at least in the sense $\lambda = 2\pi/\beta_0$, because the amplitude is attenuated so much in one of these wave lengths that even observing it becomes impossible. Of more significance, however, is the distance δ, in which the amplitude of a single wave has decreased by a factor e^{-1}:

$$\delta \equiv \frac{1}{\alpha_0} \approx \sqrt{\frac{2}{\omega\mu\sigma}} = \sqrt{\frac{1}{\pi f\mu\sigma}} \tag{8.70}$$

For copper ($\sigma \approx 5.8 \times 10^7$ mhos/m) at $f = 1$ mc, Eq. 8.70 gives $\delta \cong 0.066$ mm. At $f = 100$ mc, $\delta \cong 6.6$ microns.

The distance δ is known as the *skin depth* or *depth of penetration.* The wave impedance $\bar{\eta}$ has a magnitude $\sqrt{\omega\mu/\sigma} = \sqrt{2}(\alpha_0/\sigma) = \sqrt{2}/(\sigma\delta)$, according to Eqs. 8.69 and 8.70. For copper at 1 mc,

$|\bar{\eta}| \cong 0.37 \times 10^{-3}$ ohms, or about one millionth of $\eta_0 = \sqrt{\mu_0/\epsilon_0} =$ 377 ohms for free space. It is worth while to remark that $\bar{\eta}$ is *inductive* (with a 45° phase angle), corresponding to the fact that $\omega\epsilon$ has been neglected, leaving only σ and $j\omega\mu$.

8.2.2 An Example

Some of the implications of our preliminary calculations may be shown in terms of the example of a uniform plane wave normally incident from air upon a copper block (Fig. 8.10). The analogous

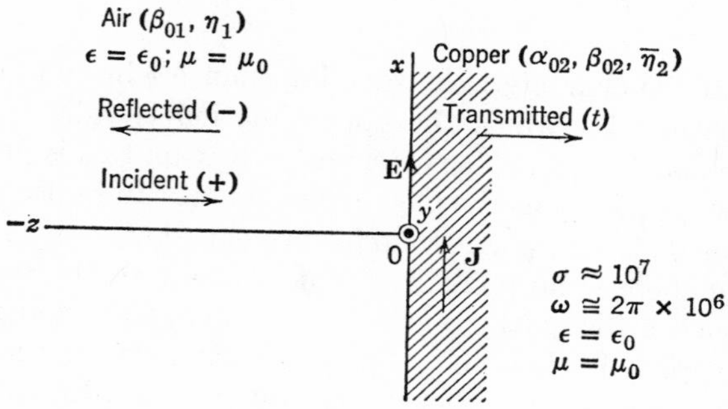

Fig. 8.10. Normal incidence of a uniform plane wave upon a metal surface.

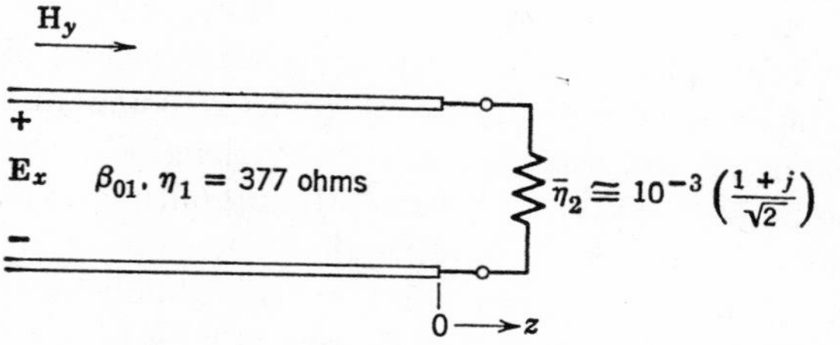

Fig. 8.11. A transmission-line analog for Fig. 8.10.

transmission-line representation appropriate for determining the reflection on the air side is shown in Fig. 8.11.

We have for the reflection coefficient at $z = 0$

$$\bar{\Gamma}_{\text{air}}(0) = \frac{\bar{\eta}_2 - \eta_1}{\bar{\eta}_2 + \eta_1} = \frac{(\bar{\eta}_2/\eta_1) - 1}{(\bar{\eta}_2/\eta_1) + 1}$$

But

$$\left|\frac{\bar{\eta}_2}{\eta_1}\right| \approx \frac{\sqrt{(\omega\mu_0)/\sigma}}{\sqrt{\mu_0/\epsilon_0}} = \sqrt{\frac{\omega\epsilon_0}{\sigma}} \ll 1$$

if $\sigma \gg \omega\epsilon_0$. We can expand the reflection coefficient to the first order in $\bar{\eta}_2/\eta_1$:

$$\begin{aligned}\bar{\Gamma}_{\mathrm{air}}(0) &= \left(\frac{\bar{\eta}_2}{\eta_1} - 1\right)\left(1 + \frac{\bar{\eta}_2}{\eta_1}\right)^{-1} \\ &= \left(\frac{\bar{\eta}_2}{\eta_1} - 1\right)\left[1 - \left(\frac{\bar{\eta}_2}{\eta_1}\right) + \left(\frac{\bar{\eta}_2}{\eta_1}\right)^2 - \left(\frac{\bar{\eta}_2}{\eta_1}\right)^3 + \cdots\right] \\ &= -1 + 2\left(\frac{\bar{\eta}_2}{\eta_1}\right) - 2\left(\frac{\bar{\eta}_2}{\eta_1}\right)^2 + 2\left(\frac{\bar{\eta}_2}{\eta_1}\right)^3 - \cdots\end{aligned}$$

or

$$\bar{\Gamma}_{\mathrm{air}}(0) \approx -1 + 2\left(\frac{\bar{\eta}_2}{\eta_1}\right) \qquad \left|\frac{\bar{\eta}_2}{\eta_1}\right| \ll 1 \tag{8.71}$$

Thus

$$\bar{\Gamma}_{\mathrm{air}}(0) \approx -1 + 2\sqrt{\frac{j\omega\epsilon_0}{\sigma}} = \left(-1 + \sqrt{\frac{2\omega\epsilon_0}{\sigma}}\right) + j\sqrt{\frac{2\omega\epsilon_0}{\sigma}} \qquad \frac{\omega\epsilon_0}{\sigma} \ll 1 \tag{8.72}$$

Accordingly, the fraction of incident power reflected is approximately

$$|\bar{\Gamma}_{\mathrm{air}}(0)|^2 \approx 1 - 4\sqrt{\frac{\omega\epsilon_0}{2\sigma}}$$

and the fraction transmitted into the metal is

$$1 - |\bar{\Gamma}_{\mathrm{air}}(0)|^2 \approx 4\sqrt{\frac{\omega\epsilon_0}{2\sigma}}$$

Therefore the time-average power $\langle P_d \rangle$ entering the metal per square meter, being the incident power times the fraction of it transmitted, is given by

$$\begin{aligned}\langle P_d \rangle_{\mathrm{sq\ m}} &= \frac{1}{2}|\mathrm{H}_{y1}{}^+|^2\eta_1(1 - |\bar{\Gamma}_{\mathrm{air}}(0)|^2) \approx \frac{1}{2}|\mathrm{H}_{y1}{}^+|^2\sqrt{\frac{\mu_0}{\epsilon_0}}\left(4\sqrt{\frac{\omega\epsilon_0}{2\sigma}}\right) \\ &= \frac{1}{2}|2\mathrm{H}_{y1}{}^+|^2\sqrt{\frac{\omega\mu_0}{2\sigma}} \approx \frac{1}{2}|2\mathrm{H}_{y1}{}^+|^2\eta_2{}' \qquad \frac{\omega\epsilon_0}{\sigma} \ll 1\end{aligned} \tag{8.73}$$

in which we have used the relation $\eta_2{}' = \mathrm{Re}\,\bar{\eta}_2$ from Eq. 8.69*c*.

The form of Eq. 8.73 suggests the following more rapid approximate method of solving this problem. The method employs an important technique that will be quite useful later on.

Because $|\bar{\eta}_2|$ is so small, it is clear that the air "line" in Fig. 8.11 is almost short-circuited by the metal. Of course, if we are interested in the transmitted wave and in the power dissipated (per unit area) in the copper, we cannot assume that the metal is really a short circuit; then the electric field would have to be zero on the interface. It is certainly not zero. No matter how small it may be, in percentage it is infinitely larger than zero! We can, however, determine the voltage across the load in Fig. 8.11 quite accurately by noting that, since $|\bar{\Gamma}_{\text{air}}(0)|$ is so nearly -1, the *current* through the load is approximately given by $2\mathrm{H}_{y1}{}^+$, even though the voltage is *not* accurately given by $|\mathrm{E}_{x1}{}^+| - |\mathrm{E}_{x1}{}^-| \approx 0$. This is merely a result of the well-known fact that the sum of two nearly equal large quantities is less sensitive to an error in either one than is their difference. But, knowing the current, we also know that the load voltage must be very nearly $2\mathrm{H}_{y1}{}^+\bar{\eta}_2$, which, because of the boundary conditions, corresponds to E_{xt} in Fig. 8.10. We can therefore write the transmitted wave field launched into the metal as

$$\mathrm{E}_{xt} \approx 2\mathrm{H}_{y1}{}^+\bar{\eta}_2 e^{-\alpha_{02}z} e^{-j\beta_{02}z} \tag{8.74a}$$

$$\mathrm{H}_{yt} = \frac{\mathrm{E}_{xt}}{\bar{\eta}_2} \approx 2\mathrm{H}_{y1}{}^+ e^{-\alpha_{02}z} e^{-j\beta_{02}z} \tag{8.74b}$$

where α_{02}, β_{02}, and $\bar{\eta}_2$ are to be taken from Eq. 8.69. The time-average power flowing into the metal per square meter of its surface at $z = 0$ is, accordingly,

$$\langle P_d \rangle_{\text{sq m}} = \tfrac{1}{2}\,\mathrm{Re}\,(\mathrm{E}_{xt}\mathrm{H}_{yt}{}^*) \approx \tfrac{1}{2}|2\mathrm{H}_{y1}{}^+|^2\eta_2'$$

which is in agreement with Eq. 8.73.

The complex current density $\mathbf{J}$ in the metal (Fig. 8.10) is the current flowing in the $+x$ direction per square meter of area parallel to the y, z-plane. It is a function of z:

$$\mathbf{J} = \boldsymbol{a}_x \mathrm{J}_x = \boldsymbol{a}_x \sigma \mathrm{E}_{xt} \approx \boldsymbol{a}_x 2\mathrm{H}_{y1}{}^+\bar{\eta}_2 \sigma e^{-\alpha_{02}z} e^{-j\beta_{02}z} \tag{8.74c}$$

Note that $|\mathrm{J}_x|$ falls by a factor e^{-1} in a distance δ from the surface. The total complex current I_0 per meter width along y of the conductor is

$$\mathrm{I}_0 \text{ per meter width} = \int_0^\infty \mathrm{J}_x\, dz \approx \frac{+2\mathrm{H}_{y1}\sigma\bar{\eta}_2}{\alpha_{02} + j\beta_{02}} \approx 2\mathrm{H}_{y1}{}^+ \tag{8.74d}$$

where again we have used Eqs. 8.69. It is noteworthy that the voltage drop per meter along x at the surface $z = 0$, being just $\mathrm{E}_{xt}|_{z=0}$, is simply $\bar{\eta}_2 I_0$ (according to Eqs. 8.74*a* and 8.74*d*). For this reason, $\bar{\eta}_2$, *measured in ohms, may also be thought of as the "surface impedance" of the metal: i.e., the ratio of the voltage drop per meter along the surface to the total current per meter width flowing inside the metal.* It is interesting that this impedance is just as much inductive as it is resistive.

There is still another aspect of Eq. 8.74*d* that deserves consideration. The quantity $2H_{y1}{}^{+}$ is exactly the magnetic field which would have existed tangential to the conductor had its conductivity σ been infinite. Under these conditions, we know that a surface current density (i.e., amperes per meter measured along y) would have existed which was just equal to $2H_{y1}{}^{+}$. According to Eqs. 8.74*d* and 8.74*c*, we see that very nearly this same total current per meter width still flows when σ is merely very large; only now the current is not all crammed onto the surface but extends into the metal with a density that diminishes rapidly ($e^{-\alpha_{02}z}$) with depth. In terms of this current, we find from Eqs. 8.74*d*, 8.73, and 8.70 that

$$\langle P_d \rangle_{\text{sq m}} \approx \frac{1}{2} |I_0|^2 (\eta_2') = \frac{1}{2} |I_0|^2 \left(\frac{1}{\sigma\delta}\right) \tag{8.75a}$$

But the resistance to x-directed direct-current flow of a slab of the metal having dimensions of (1 m along x) $\times$ (1 m along y) $\times$ (δ m along z) would be just $1/\sigma\delta$ ohms. This view of Eq. 8.75*a* tells us that, *insofar as power loss is concerned*, the current I_0 may be thought of as flowing with a *uniform* distribution along z, but confined within a depth δ. It is in terms of this interpretation of Eq. 8.75*a* that the concept of the skin depth δ derives its principal high-frequency applications. The crowding of the current near the metal surface is known as the *skin effect*.

The presence of the reactive (inductive) part η'' of the characteristic wave impedance in Eq. 8.69*c*, however, also has an effect upon the field in the air side of the interface in Fig. 8.10 or 8.11. This effect has to do with the phase angle of $\bar{\Gamma}_{\text{air}}(0)$ and does not show up well in the particular approximate view of the problem which leads through Eqs. 8.74 and 8.75. Specifically, the point in question here is the location of the first minimum of the electric field in the air.

The amplitudes of the electric field minima in the air are certainly very small because of the small magnitude of the wave impedance of the metal. Of course they occur regularly, with separations of a half the air wave length. If the impedance $\bar{\eta}_2$ were entirely real, and smaller than η_1, a minimum electric field would fall *at* the interface (the reflection coefficient there would be negative and real). The first minimum inside the air region would occur at $z = -(\lambda_{\text{air}}/2)$, or $\beta_{01}z = -\pi$.

Actually, the reflection coefficient $\bar{\Gamma}_{air}(0)$ in the air at the interface is not negative real, but is given approximately by Eq. 8.71 or 8.72. Writing

$$\bar{\Gamma}_{air}(0) = |\bar{\Gamma}_{air}(0)| e^{j\varphi}$$

we find from Eq. 8.72 that

$$\tan \varphi \approx \frac{\sqrt{(2\omega\epsilon_0)/\sigma}}{-1 + \sqrt{(2\omega\epsilon_0)/\sigma}} \approx -\sqrt{\frac{2\omega\epsilon_0}{\sigma}} \qquad \frac{\omega\epsilon_0}{\sigma} \ll 1$$

and that φ lies in the second quadrant. Thus

$$\varphi \approx \pi - \tan^{-1}\sqrt{\frac{2\omega\epsilon_0}{\sigma}} \approx \pi - \sqrt{\frac{2\omega\epsilon_0}{\sigma}} \qquad \frac{\omega\epsilon_0}{\sigma} \ll 1$$

because the tangent of a small angle equals that angle.

The position $z_{min} < 0$ at which the first minimum of the electric field occurs is then that for which

$$\measuredangle\, \bar{\Gamma}_{air}(z_{min}) = -\pi$$

inasmuch as $\bar{\Gamma}_{air}(z)$ must take increasingly *lagging* phases at points in the air increasingly remote from the interface. But since, in general,

$$\bar{\Gamma}_{air}(z) = \bar{\Gamma}_{air}(0)e^{j2\beta_{01}z} = |\bar{\Gamma}_{air}(0)| e^{j(\varphi+2\beta_{01}z)}$$

the above condition requires

$$\varphi + 2\beta_{01}z_{min} = -\pi$$

or

$$\beta_{01}z_{min} \approx -\pi + \sqrt{\frac{\omega\epsilon_0}{2\sigma}}$$

or

$$z_{min} \approx -\frac{\lambda_{air}}{2} + \sqrt{\frac{1}{2\omega\mu_0\sigma}} = -\frac{\lambda_{air}}{2} + \frac{1}{2}\delta \qquad \frac{\omega\epsilon_0}{\sigma} \ll 1 \qquad (8.75b)$$

That is, the actual interface gives rise to a standing-wave pattern in the air which looks as though it arises from incidence on a purely resistive medium whose surface lies *one half* of a skin depth ($\delta/2$) inside the real metal.

8.2.3 A Different Viewpoint, Similar Problem

One should not be misled by the preceding discussion into believing that the problem of skin depth is essentially only a very high-frequency phenomenon. We note that the approximation $\sigma \gg \omega\epsilon$ *in the metal* is certainly valid at very low frequencies, in spite of the fact that the particular approach to the problem we happen to have used in Fig. 8.10 is admittedly suggestive of high frequencies because we spoke of a uniform plane wave propagating in the *air*.

To clarify the low-frequency aspects of the field and current distributions in a metal, consider the example of Fig. 8.12. It represents a perfectly conducting parallel-plate line extending from $-\infty < y < \infty$, driven at $z = -l$ by a low-frequency distributed voltage source V_0,

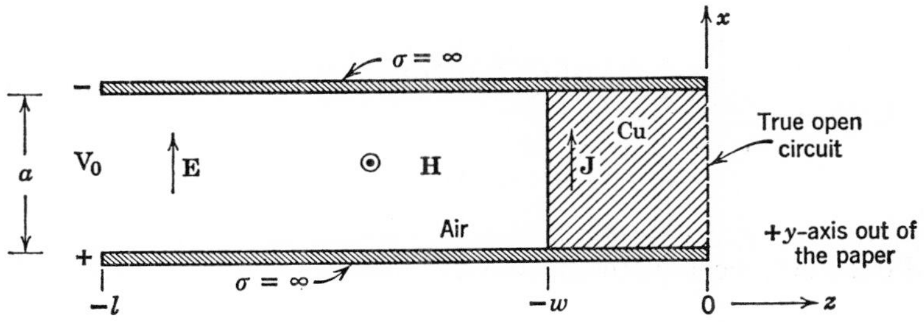

Fig. 8.12. Example for the low-frequency skin effect.

and terminated in a copper block extending along y from $-\infty$ to $+\infty$, and along z from $z = -w$ to $z = 0$. This idealized situation might represent an attempt to measure the impedance of the block between its upper and lower faces; the perfectly conducting planes serve as one method of insuring the "equipotential" character of these faces.

In order to dispose of slight difficulties with fringing fields at $z = 0$, we shall assume that this plane is a true open circuit. This is defined as the dual of a short circuit and, consequently, has the somewhat nonphysical property that tangential **H** rather than tangential **E** is zero everywhere on it. Stated in another way, we are neglecting fringing at $z = 0$, so that $\mathbf{H} = \boldsymbol{a}_y H_y$ and $\partial/\partial x \equiv \partial/\partial y \equiv 0$ in the solution. Then, if we insist that there can be no current flowing along the planes at $z = 0$, there can be no H_y at $z = 0$ for $x = 0$ and $x = a$. Since $\partial/\partial x = 0$, we therefore have $H_y = 0$ at every point in the plane $z = 0$.

For a moment let us consider the case of an applied *direct* voltage V_0. The electric field in the air *and* the copper does not vary with x, y, or z, and has the value $E_{0x} = V_0/a$ everywhere. The current density in the copper is

$$J_{0x} = \sigma E_{0x} = \sigma \left(\frac{V_0}{a}\right) \tag{8.76}$$

and the total current I_{0x} per meter along y is

$$I_{0x} = wJ_{0x} = \left(\frac{w\sigma}{a}\right) V_0 \tag{8.77}$$

which tells us that the d-c resistance of the bar is just

$$R_{\text{d-c}} = \frac{V_0}{I_{0x}} = \frac{a}{w\sigma} \qquad \text{per meter along } y \tag{8.78}$$

There will, of course, be a d-c magnetic field along y, which, from our open-circuit boundary condition, must be zero at $z = 0$. Thus, within the conductor,

$$\boldsymbol{H}_{\text{copper}} = -\boldsymbol{a}_y J_{0x} z = -\boldsymbol{a}_y \sigma \left(\frac{V_0}{a}\right) z \qquad -w \leq z \leq 0 \tag{8.79}$$

and outside the conductor (in the air)

$$\boldsymbol{H}_{\text{air}} = \boldsymbol{a}_y \sigma w \left(\frac{V_0}{a}\right) \qquad -l \leq z \leq -w \tag{8.80}$$

It is interesting to note that on the plane $z = -w$ the instantaneous (d-c) Poynting vector is $(\boldsymbol{E}_0 \times \boldsymbol{H})_{z=-w} = \boldsymbol{a}_z \sigma w (V_0/a)^2$, in the $+z$ direction (into the metal). Therefore, since no power flows out the open-circuited end at $z = 0$, the total power sent into the metal is as

$$\left(\frac{V_0}{a}\right)^2 \sigma wa = V_0^2 \left(\frac{w\sigma}{a}\right) = V_0^2/R_{\text{d-c}} \qquad \text{w/m along } y$$

it should be.

Now let us suppose that the voltage V_0 of Fig. 8.12 represents the complex amplitude of a 60-cps a-c generator located at $z = -l$. The symbols for all currents and fields will, accordingly, represent the complex amplitudes of the corresponding sinusoids. We shall suppose that the dimensions l, a, and w are all short compared with the *air* wave length (3100 miles), so that the solution for the fields *in the air* may be carried out on a quasi-static basis, starting from the foregoing d-c case.

Thus, in the air, there will be a voltage dependence on z produced by the time-rate of change of $\boldsymbol{H}_0$ such that

$$\mathrm{V}_0 - \mathrm{V}(z) \approx -j\omega[\mu_0 a(l + z)\mathrm{H}_{0y}]$$

or

$$\mathrm{V}(z) \approx \mathrm{V}_0[1 + j\omega\mu_0\sigma w(l + z)] \qquad -l \leq z \leq -w \qquad (8.81a)$$

In particular, at $z = -w$

$$\mathrm{V}(-w) \approx \mathrm{V}_0[1 + j\omega\mu_0\sigma w(l - w)] \qquad (8.81b)$$

Similarly, the magnetic field varies with z because of the displacement current of $\boldsymbol{E}_0$:

$$\mathrm{H}_y(z) - \mathrm{H}_y(-w) \approx -j\omega[\epsilon_0(z + w)\mathrm{E}_{0x}]$$

or

$$\mathrm{H}_y(z) \approx \mathrm{H}_y(-w) - j\omega\epsilon_0\left(\frac{\mathrm{V}_0}{a}\right)(z + w) \qquad -l \leq z \leq -w \qquad (8.81c)$$

At $z = -l$ specifically, therefore,

$$\mathrm{H}_y(-l) \approx \mathrm{H}_y(-w) + j\omega\epsilon_0\left(\frac{\mathrm{V}_0}{a}\right)(l - w) \qquad (8.81d)$$

Since V_0 is regarded as specified, by a voltage source at $z = -l$, Eq. 8.81*b* permits calculation of the voltage $\mathrm{V}(-w)$ [or the electric field $\mathrm{E}_x(-w) = \mathrm{V}(-w)/a$] at the air side of the copper block. Equation 8.81*d*, however, simply relates the magnetic field (or current) at the source ($z = -l$) to its value at the interface $z = -w$. The latter is not yet known; but it will be fixed shortly by conditions within the block, including the open-circuit boundary condition at $z = 0$ and the now known value of the voltage or electric field at $z = -w$ (Eq. 8.81*b*).

Inside the metal we shall have E_x and H_y independent of x and y, but dependent on z. Moreover, the variation with z will not necessarily be the same as it was in the d-c case, nor need it even be represented correctly by a first-order approximation in frequency (quasi-static solution). This is true because w is *not* necessarily short compared with the wave length (or the attenuation length) for waves *in the metal*, even though it is short compared with the wave length in *air* at 60 cps. Thus we must use the general solutions (Eqs. 8.66) for the waves in the metal, subject to the condition $\mathrm{H}_y = 0$ at $z = 0$. This condition requires $\mathrm{E}_x{}^+ = \mathrm{E}_x{}^-$; so the solutions may be written

$$\left.\begin{aligned} &\text{(a)} \qquad \mathrm{E}_x = 2\mathrm{E}_x{}^+ \cosh \bar{\gamma}_0 z \\ &\text{(b)} \qquad \mathrm{H}_y = -\frac{2\mathrm{E}_x{}^+}{\bar{\eta}} \sinh \bar{\gamma}_0 z \end{aligned}\right\} \qquad -w \leq z \leq 0 \qquad (8.82)$$

At $z = -w$, the electric field in the metal must match that in the air:

$$\frac{V(-w)}{a} = 2E_x^+ \cosh \bar{\gamma}_0 w$$

where $V(-w)$ is given by Eq. 8.81*b*. Solved for E_x^+, the above relation becomes

$$2E_x^+ = \frac{V(-w)}{a \cosh \bar{\gamma}_0 w} \tag{8.82c}$$

In view of Eq. 8.82*c*, Eqs. 8.82*a* and 8.82*b* become

$$\left.\begin{aligned} &\text{(a)} \qquad E_x = \frac{V(-w)\cosh \bar{\gamma}_0 z}{a \cosh \bar{\gamma}_0 w} \\ &\text{(b)} \qquad H_y = -\frac{V(-w)\sinh \bar{\gamma}_0 z}{\bar{\eta} a \cosh \bar{\gamma}_0 w} \end{aligned}\right\} \quad -w \leq z \leq 0 \tag{8.83}$$

Accordingly, $H_y(-w)$ is now fixed by Eq. 8.83*b*, and the boundary condition that H_y must be continuous at $z = -w$ serves only to determine the level of H_y in the air through Eq. 8.81*c*. We shall not pursue the formal substitutions for this part of the solution. It is not needed for our present purposes.

The current density distribution in the metal, however, is of immediate interest.

$$J_x = \sigma E_x = \frac{\sigma V(-w)\cosh \bar{\gamma}_0 z}{a \cosh \bar{\gamma}_0 w} \qquad -w \leq z \leq 0 \tag{8.84}$$

which we observe has the value $[\sigma V(-w)]/a$ at $z = -w$. This value is fixed completely by the source and the parallel-plate line, according to Eq. 8.81*b*. It is convenient to continue the discussion of the current distribution in terms of the ratio $J_x(z)/J_x(0)$, which is given simply by the expression

$$\frac{J_x(z)}{J_x(0)} = \cosh \bar{\gamma}_0 z \qquad -w \leq z \leq 0 \tag{8.85}$$

where we must remember that, when $\sigma \gg \omega\epsilon_0$

$$\bar{\gamma}_0 = \alpha_0 + j\beta_0 \approx \sqrt{\frac{\omega\mu_0\sigma}{2}}(1 + j) = \frac{(1 + j)}{\delta}$$

Now

$$\cosh\left(\frac{z}{\delta} + j\frac{z}{\delta}\right) = \cosh\left(\frac{z}{\delta}\right)\cos\left(\frac{z}{\delta}\right) + j\sinh\left(\frac{z}{\delta}\right)\sin\left(\frac{z}{\delta}\right)$$

So

$$|\cosh \bar{\gamma}_0 z|^2 = \cosh^2\left(\frac{z}{\delta}\right) - \sin^2\left(\frac{z}{\delta}\right) = \frac{1}{2}\left[\cosh\left(\frac{2z}{\delta}\right) + \cos\left(\frac{2z}{\delta}\right)\right]$$

Hence

$$\left|\frac{J_x(z)}{J_x(0)}\right|^2 = \frac{1}{2}\left[\cosh\left(\frac{2z}{\delta}\right) + \cos\left(\frac{2z}{\delta}\right)\right] \qquad -w \le z \le 0 \qquad (8.86)$$

A simple calculation of the first two derivatives of Eq. 8.86 with respect to z will show that for $z < 0$ the slope is always negative and the curvature always positive. Thus a plot of the relative current density versus (z/δ) appears as shown in Fig. 8.13, where it is clear that unless $w < \delta$, there is appreciable crowding of the current toward the side of the bar nearest the source.

At 60 cps, Eq. 8.70 shows that δ is about 0.85 cm in copper. The nonuniformity of the current density in large power bus bars at power frequencies may be a serious problem. For a given total current in the bar, a distribution crowded toward one surface (Fig. 8.13) causes more heating than a uniform distribution, since the heating at each point is proportional to the *square* of the current density. The result of the *skin effect* is therefore to make the a-c resistance of a metal wire larger than its d-c resistance, whenever the dimensions of the cross section are comparable to or greater than the skin depth δ.

Indeed, in trying to reduce the resistance at any given frequency above zero, there is a size of uniform solid conductor of any type beyond which it is not very profitable to go. The added cross section would not carry enough of the current to justify its weight and cost. Many schemes for stranding conductors (Litz wire), laminating them, or otherwise using different metals to produce a more uniform current distribution over the cross section have been employed to reduce conductor losses or attenuation in a variety of applications.

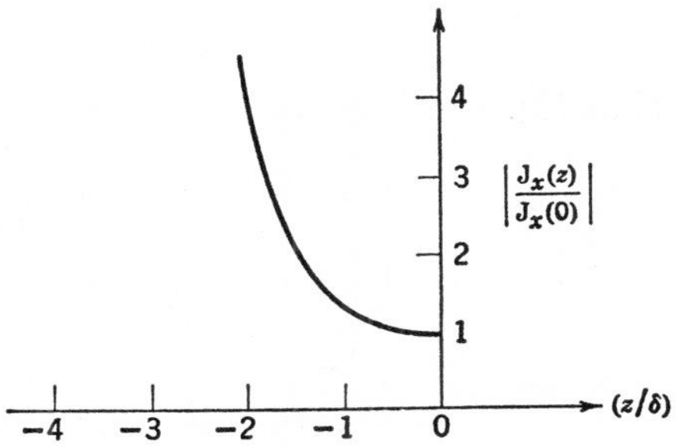

Fig. 8.13. Current density in the copper of Fig. 8.12.

We can examine more quantitatively the aforementioned increase of resistance for the bar of Fig. 8.12, and also learn something about its reactance, by computing its impedance. According to Eq. 8.84, the total current in the bar, per meter along y, is

$$\mathrm{I}_x = \int_{-w}^{0} \mathrm{J}_x\, dz = \frac{\sigma \mathrm{V}(-w)}{a \cosh \bar{\gamma}_0 w} \int_{-w}^{0} \cosh \bar{\gamma}_0 z\, dz$$

$$= \frac{w\sigma \mathrm{V}(-w)}{a} \left(\frac{\tanh \bar{\gamma}_0 w}{\bar{\gamma}_0 w} \right) \tag{8.87}$$

So its impedance is

$$\frac{\mathrm{V}(-w)}{\mathrm{I}_x} = \mathrm{Z} = \frac{a}{w\sigma} \left(\frac{\bar{\gamma}_0 w}{\tanh \bar{\gamma}_0 w} \right) \qquad \text{per meter along } y \tag{8.88}$$

Since $\bar{\gamma}_0 = \sqrt{(\omega\mu\sigma/2)}(1 + j) = (1/\delta)(1 + j)$, we see that, if the frequency is low enough (or if $w \ll \delta$), $\tanh \bar{\gamma}_0 w \to \bar{\gamma}_0 w$ and $\mathrm{Z} \to a/w\sigma$. This checks the d-c resistance of the bar given in Eq. 8.78.

For slightly higher frequencies, where w may begin to become comparable to δ, we expand the tanh in a power series

$$\tanh \bar{\gamma}_0 w = (\bar{\gamma}_0 w) - \frac{(\bar{\gamma}_0 w)^3}{3} + \frac{2(\bar{\gamma}_0 w)^5}{15} - \cdots \tag{8.89}$$

So

$$\frac{\tanh \bar{\gamma}_0 w}{\bar{\gamma}_0 w} \approx 1 - \frac{(\bar{\gamma}_0 w)^2}{3} + \frac{2(\bar{\gamma}_0 w)^4}{15}$$

and by direct long division, we find

$$\frac{\bar{\gamma}_0 w}{\tanh \bar{\gamma}_0 w} \cong \frac{1}{1 - (\bar{\gamma}_0 w)^2/3 + 2(\bar{\gamma}_0 w)^4/15}$$

$$\cong 1 + \frac{1}{3} (\bar{\gamma}_0 w)^2 - \frac{1}{45} (\bar{\gamma}_0 w)^4$$

$$= \left[1 + \frac{4}{45} \left(\frac{w}{\delta} \right)^4 \right] + j \frac{2}{3} \left(\frac{w}{\delta} \right)^2 \tag{8.90}$$

Thus Eq. 8.88 becomes

$$\mathrm{Z} \approx \frac{a}{w\sigma} \left[1 + \frac{4}{45} \left(\frac{w}{\delta} \right)^4 \right] + j \frac{2}{3} \frac{aw}{\sigma \delta^2}$$

$$= R_{\text{d-c}} \left[1 + \frac{4}{45} \left(\frac{w}{\delta} \right)^4 \right] + j\omega \left(\frac{\mu_0 a w}{3} \right)$$

$$= R_{\text{a-c}} + j\omega L_i \qquad \text{per meter along } y \tag{8.91}$$

Evidently $R_{a\text{-}c} > R_{d\text{-}c}$, and there is also an effective inductance L_i, called the *internal inductance* of the bar. The latter arises from the magnetic field in the bar. It can also be calculated to the same approximation as Eq. 8.91 from the usual first-order "flux-linkages-per-ampere" procedure, using the d-c solutions (Eqs. 8.76, 8.77, and 8.79).

At a very high frequency, or for a very large conductor, we would have $w \gg \delta$ and, therefore, $|\bar{\gamma}_0 w| \gg 1$. Then $\tanh \bar{\gamma}_0 w \to 1$, and Eq. 8.88 yields

$$Z \approx a\frac{\bar{\gamma}_0}{\sigma} = a\bar{\eta} = \frac{a}{\sigma\delta} + j\frac{a}{\sigma\delta} \qquad \text{per meter along } y \qquad (8.92)$$

which does not depend upon w. In other words, the conductor is so thick that, because of the attenuation, the end $z = 0$ cannot make itself felt at $z = -w$. Figure 8.12 is then effectively the same as Fig. 8.10. This is why when $a = 1$ m, Eq. 8.92 checks with Eq. 8.69*a*.

When $w \gg \delta$, so that Eq. 8.92 is valid, the skin effect is said to be *well developed* in the conductor. Under these circumstances, the two surfaces at $z = 0$ and $z = -w$ in Fig. 8.12 are almost completely "decoupled." We could replace the open circuit by any other termination—even another transmission line and source—without materially affecting anything on the original line. Of course, if we did use another source, there would be current flowing near the surface $z = 0$; but it would depend almost entirely upon the new source, and very little upon anything near $z = -w$ or at $z < -w$. In such a case, the shielding action of the copper for electromagnetic waves is not noticeably less effective than it would be if it were a perfect conductor.

Thus far, the skin-effect phenomenon has arisen for us only in connection with normal incidence of a uniform plane wave upon a metal surface, although we have shown that no limitation to very high frequencies is implied by this fact. The point is that "high frequency" is a relative term. In lossless media, we have seen several times that it relates to the comparison between the wave length $\lambda = 2\pi/\beta_0$ of uniform plane waves in the medium and the significant dimensions of the system under consideration. In the case of a very dissipative medium, we have just learned that the length $\delta = 1/\alpha_0$ is an equally significant quantity to be compared with these same dimensions. Since in general $\beta_0 \geq \alpha_0$, we may infer that the wave nature of the steady-state field in a given region of space becomes important when the significant dimensions of that region become comparable to or larger than $2\pi/\beta_0$. Of course, as we have seen in connection with Fig. 8.12, this condition may occur at quite different frequencies in different portions of space filled with different media. Correspondingly, at a given frequency, the field situation may be amenable to a quasi-static approximation in some portions of space, but only to the complete wave treatment in others.

We turn next to problems involving oblique incidence and nonuniform plane waves, where we shall learn more about the skin effect and encounter some new aspects of the field problem.

8.3 Oblique Incidence of Uniform Plane Waves *

Let a uniform plane wave be incident from air at an angle ϑ upon a metal boundary, as shown in Fig. 8.14. We treat the case in which the magnetic field is parallel to the boundary. The incident wave expression is therefore

$$\begin{aligned} &\text{(a)}\quad \mathbf{H}_i = \boldsymbol{a}_x \mathrm{H}_{xi} e^{-j\beta_{01} z \cos\vartheta} e^{j\beta_{01} y \sin\vartheta} \\ &\text{(b)}\quad \mathbf{E}_i = -(\boldsymbol{a}_y \eta_1 \mathrm{H}_{xi} \cos\vartheta + \boldsymbol{a}_z \eta_1 \mathrm{H}_{xi} \sin\vartheta) e^{-j\beta_{01} z \cos\vartheta} e^{j\beta_{01} y \sin\vartheta} \end{aligned} \tag{8.93}$$

where

$$\begin{aligned} &\text{(a)}\quad \beta_{01} = \omega\sqrt{\epsilon_0 \mu_0} \\ &\text{(b)}\quad \eta_1 = \sqrt{\frac{\mu_0}{\epsilon_0}} \end{aligned} \tag{8.94}$$

If there is to be any hope of meeting the continuity conditions on H_x along the whole interface, the reflected wave must also be a uniform plane wave at an angle ϑ, with magnetic field along x. It could not be a nonuniform plane wave because, if the phase were correct along the interface, there would then be attenuation of this wave along the interface. This would not match the pure phase delay along y exhibited by the incident wave in Eq. 8.93. Figure 8.14 has been drawn accordingly to show a reflected uniform plane wave. Without writing further expressions for the reflected wave, we therefore know that the transmission-line analogy for oblique incidence of uniform plane waves will be valid on the air side (compare Figs. 7.26 and 7.27*b* with Fig. 8.14). This model for the present case appears in Fig. 8.15, where Z_{z2} represents the wave impedance for the $+z$ direction in the metal.

In the metal, however, matters are more complicated than in the air. *The transmitted wave cannot in principle be a uniform plane wave.* Such a wave always has attenuation *and* phase shift along the same direction in space. If its propagation direction were chosen so the phase shift along y matched that of the incident (or reflected) wave in Eq. 8.93,

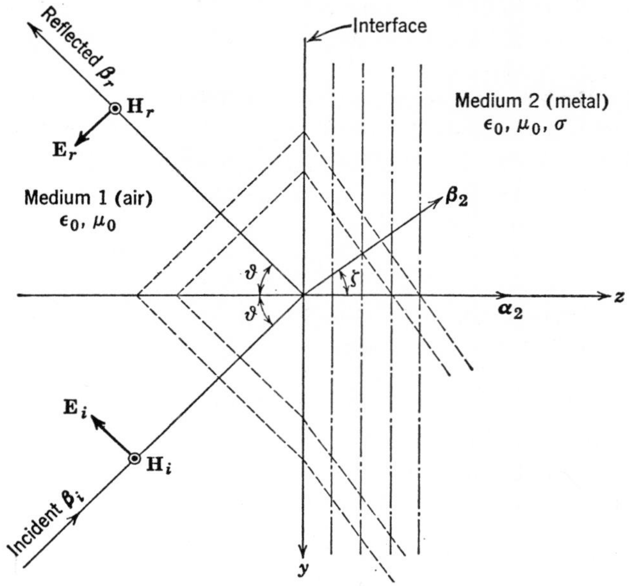

Fig. 8.14. Oblique incidence of a uniform plane wave from air upon a metal half-space. Dashed lines (– – –) are planes of constant phase; dash-dot lines (—·—·—·—) are planes of constant amplitude. Note that $\zeta < \vartheta$.

the corresponding attenuation along y of the transmitted wave would make it impossible to match the corresponding magnitudes! What is needed is a wave which has *no attenuation along the y direction* but a phase delay along $-y$ which can be made equal to that of the incident wave. The requirement for an H_x component to meet that of the other waves suggests a TM nonuniform plane wave in the metal, with no

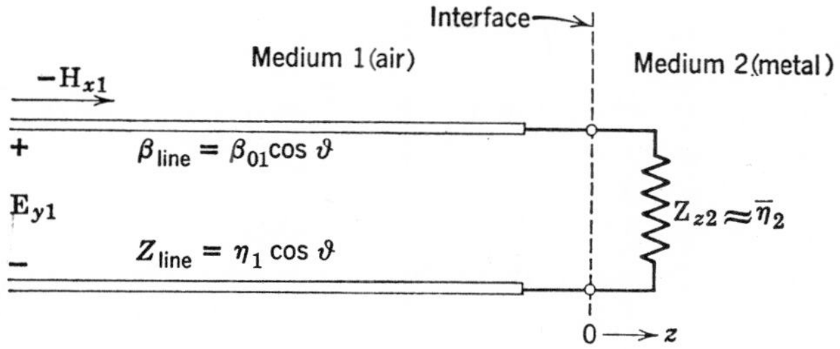

Fig. 8.15. Transmission-line analogy for x and y field components and z-variation in Fig. 8.14.

space variation along x. Moreover $\boldsymbol{\alpha}_2$ for this wave *must be normal to the interface* to avoid attenuation along that interface. The parameters α_2, β_2, and ζ must be determined by the remaining boundary conditions and properties of the nonuniform plane wave.

Thus, with reference to Fig. 8.14, the requirement that the source of power lie on the air side of the interface only, leads to the demand that the fields vanish (rather than grow exponentially) at $z = +\infty$. Since the space variation is by our convention $e^{-\bar{\boldsymbol{\gamma}}_2 \cdot \boldsymbol{r}}$, we must have

$$\boldsymbol{\alpha}_2 = \boldsymbol{a}_z \alpha_2 \qquad \alpha_2 > 0 \tag{8.95}$$

The need for the transmitted and incident waves to stay "in step" along the y-axis is expressed by the relation

$$\beta_2 \sin \zeta = \beta_{01} \sin \vartheta \tag{8.96}$$

analogous to Snell's law.

Finally, the last condition pertinent to determining $\bar{\boldsymbol{\gamma}}_2$ is the general restriction for nonuniform plane waves that

$$\bar{\boldsymbol{\gamma}}_2 \cdot \bar{\boldsymbol{\gamma}}_2 = \bar{\gamma}_{02}^2 = j\omega\mu_0(\sigma + j\omega\epsilon_0)$$

or

$$\begin{aligned} &\text{(a)} \qquad \beta_2^2 - \alpha_2^2 = \omega^2 \epsilon_0 \mu_0 = \beta_{01}^2 \\ &\text{(b)} \qquad \alpha_2 \beta_2 \cos \zeta = \frac{\omega\mu_0\sigma}{2} = \alpha_{02}\beta_{02} \end{aligned} \tag{8.97}$$

in which we have taken advantage of Eq. 8.94*a* in Eq. 8.97*a*, and written $\bar{\gamma}_{02} = \alpha_{02} + j\beta_{02}$. Observe from Eq. 8.97*a* that $\beta_2 \geq \beta_{01}$; so in Eq. 8.96, $\zeta \leq \vartheta$.

With Eqs. 8.96, 8.97*a*, and 8.97*b*, we can determine the unknown parameters α_2, β_2, and ζ. The variable ζ is eliminated first by squaring and adding Eq. 8.96 to Eq. 8.97*b* after dividing it by α_2:

$$\beta_2^2 = \beta_{01}^2 \sin^2 \vartheta + \frac{(\alpha_{02}\beta_{02})^2}{\alpha_2^2}$$

Substituting in this result the value of β_2^2 from Eq. 8.97*a* yields an equation for α_2 alone:

$$\alpha_2^4 + (\beta_{01}^2 \cos^2 \vartheta)\alpha_2^2 - (\alpha_{02}\beta_{02})^2 = 0$$

The only solution for $\alpha_2^2 > 0$ is therefore

$$\alpha_2^2 = \left(\frac{\beta_{01} \cos \vartheta}{\sqrt{2}}\right)^2 \left[\sqrt{1 + \left(\frac{\sqrt{2\alpha_{02}\beta_{02}}}{\beta_{01} \cos \vartheta}\right)^4} - 1\right]$$

But

$$\frac{\sqrt{2\alpha_{02}\beta_{02}}}{\beta_{01}} = \frac{\sqrt{2[(\omega\mu_0\sigma)/2]}}{\omega\sqrt{\epsilon_0\mu_0}} = \sqrt{\frac{\sigma}{\omega\epsilon_0}}$$

so

$$\alpha_2{}^2 = \frac{\omega\mu_0\sigma}{2}\left[\sqrt{1 + \left(\frac{\omega\epsilon_0}{\sigma}\cos^2\vartheta\right)^2} - \left(\frac{\omega\epsilon_0}{\sigma}\cos^2\vartheta\right)\right] \tag{8.98}$$

with β_2 and $\sin\zeta$ available directly from Eqs. 8.97a and 8.96 or 8.97b in that order.

Let

$$\sinh\xi = \left(\frac{\omega\epsilon_0}{\sigma}\right)\cos^2\vartheta \le \left(\frac{\omega\epsilon_0}{\sigma}\right) = \sinh\xi_0 \tag{8.99}$$

which implies that

$$\sqrt{1 + \left(\frac{\omega\epsilon_0}{\sigma}\cos^2\vartheta\right)^2} = \sqrt{1 + \sinh^2\xi} = \cosh\xi$$

and therefore

$$\alpha_2 = \sqrt{\frac{\omega\mu_0\sigma}{2}}\,(\cosh\xi - \sinh\xi)^{1/2} = e^{-\xi/2}\sqrt{\frac{\omega\mu_0\sigma}{2}} \tag{8.100}$$

Now when $\vartheta = 0$, $\alpha_2 = \alpha_{02}$, the attenuation constant for uniform plane waves, so

$$\alpha_{02} = e^{-\xi_0/2}\sqrt{\frac{\omega\mu_0\sigma}{2}}$$

Since $\xi \le \xi_0$, we have $\alpha_2 \ge \alpha_{02}$ and, consequently, from Eq. 8.97a, $\beta_2 \ge \beta_{02}$. Indeed

$$\beta_2 = \sqrt{\frac{\omega\mu_0\sigma}{2}}\left[e^{-\xi} + 2\left(\frac{\omega\epsilon_0}{\sigma}\right)\right]^{1/2} \tag{8.101}$$

and

$$\sin\zeta = \frac{\beta_{01}\sin\vartheta}{\beta_2} = \frac{\sqrt{(2\omega\epsilon_0)/\sigma}\,\sin\vartheta}{\{e^{-\xi} + 2[(\omega\epsilon_0)/\sigma]\}^{1/2}} \tag{8.102}$$

When the metal is a good conductor, we have seen that $(\omega\epsilon_0)/\sigma \lll 1$ at frequencies of usual interest. Then in Eq. 8.99 $\sinh\xi \approx \xi = (\omega\epsilon_0\cos^2\vartheta)/\sigma \lll 1$ and, computed to the "zeroth" order in $\omega\epsilon_0/\sigma$, which means neglecting $\omega\epsilon_0/\sigma$ compared to 1, Eqs. 8.100 and 8.101 show that

$$\alpha_2 \approx \alpha_{02} \approx \beta_{02} \approx \beta_2 \qquad 0 \le \vartheta \le \frac{\pi}{2} \qquad \frac{\sigma}{\omega\epsilon_0} \gg 1 \tag{8.103}$$

The nonuniform TM wave in the metal is not very nonuniform at all! The transmitted wave is altered very little by changing the angle of incidence from normal to oblique; it always remains *almost* a uniform plane wave traveling normal to the interface. The attenuation and phase vectors ($\boldsymbol{\alpha}_2$ and $\boldsymbol{\beta}_2$) have essentially the same *magnitudes* as they would have in a uniform plane wave, independent of the angle of incidence. Their relative direction ζ, however, does shift slightly with ϑ, according to Eq. 8.102:

$$\sin \zeta \approx \sqrt{\frac{2\omega\epsilon_0}{\sigma}} \sin \vartheta \qquad \frac{\sigma}{\omega\epsilon_0} \gg 1 \tag{8.104}$$

although they remain very nearly parallel in any case. Apparently $\bar{\boldsymbol{\gamma}}_2 \approx \boldsymbol{a}_z\bar{\gamma}_{02}$ to the zeroth order in $(\omega\epsilon_0)/\sigma$.

One expects from the foregoing discussion that the power dissipated in a highly conducting metal, per unit incident field strength, does not vary very much as the angle of an incident uniform plane wave is increased from the normal. The transmitted TM plane wave always seems to remain *practically* a uniform plane wave propagating (and attenuating) in a direction perpendicular to the interface. We may verify this expectation by computing first the impedance Z_{z2} (Fig. 8.15) which determines the field strength relations at the interface. At the same time, we may also examine the power flow along the interface on the metal side by considering the wave impedance Z_{-y2}. For the transmitted TM wave (see Fig. 8.14)

$$\begin{aligned}
&\text{(a)} \quad \mathbf{H}_t = \boldsymbol{a}_x \mathrm{H}_{xt} e^{-\alpha_2 z} e^{-j\beta_2 z \cos \zeta} e^{j\beta_2 y \sin \zeta} = \mathbf{H}_{02} e^{-\bar{\boldsymbol{\gamma}}_2 \cdot \boldsymbol{r}} \\
&\text{(b)} \quad \bar{\boldsymbol{\gamma}}_2 = -\boldsymbol{a}_y j\beta_2 \sin \zeta + \boldsymbol{a}_z(\alpha_2 + j\beta_2 \cos \zeta) \\
&\text{(c)} \quad \mathbf{E}_t = \frac{\mathbf{H}_t \times \bar{\boldsymbol{\gamma}}_2}{\sigma + j\omega\epsilon_0} \\
&\qquad\quad = \mathrm{H}_{xt}\left[-\mathbf{a}_y\left(\frac{\alpha_2 + j\beta_2 \cos \zeta}{\sigma + j\omega\epsilon_0}\right) - \mathbf{a}_z\left(\frac{j\beta_2 \sin \zeta}{\sigma + j\omega\epsilon_0}\right)\right] e^{-\bar{\boldsymbol{\gamma}}_2 \cdot \boldsymbol{r}}
\end{aligned} \tag{8.105}$$

Therefore

$$\begin{aligned}
&\text{(a)} \qquad Z_{z2} \equiv -\frac{\mathrm{E}_{yt}}{\mathrm{H}_{xt}} = \frac{\alpha_2 + j\beta_2 \cos \zeta}{\sigma + j\omega\epsilon_0} \\
&\text{(b)} \qquad Z_{-y2} \equiv -\frac{\mathrm{E}_{zt}}{\mathrm{H}_{xt}} = \frac{j\beta_2 \sin \zeta}{\sigma + j\omega\epsilon_0} = \frac{j\beta_{01} \sin \vartheta}{\sigma + j\omega\epsilon_0}
\end{aligned} \tag{8.106}$$

where Eq. 8.96 has been used in Eq. 8.106*b*. But by Eqs. 8.97 and 8.100

$$\beta_2 \cos \zeta = \frac{\omega\mu_0\sigma}{2\alpha_2} = \sqrt{\frac{\omega\mu_0\sigma}{2}}\, e^{\xi/2} \tag{8.107}$$

Therefore

$$Z_{z2} = \frac{\sqrt{(\omega\mu_0/2\sigma)}\,(e^{-\xi/2} + je^{\xi/2})}{1 + j(\omega\epsilon_0/\sigma)} \tag{8.108a}$$

To the zeroth order in $\omega\epsilon_0/\sigma$ then,

$$Z_{z2} \approx \sqrt{\frac{\omega\mu_0}{2\sigma}}\,(1 + j1) = \sqrt{\frac{j\omega\mu_0}{\sigma}} \approx \bar{\eta}_2 \qquad \frac{\sigma}{\omega\epsilon_0} \gg 1 \tag{8.108b}$$

as indicated in Fig. 8.15. Similarly

$$Z_{-y2} = \frac{\sqrt{(j\omega\mu_0/\sigma)}\,\sqrt{(j\omega\epsilon_0/\sigma)}\,\sin\vartheta}{1 + j(\omega\epsilon_0/\sigma)} \tag{8.109a}$$

which becomes to the zeroth order in $\omega\epsilon_0/\sigma$

$$Z_{-y2} \approx \sqrt{\frac{j\omega\mu_0}{\sigma}}\sqrt{\frac{j\omega\epsilon_0}{\sigma}}\,\sin\vartheta \approx \bar{\eta}_2\sqrt{\frac{j\omega\epsilon_0}{\sigma}}\,\sin\vartheta \qquad \frac{\sigma}{\omega\epsilon_0} \gg 1 \tag{8.109b}$$

So

$$|Z_{-y2}| \ll |Z_{z2}| \qquad \text{when } \sigma/\omega\epsilon_0 \gg 1$$

It is apparent from the above result that, in the approximation employed, the greatest part of the complex power in the metal goes in the $+z$ direction. A small amount of inductive reactive power (and no real power) goes along the interface on the metal side, as long as $\vartheta \neq 0$. Determining the real power entering Z_{z2} in Fig. 8.14 therefore suffices to find the power dissipated in the metal. This problem, however, is very similar to that of normal incidence (Fig. 8.11), except for the change of polarization and the appearance of $\eta_1 \cos\vartheta$ in place of η_1 and $\beta_{01} \cos\vartheta$ in place of β_{01}. The entire analysis and discussion comprising Eqs. 8.71 through 8.75 applies with these changes incorporated. The important results are:

$$\left.\begin{aligned}
&\text{(a)} \qquad \langle P_d\rangle_{\text{sq m}} \approx \frac{1}{2}|2H_{xi}|^2\eta_2' \approx \frac{1}{2}|2H_{xi}|^2\sqrt{\frac{\omega\mu_0}{2\sigma}} \\
&\qquad\qquad\qquad \approx \frac{1}{2}|I_{0y}|^2\frac{1}{\sigma\delta} \\
&\text{(b)} \qquad z_{\min} \approx -\frac{\lambda_{\text{air}}}{2\cos\vartheta} + \frac{1}{2}\frac{\delta}{\cos\vartheta}
\end{aligned}\right\} \quad \frac{\omega\epsilon_0}{\sigma\cos^2\vartheta} \ll 1 \tag{8.110}$$

It must be emphasized that the restriction on these results, arising from the condition $|\bar{\eta}_2/\eta_1 \cos \vartheta| \ll 1$, has a new significance. The appearance of $(\omega\epsilon_0)/(\sigma \cos^2 \vartheta) \ll 1$ instead of $\omega\epsilon_0/\sigma \ll 1$ in them means that Eqs. 8.110 *become invalid for sufficiently grazing incidence.* This happens because the characteristic impedance of the line in Fig. 8.15 gets arbitrarily small as $\cos \vartheta$ decreases, so that eventually the terminating impedance $\bar{\eta}_2$ will no longer act as a comparatively low impedance. The contrary assumptions made in connection with Eqs. 8.71 through 8.75 would then be unjustified, and a more exact calculation of the reflection coefficient at the boundary would have to be made.

One other feature of the present problem deserves discussion. Regarded from the point of view of the $-y$-axis as "longitudinal," there is certainly real power flowing in the air parallel to that axis in Fig. 8.14, just as there was in the case of a perfect conductor (Fig. 7.21) or in medium 1 of the total reflection case of Fig. 7.26. In the present situation, however, the need to supply losses in the metal requires a $+z$ component of *real power* at all points in the air. Because the $-y$-directed power *does not diminish with* y, there is no way to get power into the metal without having a source somewhere at $z < 0$. The unattenuated character of the y variation in this case, indeed, implies the existence in the air (medium 1) of a (possibly remote) source of $+z$-directed real power. Such a source extends at full strength all the way from $y = +\infty$ to $y = -\infty$. This is in addition, of course, to the source of $-y$-directed real power (possibly at $y = +\infty$) which makes possible the pure traveling-wave character of the field in the $-y$ direction. Consequently, it is not possible to use our present results to make a guided-wave or wave-guide solution by simply adding another interface parallel to the x,y-plane at an appropriate position $z < 0$, as we did previously in connection with Figs. 7.28, 7.32, or 7.35. Such a plane or interface, being passive, could not supply the necessary $+z$-directed power, unless we included provision for a (possibly remote) source in or to the left of it. In the latter event, the solution could hardly be called "guided" anyway.

Evidently the field solution of this section is simply not appropriate to describe a wave guided by two parallel interfaces, at least one of which is dissipative. The trouble is that such a guided wave would have to arise from a source at $y = \infty$, for example, and, in order to supply the metal losses, it would have to *attenuate* along $-y$. From an energy point of view, such a solution sounds possible and reasonable. Whether it will satisfy Maxwell's equations and fit the boundary conditions remains to be seen. We shall examine this question next.

8.4 Some Guided Waves *

8.4.1 Structure of the Problem *

As our principal example of a guided wave bounded by a dissipative medium, we select the arrangement shown in Fig. 8.16. This example can be viewed in three ways: First, it is an extension of the problem discussed in the last section (Fig. 8.14), made by the addition of a perfectly conducting boundary at $z = -b$. Second, it can also represent a modification of the "parallel-plate" transmission line considered

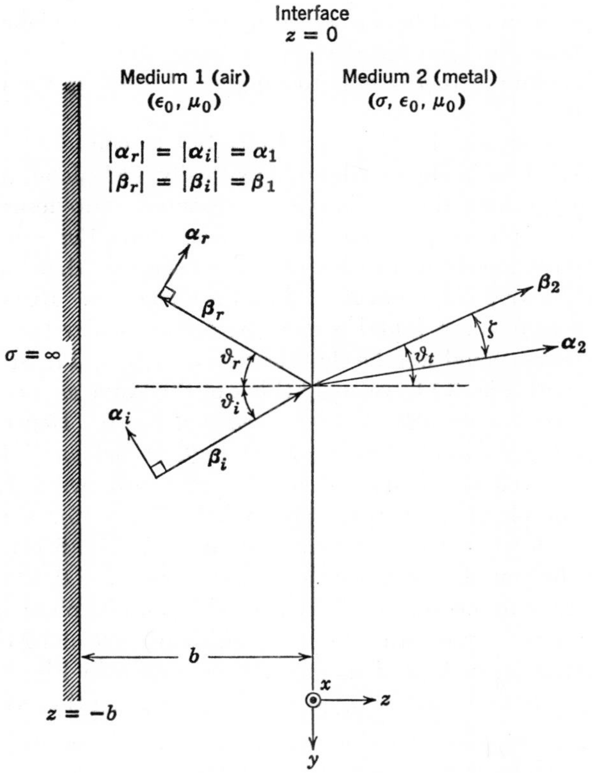

Fig. 8.16. Parallel-plate transmission structure with one dissipative boundary (medium 2). Propagation vectors of nonuniform plane waves, needed in both media to meet boundary conditions, are shown (not to scale). There is no field variation along x. The source is presumed to be at $y = +\infty$.

in Sec. 2.3. In that event, we might expect the loss in the metal (medium 2) to account at least approximately for the series resistance R per unit length of the line. The solution of this problem can then give us for the first time a field description of this resistance parameter, which we introduced rather arbitrarily in Sec. 2.4.1. Third, it provides for conductor loss in some wave-guide problems, similar to those studied in Chapter 7, leading to a discussion of attenuation in wave guides.

From all of the foregoing points of view, we would be interested in the case for which the magnetic field is parallel to the boundaries (has an x component only) and the fields do not vary with x. This is the situation we shall select here.[1] It became quite clear in the last section that this problem could *not* be solved by using *uniform* plane waves in the air. Some provision for attenuation along the $-y$ direction must be made if a source at $y = \infty$ is to supply the losses in the conductor along $z = 0$.

These requirements leave us no choice but to consider *nonuniform* plane waves in the lossless medium; specifically TM plane waves with magnetic field along the x-axis only. Moreover, our desire to have traveling-wave phase delay and attenuation along the $-y$ direction (originating perhaps from a source at $y = \infty$) suggests that these waves must have positive components of $\boldsymbol{\beta}$ and $\boldsymbol{\alpha}$ along $-y$. Recalling that in any lossless medium $\boldsymbol{\beta}$ and $\boldsymbol{\alpha}$ must be perpendicular to each other, one of the waves could well be that described by $\boldsymbol{\beta}_i$ and $\boldsymbol{\alpha}_i$ in Fig. 8.16. In analogy with Fig. 8.14, we could think of this wave as a *nonuniform plane-wave incident upon the metal boundary at* $z = 0$. Hence we have used the subscript i to identify it. The "angle of incidence" for such a wave may as well be thought of as the angle ϑ_i between $\boldsymbol{\beta}_i$ and the normal to the plane $z = 0$ (i.e., the z-axis in Fig. 8.16),[2] since this is consistent with what we have done when $\boldsymbol{\alpha}_i = 0$ (Fig. 8.14).

To meet the boundary conditions along the plane $z = 0$, a "reflected" TM plane wave in medium 1 will be needed. Both its phase delay and attenuation must agree with that of the incident wave along the y-axis, so the orientation of $\boldsymbol{\beta}_r$ and $\boldsymbol{\alpha}_r$ must be as shown in Fig. 8.16, with $|\boldsymbol{\beta}_r| = |\boldsymbol{\beta}_i| \equiv \beta_1$ and $|\boldsymbol{\alpha}_r| = |\boldsymbol{\alpha}_i| \equiv \alpha_1$. The propagation vectors for

[1] The alternate case, with the electric field parallel to the boundary, is an extension of the TE-mode wave-guide solutions considered in Sec. 7.5.1. It should be taken up as a problem.

[2] Observe the difference between this point of view and the alternative one that the *nonuniform* plane waves of Fig. 8.16 could be described instead by letting the incident angle ϑ of the *uniform* plane waves in Fig. 8.14 become complex (see Secs. 7.2.2.2 and 7.4.3.3). This approach to the solution should be taken up as a problem.

these two waves in medium 1 are therefore given by

$$\begin{aligned} &\text{(a)} \quad \bar{\boldsymbol{\gamma}}_i = -\boldsymbol{a}_y(\alpha_1 \cos \vartheta_i + j\beta_1 \sin \vartheta_i) + \boldsymbol{a}_z(-\alpha_1 \sin \vartheta_i + j\beta_1 \cos \vartheta_i) \\ &\text{(b)} \quad \bar{\boldsymbol{\gamma}}_r = -\boldsymbol{a}_y(\alpha_1 \cos \vartheta_i + j\beta_1 \sin \vartheta_i) - \boldsymbol{a}_z(-\alpha_1 \sin \vartheta_i + j\beta_1 \cos \vartheta_i) \end{aligned} \qquad (8.111)$$

where the waves behave as $e^{-\bar{\boldsymbol{\gamma}}_i \cdot \boldsymbol{r}}$ and $e^{-\bar{\boldsymbol{\gamma}}_r \cdot \boldsymbol{r}}$. It will be simpler, however, to rewrite Eqs. 8.111, abbreviating the y and z components of $\bar{\boldsymbol{\gamma}}_i$ as

$$\begin{aligned} &\text{(a)} \quad \bar{\gamma}_{y1} \equiv -\alpha_1 \cos \vartheta_i - j\beta_1 \sin \vartheta_i \\ &\text{(b)} \quad \bar{\gamma}_{z1} \equiv -\alpha_1 \sin \vartheta_i + j\beta_1 \cos \vartheta_i \end{aligned} \qquad (8.112)$$

so that Eqs. 8.111 become simply

$$\begin{aligned} &\text{(a)} \quad \bar{\boldsymbol{\gamma}}_i = \boldsymbol{a}_y\bar{\gamma}_{y1} + \boldsymbol{a}_z\bar{\gamma}_{z1} \\ &\text{(b)} \quad \bar{\boldsymbol{\gamma}}_r = \boldsymbol{a}_y\bar{\gamma}_{y1} - \boldsymbol{a}_z\bar{\gamma}_{z1} \end{aligned} \qquad (8.113)$$

Equations 8.113 are the nonuniform plane-wave analogy of the familiar law that "the angle of reflection equals the angle of incidence." In addition, these propagation vectors must also meet the general restriction on nonuniform plane waves in lossless media that

$$\bar{\boldsymbol{\gamma}}_i \cdot \bar{\boldsymbol{\gamma}}_i = \bar{\boldsymbol{\gamma}}_r \cdot \bar{\boldsymbol{\gamma}}_r = \bar{\gamma}_{y1}^2 + \bar{\gamma}_{z1}^2 = \alpha_1^2 - \beta_1^2 = -\beta_{01}^2 = -\omega^2 \epsilon_0 \mu_0 \qquad (8.114)$$

As regards the z dependence of the waves in medium 1, we notice in Eqs. 8.113 that the incident and reflected waves are characterized by *complex* propagation constants $\bar{\gamma}_{z1}$ and $-\bar{\gamma}_{z1}$ respectively, defined by Eq. 8.112*b*. This value of $\bar{\gamma}_{z1}$ may be used as the propagation constant in an equivalent transmission-line representation for the $\pm z$ directions in medium 1, which we shall use to apply the specific tangential field boundary conditions at $z = 0$ and $z = -b$. However, we need the effective characteristic impedance of the line. This may be taken as the wave impedance Z_{z1} seen looking along $+z$ for the incident TM plane wave. Inasmuch as for this wave

$$\begin{aligned} &\text{(a)} \quad \mathbf{H}_i = \boldsymbol{a}_x \mathrm{H}_{xi} e^{-\bar{\boldsymbol{\gamma}}_i \cdot \boldsymbol{r}} \\ &\text{(b)} \quad \mathbf{E}_i = \frac{\mathbf{H}_i \times \bar{\boldsymbol{\gamma}}_i}{j\omega\epsilon_0} = \left[-\boldsymbol{a}_y \left(\frac{\bar{\gamma}_{z1}}{j\omega\epsilon_0} \right) + \boldsymbol{a}_z \left(\frac{\bar{\gamma}_{y1}}{j\omega\epsilon_0} \right) \right] \mathrm{H}_{xi} e^{-\bar{\boldsymbol{\gamma}}_i \cdot \boldsymbol{r}} \end{aligned} \qquad (8.115)$$

the desired wave impedance is

$$Z_{z1} = \frac{-\mathrm{E}_{y1}}{\mathrm{H}_{x1}} = \frac{\bar{\gamma}_{z1}}{j\omega\epsilon_0} \qquad (8.116)$$

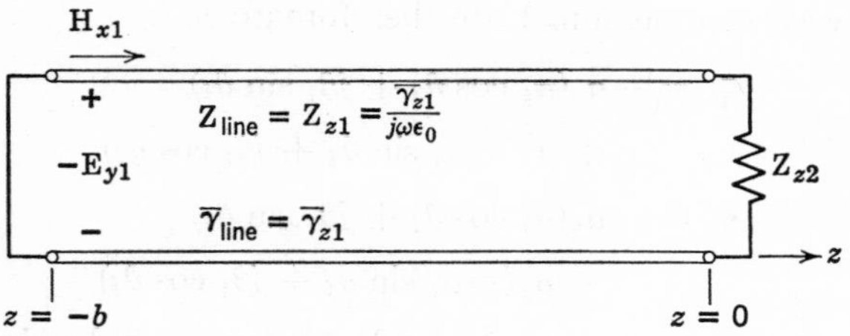

Fig. 8.17. Equivalent transmission-line representation for x and y field components and z dependence in medium 1, Fig. 8.16.

The quantities $\bar{\gamma}_{z1}$ and Z_{z1} have accordingly been set equal to the propagation constant and characteristic impedance respectively of the transmission line in Fig. 8.17.[1]

In the metal (medium 2), we may expect a "transmitted wave" in the form of another TM nonuniform plane wave. In accordance with our interest in a guided wave, the field must vanish at $z = +\infty$ (rather than grow exponentially). Unlike the situation in Fig. 8.14, however, we now require *both* attenuation and phase delay along $-y$ in Fig. 8.16. Otherwise the transmitted wave could not possibly stay "in step" with the incident (and reflected) waves along the interface $z = 0$. Thus we must allow an angle ϑ_t between the phase vector $\boldsymbol{\beta}_2$ and the $+z$-axis, and another angle ζ between the attenuation vector $\boldsymbol{\alpha}_2$ and the phase vector $\boldsymbol{\beta}_2$. Figure 8.16 shows the geometry.

Evidently with variation $e^{-\bar{\boldsymbol{\gamma}}_2 \cdot \boldsymbol{r}}$, we have

$$\begin{aligned} \bar{\boldsymbol{\gamma}}_2 &\equiv \boldsymbol{a}_y \bar{\gamma}_{y2} + \boldsymbol{a}_z \bar{\gamma}_{z2} \\ &= \boldsymbol{a}_y(-\alpha_2 \sin(\vartheta_t - \zeta) - j\beta_2 \sin\vartheta_t) \\ &\quad + \boldsymbol{a}_z(\alpha_2 \cos(\vartheta_t - \zeta) + j\beta_2 \cos\vartheta_t) \end{aligned} \tag{8.117}$$

The general restrictions for the plane wave in medium 2 are of course

$$\bar{\boldsymbol{\gamma}}_2 \cdot \bar{\boldsymbol{\gamma}}_2 \equiv \bar{\gamma}_{y2}^2 + \bar{\gamma}_{z2}^2 = -\omega^2\epsilon_0\mu_0 + j\omega\mu_0\sigma = -\beta_{01}^2 + j\omega\mu_0\sigma \tag{8.118}$$

or

$$\begin{aligned} &\text{(a)} \qquad \beta_2^2 - \alpha_2^2 = \omega^2\epsilon_0\mu_0 = \beta_{01}^2 \\ &\text{(b)} \qquad \beta_2\alpha_2 \cos\zeta = \frac{\omega\mu_0\sigma}{2} \end{aligned} \tag{8.119}$$

[1] This transmission line is unusual. Figure 8.16 and Eq. 8.112*b* show that waves on it *grow* in the direction of their phase delay. The characteristic impedance has a *negative real part.* The methods of Eq. 2.37 show that it has a *negative* series resistance per unit length. All this is reasonable. Why?

The appearance of β_{01}^2 as such in Eqs. 8.118 and 8.119*a* is special for this problem, arising from the fortuitous fact that the dielectric constant and magnetic permeability happen to be the same in both media (metal and air).

The detailed balance of attenuation and phase along the *y*-axis in media 1 and 2 leads to an extended (complex) form of Snell's law for these nonuniform plane waves

$$\bar{\gamma}_{y2} = \bar{\gamma}_{y1} \tag{8.120}$$

which implies the *two* relations

$$\begin{aligned} &\text{(a)} \qquad \beta_2 \sin \vartheta_t = \beta_1 \sin \vartheta_i \\ &\text{(b)} \qquad \alpha_2 \sin (\vartheta_t - \zeta) = \alpha_1 \cos \vartheta_i \end{aligned} \tag{8.121}$$

Equations 8.114, 8.118, and 8.120 constitute one real and two complex equations respectively for the seven real unknowns (α_1, β_1, ϑ_i, α_2, β_2, ϑ_t, ζ). One more complex equation in these same unknowns is required to complete the information. This will come from the boundary conditions on the tangential fields at $z = 0$ and $z = -b$. The equivalent line of Fig. 8.17 will provide this relation if we note first that the wave impedance Z_{z2} for medium 2 is

$$Z_{z2} = \frac{-E_{yt}}{H_{xt}} = \frac{\bar{\gamma}_{z2}}{\sigma + j\omega\epsilon_0} \tag{8.122}$$

Equation 8.122 above can be derived most easily by using the fact that it must agree with Eq. 8.116 in form, but with subscript 2 for subscript 1 and $\sigma + j\omega\epsilon_0$ for $j\omega\epsilon_0$. The impedance Z_{z2} terminates the line of Fig. 8.17 at $z = 0$.

It is quite clear that the vanishing of the tangential electric field (E_{y1}) on the perfect conductor in Fig. 8.16 requires the short-circuiting of the line in Fig. 8.17 at $z = -b$. Consequently, an effective statement of the remaining condition to be met by the system is that the impedance seen looking to the *left* at $z = 0$, as dictated by the short-circuited section of line, must be equal and opposite to the impedance Z_{z2} seen looking to *right* at $z = 0$. This condition guarantees continuity of both current and voltage on the line at $z = 0$, or of the tangential components of electric and magnetic field at $z = 0$ in the physical problem. This impedance restraint is similar to that used for resonant lines in Chapter 6. Accordingly,

$$Z_{\text{line}} \tanh (\bar{\gamma}_{\text{line}} b) = -Z_{z2}$$

or, in view of Eqs. 8.116 and 8.122,

$$\frac{\bar{\gamma}_{z1}}{j\omega\epsilon_0} \tanh \bar{\gamma}_{z1} b = -\frac{\bar{\gamma}_{z2}}{\sigma + j\omega\epsilon_0} \tag{8.123}$$

But from Eqs. 8.118 and 8.120, we have

$$\bar{\gamma}_{z2}^{\,2} = -\beta_{01}^{\,2} + j\omega\mu_0\sigma - \bar{\gamma}_{y1}^{\,2} \tag{8.124}$$

and from Eq. 8.114,

$$\bar{\gamma}_{y1}^{\,2} = -\beta_{01}^{\,2} - \bar{\gamma}_{z1}^{\,2} \tag{8.125}$$

Therefore elimination of $\bar{\gamma}_{y1}^{\,2}$ between Eqs. 8.124 and 8.125 yields another relation between $\bar{\gamma}_{z1}$ and $\bar{\gamma}_{z2}$:

$$\bar{\gamma}_{z2}^{\,2} = \bar{\gamma}_{z1}^{\,2} + j\omega\mu_0\sigma \tag{8.126}$$

Equations 8.123 and 8.126 together are, in principle, sufficient to determine completely the complex numbers $\bar{\gamma}_{z1}$ and $\bar{\gamma}_{z2}$. The rest of the problem is solved easily if we can perform the simultaneous solution of these two equations. Since, however, they are essentially transcendental equations in *complex* unknowns, direct solution in the general case is at best exceedingly laborious.

8.4.2 Approximations for Small Loss *

In view of the concluding remarks of the last section, let us limit ourselves to a simple situation which is also of great interest, namely, in which the metal (medium 2) has a very high conductivity. An extreme instance is of course the one in which medium 2 is a perfect conductor ($\sigma \to \infty$). It is profitable first to examine this case.

8.4.2.1 THE LOSSLESS CASE.* The field will not penetrate into medium 2 at all, which means $\bar{\gamma}_{z2} \to \infty$. If there is to be any nontrivial guided wave, there must be a finite (and nonzero) value of $\bar{\gamma}_{y1}$. Correspondingly, Eq. 8.125 shows that $\bar{\gamma}_{z1}$ certainly remains finite. Accordingly, $\bar{\gamma}_{z2}$ in Eq. 8.126 becomes large no faster than $\sigma^{1/2}$. This guarantees that the right-hand side of Eq. 8.123 vanishes at least as fast as $\sigma^{-1/2}$, so the allowed values of $\bar{\gamma}_{z1}$ must satisfy the relation

$$\bar{\gamma}_{z1} \tanh \bar{\gamma}_{z1} b = 0 \qquad \sigma = \infty \tag{8.127}$$

One solution is obviously

$$\bar{\gamma}_{z1} = 0 \qquad \sigma = \infty \tag{8.128a}$$

The remaining ones require

$$\tanh \bar{\gamma}_{z1} b = 0$$

or in terms of the exponentials defining the tanh

$$e^{2\bar{\gamma}_{z1} b} = 1$$

It follows that

$$2\bar{\gamma}_{z1} b = j2m\pi \qquad m = 0, 1, 2, \cdots$$

where $m \geq 0$ because $\bar{\gamma}_{z1}$ refers by definition to the incident wave, Eqs. 8.111 through 8.113. So

$$\bar{\gamma}_{z1} = j\left(\frac{m\pi}{b}\right) \qquad m = 0, 1, 2, \cdots \qquad \sigma = \infty \tag{8.128b}$$

The selection $m = 0$ in Eq. 8.128*b* actually includes the case covered by Eq. 8.128*a*, so that we have all the solutions possible in Eq. 8.128*b*.

That solution characterized by $m = 0$ in Eq. 8.128*b* has no variation of the fields along z, none along x (by assumption), and a y variation controlled by Eq. 8.125

$$\bar{\gamma}_{y1} = -j\beta_{01} \qquad \sigma = \infty \qquad m = 0 \tag{8.129}$$

indicating phase delay along $-y$. Obviously, since the magnetic field is parallel to the boundaries, and $E_{y1} = 0$ by Eq. 8.116, this solution is the simple TEM uniform plane wave between parallel, perfectly conducting plates, which we first considered in Chapter 2.

The remaining solutions are characterized by propagation along $-y$ with the propagation constant

$$\left.\begin{aligned} \bar{\gamma}_{y1} &= -\sqrt{\left(\frac{m\pi}{b}\right)^2 - \beta_{01}{}^2} \\ \text{or} \qquad \bar{\gamma}_{y1} &= -j\beta_{01}\sqrt{1 - \left(\frac{\omega_m}{\omega}\right)^2} \end{aligned}\right\} \qquad m = 1, 2, \cdots \tag{8.130a}$$

$$= -j\beta_m$$

where

$$\omega_m = \frac{m\pi}{b\sqrt{\epsilon_0\mu_0}} \qquad m = 1, 2, 3, \cdots \tag{8.130b}$$

is the cutoff frequency for the mth solution. These are wave-guide fields of the TM type between parallel planes. They are analogous to the TE waves we studied in Sec. 7.5.1, but, as suggested there, they will not meet boundary conditions in a rectangular pipe (Fig. 7.28*b*) because of the unfavorable polarization of the electric field and the fact that it does not vary with x.

We are now in a position to consider the effect (principally the attenuation) produced by a large, but not infinite, conductivity σ.

8.4.2.2 A SIMPLIFIED CALCULATION OF THE ATTENUATION.* In the problem of Sec. 8.3, we found that the skin-effect representation of the action of a good conductor could be applied in a situation of oblique incidence just as it could for normal incidence. The reason stems from the insensitivity of the form of the transmitted wave in the metal to the angle of incidence from the air side. We know from Chapter 7 and from the solution of Fig. 8.16, when the conducting boundaries are perfect, that the TM (or in the limiting case TEM) waves can be viewed as successive reflections of a uniform plane wave at oblique incidence. We might, therefore, expect skin-effect considerations to apply in the manner of Secs. 8.2 and 8.3. To proceed on that basis is, of course, to overlook the important fact developed in Sec. 8.4.1 that, with the boundary imperfectly conducting, the waves in the air are not just uniform plane waves at oblique incidence but nonuniform ones instead. Still, if the total effect is going to be small—lead to small attenuations α_r and α_i—one might argue that the change from the lossless situation will be small enough to permit calculating the loss in the metal from the assumption of oblique incidence (Fig. 8.14) in which there is *no attenuation* along $-y$. One must then compute the small, required attenuation by perturbation considerations, which equate the wall loss to power removed from the longitudinal transmission. The method is specifically the one employed in Eqs. 5.8 and 5.9, relating the attenuation to the power dissipated per unit length and the power transmitted. Such a procedure is now seen to apply to the problem of finding the "damping in space" (the attenuation), a method entirely similar to that used for high-Q systems in Chapter 6 to evaluate approximately the "damping in time" of free oscillations.

We shall return later to a more complete justification of our present procedure by solving directly Eqs. 8.123 and 8.126 under the assumption of small attenuation; but, at present, let us proceed on the already rather strong grounds outlined above.

Consider first the TEM-like wave ($m = 0$). With loss absent, let the magnetic field have magnitude $|\mathrm{H}_{x0}|$. If the metal is then given a high conductivity, the magnetic field parallel to it will still be approximately $|\mathrm{H}_{x0}|$ in strength. The power entering the metal in the z direction should be approximately

$$\langle P_{\text{diss}}\rangle_{\substack{\text{unit length}\\ \text{unit width}}} = \langle P_{\text{diss}}\rangle_{\text{sq m}} \approx \frac{1}{2}|\mathrm{H}_{x0}|^2 \operatorname{Re} \bar{\eta}_2$$

$$= \frac{1}{2}|\mathrm{H}_{x0}|^2 \sqrt{\frac{\omega\mu_0}{2\sigma}} = \frac{1}{2}|\mathrm{H}_{x0}|^2 \frac{1}{\sigma\delta} \qquad (8.131)$$

in which we are considering only a unit width (1 m) along x of the structure of Fig. 8.16, and finding the power lost per unit (differential) length along the propagation axis $(-y)$. Since the area of the inside cross section of such a unit width of line is $1 \times b = b$ sq m, and the magnetic field does not vary across this area in the lossless case, the power transmitted along $-y$ in that case is:

$$\langle P^+\rangle \approx \frac{1}{2}|\mathrm{H}_{x0}|^2\eta_1 b = \frac{1}{2}|\mathrm{H}_{x0}|^2 \sqrt{\frac{\mu_0}{\epsilon_0}}\, b \qquad (8.132)$$

Equations 8.131, 8.132, and 5.8 determine $|\alpha_y|$

$$2|\alpha_y| \approx \frac{\langle P_{\text{diss}}\rangle_{\substack{\text{unit length}\\ \text{unit width}}}}{\langle P^+\rangle}$$

or

$$|\alpha_y| \approx \frac{1}{2}\frac{1/\sigma\delta}{b\sqrt{\mu_0/\epsilon_0}} = \frac{1}{2b}\sqrt{\frac{\omega\epsilon_0}{2\sigma}} \qquad m = 0 \qquad (8.133a)$$

It follows from Eqs. 8.129 and 8.133a that

$$\bar{\gamma}_{y1} \approx -|\alpha_y| - j\beta_{01} \qquad m = 0$$

and that there must actually be a small variation of the fields with z to satisfy Eq. 8.125.

Assuming small attenuation means $|\alpha_y| \ll \beta_{01}$, and Eq. 8.125 yields accordingly

$$\bar{\gamma}_{z1}^{\,2} = -\bar{\gamma}_{y1}^{\,2} - \beta_{01}^{\,2} - 2j\beta_{01}|\alpha_y| \qquad m = 0$$

or, in view of Eq. 8.133a and the sign convention associating $\bar{\gamma}_{z1}$ with $\bar{\gamma}_{zi}$ in Fig. 8.16 and Eqs. 8.111 through 8.113,

$$\bar{\gamma}_{z1} \approx -\sqrt{(1-j)\frac{\omega\epsilon_0}{\sigma b\delta}} \qquad (8.133b)$$

This result should be compared with the one appropriate to the lossless case, Eq. 8.128a. The implication is, of course, that the plane waves in the air are actually nonuniform rather than uniform; an interesting consequence of this fact will be taken in Sec. 8.4.3.

A similar attenuation computation may be made for the TM-like solutions ($m = 1, 2, \cdots$). Again, the power lost per unit width and per unit (differential) length is

$$\langle P_{\text{diss}} \rangle_{\substack{\text{unit length}\\ \text{unit width}}} = \langle P_{\text{diss}} \rangle_{\text{sq m}} \approx \frac{1}{2} |\mathrm{H}_{x0}|^2 \operatorname{Re} \bar{\eta}_2$$

$$= \frac{1}{2} |\mathrm{H}_{x0}|^2 \sqrt{\frac{\omega \mu_0}{2\sigma}} = \frac{1}{2} |\mathrm{H}_{x0}|^2 \frac{1}{\sigma \delta} \tag{8.134}$$

but now we must insist that $|\mathrm{H}_{x0}|$ is the magnetic field strength *at the boundary* $z = 0$. This will be very nearly the same as it would have been *at this location* if σ were infinite. Note, however, that in the lossless case $|\mathrm{H}_x|$ varies with z cosinusoidally when $m \neq 0$, taking the maximum values $|\mathrm{H}_{x0}|$ at the perfectly conducting walls $z = -b$ and $z = 0$. This fact is made clear by referring to Fig. 8.17 when $\mathrm{Z}_{z2} = 0$ ($\sigma = \infty$), and observing that the TM wave solutions represent situations for which the transmission-line length b is an integer multiple m of a half wave length $\lambda_{z1}/2$. See Eq. 8.128*b* in this connection.

The power transmitted in the $-y$ direction is obtained by integration of the $-y$ component of the Poynting vector over the cross section, which we can carry out for the lossless case as a good approximation:

$$\langle P^+ \rangle = \operatorname{Re} \left[\int_{-b}^{0} \frac{(-\mathrm{E}_z)\mathrm{H}_x^*}{2} \, dz \right] \tag{8.135}$$

We may then use the wave impedance for the TM wave above cutoff in the lossless case

$$\mathrm{Z}_{-y} = \frac{-\mathrm{E}_z}{\mathrm{H}_x} = \frac{-\bar{\gamma}_{y1}}{j\omega\epsilon_0} = \frac{\beta_m}{\omega\epsilon_0} \tag{8.136}$$

to eliminate E_z from Eq. 8.135 and find

$$\langle P^+ \rangle \approx \int_{-b}^{0} \frac{1}{2} |\mathrm{H}_x|^2 \left(\frac{\beta_m}{\omega\epsilon_0} \right) dz \tag{8.137}$$

Since, when $m \neq 0$, $|\mathrm{H}_x|$ varies cosinusoidally over an integral number of half-periods, with maximum value $|\mathrm{H}_{x0}|$,

$$\int_{-b}^{0} |\mathrm{H}_x|^2 \, dz = \frac{1}{2} |\mathrm{H}_{x0}|^2 b \tag{8.138}$$

and

$$\langle P^+ \rangle = \frac{1}{4} b |\mathrm{H}_{x0}|^2 \frac{\beta_m}{\omega\epsilon_0} \tag{8.139}$$

Therefore

$$|\alpha_y| = \frac{\langle P_{\text{diss}}\rangle_{\substack{\text{unit length}\\ \text{unit width}}}}{2\langle P^+\rangle} = \frac{\omega\epsilon_0}{\beta_m b}\sqrt{\frac{\omega\mu_0}{2\sigma}} = \frac{1}{b}\sqrt{\frac{\omega\epsilon_0}{2\sigma}}\,\frac{\beta_{01}}{\beta_m}$$

$$m = 1, 2, \cdots \qquad (8.140)$$

We need not dwell on $\bar{\gamma}_{z1}$ for these cases at this point; it can be computed approximately from Eqs. 8.130a, 8.140, and 8.125 if desired.

8.4.3 Nature of the Equiphase Surfaces *

8.4.3.1 A SPECIFIC CASE.* Two nonuniform plane waves have actually been employed in the air space to solve the present problem (see Fig. 8.16), even though we have used contrary approximations to find the various propagation constants. It seems reasonable, perhaps, that the resulting total field should also be a plane wave; the phase of the magnetic field, for example, should not change along a transverse plane y = constant. That this supposition proves to be *incorrect* will not only be an interesting lesson but will also illustrate again the great variety of field solutions that can be constructed out of these nonuniform plane waves.

The point in question is the z dependence of H_{x1}, when y is held fixed. We shall treat only the approximate situation for small loss. Let $y = 0$ be the plane chosen for examination. Others would differ only by the factor $e^{-\bar{\gamma}_{y1}y}$, which we can always append later. We shall discuss the transmission-line wave $m = 0$, because it presumably relates to the situations of greatest familiarity; unusual results will therefore tend to stand out most prominently.

Referring to Eq. 8.133b, and Fig. 8.16, we see that

$$\bar{\gamma}_{z1} \approx -\sqrt{(1-j)\,\frac{\omega\epsilon_0}{\sigma b\delta}} \equiv -c + jd \qquad (8.141)$$

with

$$c = \sqrt{\frac{\sqrt{2}\,\omega\epsilon_0}{\sigma b\delta}}\cos\left(\frac{\pi}{8}\right) > 0 \qquad (8.142a)$$

$$d = \sqrt{\frac{\sqrt{2}\,\omega\epsilon_0}{\sigma b\delta}}\sin\left(\frac{\pi}{8}\right) > 0 \qquad (8.142b)$$

A sketch of the complex numbers involved in taking the root in Eq. 8.141 is helpful in establishing the algebraic signs and magnitudes of the results.

Since the transmission line of Fig. 8.17 applies to the z dependence of H_{x1} and $-E_{y1}$, we find at once that

$$H_{x1} = H_{xi}e^{-\bar{\gamma}_{z1}z} + H_{xr}e^{\bar{\gamma}_{z1}z}$$

$$-E_{y1} = Z_{z1}H_{xi}e^{-\bar{\gamma}_{z1}z} - Z_{z1}H_{xr}e^{\bar{\gamma}_{z1}z}$$

But $E_{y1} = 0$ at $z = -b$, so

$$Z_{z1}(H_{xi}e^{\bar{\gamma}_{z1}b} - H_{xr}e^{-\bar{\gamma}_{z1}b}) = 0$$

or

$$H_{xr} = H_{xi}e^{2\bar{\gamma}_{z1}b} \tag{8.142c}$$

Thus

$$\begin{aligned} H_{x1} &= H_{xi}e^{\bar{\gamma}_{z1}b}(e^{-\bar{\gamma}_{z1}(z+b)} + e^{\bar{\gamma}_{z1}(z+b)}) \\ &= H_{x0}\cosh\,[\bar{\gamma}_{z1}(z+b)] \end{aligned} \tag{8.143}$$

where H_{x0} is recognized to be the value of H_{x1} at the perfect conductor $z = -b$. Therefore

$$\begin{aligned} \frac{H_{x1}(z)}{H_{x1}(-b)} &= \cosh\,[\bar{\gamma}_{z1}(z+b)] = \cosh\,[-c(z+b) + jd(z+b)] \\ &= \cosh\,[c(z+\mathrm{b})]\cos\,[d(z+b)] \\ &\qquad -j\sinh\,[c(z+b)]\sin\,[d(z+b)] \end{aligned} \tag{8.144}$$

Now in view of the fact that in the present problem we are assuming a very large value of σ, we may restrict our interest to cases where $|\bar{\gamma}_{z1}b| \ll 1$. Since $-b \le z \le 0$, in addition to the smallness of $|\bar{\gamma}_{z1}b|$, we have then:

$$|d(z+b)| \le |bd| \le |\bar{\gamma}_{z1}b| \ll 1$$

$$|c(z+b)| \le |cb| \le |\bar{\gamma}_{z1}b| \ll 1$$

Hence

$$\frac{H_{x1}(z)}{H_{x1}(-b)} \approx 1 - jcd(b+z)^2 \tag{8.145}$$

Evaluating c and d from Eq. 8.142 leads to

$$\frac{H_{x1}(z)}{H_{x1}(-b)} \approx 1 - j\left(\frac{\omega\epsilon_0}{2\sigma}\right)\left(\frac{b}{\delta}\right)\left(1 + \frac{z}{b}\right)^2 \tag{8.146}$$

The significant feature of this result is that the phase of H_{x1} *lags* increasingly as z changes from $-b$ to 0 across the guide structure. The planes $y =$ constant are *not* equiphase surfaces for the total field.

Suppose now that we wished to find the shape of the equiphase surfaces. What we must do is to start at a point on the perfect conductor,

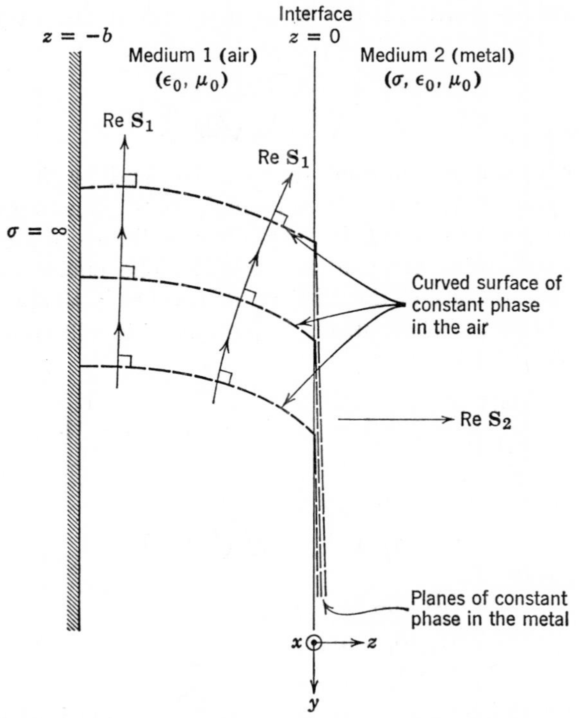

Fig. 8.18. Equiphase surfaces for a TEM-like traveling-wave solution in the parallel-plate transmission line of Fig. 8.16. The wave is delaying in phase and attenuating in the $-y$ direction. Note that Re $\mathbf{S}_1$ is exactly perpendicular to the equiphase surfaces, while Re $\mathbf{S}_2$ is approximately perpendicular to the equiphase planes and the interface.

for example, $y = 0$, $z = -b$, and determine how changes in y and z must be related to keep the total phase of H_{x1} constant. Including the y dependence, we know from Eq. 8.146 that

$$\mathrm{H}_{x1}(y, z) = \mathrm{H}_{x1}(0, -b)\left[1 - j\left(\frac{\omega\epsilon_0}{2\sigma}\frac{b}{\delta}\right)\left(1 + \frac{z}{b}\right)^2\right]e^{|\alpha_{y1}|y}e^{j|\beta_{y1}|y} \tag{8.147}$$

It is clear that y must become *more positive* as z increases if the exponential phase factor $e^{j|\beta_{y1}|y}$ is to cancel the lagging phase of the z-dependent term. A sketch appears in Fig. 8.18. The equiphase surface curves so that the wave appears to "drag" along the dissipative wall, exactly as if there were excessive friction there. The shape of the phase front is simple if the total phase shift in a plane across the guide is small.

The appropriate measure is the imaginary term in Eq. 8.146, which is small if

$$\frac{\omega\epsilon_0}{2\sigma}\frac{b}{\delta} = \sqrt{\frac{\omega\epsilon_0}{\sigma}}\left(\frac{\beta_{01}b}{2\sqrt{2}}\right) \ll 1 \tag{8.148}$$

In addition to the assumptions that $\sigma \gg \omega\epsilon_0$ and $|\bar{\gamma}_{z1}b| \ll 1$ already made, the value of $\beta_{01}b$ may not be large if Eq. 8.148 is to hold. When the TEM wave is of interest, it is usually expeditious to prevent other types of waves from being above cutoff. The choice of $\beta_{01}b < \pi$ guarantees the absence of the TM (or TE) waves in the present problem; they will all be below cutoff. Equation 8.148 will apply, and Eq. 8.146 may be written

$$\frac{\mathrm{H}_{x1}(z)}{\mathrm{H}_{x1}(-b)} \approx 1 - j\varphi \approx e^{-j\varphi} \qquad \varphi \ll 1 \tag{8.149}$$

with

$$\varphi \equiv \sqrt{\frac{\omega\epsilon_0}{\sigma}}\left(\frac{\beta_{01}b}{2\sqrt{2}}\right)\left(1 + \frac{z}{b}\right)^2 \tag{8.150}$$

Correspondingly

$$\frac{\mathrm{H}_{x1}(y, z)}{\mathrm{H}_{x1}(0, -b)} \approx e^{|\alpha_{y1}|y}e^{j(\beta_{01}y - \varphi)} \tag{8.151}$$

because $|\beta_{y1}| \approx \beta_{01}$ under our assumption of small losses. The equiphase surface which passes through ($y = 0$, $z = -b$) is characterized by the relation

$$y \approx \frac{\varphi}{\beta_{01}} = \frac{1}{2}\sqrt{\frac{\omega\epsilon_0}{2\sigma}}\, b\left(1 + \frac{z}{b}\right)^2 \tag{8.152}$$

which is a parabola. Figure 8.18 was drawn in accordance with these conditions.

8.4.3.2 A MORE GENERAL CASE.* Another view of the reasonableness of the curved phase front in the TEM-like wave stems from power considerations in the (lossless) air space between the conductors. These considerations are based specifically upon the TM character of the entire field and the absence of space variation in the direction of the magnetic field.[1]

In Fig. 8.16, the magnetic field in the air is of the form

$$\mathbf{H}_1 = \boldsymbol{a}_x \mathrm{f}(y, z) \tag{8.153}$$

where $\mathrm{f}(y, z)$ is a complex function of y and z. In order to place in

[1] A dual situation exists in lossless media for those TE waves with no space variation along the electric field direction.

evidence the equiphase (or equiamplitude) surfaces of $\mathbf{H}_1$, we write $f(y, z)$ in terms of its (complex) natural logarithm $\bar{\vartheta}(y, z)$, thus

$$f(y, z) = e^{\bar{\vartheta}(y,z)} \tag{8.154}$$

which means simply that

$$\bar{\vartheta}(y, z) \equiv \vartheta_R(y, z) + j\vartheta_I(y, z) \equiv \ln f(y, z) \tag{8.155}$$

Note that $\bar{\vartheta}(y, z)$ might be very complicated in general, and *is* complicated in the present problem (take the ln of Eq. 8.147).

The equiphase surfaces of

$$\mathbf{H}_1 = \boldsymbol{a}_x e^{\bar{\vartheta}(y,z)} = \boldsymbol{a}_x e^{\vartheta_R(y,z)} e^{j\vartheta_I(y,z)} \tag{8.156}$$

are therefore given by

$$\vartheta_I(y, z) = \text{const} \tag{8.157a}$$

and the equiamplitude surfaces by

$$\vartheta_R(y, z) = \text{const} \tag{8.157b}$$

To compute the complex Poynting vector, which will furnish information about power in this field, we need $\mathbf{E}_1$. Since the air is lossless, Maxwell's equations require

$$\mathbf{E}_1 = \frac{1}{j\omega\epsilon_0} \nabla \times \mathbf{H}_1 \tag{8.158}$$

But

$$\begin{aligned} \nabla \times \mathbf{H}_1 &= \nabla \times (\boldsymbol{a}_x e^{\bar{\vartheta}(y,z)}) = e^{\bar{\vartheta}(y,z)} \nabla \times \boldsymbol{a}_x - \boldsymbol{a}_x \times \nabla e^{\bar{\vartheta}(y,z)} \\ &= -\boldsymbol{a}_x \times [e^{\bar{\vartheta}(y,z)} \nabla\bar{\vartheta}(y, z)] \end{aligned} \tag{8.159}$$

Accordingly,

$$\begin{aligned} \mathbf{S}_1 &= \frac{1}{2}\mathbf{E}_1 \times \mathbf{H}_1^* = -\frac{1}{2j\omega\epsilon_0}[\boldsymbol{a}_x \times \nabla\bar{\vartheta}(y, z)] \times \boldsymbol{a}_x e^{2\vartheta_R(y,z)} \\ &= -\frac{e^{2\vartheta_R(y,z)}}{2j\omega\epsilon_0}\{\nabla\bar{\vartheta}(y, z) - \boldsymbol{a}_x[\boldsymbol{a}_x \cdot \nabla\bar{\vartheta}(y, z)]\} \end{aligned} \tag{8.160}$$

It happens that $\nabla\bar{\vartheta}(y, z)$ has space components only in the y, z-plane, because $\bar{\vartheta}$ is independent of x. For this reason

$$\boldsymbol{a}_x \cdot \nabla\bar{\vartheta}(y, z) \equiv 0 \tag{8.161}$$

The complex Poynting vector in Eq. 8.160 becomes finally

$$\mathbf{S}_1 = \frac{e^{2\vartheta_R(y,z)}}{2\omega\epsilon_0}[-\nabla\vartheta_I(y, z) + j\,\nabla\vartheta_R(y, z)] \tag{8.162}$$

The real component of the complex Poynting vector is perpendicular to the surfaces of constant phase, while the imaginary part is perpendicular to the surfaces of constant amplitude. In fact, Eq. 8.162 shows that the maximum real (time-average) power flows in the direction of most rapid space-rate-of-increase of phase *lag* (ϑ_I is *leading* phase, by the definition in Eq. 8.156).

These most interesting results extend considerably the validity of the similar properties, which we discovered in Chapter 7 for single, nonuniform TM and TE plane waves in lossless media. We see now that they apply in lossless media to an electromagnetic field which is made up of *any number* of TM (or TE) plane waves whose magnetic (respectively electric) fields are all parallel, but whose propagation vectors $\bar{\boldsymbol{\gamma}}$ may have different complex-vector values in the plane perpendicular to the magnetic (respectively electric) field.[1] In particular, we wish to stress here the consistency of the curved phase front (Fig. 8.18) produced by two such waves (Fig. 8.16) with the idea that real power for the entire field in the structure must not only flow in the $-y$ direction down the transmission line but also in the $+z$ direction into the metal wall. In accordance with Eq. 8.162, the phase front simply *has* to curve the way it does, if it is going to meet these power requirements.

There now emerges still another view of the function of nonuniform plane waves. These waves may arise when a region of lossless material has dissipative boundaries. Nonuniform plane waves (more than one) in the lossless region can produce a curved phase front for the purpose of supplying real power to the boundaries in one direction at the expense of power flow in another. They may, in effect, account thereby for a real attenuation (loss of real power) in one direction; although, of course, only a rerouting of power in the lossless space is actually involved. This property of two or more nonuniform plane waves is perhaps surprising at first, inasmuch as one wave alone cannot produce an attenuation of real power in any direction in a lossless medium. As illustrated further in the Problems, however, we are encountering here nothing more than a failure of superposition (direct additivity) of *power*—a lack of orthogonality between different waves. Actually, this should probably be regarded as the rule rather than the exception.

[1] They also apply to TE (but not TM) waves meeting these conditions in a dissipative medium ($\sigma \neq 0$).

8.4.4 Revaluation of Approximations for Small Loss *[1]

It is appropriate now to re-examine the basis for the approximations leading to Eqs. 8.133*a*, 8.133*b*, and 8.140. We shall do so by solving Eqs. 8.123 and 8.126, introducing the appropriate assumptions as they arise.

The periodic properties of the tanh function in Eq. 8.123 tell us that there will be an infinite set of solutions again. Physically, the phase constants along $-y$ of these solutions must correspond roughly to those of the TEM and TM solutions in the lossless case but they must attenuate along $-y$. That is, $\bar{\gamma}_{y1}$ must definitely be complex (as must $\bar{\gamma}_{z1}$). Let us therefore look for solutions to Eq. 8.123 in which $\bar{\gamma}_{z1}$ is very nearly equal to $j(m\pi/b)$, its value in the lossless case ($\sigma = \infty$), differing from it by only a small complex amount $\bar{\nu}_{z1}$

$$\bar{\gamma}_{z1} = j\frac{m\pi}{b} + \bar{\nu}_{z1} \tag{8.163a}$$

such that

$$|\bar{\nu}_{z1}b| \ll 1 \tag{8.163b}$$

But since

$$\tanh(jm\pi) = 0$$

we have by a trigonometric identity and a subsequent Taylor expansion

$$\begin{aligned}\tanh(jm\pi + \bar{\nu}_{z1}b) &= \frac{\tanh(jm\pi) + \tanh(\bar{\nu}_{z1}b)}{1 + \tanh(jm\pi)\tanh(\bar{\nu}_{z1}b)} \\ &= \tanh(\bar{\nu}_{z1}b) \approx \bar{\nu}_{z1}b \qquad |\bar{\nu}_{z1}b| \ll 1\end{aligned} \tag{8.164}$$

Thus if $\sigma \gg \omega\epsilon_0$, Eq. 8.123 becomes approximately

$$\left(j\frac{m\pi}{b} + \bar{\nu}_{z1}\right)\bar{\nu}_{z1}b \approx -\frac{j\omega\epsilon_0}{\sigma}\bar{\gamma}_{z2} \qquad \text{if } |\bar{\nu}_{z1}b| \ll 1 \quad \text{and} \quad \frac{\omega\epsilon_0}{\sigma} \ll 1 \tag{8.165}$$

According to Eq. 8.126, however,

$$\bar{\gamma}_{z2}^{\,2} = j\omega\mu_0\sigma\left(1 + \frac{\bar{\gamma}_{z1}^{\,2}}{j\omega\mu_0\sigma}\right)$$

But from Eq. 8.128*b* we know that $|\bar{\gamma}_{z1}|$ is finite when $\sigma = \infty$; so we expect that, for σ large enough to make the effect of the loss small, we may set

$$|\bar{\gamma}_{z1}|^2 \ll \omega\mu_0\sigma \tag{8.166}$$

[1] This section may be omitted without disrupting the continuity of other starred sections.

On this basis, we learn by a binomial expansion that

$$\bar{\gamma}_{z2} \approx \sqrt{j\omega\mu_0\sigma}\left[1 + \frac{1}{2}\frac{\left(j\dfrac{m\pi}{b} + \bar{\nu}_{z1}\right)^2}{j\omega\mu_0\sigma}\right]$$

$$= \sqrt{j\omega\mu_0\sigma} - \frac{1}{2}\left(\frac{m\pi}{b}\right)^2\frac{1}{\sqrt{j\omega\mu_0\sigma}} + \frac{j\dfrac{m\pi}{b}\bar{\nu}_{z1}}{\sqrt{j\omega\mu_0\sigma}} + \frac{1}{2}\frac{\bar{\nu}_{z1}{}^2}{\sqrt{j\omega\mu_0\sigma}} \qquad (8.167)$$

Equation 8.167 substituted in Eq. 8.165 yields

$$\bar{\nu}_{z1}{}^2\left(1 + \frac{j\omega\epsilon_0}{2\sigma b\sqrt{j\omega\mu_0\sigma}}\right) + \bar{\nu}_{z1}\left[\frac{jm\pi}{b}\left(1 + \frac{j\omega\epsilon_0}{\sigma b\sqrt{j\omega\mu_0\sigma}}\right)\right]$$

$$+ \frac{j\omega\epsilon_0}{b\sigma}\sqrt{j\omega\mu_0\sigma}\left[1 - \frac{1}{2}\frac{(m\pi/b)^2}{j\omega\mu_0\sigma}\right]$$

$$= 0 \qquad |\bar{\gamma}_{z1}|^2 \ll \omega\mu_0\sigma \qquad |\bar{\nu}_z b| \ll 1 \qquad \frac{\omega\epsilon_0}{\sigma} \ll 1 \qquad (8.168)$$

Now Eq. 8.166 needs more interpretation. Written out in terms of $\bar{\nu}_{z1} = \Delta\alpha_{z1} + j\,\Delta\beta_{z1}$, it reads

$$(\Delta\alpha_{z1})^2 + \left(\frac{m\pi}{b} + \Delta\beta_{z1}\right)^2 \ll \omega\mu_0\sigma$$

or

$$|\bar{\nu}_{z1}|^2 + \left(\frac{m\pi}{b}\right)^2 + 2\left(\frac{m\pi}{b}\right)(\Delta\beta_{z1}) \ll \omega\mu_0\sigma$$

or

$$|\bar{\nu}_{z1}b|^2 + (m\pi)^2 + 2m\pi(b\,\Delta\beta_{z1}) \ll \omega\mu_0\sigma b^2 = 2\left(\frac{b}{\delta}\right)^2 \qquad (8.169)$$

where $\delta = (2/\omega\mu_0\sigma)^{1/2}$ is the skin depth in the metal (medium 2). Inasmuch as we are already assuming that $|\bar{\nu}_{z1}b| \ll 1$, it must also be true that $|b\,\Delta\beta_{z1}| \ll 1$. The essential new assumption in Eq. 8.169 is therefore

$$\text{(a)} \qquad \frac{\sqrt{2}\,b}{\delta} = b\sqrt{\omega\mu_0\sigma} \gg |\bar{\nu}_{z1}b| \qquad m = 0 \text{ (TEM-like wave)}$$

$$(8.170)$$

$$\text{(b)} \qquad \frac{\sqrt{2}\,b}{\delta} = b\sqrt{\omega\mu_0\sigma} \gg m\pi \qquad m = 1, 2, \cdots \text{ (TM waves)}$$

It follows that the subtractive term in the last bracket on the left side of Eq. 8.168 may be neglected for those allowed values of m that are not too large. Moreover, the additive terms in the other two brackets in Eq. 8.168 will also be negligible if

$$\frac{\omega\epsilon_0}{\sigma} \ll \frac{\sqrt{2}\,b}{\delta} \tag{8.171}$$

Since $(\omega\epsilon_0/\sigma) \ll 1$ anyway, by previous assumptions, Eq. 8.170b for b/δ is stronger than Eq. 8.171, when $m \neq 0$. When $m = 0$, it turns out that Eq. 8.171 duplicates Eq. 8.170a, although this fact is only apparent when the size of $\bar{\nu}_{z1}$ is known. We could check it now by using Eq. 8.141, but prefer to do so later after the present analysis is completed.

On the basis of the assumptions stated in Eqs. 8.165, 8.170b, and either 8.171 or 8.170a, Eq. 8.168 simplifies to

$$\bar{\nu}_{z1}^2 + j\left(\frac{m\pi}{b}\right)\bar{\nu}_{z1} - (1 - j)\left(\frac{\omega\epsilon_0}{\sigma b\delta}\right) = 0 \tag{8.172}$$

The solution is

$$\bar{\nu}_{z1} = -j\frac{m\pi}{2b} \pm \sqrt{\left(j\frac{m\pi}{2b}\right)^2 + (1 - j)\left(\frac{\omega\epsilon_0}{\sigma b\delta}\right)} \tag{8.173}$$

Only the (+) root is acceptable when $m > 0$, the other one being rejected because $\bar{\nu}_{z1} \to 0$ when $\sigma \to \infty$ is required by the definition of $\bar{\nu}_{z1}$. Since we are interested only in solutions for which $|\bar{\nu}_{z1}b| \ll 1$, the second term under the root in Eq. 8.173 is small compared to the first when $m \neq 0$. A binomial expansion of the root then yields to the first order

$$\bar{\nu}_{z1} \approx +\frac{jm\pi}{2b}\left\{-1 + \left[1 + \frac{(1 - j)\omega\epsilon_0}{2\sigma b\delta[j(m\pi/2b)]^2}\right]\right\}$$

$$\approx -\frac{(1 + j)\omega\epsilon_0}{\sigma\delta m\pi} \qquad m \neq 0 \tag{8.174a}$$

If $m = 0$, however, Eq. 8.173 becomes simply

$$\bar{\gamma}_{z1} = \bar{\nu}_{z1} \approx -\sqrt{(1 - j)\frac{\omega\epsilon_0}{\sigma b\delta}} \qquad m = 0 \tag{8.174b}$$

The choice of the (−) sign now simply means that $\bar{\nu}_{z1}$ refers to the "incident" wave, which, by definition, lags in phase along $+z$ (Fig. 8.16).

It will be desirable to reconcile Eqs. 8.174*a* and 8.174*b* with the various assumptions we have made in deriving them, before continuing our solution further. The criterion $|\bar{\nu}_{z1}b| \ll 1$ becomes

$$\text{(a)} \qquad \frac{\omega\epsilon_0}{\sigma} \ll m\pi\left(\frac{\delta}{\sqrt{2}\,b}\right) \qquad m \neq 0 \text{ (TM waves)}$$
$$\text{(b)} \qquad \frac{\omega\epsilon_0}{\sigma} \ll \left(\frac{\delta}{\sqrt{2}\,b}\right) \qquad m = 0 \text{ (TEM-like wave)} \tag{8.175}$$

Meeting inequalities, Eqs. 8.175*a* and 8.170*b*, when $m \neq 0$ is certainly possible with $(\omega\epsilon_0)/\sigma \ll 1$, but they are actually somewhat more stringent. Specifically, they demand

$$\frac{\omega\epsilon_0}{\sigma} \ll m\pi\left(\frac{\delta}{\sqrt{2}b}\right) \ll 1 \qquad m \neq 0 \text{ (TM waves)} \tag{8.176}$$

This result can nevertheless be achieved with substantial freedom for large enough values of σ, because $\delta \sim \sigma^{-(1/2)}$.

Next, Eqs. 8.171 and 8.175*b* bracket the value of b/δ differently:

$$\frac{\omega\epsilon_0}{\sigma} \ll \frac{\sqrt{2}\,b}{\delta} \ll \frac{\sigma}{\omega\epsilon_0} \qquad m = 0 \qquad \frac{\omega\epsilon_0}{\sigma} \ll 1 \text{ (TEM-like wave)} \tag{8.177}$$

but this is again a broad range when σ is large.

Finally, to check the inequality of Eq. 8.170*a*, we substitute Eq. 8.174*b* into it. There results

$$\frac{\sqrt{2}\,b}{\delta} \gg \sqrt{\frac{\omega\epsilon_0}{\sigma}\left(\frac{b}{\delta}\right)\sqrt{2}}$$
or
$$\frac{\sqrt{2}\,b}{\delta} \gg \frac{\omega\epsilon_0}{\sigma} \tag{8.178}$$

which is covered already by Eq. 8.177 or 8.171, as we suggested would be the case. Note the agreement between Eqs. 8.174*b* and 8.141.

Accordingly, a summary of our results and corresponding approximations is:

$$\left.\begin{array}{l} \bar{\nu}_{z1} \approx -\sqrt{(1-j)\dfrac{\omega\epsilon_0}{\sigma b\delta}} \\ \text{if} \\ \dfrac{\omega\epsilon_0}{\sigma} \ll 1 \\ \text{and} \\ \dfrac{\omega\epsilon_0}{\sigma} \ll \dfrac{\sqrt{2}b}{\delta} \ll \dfrac{\sigma}{\omega\epsilon_0} \end{array}\right\} \quad m = 0, \text{(TEM-like wave)} \tag{8.179a}$$

$$\left.\begin{array}{l} \bar{\nu}_{z1} \approx -\dfrac{(1+j)\omega\epsilon_0}{\sigma\delta m\pi} \\ \text{if} \\ \dfrac{\omega\epsilon_0}{\sigma} \ll m\pi \dfrac{\delta}{\sqrt{2}b} \ll 1 \end{array}\right\} \quad m = 1, 2, \cdots \text{(TM waves)} \tag{8.179b}$$

The way in which the propagation along the y direction is affected by the loss is contained in $\bar{\gamma}_{y1}$. In the lossless case, for the TEM solution, we had $\bar{\gamma}_{y1} = -j\beta_{01}$ (Eq. 8.129). Since $m = 0$ for this case, we find from Eqs. 8.125, 8.163, and 8.179a that, when the loss is present,

$$\bar{\gamma}_{y1}^{\,2} \approx -\beta_{01}^{\,2} - (1-j)\frac{\omega\epsilon_0}{\sigma b\delta} \tag{8.180}$$

The simplest and most interesting results arise when the attenuation constant α_{y1} is much smaller in magnitude than that of the phase constant β_{y1}. Then the second term of Eq. 8.180 is much smaller than the first, and the square root can be taken by the binomial theorem, retaining only the leading terms:

$$\begin{aligned} \bar{\gamma}_{y1} &\approx -j\beta_{01} + \frac{1}{-j\beta_{01}}\left[-(1-j)\frac{\omega\epsilon_0}{2\sigma b\delta}\right] \\ &= -j\left(\beta_{01} + \frac{\omega\epsilon_0}{2\beta_{01}\sigma b\epsilon}\right) - \frac{\omega\epsilon_0}{2\beta_{01}\sigma b\delta} \\ &= -j\left(\beta_{01} + \frac{1}{2b}\sqrt{\frac{\omega\epsilon_0}{2\sigma}}\right) - \frac{1}{2b}\sqrt{\frac{\omega\epsilon_0}{2\sigma}} \equiv \alpha_{y1} + j\beta_{y1} \end{aligned} \tag{8.181}$$

provided that

$$|\alpha_{y1}| \ll |\beta_{y1}|$$

or

$$\frac{1}{2b}\sqrt{\frac{\omega\epsilon_0}{2\sigma}} \ll \omega\sqrt{\epsilon_0\mu_0}$$

or

$$\frac{1}{4}\frac{\delta}{b} \ll 1 \tag{8.182}$$

The last condition, Eq. 8.182, is consistent with that in Eq. 8.179*a*, but it is an additional constraint upon our results. Thus, for the propagation constant along the conductors of Fig. 8.16, Eqs. 8.181, 8.179*a*, and 8.182 yield finally

$$\left.\begin{aligned}
\alpha_{y1} &\approx -\frac{1}{2b}\sqrt{\frac{\omega\epsilon_0}{2\sigma}} = -\frac{1}{4}\frac{\delta}{b}\beta_{01} \\
\beta_{y1} &\approx -\left(\beta_{01} + \frac{1}{2b}\sqrt{\frac{\omega\epsilon_0}{2\sigma}}\right) = -\beta_{01}\left(1 + \frac{1}{4}\frac{\delta}{b}\right) \\
&\approx -\beta_{01} \\
\text{if } &\frac{\omega\epsilon_0}{\sigma} \ll 1 \ll \frac{b}{\delta} \ll \frac{\sigma}{\omega\epsilon_0}
\end{aligned}\right\} \quad \begin{gathered} m = 0, \\ \text{(TEM-like wave)} \end{gathered} \tag{8.183}$$

where factors of $\sqrt{2}$ or 4 which should strictly appear in the inequalities have been absorbed in the symbol ($\ll$). It turns out in most practical transmission-line or wave-guide examples that, if the restrictions can be met at all, they can be met by factors of at least 10^2. Note that Eq. 8.183 agrees with Eq. 8.133*a* and the assumption $\beta_{y1} \approx -\beta_{01}$ made to obtain the latter.

The waves with $m > 0$ are described by

$$\bar{\gamma}_{y1}^2 = -\beta_{01}^2 - \left(j\frac{m\pi}{b} + \bar{\nu}_{z1}\right)^2 = \left(\frac{m\pi}{b}\right)^2 - \beta_{01}^2 - 2j\frac{m\pi}{b}\bar{\nu}_{z1} - \bar{\nu}_{z1}^2 \tag{8.184}$$

but since $|\bar{\nu}_z b| \ll 1$ the term in $\bar{\nu}_{z1}^2$ is negligible. That is,

$$\bar{\gamma}_{y1}^2 \approx \bar{\gamma}_m^2 - 2j\frac{m\pi}{b}\bar{\nu}_{z1} \tag{8.185}$$

where

$$\bar{\gamma}_m = -\sqrt{\left(\frac{m\pi}{b}\right)^2 - \beta_{01}^2} \tag{8.186}$$

is the propagation constant in the lossless case (see Eq. 8.130*a*). Of course $\bar{\gamma}_m$ is *real* below the cutoff frequency, zero at the cutoff frequency, and *imaginary* above the cutoff frequency. Treatment of Eq. 8.185 differs in the three cases. We will consider only the case above cutoff, for which

$$\bar{\gamma}_m = -j\beta_m = -j\sqrt{\beta_{01}{}^2 - \left(\frac{m\pi}{b}\right)^2} \tag{8.187}$$

If the attenuation in a guide wave length $(2\pi/\beta_m)$ is to be small again, the second term of Eq. 8.185 must be small compared to the first. Binomial expansion on this basis yields

$$\bar{\gamma}_{y1} \approx -j\beta_m + \frac{1}{-j2\beta_m}\left(-2j\frac{m\pi}{b}\right)\left[-\frac{(1+j)\omega\epsilon_0}{\sigma\delta m\pi}\right]$$

$$= -j\beta_m - \frac{(1+j)\omega\epsilon_0}{\beta_m b\sigma\delta} = -j\beta_m - \frac{(1+j)(\sqrt{\omega\epsilon_0/2\sigma})\beta_{01}}{b\beta_m}$$

or

$$\bar{\gamma}_{y1} = -j\left[\beta_m + \frac{(\sqrt{\omega\epsilon_0/2\sigma})}{b}\frac{\beta_{01}}{\beta_m}\right] - \sqrt{\frac{\omega\epsilon_0}{2\sigma}}\frac{\beta_{01}}{b\beta_m} \qquad m > 0 \tag{8.188}$$

provided

$$|\alpha_{y1}| \ll |\beta_{y1}|$$

or

$$\frac{1}{2}\frac{\delta}{b} \ll \left(\frac{\beta_m}{\beta_{01}}\right)^2 = 1 - \left(\frac{m\pi}{b\beta_{01}}\right)^2 \tag{8.189}$$

The demands of Eq. 8.189 cannot be met in the immediate neighborhood of the cutoff frequency. Well above cutoff, however, this equation becomes weaker with respect to δ/b than Eq. 8.179. In summary:

$$\left.\begin{aligned}
\alpha_{y1} &\approx -\frac{1}{b}\sqrt{\frac{\omega\epsilon_0}{2\sigma}}\frac{\beta_{01}}{\beta_m} = -\frac{1}{2}\frac{\delta}{b}\left(\frac{\beta_{01}}{\beta_m}\right)^2\beta_m \\
\beta_{y1} &\approx -\left(\beta_m + \frac{1}{b}\sqrt{\frac{\omega\epsilon_0}{2\sigma}}\frac{\beta_{01}}{\beta_m}\right) \\
&= -\beta_m\left[1 + \frac{1}{2}\frac{\delta}{b}\left(\frac{\beta_{01}}{\beta_m}\right)^2\right] \approx -\beta_m \\
\text{if} \quad & \frac{1}{2}\frac{\delta}{b} \ll \left(\frac{\beta_m}{\beta_{01}}\right)^2 = 1 - \left(\frac{m\pi}{b\beta_{01}}\right)^2 > 0 \\
\text{and} \quad & \frac{\omega\epsilon_0}{\sigma} \ll \frac{m\pi}{\sqrt{2}}\frac{\delta}{b} \ll 1
\end{aligned}\right\} \quad \begin{aligned} &m = 1, 2, \cdots \\ &\text{(TM waves)}\end{aligned} \tag{8.190}$$

The above value of α_{y1} should be compared with that in Eq. 8.140, which was derived on the assumption that $\beta_{y1} \approx -\beta_m$, as in the lossless case.

We have found that the notion of skin effect or surface impedance for a highly conducting metal should be valid for the *approximate* computation of losses in three different situations:

1. Normal incidence of a uniform plane wave.
2. Oblique incidence of a uniform plane.
3. Oblique incidence of a nonuniform plane wave (or the case of a guided wave).

Of course the detailed approximations are not the same in each case, more complicated conditions being imposed successively down the list. Briefly, though, some careful consideration shows that they reduce in each example to two major ideas: First, the propagation constant for the wave in the metal is approximately given by

$$\bar{\gamma}_{\text{metal}} \approx \sqrt{j\omega\mu_{\text{metal}}\sigma_{\text{metal}}}$$

This simply says that $\sigma \gg \omega\epsilon_0$, and that the wave in the metal is very nearly a uniform plane wave traveling normal to the surface. Second, some more involved conditions are imposed which amount to the idea that the presence of the imperfect conductor, instead of a perfect one, must produce only small changes in the field *elsewhere*, if the skin-effect treatment of the conductor itself is to be valid.

8.4.5 Transmission-Line Representation *

We mentioned in Sec. 2.4.1 that, although the transmission-line formulation of the behavior of a structure like that in Fig. 8.16 could be carried out exactly if both conductors were perfect, approximations became necessary when they were not. We had in mind, of course, the problem of defining the parameters R, L, C, and G, starting from a field solution in the structure. These remarks need clarification. We are now in a position to clarify them because the field solution is available.

We shall eventually limit ourselves to a study of the TEM ($m = 0$) and TEM-like waves, because these solutions correspond to use of the structure as a conventional transmission line rather than as a wave guide. It will be instructive however to consider all the solutions ($m = 0, 1, 2, \cdots$) together, until specialization becomes absolutely necessary.

8.4.5.1 THE FIELDS.* First we must examine the fields a little more carefully. Corresponding to the nonuniform plane waves shown in Fig. 8.16, we have in the air, medium 1,

$$\text{(a)} \qquad \mathbf{H}_1^- = \boldsymbol{a}_x(\mathrm{H}_{xi}^- e^{-\bar{\gamma}_{z1}z} + \mathrm{H}_{xr}^- e^{\bar{\gamma}_{z1}z})e^{\bar{\gamma}_y y}$$

$$\text{(b)} \qquad \mathbf{E}_1^- = \left[\left(-\boldsymbol{a}_y \frac{\bar{\gamma}_{z1}}{j\omega\epsilon_0} - \boldsymbol{a}_z \frac{\bar{\gamma}_y}{j\omega\epsilon_0}\right)\mathrm{H}_{xi}^- e^{-\bar{\gamma}_{z1}z} + \left(\boldsymbol{a}_y \frac{\bar{\gamma}_{z1}}{j\omega\epsilon_0} - \boldsymbol{a}_z \frac{\bar{\gamma}_y}{j\omega\epsilon_0}\right)\mathrm{H}_{xr}^- e^{\bar{\gamma}_{z1}z}\right]e^{\bar{\gamma}_y y} \qquad (8.191)$$

where we have set $-\bar{\gamma}_{y1} = -\bar{\gamma}_{y2} \equiv \bar{\gamma}_y$. The form in which Eq. 8.191 has been written and the use of a superscript $(-)$ in E_1 and $\mathbf{H}_1$ are designed to emphasize that the entire field defines a traveling wave along the $-y$ direction. Similarly in the metal, medium 2,

$$\text{(a)} \qquad \mathbf{H}_2^- = \boldsymbol{a}_x(\mathrm{H}_{xt}^- e^{-\bar{\gamma}_{z2}z})e^{\bar{\gamma}_y y}$$

$$\text{(b)} \qquad \mathbf{E}_2^- = -\left[\mathrm{H}_{xt}^-\left(\boldsymbol{a}_y \frac{\bar{\gamma}_{z2}}{\sigma + j\omega\epsilon_0} + \boldsymbol{a}_z \frac{\bar{\gamma}_y}{\sigma + j\omega\epsilon_0}\right)e^{-\bar{\gamma}_{z2}z}\right]e^{\bar{\gamma}_y y} \qquad (8.192)$$

There are boundary conditions still to be applied to Eqs. 8.191 and 8.192. First, there is Eq. 8.142c relating H_{xi}^- and H_{xr}^- at the short-circuit plane $z = -b$. Then there remains the continuity condition on tangential **H** at $z = 0$:

$$\mathrm{H}_{xi}^- + \mathrm{H}_{xr}^- = \mathrm{H}_{xt}^- \qquad (8.193)$$

There is no need to rewrite the continuity of E_y at $z = 0$ because this is already covered by the wave-impedance continuity Eq. 8.123, together with Eq. 8.193. We have already discussed the effect of the boundary conditions upon the propagation constants $\bar{\gamma}_{z1}$, $\bar{\gamma}_{z2}$, and $\bar{\gamma}_y$.

Since we are interested in the behavior of the system as regards the longitudinal (y) axis, the field components of importance are H_x and E_z. Inserting the Eqs. 8.142c and 8.193 into Eqs. 8.191 and 8.192, we find for these components in medium 1:

$$\left.\begin{aligned} &\text{(a)} \quad \mathrm{H}_{x1}^- = \mathrm{H}_{x0}^-\{\cosh[\bar{\gamma}_{z1}(z+b)]\}e^{\bar{\gamma}_y y} \\ &\text{(b)} \quad \mathrm{E}_{z1}^- = -\left(\frac{\bar{\gamma}_y}{j\omega\epsilon_0}\right)\mathrm{H}_{x1}^- \end{aligned}\right\} \quad -b \le z \le 0 \qquad (8.194)$$

and in medium 2:

$$\left.\begin{array}{ll}\text{(a)} & H_{x2}^{-} = H_{x0}^{-}\{[\cosh(\bar{\gamma}_{z1}b)]e^{-\bar{\gamma}_{z2}z}\}e^{\bar{\gamma}_y y} \\ \text{(b)} & E_{z2}^{-} = -\left(\dfrac{\bar{\gamma}_y}{\sigma + j\omega\epsilon_0}\right)H_{x2}^{-}\end{array}\right\} \quad z \geq 0 \qquad (8.195)$$

Observe that there is only one unknown complex constant left in the solution. We have chosen it to be H_{x0}^{-}, the magnetic field strength at $y = 0$, $z = -b$. This constant sets the entire level of excitation in both media of the field traveling in the $-y$ direction. In terms of transmission-line thinking, we have here a $(-)$ wave whose amplitude is proportional to H_{x0}^{-}.

Now in considering the situation of Fig. 8.16, we could have started originally with the incident wave coming from the upper left instead of from the lower left. The entire solution would have looked very similar to the present one, except that phase delay and attenuation would have taken place along $+y$ instead of $-y$. That is, we would simply have $\bar{\gamma}_y$ reversed in algebraic sign everywhere it appears. This means there exists a solution to our problem which corresponds to a transmission-line wave along the $+y$ direction. Its amplitude need not be the same as that of the $(-)$ wave in Eqs. 8.194 and 8.195, so it may be denoted by H_{x0}^{+}. Making the replacements $-\bar{\gamma}_y$ for $\bar{\gamma}_y$, and H_{x0}^{+} for H_{x0}^{-} in Eqs. 8.194 and 8.195, we find in medium 1:

$$\left.\begin{array}{ll}\text{(a)} & H_{x1}^{+} = H_{x0}^{+}\{\cosh[\bar{\gamma}_{z1}(z+b)]\}e^{-\bar{\gamma}_y y} \\ \text{(b)} & E_{z1}^{+} = \left(\dfrac{\bar{\gamma}_y}{j\omega\epsilon_0}\right)H_{x1}^{+}\end{array}\right\} \quad -b \leq z \leq 0 \qquad (8.196)$$

and in medium 2:

$$\left.\begin{array}{ll}\text{(a)} & H_{x2}^{+} = H_{x0}^{+}[(\cosh\bar{\gamma}_{z1}b)e^{-\bar{\gamma}_{z2}z}]e^{-\bar{\gamma}_y y} \\ \text{(b)} & E_{z2}^{+} = \left(\dfrac{\bar{\gamma}_y}{\sigma + j\omega\epsilon_0}\right)H_{x2}^{+}\end{array}\right\} \quad z \geq 0 \qquad (8.197)$$

Comparison of Eqs. 8.194 and 8.196 on one hand, and Eqs. 8.195 and 8.197 on the other, shows that, if we choose a fixed value of z, a transmission-line representation for the y dependence is certainly possible. As illustrated in Fig. 8.19, H_x is like a current and E_z is like a voltage. The propagation constant is $\bar{\gamma}_y$, while the characteristic impedance is either $\bar{\gamma}_y/(j\omega\epsilon_0)$ or $\bar{\gamma}_y/(\sigma + j\omega\epsilon_0)$, depending upon whether z lies in medium 1 or medium 2 respectively. This situation is,

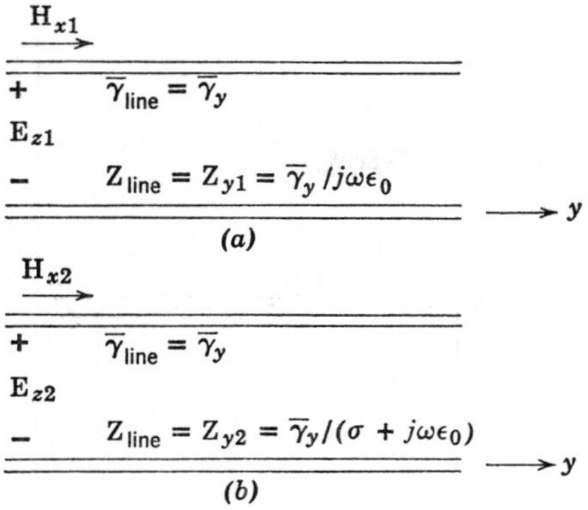

Fig. 8.19. Transmission-line representations for x and z field components and y dependence of waves in the structure of Fig. 8.16. (*a*) Applies for medium 1, $-b \leq z \leq 0$; (*b*) applies for medium 2, $z \geq 0$. These representations are valid in any longitudinal plane z = const.

of course, similar to that in our earlier transmission-line representations of plane-wave reflection problems, where only *some* of the field components and *some* of the space dependence are accounted for by the representation. See, for example, Figs. 8.17, 8.15, and those for oblique incidence in Chapter 7.

8.4.5.2 VOLTAGE AND CURRENT.* In the present case we would like to treat a unit width of the structure along x in terms of variables which depend upon y only, absorbing the z dependence in a "voltage" and a "current" which are proportional to the field strengths, but not equal to them. Usually we accomplish this reduction by: (1) defining voltage as the line integral of the electric field between two conductors, carrying out the integral in a transverse plane of the structure; and (2) defining current as that flowing in one of the conductors at the position of the voltage plane. Exactly the same value of current may also be regarded as obtained from a closed line integral of the magnetic field, chosen in such a way that the contour of integration lies in the transverse plane and encircles the chosen conductor. We employed these procedures in Sec. 2.3, and it would seem convenient to do so here.

In particular, the line integral of E_z between the conductors was a successful voltage definition when medium 2 was a perfect conductor, so perhaps it should be retained even when that medium has finite conductivity. The trouble is that, on such a basis, the voltage will not involve any of the electric field in medium 2, and this omission looks somehow unfair. Possibly it would be wiser to integrate all the way from $z = -b$ to $z = +\infty$, thereby including all the field. If medium 2 were a perfect conductor, it would not matter which choice were made; but when it is not, the choice seems difficult. Maybe the current definition will help to resolve the dilemma.

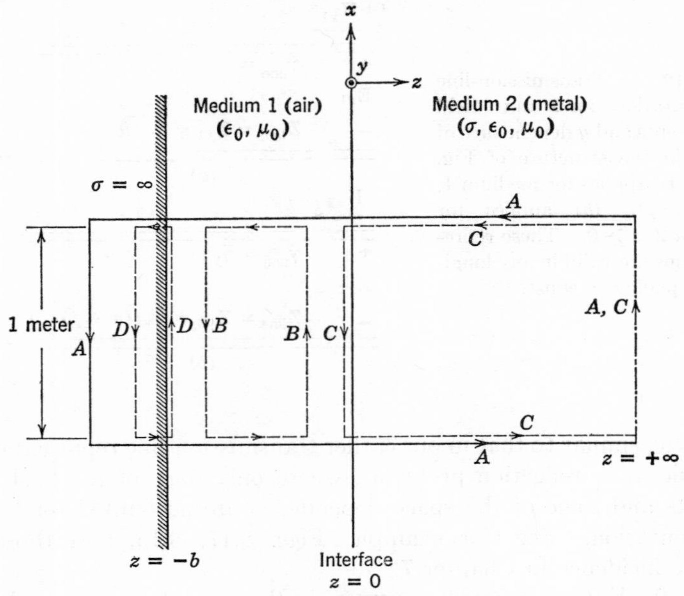

Fig. 8.20. Various closed contours for the definition of transmission-line current as a line integral of magnetic field in the parallel-plate transmission structure. The magnetic field is of the form $\mathbf{H} = \boldsymbol{a}_x \mathbf{H}_x(y, z)$, and these contours make current positive in the $+y$ direction.

The current definition is not easier in this problem. It is harder than the voltage definition. There are several components of current in the longitudinal (y) direction, any one of which might be taken as the "current" for the transmission line. First there is the longitudinal conduction current on the surface of the perfect conductor at $z = -b$. Then there is the longitudinal displacement current produced by E_{y1} in the air space. Finally there is both a conduction current and a displacement current in the metal produced by E_{y2}. The sum of all these currents is zero. We know it is zero because the magnetic field vanishes inside the perfect conductor ($z < -b$) and at $z = +\infty$, while it has only an x-component elsewhere. Accordingly, a contour integral of $\mathbf{H}$ about the path A shown in Fig. 8.20 must vanish, which implies no net current in a longitudinal direction. Thus we could, in fact, define the "transmission-line current" as that enclosed by any path in the transverse plane, such as path B in Fig. 8.20, with the assurance that the total remaining current outside this path, and within a unit width of the plate structure, would correctly be equal and opposite to it in value.

This remaining current would constitute the "return" current of the transmission-line representation.

In the general case, then, it seems that there are no conventional line-integral definitions of voltage and current that seem especially appropriate for our problem. Apparently, we could choose them in a variety of ways. So why not be entirely arbitrary about it, and select a pair at random? The only reason for being a little careful here is that we probably would like the product of current and voltage to give correctly the total power flowing in the longitudinal direction. It is doubtful that this condition would be met by completely arbitrary choices. This is what really lies behind our worries that "all the field" might not be involved in our voltage definition, and that there seemed to be no particularly "appropriate" current definition.

The obvious way out of our troubles is now seen to be the *arbitrary* choice of a definition for *either* the voltage V *or* the current I, making it proportional to either the electric or magnetic field strength respectively. Having defined one, the other can simply be forced to make $\frac{1}{2}$ VI* equal to the complex power (which is computed from the exact field solution by integrating the complex Poynting vector over the entire cross section). This is the way one often handles wave-guide modes, similar to our solutions with $m > 0$. In fact, these modes must be handled in some such manner even when the boundaries are perfectly conducting, because there is always a longitudinal component of either the electric or magnetic field (or both) to obscure one (or both) of the definitions of current and voltage. In principle, the same thing must also be done for the TEM-like (actually TM) solution that occurs with a dissipative boundary.

Fortunately, when the conductivity of the metal boundary is high, the difficulties discussed in principle above do not amount to much numerically for the TEM-like wave. First of all, the displacement current in the metal is negligible. Second, the magnetic field changes very little with position z in the air, as long as $|\bar{\gamma}_{z1}b| \ll 1$. This is clear from the hyperbolic functions in Eqs. 8.194 and 8.196. It follows that the longitudinal current in the metal, obtained by integration of the magnetic field around the contour C of Fig. 8.20, must be very nearly equal and opposite in sign to that on the surface of the perfect conductor at $z = -b$. Another way of saying the same thing is to point out that the total longitudinal displacement current in the air is very small compared to the conduction currents in the metal boundaries. We can therefore safely take as our transmission-line current I(y) either the $+y$-directed current or the tangential magnetic field H_x, on the surface of the perfect conductor. Incidentally, the corresponding closed contour for the latter is the one marked D in Fig. 8.20, so the line integral is numerically equal to H_x.

Observe that the magnetic field is continuous at $z = 0$, but drops off rapidly in the metal because of the factor $e^{-\tilde{\gamma}_{z2}z}$.

Similarly, the transverse electric field E_z is almost constant across the air space. It drops suddenly by approximately the ratio $(\omega\epsilon_0/\sigma)$ just inside the metal and then falls rapidly as $e^{-\tilde{\gamma}_{z2}z}$ thereafter. Because of the sudden decrease of E_z in going from air into the metal as well as of the small penetration distance (skin depth) of the entire field in the metal, the total power flowing longitudinally (in the y-direction) inside the metal will be much smaller than that flowing in the air. Indeed, it is usually entirely negligible by comparison.

8.4.5.3 Transmission-Line Parameters.* Let us therefore first define the transmission-line voltage by the relation

$$V(y) = \int_{-b}^{0} E_z\, dz \bigg|_{y=\text{const}} \tag{8.198}$$

chosen to make V positive when the perfectly conducting surface at $z = -b$ is at higher potential than the metal one at $z = 0$. This choice of sign is consistent with the previous current definition as regards our usual transmission-line notation. That is, when V and I are positive, power will flow in the positive direction along the line ($+y$ here).

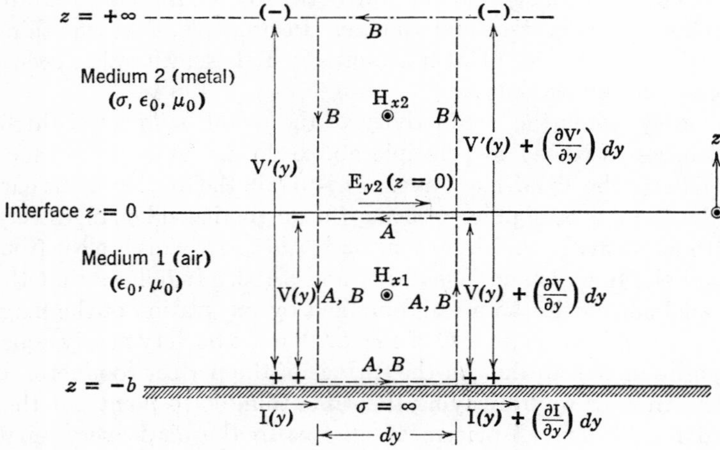

Fig. 8.21. Voltage definitions for the parallel-plate transmission line of Fig. 8.16, showing also contours A and B for writing the voltage-balance differential equation. Observe that the current $I(y)$ is defined by contour D of Fig. 8.20, and flows on the surface $z = -b$ of the perfect conductor.

We may now construct the transmission-line differential equations for a unit width of the planes in the x direction by inspection of Fig. 8.21, keeping in mind the approximations we have justified from the field solution. The rise in line voltage in a distance dy is $(\partial \mathrm{V}/\partial y)\, dy$. It must be accounted for in part by the negative time-rate-of-change of flux in the air through the contour A in Fig. 8.21,

$$-j\omega\mu_0(dy)\int_{-b}^{0} \mathrm{H}_x\, dz \approx -j\omega\mu_0(dy)(b\mathrm{I}) \tag{8.199}$$

and in part by the line integrals of the electric field component E_y along distances dy on the surfaces at $z = 0$ and $z = -b$. But $\mathrm{E}_y = 0$ at $z = -b$, because of the perfect conductor. At $z = 0$, E_y is related to H_x by the very nearly uniform plane-wave character of the field in the metal

$$-\mathrm{E}_{y1}(z = 0) = -\mathrm{E}_{y2}(z = 0) \approx \bar{\eta}_2\mathrm{H}_x(z = 0) \approx \mathrm{I}\sqrt{\frac{j\omega\mu_0}{\sigma}} \tag{8.200}$$

and this produces a contribution to the voltage rise of $(dy)\mathrm{E}_y(z = 0)$. The voltage balance then reads

$$\frac{\partial \mathrm{V}}{\partial y} \approx -j\omega\mu_0 b\mathrm{I} - (1 + j)\sqrt{\frac{\omega\mu_0}{2\sigma}}\,\mathrm{I}$$

$$= -j\left(\omega\mu_0 b + \sqrt{\frac{\omega\mu_0}{2\sigma}}\right)\mathrm{I} - \sqrt{\frac{\omega\mu_0}{2\sigma}}\,\mathrm{I}$$

or

$$\frac{\partial \mathrm{V}}{\partial y} \approx -j\omega L\mathrm{I} - j(\omega L_i)\mathrm{I} - R\mathrm{I} \tag{8.201}$$

where

$$\omega L_i = \sqrt{\frac{\omega\mu_0}{2\sigma}} = \frac{1}{\sigma\delta} \tag{8.202}$$

and

$$R = \sqrt{\frac{\omega\mu_0}{2\sigma}} = \frac{1}{\sigma\delta} \tag{8.203}$$

and

$$L = \mu_0 b \tag{8.204}$$

The resistance R is the same as the d-c resistance of a sheet of metal of conductivity σ, having length and width of 1 m each and a thickness equal to the skin depth δ. The inductance L is associated with flux in the air between the plates, and is the same as was found for the lossless case (Chapter 2). The reactance ωL_i is associated with the reactive part of the surface "impedance drop" along the metal. Whereas R accounts for real power dissipated in the metal, ωL_i accounts for the (inductive) reactive power furnished to it. Sometimes called the *internal inductance* of the metal in the line structure, in contrast with the *external inductance* L, L_i is frequency-dependent ($\sim\omega^{-1/2}$). The resistance R has the same frequency dependence as the whole reactance ωL_i. These circumstances support the contention made in Sec. 2.4.1 regarding frequency dependence of the line parameters produced by losses in the metal walls of a transmission line.

As regards the other transmission-line equation, we must consider $(\partial \mathrm{I}/\partial y)\,dy$ in Fig. 8.21. Recalling that I is the longitudinal current in the $+y$ direction on the surface of the perfect conductor, where $\mathrm{E}_y = 0$, we see from the figure that the increase in I with y must be accounted for by the displacement current arriving perpendicular to the surface $z = -b$. This will be equal to $-j\omega\epsilon_0 \mathrm{E}_z(z = -b)\,dy$. But to the extent that E_z does not vary with z across the air space, we have

$$\mathrm{E}_z(z = -b) \approx \frac{\mathrm{V}}{b} \tag{8.205}$$

Therefore

$$\frac{\partial \mathrm{I}}{\partial y} \approx -j\omega\left(\frac{\epsilon_0}{b}\right)\mathrm{V} = -j\omega C\mathrm{V} \tag{8.206}$$

where

$$C = \frac{\epsilon_0}{b} \tag{8.207}$$

as it was in the lossless case (Chapter 2).

It is interesting to note that the internal reactance ωL_i accounts for the correction to the phase constant which appears in Eq. 8.183. If the attenuation is small, we know for the line that

$$\begin{aligned} |\beta_y| = |\beta_{y1}| = |\beta_{y2}| &\approx \omega\sqrt{C(L + L_i)} \\ &= \omega\sqrt{LC}\sqrt{1 + \frac{L_i}{L}} \\ &\approx \beta_{01}\left(1 + \frac{1}{2}\frac{L_i}{L}\right) = \beta_{01}\left(1 + \frac{1}{4}\frac{\delta}{b}\right) \end{aligned} \tag{8.208}$$

where we have used Eqs. 8.204, 8.202, and 8.207 for L, L_i, and C. Similarly, the resistance R accounts for the attenuation α_y. For small attenuation, we know from Chapter 5 that

$$|\alpha_y| \approx \frac{R}{2Z_0} \tag{8.209}$$

The differential Eqs. 8.201 and 8.206, however, lead to

$$Z_0 = \sqrt{\frac{(R + j\omega L_i) + j\omega L}{j\omega C}} \approx \sqrt{\frac{L}{C}} = b\sqrt{\frac{\mu_0}{\epsilon_0}} \tag{8.210}$$

if

$$R = \omega L_i \ll \omega L \tag{8.211}$$

The approximation (Eq. 8.211) is necessary to conform with the approximations in Eq. 8.209, which applies only when Z_0 is nearly real. From Eqs. 8.203, 8.210, and 8.209, therefore, we find

$$|\alpha_y| \approx \frac{1}{2\sigma\delta b}\sqrt{\frac{\epsilon_0}{\mu_0}} = \frac{1}{2b}\sqrt{\frac{\omega\epsilon_0}{2\sigma}} \tag{8.212}$$

in agreement with Eq. 8.183.

We have suggested that, because of the validity of our approximations when σ is large, variations in the definitions of voltage or current will not alter very much the transmission-line description of Fig. 8.16 for the TEM-like waves. As an example, try a second definition of voltage:

$$\mathrm{V}'(y) = \int_{-b}^{+\infty} \mathrm{E}_z\, dz \tag{8.213}$$

The appropriate contour for voltage balance (contour B, Fig. 8.21) now stretches out to $z = +\infty$, where all fields vanish. In this situation, the voltage rise along y, $\partial\mathrm{V}'/\partial y$, must be balanced *entirely* by flux linkages through the contour all the way from $z = -b$ to $z = +\infty$. Between $z = -b$ and $z = 0$ we have those in air, $-j\omega\mu_0 b\mathrm{I}$ as before. This time, however, we must account for the component of the magnetic field in the metal, which is of the form $\mathrm{H}_x(z = 0)e^{-\bar{\gamma}_{z2}z}$. Thus the flux linkages of the contour in the metal are given by

$$-j\omega\mu_0 \int_0^\infty \mathrm{H}_x(z = 0)e^{-\bar{\gamma}_{z2}z}\, dz = -\frac{j\omega\mu_0}{\bar{\gamma}_{z2}}\mathrm{H}_x(z = 0) \approx \frac{-j\omega\mu_0}{\bar{\gamma}_{z2}}\mathrm{I} \tag{8.214}$$

because $\mathrm{H}_x(z = 0) \approx \mathrm{H}_x(z = -b)$. We have indicated that $\bar{\gamma}_{z2} \approx \sqrt{j\omega\mu_0\sigma}$, so the voltage balance equation becomes

$$\begin{aligned} \frac{\partial\mathrm{V}'}{\partial y} &= -j\omega\mu_0 b\mathrm{I} - \sqrt{\frac{j\omega\mu_0}{\sigma}}\,\mathrm{I} \\ &= -j\omega L\mathrm{I} - j(\omega L_i)\mathrm{I} - R\mathrm{I} \end{aligned} \tag{8.215}$$

a result identical with Eq. 8.201.

Here the resistance R and the internal reactance ωL_i show up as the real and imaginary parts respectively of the flux linkages in the metal. On the basis of this particular physical interpretation, incidentally, we see, rather more easily perhaps than we could from our previous point of view, some circumstances in which the internal reactance may become quite important. An increase of the permeability of the metal (medium 2) certainly increases the flux linkages in that region, as compared to those in the air. A decrease of the spacing b has the same net effect, but does so by reducing flux linkages in the air. Finally, a reduction in frequency increases the skin depth $\delta(\sim \omega^{-1/2})$, thereby increasing the effective penetration distance of the field into medium 2, as compared with dimension b, and increasing the ratio of L_i to L. The use of copper-coated steel wire to achieve mechanical strength in some power-transmission lines may give rise to just such a combination of circumstances.

Returning to the transmission-line equations, the second one involves the transverse displacement current. In the present case, we need to know not only that E_z varies very little across the air space but also that it is very small in the metal (medium 2). These facts are both included in our approximations when σ is large, so we may write

$$\mathrm{E}_z(z = -b) \approx \frac{\mathrm{V}'}{b} \tag{8.216}$$

just as we did in Eq. 8.205. Therefore by the same reasoning which leads to Eq. 8.206 from Eq. 8.205, we now find

$$\frac{\partial \mathrm{I}}{\partial y} \approx -j\omega\left(\frac{\epsilon_0}{b}\right)\mathrm{V}' = -j\omega C\mathrm{V}' \tag{8.217}$$

Equations 8.215 and 8.217 show that the new voltage definition (Eq. 8.213) produces no change in the transmission-line description of the structure within the allowable approximations for high conductivity of medium 2. Broadly speaking, one may expect to find difficulties with these approximations only when the attenuation of the line structure in a distance of one wave length is appreciable. In terms of our present example, this means the assumption $|\alpha_y| \ll |\beta_y|$ would not be valid, which means, in turn, that the surface impedance $\bar{\eta}_2$ of the metal must become comparable to the inductive reactance per unit length (ωL) that the structure would have if the metal were a perfect conductor.

8.4.5.4 Power. As a last check upon our definitions of current and voltage, we should compare the longitudinal power to which they lead with the exact value computed from the field expressions. For this purpose, we need to deal only with a +wave (along $+y$), given by Eqs. 8.196 and 8.197, and with the first voltage definition given in Eq. 8.198.

Accordingly, by our definition of current as that on the surface $z = -b$, we have

$$\mathrm{I}^+(y) \equiv \mathrm{H}_x^{\,+}(z = -b) = \mathrm{H}_{x0}^{\,+}e^{-\bar{\gamma}_y y} \tag{8.218}$$

from Eq. 8.196*a*. From Eqs. 8.198 and 8.196*b*, the voltage becomes

$$\begin{aligned} \mathrm{V}^+(y) &= \int_{-b}^{0} \mathrm{E}_{z1}^{\,+}\,dz = \frac{\bar{\gamma}_y}{j\omega\epsilon_0}\,\mathrm{H}_{x0}^{\,+}e^{-\bar{\gamma}_y y}\int_{-b}^{0} \cosh\,[\bar{\gamma}_{z1}(z+b)]\,dz \\ &= \left(\frac{\bar{\gamma}_y b}{j\omega\epsilon_0}\right)\mathrm{H}_{x0}^{\,+}e^{-\bar{\gamma}_y y}\left[\frac{\sinh\,(\bar{\gamma}_{z1}b)}{(\bar{\gamma}_{z1}b)}\right] \end{aligned} \tag{8.219}$$

Thus the complex power flowing in the $+y$ direction would be given from the voltage and current as

$$\langle P^+\rangle + jQ^+ = \frac{1}{2}\,\mathrm{V}^+\mathrm{I}^{+*} = \left(\frac{\bar{\gamma}_y b}{2j\omega\epsilon_0}\right)|\mathrm{H}_{x0}^{\,+}|^2 e^{-2\alpha_y y}\left[\frac{\sinh\,(\bar{\gamma}_{z1}b)}{(\bar{\gamma}_{z1}b)}\right] \tag{8.220}$$

For the calculation of the true power we must find the longitudinal power using the field expressions in both medium 1 and medium 2. In the first medium

$$\begin{aligned} \langle P_1^{\,+}\rangle + jQ_1^{\,+} &= \frac{1}{2}\int_{-b}^{0} \mathrm{E}_{z1}^{\,+}\mathrm{H}_{x1}^{\,+*}\,dz \\ &= \frac{1}{2}\,\frac{\bar{\gamma}_y}{j\omega\epsilon_0}|\mathrm{H}_{x0}^{\,+}|^2 e^{-2\alpha_y y}\int_{-b}^{0} |\cosh\,[\bar{\gamma}_{z1}(z+b)]|^2\,dz \end{aligned} \tag{8.221}$$

But in general

$$\begin{aligned} |\cosh\bar{\vartheta}|^2 &= \cosh\bar{\vartheta}\,\cosh\bar{\vartheta}^* = \tfrac{1}{4}(e^{\bar{\vartheta}} + e^{-\bar{\vartheta}})(e^{\bar{\vartheta}^*} + e^{-\bar{\vartheta}^*}) \\ &= \tfrac{1}{2}[\cosh\,(\bar{\vartheta} + \bar{\vartheta}^*) + \cosh\,(\bar{\vartheta} - \bar{\vartheta}^*)] \end{aligned}$$

where $\bar{\vartheta} + \bar{\vartheta}^* = 2\,\mathrm{Re}\,\bar{\vartheta}$, which is pure real, and $\bar{\vartheta} - \bar{\vartheta}^* = 2j\,\mathrm{Im}\,\bar{\vartheta}$, which is pure imaginary. Thus

$$|\cosh\,[\bar{\gamma}_{z1}(z+b)]|^2 = \tfrac{1}{2}[\cosh 2\alpha_{z1}(z+b) + \cos 2\beta_{z1}(z+b)] \tag{8.222}$$

The integration in Eq. 8.221 is now easy with the use of Eq. 8.222. We find in medium 1

$$\langle P_1{}^+\rangle + jQ_1{}^+ = \left(\frac{\bar{\gamma}_y b}{2j\omega\epsilon_0}\right)|\mathrm{H}_{x0}{}^+|^2 e^{-2\alpha_y y}\left\{\frac{1}{2}\left[\left(\frac{\sinh 2\alpha_{z1}b}{2\alpha_{z1}b}\right) + \left(\frac{\sin 2\beta_{z1}b}{2\beta_{z1}b}\right)\right]\right\} \quad (8.223)$$

In medium 2, from Eq. 8.197, there results

$$\begin{aligned}\langle P_2{}^+\rangle + jQ_2{}^+ &= \frac{1}{2}\int_0^\infty \mathrm{E}_{z2}{}^+\mathrm{H}_{x2}{}^{+*}\,dz \\ &= \frac{\bar{\gamma}_y}{2(\sigma + j\omega\epsilon_0)}|\mathrm{H}_{x0}{}^+|^2|\cosh\bar{\gamma}_{z1}b|^2 e^{-2\alpha_y y}\int_0^\infty e^{-2\alpha_{z2}z}\,dz \\ &= \frac{\bar{\gamma}_y}{4\alpha_{z2}(\sigma + j\omega\epsilon_0)}|\mathrm{H}_{x0}{}^+|^2|\cosh\bar{\gamma}_{z1}b|^2 e^{-2\alpha_y y} \qquad (8.224)\end{aligned}$$

Finally, adding Eqs. 8.223 and 8.224, we have

$$\begin{aligned}\langle P^+\rangle + jQ^+ &= (\langle P_1{}^+\rangle + jQ_1{}^+) + (\langle P_2{}^+\rangle + jQ_2{}^+) \\ &= \left(\frac{\bar{\gamma}_y b}{2j\omega\epsilon_0}\right)|\mathrm{H}_{x0}{}^+|^2 e^{-2\alpha_y y}\left[\frac{1}{2}\left(\frac{\sinh 2\alpha_{z1}b}{2\alpha_{z1}b} + \frac{\sin 2\beta_{z1}b}{2\beta_{z1}b}\right)\right. \\ &\qquad \left. + \left(\frac{j\omega\epsilon_0}{\sigma + j\omega\epsilon_0}\right)\frac{|\cosh\bar{\gamma}_{z1}b|^2}{2\alpha_{z2}b}\right] \qquad (8.225)\end{aligned}$$

It is obvious that Eqs. 8.225 and 8.220 do *not* agree exactly. On the other hand, subject to the approximations we have found valid when σ is large (Secs. 8.4.2 and 8.4.4), they are *very close indeed.* Specifically, on account of the valid condition $|\bar{\gamma}_{z1}b| \ll 1$, we know that $\alpha_{z1}b \ll 1$ and $\beta_{z1}b \ll 1$. Therefore

$$\left.\begin{array}{ll}\text{(a)} & \dfrac{\sinh\bar{\gamma}_{z1}b}{\bar{\gamma}_{z1}b} \approx 1 \\[2ex] \text{(b)} & \dfrac{\sinh 2\alpha_{z1}b}{2\alpha_{z1}b} \approx 1 \\[2ex] \text{(c)} & \dfrac{\sin 2\beta_{z1}b}{2\beta_{z1}b} \approx 1 \\[2ex] \text{(d)} & \cosh\bar{\gamma}_{z1}b \approx 1\end{array}\right\} \quad \text{if } |\bar{\gamma}_{z1}b| \ll 1 \qquad (8.226)$$

These results mean that the power given by the voltage and current in Eq. 8.220 is very nearly equal to the exact power in medium 1, given by Eq. 8.223. Also in view of the additional valid conditions $\delta/b \ll 1$ and $(\omega\epsilon_0/\sigma) \ll 1$, the last term in the brackets of Eq. 8.225 is very small compared to the first one:

$$\frac{1}{2}\left(\frac{\sinh 2\alpha_{z1}b}{2\alpha_{z1}b} + \frac{\sin 2\beta_{z1}b}{2\beta_{z1}b}\right) \approx 1 \ggg \frac{1}{2}\left(\frac{\omega\epsilon_0}{\sigma}\right)\frac{|\cosh \bar{\gamma}_{z1}b|^2}{\alpha_{z2}b} \approx \frac{1}{2}\left(\frac{\omega\epsilon_0}{\sigma}\right)\frac{\delta}{b} \tag{8.227}$$

where we have used Eqs. 8.226 and the fact that $\delta \approx 1/\alpha_{z2}$. The longitudinal power flowing in medium 2 is much, much less than that flowing in the air space.

8.4.6 Generalization to Two Imperfect Metal Boundaries *

One obvious question is in order regarding the comparison between Fig. 8.16 and the lossless parallel-plate line considered in Chapter 2. It hardly seems realistic to allow for finite conductivity in only *one* of the two plates; we should have allowed it in both of them. This modification of the problem is not too difficult to make. We do so in the following manner.

Consider the situation shown in Fig. 8.22*a*, comprising a perfectly conducting, infinitesimally thin sheet occupying the plane $z = 0$, and two identical metal half-spaces occupying the regions $z \geq b$ and $z \leq -b$ respectively. These metals have parameters σ, ϵ_0, μ_0. Now, obviously, the arrangement for $z \geq 0$ is identical with that of Fig. 8.16, except for the unimportant shift of origin on the z-axis to the left by b units. Therefore one of the field solutions we have found for Fig. 8.16 certainly fits in this region. Let it be denoted by $\mathbf{E}_a$, $\mathbf{H}_a$. Note that $\mathrm{E}_{ya}(z = 0) = 0$, and let $\mathrm{H}_{xa}(z = 0)$ and $\mathrm{E}_{za}(z = 0)$ be the values of the tangential magnetic field and transverse electric field of this solution on the right side of the perfect conductor.

Next, observing that the region $z \leq 0$ is really identical physically to that for $z \geq 0$, we recognize that a field $\mathbf{E}_b$, $\mathbf{H}_b$ defined by the relations

$$\left.\begin{aligned} &\text{(a)} \qquad \mathrm{H}_{xb}(-z) = +\mathrm{H}_{xa}(z) \\ &\text{(b)} \qquad \mathrm{E}_{yb}(-z) = -\mathrm{E}_{ya}(z) \\ &\text{(c)} \qquad \mathrm{E}_{zb}(-z) = +\mathrm{E}_{za}(z) \end{aligned}\right\} \quad \text{for } all\ (x, y) \tag{8.228}$$

surely meets all the boundary conditions in $z \leq 0$. The only questions are whether or not the field is a solution of Maxwell's equations, and

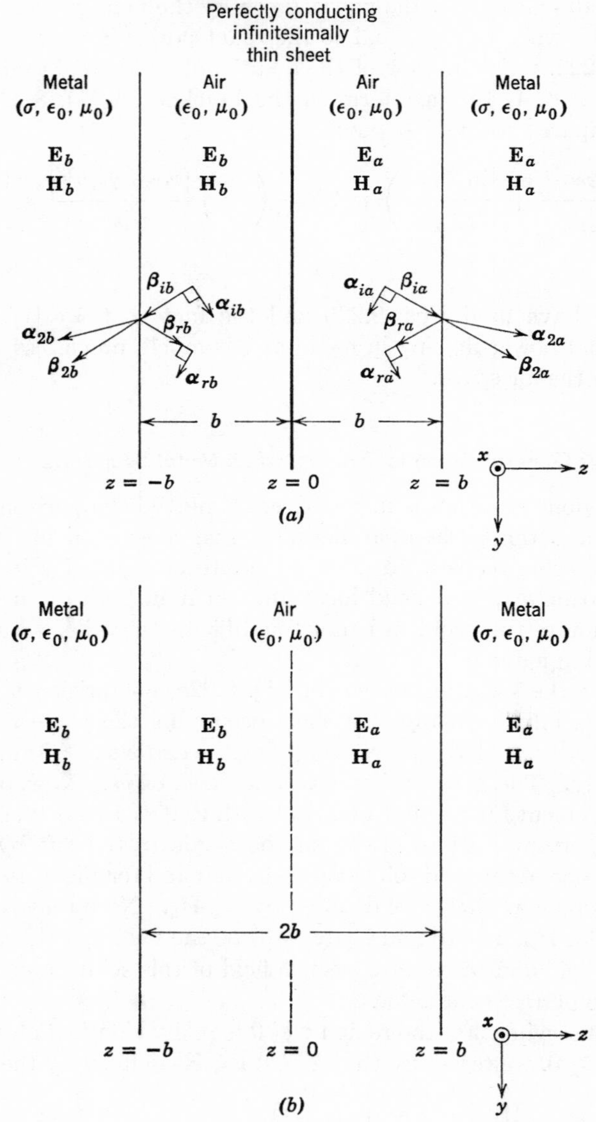

Fig. 8.22. Extension of the solution of Fig. 8.16 to apply to a parallel-plate transmission system with two imperfectly conducting walls. (*a*) First step in the extension, showing propagation vectors for a $+y$-wave solution; (*b*) removal of perfectly conducting plane at $z = 0$ to complete the extension.

whether the center conductor may be removed. That it is a solution to the field equations we can see from the following argument.

We should have for the field $\mathbf{E}_b$, $\mathbf{H}_b$

$$\mathbf{E}_b = \frac{1}{\sigma + j\omega\epsilon_0} \nabla \times \mathbf{H}_b \tag{8.229}$$

if it is to obey one of Maxwell's equations ($\sigma = 0$ in $-b < z \leq 0$, and $\sigma \neq 0$ in $z \leq -b$). But since $\mathbf{H}_b = \boldsymbol{a}_x \mathrm{H}_{xb}(y, z)$

$$\nabla \times \mathbf{H}_b = \boldsymbol{a}_y \frac{\partial \mathrm{H}_{xb}}{\partial z} - \boldsymbol{a}_z \frac{\partial \mathrm{H}_{xb}}{\partial y} \tag{8.230}$$

According to Eq. 8.228*a*, however,

$$\begin{aligned} &\text{(a)} \qquad \frac{\partial \mathrm{H}_{xb}}{\partial z} = -\frac{\partial \mathrm{H}_{xa}}{\partial z} \\ &\text{(b)} \qquad \frac{\partial \mathrm{H}_{xb}}{\partial y} = +\frac{\partial \mathrm{H}_{xa}}{\partial y} \end{aligned} \tag{8.231}$$

Thus $\mathbf{E}_b$, $\mathbf{H}_b$ will be a solution to Eq. 8.229 if

$$\begin{aligned} &\text{(a)} \qquad (\sigma + j\omega\epsilon_0)\mathrm{E}_{yb} = \frac{\partial \mathrm{H}_{xb}}{\partial z} = -\frac{\partial \mathrm{H}_{xa}}{\partial z} = -\mathrm{E}_{ya}(\sigma + j\omega\epsilon_0) \\ &\text{(b)} \qquad (\sigma + j\omega\epsilon_0)\mathrm{E}_{zb} = -\frac{\partial \mathrm{H}_{xb}}{\partial y} = -\frac{\partial \mathrm{H}_{xa}}{\partial y} = \mathrm{E}_{za}(\sigma + j\omega\epsilon_0) \end{aligned} \tag{8.232}$$

because $\mathbf{E}_a$, $\mathbf{H}_a$ *is* a solution to Maxwell's equations. But Eqs. 8.232*a* and 8.232*b* are, in fact, met by Eqs. 8.228*b* and 8.228*c* respectively, so $\mathbf{E}_b$, $\mathbf{H}_b$, as defined by Eqs. 8.228, does satisfy Eq. 8.229.

An exactly similar argument shows that $\mathbf{E}_b$, $\mathbf{H}_b$ also satisfies the other Maxwell equation, $\nabla \times \mathbf{E}_b = -j\omega\mu_0\mathbf{H}_b$. Indeed the relation between the (*b*) and (*a*) solutions is perhaps best illustrated in terms of the nonuniform plane waves from which they may be constructed. For the case where the total solution represents a wave propagating in the $+y$ direction, the propagation vectors of these component waves are shown in Fig. 8.22*a*. This figure should be compared with Fig. 8.16 to make clear the simplicity of the actual "mirror" relation between the fields, which is so abstractly described by Eq. 8.228.

Noting that, in fact,

$$\begin{aligned} \mathrm{E}_{ya}(z = 0) &= 0 = -\mathrm{E}_{yb}(z = 0) \\ \mathrm{H}_{xa}(z = 0) &= \mathrm{H}_{xb}(z = 0) \\ \mathrm{E}_{za}(z = 0) &= \mathrm{E}_{zb}(z = 0) \end{aligned}$$

it is clear that the electric and magnetic fields are actually *continuous* across the perfectly conducting plane $z = 0$. There are neither *net* charges nor currents on this plane. Hence the plane may be removed entirely without disturbing the fields, and we have a solution to the problem of two dissipative planes bounding an air region of width $2b$ as shown in Fig. 8.22*b*. Every solution to the situation of Fig. 8.16 may, by inserting its "image field" given by Eqs. 8.228, generate a solution to the structure of Fig. 8.22*b*.

The field solutions for Fig. 8.22*b*, generated in the foregoing way from those for Fig. 8.16, are the ones in which the *transverse* field components (along x and z) are *even* functions of z, and the *longitudinal* field components (along y) are *odd* functions of z. There will, however, be other solutions for this structure in which the *transverse* fields are *odd* in z, and the *longitudinal* fields are *even* in z. For these, $H_x(z = 0) = 0 = E_y(z = 0)$ and $E_z(z = 0) \neq 0$. We did not find them from Fig. 8.16 because they would have required an "open-circuit" along $z = -b$ rather than a short circuit (see the transmission-line representation in Fig. 8.17). The important point for our purposes here is that the TEM-like solution for Fig. 8.22*b* is included in the set generated by Eqs. 8.228, using the TEM-like solution for Fig. 8.16. This is quite obvious from the fact that this solution for Fig. 8.22*b* must have longitudinal currents going in *opposite* directions in the metals at $z = b$ and $z = -b$, and it must be a solution which goes smoothly over to the TEM wave of Chapter 2 when $\sigma \to \infty$. It has the required symmetry.

Before closing the discussion of Fig. 8.22, it is essential to point out that a parallel-plate configuration with two dissipative boundaries, spaced by b meters (not $2b$ meters), will have *twice* the losses and *twice* the internal reactance drop of that in Fig. 8.16 for the same peak field strengths. Its per unit length series resistance R and internal inductance L_i (but not its external inductance L) will be twice the values given for Fig. 8.16. Correspondingly, its attenuation constant will also be twice as great. Let this be a general reminder that in considering physical two-wire transmission lines, the distributed equivalent-circuit series resistance R per unit length and internal inductance L_i per unit length *must include the effects of a unit length of both wires.*

PROBLEMS

Problem 8.1. (a) For the following materials, evaluate the intrinsic impedance $\bar{\eta}$, the phase velocity v_p, the wave length λ, and the skin depth δ at frequencies of 10^2, 10^6, and 10^{10} cps. (i) Lake water: $\sigma = 5 \times 10^{-3}$ mho/m; $\mu = \mu_0$; $\epsilon = 81\ \epsilon_0$;

(ii) Sea water: $\sigma = 5$ mhos/m; $\mu = \mu_0$; $\epsilon = 81\ \epsilon_0$; (iii) Iron: $\sigma = 10^7$ mhos/m; $\mu = 1000\mu_0$; $\epsilon = \epsilon_0$; (iv) Copper: $\sigma = 5 \times 10^7$ mhos/m; $\mu = \mu_0$; $\epsilon = \epsilon_0$. (b) In each material, at what frequency is the phase velocity of an electromagnetic plane wave equal to the velocity of sound in the same material? (c) For the same materials, evaluate the frequency at which the conduction and displacement current densities are equal. (d) At the frequency specified in (c), what is the ratio α/β and what is the attenuation in decibels per wave length? (e) Compute and discuss the significance of the group velocity in each case under (a) and (c).

Problem 8.2. (a) With reference to Prob. 8.1, for sea water, comment on the possibility of radio communication between submerged submarines, and discuss the choice of frequency for this purpose. (b) Discuss the added features of the problem of radio communication between craft above the surface and submerged submarines, using some numerical field-strength values.

Problem 8.3. Consider a single traveling uniform plane wave of frequency ω in a dissipative medium ($0 < \sigma < \infty$). (a) If $\psi(t)$ is the angle made by $\boldsymbol{E}(z, t)$ with a fixed x-axis normal to the propagation direction (z), show that the angular speed of rotation of the electric field is $d\psi/dt = |\boldsymbol{E} \times (d\boldsymbol{E}/dt)|/|\boldsymbol{E}|^2$ and evaluate it as a function of time. (b) Determine all possible states of polarization of this wave for which the space angle between $\boldsymbol{E}(z, t)$ and $\boldsymbol{H}(z, t)$ is independent of time.

Problem 8.4. Two steady-state uniform plane waves propagate in the $+z$-direction in a dissipative medium. (a) Find the necessary and sufficient condition on their complex amplitudes which will make them orthogonal in complex power. Interpret in terms of polarization resitrictions in the time domain. (b) Find the necessary and sufficient conditions on the second wave to make the two orthogonal in instantaneous power, if the first wave is: (i) linearly polarized; (ii) circularly polarized; (iii) elliptically polarized.

Problem 8.5. A uniform plane wave propagates in a good conductor ($\sigma/\omega\epsilon \gg 1$). The only component of electric field is given by $E_x = \mathrm{Re}\,(\mathrm{E}_{x0}e^{-\alpha z}e^{-j\beta z}e^{j\omega t})$. By formal integration, find the magnitude and relative phase of the total current passing through a 1-m width of the y, z-plane extending from $z = 0$ to $z = \infty$. Express your results in terms of E_{x0}, σ, and α.

Problem 8.6. A uniform plane wave traveling in the $+z$-direction is incident normally from air upon a dissipative dielectric with parameters $\epsilon = 8\epsilon_0$, $\mu = \mu_0$, $\sigma = 2\sqrt{3}$ mhos/m. The frequency is 4500 mc. The magnetic field is in the $+x$-direction. (a) At what depth into the dielectric is the magnetic field strength ¼ of its strength at the interface? (b) What per cent of the incident power is dissipated in the dielectric?

Problem 8.7. Uniform plane electromagnetic waves in air are incident normally on a copper sheet of thickness t. The conductor has permeability μ and conductivity $\sigma \gg \omega\epsilon$. (a) Show that the ratio of the complex electric field strength $\mathbf{E}_2$ in the transmitted wave at the second surface to the total complex electric field strength $\mathbf{E}_1$ at the first surface is approximately equal to sech $\bar{\gamma}t$, where $\bar{\gamma} = \sqrt{(\omega\mu\sigma/2)}\,(1 + j)$. (b) For $\sigma = 10^7$ mhos/m, $\mu = \mu_0$, and $t = 1$ mm, what is the transmission loss in decibels $[10 \log (\|\mathbf{E}_1\|^2/\|\mathbf{E}_2\|^2)]$ at: (i) 100 cps? (ii) 10^6 cps? (iii) 10^{10} cps? (c) Consider the possibility of uniform plane waves incident upon both sides of the sheet. Regarding the two surfaces as "terminal pairs," make a two-terminal-pair network equivalent of the transmission-line representation of the sheet. (d) Under what conditions does the sheet act like a lumped resistor across an air transmission line?

Problem 8.8. A thin mylar sphere is ejected from a rocket and expands to a large radius (≈ 30 ft). The plastic is to be coated with a thin layer of aluminum ($\sigma = 3.72 \times 10^7$ mhos/m) so that the sphere may be used as a radar target. Approximately what thickness of aluminum is required if, at a frequency of 400 mc, 95% of the incident power per unit area is to be reflected at the surface?

Problem 8.9. In an attempt to make a ship invisible to enemy radar, it is proposed to coat the hull with a sheet of lossless, nonmagnetic dielectric having $\epsilon/\epsilon_0 = 9.0$ and a thickness of 1 cm. The hull may be regarded as perfectly conducting, and the outer surface of the dielectric will have a thin carbon coating of sheet resistance 377 ohms/square. (a) Determine the lowest frequency at which reflection would be eliminated entirely if the hull were a plane surface and the radar signal were a uniform plane wave at normal incidence. (b) Under conditions (a), what variation of frequency could be made without producing a reflected field strength more than 30% of that incident ($|\bar{\Gamma}| \le 0.30$)?

Problem 8.10. An idealized resonant cavity consists of a 1-m square area of two infinite, parallel, thick metal plates separated by an air space of l m. The fields between the plates have no components normal to them. The lowest natural frequency is $f = 3 \times 10^8$ cps. (a) Find l. (b) Determine the depth of penetration δ for this frequency, if the resistivity of the metal is 1.7×10^{-8} ohm-m. (c) Calculate an approximate value of Q for the system at the above frequency.

Problem 8.11. Consider the problem of normal incidence of a uniform plane wave from air upon a magnetic material for which $\epsilon \approx \epsilon_0$, $\sigma \gg \omega\epsilon$, and $\mu \gg \mu_0$, such that $\sigma/\omega\epsilon$ is comparable in value with μ/μ_0. (a) Can the usual "skin depth" calculation be employed for the effective resistance and internal reactance of the medium, using twice the incident magnetic field to define the total current? If so, explain in detail. If not, develop an alternative procedure. (b) Discuss the relationship between the "skin effect" in metals and the "eddy-current crowding" of flux in magnetic materials. Comment carefully on the advantages, if any, of laminating magnetic cores.

Problem 8.12. In connection with more than one nonuniform plane wave simultaneously present in a *lossless* medium, discuss how time-average power can be carried along a direction of attenuation, as in Figs. 8.16 or 8.18. Refer to Probs. 3.49, 7.30, and 7.31 for helpful "transmission-line" analogies. Comment on the idea that these results are intimately connected with failure of power orthogonality. Apply the above reasoning analytically to explain why light reflected internally from one face of a glass prism passes instead through that face when a second prism is brought up close to (but not touching) the one in question.

Problem 8.13. (a) A uniform plane wave in air is reflected obliquely from a perfect conductor. Describe the resulting surfaces of constant phase in the air. (b) A uniform plane wave suffers total reflection at an interface between *lossless* dielectrics. Describe the equiphase surfaces in both media. (c) Repeat (a) if the conductor is not quite perfect, and extend the equiphase surfaces into the conductor.

Problem 8.14. Solve the problem of oblique incidence of a uniform plane wave from air onto an imperfect conductor, for the case where $\boldsymbol{E}$ is parallel to the boundary (see Fig. 8.14).

Problem 8.15. Consider oblique incidence of a uniform plane wave *from* an imperfect conductor onto an air interface. Treat both elementary polarizations. Discuss the physical interpretation of this problem and its results.

Problem 8.16. Solve the problem of Fig. 8.16 for $\boldsymbol{E}$ parallel to the walls.

Problem 8.17. Solve the problem of Fig. 8.16 by starting from the *solution* for Fig. 8.14 and letting ϑ become complex, $\bar{\vartheta} = \vartheta_R + j\vartheta_I$, to fit the extra boundary condition at $z = -b$. Carry out the work for both elementary polarizations (see Prob. 8.14).

Problem 8.18. Without approximating $\bar{\gamma}_{y1}$, but using the same approximation for $\bar{\gamma}_{z1}$ used in the text, plot $\bar{\gamma}_{y1}$ versus ω for a $\mathrm{TM}_{1,0}$ wave in the structure of Fig. 8.16. Choose b and σ appropriately, and extend the plot to values of ω that would be well below the cutoff frequency for this wave if loss were absent.

Problem 8.19. (a) Find approximately the attenuation constant for $\mathrm{TE}_{m,0}$ waves in a rectangular wave guide with imperfectly conducting walls. (b) In case the wide faces are perfectly conducting, show that your result (a) satisfies the transcendental equation for Fig. 8.16 with appropriate definition of the guide dimensions.

Problem 8.20. Work out and plot the equiphase surface for the TEM-like wave in Figs. 8.16 and 8.18 under conditions $\omega\epsilon_0/\sigma \gg 1$ and $\beta_{01}b > 1$.

Problem 8.21. Consider the parallel-plate line of Figs. 8.16 and 8.21 operating in the TEM-like mode. (a) Define voltage and current for a (+) wave as is done in Eqs. 8.218 and 8.219. Noting that $Z_0 = V_+/I_+$ for a transmission line and that $\bar{\gamma}_y$ is its propagation constant, find *exact* expressions for the series impedance Z_S and shunt admittance Y_P per unit length of the line. Is it *exactly* true that $G_{\text{line}} = 0$? Explain. (b) Is it possible to redefine voltage and current for this structure so that the two conditions $G_{\text{line}} = 0$, $\frac{1}{2}VI^* =$ total longitudinal complex power will be *exactly* valid? (c) Show that if the current is defined as in Eq. 8.218, and the voltage is defined as twice the complex power carried longitudinally *in the air space only* (Eq. 8.223), per unit line current, then $G_{\text{line}} \equiv 0$. (d) Show by direct exact calculation and integration that the total longitudinal conduction and displacement current in medium 2 of Fig. 8.16 is very nearly equal and opposite to the surface current on the perfect conductor, provided only that $|\bar{\gamma}_{z1}b| \ll 1$.

Problem 8.22. Solve the problem of guided waves in Fig. 8.16 if medium 1 has σ, ϵ_0, μ_0 and medium 2 has $\sigma = 0$, ϵ_0, μ_0.

(a) Find $\bar{\gamma}_y$ for guided TM waves when: (i) $\sigma \ll \omega\epsilon_0$; (ii) $\sigma \gg \omega\epsilon_0$. Look for a guided wave which persists down to $\omega = 0$, and compare with the results of Sec. 7.5.3.
(b) Repeat the above for TE waves.

Problem 8.23. Consider oblique incidence of uniform plane waves from air onto a medium with σ, ϵ_0, $\mu \gg \mu_0$. Treat both polarizations, and include cases with $\sigma \gg \omega\epsilon_0$ and $\sigma \ll \omega\epsilon_0$. Compare with Prob. 8.11.

Problem 8.24. In a certain medium having constant parameters σ, ϵ, μ, let $\boldsymbol{E} = \mathrm{Re}\,[\mathbf{E}e^{st}]$, $\boldsymbol{H} = \mathrm{Re}\,[\mathbf{H}e^{st}]$, and *force* $\mathbf{E} = \mathbf{E}_0 e^{-j\boldsymbol{\beta}\cdot\boldsymbol{r}}$, $\mathbf{H} = \mathbf{H}_0 e^{-j\boldsymbol{\beta}\cdot\boldsymbol{r}}$, with $\boldsymbol{\beta}$ a real vector.

(a) In terms of $\boldsymbol{\beta}$, find the allowed values of s for such waves and interpret the corresponding solutions. Consider their application to Fourier expansions in *space*, for resonance problems, etc. (b) Repeat (a) if we *force* $\mathbf{E} = \mathbf{E}_0 e^{-\boldsymbol{\alpha}\cdot\mathbf{r}}$, $\mathbf{H} = \mathbf{H}_0 e^{-\boldsymbol{\alpha}\cdot\mathbf{r}}$, with $\boldsymbol{\alpha}$ a real vector. Compare your results for the case $\sigma = 0$ with those found in Prob. 7.28.

Problem 8.25. (a) The electric properties of an isotropic material may be described in terms of its dielectric constant or electric susceptibility. Consider a uniform medium containing N mobile electrons per unit volume which are executing vibrations under the influence of an impressed uniform plane wave of frequency ω. Neglect the possibility of collisions of electrons either with themselves or with ions or neutral molecules. Neglect also the forces arising from space charge, and those arising from the velocity of the electrons and the impressed magnetic field. Write and solve the equations of steady-state motion of the electrons. From the relation between the complex polarization vector $\mathbf{P}$ (i.e., the dipole moment per unit volume) and the complex displacement $\mathbf{r}$ of an electron, due to the impressed field, derive an expression for the electric susceptibility χ_e of the medium. From χ_e, determine the dielectric constant ϵ of the medium. (b) Discuss the validity of neglecting the interaction with the magnetic field, i.e., compare the electric and magnetic forces. (c) The effect of collisions (primarily with ions and neutral molecules) constitutes an energy loss. The usual approach is to treat this loss as a drag similar to viscous friction, i.e., in terms of a retarding force proportional to the velocity. The proportionality constant is normally written as $m\nu$, where m is the electron mass. What are the dimensions of ν? (It is possible to identify ν as the collision frequency, which is the average number of collisions per second.) (d) Modify the results of (a) to include the effects of collisions. For convenience, the following abbreviations are suggested: $g = \nu/\omega$ and $\mathrm{u} = 1 - jg$. What is the effective "conductivity" σ of the medium? (e) What are the phase constant β and attenuation constant α for a uniform plane wave in the medium? When collisions are neglected, what is the "critical frequency," ω_p, at which β vanishes? With reference to Prob. 7.32, discuss the significance of this frequency. If χ_0 is used to denote the susceptibility when collisions are neglected, write χ_0 as a function of ω_p and ω. Finally, discuss the effect of collisions on the "critical frequency" phenomenon.

CHAPTER NINE

Transverse Electromagnetic Waves

In our initial consideration of transmission lines, we assumed that the product of the voltage and current accounted for the power flowing along the line, and that, in terms of the fields, the voltage and current otherwise retained essentially their static or d-c steady-state significance. Indeed in Chapter 2 we showed, on the one hand, that a particular physical structure could support a time-varying field which met all these conditions and led without approximation to the distributed parameters of transmission-line theory. On the other hand, in Chapter 8 it became clear that the presence of longitudinal field components could alter in various degrees the whole idea of voltage and current, as well as the associated distributed parameters.

Our present task is to show that, by exercising some care in the definitions, something very like the familiar static concepts of voltage and current can always be made *precise* for one class of wave solutions to the field equations. These are the *Transverse Electromagnetic* (*or TEM*) *waves*, of which we have so far encountered only one special case—the uniform plane wave. We shall now study the general form of such TEM waves, the type of structure which can support them, and the corresponding transmission-line description to which they lead.

9.1 The TEM Form of Maxwell's Equations (Time Domain)

We wish to investigate solutions of the field equations appropriate to the spaces between conductors of a reasonably general transmission line. The line is visualized as an arrangement of *perfectly conducting* cylinders or sheets, all parallel to the z-axis. There may be many conductors, and some may be concentric with others of tubular character; but the cross section of each conductor and the relative arrangement of all the conductors do not vary with z; i.e., all planes perpendicular to the z-axis produce cross sections of the whole system that are identical. Since we shall be dealing in the time domain for the present, the medium in which the conductors are imbedded is assumed to be characterized everywhere by *constant* parameters ϵ, μ, and σ.

We expect to be interested in electric and magnetic fields which, in general, will vary with position both over the cross section and along z, but which can carry active power only along $\pm z$. A field solution in which neither $\boldsymbol{E}$ nor $\boldsymbol{H}$ has a z component would meet this power condition ideally since the only component of the Poynting vector would have to be along $\pm z$. Moreover, the perfectly conducting boundaries of the system require a zero tangential component of the electric field upon them. The assumption

$$E_z \equiv H_z \equiv 0 \tag{9.1}$$

is consistent with this requirement, although we have seen in Chapters 7 and 8 that it is neither necessary nor sufficient. Another justification for assuming Eq. 9.1 is our desire to find a "transmission line" solution, which must remain valid at all frequencies, including direct current, and our knowledge that the d-c fields on the type of perfectly conducting structure in question would have this property.

For the sake of concrete illustration, then, consider the two-conductor configuration and coordinate axes shown in Fig. 9.1. In accordance with Eq. 9.1, we wish a field solution for which

$$\begin{aligned} &\text{(a)} \qquad \boldsymbol{E} = \boldsymbol{a}_x E_x + \boldsymbol{a}_y E_y = \boldsymbol{E}_T \\ &\text{(b)} \qquad \boldsymbol{H} = \boldsymbol{a}_x H_x + \boldsymbol{a}_y H_y = \boldsymbol{H}_T \end{aligned} \tag{9.2}$$

where the subscript T reminds us that the fields lie entirely in the *transverse* plane of the system. This is the reason for naming our solu-

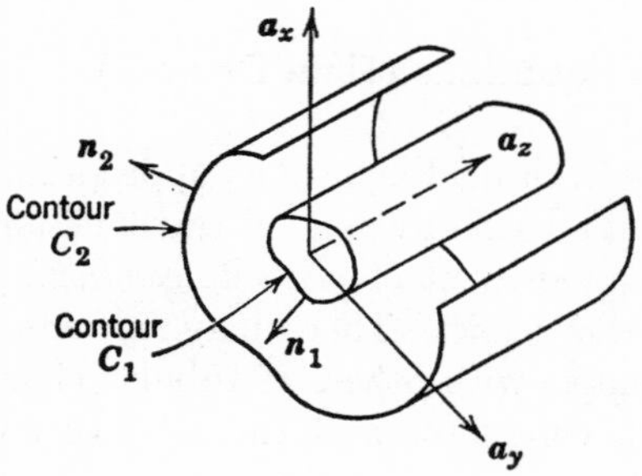

Fig. 9.1. Coordinate system and conductors for TEM field solution.

tion the TEM field. Observe, however, that $\boldsymbol{E}_T$ and $\boldsymbol{H}_T$ are *functions* of x, y, z, and t in general.

In view of Eqs. 9.1 and 9.2, and the assumption of constant medium parameters, Maxwell's equations become

$$\begin{aligned} &\text{(a)} \qquad \nabla \times \boldsymbol{E}_T = -\mu \frac{\partial \boldsymbol{H}_T}{\partial t} \\ &\text{(b)} \qquad \nabla \times \boldsymbol{H}_T = \sigma \boldsymbol{E}_T + \epsilon \frac{\partial \boldsymbol{E}_T}{\partial t} \end{aligned} \tag{9.3}$$

Now the vector operator ∇ can be split into two parts: one representing differentiation with respect to the coordinates in the transverse plane (x and y, for example), and one representing differentiation with respect to the longitudinal or axial coordinate z. Thus

$$\nabla \equiv \left(\boldsymbol{a}_x \frac{\partial}{\partial x} + \boldsymbol{a}_y \frac{\partial}{\partial y}\right) + \boldsymbol{a}_z \frac{\partial}{\partial z} \equiv \nabla_T + \boldsymbol{a}_z \frac{\partial}{\partial z} \tag{9.4}$$

where ∇_T represents the transverse components of ∇, and differentiates only with respect to the transverse coordinates. Note the fact that this split separates not only transverse and longitudinal *space dependence* but also transverse and longitudinal *vector components*. Using Eq. 9.4, we find that Eqs. 9.3 become

$$\begin{aligned} &\text{(a)} \quad \left(\nabla_T + \boldsymbol{a}_z \frac{\partial}{\partial z}\right) \times \boldsymbol{E}_T = \nabla_T \times \boldsymbol{E}_T + \frac{\partial}{\partial z}(\boldsymbol{a}_z \times \boldsymbol{E}_T) = -\mu \frac{\partial \boldsymbol{H}_T}{\partial t} \\ &\text{(b)} \quad \left(\nabla_T + \boldsymbol{a}_z \frac{\partial}{\partial z}\right) \times \boldsymbol{H}_T = \nabla_T \times \boldsymbol{H}_T + \frac{\partial}{\partial z}(\boldsymbol{a}_z \times \boldsymbol{H}_T) \\ &\qquad\qquad = \sigma \boldsymbol{E}_T + \epsilon \frac{\partial \boldsymbol{E}_T}{\partial t} \end{aligned} \tag{9.5}$$

in which we have also used the fact that $\boldsymbol{a}_z$ is a constant vector in both magnitude and direction, and may therefore be moved inside the z derivative without altering the results.

In Eq. 9.5*b*, the term $\nabla_T \times \boldsymbol{H}_T$ is a vector along the z axis because it involves cross products of unit vectors in ∇_T and $\boldsymbol{H}_T$, both of which have components only in the transverse x, y-plane. The remaining terms, however, are all vectors in the transverse plane. Therefore, in order for the entire vector equation to balance in all three space directions, we must have

$$\text{(a)} \qquad \nabla_T \times \boldsymbol{H}_T = 0$$
$$\text{(b)} \qquad \frac{\partial}{\partial z}(\boldsymbol{a}_z \times \boldsymbol{H}_T) = \sigma \boldsymbol{E}_T + \epsilon \frac{\partial \boldsymbol{E}_T}{\partial t} \tag{9.6}$$

Similarly for Eq. 9.5*a* we must have

$$\nabla_T \times \boldsymbol{E}_T = 0 \tag{9.7a}$$

and

$$\frac{\partial}{\partial z}(\boldsymbol{a}_z \times \boldsymbol{E}_T) = -\mu \frac{\partial \boldsymbol{H}_T}{\partial t}$$

The last equation may be related more conveniently to Eq. 9.6*b* by taking the cross product of both sides with the constant unit vector $\boldsymbol{a}_z$. Thus

$$\frac{\partial \boldsymbol{E}_T}{\partial z} = \mu \frac{\partial}{\partial t}(\boldsymbol{a}_z \times \boldsymbol{H}_T) \tag{9.7b}$$

where we have noted that

$$(\boldsymbol{a}_z \times \boldsymbol{E}_T) \times \boldsymbol{a}_z = \boldsymbol{E}_T - (\boldsymbol{a}_z \cdot \boldsymbol{E}_T)\boldsymbol{a}_z = \boldsymbol{E}_T$$

because $\boldsymbol{a}_z$ and $\boldsymbol{E}_T$ are perpendicular in space.

The implications of these results are interesting. Consider first the fact that

$$\nabla \cdot \boldsymbol{B} = 0 \tag{9.8}$$

as we pointed out in Eqs. 1.15 and 1.19. In our present case, $\boldsymbol{H}$ is entirely in the transverse plane ($\boldsymbol{H} \equiv \boldsymbol{H}_T$) and μ is a constant; so Eq. 9.8 becomes

$$\nabla \cdot \boldsymbol{H}_T = \nabla_T \cdot \boldsymbol{H}_T = 0 \tag{9.9}$$

Now Eqs. 9.6*a* and 9.9 together tell us that, for any value of z or t, *the vector* $\boldsymbol{H}_T$ *in the* x, y*-plane obeys the equations of magnetostatics in the two-dimensional space between the line conductors.*

Moreover, the boundary condition on $\boldsymbol{H}_T$ at the conductor surfaces, defined in the transverse plane by contours C_1 and C_2 in Fig. 9.1, is that there shall be no component of $\boldsymbol{H}_T$ *normal* to these surfaces. This condition follows analytically from the limiting form of Eq. 9.8 applied at the interface between two media, one of which is a perfect conductor and can, therefore, support no time-varying field at all inside it. (For further elaboration of this point, refer to the discussions concerning Eqs. 1.22 through 1.25 and consult the reference cited at the beginning of Chapter 1). Physically, however, the meaning of this boundary condition is quite simple. Any attempt to establish a time-varying $\boldsymbol{H}$

field normal to a perfect conductor results in tremendously efficient eddy currents on the surface, which produce a back magnetomotive force just sufficient to cancel out the applied field. If this were not so, there would remain a time-varying $\boldsymbol{H}$ normal to the conductor, resulting in an electric field tangential to the conductor (by Faraday's law). Such an electric field would violate our primary boundary condition (Eq. 1.25), and, incidentally, would do so by producing an infinite *total* current in the conductor.

Thus the magnetic field $\boldsymbol{H}_T$ obeys the boundary conditions

$$\left.\begin{aligned} \boldsymbol{n}_1 \cdot \boldsymbol{H}_T &= 0 \qquad \text{on } C_1 \\ \boldsymbol{n}_2 \cdot \boldsymbol{H}_T &= 0 \qquad \text{on } C_2 \end{aligned}\right\} \quad \text{for all } t \text{ and } z \tag{9.10}$$

where $\boldsymbol{n}_1$ and $\boldsymbol{n}_2$ are unit vectors normal to the respective contours, as shown in Fig. 9.1. These are, incidentally, the same conditions we would have if we passed direct current longitudinally through the conductors.

Therefore we conclude from Eqs. 9.6*a*, 9.9, and 9.10 that, *between perfect conductors, the (transverse) magnetic field of a TEM wave has at any moment of time the same space distribution over any cross section of the system as would the two-dimensional magnetostatic field arising there from appropriate direct currents flowing along the conductors.*

Considerations for the electric field $\boldsymbol{E}_T$ are similar, except for one slight difficulty. In general, for time-varying fields, we have the complete form of the extended Ampere law in a source-free conducting medium

$$\nabla \times \boldsymbol{H} = \sigma \boldsymbol{E} + \epsilon \frac{\partial \boldsymbol{E}}{\partial t} \tag{9.11}$$

and we also have Gauss' law

$$\nabla \cdot \boldsymbol{D} = \rho$$

or, since ϵ is constant,

$$\nabla \cdot \boldsymbol{E} = \frac{\rho}{\epsilon} \tag{9.12}$$

Taking the divergence of Eq. 9.11, noting that $\nabla \cdot (\nabla \times \boldsymbol{H}) \equiv 0$, we find that, because σ and ϵ are constants,

$$\sigma(\nabla \cdot \boldsymbol{E}) + \epsilon \frac{\partial}{\partial t} (\nabla \cdot \boldsymbol{E}) = 0 \tag{9.13}$$

or, with Eq. 9.12,

$$\frac{\partial \rho}{\partial t} + \frac{\sigma}{\epsilon} \rho = 0 \tag{9.14}$$

In other words, $\nabla \cdot \boldsymbol{E}$ (or ρ) does not appear offhand to be zero!

However, Eq. 9.14 has an easy general solution

$$\rho(x, y, z; t) = \rho_0(x, y, z)e^{-\sigma t/\epsilon} \tag{9.15}$$

where $\rho_0(x, y, z)$ is the charge density at any point at time $t = 0$ [i.e., $\rho(x, y, z; 0)$]. Equation 9.15 says that any charge density present at $t = 0$, wherever located in space, vanishes with time, absolutely. At every point in the space characterized by uniform values of ϵ and σ, the charge density *disappears* with a time constant ϵ/σ—never to return. Where does it go? It goes to places where ϵ and σ are *not* constant: i.e., to metal boundaries, interfaces between different dielectrics, etc. In our transmission-line problem, it goes to the metal conductors. Of course, if at $t = 0$ there is no charge density anywhere between the conductors, then $\rho_0(x, y, z) \equiv 0$ everywhere. Hence, by Eq. 9.15, $\rho(x, y, z; t) \equiv 0$ everywhere for $t \geq 0$.

What we have before us is simply a generalized leaky capacitor. If we agree not to throw any free charge into the dielectric *inside*, but only to pass equal and opposite currents through the metal terminals, then there will *never* be any free charge inside. If, however, someone does throw some free charge inside at $t = 0$, we have only to wait a few time constants (ϵ/σ) for it to leak out to the terminals.

In any case, we are at present interested in the behavior of the system used as a transmission line, when we apply voltages or currents to the conductors, and not in its behavior for the rare case of an internally charged dielectric medium. Hence we shall ask only for those solutions in which

$$\nabla \cdot \boldsymbol{E} = \nabla_T \cdot \boldsymbol{E}_T = 0 \tag{9.16}$$

The boundary conditions on $\boldsymbol{E}_T$ are that its tangential components must vanish on the perfect conductors

$$\left.\begin{aligned} \boldsymbol{n}_1 \times \boldsymbol{E}_T &= 0 \qquad \text{on } C_1 \\ \boldsymbol{n}_2 \times \boldsymbol{E}_T &= 0 \qquad \text{on } C_2 \end{aligned}\right\} \quad \text{for all } t \text{ and } z \tag{9.17}$$

Thus Eqs. 9.17, 9.16, and 9.7*a* show that *for any z and t the (transverse) electric field of a TEM wave is distributed over the cross section just like the two-dimensional source-free electrostatic field that would arise there from appropriate direct voltages applied to the conductors.*

It is an interesting consequence of the conclusion following Eq. 9.17 that no TEM solution is possible inside a hollow metal pipe (wave guide). This result stems from the fact that, as a two-dimensional, electrostatic problem, a single, closed, perfectly conducting tube is a *complete electrostatic shield* for the source-free region inside it. Therefore $\boldsymbol{E}_T$ for a TEM wave cannot exist. With regard to $\boldsymbol{H}_T$, Eqs. 9.6*b* and 9.7*b* show

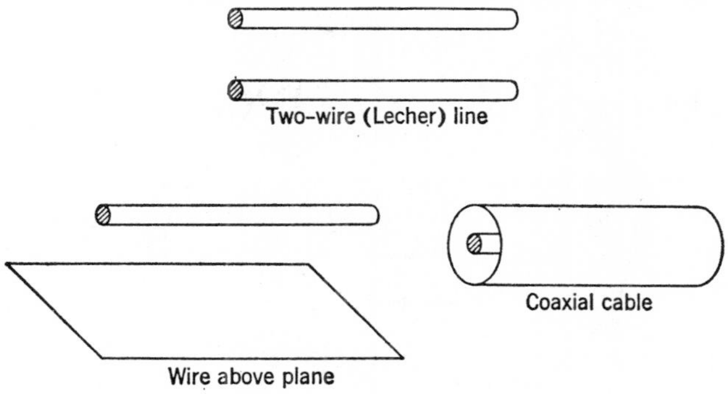

Fig. 9.2. Arrangements of two conductors for TEM propagation.

that it does not vary with either z or t under these conditions. There could be a real source-free *magnetostatic* field in the tube, but this is hardly surprising or interesting (refer to the discussion following Eq. 2.28). If we wish to have a time-varying TEM solution, we must have at least two conductors.[1] Some typical two-conductor arrangements are shown in Fig. 9.2.

We are now prepared to examine the general role of the TEM wave in a two-wire transmission line comprised of perfect conductors imbedded in a medium having constant parameters.

9.2 Transmission-Line Concepts for Two-Conductor Lines (Time Domain)

Consider the two-conductor system of Fig. 9.1, shown at a cross section z = constant in Fig. 9.3. Imagine the field in the space between conductors at a given instant of time t, so that our interest centers upon the behavior of $\boldsymbol{E}_T$ and $\boldsymbol{H}_T$ with respect to the transverse coordinates (x, y) at *fixed* values of z and t.

9.2.1 Voltage, Charge, and Capacitance

First let us examine the electric-field distribution, which satisfies Eqs. 9.7a, 9.16, and 9.17. As we have mentioned before, it is a two-dimensional electrostatic type of field between perfect conductors. It can be derived from a two-dimensional scalar potential because of

[1] The uniform plane wave may seem at first glance to contradict this statement.

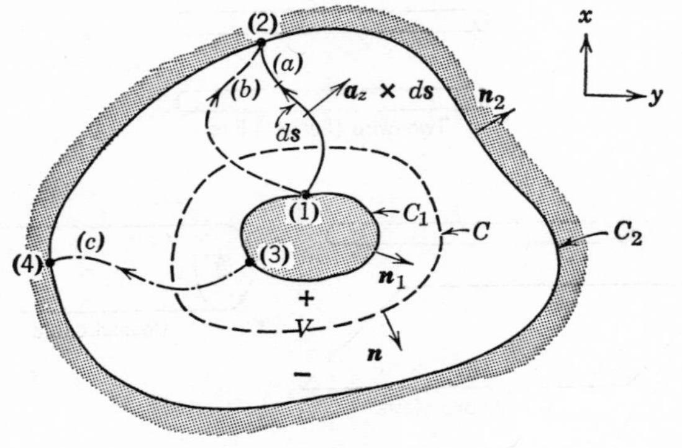

Fig. 9.3. Geometry for definitions of voltage, charge, and capacitance.

Eq. 9.7*a*. Indeed, as far as the x and y dependence of $\boldsymbol{E}_T$ is concerned, the mathematical problem posed by Eqs. 9.7*a*, 9.16, and 9.17 is exactly the same as we would encounter if we applied a direct voltage between the inner and outer conductors and looked for the two-dimensional electrostatic field between them (independent of z and t). Therefore we can define a voltage with the polarity shown in Fig. 9.3 by the relation

$$V(z, t) = \int_{(1)}^{(2)} \boldsymbol{E}_T \cdot d\boldsymbol{s} \bigg|_{\text{along } (a)} \tag{9.18}$$

in which the integration is carried out with z and t held constant. Nevertheless, the voltage will be a function of z and t because our actual field $\boldsymbol{E}_T$ does depend upon these coordinates. Mathematically, this situation is described by saying that integrals like that in Eq. 9.18 are carried out upon the two transverse variables x and y, with the remaining variables z and t as *parameters*.

The next important point is that, *given a value* z_0 *of* z *and* t_0 *of* t, we could just as well have integrated along paths (b) or (c) of Fig. 9.3 in setting up Eq. 9.18 because these paths must all give the *same* answer for $V(z_0, t_0)$. This fact follows from Stokes' theorem applied to Eq. 9.7*a* in the two-dimensional region between the conductors at a given z_0 and t_0:

$$\int_{\substack{\text{between paths} \\ (a) \text{ and } (b)}} \boldsymbol{a}_z \cdot \nabla_T \times \boldsymbol{E}_T \, da = \int_{(1)}^{(2)} \boldsymbol{E}_T \cdot d\boldsymbol{s} \bigg|_{\text{along } (a)} + \int_{(2)}^{(1)} \boldsymbol{E}_T \cdot d\boldsymbol{s} \bigg|_{\text{along } (b)} = 0 \tag{9.19}$$

where da is a scalar element of area between the paths (a) and (b).

Therefore

$$\int_{(1)}^{(2)} \boldsymbol{E}_T \cdot d\boldsymbol{s}\bigg|_{\text{along }(b)} = -\int_{(2)}^{(1)} \boldsymbol{E}_T \cdot d\boldsymbol{s}\bigg|_{\text{along }(b)}$$

$$= \int_{(1)}^{(2)} \boldsymbol{E}_T \cdot d\boldsymbol{s}\bigg|_{\text{along }(a)} = V(z_0, t_0) \qquad (9.20)$$

Similarly, considering the paths (a) and (c), we have by Stokes' theorem

$$\int\limits_{\substack{\text{between paths (1),}\\ \text{(2) and (4), (3)}}} \boldsymbol{a}_z \cdot \nabla_T \times \boldsymbol{E}_T\, da = \int_{(1)}^{(2)} \boldsymbol{E}_T \cdot d\boldsymbol{s}\bigg|_{\text{along }(a)} + \int_{(2)}^{(4)} \boldsymbol{E}_T \cdot d\boldsymbol{s}\bigg|_{\text{along }(C_2)}$$

$$+ \int_{(4)}^{(3)} \boldsymbol{E}_T \cdot d\boldsymbol{s}\bigg|_{\text{along }(c)}$$

$$+ \int_{(3)}^{(1)} \boldsymbol{E}_T \cdot d\boldsymbol{s}\bigg|_{\text{along }(C_1)} = 0 \quad (9.21)$$

But on account of the boundary conditions (Eq. 9.17), the second and fourth integrals in Eq. 9.21 must vanish; so

$$\int_{(3)}^{(4)} \boldsymbol{E}_T \cdot d\boldsymbol{s}\bigg|_{\text{along }(c)} = -\int_{(4)}^{(3)} \boldsymbol{E}_T \cdot d\boldsymbol{s}\bigg|_{\text{along }(c)}$$

$$= \int_{(1)}^{(2)} \boldsymbol{E}_T \cdot d\boldsymbol{s}\bigg|_{\text{along }(a)} = V(z_0, t_0) \quad (9.22)$$

In the sense that the choice of path between conductors (1) and (2) does not alter the result, *as long as the entire path lies in a transverse plane and is traversed at a single moment,* we say that the voltage $V(z, t)$ is defined *uniquely* as the line integral of the electric field. Physically, this fact is a rather obvious consequence of the absence of any longitudinal magnetic field component H_z.

Now the charge q_1 per unit (differential) length of the inner conductor resides on its outer surface, because the conductivity of the conductors is infinite. The surface charge density is equal to the normal electric-displacement field; so at a given (z, t)

$$q_1(z, t) = \epsilon \oint_{C_1} \boldsymbol{n}_1 \cdot \boldsymbol{E}_T\, dl \qquad (9.23)$$

where dl is an arc of the contour C_1 in Fig. 9.3. Because of Eq. 9.16, however, Guass' theorem in two dimensions applied to the region between C_1 and C yields

$$\int_{\substack{\text{between} \\ C_1 \text{ and } C}} \nabla_T \cdot \boldsymbol{E}_T \, da = \oint_C \boldsymbol{n} \cdot \boldsymbol{E}_T \, dl - \oint_{C_1} \boldsymbol{n}_1 \cdot \boldsymbol{E}_T \, dl = 0 \tag{9.24}$$

in which C is any contour that lies in the region between conductors and encloses the inner one, and $\boldsymbol{n}$ is a unit vector normal to C in the sense shown in the figure. In particular, choosing C_2 for C, we have by Eqs. 9.24 and 9.23

$$\epsilon \oint_{C_2} \boldsymbol{n}_2 \cdot \boldsymbol{E}_T \, dl = q_1(z, t) \tag{9.25}$$

Noting that the actual charge $q_2(z, t)$ per unit (differential) length residing on the inner surface of the outer conductor is the normal displacement field integrated over the conductor surface,

$$q_2(z, t) = -\epsilon \oint_{C_2} \boldsymbol{n}_2 \cdot \boldsymbol{E}_T \, dl = -q_1(z, t) \tag{9.26}$$

we see that the charges on the two conductors turn out to be equal and opposite. This is to be expected since there are only two conductors (and the dielectric is assumed to be charge-free).

It follows from Eqs. 9.18, 9.23, and the essentially static form of $\boldsymbol{E}_T$ at fixed (z, t), that a capacitance C per unit (differential) length can be defined in the form

$$C \equiv \frac{q_1}{V} = \frac{\epsilon \oint_{C_1} \boldsymbol{E}_T \cdot \boldsymbol{n}_1 \, dl}{\int_{(1)}^{(2)} \boldsymbol{E}_T \cdot d\boldsymbol{s}} \tag{9.27}$$

which depends *only on the value of the constant* ϵ *and the geometry of the conductors.* Moreover, it is the *same* capacitance per unit length that we would calculate for these conductors from electrostatics.

9.2.2 Current, Flux Linkage, and Inductance

For the magnetic field, we refer to Fig. 9.4 and Eqs. 9.6*a*, 9.9, and 9.10. Again, given z and t, the problem is *formally* a magnetostatic one. The current flows on the outer surface of the inner conductor, and on

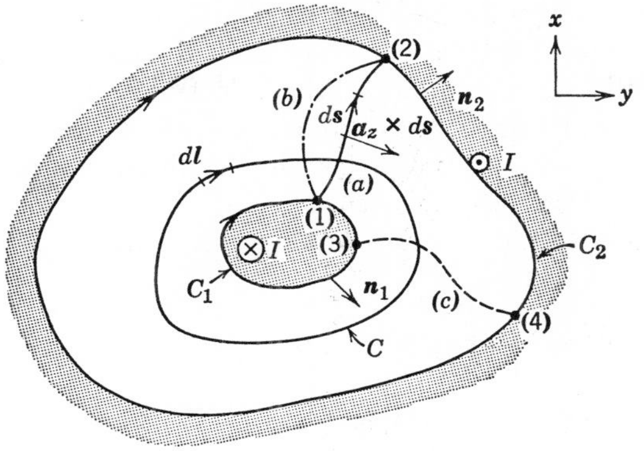

Fig. 9.4. Geometry for definitions of current, flux linkage, and inductance.

the inner surface of the outer one. We define for the inner conductor

$$I_1(z, t) = \oint_{C_1} \boldsymbol{H}_T \cdot d\boldsymbol{l} \tag{9.28}$$

which is positive in the $+z$ direction, consistent for transmission-line notation with the choice of voltage polarity in Fig. 9.3 and Eq. 9.18.

Note next that by Stokes' theorem and Eq. 9.6*a*

$$\int_{\substack{\text{between} \\ C_1 \text{ and } C}} \boldsymbol{a}_z \cdot \nabla_T \times \boldsymbol{H}_T \, da = \oint_C \boldsymbol{H}_T \cdot d\boldsymbol{l} - \oint_{C_1} \boldsymbol{H}_T \cdot d\boldsymbol{l} = 0 \tag{9.29}$$

where C is any contour that lies between the two conductors and surrounds the inner one. Observing that the surface-current density on the inside surface of the outer conductor is $\boldsymbol{a}_z K_{2z} = -\boldsymbol{n}_2 \times \boldsymbol{H}_T$, we have for the $+z$-directed current on the outer conductor,

$$I_2(z, t) = \oint_{C_2} K_{2z} \, dl = \oint_{C_2} K_{2z}(\boldsymbol{a}_z \times \boldsymbol{n}_2) \cdot d\boldsymbol{l} = -\oint_{C_2} \boldsymbol{H}_T \cdot d\boldsymbol{l} \tag{9.30}$$

where we have used the boundary condition (Eq. 9.10) in the last step. Therefore, by Eqs. 9.30 and 9.29, choosing $C = C_2$, we find

$$I_2(z, t) = -\oint_{C_1} \boldsymbol{H}_T \cdot d\boldsymbol{l} = -I_1(z, t) \tag{9.31}$$

The currents on the inner and outer conductors turn out to be equal and opposite. For several reasons this too is to be expected. First, the absence of magnetic field just *outside* C_2 (in a perfect conductor) means that no total current can be contained within a contour located there. Second, there is no longitudinal electric field E_z to make either longitudinal displacement or conduction current flow in the interior dielectric. Accordingly we say that the definition of current is *unique*, on the basis of Eqs. 9.29 and 9.31.

We can now define the flux linkage $\Lambda(z, t)$ per unit (differential) length as

$$\Lambda(z, t) = \int_{(1)}^{(2)} \mu \boldsymbol{H}_T \cdot (\boldsymbol{a}_z \times d\boldsymbol{s}) \Big|_{\substack{\text{across paths } (a), \\ (b), \text{ or } (c)}} \tag{9.32}$$

The fact that surfaces (*a*), (*b*), and (*c*) in Fig. 9.4 are equally good for the definition stems from Gauss' theorem in two dimensions applied to Eq. 9.9. For example:

$$\int_{\substack{\text{between} \\ (a) \text{ and } (c)}} \nabla \cdot \boldsymbol{H}_T \, da = \int_{(3)}^{(4)} \boldsymbol{H}_T \cdot (\boldsymbol{a}_z \times d\boldsymbol{s}) \Big|_{\text{across } (c)} + \int_{(4)}^{(2)} \boldsymbol{n}_2 \cdot \boldsymbol{H}_T \, dl \Big|_{\text{across } C_2}$$
$$+ \int_{(1)}^{(2)} \boldsymbol{H}_T \cdot (d\boldsymbol{s} \times \boldsymbol{a}_z) \Big|_{\text{across } (a)}$$
$$- \int_{(1)}^{(3)} \boldsymbol{n}_1 \cdot \boldsymbol{H}_T \, dl \Big|_{\text{across } C_1} = 0 \tag{9.33}$$

The boundary conditions (Eq. 9.10) eliminate the second and fourth integrals in Eq. 9.33, leaving the result

$$\int_{(3)}^{(4)} \mu \boldsymbol{H}_T \cdot (\boldsymbol{a}_z \times d\boldsymbol{s}) \Big|_{\text{across } (c)} = \int_{(1)}^{(2)} \mu \boldsymbol{H}_T \cdot (\boldsymbol{a}_z \times d\boldsymbol{s}) \Big|_{\text{across } (a)}$$
$$= \Lambda(z, t) \tag{9.34}$$

Equations 9.28 and 9.32 lead to an inductance L per unit (differential) length

$$L \equiv \frac{\Lambda}{I_1} = \frac{\mu \displaystyle\int_{(1)}^{(2)} \boldsymbol{H}_T \cdot (\boldsymbol{a}_z \times d\boldsymbol{s})}{\displaystyle\oint_{C_1} \boldsymbol{H}_T \cdot d\boldsymbol{l}} \tag{9.35}$$

which depends *only upon the geometry of the conductors and the value of the constant* μ. It is the same as the d-c inductance per unit length for the same conductors, provided we note that the current in *perfect conductors* is assumed to flow entirely on the surface even in the limiting case of direct current (see the discussion preceding Eq. 1.25).

9.2.3 Shunt Conductance

Finally, since the conduction current density flowing from one conductor to the other, through the medium between them, is $\boldsymbol{J}_T = \sigma \boldsymbol{E}_T$, it is easy to show from our discussion of the electric field and the capacitance that

$$G \equiv \frac{\sigma \oint_{C_1} \boldsymbol{E}_T \cdot \boldsymbol{n}_1 \, dl}{\int_{(1)}^{(2)} \boldsymbol{E}_T \cdot d\boldsymbol{s}} = \left(\frac{\sigma}{\epsilon}\right) C \tag{9.36}$$

is the shunt conductance per unit (differential) length, and is the same as we would obtain by applying a direct voltage between the conductors.

9.2.4 The Transmission-Line Equations

That the field equations for TEM waves lead directly to the same transmission-line equations we have used so often in previous chapters follows from our developments above regarding two conductors taken in conjunction with the rest of Maxwell's equations (Eqs. 9.7*b* and 9.6*b*). We proceed as follows.

With a view toward introducing the voltage V, let us integrate Eq. 9.7*b* along a path like (*a*) in Fig. 9.3. Noting again that the integration defining V in Eq. 9.18 is performed only upon x and y, with z and t as parameters, we have from Eqs. 9.7*b* and a permissible interchange of integration and differentiation operations

$$\frac{\partial}{\partial z}\left(\int_{(1)}^{(2)} \boldsymbol{E}_T \cdot d\boldsymbol{s}\right) = \mu \frac{\partial}{\partial t}\left[\int_{(1)}^{(2)} (\boldsymbol{a}_z \times \boldsymbol{H}_T) \cdot d\boldsymbol{s}\right] \tag{9.37}$$

Use of Eq. 9.18 and of the vector identity $(\boldsymbol{a}_z \times \boldsymbol{H}_T) \cdot d\boldsymbol{s} = -\boldsymbol{H}_T \cdot (\boldsymbol{a}_z \times d\boldsymbol{s})$ yields

$$\frac{\partial V}{\partial z} = -\frac{\partial}{\partial t}\left[\mu \int_{(1)}^{(2)} \boldsymbol{H}_T \cdot (\boldsymbol{a}_z \times d\boldsymbol{s})\right] \tag{9.38}$$

But according to Eqs. 9.35 and 9.28, the term in brackets on the right is just the flux linkage $\Lambda = LI$, where I is now being written for the current in the inner conductor flowing in the $+z$ direction (previously called I_1). Equation 9.38 becomes

$$\frac{\partial V}{\partial z} = -L \frac{\partial I}{\partial t} \tag{9.39}$$

which is one of the transmission-line equations.

There is even more to learn from Eq. 9.7*b*, however, if we integrate it around the contour C_1 in Fig. 9.4 with a view toward introducing the current on the right-hand side. Specifically, by arguments similar to those used above

$$\frac{\partial}{\partial z}\left(\oint_{C_1} \boldsymbol{n}_1 \cdot \boldsymbol{E}_T \, dl\right) = \mu \frac{\partial}{\partial t}\left[\oint_{C_1} \boldsymbol{n}_1 \cdot (\boldsymbol{a}_z \times \boldsymbol{H}_T) \, dl\right] \tag{9.40}$$

But, by some vector identities and Eq. 9.28, we have

$$\oint_{C_1} \boldsymbol{n}_1 \cdot (\boldsymbol{a}_z \times \boldsymbol{H}_T) \, dl = \oint_{C_1} (\boldsymbol{n}_1 \times \boldsymbol{a}_z) \cdot \boldsymbol{H}_T \, dl = -\oint_{C_1} \boldsymbol{H}_T \cdot \boldsymbol{dl} = -I \tag{9.41}$$

and by Eqs. 9.27 and 9.18,

$$\oint_{C_1} \boldsymbol{n}_1 \cdot \boldsymbol{E}_T \, dl = \frac{C}{\epsilon} V \tag{9.42}$$

On this basis, Eq. 9.40 reads

$$\frac{\partial V}{\partial z} = -\left(\frac{\epsilon\mu}{C}\right)\frac{\partial I}{\partial t} \tag{9.43}$$

From a comparison of Eqs. 9.39 and 9.43, we learn the remarkable fact that

$$LC = \epsilon\mu \tag{9.44}$$

which appears to be a very surprising relation indeed between the d-c inductance and capacitance per unit length of *any two parallel perfect conductors* immersed in a dielectric with constant parameters ϵ and μ! Apparently, no matter what the shape of the conductor cross sections may be, or what the spacing between them may be, the LC product remains fixed, *by the parameters of the surrounding medium*, at the value $\epsilon\mu$. We shall give a further discussion of this matter later on.

For the moment, let us treat the remaining field equation, Eq. 9.6*b*, in the manner we employed in passing from Eq. 9.7*b* to Eq. 9.37.

$$\frac{\partial}{\partial z}\left[\int_{(1)}^{(2)} (\boldsymbol{a}_z \times \boldsymbol{H}_T) \cdot \boldsymbol{ds}\right] = \sigma V + \epsilon \frac{\partial V}{\partial t} \tag{9.45}$$

By steps similar to those leading to Eq. 9.39 from Eq. 9.37, we can rewrite Eq. 9.45 in the form

$$-\frac{\partial I}{\partial z} = \left(\frac{\mu\sigma}{L}\right) V + \left(\frac{\mu\epsilon}{L}\right)\frac{\partial V}{\partial t} \tag{9.46}$$

Equations 9.44 and 9.36, however, allow us to conclude that

$$\frac{\partial I}{\partial z} = -\left(GV + C\frac{\partial V}{\partial t}\right) \tag{9.47}$$

which is the second transmission-line equation.

We must now prove the last important transmission-line property of TEM waves for a structure comprising two perfect conductors surrounded by a medium with constant parameters: The power carried across a plane z = constant, as given by the product VI, checks exactly with the integral of the z component of the Poynting vector over the cross section of the line.

9.2.5 Voltage, Current, and Power

To carry out the proof of the statement above, let us compute the instantaneous power from the fields,

$$P(z, t) = \int_A \boldsymbol{a}_z \cdot (\boldsymbol{E}_T \times \boldsymbol{H}_T)\, da = \int_A \boldsymbol{E}_T \cdot (\boldsymbol{H}_T \times \boldsymbol{a}_z)\, da \tag{9.48}$$

where A is the area of the cross section between conductors, as shown in Fig. 9.5, and the integration is carried out with z and t held constant.

On account of Eq. 9.7a, a scalar potential $\phi(x, y, z; t)$ can be defined, such that

$$\boldsymbol{E}_T = -\nabla_T \phi \tag{9.49}$$

in view of which

$$P(z, t) = \int_A (\boldsymbol{a}_z \times \boldsymbol{H}_T) \cdot \nabla_T \phi\, da \tag{9.50}$$

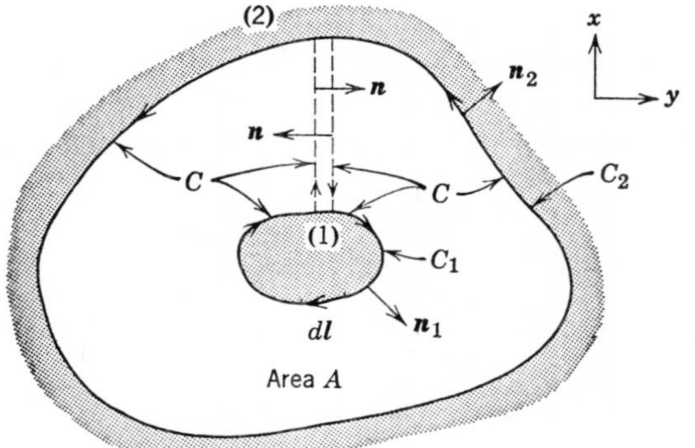

Fig. 9.5. Geometry for power calculation.

Now, by Gauss' theorem in two dimensions, with reference to the area A and the bent contour C bounding it (Fig. 9.5), we have the formal identity

$$\int_A \nabla_T \cdot [\phi(\boldsymbol{a}_z \times \boldsymbol{H}_T)]\, da = \oint_C \phi \boldsymbol{n} \cdot (\boldsymbol{a}_z \times \boldsymbol{H}_T)\, dl \tag{9.51}$$

where $\boldsymbol{n}$ is the *outward* normal to the contour C of area A. Since the dotted portions of C have opposed normals, the integrations over these lines cancel out. Therefore

$$\oint_C \phi(\boldsymbol{n} \times \boldsymbol{a}_z) \cdot \boldsymbol{H}_T\, dl = \oint_{C_2} \phi \boldsymbol{H}_T \cdot (\boldsymbol{n}_2 \times \boldsymbol{a}_z)\, dl - \oint_{C_1} \phi \boldsymbol{H}_T \cdot (\boldsymbol{n}_1 \times \boldsymbol{a}_z)\, dl \tag{9.52}$$

The minus sign arises because $\boldsymbol{n}_1$ is defined in the figure as the *inward* normal to A.

The conductor surfaces are equipotentials because of boundary condition (Eq. 9.17), and we can always choose one of them as the zero of potential. Let this one be the outer conductor (2) in our case. Then the inner conductor (1) will be at potential V, according to Eqs. 9.49 and 9.18. Since V is a function of only z and t,

$$\begin{aligned} \phi &= 0 \qquad \text{on } C_2 \\ \phi &= V \qquad \text{on } C_1 \end{aligned} \tag{9.53}$$

and we have, for Eq. 9.52,

$$\oint_C \phi(\boldsymbol{n} \times \boldsymbol{a}_z) \cdot \boldsymbol{H}_T\, dl = +V \oint_{C_1} \boldsymbol{H}_T \cdot d\boldsymbol{l} = VI \tag{9.54}$$

in which we have used Eq. 9.28 and written I for I_1 (as we did in Eq. 9.39, etc.).

In the light of Eq. 9.54, Eq. 9.51 becomes

$$\int_A \nabla_T \cdot [\phi(\boldsymbol{a}_z \times \boldsymbol{H}_T)]\, da = VI \tag{9.55}$$

Direct expansion of the integrand in Eq. 9.55, however, shows that

$$\int_A \nabla_T \cdot [\phi(\boldsymbol{a}_z \times \boldsymbol{H}_T)]\, da = \int_A (\boldsymbol{a}_z \times \boldsymbol{H}_T) \cdot \nabla_T \phi\, da + \int_A \phi \nabla_T \cdot (\boldsymbol{a}_z \times \boldsymbol{H}_T)\, da \tag{9.56}$$

But, by further vector expansion,

$$\nabla_T \cdot (\boldsymbol{a}_z \times \boldsymbol{H}_T) = \boldsymbol{H}_T \cdot (\nabla_T \times \boldsymbol{a}_z) - \boldsymbol{a}_z \cdot (\nabla_T \times \boldsymbol{H}_T) = 0 \quad (9.57)$$

because $\boldsymbol{a}_z$ is a constant vector and $\nabla_T \times \boldsymbol{H}_T = 0$, by Eq. 9.6*a*. It follows from Eqs. 9.55, 9.56, and 9.57 that

$$\int_A (\boldsymbol{a}_z \times \boldsymbol{H}_T) \cdot \nabla_T \phi \, da = VI \quad (9.58)$$

or, in view of Eq. 9.50,

$$P(z, t) = V(z, t) I(z, t) \quad (9.59)$$

We conclude that, for TEM waves guided by two perfect conductors in a medium with constant parameters, the field definitions of voltage V and current I employed in Eqs. 9.18 and 9.28 are consistent with the power flow, in the sense that VI gives correctly the same result as integrating the Poynting vector over the cross section of the line.

9.3 Some General Features of TEM Waves in the Sinusoidal Steady State*[1]

Although we could proceed further with our analysis in the time domain if the medium under discussion were lossless ($\sigma = 0$) and had constant parameters, we prefer to retain the possibility of having dissipative media ($\sigma \neq 0$). This preference alone makes it necessary to switch our discussion to the sinusoidal steady state, or frequency domain. An additional advantage of doing so is that we need no longer assume that σ, ϵ, and μ are constant; i.e., *σ, ϵ, and μ may now be considered as functions of frequency, although not of either position, field strength, or field direction.* The medium is accordingly homogeneous, linear, and isotropic; but the connections between $\boldsymbol{D}$ and $\boldsymbol{E}$ or $\boldsymbol{B}$ and $\boldsymbol{H}$ in the time domain may be linear integrodifferential equations instead of just constants [this is what makes $\sigma(\omega)$, $\epsilon(\omega)$, and $\mu(\omega)$ frequency-dependent].

To introduce the steady state, we set

$$\boldsymbol{E}_T = \text{Re}\,(\mathbf{E}_T e^{j\omega t})$$

and (9.60)

$$\boldsymbol{H}_T = \text{Re}\,(\mathbf{H}_T e^{j\omega t})$$

where $\mathbf{E}_T$ and $\mathbf{H}_T$ are complex vectors with only transverse space com-

[1] In a minimum treatment, the pertinent parts of Secs. 9.3 through 9.5 should not be omitted entirely, but can be covered very simply by restricting consideration to cases of constant ϵ, μ, and $\sigma = 0$. Complete freedom to switch between time and frequency domains, if and when desired, is then available.

ponents (they are functions of x, y, *and* z, but not of t). Dropping the real-part restriction until after solution, we find for Eqs. 9.6*b* and 9.7*b* respectively

$$\text{(a)} \qquad \frac{\partial}{\partial z}(\boldsymbol{a}_z \times \mathbf{H}_T) = (\sigma + j\omega\epsilon)\mathbf{E}_T$$

$$\text{(b)} \qquad \frac{\partial \mathbf{E}_T}{\partial z} = j\omega\mu(\boldsymbol{a}_z \times \mathbf{H}_T) \tag{9.61}$$

We can eliminate $\boldsymbol{a}_z \times \mathbf{H}_T$ by substituting its value from Eq. 9.61*b* into Eq. 9.61*a*, with the result

$$\frac{\partial^2 \mathbf{E}_T}{\partial z^2} = j\omega\mu(\sigma + j\omega\epsilon)\mathbf{E}_T \tag{9.62}$$

For any given values of x and y, Eq. 9.62 tells us how $\mathbf{E}_T$ depends upon z. Indeed, the solution is

$$\mathbf{E}_T = \hat{\mathbf{E}}_T{}^{+}e^{-\bar{\gamma}_0 z} + \hat{\mathbf{E}}_T{}^{-}e^{\bar{\gamma}_0 z} \tag{9.63}$$

with

$$\bar{\gamma}_0 = \sqrt{j\omega\mu(\sigma + j\omega\epsilon)} = \alpha_0 + j\beta_0 \tag{9.64}$$

as in Eq. 8.6. Observe carefully that $\hat{\mathbf{E}}_T{}^{+}$ and $\hat{\mathbf{E}}_T{}^{-}$ are complex vectors with only transverse components (x and y). They may be functions of the transverse coordinates x and y, but they do not depend upon z. *We have used the mark* (^) *over the complex-vector coefficients* $\hat{\mathbf{E}}_T{}^{+}$ *and* $\hat{\mathbf{E}}_T{}^{-}$ *to remind us that they are functions only of the transverse coordinates* (x, y).

We can now find $\mathbf{H}_T$ from Eq. 9.61*b*.

$$\mathbf{H}_T = -\frac{1}{j\omega\mu}\frac{\partial}{\partial z}(\boldsymbol{a}_z \times \mathbf{E}_T)$$

$$= \frac{\bar{\gamma}_0}{j\omega\mu}[(\boldsymbol{a}_z \times \hat{\mathbf{E}}_T{}^{+})e^{-\bar{\gamma}_0 z} - (\boldsymbol{a}_z \times \hat{\mathbf{E}}_T{}^{-})e^{\bar{\gamma}_0 z}] \tag{9.65}$$

This is of the form

$$\mathbf{H}_T = \hat{\mathbf{H}}_T{}^{+}e^{-\bar{\gamma}_0 z} + \hat{\mathbf{H}}_T{}^{-}e^{\bar{\gamma}_0 z} \tag{9.66}$$

which we could also have obtained by eliminating $\mathbf{E}_T$ from Eqs. 9.61 and finding the same differential equation (Eq. 9.62) for $\mathbf{H}_T$ as we did for $\mathbf{E}_T$.

If we now make use of the characteristic wave impedance of the medium between the conductors,

$$\bar{\eta} \equiv \frac{j\omega\mu}{\bar{\gamma}_0} = \sqrt{\frac{j\omega\mu}{\sigma + j\omega\epsilon}} \tag{9.67}$$

from Eq. 8.23, we find from Eqs. 9.65, 9.66, and 9.67 that

$$\text{(a)} \qquad \hat{\mathbf{H}}_T^+ = \frac{1}{\bar{\eta}} (\boldsymbol{a}_z \times \hat{\mathbf{E}}_T^+)$$
$$\text{(b)} \qquad \hat{\mathbf{H}}_T^- = -\frac{1}{\bar{\eta}} (\boldsymbol{a}_z \times \hat{\mathbf{E}}_T^-) \qquad (9.68)$$

Formally, as far as the behavior of the fields $\mathbf{E}_T$ and $\mathbf{H}_T$ with z is concerned, Eqs. 9.63, 9.66, and 9.68 look very much like the Eqs. 8.22 and 8.19 for uniform plane waves in a dissipative medium. The difference is that $\mathbf{E}_T$, for example, is a function of x and y through $\hat{\mathbf{E}}_T^+$ and $\hat{\mathbf{E}}_T^-$. Similarly, $\mathbf{H}_T$ is also a function of the transverse coordinates.

According to Eqs. 9.63 and 9.66, the fields are composed of attenuated waves moving in the $\pm z$ directions. Let us examine only the (+) wave:

$$\text{(a)} \qquad \mathbf{E}_T^+ = \hat{\mathbf{E}}_T^+ e^{-\alpha_0 z} e^{-j\beta_0 z}$$
$$\text{(b)} \qquad \mathbf{H}_T^+ = \frac{1}{\bar{\eta}} (\boldsymbol{a}_z \times \hat{\mathbf{E}}_T^+) e^{-\alpha_0 z} e^{-j\beta_0 z} \qquad (9.69)$$

similar to Eqs. 8.22. Now the actual (+) wave fields, as real functions of time and position, are:

$$\boldsymbol{E}_T^+ = \text{Re}\,(\mathbf{E}_T^+ e^{j\omega t}) = e^{-\alpha_0 z}\,\text{Re}\,(\hat{\mathbf{E}}_T^+ e^{j(\omega t - \beta_0 z)}) \qquad (9.70)$$

$$\boldsymbol{H}_T^+ = \text{Re}\,(\mathbf{H}_T^+ e^{j\omega t}) = e^{-\alpha_0 z}\,\text{Re}\left(\frac{\boldsymbol{a}_z \times \hat{\mathbf{E}}_T^+}{\bar{\eta}} e^{j(\omega t - \beta_0 z)}\right)$$
$$= \frac{e^{-\alpha_0 z}}{|\bar{\eta}|} \text{Re}\,[(\boldsymbol{a}_z \times \hat{\mathbf{E}}_T^+) e^{j(\omega t - \beta_0 z - \varphi)}] \qquad (9.71)$$

where

$$\bar{\eta} = |\bar{\eta}| e^{j\varphi} \qquad 0 \le \varphi \le \frac{\pi}{4} \qquad (9.72)$$

as in Eq. 8.26.

We wish to examine the relative space orientation of the real vectors $\boldsymbol{E}_T^+$ and $\boldsymbol{H}_T^+$, both of which always lie in the x, y- (or transverse) plane. To do this, in a manner similar to that of Fig. 8.3, we may express the complex vector $\hat{\mathbf{E}}_T^+$ in terms of its real and imaginary parts

$$\hat{\mathbf{E}}_T^+ = \hat{\boldsymbol{E}}_{Tr}^+ + j\hat{\boldsymbol{E}}_{Ti}^+ \qquad (9.73)$$

where it must be kept in mind that $\hat{\boldsymbol{E}}_{Tr}^+$ and $\hat{\boldsymbol{E}}_{Ti}^+$ are functions of (x, y) only. Accordingly, Eqs. 9.70 and 9.71 become

$$\boldsymbol{E}_T^+ = e^{-\alpha_0 z}[\hat{\boldsymbol{E}}_{Tr}^+ \cos(\omega t - \beta_0 z) - \hat{\boldsymbol{E}}_{Ti}^+ \sin(\omega t - \beta_0 z)] \qquad (9.74)$$

$$\boldsymbol{H}_T{}^+ = \frac{e^{-\alpha_0 z}}{|\bar{\eta}|} [(\boldsymbol{a}_z \times \hat{\boldsymbol{E}}_{Tr}{}^+) \cos(\omega t - \beta_0 z - \varphi) - (\boldsymbol{a}_z \times \hat{\boldsymbol{E}}_{Ti}{}^+) \sin(\omega t - \beta_0 z - \varphi)] \quad (9.75)$$

similar to Eqs. 8.27. In general, vectors $\hat{\boldsymbol{E}}_{Tr}{}^+$ and $\hat{\boldsymbol{E}}_{Ti}{}^+$ in the x, y-plane may have any magnitudes and directions we wish at *one* given point (x_0, y_0) in that plane—at least as far as Maxwell's equations are concerned. This follows from Eqs. 9.62 and 9.63, in which $\hat{\mathbf{E}}_T{}^+$ appears as an arbitrary "constant" in the solution. The *variations* of $\hat{\boldsymbol{E}}_{Tr}{}^+$ and $\hat{\boldsymbol{E}}_{Ti}{}^+$ with x and y are, of course, restricted by Eqs. 9.7a and 9.16, which are part of Maxwell's equations; but we shall see shortly that, unless the boundary conditions (Eq. 9.17) are taken into account, these equations do not force any particular relation between $\hat{\boldsymbol{E}}_{Tr}{}^+(x_0, y_0)$ and $\hat{\boldsymbol{E}}_{Ti}{}^+(x_0, y_0)$.

Aside from boundary conditions, then, the fields $\boldsymbol{E}_T{}^+$ and $\boldsymbol{H}_T{}^+$ might look like those presented in Fig. 9.6, which is a picture of Eqs. 9.74 and 9.75, drawn for $z = 0$, $x = x_0$, $y = y_0$, and for $\varphi = \pi/4$ and $\varphi = 0$. One complete period of time variation $(0 \leq t \leq 2\pi/\omega)$ is shown. The principal features of this figure are the same as those of Fig. 8.3, and the discussion of the latter applies to the former. In particular, *as long as the medium is dissipative* $(\sigma > 0, \varphi > 0)$, *and the polarization is other than linear, the space angle between* $\boldsymbol{E}_T{}^+$ *and* $\boldsymbol{H}_T{}^+$ *is not 90°; in fact, it varies with time.*

Only if the medium is lossless ($\sigma = 0$ and $\varphi = 0$ in Fig. 9.6 or Eq. 9.75), will $\boldsymbol{H}_T{}^+$ trace its ellipse in a manner exactly similar to $\boldsymbol{E}_T{}^+$ when the latter is not linearly polarized. *In the lossless case,* $\boldsymbol{E}_T{}^+$ *and* $\boldsymbol{H}_T{}^+$ *will not only have a constant space angle between them, but this angle will be 90°.* We found this to be true of uniform plane waves, and now reach the same conclusion for any TEM wave. To summarize this important special (lossless) situation, we rewrite Eqs. 9.74, 9.75, 9.67, and 9.64 as they will appear in a lossless medium:

$$\left.\begin{aligned}
&\text{(a)} \quad \boldsymbol{E}_T{}^+ = \hat{\boldsymbol{E}}_{Tr}{}^+ \cos(\omega t - \beta_0 z) - \hat{\boldsymbol{E}}_{Ti}{}^+ \sin(\omega t - \beta_0 z) \\
&\text{(b)} \quad \boldsymbol{H}_T{}^+ = \sqrt{\frac{\epsilon}{\mu}}\, \boldsymbol{a}_z \times \boldsymbol{E}_T{}^+ \\
&\text{(c)} \quad \eta = \sqrt{\frac{\mu}{\epsilon}} \\
&\text{(d)} \quad \bar{\gamma}_0 = j\beta_0 = j\omega\sqrt{\epsilon\mu}
\end{aligned}\right\} \begin{array}{l}\text{for a lossless medium} \\ (\sigma = 0)\end{array} \quad (9.76)$$

From Eqs. 9.63, 9.66, and 9.68, it is clear that, *in any case*, the properties of the $(-)$ wave alone can be obtained from those of the $(+)$

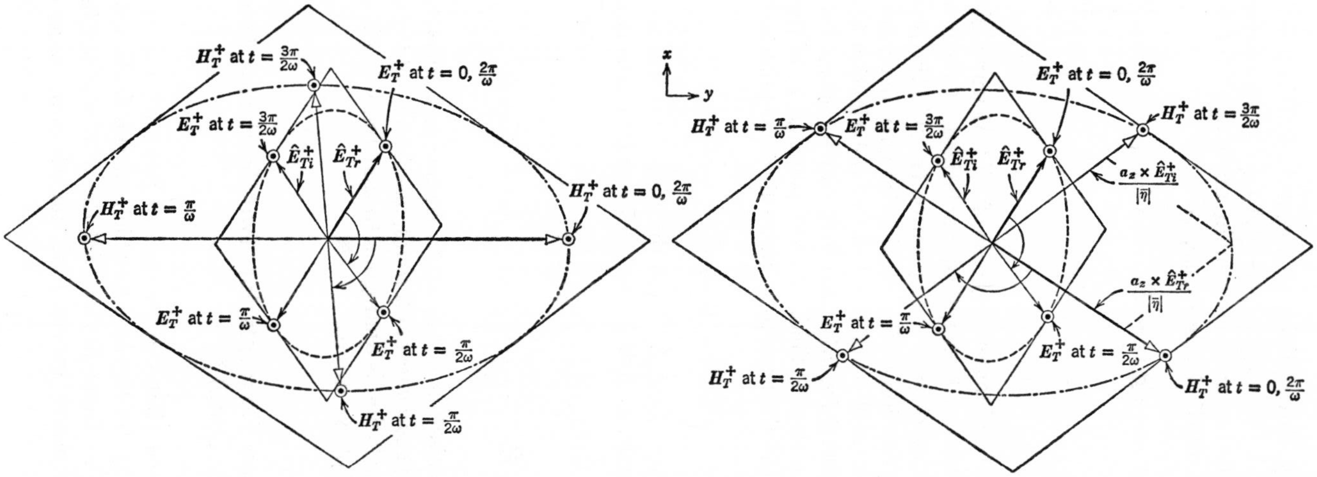

$\varphi = \frac{\pi}{4}$
(a)

$\varphi = 0$
(b)

Fig. 9.6. Loci of instantaneous electric and magnetic fields $\boldsymbol{E}_T{}^+$ and $\boldsymbol{H}_T{}^+$ for a TEM wave, at a point $x = x_0$, $y = y_0$, $z = 0$, with sinusoidal time dependence (ωt) and $0 \leq t \leq 2\pi/\omega$. Fields are shown in ($a$) for a highly dissipative medium ($\varphi = \pi/4$), and in (b) for a lossless medium ($\varphi = 0$). Compare with Fig. 8.3.

wave by making only three slight changes: 1. Putting $-z$ for z; 2. putting $-\boldsymbol{a}_z$ for $\boldsymbol{a}_z$; 3. changing superscript (+) to superscript (−). All of our other detailed comments remain unchanged for the (−) wave; so for it we find, in the lossless case,

$$\left.\begin{aligned} &\text{(a)} \quad \boldsymbol{E}_T^- = \hat{\boldsymbol{E}}_{Tr}^- \cos(\omega t + \beta_0 z) - \hat{\boldsymbol{E}}_{Ti}^- \sin(\omega t + \beta_0 z) \\ &\text{(b)} \quad \boldsymbol{H}_T^- = -\sqrt{\frac{\epsilon}{\mu}}\, \boldsymbol{a}_z \times \boldsymbol{E}_T^- \end{aligned}\right\} \begin{matrix}\text{for a}\\ \text{lossless}\\ \text{medium}\\ (\sigma \equiv 0)\end{matrix} \tag{9.77}$$

In general the electric and magnetic fields of a TEM wave, like those of a uniform plane wave, may be elliptically polarized. Of course, since $\hat{\boldsymbol{E}}_{Tr}^+$ and $\hat{\boldsymbol{E}}_{Ti}^+$ (for example) vary with position (x, y) in the transverse plane, the state of polarization of $\boldsymbol{E}_T^+$ could also vary over this plane. Whether it does or not, and what the state of polarization is at any point, depends upon Eqs. 9.7*a*, 9.16, and 9.17 which control the transverse dependence of the electric field. At any point (x, y) of the transverse plane at which the field happens to be linearly polarized, however, we must have

$$\hat{\mathbf{E}}_T^+ = \hat{\boldsymbol{E}}_T^+ e^{j\psi} \tag{9.78}$$

where $\hat{\boldsymbol{E}}_T^+$ is a *real vector*. Both $\hat{\boldsymbol{E}}_T^+$ and the angle ψ may depend upon x and y. If Eq. 9.78 holds, we find for Eqs. 9.74 and 9.75 the new forms

$$\left.\begin{aligned} &\text{(a)} \quad \boldsymbol{E}_T^+ = \hat{\boldsymbol{E}}_T^+ e^{-\alpha_0 z} \cos(\omega t - \beta_0 z + \psi) \\ &\text{(b)} \quad \boldsymbol{H}_T^+ = (\boldsymbol{a}_z \times \hat{\boldsymbol{E}}_T^+) \frac{e^{-\alpha_0 z}}{|\bar{\eta}|} \cos(\omega t - \beta_0 z + \psi - \varphi) \end{aligned}\right\} \begin{matrix}\text{linear}\\ \text{polari-}\\ \text{zation}\end{matrix} \tag{9.79}$$

Note that when the polarization is linear, the electric and magnetic fields of a TEM wave are perpendicular at all times, whether or not the medium contains loss. This conclusion is analogous to that following Eqs. 8.28 and 8.29 for uniform plane waves.

9.4 Transmission-Line Concepts for Two-Conductor Lines (Sinusoidal Steady State) *[1]

Imagine for a moment that the analysis of Secs. 9.1 and 9.2 were repeated, but specifically for the case in which all the fields had sinusoidal time dependence at frequency ω. All the results given in those sections would remain intact, except that the relevant values of the

[1] See footnote, Sec. 9.3.

medium parameters would simply have to be those valid only at the particular frequency ω involved in the time dependence.

In particular we would find for Eq. 9.27 a slightly modified result

$$C = \frac{\epsilon(\omega)}{K_1} \tag{9.80a}$$

where

$$K_1 = \frac{\displaystyle\int_{(1)}^{(2)} \boldsymbol{E}_T \cdot d\boldsymbol{s}}{\displaystyle\oint_{C_1} \boldsymbol{E}_T \cdot \boldsymbol{n}_1 \, dl} \tag{9.80b}$$

is again a purely geometrical factor. It can always be calculated from a consideration of the d-c electric field in the structure, because the dependence of $\boldsymbol{E}_T$ upon the transverse coordinates is just like the electrostatic field. Similarly, in place of Eq. 9.35 we would have

$$L = K_2 \mu(\omega) \tag{9.80c}$$

where

$$K_2 = \frac{\displaystyle\int_{(1)}^{(2)} \boldsymbol{H}_T \cdot (\boldsymbol{a}_z \times d\boldsymbol{s})}{\displaystyle\oint_{C_1} \boldsymbol{H}_T \cdot d\boldsymbol{l}} \tag{9.80d}$$

is also a purely geometrical factor, calculable from considerations of the d-c magnetic field in the structure. Finally,

$$G = \left[\frac{\sigma(\omega)}{\epsilon(\omega)}\right] C \tag{9.80e}$$

replaces Eq. 9.36.

It follows that even with perfectly conducting boundaries, the transmission-line parameters L, C, and G will be functions of frequency if the medium parameters μ, ϵ, and σ are functions of frequency. For a TEM wave, in fact, the transmission-line parameters are directly proportional to the corresponding parameters of the medium, with proportionality factors involving only the geometry of the cross section of the conductor configuration. This circumstance should be compared with the frequency dependence produced by conductor losses, through the action of skin effect, which takes place even if σ, ϵ, and μ are, in fact, independent of frequency in all the media comprising the system (see Sec. 8.4.5).

Continuing our discussion of the modification of previous results, it is clear that for sinusoidal time dependence Eqs. 9.39 and 9.47 could be written as the real parts of their frequency-domain counterparts. After canceling $e^{j\omega t}$, we have

$$\text{(a)} \qquad \frac{\partial \mathrm{V}}{\partial z} = -j\omega L\mathrm{I}$$

$$\text{(b)} \qquad \frac{\partial \mathrm{I}}{\partial z} = (G + j\omega C)\mathrm{V} \tag{9.81}$$

where L, C, and G are those given by Eqs. 9.80. Evidently, Eqs. 9.81 lead to the familiar complex voltage and current transmission-line waves. The propagation constant is of course

$$\bar{\gamma} = \sqrt{j\omega L(G + j\omega C)}$$

But the propagation constant of *any* TEM wave must be that given by Eq. 9.64

$$\bar{\gamma}_0 = \sqrt{j\omega\mu(\sigma + j\omega\epsilon)}$$

and if the transmission-line voltage and current waves are to be accounted for by a TEM wave, it must be true that

$$\bar{\gamma} = \bar{\gamma}_0 \tag{9.82a}$$

Writing out Eq. 9.82*a* and equating real and imaginary parts lead to the relations

$$LC = \epsilon\mu \tag{9.82b}$$

and

$$LG = \sigma\mu \tag{9.82c}$$

Equation 9.82*a* is again the result of Eq. 9.44, but without the restriction contained in the latter that ϵ and μ must be constant. Similarly, the quotient of Eqs. 9.82*b* and 9.82*c* repeats the conclusion of Eq. 9.80*e*.

It follows from the application of Eq. 9.82*b* to the product of Eqs. 9.80*a* and 9.80*c* that

$$K_1 = K_2 \equiv K \tag{9.83}$$

a conclusion which is hardly obvious from either the geometry, or the equivalent static-field calculations of the inductance and capacitance per unit length of an arbitrary two-conductor line configuration!

We can perhaps see now that the remarkable results of Eqs. 9.44, 9.82*b*, or 9.83 are connected with the fact that, for the type of transmission line we have just been discussing, the entire phenomenon of energy storage and propagation takes place in the medium *outside* the conductors. The flow of current on the conductors themselves actually has nothing to do with the transmission of energy, but serves only to *guide* it along the line. For this reason, the velocity of propagation (and the attenuation, if any) *must* be controlled, at least in part, by that of the waves in the medium surrounding the conductors. In the special case of TEM waves this is the velocity of light (and its attenuation, if any), which happens to be independent of the particular field distribution over the cross section. That is to say, the propagation constant of a TEM wave is independent of the geometry. Thus any two-conductor system which can support a true TEM wave must end up with transmission-line parameters that slave to the propagation requirements of such a wave.

Unlike the propagation constant, the characteristic impedance of the line does depend upon its geometry. From Eqs. 9.81 we know that the characteristic impedance must be given by

$$\mathrm{Z}_0 = \frac{j\omega L}{\bar{\gamma}}$$

But on account of Eqs. 9.82*a*, 9.80*c*, and 9.83

$$\mathrm{Z}_0 = \frac{j\omega L}{\bar{\gamma}} = \frac{j\omega L}{\bar{\gamma}_0} = \frac{j\omega K\mu}{\sqrt{j\omega\mu(\sigma + j\omega\epsilon)}} = K\sqrt{\frac{j\omega\mu}{\sigma + j\omega\epsilon}} = K\bar{\eta} \quad (9.84)$$

where $\bar{\eta}$ is the characteristic wave impedance of the medium at frequency ω. Clearly, however, the entire set of transmission-line parameters depends upon the geometry only to the extent of the single factor K. This factor may be found, for example, from a static-field solution for the capacitance per unit length, the inductance per unit length, or the conductance per unit length; or from a knowledge of the characteristic impedance of the line at some frequency. Obviously one chooses the easiest of these methods, depending upon the situation. In addition, of course, the values of the medium parameters σ, ϵ, and μ must be known at all frequencies of interest.

It is important to understand at this point that our results are exactly correct only if the transmission-line fields are TEM. Any condition which prevents this circumstance may upset the argument to a numerically large or small extent. For example, if the conductors are not perfect, we have seen in Chapter 8 that there must be a longitudinal electric field on their surfaces, and therefore also in the space outside them. E_z is not zero, and the entire field problem is not TEM. The voltage and current definitions we have used no longer have the uniqueness or power-flow properties described in our discussion, and any *exact* transmission-line formulation must be carried out on quite a different basis (if, indeed, it is worth doing at all). When the conductor loss is small, however, we have also seen that the *numerical* results will not be altered very much if we stick to the definitions of voltage and current employed here—but we are then getting along on an approximation (however accurate).

A more extreme example is that of so-called continuously loaded cable, in which our "inner conductor" is replaced by a thinly insulated wire wrapped helically around a cylindrical (possibly ferromagnetic) form. Even if the wire is a perfect conductor, the winding pitch produces an E_z and an H_z. The field is not TEM, and $LC \neq \epsilon\mu$. Nevertheless, it is possible to define a voltage and a current as line integrals in the transverse plane; these turn out to represent the behavior of the system accurately enough for practical purposes, even though they do not have exactly the uniqueness and power-flow properties discussed above, and which we usually assume for them when we interconnect such lines with other circuit elements. As one might expect, the equivalent transmission-line inductance L per unit length assigned on this basis is raised greatly above that associated with flux in the space between conductors (which is all that is accounted for by Eq. 9.80*c*), without the corresponding decrease in C required if Eq. 9.82*b* still applied.

Situations like those mentioned above, in which transmission-line equations may be applied with our relatively conventional interpretations for voltage and current but only as (very good) approximations, are, of course, the rule rather than the exception in practical two-wire lines.

Also exceedingly common are applications of transmission-line representations to describe wave-guide behavior. As indicated in Chapter 8, however, the fields in these structures are very far from TEM. The assignment of "voltage" and "current" is really only an analytical device; it is carried out almost arbitrarily. The very absence of terminals in the configuration itself precludes trying to interconnect wave guides with ordinary circuit elements on the basis of a blind faith in voltage or current, so the whole question of how to define them "best" just does not turn out to be important.

The point is, transmission-line thinking is important in a good many places, whether or not anything like the *lumped-circuit* notions of voltage and current remain valid. Perhaps the best reason for dwelling on the TEM waves, in the transverse plane of which these notions do apply exactly, is to provide an idealized model which we can understand so well that departures from it will be easy to recognize and evaluate.

The feature which thus distinguishes the ideal TEM transmission-line case from all the others is that it is the *only* one for which the following *three* conditions are all *simultaneously* and *exactly* valid:

1. Voltage $V(z, t)$ is defined *uniquely* by a transverse-plane line integral of the electric field along *any* path joining the conductors.
2. Current $I(z, t)$ is defined *uniquely* by a transverse-plane line integral of the magnetic field around *any* contour surrounding one of the conductors, and is, in fact, the current carried by that conductor.
3. The power carried across a plane z = constant, as given by the product VI, checks with the integral of the z component of the Poynting vector over the cross section of the line.

9.5 More About the TEM Field in a Two-Conductor Line *[1]

It is clear from the analysis so far that, if the field in our two-wire line is of TEM character, the regular transmission-line equations and power considerations will apply in terms of a reasonably familiar voltage and current. We have not yet looked directly at the behavior of the field solution itself in the transverse plane. Let us do so first in the frequency domain, and draw time-domain conclusions afterwards, if possible.

We suppose the line is operating with complex voltage $\mathrm{V}(z)$ and current $\mathrm{I}(z)$; the voltage being reckoned (+) on the inner conductor, as in Fig. 9.3, and the current being that flowing in the $+z$ direction on that conductor, as in Fig. 9.4. The fields will be TEM, which in the most general case conform to $\mathbf{E}_T$ and $\mathbf{H}_T$ of Eqs. 9.63 through 9.69. From transmission-line theory we know that $\mathrm{V}(z) = \mathrm{V}_+ e^{-\bar{\gamma}_0 z} + \mathrm{V}_- e^{\bar{\gamma}_0 z}$; whereas from Eq. 9.63 for the TEM field, $\mathbf{E}_T = \hat{\mathbf{E}}_T{}^+ e^{-\bar{\gamma}_0 z} + \hat{\mathbf{E}}_T{}^- e^{\bar{\gamma}_0 z}$. The same value of $\bar{\gamma}_0$ appears in both expressions, as indicated by Eq. 8.82*a*. If, then, the relation (Eq. 9.18) between voltage and electric field is to hold for *all* values of z and t, we require separately

$$\text{(a)} \qquad \mathrm{V}_+ = \int_{(1)}^{(2)} \hat{\mathbf{E}}_T{}^+ \cdot d\mathbf{s} \qquad (9.85)$$

$$\text{(b)} \qquad \mathrm{V}_- = \int_{(1)}^{(2)} \hat{\mathbf{E}}_T{}^- \cdot d\mathbf{s}$$

[1] See footnote. Sec. 9.3.

Now, in the frequency domain, Eqs. 9.7*a* and 9.16 become respectively

$$\left.\begin{aligned} &\text{(a)} \quad \nabla_T \times (\hat{\mathbf{E}}_T{}^+ e^{-\bar{\gamma}_0 z} + \hat{\mathbf{E}}_T{}^- e^{\bar{\gamma}_0 z}) = 0 \\ &\text{(b)} \quad \nabla_T \cdot (\hat{\mathbf{E}}_T{}^+ e^{-\bar{\gamma}_0 z} + \hat{\mathbf{E}}_T{}^- e^{\bar{\gamma}_0 z}) = 0 \end{aligned}\right\} \quad \text{for } all\ z \tag{9.86}$$

yielding separately

$$\begin{aligned} &\text{(a)} \quad \nabla_T \times \hat{\mathbf{E}}_T{}^+ = 0 \\ &\qquad\ \nabla_T \cdot \hat{\mathbf{E}}_T{}^+ = 0 \\ &\text{(b)} \quad \nabla_T \times \hat{\mathbf{E}}_T{}^- = 0 \\ &\qquad\ \nabla_T \cdot \hat{\mathbf{E}}_T{}^- = 0 \end{aligned} \tag{9.87}$$

The boundary conditions (Eq. 9.17) may be split similarly:

$$\begin{aligned} &\text{(a)} \quad \boldsymbol{n}_1 \times \hat{\mathbf{E}}_T{}^+ = 0 \quad \text{on } C_1 \\ &\qquad\ \boldsymbol{n}_2 \times \hat{\mathbf{E}}_T{}^+ = 0 \quad \text{on } C_2 \\ &\text{(b)} \quad \boldsymbol{n}_1 \times \hat{\mathbf{E}}_T{}^- = 0 \quad \text{on } C_1 \\ &\qquad\ \boldsymbol{n}_2 \times \hat{\mathbf{E}}_T{}^- = 0 \quad \text{on } C_2 \end{aligned} \tag{9.88}$$

because they must hold for all z.

It is sufficient, therefore, to consider only the (+) wave, since Eqs. 9.85, 9.87, and 9.88 are identical for both waves. This is reasonable, because turning the line end for end should not affect the general solution.

For the (+) wave in the frequency domain, we can, by Eq. 9.87*a*, define a *complex* scalar potential $\Phi^+(x, y)$ such that

$$\hat{\mathbf{E}}_T{}^+ = -\nabla_T \Phi^+ \tag{9.89}$$

which is also a consequence of Eq. 9.49. Then

$$\nabla_T \cdot \hat{\mathbf{E}}_T{}^+ = \nabla_T{}^2 \Phi^+ = 0 \tag{9.90}$$

and from Eqs. 9.85*a* and 9.88*a*, the boundary conditions on Φ^+ may be taken to be

$$\begin{aligned} &\text{(a)} \quad \Phi^+ = 0 \quad \text{on } C_2 \\ &\text{(b)} \quad \Phi^+ = \mathrm{V}_+ \quad \text{on } C_1 \end{aligned} \tag{9.91}$$

This also checks with Eq. 9.53.

Equations 9.90 and 9.91 are sufficient to determine $\Phi^+(x, y)$, given V_+. In general, since V_+ is complex, we expect Φ^+ to be complex.

Suppose, therefore, we break up V_+ and Φ^+ into their real and imaginary parts:

$$\begin{aligned} &\text{(a)} \qquad \mathrm{V}_+ = V_{+r} + jV_{+i} \\ &\text{(b)} \qquad \Phi^+(x, y) = \phi_r^{\ +}(x, y) + j\phi_i^{\ +}(x, y) \end{aligned} \tag{9.92}$$

Equations 9.90 and 9.91 can then also be separated into real and imaginary parts, yielding two sets of results:

$$\begin{aligned} &\text{(a)} \qquad \nabla_T^{\,2}\phi_r^{\ +} = 0 \\ &\qquad\qquad \phi_r^{\ +} = V_{+r} \qquad \text{on } C_1 \\ &\qquad\qquad \phi_r^{\ +} = 0 \qquad \text{on } C_2 \\ &\text{(b)} \qquad \nabla_T^{\,2}\phi_i^{\ +} = 0 \\ &\qquad\qquad \phi_i^{\ +} = V_{+i} \qquad \text{on } C_1 \\ &\qquad\qquad \phi_i^{\ +} = 0 \qquad \text{on } C_2 \end{aligned} \tag{9.93}$$

Observe that Eq. 9.93*a* really defines a completely electrostatic problem. It is the problem of finding the (real) scalar potential $\phi_r^{\ +}$ as a function of position in two dimensions (x, y), produced when one conductor is held at ground and the other at a direct voltage V_{+r}.

Now, *for a given value of* V_{+r}, there is only *one* function $\phi_r^{\ +}(x, y)$ which satisfies Eqs. 9.93*a*. This is a consequence of the uniqueness theorem for solutions to Laplace's equation. If we specify the potentials of all conductors of a system, and the condition at infinity if necessary, the potential at all points in space is determined completely. This fact is very reasonable on physical grounds—there should not be more than one answer. Still it is worth proving analytically, because we shall need to rely quite heavily on this conclusion.

Assume that there exist *two* solutions $\phi_1(x, y)$ and $\phi_2(x, y)$ which satisfy *all* of Eqs. 9.93*a*. Then their difference $\phi(x, y) \equiv \phi_1(x, y) - \phi_2(x, y)$ satisfies the equations

$$\begin{aligned} &\text{(a)} \qquad \nabla_T^{\,2}\phi = 0 \\ &\text{(b)} \qquad \phi = 0 \qquad \text{on } C_1 \textit{ and } C_2 \end{aligned} \tag{9.94}$$

With reference to Fig. 9.5, we can again apply Gauss' theorem in two dimensions, in the manner of Eq. 9.51, but this time to the function $(\phi\nabla_T\phi)$

$$\int_A \nabla_T \cdot (\phi\nabla_T\phi)\, da = \oint_C \phi(\boldsymbol{n}\cdot\nabla_T\phi)\, dl = 0 \tag{9.95}$$

The second integral is set equal to zero, because ϕ is zero on C_1 and C_2 (Eq. 9.94*b*), and the integrations over the dotted lines cancel out on account of the oppositely directed normals. Expansion of the left side of Eq. 9.95 then yields

$$\int_A \phi \nabla_T{}^2 \phi \, da + \int_A (\nabla_T \phi) \cdot (\nabla_T \phi) \, da = 0$$

or, in view of Eq. 9.94*a*,

$$\int_A |\nabla_T \phi|^2 \, da = 0 \tag{9.96}$$

The condition of Eq. 9.96 cannot be met unless $\nabla_T \phi \equiv 0$ over A, because the integrand is never negative. Thus $\phi =$ constant looks like a possible solution, but, on account of the boundary conditions (Eq. 9.94*b*), $\phi = 0$ is the only allowable answer. The "two" supposed solutions ϕ_1 and ϕ_2 are one and the same solution!

Collaterally, it is interesting to note that we have proved that the only electrostatic potential that can exist in a source-free region, surrounded *completely* by conductors which are all at zero potential, is the solution $\phi = 0$. This is the purely analytical basis for our earlier statement that, on physical grounds, no TEM solution can exist inside a hollow, perfectly conducting pipe.

Our present need of the uniqueness theorem proved above, however, arises in connection with Eqs. 9.93*b*. If we had found a solution for $\phi_r{}^+(x, y)$ which met the conditions of Eq. 9.93*a*, then one obvious solution for $\phi_i{}^+(x, y)$ in Eqs. 9.93*b* would be

$$\phi_i{}^+(x, y) = \left(\frac{V_{+i}}{V_{+r}}\right) \phi_r{}^+(x, y) = \text{constant} \times \phi_r{}^+(x, y) \tag{9.97}$$

But, because of the uniqueness theorem, Eq. 9.97 is the *only* solution to Eqs. 9.93*b* for a given value of V_{+i}. Thus the functions $\phi_i{}^+(x, y)$ and $\phi_r{}^+(x, y)$ *are not really different.* They differ only by a constant (real) multiplier, which means that they are linearly dependent solutions. We can therefore always write the complex function $\Phi^+(x, y)$ in the form

$$\Phi^+(x, y) = \phi^+(x, y) e^{j\psi} \tag{9.98}$$

where $\phi^+(x, y)$ is a *real* function, and the angle ψ is *not* a function of x and y. Indeed, if $\mathrm{V}_+ = |\mathrm{V}_+| e^{j\psi}$, then $\phi^+(x, y)$ satisfies the equations

$$\begin{aligned} &\text{(a)} \quad && \nabla_T{}^2 \phi^+ = 0 && \\ &\text{(b)} && \phi^+ = |\mathrm{V}_+| && \text{on } C_1 \\ &\text{(c)} && \phi^+ = 0 && \text{on } C_2 \end{aligned} \tag{9.99}$$

and $\phi^+(x, y)$ is again simply the *electrostatic* potential between the conductors due to a d-c potential difference $|V_+|$ applied to them.

The electric field in Eq. 9.89 now becomes

$$\hat{\mathbf{E}}_T{}^+ = (-\nabla_T\phi^+)e^{j\psi} = \hat{\boldsymbol{E}}_T{}^+e^{j\psi} \tag{9.100}$$

in which $\hat{\boldsymbol{E}}_T{}^+$ is exactly the same as the *static* electric field between the conductors for a voltage $|V_+|$ between them. The instantaneous (+) wave electric field $\boldsymbol{E}_T{}^+$ in the sinusoidal steady state is *linearly polarized* (Eqs. 9.100 and 9.78). Moreover, at any instant, its distribution in a given transverse plane looks exactly like the *electrostatic* field. As time goes on, however, the field strength (Eq. 9.79*a*) everywhere in the plane waxes, wanes, and reverses periodically because it is actually an alternating field. Comparing this time variation at different cross sections, Eq. 9.79*a* also shows that its phase lags along $+z$, and its amplitude attenuates in the same direction if there is loss in the medium ($\sigma \neq 0$).

The instantaneous magnetic field $\boldsymbol{H}_T$ of the (+) wave is everywhere at right angles to the electric field in space (Eq. 9.79), but differs in strength and time phase according to the wave impedance $\bar{\eta} = |\bar{\eta}|e^{j\varphi}$. The relative directions of $\boldsymbol{E}_T{}^+$ and $\boldsymbol{H}_T{}^+$ are such as to yield power flow $\boldsymbol{S}^+ = \boldsymbol{E}_T{}^+ \times \boldsymbol{H}_T{}^+$ with average value in the $+z$ direction.

In view of Eqs. 9.6*a*, 9.9, and 9.28, along with the fact that $\boldsymbol{H}_T{}^+$ is linearly polarized according to Eq. 9.79, we can see that the magnetic field also has exactly the same dependence upon x and y as would the magnetostatic field produced by equal and opposite direct currents in the two conductors.

Entirely similar comments could be made about the (−) wave, which has precisely the same configurations of electric and magnetic fields over the cross section as does the (+) wave. The only difference is that the relative directions of $\boldsymbol{E}_T{}^-$ and $\boldsymbol{H}_T{}^-$ must be such as to make $\boldsymbol{S}^- = \boldsymbol{E}_T{}^- \times \boldsymbol{H}_T{}^-$ a vector for which the average value points in the $-z$ direction. This means that, if we compared a (+) wave and a (−) wave for which the $\boldsymbol{E}_T$ happened to be the same at some (x, y, z, t), the magnetic fields would simply be oppositely directed in the two waves $(\boldsymbol{H}_T{}^- = -\boldsymbol{H}_T{}^+)$ at the same (x, y, z, t).

It follows that in a general standing wave on the two-conductor line, i.e., an arbitrary combination of (+) and (−) waves, the electric and magnetic fields at any given values of z_0 and t_0 are still distributed over the cross section just as they would be if a direct current and a direct voltage were impressed, equal respectively to the existing $I(z_0, t_0)$ and $V(z_0, t_0)$ on the line. The behavior of the fields $\boldsymbol{E}_T$ and $\boldsymbol{H}_T$ with z and t, however, is just that of $V(z, t)$ and $I(z, t)$ respectively.

In short, the electric and magnetic fields in any cross section are proportional, respectively, to the instantaneous voltage and current at that cross section, and they vary over the cross section exactly as if the same voltage and current existed statically.

The conclusions in the last paragraph have been drawn from the sinusoidal steady-state behavior of the fields, according to Eqs. 9.99, 9.100, and 9.79. In fact, however, they are true for *any* time dependence, not just for sinusoidal time variation. We can appreciate this fact by considering a Fourier expansion of an arbitrary wave form in time. Treating each frequency component separately, we observe that Eqs. 9.99 and 9.100 do not depend upon the frequency but only upon the amplitude and phase of such a component. Thus the pattern of the field over a given cross section is the *same* static pattern for each frequency component, and, accordingly, is the same for the sum of them all. The various amplitudes and phases of the components show up only in the z and t dependence, and this is contained completely in the resulting $I(z, t)$ and $V(z, t)$ functions.

If the medium is dissipative ($\sigma \neq 0$), the attenuation and phase constants α_0 and β_0 in Eq. 9.79 are not simple functions of frequency (see Chapter 8). Each frequency component is attenuated and delayed differently by the line, even though its transverse-field pattern is always the same. The wave form in time or longitudinal distance distorts progressively.

If the medium is lossless ($\sigma = 0$), $\alpha_0 = 0$ and $\beta_0 = \omega\sqrt{\epsilon\mu}$, and the relative phase and amplitude of the various frequency components are not altered *by the line*. The line is not dispersive. Wave shapes in time or longitudinal distance do not change as they move on the line.

These comments, however, relate to various possibilities for $I(z, t)$ and $V(z, t)$ only and not to the (x, y) dependence. The latter is always the static pattern. This implies, for example, that *in a two-wire line the field vectors of a TEM wave cannot rotate, no matter what the time dependence may be.*

9.6 Some Examples

9.6.1 Coaxial Cable

9.6.1.1 TEM FIELD. To illustrate the general remarks of the preceding sections, let us first determine the field in the (+) wave and the transmission-line constants that characterize a coaxial line with perfectly conducting boundaries. The structure appears in Fig. 9.7*a*, and the medium between the (perfect) inner and outer conductors has constant parameters ϵ, μ, and σ. The shape of the conducting boundaries indicates the use of polar coordinates.

Ordinarily, as a first step, we would solve Laplace's equation (here in polar coordinates) to find the potential $\phi(r, \vartheta)$ caused by applying $+V$ volts (d-c) to the inner conductor, with the outer conductor at

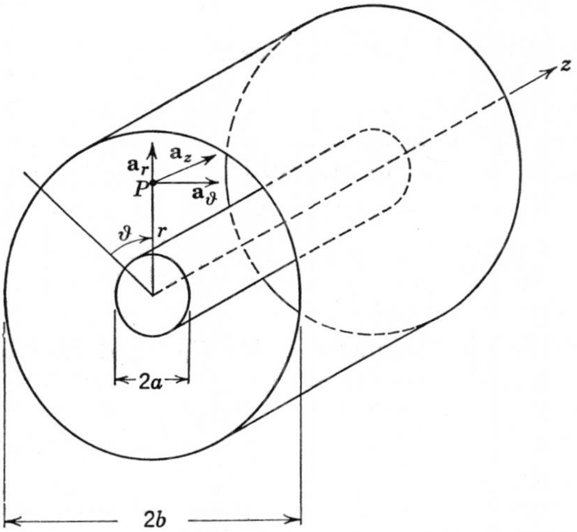

Fig. 9.7a. Coordinates and geometry for a coaxial line.

zero volts. From the resulting potential function, we could then find the static field.

In the present problem, however, the circular symmetry allows us to find the field more simply, directly from Gauss' theorem. Imagine a static charge $+q$ per unit length on the outer surface of the inner conductor (and $-q$ per unit length on the inner surface of the outer conductor). The electric field is independent of ϑ, by reason of symmetry.

$$\hat{\boldsymbol{E}}_T = \boldsymbol{a}_r E_r(r, \vartheta) = \boldsymbol{a}_r E_r(r)$$

By Gauss' theorem

$$2\pi r \epsilon E_r = q$$

or

$$E_r = \frac{q}{2\pi \epsilon r} \tag{9.101}$$

The voltage V on the inner conductor (with the outer one at zero volts) is then

$$V = \int_a^b E_r\, dr = \frac{q}{2\pi\epsilon} \ln \left(\frac{b}{a}\right) \tag{9.102}$$

yielding, from Eq. 9.101,

$$E_r = \frac{V}{r \ln (b/a)} \tag{9.103}$$

Thus, for a (+) wave in the frequency domain,

$$\mathbf{E}_T^{+} = \boldsymbol{a}_r \left[\frac{\mathrm{V}_+ e^{-\bar{\gamma}_0 z}}{r \ln (b/a)}\right] \tag{9.104a}$$

$$\mathbf{H}_T^{+} = \frac{1}{\bar{\eta}} \boldsymbol{a}_z \times \mathbf{E}_T^{+} = \boldsymbol{a}_\vartheta \left[\frac{\mathrm{V}_+ e^{-\bar{\gamma}_0 z}}{\bar{\eta} r \ln (b/a)}\right] \tag{9.104b}$$

The corresponding instantaneous fields $\boldsymbol{E}_T^{+}$ and $\boldsymbol{H}_T^{+}$ are shown in Fig. 9.7*b* for the lossless case ($\sigma = 0$).

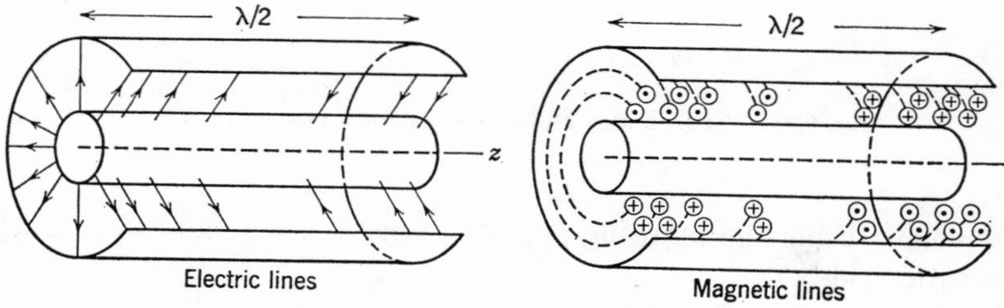

Fig. 9.7b. Instantaneous (+) wave electric and magnetic field lines in a lossless coaxial cable.

Observe that by definition, on the inner conductor,

$$\mathrm{I}_+ \equiv \int_0^{2\pi} \hat{\mathbf{H}}_T^{+}(a) \cdot \boldsymbol{a}_\vartheta a\, d\vartheta = \frac{2\pi \mathrm{V}_+}{\bar{\eta} \ln (b/a)} \tag{9.105}$$

which implies

$$\mathrm{Z}_0 = \frac{\mathrm{V}_+}{\mathrm{I}_+} = \bar{\eta} \left[\frac{\ln (b/a)}{2\pi}\right] \tag{9.106}$$

That Eq. 9.106 is correct also follows from a transmission-line point of view. According to Eq. 9.102,

$$C = \frac{q}{V} = \frac{2\pi\epsilon}{\ln (b/a)} \tag{9.107a}$$

so that

$$L = \frac{\epsilon\mu}{C} = \frac{\mu \ln (b/a)}{2\pi} \tag{9.107b}$$

and

$$G = \frac{\sigma}{\epsilon} C = \frac{2\pi\sigma}{\ln (b/a)} \tag{9.107c}$$

Therefore

$$Z_0 = \sqrt{\frac{j\omega L}{G + j\omega C}} = \frac{\ln (b/a)}{2\pi} \sqrt{\frac{j\omega\mu}{\sigma + j\omega\epsilon}} = \bar{\eta} \left[\frac{\ln (b/a)}{2\pi}\right] \tag{9.107d}$$

As a final check, let us express $\mathbf{H}_T{}^+$ in terms of I_+ by substituting V_+ from Eq. 9.106 into Eq. 9.104*b*. We find

$$\mathbf{H}_T{}^+ = \boldsymbol{a}_\vartheta \frac{I_+ e^{-\bar{\gamma}_0 z}}{2\pi r} \tag{9.108}$$

which shows that the dependence of $\mathbf{H}_T{}^+$ upon the transverse variables (r, ϑ) is exactly what it would be if it were to arise from a current I_+ in the inner conductor. Incidentally, the inductance per unit length on the basis of Eq. 9.108 is

$$L = \frac{\Lambda}{I_+}\bigg|_{z=0} = \frac{\mu}{I_+} \int_a^b \mathbf{H}_T \cdot \boldsymbol{a}_\vartheta \, dr = \frac{\mu}{2\pi} \ln \left(\frac{b}{a}\right)$$

in agreement with Eq. 9.107*b*.

The (−) wave field is exactly like our results for the (+) wave, except for three minor modifications:

1. All (+) sub- or superscripts become (−).
2. $\bar{\gamma}_0$ is replaced by $-\bar{\gamma}_0$.
3. The signs of I and $\mathbf{H}_T$ are each reversed.

9.6.1.2 Approximate Effect of Imperfectly Conducting Walls. If the conductors comprising the coaxial line of Fig. 9.7*a* have finite conductivity, the electric field must have a longitudinal component accompanying the longitudinal current flow. To make an exact calculation of the fields in the dielectric and the metals, a solution of the type used for the parallel-plate lines of Figs. 8.16 and 8.22 would have to be carried out. This time, however, it would have to be done in the circular geometry of Fig. 9.7*a*, which makes the problem more involved.

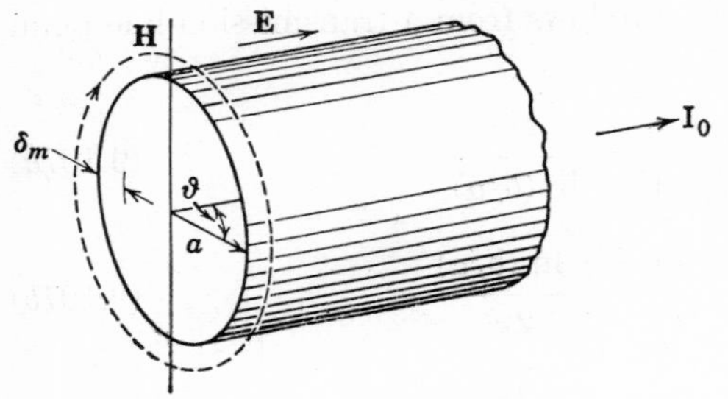

Fig. 9.7c. Skin effect in a round wire.

The philosophy outlined in Secs. 8.2.2 (Eqs. 8.73 and 8.74) and 8.4.2.2 may be employed to simplify the present situation, provided certain assumptions are permissible. Figure 9.7*c* shows most of the important points as they apply, for instance, to the inner conductor. First, if the conductivity of the metals is high enough, the magnetic field on their surfaces will not be much different from its value for perfect conductors (for the same total conductor current I_0). Next, the field penetrating the conductor will be very much like that in a parallel-plate line (Fig. 8.16 or 8.22), provided the skin depth δ_m in the metal is small compared to the radius a of the wire ($\delta_m \ll a$). This condition means that the penetrating fields die out before the change in curvature of the geometry with radial distance becomes effective enough to modify the solution. Neglecting curvature underestimates losses. An exact calculation shows that $a > 5\delta_m$ is required to hold the errors within 5%, and that errors become extremely serious when $a \leq \sqrt{2}\delta_m$.

The two approximations discussed above permit us to apply the result of Sec. 8.4.2.2 that

$$\langle P_{\text{diss}}\rangle_{\text{sq m}} \approx \frac{1}{2}|\mathrm{H}_\vartheta|^2\eta_m' = \frac{1}{2}|\mathrm{H}_\vartheta|^2\frac{1}{\sigma_m\delta_m}$$

where H_ϑ is the tangential magnetic field present *when the conductor is perfect*, η_m' is the real part of the characteristic wave impedance of the metal, and σ_m is the conductivity of the metal. By reason of the circular symmetry in the present problem

$$\mathrm{H}_\vartheta = \frac{\mathrm{I}_0}{2\pi a}$$

so

$$\langle P_{\text{diss}}\rangle_{\text{meter}} \approx \frac{1}{2}\frac{|\mathrm{I}_0|^2}{4\pi a^2}\eta_m'(2\pi a)$$

$$= \frac{1}{2}\frac{|\mathrm{I}_0|^2}{2\pi a\sigma_m\delta_m}$$

The effective series resistance per unit length of this conductor is therefore

$$R_0 \approx \frac{1}{2\pi a \sigma_m \delta_m} \qquad \text{ohms/m (of wire)} \qquad (9.109a)$$

Again it is equal to the d-c resistance per unit length which would arise if the current I_0 were uniformly distributed over the depth δ_m (provided $\delta_m \ll a$).

In the coaxial line, we must account for losses in both conductors. Certainly $\delta_m \ll a$ guarantees $\delta_m \ll b$, so applying the reasoning of Eq. 9.109*a* to each conductor, we find

$$R \approx \frac{1}{2\pi \sigma_m \delta_m}\left(\frac{1}{a} + \frac{1}{b}\right) \qquad \text{ohms/m (of line)} \qquad (9.109b)$$

To a first approximation, the presence of series resistance (Eq. 9.109*b*) does not alter the other line parameters (Eq. 9.107). The reasons for this fact when the dielectric is lossless ($\sigma = 0$) are contained in the example of Sec. 8.4.5. They may be carried over to the present case as long as the conductivity of the metal is much greater than that of the dielectric ($\sigma_m \gg \sigma$). Needless to say, this restriction is no problem in practice!

There is, however, the possibility that the internal inductance may become important enough to consider. If so, it may easily be calculated from the resistance within the approximations we have been making. Using the results from Sec. 8.4.5, we have

$$\omega L_i \approx R \qquad (9.109c)$$

Sometimes Eq. 9.109*a* is accepted as giving the "a-c resistance" per unit length of *any* piece of wire with a well-developed skin effect ($\delta_m \ll a$). The underlying assumption of axial symmetry actually involved in that equation is forgotten. Often, as a matter of fact, fields from other nearby current-carrying conductors do upset this symmetry, thereby invalidating Eq. 9.109*a*. The current in the conductor of interest not only crowds toward its surface but also crowds in the ϑ-coordinate toward the region (or regions) nearest the neighboring wires. This *proximity effect* raises the effective resistance above the value given by Eq. 9.109*a*, for the same reason that crowding of current toward the surface raises the resistance above the d-c value. The restriction of a given total longitudinal current to a smaller cross-sectional area increases proportionately the current density and reduces proportionately the volume in which dissipation takes place; but it increases by the *square* of the area ratio the volume density of dissipation (which depends upon the *square* of the current density), thereby increasing total losses. Proximity effect is often important in the turns of high-frequency inductors, and it may be important in some open-wire transmission lines. Indeed the open-wire line furnishes such an excellent analyzable illustration, not only of proximity effect but also of several detailed matters relating to the calculation of the line parameters, that we shall now take it up.

9.6.2 Open-Wire Line *

The transmission line of interest here is the two-wire (Lecher) line of Fig. 9.2. Let the spacing between perfectly conducting wires be $2a$ on centers, and let their radii be $b \leq a$, as shown in detail on Fig. 9.8*a*. We shall calculate the capacitance per unit length between the wires. This is a two-dimensional electrostatic problem.

If a charge of $+1$ coulomb/m is placed on one conductor, and -1 coulomb/m on the other, the potential difference between them will by definition be the reciprocal of their capacitance per unit length C. The periphery of each conductor must be an equipotential, and it is obvious from symmetry that the mid-plane between them is also an equipotential. Take the zero of potential ϕ to be on this mid-plane, so that one conductor [marked $(+)$] is at potential $+(1/2C)$ and the other [marked $(-)$] at $-(1/2C)$. These potentials are what we must find.

Now it happens that the equipotential surfaces produced by two equal and opposite parallel line charges, spaced by distance $2d$, are circular cylinders whose cross sections form a set of "bipolar circles." These surfaces are defined by the relation (Fig. 9.8*b*)

$$\phi = \frac{1}{2\pi\epsilon}[-\ln r_1 + \ln r_2] = \frac{1}{2\pi\epsilon}\ln\frac{r_2}{r_1} = \text{constant}$$

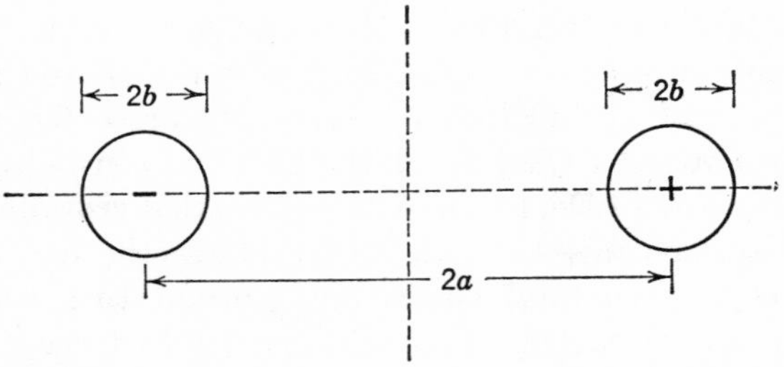

Fig. 9.8a. Geometry of two-wire line.

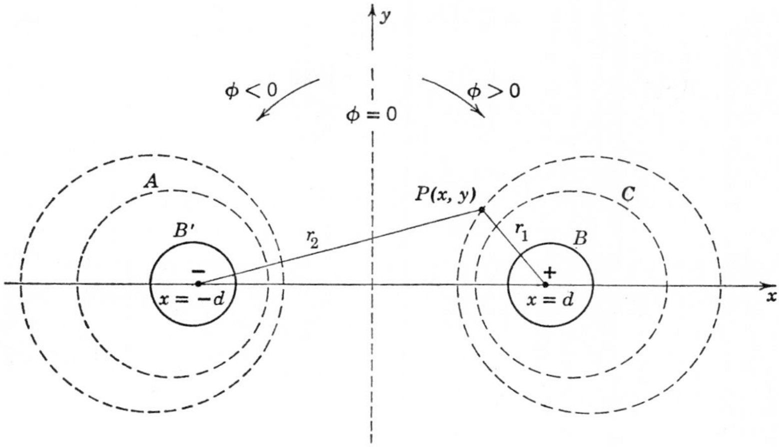

Fig. 9.8b. Equipotentials of (+) and (−) line charges at $x = \pm d$.

provided each line charge has a strength of 1 coulomb/m. It follows that the equipotentials have the equation

$$\left(\frac{r_2}{r_1}\right)^2 = e^{4\pi\epsilon\phi} \equiv K^2 \qquad K \gtrless 1 \text{ as } \phi \gtrless 0 \tag{9.110a}$$

or by the geometry of Fig. 9.8*b*

$$\frac{(x + d)^2 + y^2}{(x - d)^2 + y^2} = K^2$$

Multiplying out the last relation, and completing the square to obtain a standard algebraic form, yields

$$\left[x - d\left(\frac{K^2 + 1}{K^2 - 1}\right)\right]^2 + y^2 = \left(\frac{2dK}{K^2 - 1}\right)^2$$

describing a circle with center at

$$x_c = d\left(\frac{K^2 + 1}{K^2 - 1}\right) \qquad y = 0 \tag{9.110b}$$

and radius

$$\rho = \frac{2dK}{|K^2 - 1|} \tag{9.110c}$$

Observe from Eq. 9.110*b* that $|x_c| > d$ and that x_c has equal and opposite values for K and K^{-1}, which corresponds to changing from potential ϕ to $-\phi$ in Eq. 9.110*a*. The circles in Fig. 9.8*b* are some of the equipotentials.

Evidently, the potential field of two line charges can represent the potential field outside the wires of our transmission line. We must only arrange that the spacing and radii of the wires place their surfaces on a pair of equipotentials (which are *not* concentric with the line charges). As a matter of fact, it is clear that we could solve the problem with conductors of unequal size (circles A and B), and even with one conductor inside another, with an offset between their centers (circles C and B). Here, however, we shall take only the symmetrical case (solid circles B' and B), whose geometry must be arranged to agree with Fig. 9.8*a*.

Equating the position x_{c_B} of the center of B in Fig. 9.8*b* and Eq. 9.110*b* with that of the (+) conductor in Fig. 9.8*a* yields

$$d\left(\frac{K^2+1}{K^2-1}\right) = a \tag{9.110d}$$

Equating radii (Eq. 9.110*c*) yields

$$\frac{2dK}{K^2-1} = b \tag{9.110e}$$

Elimination of d by division of Eqs. 9.110*d* and 9.110*e* results in

$$\frac{K^2-1}{2K} = \frac{a}{b}$$

of which the only solution with $K > 1$ ($\phi > 0$) is

$$K = \frac{a+\sqrt{a^2-b^2}}{b} \tag{9.110f}$$

Each conductor has charge 1 coulomb/m because the line sources creating the potential field we are using had this amount of charge. The positive conductor therefore has potential $\phi = +(1/2C)$, so in view of the definition of K in Eq. 9.110*a* and Eq. 9.110*f*

$$e^{\pi\epsilon/C} = \frac{a+\sqrt{a^2-b^2}}{b}$$

or

$$C = \frac{\pi\epsilon}{\ln\left[(a+\sqrt{a^2-b^2})/b\right]} \tag{9.111a}$$

Because $LC = \mu\epsilon$

$$L = \frac{\mu}{\pi} \ln \left(\frac{a + \sqrt{a^2 - b^2}}{b} \right) \tag{9.111b}$$

and because $G/C = \sigma/\epsilon$

$$G = \frac{\pi\sigma}{\ln [(a + \sqrt{a^2 - b^2})/b]} \tag{9.111c}$$

It is obvious from the spacing of the equipotentials, or from the eccentric position of the equivalent line sources with respect to the actual conductors, that the normal electric field (or surface-charge density) is strongest on the portions of the conductor surfaces that come closest together. In our TEM wave, the magnetic field is perpendicular to, and proportional to, the electric field. Therefore it and the longitudinal surface-current density are also strongest on the mutually nearest portions of the conductor surfaces.

To compute approximately the effective series resistance per unit length of the line produced by imperfectly conducting wires, assuming a high conductivity metal in which $\delta_m \ll b$, we must evaluate

$$\langle P_{\text{diss}} \rangle_{\text{meter}} \approx 2 \left(\frac{\eta_m'}{2} \oint_{\text{circle } B} |\mathrm{H}_{\text{tang}}|^2 \, dl \right)$$

where H_{tang} applies to the situation with perfectly conducting wires and is essentially a two-dimensional magnetostatic field. The equation above includes *both* conductors because of the symmetry of this particular problem and the extra factor of 2. The integral involved is not easy to carry out in a straightforward manner, because the expression for H_{tang} (even in the perfectly conducting case) is not simple, and the path of integration is awkward. There are ways of doing it more or less directly, but there is a surprisingly simple indirect method which relates the integral in question to the inductance L per unit length!

The static magnetic energy stored per unit length is given from two points of view by the relation

$$u_m = \tfrac{1}{2} L I^2 = \tfrac{1}{2} \mu \int_A |\boldsymbol{H}|^2 \, da$$

where A designates the whole region of the cross section outside the perfect conductors. Suppose now we envision a slight expansion Δb_+ in the radius of just the (+) conductor in Fig. 9.8*a*, keeping its center fixed, and holding the total current constant. It is clear that the change (actually decrease in this case) of magnetic stored energy is just what is "wiped out" by the expansion of the perfect conductor, namely,

$$-\Delta u_m = (\tfrac{1}{2}\mu \oint_{\text{circle } B} |\boldsymbol{H}|^2\, dl)\, \Delta b_+$$

$$= (\tfrac{1}{2}\mu \oint_{\text{circle } B} H_{\text{tang}}^2\, dl)\, \Delta b_+$$

since $\boldsymbol{H}$ is entirely tangent to a perfect conductor anyway. But also, at constant current,

$$-\Delta u_m = -\tfrac{1}{2} I^2\, \Delta L$$

where ΔL is the change of *system* inductance caused by changing *only the radius of one conductor* (the positive one in this case). So

$$\oint_{\text{circle } B} H_{\text{tang}}^2\, dl = -\frac{I^2}{\mu}\left(\frac{\partial L}{\partial b_+}\right)$$

It follows that the a-c series resistance for both conductors together is given in this symmetrical situation by

$$\frac{1}{2}|\mathrm{I}|^2 R = -\eta_m{}' \frac{|\mathrm{I}|^2}{\mu}\left(\frac{\partial L}{\partial b_+}\right)$$

or

$$R = \frac{-2\eta_m{}'}{\mu}\left(\frac{\partial L}{\partial b_+}\right)$$

Using Eq. 9.111*b*, however, we find by direct differentiation that

$$\frac{\partial L}{\partial b} = -\frac{\mu}{\pi}\left[\frac{1}{a + \sqrt{a^2 - b^2}}\left(\frac{b}{\sqrt{a^2 - b^2}}\right) + \frac{1}{b}\right]$$

$$= -\frac{\mu a}{\pi b\sqrt{a^2 - b^2}} = \frac{-\mu}{\pi b\sqrt{1 - (b/a)^2}}$$

This result applies to the change of system inductance when the radius b of *both* conductors is changed at the same time, since Eq. 9.111*b* for L automatically includes the fact that both conductors happen to have the same radius b. Had we solved the original problem for different conductor radii, say b_+ and b_-, we would find both b_+ and b_- entering Eq. 9.111*b* for L. We would then need $(\partial L/\partial b_+)$ for the computation of the magnetic field integral entering that part of R $(=R_+)$ contributed by the $(+)$ conductor, and, of course, $(\partial L/\partial b_-)$ for that part (R_-)

contributed by the (−) conductor. In the present symmetrical situation, it is obvious that $R_- = R_+ = \frac{1}{2}R$ and that

$$\frac{\partial L}{\partial b_-} = \frac{\partial L}{\partial b_+} = \frac{1}{2}\frac{\partial L}{\partial b} = \frac{-\mu}{2\pi b\sqrt{1-(b/a)^2}}$$

Therefore

$$R = \frac{\eta_m{}'}{\pi b\sqrt{1-(b/a)^2}} = 2\left[\frac{1}{2\pi b\sigma_m\delta_m\sqrt{1-(b/a)^2}}\right] \qquad (9.112)$$

From a comparison of Eqs. 9.112 and 9.109*a*, it is clear that only for large spacings a may we treat the resistance of a two-wire line, with well-developed skin effect, by simply using Eq. 9.109*a* for the contribution of each conductor. The proximity effect acts through the term $[1 - (b/a)^2]^{-1/2}$ in Eq. 9.112 to increase the effective resistance above what it would be if each conductor were in a circularly symmetric electrical environment. The magnitude of this effect may not be great enough to worry about if $b/a \ll 1$. For example, if $a \geq 10b$, so that the conductor spacing on centers ($2a$) is at least 10 times the *diameter* ($2b$) of each one, the enhancement of series resistance, above that given by applying Eq. 9.109*a*, cannot amount to more than ½%. If however $a = 2b$, so that the shortest distance between the conductor surfaces is equal to the diameter of either conductor, the enhancement is about 12.5%.

9.7 Transmission-Line Concepts for Multiconductor Lines *

We found in Sec. 9.5 that a two-conductor line in the sinusoidal steady state always gives rise to linearly polarized TEM waves, and is, therefore, incapable of supporting the most general type of TEM solutions to Maxwell's equations. This limitation is removed when more than two conductors are used to guide the electromagnetic energy.

Let us explore the situation when TEM waves are guided by three perfect conductors. This case contains nearly all the features of the most general one (n conductors), and is also a practical example. In electric-power transmission it is used for some three-phase systems, and in communications it often occurs when a "ground" forms the unavoidable third conductor in what is otherwise supposed to be a two-wire system.

With reference to Fig. 9.9, we can still define the voltage between any two of the conductors in the manner of Eqs. 9.18 through 9.22. Considering a closed path touching all *three* conductors, however, leads us to the conclusion that there are actually only *two independent*

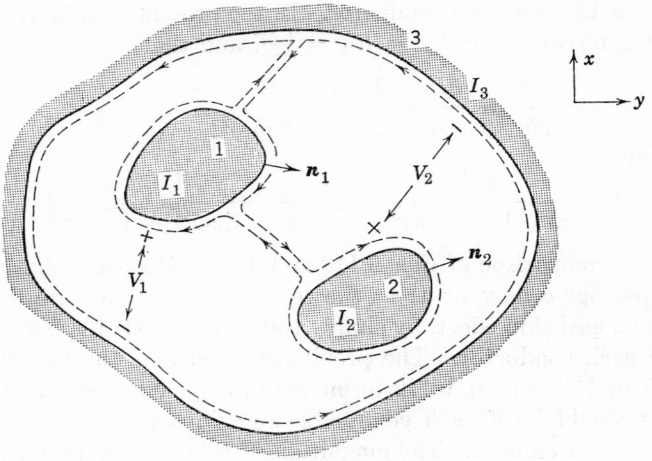

Fig. 9.9. A three-conductor line. The currents are all chosen positive in the $+z$ direction.

voltages; i.e., one conductor can always be selected as the potential reference. We have chosen the voltages indicated by V_1 and V_2.

The definition (Eq. 9.28) of current in each conductor in terms of the magnetic field can again be used, but application of the reasoning in Eqs. 9.29 through 9.31 to the dotted contour shown in Fig. 9.9 proves that $I_1 + I_2 + I_3 = 0$. Thus only *two* currents are independent. We shall let these be I_1 and I_2 as illustrated.

Transferring our thinking to the frequency domain leads us to consider the general TEM wave solution given by Eqs. 9.63 through 9.69 as being guided by the three conductors. In particular, depending upon the terminations, there will usually be a $(+)$ and a $(-)$ wave on the line. We can first consider these separately, choosing the $(+)$ wave for discussion (Eqs. 9.69). The steps involved in Eqs. 9.85 through 9.90 can be retraced almost exactly in the present case; except that the boundary conditions (Eq. 9.88*a*) must now be applied on all *three* conductors, and *two* voltages V_{1+} and V_{2+} must be defined by modifying Eq. 9.85*a* according to the voltage definitions shown in Fig. 9.9.

The complex scalar potential $\Phi^+(x, y)$ obeying Eqs. 9.89 and 9.90 must now satisfy the following boundary conditions:

$$\begin{aligned} &(a) \qquad \Phi^+ = \mathrm{V}_{1+} \quad \text{on } C_1 \\ &(b) \qquad \Phi^+ = \mathrm{V}_{2+} \quad \text{on } C_2 \qquad (9.113) \\ &(c) \qquad \Phi^+ = 0 \quad \text{on } C_3 \end{aligned}$$

However, since Eq. 9.90 is a linear differential equation, our solution Φ^+ can always be thought of as the superposition of two separate solutions

$$\Phi^+ = \Phi_a^+ + \Phi_b^+ \tag{9.114}$$

where

$$\text{(a)}\quad \begin{aligned} &\nabla_T^2\Phi_a^+ = 0 \\ &\Phi_a^+ = \mathrm{V}_{1+} \quad \text{on } C_1 \\ &\Phi_a^+ = 0 \quad \text{on } C_2 \\ &\Phi_a^+ = 0 \quad \text{on } C_3 \end{aligned}$$

$$\text{(b)}\quad \begin{aligned} &\nabla_T^2\Phi_b^+ = 0 \\ &\Phi_b^+ = 0 \quad \text{on } C_1 \\ &\Phi_b^+ = \mathrm{V}_{2+} \quad \text{on } C_2 \\ &\Phi_b^+ = 0 \quad \text{on } C_3 \end{aligned} \tag{9.115}$$

Certainly the function $(\Phi_a^+ + \Phi_b^+)$ is a solution of Eq. 9.90; and certainly it also meets the boundary conditions (Eqs. 9.113) because of Eqs. 9.115. But then, *by the uniqueness theorem*, it is the *only* solution for Φ^+, given V_{1+} and V_{2+}.

Observe that the solution Φ_a^+ obeys boundary conditions for what amounts to a *two-wire* line. It is formed by applying V_{1+} to conductor 1, and holding conductors 2 and 3 at the *same* potential (zero), thus making them *electrically* (but not physically) equivalent to a single conductor. The reasoning in Eqs. 9.91 through 9.100 therefore applies directly, treating C_2 and C_3 together as the "return" side of the line. It follows that, if $\mathrm{V}_{1+} = |\mathrm{V}_{1+}|e^{j\psi}$, we shall have $\Phi_a^+(x, y) = \phi_a^+(x, y)e^{j\psi}$, with $\phi_a^+(x, y)$ the (real) d-c potential produced by $|\mathrm{V}_{1+}|$. Consequently, the electric field of this solution is linearly polarized, of the form (Eq. 9.100)

$$\hat{\mathbf{E}}_{Ta}^+ = \hat{\boldsymbol{E}}_{Ta}^+ e^{j\psi} \tag{9.116}$$

where $\hat{\boldsymbol{E}}_{Ta}^+$ is the (real) *electrostatic* field in the cross section produced by a direct voltage $|\mathrm{V}_{1+}|$ applied to conductor 1, with the other conductors grounded. We can actually visualize this d-c field in terms of Fig. 9.9. The electric lines will emanate from conductor 1 and terminate *partly* on conductor 2 and *partly* on conductor 3. The division of the lines between the latter conductors will be governed by the charge distribution.

As a two-dimensional electrostatic problem, we know that the charge q_{1a}^+ on conductor 1 (per unit length) produced by a voltage $|\mathrm{V}_{1+}|$ on this conductor, and by zero volts on the others, is given by

$$q_{1a}^{+} = C_{11}|\mathrm{V}_{1+}| \tag{9.117}$$

in which C_{11} is a (real) "self-capacitance" coefficient depending *only* upon ϵ and the geometry. For simple conductor geometries, this capacitance can be calculated analytically by well-known methods of electrostatics. We need not dwell upon them here. In any case, C_{11} can always be measured.

Similarly, the charge q_{2a}^{+} per unit length "induced" on conductor 2 by a potential $|\mathrm{V}_{1+}|$ on conductor 1, when conductors 2 and 3 are at zero volts, is

$$q_{2a}^{+} = C_{21}|\mathrm{V}_{1+}| \tag{9.118}$$

with C_{21} a (real) "mutual capacitance" coefficient determined from electrostatic calculations or measurements. The charge on conductor 3 is not independent of the others, since the algebraic sum of all the charges must vanish (see the reasoning in Eqs. 9.23 through 9.26).

Returning to the frequency domain, we find the *complex* charges q_{1a}^{+} and q_{2a}^{+} per unit length for the TEM wave to be given by Eqs. 9.23, 9.116, 9.117, and 9.118, as

$$\begin{aligned} &\text{(a)} \qquad \mathrm{q}_{1a}^{+} = q_{1a}^{+}e^{j\psi} = C_{11}\mathrm{V}_{1+} \\ &\text{(b)} \qquad \mathrm{q}_{2a}^{+} = q_{2a}^{+}e^{j\psi} = C_{21}\mathrm{V}_{1+} \end{aligned} \tag{9.119}$$

The factor $e^{-\bar{\gamma}_0 z}$ has been omitted; i.e., the actual charges per unit length in this solution are $\mathrm{q}_{1a}^{+}e^{-\bar{\gamma}_0 z}$ and $\mathrm{q}_{2a}^{+}e^{-\bar{\gamma}_0 z}$.

The solution Φ_b^{+} in Eqs. 9.115 is now seen to be another "two-wire" solution, this time with conductor 2 at voltage V_{2+}, and conductors 1 and 3 together at ground. This solution is quite different looking from the previous one (see Fig. 9.9, and visualize the corresponding electrostatic field). It is *not* just a complex constant times Φ_a^{+}. If it were, its boundary conditions would also have to be simply a complex constant times those for Φ_a^{+} (by the uniqueness theorem again!). It is evident from Eqs. 9.115*a* and 9.115*b* that the boundary conditions for the latter cannot be obtained from that of the former by multiplying all the conductor voltages by a single complex constant. Hence Φ_b^{+} and Φ_a^{+} are linearly independent functions of (x, y).

Corresponding to the solution Φ_b^{+}, there is an electric field due to $\mathrm{V}_{2+} = |\mathrm{V}_{2+}|e^{j\psi'}$, given by

$$\hat{\mathbf{E}}_{Tb}^{+} = \hat{\boldsymbol{E}}_{Tb}^{+}e^{j\psi'} \tag{9.120}$$

and complex charges per unit length on conductors 1 and 2 given by

$$\begin{aligned} &\text{(a)} \qquad q_{2b}{}^{+} = C_{22}V_{2+} \\ &\text{(b)} \qquad q_{1b}{}^{+} = C_{12}V_{2+} = C_{21}V_{2+} \end{aligned} \tag{9.121}$$

The fact that $C_{12} = C_{21}$ (Eqs. 9.119*b* and 9.121*b*) stems from the reciprocity theorem (for capacitance networks, in this case). The charge induced on conductor 2 (at ground potential) by 1 v applied to conductor 1 must equal the charge produced on conductor 1 (at ground potential) by 1 v applied to conductor 2, all other conductors in the system being held at ground potential in both cases.

The capacitance coefficient C_{22} need not, of course, be the same as C_{11}, although it may be if the conductor arrangement had some special symmetry. It is also possible for $C_{12} = C_{21} = 0$, if the third conductor actually *shields* the other two completely from each other. This would happen in Fig. 9.9, for example, if conductor 3 included a plane running right across from one side to the other, between conductors 1 and 2. Such cases represent trivial examples of three-conductor lines, however, since these systems really become two completely separate two-wire lines.

There remains only one other general restriction among the capacitance coefficients C_{11}, C_{12}, and C_{22}. If we apply direct voltages V_1 and V_2 to conductors 1 and 2 (with respect to 3), the total electric stored energy will be given by $2W_e = C_{11}V_1{}^2 + 2C_{12}V_1V_2 + C_{22}V_2{}^2$. This energy *must* be a positive quantity for *all* (real) choices of V_1 and V_2. Thus the equation $W_e = 0$ cannot have any real roots for V_1 and V_2 other than $V_1 = V_2 = 0$. This requires the discriminant of the quadratic equation $W_e = 0$ to be negative (or at most, zero); i.e.,

$$\frac{C_{12}}{\sqrt{C_{11}C_{22}}} \leq 1 \tag{9.122}$$

Equation 9.122 is a general restriction on the "coupling coefficient" of a two-terminal-pair capacitance network, analogous to the well-known dual situation with inductances and mutual inductances.

Returning now to the TEM wave problem, we note that the general (+) wave is the superposition of the (+, *a*) and (+, *b*) solutions; so, for the charges on conductors 1 and 2, we have

$$\begin{aligned} &\text{(a)} \qquad q_{1+} = q_{1a}{}^{+} + q_{1b}{}^{+} = C_{11}V_{1+} + C_{12}V_{2+} \\ &\text{(b)} \qquad q_{2+} = q_{2a}{}^{+} + q_{2b}{}^{+} = C_{12}V_{1+} + C_{22}V_{2+} \end{aligned} \tag{9.123}$$

But on either conductor in Fig. 9.9, from Eq. 9.25,

$$q_{1,2}{}^{+} = \epsilon \oint_{C_{1,2}} \boldsymbol{n}_{1,2} \cdot \hat{\mathbf{E}}_T{}^{+} \, dl \tag{9.124}$$

and by Eq. 9.69*b* for a TEM wave,

$$\hat{\mathbf{E}}_T{}^{+} = \bar{\eta} \hat{\mathbf{H}}_T{}^{+} \times \boldsymbol{a}_z \tag{9.125}$$

So

$$\begin{aligned} q_{1,2}{}^{+} &= \epsilon\bar{\eta} \oint_{C_{1,2}} \boldsymbol{n}_{1,2} \cdot (\hat{\mathbf{H}}_T{}^{+} \times \boldsymbol{a}_z) \, dl \\ &= \epsilon\bar{\eta} \oint_{C_{1,2}} \boldsymbol{a}_z \cdot (\boldsymbol{n}_{1,2} \times \hat{\mathbf{H}}_T{}^{+}) \, dl \\ &= \epsilon\bar{\eta} \oint_{C_{1,2}} \hat{\mathbf{H}}_T{}^{+} \cdot d\boldsymbol{l} = \epsilon\bar{\eta} \mathrm{I}_{1,2+} \end{aligned} \tag{9.126}$$

Therefore, using Eq. 9.126 to eliminate the q from Eq. 9.123, we have

$$\begin{aligned} &\text{(a)} \quad \mathrm{I}_{1+} = \left(\frac{C_{11}}{\epsilon\bar{\eta}}\right) \mathrm{V}_{1+} + \left(\frac{C_{12}}{\epsilon\bar{\eta}}\right) \mathrm{V}_{2+} = \mathrm{Y}_{11}\mathrm{V}_{1+} + \mathrm{Y}_{12}\mathrm{V}_{2+} \\ &\text{(b)} \quad \mathrm{I}_{2+} = \left(\frac{C_{12}}{\epsilon\bar{\eta}}\right) \mathrm{V}_{1+} + \left(\frac{C_{22}}{\epsilon\bar{\eta}}\right) \mathrm{V}_{2+} = \mathrm{Y}_{12}\mathrm{V}_{1+} + \mathrm{Y}_{22}\mathrm{V}_{2+} \end{aligned} \tag{9.127}$$

We find that, at any value of z, the (+) wave conductor currents are related to the (+) wave voltages by a two-terminal-pair network equation (Eq. 9.127). The (+) wave described by (I_{1+}, V_{1+}), which would exist by itself only when $\mathrm{V}_{2+} = 0$, is said to be *coupled* to the (+) wave described by (I_{2+}, V_{2+}), which would exist by itself only when $\mathrm{V}_{1+} = 0$. From this point of view, the three-wire line behaves like a *pair* of *coupled* two-wire lines, the two-wire lines in this case being formed by choosing either conductor 1 or 2 as one wire, and grounding the remaining two together. Observe that all of the admittances Y have the *same* complex angle (that of $\bar{\eta}^{-1}$); so the "coupling admittance" Y_{12} is *not* just inductive, capacitive, or conductive, but includes all three types of elements. Even if the medium is lossless ($\sigma = 0$, $\bar{\eta} = \sqrt{\mu/\epsilon}$), the coupling admittance Y_{12} must contain *both* inductive and capacitive coupling because it is a *real* number. This is reasonable, because physically both the electric and the magnetic fields produced by one conductor will generally induce charges and currents in the other.

For (−) waves, the discussion would follow that of the (+) wave, except that in Eqs. 9.125 and 9.126 the signs would change. This occurs because of the difference in relative directions of $\hat{\mathbf{E}}_T{}^{-}$ and $\hat{\mathbf{H}}_T{}^{-}$

compared to $\hat{\mathbf{E}}_T^+$ and $\hat{\mathbf{H}}_T^+$ (see Eq. 9.68), and the fact that we nevertheless retain the *same* reference directions for conductor currents and voltages. Thus

$$\text{(a)} \quad I_{1-} = -(Y_{11}V_{1-} + Y_{12}V_{2-})$$
$$\text{(b)} \quad I_{2-} = -(Y_{12}V_{1-} + Y_{22}V_{2-}) \tag{9.128}$$

Consequently, the complete solutions for voltage and current are

$$\text{(a)} \quad \begin{aligned} V_1 &= V_{1+}e^{-\bar{\gamma}_0 z} + V_{1-}e^{\bar{\gamma}_0 z} \\ V_2 &= V_{2+}e^{-\bar{\gamma}_0 z} + V_{2-}e^{\bar{\gamma}_0 z} \end{aligned}$$

$$\text{(b)} \quad \begin{aligned} I_1 &= (Y_{11}V_{1+} + Y_{12}V_{2+})e^{-\bar{\gamma}_0 z} \\ &\quad - (Y_{11}V_{1-} + Y_{12}V_{2-})e^{\bar{\gamma}_0 z} \\ I_2 &= (Y_{12}V_{1+} + Y_{22}V_{2+})e^{-\bar{\gamma}_0 z} \\ &\quad - (Y_{12}V_{1-} + Y_{22}V_{2-})e^{\bar{\gamma}_0 z} \end{aligned} \tag{9.129}$$

These relationships show that there are two traveling waves (+ and −) on the line, and that the complex amplitude of *each* of them depends upon the voltages applied to *both* conductors.

In view of these two waves, which are simply the TEM *field configurations* traveling in the $+z$ and $-z$ directions, we are led to try to think about the voltages and currents in the system in terms of some new concepts which include both line voltages *together* as a single "voltage" unit, and both line currents *together* as a single "current" unit. Matrices are just such concepts.

Thus, we may define a "(+) wave voltage matrix" (a column matrix)

$$\mathscr{v}_+ = \begin{pmatrix} V_{1+} \\ V_{2+} \end{pmatrix} \tag{9.130a}$$

and a "(+) wave current matrix"

$$i_+ = \begin{pmatrix} I_{1+} \\ I_{2+} \end{pmatrix} \tag{9.130b}$$

which together define completely the electric and magnetic fields respectively of the (+) TEM wave on a given three-wire system. Similarly, for the (−) wave we define

$$\text{(a)} \quad \mathscr{v}_- = \begin{pmatrix} V_{1-} \\ V_{2-} \end{pmatrix}$$

$$\text{(b)} \quad i_- = \begin{pmatrix} I_{1-} \\ I_{2-} \end{pmatrix} \tag{9.131}$$

and for the total voltage and current

$$\text{(a)} \qquad \mathscr{v} = \begin{pmatrix} V_1 \\ V_2 \end{pmatrix}$$
$$\text{(b)} \qquad \mathscr{i} = \begin{pmatrix} I_1 \\ I_2 \end{pmatrix} \tag{9.132}$$

Characterizing the line itself, we then have the square "admittance-per-unit-length matrix"

$$\mathcal{Y}_0 = \begin{pmatrix} Y_{11} & Y_{12} \\ Y_{12} & Y_{22} \end{pmatrix} \tag{9.133}$$

Accordingly, Eqs. 9.127 to 9.129 become simply

$$\text{(a)} \qquad \mathscr{i}_+ = \mathcal{Y}_0 \mathscr{v}_+$$
$$\text{(b)} \qquad \mathscr{i}_- = -\mathcal{Y}_0 \mathscr{v}_-$$
$$\text{(c)} \qquad \mathscr{v} = \mathscr{v}_+ e^{-\bar{\gamma}_0 z} + \mathscr{v}_- e^{\bar{\gamma}_0 z} \tag{9.134}$$
$$\text{(d)} \qquad \mathscr{i} = \mathscr{i}_+ e^{-\bar{\gamma}_0 z} + \mathscr{i}_- e^{\bar{\gamma}_0 z} = \mathcal{Y}_0(\mathscr{v}_+ e^{-\bar{\gamma}_0 z} - \mathscr{v}_- e^{\bar{\gamma}_0 z})$$

Except for the fact that they are matrix equations, so that (among other things) the order of multiplication is important, Eqs. 9.134 are identical in *form* with those describing a two-wire line. This emphasizes the fact that the essential wave character of the situation is unchanged; but we must add that a good many of the detailed properties of the system are hidden by the condensed matrix notation, and they may turn out to be quite new in specific problems.

To proceed further with our immediate discussion, we must consider the terminations of the line. In Fig. 9.10 appears a three-terminal passive network which can serve as a load at $z = 0$. We describe this

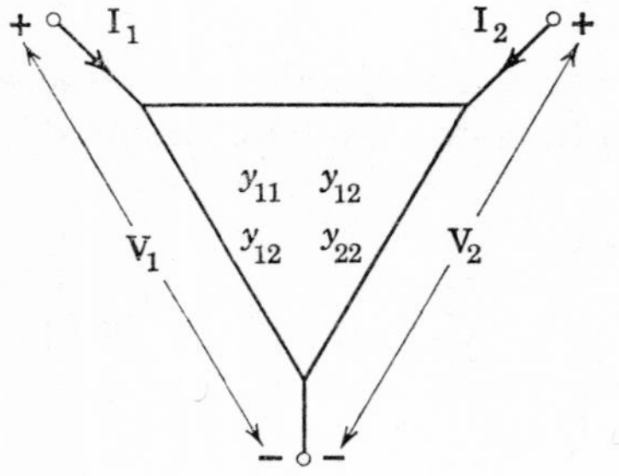

Fig. 9.10. Passive reciprocal three-terminal load network for a three-wire line.

load by the two-terminal-pair $\mathcal{Y}$ matrix

$$\mathcal{Y} = \begin{pmatrix} y_{11} & y_{12} \\ y_{12} & y_{22} \end{pmatrix} \tag{9.135a}$$

so that

$$\boldsymbol{\imath} = \mathcal{Y}\boldsymbol{\nu} \qquad \text{at } z = 0 \tag{9.135b}$$

is the load boundary condition for the line.

Applying Eq. 9.135*b* to the line equations (Eqs. 9.134*c* and 9.134*d*) at $z = 0$, we find

$$\mathcal{Y}_0\boldsymbol{\nu}_+ - \mathcal{Y}_0\boldsymbol{\nu}_- = \mathcal{Y}\boldsymbol{\nu}_+ + \mathcal{Y}\boldsymbol{\nu}_-$$

or

$$(\mathcal{Y}_0 + \mathcal{Y})\boldsymbol{\nu}_- = (\mathcal{Y}_0 - \mathcal{Y})\boldsymbol{\nu}_+$$

or

$$\boldsymbol{\nu}_- = (\mathcal{Y}_0 + \mathcal{Y})^{-1}(\mathcal{Y}_0 - \mathcal{Y})\boldsymbol{\nu}_+ \tag{9.136}$$

By analogy with two-wire lines, we can define a *matrix reflection coefficient* $\mathcal{R}$

$$\mathcal{R} \equiv (\mathcal{Y}_0 + \mathcal{Y})^{-1}(\mathcal{Y}_0 - \mathcal{Y}) = \begin{pmatrix} \bar{\Gamma}_{11} & \bar{\Gamma}_{12} \\ \bar{\Gamma}_{21} & \bar{\Gamma}_{22} \end{pmatrix} \tag{9.137}$$

which reduces Eq. 9.136 to

$$\boldsymbol{\nu}_- = \mathcal{R}\boldsymbol{\nu}_+ \tag{9.138}$$

Observe that $\bar{\Gamma}_{21}$ and $\bar{\Gamma}_{12}$ are *not* equal in general (even though $Y_{12} = Y_{21}$ and $y_{12} = y_{21}$). One special situation in which they are equal is when both $Y_{11} = Y_{22}$ *and* $y_{11} = y_{22}$. But then $\bar{\Gamma}_{11} = \bar{\Gamma}_{22}$ also. These conditions occur if the line conductors 1 and 2 as well as the load network are *electrically* symmetrical with respect to the reference conductor 3. This usually (but not always) means that they are also *physically* symmetrical with respect to conductor 3. We shall refer to such a system, in which $Y_{11} = Y_{22}$ and $y_{11} = y_{22}$, as being *mirror-symmetric.*

An even higher symmetry is possible, of course, if the line and load are symmetrical about a longitudinal axis. They would be, for instance, if the three conductors viewed in a cross section were identical, and disposed at the corners of an equilateral triangle, and the load comprised three equal resistances connected, respectively, between conductors 1 and 2, 2 and 3, and 3 and 1. In such cases, the system may be called *rotation-symmetric.* The electrical conditions for rotation symmetry are evidently $Y_{11} = Y_{22} = 2Y_{12} = Y$, and $y_{11} = y_{22} = 2y_{12} = y$, for which case Eq. 9.137 shows that $\bar{\Gamma}_{12} = \bar{\Gamma}_{21} = 0$, and $\bar{\Gamma}_{11} = \bar{\Gamma}_{22} = (Y - y)/(Y + y)$.

Equations 9.137 and 9.138 show that reflection can only be eliminated *entirely* (i.e., $V_{1-} = V_{2-} = 0$ for *all* values of V_{1+} and V_{2+}) if $\mathcal{R} = 0$, or

$$\mathcal{Y} = \mathcal{Y}_0 \tag{9.139a}$$

In detail

$$\begin{aligned} y_{11} &= Y_{11} \\ y_{12} &= Y_{12} \\ y_{22} &= Y_{22} \end{aligned} \tag{9.139b}$$

so that the three-terminal load network must look electrically exactly as the line does (per unit length). This constitutes a "matched" condition.

On the other hand, if all three wires are left disconnected at $z = 0$, the "load network" has $y_{11} = y_{12} = y_{22} = 0$; and by Eq. 9.137,

$$\mathcal{R}_{oc} = \mathcal{Y}_0^{-1}\mathcal{Y}_0 = \begin{pmatrix} 1 & 0 \\ 0 & 1 \end{pmatrix} \tag{9.140a}$$

This means, in Eq. 9.138,

$$\mathcal{V}_- = \mathcal{V}_+ \qquad \text{on open circuit} \tag{9.140b}$$

or

$$\left.\begin{aligned} V_{1-} &= V_{1+} \\ V_{2-} &= V_{2+} \end{aligned}\right\} \qquad \text{on open circuit} \tag{9.140c}$$

so that the complete voltage and current in Eq. 9.134 represent a pure standing wave (in the lossless case).

Similarly, if we short-circuit all the lines together, $y_{11} = y_{12} = y_{22} = \infty$, and the limiting form of Eq. 9.137 yields

$$\mathcal{R}_{sc} = -\mathcal{Y}^{-1}\mathcal{Y} = \begin{pmatrix} -1 & 0 \\ 0 & -1 \end{pmatrix} \tag{9.141a}$$

leading again to a complete reflection

$$\mathcal{V}_- = -\mathcal{V}_+ \qquad \text{on short circuit} \tag{9.141b}$$

or

$$\left.\begin{aligned} V_{1-} &= -V_{1+} \\ V_{2-} &= -V_{2+} \end{aligned}\right\} \qquad \text{on short circuit} \tag{9.141c}$$

We apparently find once more, in the foregoing results, a rather complete analogy with pure traveling and standing waves on a two-wire line. Note, however, that these situations require very special forms of the load network, viewed as a *two-terminal-pair* device. In general, the load boundary condition in Eq. 9.138 does *not* fix the character of the "standing-wave" pattern on the line, as it would in a two-wire line. We can emphasize this fact by writing out Eq. 9.138 in the form,

$$\frac{V_{1-}}{V_{1+}} = \bar{\Gamma}_{11} + \bar{\Gamma}_{12}\frac{V_{2+}}{V_{1+}}$$
$$\frac{V_{2-}}{V_{2+}} = \bar{\Gamma}_{21}\frac{V_{1+}}{V_{2+}} + \bar{\Gamma}_{22} \qquad (9.142)$$

and observing that the load does not determine the ratios V_{1-}/V_{1+} and V_{2-}/V_{2+}, unless the ratio V_{2+}/V_{1+} is known. Thus the load alone does not determine the general form of the voltage on either conductor (Eqs. 9.129*a*). To be able to determine the form, *the character of the source must also be specified*—and this will, in fact, fix all four constants V_{1+}, V_{1-}, V_{2+}, and V_{2-}. Indeed, we can conclude from Eq. 9.142 that the "apparent" voltage reflection coefficients V_{1-}/V_{1+} and V_{2-}/V_{2+} observed by voltage standing-wave measurements on the two conductors will *generally be different!*

One case in which the standing-wave ratios will be the same occurs for a *mirror-symmetrical* line and load ($\bar{\Gamma}_{11} = \bar{\Gamma}_{22}$, $\bar{\Gamma}_{12} = \bar{\Gamma}_{21}$), driven by a *symmetrical* (push-push) source, so that $V_{1+} = V_{2+}$ by symmetry. Another case is that of a mirror-symmetrical line and load driven by an *antisymmetrical* (push-pull) source, so that $V_{1+} = -V_{2+}$ by symmetry. In both of these cases, however, the three-wire line is actually being used as a two-wire line: In the first, conductors 1 and 2 together form a "single wire," with conductor 3 the "ground return"; in the second, conductors 1 and 2 form the two-wire line, with conductor 3 carrying no current, i.e., we have a two-wire "balanced-to-ground" line. The existence of only one standing-wave ratio is hardly surprising. Of course, any failure of the symmetry of the *line*, *load*, or *source* will upset these balanced situations and lead to typical three-wire behavior instead of two-wire results.

When the line and load system is *rotation-symmetric*, we have seen that $\bar{\Gamma}_{12} = \bar{\Gamma}_{21} = 0$ and $\bar{\Gamma}_{11} = \bar{\Gamma}_{22}$. Then Eq. 9.142 does determine the ratios V_{1-}/V_{1+} and V_{2-}/V_{2+}, *regardless* of the values of V_{2+} and V_{1+}. Moreover, these ratios are equal for all types of driving source. This conclusion is subject to the caution that three conductors can never have rotation symmetry unless they form an open-wire line, and then the external world (or "ground") forms an unavoidable fourth conductor which may or may not destroy the rotation symmetry of the system. It usually will destroy the symmetry to some extent, unless it actually encloses quite closely the three original conductors.[1] In that event, it certainly forms a very definite fourth conductor, which ought to be included explicitly in the discussion.

It is interesting to learn now that, by a change of current and voltage variables, the line equations (Eqs. 9.129) can always be brought into "uncoupled" form; i.e., new variables $I_{1,2}'$ and $V_{1,2}'$ can be chosen so that Eq. 9.127 (and similarly also Eq. 9.128) becomes

$$\begin{aligned} &(a) \qquad I_{1+}' = Y_{11}'V_{1+}' \\ &(b) \qquad I_{2+}' = Y_{22}'V_{2+}' \end{aligned} \tag{9.143}$$

while the complex power remains invariant in form and value,

$$\tfrac{1}{2}(V_{1+}'I_{1+}'^* + V_{2+}'I_{2+}'^*) = \tfrac{1}{2}(V_{1+}I_{1+}^* + V_{2+}I_{2+}^*) \tag{9.144}$$

As a result of these conditions, the total complex power may be computed as the sum of that carried separately by each of the two new independent waves (Eqs. 9.143). The change of variable makes the solutions *orthogonal as regards complex power*, which is not true of those in Eq. 9.127. We shall illustrate this reduction to uncoupled form only for a mirror-symmetric line, having $Y_{11} = Y_{22}$, although it can be done in general (with more algebra).

The simplifying clue for the present case lies in our earlier comments about the symmetrical and antisymmetrical drives, each leading to essentially two-wire behavior on a mirror-symmetric three-wire system. Since Eqs. 9.143 indicate essentially such two-wire behavior for each of the new (+) waves (I_{1+}', V_{1+}') and (I_{2+}', V_{2+}'), we try the sum (push-push) and difference (push-pull) variables [2]

$$\begin{aligned} &(a) \qquad I_{1+}' = \frac{I_{1+} + I_{2+}}{\sqrt{2}} \qquad V_{1+}' = \frac{V_{1+} + V_{2+}}{\sqrt{2}} \\ &(b) \qquad I_{2+}' = \frac{I_{1+} - I_{2+}}{\sqrt{2}} \qquad V_{2+}' = \frac{V_{1+} - V_{2+}}{\sqrt{2}} \end{aligned} \tag{9.145}$$

The $\sqrt{2}$ in Eqs. 9.145 is used to meet Eq. 9.144, and, at the same time,

[1] The practice of transposing power lines, which are geometrically long and electrically quite short, does reduce the unsymmetric effect of ground on the average over the run. This technique is very difficult to apply at much higher frequencies where lines become shorter geometrically and longer electrically. In any case, a transposed system does not fit strictly into the longitudinally uniform systems assumed here.

[2] This transformation is essentially the same as the Clark transformation in the theory of polyphase machines and power-transmission systems. See White and Woodson, *loc. cit.*

to have the voltage and current transformations identical. In matrix form, Eqs. 9.145 become

$$\text{(a)} \qquad i_+' = \begin{pmatrix} \frac{1}{\sqrt{2}} & \frac{1}{\sqrt{2}} \\ \frac{1}{\sqrt{2}} & -\frac{1}{\sqrt{2}} \end{pmatrix} i_+ = \mathfrak{J} i_+$$

$$\text{(b)} \qquad v_+' = \begin{pmatrix} \frac{1}{\sqrt{2}} & \frac{1}{\sqrt{2}} \\ \frac{1}{\sqrt{2}} & -\frac{1}{\sqrt{2}} \end{pmatrix} v_+ = \mathfrak{J} v_+ \qquad (9.146)$$

and we observe that the determinant of the transformation matrix $\mathfrak{J}$ is $\det \mathfrak{J} = -1$. Solution of Eqs. 9.146*a* and 9.146*b* for i_+ and v_+ respectively yields

$$\text{(a)} \qquad i_+ = \mathfrak{J}^{-1} i_+' = \mathfrak{J} i_+'$$

$$\text{(b)} \qquad v_+ = \mathfrak{J}^{-1} v_+' = \mathfrak{J} v_+' \qquad (9.147)$$

because $\mathfrak{J}^{-1} = \mathfrak{J}$. With Eq. 9.147, Eq. 9.134*a* becomes

$$\mathfrak{J}^{-1} i_+' = \mathcal{Y}_0 \mathfrak{J} v_+'$$

or

$$i_+' = (\mathfrak{J} \mathcal{Y}_0 \mathfrak{J}) v_+' = \mathcal{Y}_0' v_+' \qquad (9.148)$$

But with $Y_{22} = Y_{11}$,

$$\mathcal{Y}_0' = \mathfrak{J}\mathcal{Y}_0\mathfrak{J} = \frac{1}{2}\begin{pmatrix} 1 & 1 \\ 1 & -1 \end{pmatrix}\begin{pmatrix} Y_{11} & Y_{12} \\ Y_{12} & Y_{11} \end{pmatrix}\begin{pmatrix} 1 & 1 \\ 1 & -1 \end{pmatrix}$$

$$= \frac{1}{2}\begin{pmatrix} 1 & 1 \\ 1 & -1 \end{pmatrix}\begin{pmatrix} Y_{11} + Y_{12} & Y_{11} - Y_{12} \\ Y_{11} + Y_{12} & Y_{12} - Y_{11} \end{pmatrix}$$

$$= \begin{pmatrix} Y_{11} + Y_{12} & 0 \\ 0 & Y_{11} - Y_{12} \end{pmatrix} \qquad (9.149)$$

which, indeed, brings Eq. 9.148 into the form of Eq. 9.143. Specifically,

$$\text{(a)} \qquad Y_{11}' = Y_{11} + Y_{12}$$

$$\text{(b)} \qquad Y_{22}' = Y_{11} - Y_{12} \qquad (9.150)$$

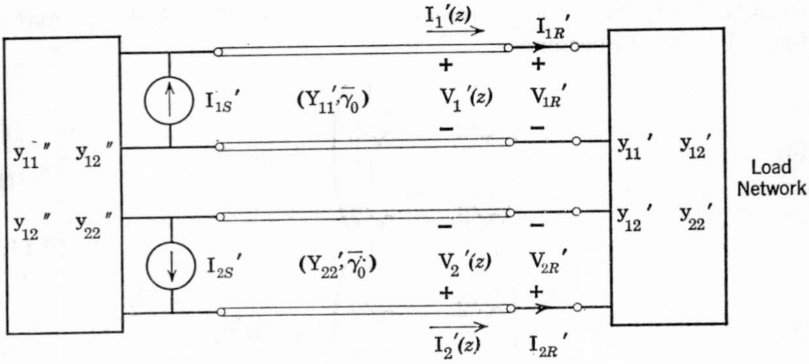

Fig. 9.11. Transformation of a three-wire line into two separate lines operating between transformed source and load networks.

Since we apply the *same* transformation $\mathfrak{I}$ to *all* current and voltage variables [(+), (−), and totals], Eqs. 9.129 become, via Eq. 9.134 transformed,

$$
\begin{aligned}
&\text{(a)} \qquad V_1' = V_{1+}'e^{-\bar{\gamma}_0 z} + V_{1-}'e^{\bar{\gamma}_0 z} \\
&\text{(b)} \qquad I_1' = Y_{11}'(V_{1+}'e^{-\bar{\gamma}_0 z} - V_{1-}'e^{\bar{\gamma}_0 z}) \\
&\text{(c)} \qquad V_2' = V_{2+}'e^{-\bar{\gamma}_0 z} + V_{2-}'e^{\bar{\gamma}_0 z} \\
&\text{(d)} \qquad I_2' = Y_{22}'(V_{2+}'e^{-\bar{\gamma}_0 z} - V_{2-}'e^{\bar{\gamma}_0 z})
\end{aligned}
\qquad (9.151)
$$

In the new variables, our three-wire line becomes *two separate* lines; they are no longer coupled along z. However, the source and load network boundary conditions must also be rephrased in terms of the new variables, and the result will usually tie the two "new" transmission lines together at their ends in some such manner as illustrated in Fig. 9.11. This problem may, nevertheless, be much simpler to handle than the original one.

Only if the source and load *networks* are also mirror-symmetric for the particular case we chose to deal with here (although the current sources themselves need *not* be), does the entire problem in the new variables separate into two *completely* uncoupled transmission-line systems. Actually this circumstance is apt to arise a little more often in practice than we might offhand expect, because multiwire lines tend to be associated with symmetrical terminal equipment, such as motors and generators. Nevertheless, more complicated cases are very common and cannot be dismissed by any means.

PROBLEMS

Problem 9.1. Measurements in an air-dielectric, lossless line with $Z_0 = 100$ ohms give the following results: VSWR $= 4$; distance between successive voltage minima $= 16$ cm; distance from termination to first current minimum $= 2$ cm. Find: (a) The frequency of operation. (b) The load impedance. To match this load, a quarter-wave-length lossless dielectric-filled line of the same geometry is to be inserted as a matching section. Assume the dielectric has $\mu = \mu_0$. Find: (c) The minimum distance between the load and the matching section. (d) The required dielectric constant of the added section. (e) The length (in cm) of the matching section.

Problem 9.2. A parallel-plate transmission line has perfect conductors and the medium inside has constant parameters ϵ, μ, σ ($\neq 0$). (a) Find all possible d-c solutions for $V(z)$ and $I(z)$ with voltage V_0 applied between the conductors. (b) Repeat (a), but with $I(z) = I_0$ constant. (c) Compare the field distributions in the cross section corresponding to conditions (a) and (b) with the comments made following Eqs. 9.9, 9.10, 9.16, and 9.17. (d) Develop the d-c circuit theory of this transmission line with arbitrary source and load resistances.

Problem 9.3. A lossless air-filled coaxial line has an outer conductor of $1\frac{1}{4}$-in. outside diameter and a wall thickness of 49 mils. The diameter of the inner conductor is $\frac{1}{2}$ in. The line is 3 m long. It is driven by a pulse voltage source in series with 100 ohms, and it is loaded with 100 ohms. If the voltage source generates a single 1000-v rectangular pulse of duration of 20 mμsec: (a) Find the maximum value of the instantaneous electric field that occurs anywhere within the line. (b) Write expressions for the instantaneous vector electric and magnetic fields in the line, as functions of time and position. (c) Repeat (a) and (b) for pulse durations of 30, 100, and ∞ mμsec.

Problem 9.4. (a) Find the approximate resistance and internal inductance per unit length of the coaxial line of Fig. 9.7 by considering the series voltage drop produced by the current and the complex surface impedance of the metal. State the important assumptions. (b) Repeat (a), but from the point of view of "complex flux linkages." (c) Find the series impedance per unit length of the coaxial line by the method used to obtain Eq. 9.112, and discuss the assumptions involved.

Problem 9.5. Consider the possibility of waves other than TEM being "guided" within the annular ring of a lossless coaxial line. (a) Qualitatively, how might such waves progress by (i) spiraling or (ii) bouncing, within the available space? Which type of propagation (other than TEM) would you expect to have the lowest cutoff frequency? (b) In case $(b - a) \ll (b + a)/2$ in Fig. 9.7*a*, one might expect the annular space to permit waves to propagate in about the same way as it would if it were "uncurled" to become a rectangular or parallel-plate wave guide. The boundary conditions at the "sides" would be different, however. Why and how? (c) Making use of the ideas suggested in (b), evaluate approximately the lowest cutoff frequency of "higher modes" in a coaxial line and, hence, the limitations on dimensions which must be observed if only the TEM wave is to propagate. (d) Discuss the possibility of "higher modes" on an open two-wire line. In view of your answer, is there any reason why the dimensions of such a line cannot be made arbitrarily large at a given frequency? See Chapter 10.

Problem 9.6. Assuming a well-developed skin effect, small attenuation, and a loss-free dielectric, calculate and plot Z_0 and α versus b/a, for various values of b, in: (a) A coaxial line (Fig. 9.7*a*). (b) A symmetrical open two-wire line (Fig. 9.8*a*). (c) What are the best methods for achieving low-attenuation α with a given value of Z_0 in each case? (d) In view of Prob. 9.5, what factors limit our ability to make an arbitrarily low value of α with a fixed value of Z_0 for each type of line? (e) For a coaxial line with outer radius b fixed, find the inner radius a yielding minimum attenuation constant α. Find also the corresponding characteristic impedance. (f) Is there an "optimum design" like that of (e) for a symmetrical two-wire line? Explain.

Problem 9.7. In an air-filled lossless coaxial line, if the instantaneous electric field anywhere ever exceeds E_{max} ($\approx 3 \times 10^6$) v/m, breakdown will occur. (a) In terms of E_{max} and the line geometry, find the maximum time-averaged power $\langle P \rangle_{max}$ which can be delivered to a *matched* load without causing breakdown. (b) Is there an optimum choice of inner radius, given the outer radius, which maximizes $(\langle P \rangle_{max}/E_{max})$? Find the corresponding value of Z_0 and compare with Prob. 9.6. (c) Answer (a) and (b) for a symmetrical two-wire line. (d) In terms of $\langle P \rangle_{max}$ found above, determine the maximum time-averaged power $\langle P' \rangle_{max}$ deliverable without causing breakdown to a *mismatched* load which produces a VSWR equal to s.

Problem 9.8. A certain air-filled coaxial line has thick metal walls. Its attenuation constant α is 3×10^{-3} neper/m at 300 mc. (a) What is its attenuation constant at 600 mc? (b) If all its cross-sectional linear dimensions are doubled, what is its attenuation constant α at 300 mc? (c) Suppose the *original* line is filled with a dielectric for which $\epsilon = 2\epsilon_0$ and the loss tangent $\sigma/\omega\epsilon$ is $(15/\pi) \times 10^{-4}$ at 300 mc. Find the attenuation constant α at 300 mc. (d) Answer (c) at 600 mc if the loss tangent of the dielectric does not change.

Problem 9.9. A single round wire of radius b parallel to an infinite metal ground at distance a constitutes a two-wire transmission line. (a) Find exactly the parameters of the line if the system is lossless. (b) Find approximately the series resistance R per unit length if both conductors are good, but imperfect. What equivalent width of the ground plane would account for its contribution to R if the current distribution across its surface were uniform in the transverse plane?

Problem 9.10. As a result of an error in manufacture, the center conductor of an air-dielectric coaxial line is placed considerably off center. The measured capacitance C per unit length consequently varies from the design specifications by 5%. (a) Is C larger or smaller than it should be? (b) What is the percentage of error in the characteristic impedance Z_0 of the line? (c) If the line normally has small conductor losses, in which direction is its attenuation constant α changed? Outline qualitatively but carefully all the factors upon which you base your answer. (d) Analyze exactly how the capacitance varies with eccentricity of the center conductor. What choice of dimensions tends to minimize errors caused by this centering problem? Compare with Probs. 9.7 and 9.6. (e) Determine a quantitative answer to (c) by plotting attenuation against eccentricity. Use the method leading to Eq. 9.112.

Problem 9.11. The structure shown in cross section in Fig. 9.12 is an air-filled perfectly conducting two-wire transmission line. Assume that $a \ll d \sin(\vartheta/2)$.

(a) For what values of ϑ will the method of images suffice to determine the capacitance C per unit length? (b) Suppose the center conductor is *not* on the bisector of ϑ, but is nearer to one side. For what values of ϑ will the image method suffice to determine C? (Assume that a is still much smaller than the distance between the center conductor and the nearest part of the outer one.) (c) Find C for $\vartheta = 90°$, when (i) the center conductor is on the bisector; (ii) the center conductor is twice as near to one side as the other. (d) If in fact the center conductor is too large to satisfy our assumption above, will the actual value of C be larger or smaller than we calculated? (e) In (c), calculate the maximum electric field strength if the peak line voltage is V_0. (f) In (c), calculate the characteristic impedance. (g) In (c), calculate the approximate attenuation constant α which would be caused by well-developed skin effect in a real metal inner conductor, neglecting losses in the outer conductor. Would this be a reasonable approximation to the actual attenuation constant of a real metal line? Justify your answer. See Prob. 9.9.

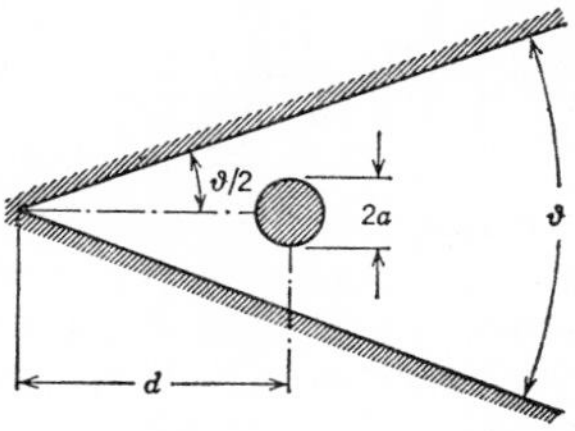

Fig. 9.12. Problem 9.11.

Problem 9.12. A "strip line" (Fig. 9.13) is made by laying a strip of polystyrene tape ½ in. wide and 0.020 in. thick on a metal ground plane. On this is laid a ¼ in. wide strip of copper sheet 0.010 in. thick. The dielectric constant of polystyrene is about 2.25. Neglect dielectric losses. Assume perfect conductors. (a) Does this structure propagate a TEM wave? If not, what kind of wave does it propagate when used as a two-wire transmission line? (b) Give simple upper and lower limits to the characteristic impedance of this line. (c) What do you think controls the phase velocity of transmission-line waves on this structure?

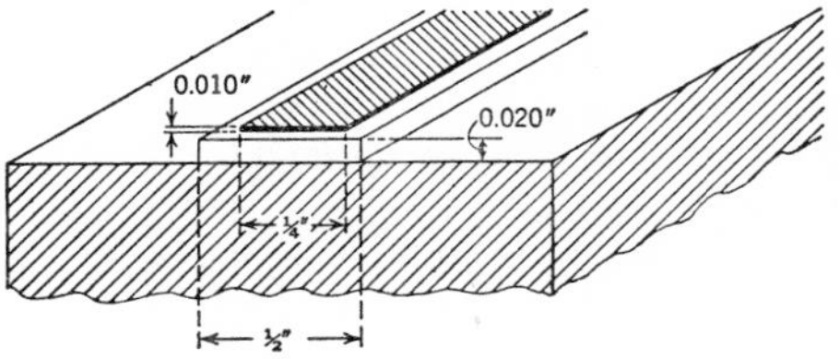

Fig. 9.13. Problem 9.12.

Why? (d) Comment on the validity of measuring the capacitance C and inductance L per unit length at low frequencies, and computing $Z_0 = \sqrt{L/C}$ and $v = 1/\sqrt{LC}$.

Problem 9.13. A thick-walled section of brass coaxial line is "short circuited" with brass plates of similar thickness, forming a closed air-filled resonator. The ratio of conductor radii is $e = 2.718 \cdots$, and the resonator has a Q of 2000 at its lowest natural frequency of 2000 mc. (a) If the structure is filled with lossless dielectric having $\mu = \mu_0$, $\epsilon = 4\epsilon_0$, find the new lowest nonzero natural frequency and the corresponding Q. (b) Repeat (a) if $\mu = 4\mu_0$ and $\epsilon = \epsilon_0$. (c) Repeat (a) if $\mu = 2\mu_0$ and $\epsilon = 2\epsilon_0$. (d) Repeat (a) if the medium has a loss, $\sigma = 10^{-4}$ mho/m, $\epsilon = 4\epsilon_0$, and $\mu = \mu_0$. (e) If all the linear dimensions of the original air-filled resonator are increased by a factor of 3, find the new lowest nonzero frequency and corresponding Q. (f) For the original air-filled structure, find the second lowest nonzero natural frequency and corresponding Q. (g) If in the original air-filled resonator a pair of terminals is connected in parallel with the line at its center, find the impedance seen at these terminals at the frequency 2000 mc.

Problem 9.14. One of the end thirds of a lossless coaxial line short-circuited at both ends is filled with a lossless dielectric having $\mu = \mu_0$ and $\epsilon = 4\epsilon_0$, while the rest is air-filled. (a) If the total length of the resonator is 30 cm, determine its two lowest nonzero natural frequencies. (b) At each of the above natural frequencies, find the ratio of the time-average magnetic energy stored in the dielectric-filled section to that in the air-filled section. (Check your result by considering the problem two ways: first by equivalent transmission-line methods, and then directly in terms of the electromagnetic fields themselves.) (c) If the radius of the inner conductor is 1 cm, that of the outer is e (= 2.718···) cm, the conductivity of the copper is 2×10^7 mhos/m, and the dielectric has some loss in the form of a conductivity $10^{-4}/9\pi$ mho/m, find the Q's of the system at the natural frequencies considered in (a) and (b). Assume all walls are thick enough for the skin effect to be well developed.

Problem 9.15. (a) Find the lowest natural frequency of the cylindrical re-entrant cavity shown in cross section in Fig. 9.14a. Justify a transmission-line equivalent

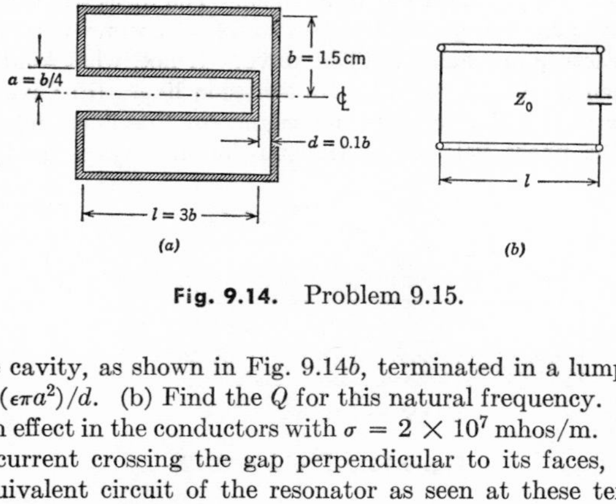

Fig. 9.14. Problem 9.15.

circuit for the cavity, as shown in Fig. 9.14b, terminated in a lumped capacitance equal to $C = (\epsilon\pi a^2)/d$. (b) Find the Q for this natural frequency. Assume a well-developed skin effect in the conductors with $\sigma = 2 \times 10^7$ mhos/m. (c) If the cavity is driven by current crossing the gap perpendicular to its faces, give a lumped-parameter equivalent circuit of the resonator as seen at these terminals for frequencies near the lowest resonance.

Repeat (a), (b), and (c) under the following new conditions (it should not be necessary to repeat all the work in detail): (d) All linear dimensions are increased by a factor of 2. (e) The original dimensions are restored, but the resonator is completely filled with a loss-free dielectric having $\epsilon/\epsilon_0 = 4$. (f) The conductivity of boundary conductor material in the original resonator is increased by a factor of 2, everything else remaining as it was originally.

Problem 9.16. The two coaxial lines in Fig. 9.15 have solid outer (and inner) conductors, and are lossless. They are of equal physical length, but the electrically long one ($\lambda/8$) is filled with a dielectric having $\epsilon > \epsilon_0$. The diameters are, however, adjusted so that the characteristic impedances of both lines are 50 ohms. Various physical connections between the two lines are made as illustrated in the figure and described in (b). (a) Noting that the space *outside* the two *outer* conductors forms a third two-wire transmission line, make sketches showing how the *three* lines are electrically interconnected by making the physical connections in each case described in (b). Note that no other conductors or "grounds" are assumed to be present, so that the outside surfaces of the two outer conductors must necessarily

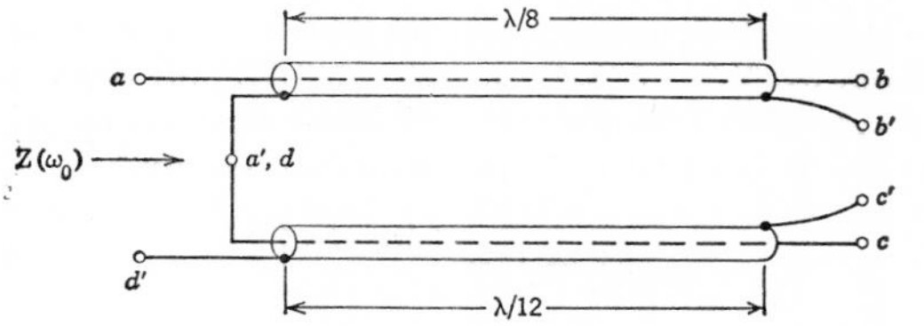

Fig. 9.15. Problem 9.16.

carry equal and opposite currents. Describe the conditions under which the electrical effect of the third line may be *neglected* in each case, and answer (b) assuming these conditions are met. (b) Find the input impedance $Z(\omega_0)$ at ad': (i) if the lines are connected as in the figure, with b, b', c, c', left open-circuited; (ii) if b' is short-circuited to c; (iii) if b' is short-circuited to c'; (iv) if b' is short-circuited to c' and an impedance $Z_L = j50(\sqrt{3} - 1)$ ohms is connected from b to c.

Problem 9.17. Calculate the constants C, G, and L of the transmission lines shown in Figs. 9.16*a*, 9.16*b*, and 9.16*c*. The radius of the cylindrical conductors is a,

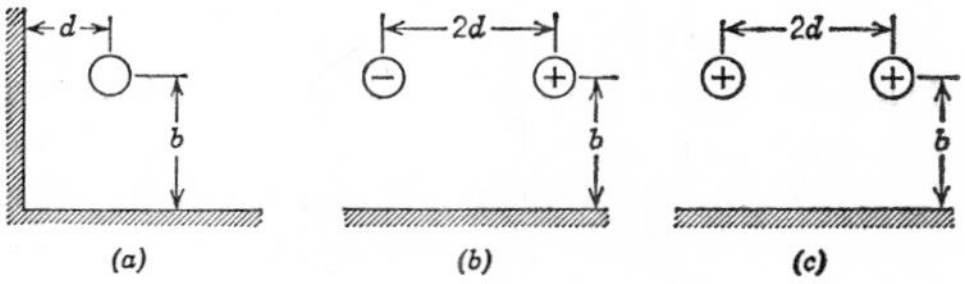

Fig. 9.16. Problem 9.17.

which is small compared to b and d. The conductors and ground planes are perfect. The (+) and (−) signs indicate "push-pull" or "push-push" constraints.

Problem 9.18. Fill in the details of the following outline of a general method for defining new variables to decouple the multiconductor transmission-line equations. The superscript dagger (†) means "the complex conjugate of the transpose" of any matrix to which it is applied. (a) If the transformation of both current and voltage involves the same matrix $\mathfrak{I}$, in the form: $i_+' = \mathfrak{I}i_+$, $v_+' = \mathfrak{I}v_+$, invariance of the complex power expression for all v_+ or i_+ requires $i_+^\dagger v_+ = i_+'^\dagger v_+' = i_+^\dagger \mathfrak{I}^\dagger \mathfrak{I} v_+$, or

$$\mathfrak{I}^\dagger \mathfrak{I} = \mathcal{I} \equiv \begin{bmatrix} 1 & 0 & \cdots & 0 \\ 0 & 1 & \cdots & 0 \\ & & \cdots & \\ 0 & 0 & \cdots & 1 \end{bmatrix}$$

(b) Thus for three conductors, the most general permitted form of $\mathfrak{J}$ is

$$\mathfrak{J} = \begin{bmatrix} \alpha e^{j\psi} & \sqrt{1-\alpha^2}\, e^{jv} \\ \sqrt{1-\alpha^2}\, e^{j\varphi} & -\alpha e^{j(\varphi+v-\psi)} \end{bmatrix}$$

where $\alpha \geq 0$ and ψ, φ, v lie between π and $-\pi$. Check the transformation used in Eqs. 9.145–9.147 against this result. Note that the determinant det $\mathfrak{J}$ may not be more general than det $\mathfrak{J} = -e^{j(\varphi+v)}$. (c) The new admittance-per-unit-length matrix $\mathcal{Y}_0'$ is given by $\mathcal{Y}_0' = \mathfrak{J}\mathcal{Y}_0\mathfrak{J}^{-1}$ or $\mathcal{Y}_0' = \mathfrak{J}\mathcal{Y}_0\mathfrak{J}\dagger$. (d) Since $\mathcal{Y}_0$ is symmetric, there is in most cases only one matrix $\mathfrak{J}$ which not only makes $\mathcal{Y}_0'$ diagonal in (c), but also meets the conditions (a).[1] This matrix is given by

$$\mathfrak{J}\dagger \equiv \mathfrak{L} = \begin{bmatrix} l_{11} & l_{12} \\ l_{21} & l_{22} \end{bmatrix}$$

where $\mathcal{Y}_0\mathfrak{L} = \mathfrak{L}\mathfrak{D}$, with

$$\mathfrak{D} = \begin{bmatrix} \lambda_1 & 0 \\ 0 & \lambda_2 \end{bmatrix}$$

and where $\lambda_{1,2}$ are the two roots of the equation

$$\det(\mathcal{Y}_0 - \lambda\mathfrak{g}) = \begin{vmatrix} (Y_{11} - \lambda) & Y_{12} \\ Y_{12} & (Y_{22} - \lambda) \end{vmatrix} = 0$$

To find l_{ij}, one solves the homogeneous equations

$$(\mathcal{Y}_0 - \lambda_1\mathfrak{g})\boldsymbol{\ell}_1 = 0 \qquad \boldsymbol{\ell}_1 = \begin{bmatrix} l_{11} \\ l_{21} \end{bmatrix}$$

$$(\mathcal{Y}_0 - \lambda_2\mathfrak{g})\boldsymbol{\ell}_2 = 0 \qquad \boldsymbol{\ell}_2 = \begin{bmatrix} l_{12} \\ l_{22} \end{bmatrix}$$

in a way that makes $\boldsymbol{\ell}_{1,2}{}^\dagger\boldsymbol{\ell}_{1,2} = 1$. Check by this method the results found in the text for the mirror-symmetric case, and then extend the work to the more general situation $Y_{11} \neq Y_{22}$. (e) Discuss the three-conductor line for the case where one conductor carries no current. Does the solution behave properly like that of a two-conductor line?

Problem 9.19. (a) Develop the theory of three-phase four-wire lines, comprising three conductors and a neutral. Impose any interesting symmetries. What happens if there is no current in the neutral? Compare the result with the three-wire line. (b) Develop the theory of four-wire lines based upon a viewpoint which treats them as a pair of coupled two-wire lines. Impose any necessary constraints and symmetries. (c) For polyphase systems, a change of variable to "symmetrical components" is often made. Is this change of variable included among those discussed in the text and in Prob. 9.18? Explain your reasoning, and find the matrix $\mathfrak{J}$ that does describe the "symmetrical components" transformation for three- and four-wire lines. See White and Woodson, *loc. cit.*

[1] See, for example, E. A. Guillemin, *The Mathematics of Circuit Analysis*, Technology Press, M.I.T., 1949, Chapter 3.

C H A P T E R   T E N

Elements of Radiation

The *possibility* of radiation is not something we are going to take up for the first time in this chapter; on the contrary, we have already accepted this possibility with the truth of Maxwell's equations. It is important to understand from the beginning that radiation is not a phenomenon essentially different from those we have been considering up to now. Actually, like the other waves we have studied, it is nothing more than a class of solutions to Maxwell's equations, arrived at under certain rather unique types of boundary conditions. The only new questions we must face therefore relate to: (1) Defining a class of problems, including especially the boundary conditions, representing the phenomenon of radiation; (2) obtaining solutions to that class of problems; and (3) studying the relationship between the source and the field solution, for some very simple cases, to achieve a better understanding of the conditions under which radiation can be maximized, minimized, or otherwise controlled.

10.1 Definition of the Problem

If we ask what are the essential differences between a radiation situation and those we have considered previously in connection with transmission lines or plane waves, we will discover that the answer is surprisingly elusive. One could argue, for example, that a TEM wave actually "radiates" from the source to a load, but happens to be *guided* during the journey by the wires of the transmission line along which it is propagating. Even more convincingly, perhaps, the waveguide modes represent the radiation of an electromagnetic field from a source to a load, along a zig-zag path within some material boundaries. There is, indeed, no doubt that these situations do have one essential characteristic we associate with the very idea of radiation, namely, that of a traveling wave. But it is probably also true that the most clear-cut radiation occurs when no material body extends physically from the source to the receiver, and when the source itself does not extend to infinity but is confined to a finite volume of space. That is why the transmission-line and wave-guide waves, on one hand, and the various types of plane waves which we have discussed, on the other hand, seem at best to be only marginal examples of radiation.

We therefore insist upon discussing the relationship between a traveling wave in space and a "source" of finite dimensions. But in so doing we unfortunately raise some very severe questions having to do with the nature of a "source." Their difficulty is attested to by the fifty-year gap that intervened between Maxwell's discovery that the electromagnetic field obeyed a wave equation involving the velocity of light and Hertz's construction of generating and detecting equipment for other than optical electromagnetic waves.

Nevertheless, the concept of the electric source is one we have seen before in circuit theory. There, in addition to the Kirchhoff laws, and to the volt-ampere characteristics of the elements R, L, and C, we introduced the notion of voltage and current sources. These sources have the property that their essential characteristics (voltage for the voltage source, and current for the current source) do not vary with the loading conditions imposed upon them. They are able to produce electric energy indefinitely, and we do not inquire into the internal mechanism that makes this possible. That is to say, for the purposes of network analysis and discussion, it is not necessary to consider in detail the energy-conversion processes that result in the transformation from the nonelectric into electric form. We simply consider the relationship between the system response and the excitation produced by the idealized sources. We also recognize, however, that a number of real physical network sources can be represented electrically by a combination of the idealized current or voltage sources with passive network components. It is this latter feature, of course, that makes it possible to relate the network theory to the physical world.

The essential problem facing us now is seen to be the identification of appropriate idealized sources of the electromagnetic field, in terms of which we may not only calculate the field in space but also describe in a reasonable way the characteristics of physical sources of interest. Of course, in the present case, we must acknowledge the three-dimensional character of the entire situation. In this particular respect, our experience with lumped networks is much less helpful than our experience with electrostatics and magnetostatics. In electrostatics, for example, the use of the point-charge idealization is suggestive. Similarly, in magnetostatics, the circulating filamentary current is a useful source idealization. Although these two quantities represent sources, in the sense that we imagine them to be prescribed, independent of external conditions, they do not have the other feature generally associated with the source concept; namely, they do not produce any electric power. Nevertheless, it is clear that their failure to do so is connected only with the static character of the problems to which each is appropriate rather than with any fault in the spacial or electrical character of the idealization itself.

In the case of time-varying electromagnetic fields, even the development of idealized sources must conform to the general requirement of the conservation of charge. It is therefore not possible, for example, to consider isolated charges whose value changes with time. Whenever we have time-dependent charges, we must allow either a conduction or displacement current to connect them. It is possible, of course, to think of time-varying currents without time-varying charges, but only if these currents are without divergence. It is evident, therefore, that the generalization to a time-varying form of the *single* point charge of electrostatics is unreasonable.

Two point charges of equal magnitude and opposite sign may however be generalized from the static to the dynamic case, if we will permit a conduction and/or displacement current to connect them. This same configuration may, interestingly enough, also be arrived at from another point of view. Often in magnetostatics, the concept of a "current element," or a short piece of filamentry current, is considered. Actually, in the static case, such a current element is nonphysical. Its existence would imply in the d-c steady state a continuous building up of positive and negative charges at its ends. On the other hand, if we consider such a current element in a time-varying situation like the sinusoidal steady state, the idealization becomes much more attractive than before. Such a time-varying current may then appropriately be terminated at its ends in time-varying charges, and we are led back to the previous *dipole* arrangement.

If therefore we insist upon having not only an electric source but also one which (as in statics) occupies essentially a point in space, the notion of an *electric dipole and associated current element* becomes extremely attractive. Clearly, any distribution of currents that satisfies the law of conservation of charge can be made up from an appropriate sequence of these dipole charge-current elements placed end to end and side by side. In other words, if we knew the electromagnetic field produced by such an element, we could, by integration, determine the field produced by any prescribed current distribution obeying the conservation of charge.

One might now raise the objection that, in practical radiation problems, one does not necessarily know a distribution of current in terms of which to calculate the electromagnetic field. This objection however is no more fundamental in the present case than the one which maintains that Coulomb's law is useless because electrostatic problems do not always come to us in the form of a prescribed charge distribution from which we are requested to find the electric field. In point of fact, in electrostatic problems, if metal bodies at different prescribed potentials represent the actual physical problem, we can still begin the solution from the field of a point-charge (Coulomb's law). The method, of course, is first to assume momentarily that we know the (actually unknown) charge distributions on the metal bodies whose potential is specified. We proceed next to compute the field at an arbitrary point in space by using Coulomb's law and the assumed charge distribution, thereby obtaining a field expressed as an integral over the unknown charge density. If in terms of this integral expression for the field we now apply the boundary condition that the metal bodies be equipotentials or, what is the same thing, that the tangential component of the electric field vanish over their surfaces, we shall have an integral equation to solve for the unknown charge distribution. Having solved said equation, the fields will follow from Coulomb's law. Although this approach may or may not be particularly convenient for the solution of a given problem, the point to be made here is the sufficiency of the method in principle.

To see how the same ideas apply to the relationship between the dipole current-element solution and the radiation from a more nearly physical source, let us examine one such typical situation shown in Fig. 10.1. The configuration represents a perfectly conducting coaxial line feeding a very simple electric-dipole antenna. Somewhere to the left of the figure, and not shown, is a well-shielded signal generator. That is, the generator is assumed to be enclosed in a metal box, completely contiguous with the outer conductor of the coaxial cable, and its output lead presumably comes through a coaxial jack whose center conductor connects with that of the coaxial line. Thus the generator does not produce directly any field *outside* the outer conductor of the line.

We shall suppose that, at a large enough distance back from the end of the line connected to the antenna wires in Fig. 10.1, we have a valid terminal pair; i.e., we have a position in which the total field within the coaxial line structure contains only the TEM coaxial line waves.

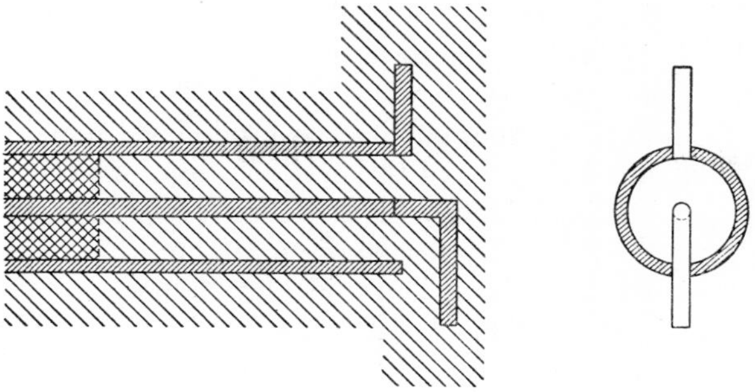

Fig. 10.1. A physical arrangement for radiating electromagnetic waves.

Fringing fields from the generator or load ends have vanished. The dotted plane in the figure shows a position at which this condition presumably applies. Observe that we may only assume this simple form of TEM field solution in the space between the surface of the inner conductor, and the inner surface of the outer conductor.

There will naturally be some current on the outer surface of the outer conductor. Since, as pointed out above, this current does not originate from the generator, it may be thought of as arising either from the leakage of magnetic field out the open end, from the magnetic field of the currents on the dipole arms, or to interconnect the time-varying charges on the outer conductor. Some of these charges are needed to terminate the fringing electric field from the open end and the attached dipole-arm wires.

The specific steady-state problem that we might have to solve involves: (1) The specification of either the voltage or the current at the dotted terminal plane; (2) the condition that the conductors in the problem be perfect; and (3) a statement about the nature of the field solution at points distant from the open end, but residing in the space outside the outer conductor.

The latter statement is, as we have pointed out before, most important in defining the radiation problem, because it must be the one that tells us of the essentially traveling-wave (rather than standing-wave) character of the solution in the open space surrounding the source.

The specification of the voltage at the dotted terminal plane, as listed above, is equivalent to specifying the electric field intensity across the whole cross section defined by this plane. This is true because the electric field configuration is fixed in all details except amplitude level by the nature of the TEM wave in the coaxial structure. The amplitude is fixed by the assumed value of the voltage across the line at the plane. Of course, we also know the configuration of the magnetic field at this plane; but its amplitude level remains unknown until we have found the rest of the field solution. In other words, we cannot predict the impedance seen at the reference terminal pair until after we know the field. By reason of our assumptions about the field at the dotted terminal plane, it would only be necessary to solve the field problem in the region of space indicated by the light, diagonal cross-hatching; the solution in the region shown in double cross hatch would then follow from transmission-line principles.

If we knew the distribution of currents on all conductor surfaces bounding the region of light cross-hatching in Fig. 10.1 (the corresponding charges need not be known separately because of the conservation-of-charge requirement), we could decompose this current distribution into small sections, each of which would represent a current element with appropriate charges at its ends. If we knew the field solution for one such current element, we could then express the field at any point in the lightly cross-hatched region of space, as well as at all points on the conductor surfaces and across the terminal plane, in terms of an integral over the unknown current-element distribution. We would then have to impose the conditions upon this integral that: (1) it reduce to the given electric field across the dotted terminal plane; (2) the tangential electric field vanish over all sections of the perfect conductor bounding the lightly crossed-hatched area of interest; and (3) the field solution vanish at large distances from the open end in such a way as to represent outgoing waves from that end. All this would lead us to a very complicated integral equation for the unknown current distribution. It is obvious that in the present geometry the procedure outlined above, although possible, is hardly practical.

An alternative procedure, equally difficult and impractical in the case of Fig. 10.1, would be to search for solutions to Maxwell's equations that are valid *in the free space* defined by the lightly cross-hatched area, and use a (possibly infinite) superposition of these free-space solutions to match all of the boundary conditions referred to above. This would be the same kind of approach commonly used in electrostatics, for example, in which the various free-space solutions of Laplace's equation in appropriate coordinate systems are arrived at by the separation of variables, and are then superposed to meet specified boundary conditions in space. Evidently, the difficulty is the lack of any convenient coordinate system in terms of which one could carry out this type of solution for the geometry of Fig. 10.1.

There are two major points to be made from the discussion pertinent to Fig. 10.1. The first one relates to the fact that practical "antenna" problems are apt to involve exceedingly complicated geometries for which exact solution will not, in general, be feasible. It is therefore more important to understand the general nature of the solutions than it is to become too deeply immersed in the details of the calculation for a specific complicated situation. The second point is that, in principle, the dipole current-element solution of which we have spoken is sufficient to solve *any* steady-state problem, however complicated. Accordingly, if we understand well the nature of the field solution appropriate to such a dipole current element, we shall have the starting point of a general understanding of the field solution in more complicated cases.

The question now is, what are the conditions that must be applied to determine the steady-state radiation field solution arising from an element of charge and current of the kind described in the foregoing? We have tried to make it clear that what we desire is a "point source" solution, in the sense that the electromagnetic field involved will be a solution to the complex Maxwell equations *in free space* at all points except a single one, where the source is located. This is, after all, the same kind of thing we mean by the electrostatic field of a point charge. A point charge or "point" current element, however, is always the result of a limiting process applied to something finite, and it results in singularities of the solution at its location. In the case of a point charge, we might start from the notion of a finite, spherical charge distribution. The radius of the sphere is then allowed to approach zero while the charge density at all points within it increases in such a manner that the total charge contained within the sphere remains finite. The electric field and potential become infinite at the point charge. Similarly, in the case of our current element or electric dipole, we shall wish to consider the limiting case of a source roughly like that described in Fig. 10.2.

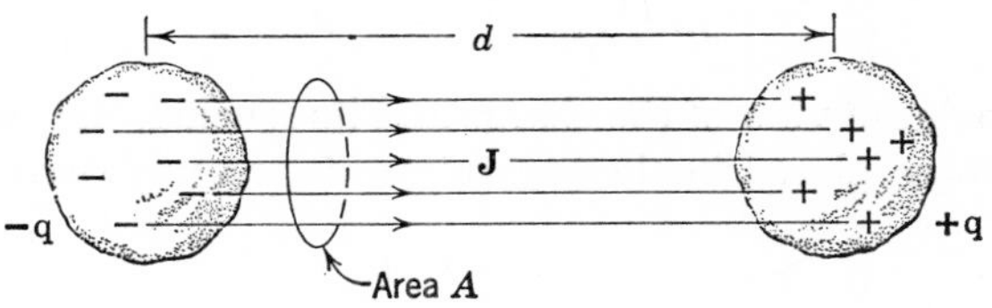

Fig. 10.2. A configuration leading in the limit of small size to an electric-dipole or current-element "point source," for which $(1/j\omega) \lim_{\substack{A\to 0\\ d\to 0\\ \mathbf{J}\to\infty}} \mathbf{J}Ad = \lim_{\substack{A\to 0\\ d\to 0\\ \mathrm{q}\to\infty}} \mathrm{q}\boldsymbol{d} = \text{const} = \mathbf{p}.$

The kind of limit process we have in mind is one in which we shrink the dipole structure down to a point in the following manner:

$$\lim_{\substack{J\to\infty \\ A\to 0 \\ d\to 0}} JAd = \lim_{\substack{I\to\infty \\ d\to 0}} Id = \lim_{\substack{J\to\infty \\ \Delta V\to 0}} J\Delta V = \text{constant} \tag{10.1}$$

where, as illustrated in Fig. 10.2, J is a *conduction current density*, A is the cross-sectional area and d is the length of the current element. Obviously $\Delta V = Ad$ is the volume of the current element. Of course, to discuss the conservation of charge, we must draw some surfaces that intercept the current element and enclose all the like-signed charges, as shown in the shaded parts of Fig. 10.2. The conservation of charge requires that

(a) $$JA = I = j\omega q$$

(b) $$\lim_{\substack{J\to\infty \\ A\to 0 \\ d\to 0}} JAd = \lim_{\substack{I\to\infty \\ d\to 0}} Id = j\omega \lim_{\substack{q\to\infty \\ d\to 0}} qd \tag{10.2}$$

which implies existence of a complex constant P, such that

$$p = \lim_{\substack{q\to\infty \\ d\to 0}} qd = \frac{1}{j\omega} \lim_{\substack{J\to\infty \\ A\to 0 \\ d\to 0}} JAd = \frac{1}{j\omega} \lim_{\substack{I\to\infty \\ d\to 0}} Id \tag{10.3}$$

Thus far, our dipole- or current-element "point source" is characterized by only a single complex parameter p. But this parameter is not quite sufficient to describe the entire effect of this source. In addition it is necessary to describe its orientation. This means that we must define a *vector* whose magnitude and time phase are contained in p of Eq. 10.3 and whose space direction indicates that of the flow of conduction current going from the negatively to the positively charged end of the dipole. Thus

$$\mathbf{p} = \frac{1}{j\omega} \lim_{\substack{JA = I\to\infty \\ d\to 0}} \mathbf{J}Ad = \frac{1}{j\omega} \lim_{\substack{I\to\infty \\ d\to 0}} \mathbf{I}d \tag{10.4}$$

From the point of view of the charge and its spacing, the vector character of $\mathbf{p}$ can be placed in the separation $\boldsymbol{d}$, pointing from $(-)$ to $(+)$ charge:

$$\mathbf{p} = \lim_{\substack{q\to\infty \\ d\to 0}} q\boldsymbol{d} = \frac{1}{j\omega} \lim_{\substack{I\to\infty \\ d\to 0}} \mathbf{I}d \tag{10.5}$$

In contrast with the point-charge source, or monopole, of electrostatics, which is a scalar quantity, we have here for the dynamic case a *vector* "point source" characterized by the *vector dipole moment* $\mathbf{p}$. It may be visualized, as in magnetostatics, in terms of a (vector) current times a length (but now with the intervention of the frequency ω), or as in electrostatics, in terms of a charge times a (vector) distance. In any case, it is most important to bear in mind that, in so far as our present definitions are concerned, the dipole moment vector $\mathbf{p}$ describes the character of an *infinitesimally small* "point source." In particular, because the charge q and the current I become infinite in the limiting processes under consideration, we should expect the electromagnetic fields to become singular at the point of location of this source. Such a situation certainly arises in the electrostatic case of a point charge and in the magnetostatic case of a current element. It could hardly be expected to be absent in the presence of their combination.

10.2 Spherical Coordinates

The problem before us now is to find a solution to the complex Maxwell equations. The behavior of this solution at one point in space must correspond to the presence there of a dipole point source, and the behavior of the solution at infinity must correspond to outgoing waves emanating from the source point.

To solve this problem, it proves convenient to locate the point dipole at the origin of a set of spherical coordinates, and to discuss the solutions to Maxwell's equations in these coordinates. For simplicity we choose to orient the dipole vector moment $\mathbf{p}$ or $\mathbf{I}d$ along the polar or z-axis of the spherical system. Our method of approach will then be a special case of the second one discussed in connection with solving a general radiation problem. Specifically, from an examination of the solutions to Maxwell's equations *in free space* and in spherical coordinates, we shall try to choose that one (or linear combination) which meets all the necessary boundary conditions.

For the problem at hand, we shall designate a point P in space by its spherical coordinates (r, ϑ, φ), as shown in Fig. 10.3. The radial coordinate r is the scalar distance measured from the origin to the point under consideration. A unit vector $\boldsymbol{a}_r$ is defined as one which has a unit magnitude and a direction radially outward. The angle ϑ is measured from a fixed axis in space, usually chosen as the $+z$-axis of the Cartesian system, to the radial vector $\boldsymbol{a}_r r$. The unit vector $\boldsymbol{a}_\vartheta$ is then in the direction of increasing ϑ. The projection of the radial vector in the x, y-plane defines the angle φ measured from the $+x$-axis, and the unit vector $\boldsymbol{a}_\varphi$ in the direction of increasing φ. The three unit

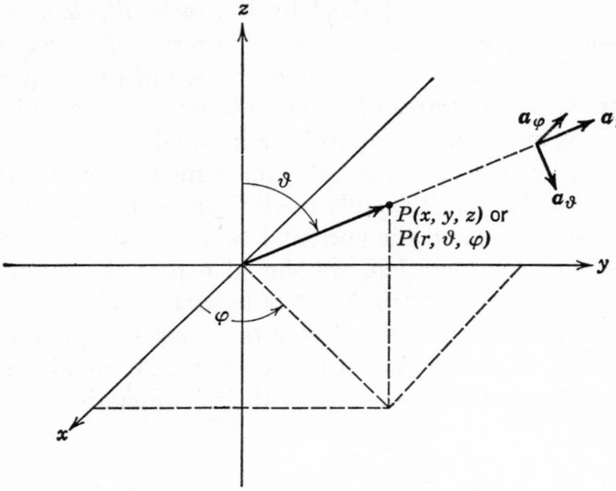

Fig. 10.3. Spherical coordinates.

vectors form an orthogonal set, in the sense that they are perpendicular to one another. It is a right-handed set, since the cross product of any two of them in the order r, ϑ, φ, is the third unit vector (Fig. 10.3). In terms of the conventional Cartesian system (x, y, z), we have directly by geometry

$$r = \sqrt{x^2 + y^2 + z^2}$$

$$\cos \vartheta = \frac{z}{\sqrt{x^2 + y^2 + z^2}} \qquad \cos \varphi = \frac{x}{\sqrt{x^2 + y^2}} \tag{10.6}$$

$$x = r \sin \vartheta \cos \varphi \qquad y = r \sin \vartheta \sin \varphi \qquad z = r \cos \vartheta \tag{10.7}$$

$$\begin{aligned} \boldsymbol{a}_r &= \boldsymbol{a}_x \sin \vartheta \cos \varphi + \boldsymbol{a}_y \sin \vartheta \sin \varphi + \boldsymbol{a}_z \cos \vartheta \\ \boldsymbol{a}_\vartheta &= \boldsymbol{a}_x \cos \vartheta \cos \varphi + \boldsymbol{a}_y \cos \vartheta \sin \varphi - \boldsymbol{a}_z \sin \vartheta \\ \boldsymbol{a}_\varphi &= -\boldsymbol{a}_x \sin \varphi + \boldsymbol{a}_y \cos \varphi \end{aligned} \tag{10.8}$$

and

$$\begin{aligned} \boldsymbol{a}_x &= \boldsymbol{a}_r \sin \vartheta \cos \varphi + \boldsymbol{a}_\vartheta \cos \vartheta \cos \varphi - \boldsymbol{a}_\varphi \sin \varphi \\ \boldsymbol{a}_y &= \boldsymbol{a}_r \sin \vartheta \sin \varphi + \boldsymbol{a}_\vartheta \cos \vartheta \sin \varphi + \boldsymbol{a}_\varphi \cos \varphi \\ \boldsymbol{a}_z &= \boldsymbol{a}_r \cos \vartheta - \boldsymbol{a}_\vartheta \sin \vartheta \end{aligned} \tag{10.9}$$

The incremental vector distance $\boldsymbol{ds}$ in space can be expressed in terms of the other differentials as follows:

$$\boldsymbol{ds} = \boldsymbol{a}_r \, dr + \boldsymbol{a}_\vartheta r \, d\vartheta + \boldsymbol{a}_\varphi r \sin \vartheta \, d\varphi \tag{10.10}$$

The scalar coefficients $(1, r, r \sin \vartheta)$ multiplying the differentials themselves are called *metric coefficients* of the spherical coordinates. They convert coordinate changes into *distance* changes. Thus, according to the definitions of gradient, divergence, and curl in vector analysis, we have directly

$$\nabla\psi = \left(\boldsymbol{a}_r \frac{\partial}{\partial r} + \boldsymbol{a}_\vartheta \frac{1}{r}\frac{\partial}{\partial \vartheta} + \boldsymbol{a}_\varphi \frac{1}{r \sin\vartheta}\frac{\partial}{\partial \varphi}\right)\psi \tag{10.11}$$

$$\nabla \cdot \boldsymbol{A} = \frac{1}{r^2 \sin\vartheta}\left[\frac{\partial}{\partial r}(r^2 \sin\vartheta A_r) + \frac{\partial}{\partial \vartheta}(r \sin\vartheta A_\vartheta) + \frac{\partial}{\partial \varphi}(r A_\varphi)\right]$$

$$= \frac{1}{r^2}\frac{\partial}{\partial r}(r^2 A_r) + \frac{1}{r\sin\vartheta}\frac{\partial}{\partial\vartheta}(\sin\vartheta A_\vartheta) + \frac{1}{r \sin\vartheta}\frac{\partial}{\partial\varphi}A_\varphi \tag{10.12}$$

$$\nabla \times \boldsymbol{A} = \frac{1}{r^2 \sin\vartheta}\begin{vmatrix} \boldsymbol{a}_r & \boldsymbol{a}_\vartheta r & \boldsymbol{a}_\varphi r \sin\vartheta \\ \dfrac{\partial}{\partial r} & \dfrac{\partial}{\partial \vartheta} & \dfrac{\partial}{\partial \varphi} \\ A_r & r A_\vartheta & r \sin\vartheta A_\varphi \end{vmatrix} \tag{10.13}$$

where A_r, A_ϑ, and A_φ are the spherical components of a vector $\boldsymbol{A}$.

We shall start from the sinusoidal steady-state form of Maxwell equations, and attempt to obtain a solution in the charge-free and current-free space beyond the immediate neighborhood of the origin. The solution will have a discontinuity of the field at the origin. The nature of the discontinuity must be suitable to describe an electric dipole point source whose complex vector moment $\mathbf{p}$ lies along $+z$.

It is clear that the field will have circular symmetry about the z-axis. This means that the field is independent of φ; so the partial derivatives with respect to φ are identically zero. With this simplification, the two Maxwell equations in complex form may be expanded in the spherical coordinates (r, ϑ, φ), to become

$$\text{(a)} \qquad -j\omega\mu \mathrm{H}_\varphi = \frac{1}{r}\frac{\partial}{\partial r}(r\mathrm{E}_\vartheta) - \frac{1}{r}\frac{\partial}{\partial\vartheta}\mathrm{E}_r$$

$$\text{(b)} \qquad j\omega\epsilon \mathrm{E}_r = \frac{1}{r \sin\vartheta}\frac{\partial}{\partial\vartheta}(\sin\vartheta \mathrm{H}_\varphi) \tag{10.14}$$

$$\text{(c)} \qquad j\omega\epsilon \mathrm{E}_\vartheta = -\frac{1}{r}\frac{\partial}{\partial r}(r\mathrm{H}_\varphi)$$

$$\text{(a)} \qquad j\omega\epsilon \mathrm{E}_\varphi = \frac{1}{r}\frac{\partial}{\partial r}(r\mathrm{H}_\vartheta) - \frac{1}{r}\frac{\partial}{\partial \vartheta}\mathrm{H}_r$$

$$\text{(b)} \qquad -j\omega\mu \mathrm{H}_r = \frac{1}{r \sin \vartheta}\left(\frac{\partial}{\partial \vartheta} \sin \vartheta \mathrm{E}_\varphi\right) \qquad (10.15)$$

$$\text{(c)} \qquad -j\omega\mu \mathrm{H}_\vartheta = -\frac{1}{r}\frac{\partial}{\partial r}(r\mathrm{E}_\varphi)$$

where H_r, H_φ, H_ϑ, E_r, E_ϑ, and E_φ are the components of the vector fields **H** and **E**, and we have regrouped the equations into a particularly convenient form.

10.3 Solution of Maxwell's Equations

The field equations found in the previous section fall into two groups. The first three equations, Eqs. 10.14, involve H_φ, E_r, and E_ϑ *only*, and the second group, Eqs. 10.15, is the dual of the first. The two groups of equations are *completely independent*. If the radial direction is considered as the "longitudinal" direction, or the direction of propagation, and the spherical surfaces are considered as "transverse," the solutions of the first group of equations will have only transverse components (H_φ) of the magnetic field. These waves will be called transverse magnetic waves, or *TM waves*. The waves associated with the second group will be called the transverse electric waves, or *TE waves*. We need consider the first group (TM waves) only, since a current element along z definitely requires H_φ and nothing in this source requires H_ϑ or H_r.

We shall first eliminate E_r and E_ϑ from the first group of equations by substituting E_r and E_ϑ from Eqs. 10.14*b* and 10.14*c* into Eq. 10.14*a*. We obtain then a partial-differential equation for H_φ.

$$\frac{1}{r}\frac{\partial^2}{\partial r^2}(r\mathrm{H}\ \ ) + \frac{1}{r^2}\frac{\partial}{\partial \vartheta}\left[\frac{1}{\sin \vartheta}\frac{\partial}{\partial \vartheta}(\sin \vartheta \mathrm{H}_\varphi)\right] + \beta_0{}^2\mathrm{H}_\varphi = 0 \qquad (10.16)$$

where $\beta_0 = \omega\sqrt{\epsilon\mu}$. This is the *wave equation* for H_φ. Once the solution for H_φ is obtained, E_r and E_ϑ can be determined from Eqs. 10.14*b* and 10.14*c*.

One of the standard techniques of finding solutions to a second-order partial-differential equation is the separation of variables. In this method, we arbitrarily look for solutions of the product form

$$\mathrm{H}_\varphi = \mathrm{R}(r)\Theta(\vartheta) \qquad (10.17)$$

where the function R is a complex function of r *only*, and the function

Θ is a complex function of ϑ *only*. Substituting Eq. 10.17 into Eq. 10.16, and multiplying through by $r^2/(R\Theta)$ to "separate the variables," we find:

$$\frac{r}{R}\frac{\partial^2}{\partial r^2}(rR) + \frac{1}{\Theta}\frac{\partial}{\partial \vartheta}\left[\frac{1}{\sin\vartheta}\frac{\partial}{\partial\vartheta}(\Theta\sin\vartheta)\right] + \beta_0{}^2 r^2 = 0 \quad (10.18)$$

We notice that the middle term of Eq. 10.18 is a function of ϑ only, and the others are functions of r only. In order to have such an equation valid for *all* values of ϑ and r, which are completely *independent* variables, the second term must be equal to a constant C and the remaining terms must equal $-$C. Thus

$$\frac{d}{d\vartheta}\left[\frac{1}{\sin\vartheta}\frac{d}{d\vartheta}(\Theta\sin\vartheta)\right] = C\Theta \quad (10.19)$$

and, from Eqs. 10.18 and 10.19,

$$r\frac{d^2}{dr^2}(rR) + CR + \beta_0{}^2 r^2 R = 0 \quad (10.20)$$

The partial derivatives have been replaced by the ordinary derivatives, since R and Θ are each functions of one variable.

It is important to recognize that the radial dependence of H_φ in Eqs. 10.16 or 10.20 is a function of $\beta_0 r = 2\pi r/\lambda$, rather than just r alone. In other words, changing the independent variable from r to $r' = \beta_0 r$ in Eq. 10.20 yields

$$r'\frac{d^2(r'R)}{dr'^2} + CR + r'^2 R = 0 \quad (10.21)$$

in which now none of the coefficients changes with frequency. This means that the behavior of the solution at low frequencies and its behavior at short distances (small r) are equivalent. *The field at distances from the origin that are very short compared to a wave length must approach the static configuration.*

Those who are quite familiar with differential equations will recognize Eq. 10.19 as the Legendre equation, whose solutions are the associated Legendre functions for a set of discrete C values. The value of C is undetermined in our problem so far. One of the simple solutions of Eq. 10.19 is obviously $1/\sin\vartheta$, for which C must be zero. The first operation of the differentiation is then identically zero, since $\Theta\sin\vartheta$ is unity. Unfortunately, $1/\sin\vartheta$ becomes infinite when ϑ becomes 0 or π. This will force H_φ to be infinite along the entire z-axis. It is a solution which is not appropriate for our point source of radiation, because it implies a discontinuity of the field that extends at full strength along the z-axis to ∞.

We are searching for another solution, in which H_φ will look like the field of a static current element for small r or β_0 (i.e., for small $r' = \beta_0 r$). Since we know from the Biot-Savart law that $H_\varphi \sim \sin\vartheta$ at low frequencies, and since Θ in the present separable solution remains the same for all values of r', we should try

$$\Theta = \sin\vartheta \tag{10.22}$$

We find that Eq. 10.19 will be satisfied for $C = -2$. Thus Eq. 10.22 is a solution of the equation

$$\frac{d}{d\vartheta}\left[\frac{1}{\sin\vartheta}\frac{d}{d\vartheta}(\Theta\sin\vartheta)\right] + 2\Theta = 0 \tag{10.23}$$

and it is the solution we want.

The remaining equation to be solved is Eq. 10.21 with $C = -2$. It becomes

$$\frac{d^2}{dr'^2}(r'R) - \frac{2}{r'}R + r'R = 0 \tag{10.24}$$

If $r' = \beta_0 r \gg 1$, the second term of Eq. 10.24 is small compared with either one of the others. The remainder equation for $(r'R)$ is that of a simple harmonic function, and, therefore,

$$R \cong \frac{A'}{r'}e^{\pm jr'} = \frac{A'}{\beta_0 r}e^{\pm j\beta_0 r} \tag{10.25}$$

For radiation problems where the direction of propagation is radially outward, with a time variation $e^{j\omega t}$, the negative sign in the exponent must be chosen. Here we are using the "outgoing wave" boundary condition at ∞ which we have discussed before. The complete solution for H_φ will then be given, from Eqs. 10.22 and 10.25, as

$$H_\varphi \cong \frac{A}{j\beta_0 r}e^{-j\beta_0 r}\sin\vartheta \tag{10.26}$$

where we have written A/j for A'. From Eqs. 10.14*b* and 10.14*c* we find that when $r' = \beta_0 r \gg 1$

$$\begin{aligned} &\text{(a)} \qquad E_\vartheta \cong \sqrt{\frac{\mu}{\epsilon}}\frac{A}{j\beta_0 r}e^{-j\beta_0 r}\sin\vartheta \\ &\text{(b)} \qquad E_r \cong 0 \end{aligned} \tag{10.27}$$

The above approximate solution of the Maxwell equations has an angular variation $\sin \vartheta$. If we consider a small volume of the space which is far away from the origin and take the *linear dimensions* of the volume to be small compared with its mean distance from the origin, we can neglect the variation of the amplitude factor $\sin \vartheta/\beta_0 r$. The solution is then that of a uniform plane wave along r!

If $r' = \beta_0 r \ll 1$, the third term of Eq. 10.24 can be neglected. The substitution $u = \mathrm{R}/r'$ reduces the resulting equation to a separable first-order one in du/dr'. Solution and subsequent integration then yield either

$$\mathrm{R} \cong \frac{\mathrm{B}}{(\beta_0 r)^2} \qquad \text{or} \qquad \mathrm{R} \cong \mathrm{D}\beta_0 r \tag{10.28a}$$

The second solution vanishes at the origin, which is not by itself appropriate for the singular source there. Still, it is worth looking at for a moment before going on. On the basis of it, the complete solution for H_φ would be

$$\mathrm{H}_\varphi \cong \mathrm{D}\beta_0 r \sin \vartheta \tag{10.28b}$$

We recognize that $r \sin \vartheta$ is the radial coordinate of a *cylindrical* coordinate system. The corresponding solutions for the electric field components are

$$\mathrm{E}_r \cong -j\mathrm{D}\frac{\beta_0}{\omega\epsilon} 2 \cos \vartheta \tag{10.28c}$$

$$\mathrm{E}_\vartheta \cong j\mathrm{D}\frac{\beta_0}{\omega\epsilon} 2 \sin \vartheta \tag{10.28d}$$

which can be combined to be written simply

$$\begin{aligned} \mathrm{E}_z &= -j\mathrm{D}\frac{\beta_0}{\omega\epsilon} \\ &= -j\mathrm{D}\sqrt{\frac{\mu}{\epsilon}} \end{aligned} \tag{10.28e}$$

This is the static electric field between two parallel conducting plates, and Eq. 10.28*b* is the *quasi-static* magnetic field induced by it. This field has no discontinuity at or near the origin, as we mentioned above, and therefore no current or charges there.

It is interesting to observe that Eq. 10.28*b*, valid when $r' = \beta_0 r \ll 1$, also happens to be hidden already in Eq. 10.26, which was originally derived under the condition $r' = \beta_0 r \gg 1$. We see this by expanding

$e^{-j\beta_0 r}$ into its power series, and taking only three terms when $\beta_0 r \ll 1$:

$$\mathrm{H}_\varphi = \frac{\mathrm{A}}{j\beta_0 r}\sin\vartheta\left[1 - j\beta_0 r + \frac{1}{2}(j\beta_0 r)^2 - \cdots\right]$$

$$= \frac{\mathrm{A}}{j\beta_0 r}\sin\vartheta - \mathrm{A}\sin\vartheta + \underbrace{\frac{1}{2}\mathrm{A}(j\beta_0 r)\sin\vartheta}_{\mathrm{D}\beta_0 r\sin\vartheta} - \cdots$$

There remains the first solution in Eq. 10.28. It leads to the following fields.

$$\mathrm{H}_\varphi \cong \frac{\mathrm{B}}{(\beta_0 r)^2}\sin\vartheta$$

$$\mathrm{E}_r \cong -j\sqrt{\frac{\mu}{\epsilon}}\frac{2B}{(\beta_0 r)^3}\cos\vartheta \tag{10.29}$$

$$\mathrm{E}_\vartheta \cong -j\sqrt{\frac{\mu}{\epsilon}}\frac{\mathrm{B}}{(\beta_0 r)^3}\sin\vartheta$$

The magnetic field can be recognized as that of an infinitesimally small current element oriented along the z-axis and located at the origin. The electric field has the same space variation as that of a similarly oriented and located electrostatic dipole. Indeed, a little more study will show us later that the current and charges obey the equation of continuity (Eq. 10.2), and, accordingly, do meet the boundary conditions for the point source we have in mind.

The approximate solutions derived so far suggest that the complete outgoing wave solution of Eq. 10.24 for R is of the form

$$\mathrm{R} = \left[\frac{\mathrm{A}}{jr'} - \frac{\mathrm{B}}{(jr')^2}\right]e^{-jr'} = \left(\frac{\mathrm{A}}{j\beta_0 r} - \frac{\mathrm{B}}{(j\beta_0 r)^2}\right)e^{-j\beta_0 r}$$

By direct substitution of the above function into Eq. 10.24, we find that it will be a solution, provided the constants A and B are the negatives of each other. Thus we have for the fields

$$\text{(a)} \qquad \mathrm{H}_\varphi = \mathrm{A}\left[\frac{1}{j\beta_0 r} + \frac{1}{(j\beta_0 r)^2}\right]e^{-j\beta_0 r}\sin\vartheta$$

$$\text{(b)} \qquad \mathrm{E}_\vartheta = \sqrt{\frac{\mu}{\epsilon}}\mathrm{A}\left[\frac{1}{j\beta_0 r} + \frac{1}{(j\beta_0 r)^2} + \frac{1}{(j\beta_0 r)^3}\right]e^{-j\beta_0 r}\sin\vartheta \tag{10.30}$$

$$\text{(c)} \qquad \mathrm{E}_r = 2\sqrt{\frac{\mu}{\epsilon}}\mathrm{A}\left[\frac{1}{(j\beta_0 r)^2} + \frac{1}{(j\beta_0 r)^3}\right]e^{-j\beta_0 r}\cos\vartheta$$

The expressions for E_ϑ and E_r were obtained, as previously, from Eqs. 10.14*b* and 10.14*c*.

Equation 10.24 for the radial variation of the field is a second-order ordinary differential equation with real coefficients. It has two independent solutions. Since the coefficients are real, the conjugate of the complex solution (Eq. 10.30*a*) will be another solution to Eq. 10.24, and also to the wave equation (Eq. 10.16). The second solution represents a wave propagating radially *inward*, on account of the exponential factor $e^{+j\beta_0 r}$. These two solutions are analogous to the two traveling waves along a transmission line. Also, the real and imaginary parts of H_φ satisfy the wave equation. They contain $\cos \beta_0 r$ and $\sin \beta_0 r$, and are analogous to the two possible complete standing waves along a transmission line. Since we are primarily interested in the outgoing radiation phenomenon for the time being, as we have remarked before, the solution given by Eqs. 10.30 is the appropriate one.

This solution has a second-order singularity at the origin for the magnetic field, and a third-order one for the electric field. The singularities correspond to discontinuities of some of the field components at the origin, which, in turn, indicate the presence of charges and currents there. Since we started from the Maxwell equations in a current-free and charge-free space, the solution is not valid *at* the origin. However, we can investigate the nature of the discontinuities there by exploring the field *near* the origin, and verify that in the limit it represents our "point source" properly.

If the variable $r' = \beta_0 r \ll 1$, we can neglect the variation of the exponential factor and drop all the terms other than the highest power term of $1/j\beta_0 r$ in each of the Eqs. 10.30. Thus Eqs. 10.30 become

$$\begin{aligned} H_\varphi &\cong A \frac{\sin \vartheta}{(j\beta_0 r)^2} \\ E_\vartheta &\cong \sqrt{\frac{\mu}{\epsilon}}\, A \frac{\sin \vartheta}{(j\beta_0 r)^3} \\ E_r &\cong 2 \sqrt{\frac{\mu}{\epsilon}}\, A \frac{\cos \vartheta}{(j\beta_0 r)^3} \end{aligned} \tag{10.31}$$

for $\beta_0 r \ll 1$. It is clear that H_φ reverses its direction from one side of the z-axis to the other, and that the line integral of H_φ around the z-axis is not zero. The line integral $\int H_\varphi r \sin \vartheta \, d\varphi$ is equal to the total current (including *both* conduction and displacement currents) enclosed by the contour.

To determine the nature of the discontinuity of the fields more precisely, it is convenient to express them in the *cylindrical* coordinates (z, ρ, φ):

$$r = +\sqrt{z^2 + \rho^2}$$

$$\sin \vartheta = \frac{\rho}{\sqrt{z^2 + \rho^2}} \tag{10.32}$$

$$\cos \vartheta = \frac{z}{\sqrt{z^2 + \rho^2}}$$

$$\boldsymbol{a}_z = \boldsymbol{a}_r \cos \vartheta - \boldsymbol{a}_\vartheta \sin \vartheta$$

$$\boldsymbol{a}_\rho = \boldsymbol{a}_r \sin \vartheta + \boldsymbol{a}_\vartheta \cos \vartheta \tag{10.33}$$

$$\boldsymbol{a}_\varphi = \boldsymbol{a}_\varphi$$

After the change of variables and rearrangement, we have for Eqs. 10.31

$$\mathrm{H}_\varphi \cong -\frac{\mathrm{A}}{\beta_0{}^2} \frac{\rho}{(z^2 + \rho^2)^{3/2}}$$

$$\mathrm{E}_z \cong \sqrt{\frac{\mu}{\epsilon}} \frac{j\mathrm{A}}{\beta_0{}^3} \frac{2z^2 - \rho^2}{(z^2 + \rho^2)^{5/2}} \tag{10.34}$$

$$\mathrm{E}_\rho \cong \sqrt{\frac{\mu}{\epsilon}} \frac{j\mathrm{A}}{\beta_0{}^3} \frac{3\rho z}{(z^2 + \rho^2)^{5/2}}$$

The component E_z is an even function of z and ρ. For a finite value of ρ, E_z changes its sign twice as z varies from $-\infty$ to $+\infty$. For constant values of z and ρ, the E_z function remains constant for all values of φ. Although the function becomes infinite at the origin ($\rho = 0$, $z = 0$), it is continuous through the z-axis elsewhere. The component E_ρ is an odd function of z as well as of ρ. In the cylindrical coordinate system, however, the radial variable ρ is always considered *positive*. Nevertheless, the vector $\boldsymbol{a}_\rho \mathrm{E}_\rho$ reverses its direction as we pass through the z-axis. This behavior of $\boldsymbol{a}_\rho \mathrm{E}_\rho$ suggests a charge distribution along the z-axis, and the behavior of H_φ suggests a current along the z-axis.

From the point of view of the electromagnetic field, we can only detect *total* current, including both the displacement and conduction currents. We can identify their sum in terms of the line integral of the magnetic field around a contour through which the total current of interest is presumed to flow. Similarly, our identification of charge from the field viewpoint rests upon the surface integral of the normal component of an electric field. Therefore, to identify the nature of the singular charge and current distribution implied by the field solution (Eq. 10.34), and to compare it with the limiting notions contained in our original description appropriate to Fig. 10.2, we shall investigate the charge and current contained within a cylinder of radius ρ drawn about the z-axis of our coordinate system.

In the discussion to follow it will be important to keep in mind the fact that our interests will be in the limiting condition when ρ approaches zero. Accordingly the total current I_z flowing through the cylinder of radius ρ in question is

$$\begin{aligned} I_z &\equiv \int_0^{2\pi} H_\varphi \rho \, d\varphi \\ &= -\frac{2\pi A}{\beta_0^2} \frac{\rho^2}{(z^2 + \rho^2)^{3/2}} \end{aligned} \tag{10.35}$$

The equivalent charge distribution may be based upon only the radial component E_ρ of the electric field emerging from the cylinder. This fact is reasonably clear in terms of our eventual interest in the limit $\rho \to 0$, but the importance of defining the charge in this way is more strongly connected with our wish to obtain a continuity relationship similar to Eq. 10.2 between the current and charges we are defining. The point is in this connection that Maxwell's equations guarantee for us that the *total* current, including both displacement and conduction components, has no divergence. Since I_z defined in Eq. 10.35 contains the z components of both the conduction and displacement currents within the cylinder, it is the radial displacement current leaving the cylinder that must be equal and opposite to I_z. Therefore, if the effective charge density $\bar{\eta}$ per unit length along z contained within the cylinder is taken to be

$$\begin{aligned} \bar{\eta} &= \int_0^{2\pi} \epsilon E_\rho \rho \, d\varphi \\ &= \sqrt{\mu\epsilon}\, \frac{j2\pi A}{\beta_0^3} \frac{3\rho^2 z}{(z^2 + \rho^2)^{5/2}} \end{aligned} \tag{10.36}$$

then, by this definition, that of I_z, and the argument just presented, we must necessarily find that the relation

$$\frac{\partial I_z}{\partial z} = -j\omega\bar{\eta} \tag{10.37}$$

is obeyed. It is indeed easy to show directly from the results of Eqs.

10.35 and 10.36 that Eq. 10.37 is satisfied, and this means that, notwithstanding its development, I_z *may* always be thought of as a pure *conduction* current line source, accompanied by a corresponding line charge $\bar{\eta}(z)$ per unit length.

Notice in Eqs. 10.35 and 10.36 that at all points other than $z = 0$, both I_z and $\bar{\eta}$ approach zero as ρ approaches zero. At the point $z = 0$, however, I_z becomes infinite as ρ approaches zero. The nature of these functions of the two variables z and ρ is best illustrated in the sketches of Fig. 10.4 which show both I_z and $\bar{\eta}$ plotted versus z, with decreasing values of ρ as a parameter. It is quite clear from this figure that, as the radius becomes smaller, the effective current I_z behaves more and more like an impulse located at the origin. Similarly, the charge density $\bar{\eta}$ looks more and more like an electric dipole in which the charges of opposite sign are becoming large and their spacing small.

Specifically, if we investigate the line integral of I_z along the z-axis, we find

$$\begin{aligned}\int_{-\infty}^{\infty} \mathrm{I}_z\, dz &= 2\int_0^{\infty} -\frac{2\pi \mathrm{A}}{\beta_0{}^2} \frac{\rho^2\, dz}{(z^2+\rho^2)^{3/2}} \\ &= \frac{4\pi \mathrm{A}}{\beta_0{}^2} \frac{z/\rho}{\sqrt{(z/\rho)^2+1}}\bigg|_0^{\infty} \\ &= -\frac{4\pi \mathrm{A}}{\beta_0{}^2} \equiv \mathrm{I}d \qquad (10.38)\end{aligned}$$

The important point is, of course, that the line integral of the current along the z-axis is independent of the radius ρ as the radius becomes small. In other words, the area under the curve in Fig. 10.4*a* approaches a definite limit as the cylinder radius approaches zero. The sense of Eq. 10.38 is that this limiting value may be called $\mathrm{I}d$, simply on the basis of its dimensional character as current times distance. At this particular point in the discussion we are not trying to identify either I or d separately, but only to recognize their product as a value which makes sense in the limit of small radii.

The charge distribution $\bar{\eta}$ is an odd function of z. The total charge is therefore zero. We know however that it is the charge times distance in which we shall be interested, and accordingly we investigate the first moment of the charge distribution along the z-axis. This moment is given by

$$\begin{aligned}\int_{-\infty}^{\infty} \bar{\eta} z\, dz &= j\frac{6\pi \mathrm{A}}{\omega\beta_0{}^2}\int_{-\infty}^{\infty} \frac{\rho^2 z^2\, dz}{(z^2+\rho^2)^{5/2}} \\ &= j\frac{4\pi \mathrm{A}}{\omega\beta_0{}^2} \equiv \mathrm{q}d \qquad (10.39)\end{aligned}$$

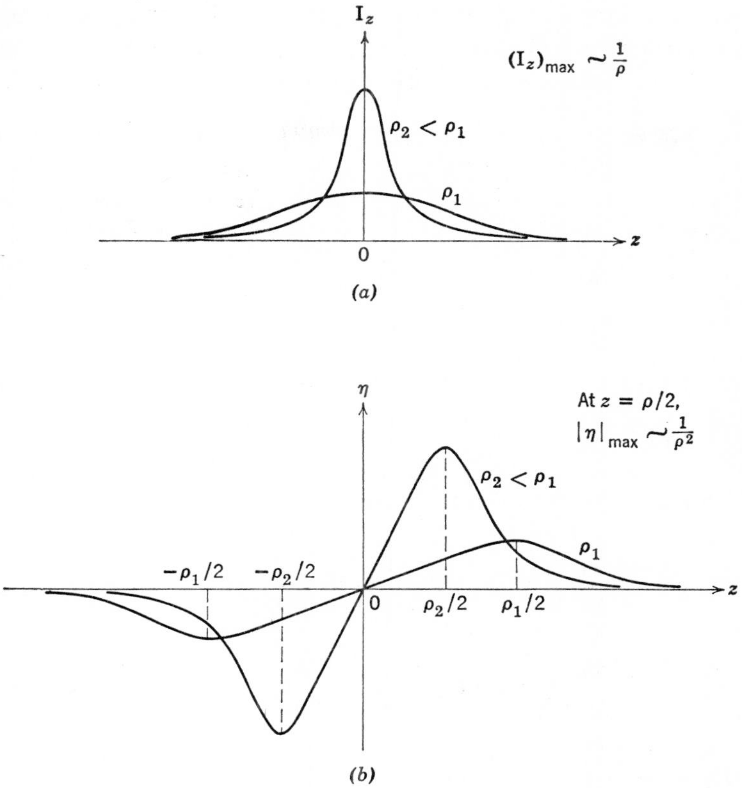

Fig. 10.4. (*a*) Effective line current and (*b*) line charge distributions defined by Eqs. 10.35 and 10.36.

In our result, there are two significant points to emphasize. One is that the charge moment is finite and independent of the radius ρ for small ρ. The second point is that this moment defines a number with the dimensions of charge times distance, which we can, if we wish, call the product of a charge q' and a distance d'. There is of course nothing in Eq. 10.39 to suggest a separate identity for either the charge q' or the distance d', only their product $q'd'$ being defined by the integral.

Nevertheless, in view of Eq. 10.38, we are at liberty to *define* the distance d' arising from the charge moment in Eq. 10.39 as the *same one* (d) that appears in the current moment integration in Eq. 10.38. Our desire to do this has in fact led us to use the same d in Eqs. 10.39 and 10.38. With this assumption, a comparison of Eqs. 10.38 and 10.39

shows that the conservation of charge then holds between the current I and the charge q. That is

$$j\omega \mathrm{q} = \mathrm{I} \tag{10.40}$$

which agrees with Eq. 10.2*a*. Thus I *may* always be thought of as a pure conduction current, and q as a free charge terminating it, even though, prior to the limiting process, we did not define them this way. Actually, at this point, Eq. 10.40 is less important than the fact that the current moment defined by Eq. 10.38 and the charge moment defined by Eq. 10.39 for the field solution (Eq. 10.30) are, in fact, related by the factor $j\omega$ characteristic of the limiting point source described by Eq. 10.2*b*. This means that our field solution may indeed be produced by an infinitesimal dipole source, and we shall so regard it for the time being.

The complete expression of the field (Eq. 10.30), called the field of an *electric dipole*, can now be rewritten in terms of the current moment, by using Eq. 10.38 to eliminate the constant A.

$$\text{(a)}\quad \mathrm{E}_r = -\frac{\mathrm{I}d\beta_0^2}{2\pi}\sqrt{\frac{\mu}{\epsilon}}\cos\vartheta\left[\frac{1}{(j\beta_0 r)^2}+\frac{1}{(j\beta_0 r)^3}\right]e^{-j\beta_0 r}$$

$$\text{(b)}\quad \mathrm{E}_\vartheta = -\frac{\mathrm{I}d\beta_0^2}{4\pi}\sqrt{\frac{\mu}{\epsilon}}\sin\vartheta\left[\frac{1}{j\beta_0 r}+\frac{1}{(j\beta_0 r)^2}+\frac{1}{(j\beta_0 r)^3}\right]e^{-j\beta_0 r} \tag{10.41}$$

$$\text{(c)}\quad \mathrm{H}_\varphi = -\frac{\mathrm{I}d\beta_0^2}{4\pi}\sin\vartheta\left[\frac{1}{j\beta_0 r}+\frac{1}{(j\beta_0 r)^2}\right]e^{-j\beta_0 r}$$

In choosing the above solution as the particular solution of the Maxwell equations, we have used the "radiation condition" or the condition of outgoing waves at infinity. We find that both E_ϑ and H_φ will vanish as $1/\beta_0 r$, and both will vary exponentially as $e^{-j\beta_0 r}$, representing a wave propagating outward. The Poynting vector under this condition is real and varies as $(1/\beta_0 r)^2$, *such that the total power radiated through a complete sphere is constant.* In the case of a lossless transmission line in the steady state, the particular solution to be used depends upon the terminal conditions at both ends. For a semi-infinite line, it is tacitly assumed that the end of the line at infinity is, in effect, matched to the characteristic impedance to allow the existence of just a single traveling wave along the line. In the frequency-domain solution, the terminal condition at infinity influences the behavior of the line solution near the input. Similarly, the boundary condition at infinity influences the dipole solution at all points in space.

10.4 Wave Impedance

The ratios of the components of the field vectors occur frequently in calculations. For plane waves we defined various wave impedances which we use to solve transmission and reflection problems. In spherical coordinates, the wave impedances are defined as:

$$\begin{aligned} Z_r &= \frac{E_\vartheta}{H_\varphi} \quad \text{or} \quad -\frac{E_\varphi}{H_\vartheta} \\ Z_\vartheta &= \frac{E_\varphi}{H_r} \quad \text{or} \quad -\frac{E_r}{H_\varphi} \\ Z_\varphi &= \frac{E_r}{H_\vartheta} \quad \text{or} \quad -\frac{E_\vartheta}{H_r} \end{aligned} \tag{10.42}$$

In the above definitions, the impedances are defined with respect to power flowing in the directions of the unit vectors $\boldsymbol{a}_r$, $\boldsymbol{a}_\vartheta$, $\boldsymbol{a}_\varphi$. The signs are fixed by the right-hand rule, in the same way as the components of Poynting's vector in spherical coordinates. It should be pointed out that the two wave impedances of the same subscript in Eq. 10.42 are not necessarily equal to each other.

For our electric dipole field (Eq. 10.41), there are only two wave impedances

$$\text{(a)} \qquad Z_r = \frac{E_\vartheta}{H_\varphi} = \sqrt{\frac{\mu}{\epsilon}}\,\frac{1 + j\beta_0 r + (j\beta_0 r)^2}{j\beta_0 r + (j\beta_0 r)^2}$$

$$\text{(b)} \qquad Z_\vartheta = \frac{-E_r}{H_\varphi} = -2\sqrt{\frac{\mu}{\epsilon}}\,\frac{\cot\vartheta}{j\beta_0 r} \tag{10.43}$$

The angular wave impedance is capacitive, since β is proportional to ω. The negative sign is the result of the defined direction of the wave impedance. There is no time-average power flow in the angular direction.

The radial wave impedance is a complex function of $\beta_0 r$ and independent of ϑ. A part of the electric field E_ϑ is in phase with the magnetic field H_φ to supply the radial power flow. Equation 10.43*a* can be developed into the following continuous fraction:

$$Z_r = \frac{1}{j\omega\epsilon r} + \frac{1}{1/(j\omega\mu r) + \sqrt{\epsilon/\mu}} \tag{10.44}$$

This particular form of the expression suggests an "equivalent circuit"

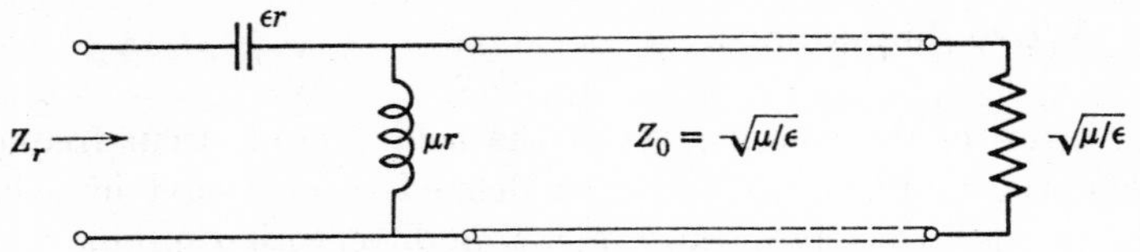

Fig. 10.5. The radial wave impedance of the field of an electric dipole.

for Z_r, as shown in Fig. 10.5. The circuit shows what the wave at radius r "sees," radially outward: a capacitance of ϵr farad, in series with a matched line of characteristic impedance $\sqrt{\mu/\epsilon}$ which is shunted at the input by an inductance of μr henrys. At a low frequency or a small radius, $\beta_0 r \ll 1$, the characteristic resistance of the transmission line, representing the real power radiated is practically short-circuited by the inductance, and Z_r is substantially capacitive. At a high frequency or a large radius, $\beta_0 r \gg 1$, the capacitance is practically a short circuit, and the inductance an open circuit. The wave impedance is resistive and equal to the intrinsic impedance of the free space. As a whole, the wave impedance has the characteristics of a high-pass filter. In spite of the presence of the inductor and the capacitor in Fig. 10.5, the circuit does not resonate at any finite frequency. The imaginary part of Z_r varies monotonically with $r' = \beta_0 r$ *but always remains capacitive.*

10.5 Complex Power

The complex Poynting vector of the dipole field is, after manipulation,

$$\mathbf{S} = \frac{1}{2}\left(\frac{|\mathrm{I}|\,d\beta_0{}^2}{4\pi}\right)^2 \sqrt{\frac{\mu}{\epsilon}}\left\{\boldsymbol{a}_r \sin^2\vartheta\left[\frac{1}{(\beta_0 r)^2} + \frac{1}{j(\beta_0 r)^5}\right]\right.$$
$$\left. + \boldsymbol{a}_\vartheta j \sin 2\vartheta\left[\frac{1}{(\beta_0 r)^3} + \frac{1}{(\beta_0 r)^5}\right]\right\} \qquad (10.45)$$

The angular component of the complex Poynting vector is pure imaginary and is consistent with the reactive angular wave impedance. The radial component can be written as

$$\mathrm{S}_r = \frac{(|\mathrm{I}|\,d)^2}{32\pi^2}\sqrt{\frac{\mu}{\epsilon}}\sin^2\vartheta\left(\frac{\beta_0{}^2}{r^2} + \frac{1}{j\beta_0 r^5}\right) \qquad (10.46)$$

The surface integral of S_r over a sphere with the dipole at the center is the total complex power radiated. If for a vacuum we take the intrinsic impedance of free space as $\sqrt{\mu/\epsilon} = 120\pi$, the complex power radiated is

$$\text{(a)} \qquad \langle P_r \rangle + jQ_r = r^2 \int_0^{2\pi} d\varphi \int_0^{\pi} S_r \sin\vartheta \, d\vartheta$$

$$= |I|^2 \frac{\sqrt{\mu/\epsilon}}{12\pi} \left[(\beta_0 d)^2 + \frac{d^2}{j\beta_0 r^3} \right] \qquad (10.47)$$

$$\text{(b)} \qquad \langle P_r \rangle + jQ_r = \frac{1}{2} |I|^2 \left[20(\beta_0 d)^2 + \frac{20d^2}{j\beta_0 r^3} \right]$$

The real part (first term) is independent of the radius. For a constant *current* moment $|I|d$, the real part is proportional to the square of frequency. The imaginary part (second term) has the appearance of the reactive power associated with a capacitance.

We shall come back to the radiation aspects of the electric-dipole-field problem later. For the time being, we should like to discuss the field of sources that correspond more nearly to physical ones than the point-source electric dipole we have considered thus far.

10.6 The Physical Electric Dipole

As we have pointed out in our earlier discussions, the function of the point-source electric-dipole solution is to act as an elementary electromagnetic field which can be used to compute that of any given current distribution. Of course, if we happen to know the current distribution, only the integration of the point-source solution is required to calculate the field. If the current distribution in question is not known, the integration does not actually give the answer. It supplies an integral form of the field which, upon application of the boundary conditions, will yield an integral equation. In terms of this integral equation the problem can, in principle at least, be solved.

There is one case in which only a single parameter of the current distribution is required for the purpose of calculating the principal features of the radiation. This is the case of the very short antenna. An example is shown in Fig. 10.6*a*, which represents two conductors forming the arms of a physical dipole. The problem of making terminal connections to this dipole leads us into complications similar to those of Fig. 10.1, and we avoid these difficulties by assuming that the two con-

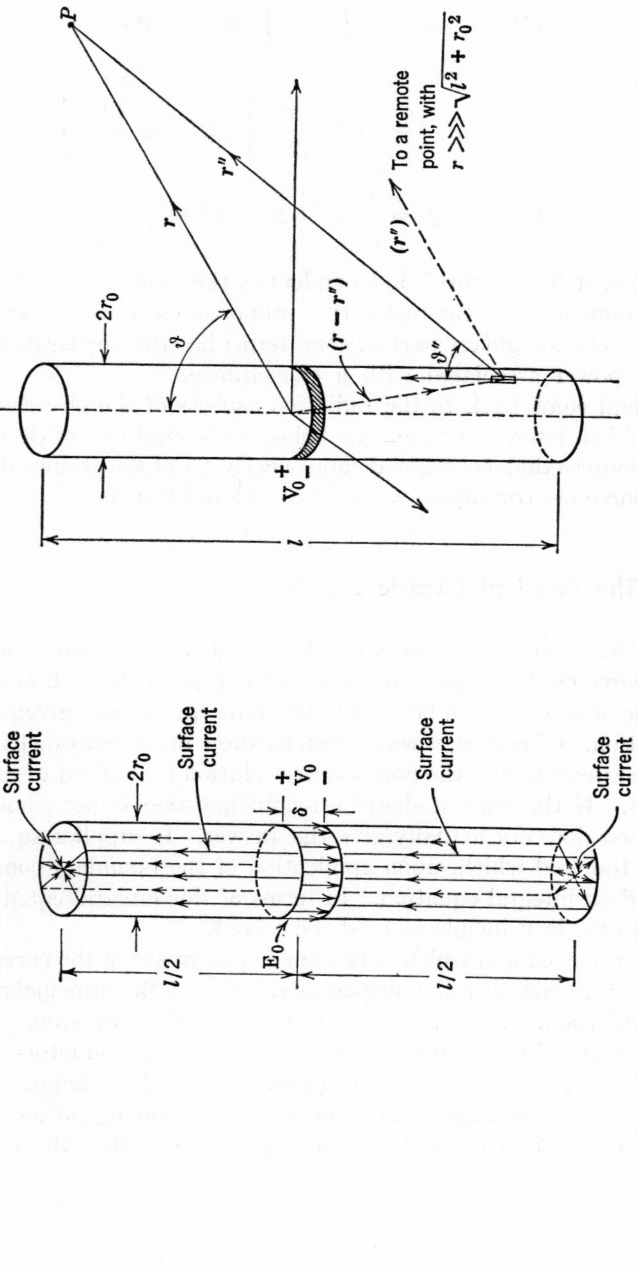

Fig. 10.6. Physical electric dipole.

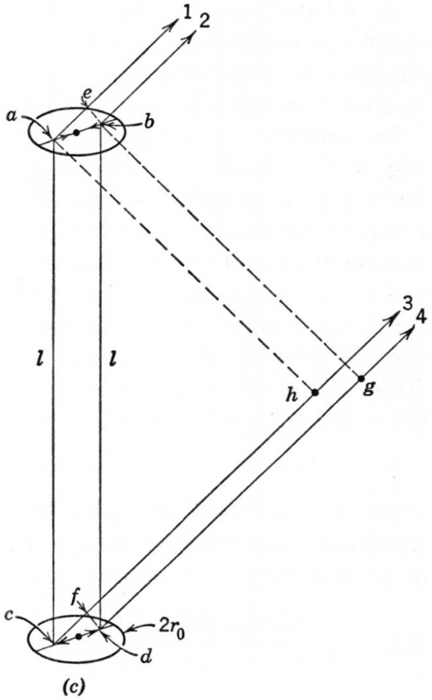

Fig. 10.6. (*Continued*)

ductors are driven from a very idealized "gap source" indicated at the center of the figure. Such a gap source may be thought of as located inside the gap and producing a strong uniform electric field E_0 across the gap surface, between the contiguous ends of the two conductors, extending over the distance δ shown. The source actually intended here is the limiting case in which δ becomes vanishingly small while E_0 becomes infinitely large, in such a manner that the product $E_0\delta$ represents a finite voltage V_0. It is assumed that the system is circularly symmetric, so that none of the electromagnetic quantities vary with the spherical longitude angle φ.

In particular the current on the surface of the cylinder extending from the center to the ends may be denoted by $\mathbf{K}_s$ and, in the notation of Fig. 10.6*b*, depends only upon z, the position on the longitudinal axis of the cylinder. We may therefore write

$$\mathbf{K}_s = \boldsymbol{a}_z K_s(z) \qquad -\frac{l}{2} < z < \frac{l}{2} \tag{10.48}$$

In order to calculate the electromagnetic field at some point P in the space surrounding the cylinder, it will be sufficient to calculate either the electric or the magnetic field at that point. The other field may be found by differentiation from Maxwell's equations.

In terms of the spherical coordinates (r, ϑ, φ) of the point P in Fig. 10.6*b*, it is clear that the longitudinal surface currents on the cylinder will produce only φ components of the magnetic field. Let us first compute the amount of this component at point P which is so produced, leaving until later the discussion of all the effects of the currents on the end caps of the cylinder (Fig. 10.6*a*). Accordingly, we must calculate the effect at point P of the field caused by a small current element like the one shown in the lower right-hand portion of the cylinder in Fig. 10.6*b*. Since the width of this element is $r_0\, d\varphi$, its current is $\mathbf{K}_s(z) r_0\, d\varphi$. Since its length is dz, its vector current moment is given by

$$d(\mathbf{I}d) = \boldsymbol{a}_z \mathrm{K}_s(z') r_0\, d\varphi'\, dz' \tag{10.49}$$

where we have used primes to distinguish the position of the element from that of the observation point P. Thus the magnetic field produced at point P may be computed from Eq. 10.41

$$d\mathrm{H}_\varphi = -\frac{\mathrm{K}_s(z') r_0 \beta_0^2\, d\varphi'\, dz'}{4\pi} \sin\vartheta'' \left[\frac{1}{j\beta_0 r''} + \frac{1}{(j\beta_0 r'')^2}\right] e^{-j\beta_0 r''} \tag{10.50}$$

Adding up the effects at point P of the φ components of the magnetic fields produced by little elements of area from all parts of the cylindrical surface (except the end caps), we find by an integration over all values of φ' and z' covering the cylinder surface

$$\mathrm{H}_\varphi = -\int_0^{2\pi}\int_{-l/2}^{l/2} \frac{\mathrm{K}_s(z') r_0\, d\varphi'\, dz'}{4\pi} \beta_0^2 \sin\vartheta'' \left[\frac{1}{j\beta_0 r''} + \frac{1}{(j\beta_0 r'')^2}\right] e^{-j\beta_0 r''} \tag{10.51}$$

Note that for a given point P, both ϑ'' and r'' are functions of z' and φ', the location coordinates of the area element.

We wish now to make a series of approximations in Eq. 10.51 that will allow us to deal in a simple way with the field produced in certain parts of space by a physical radiator, all dimensions of which are very short compared to a wave length. The following discussion will define the appropriate regions of space and develop the approximations required.

It is clear from Fig. 10.6*b* that

$$|\boldsymbol{r} - \boldsymbol{r}''| \le \sqrt{(l/2)^2 + r_0^2} \tag{10.52}$$

Let us agree to *consider only points P which are very far away from the radiator in comparison to its dimensions.* We shall then be assuming that

$$r \gg \sqrt{l^2 + 4r_0^2} \tag{10.53}$$

Under these conditions, visualization of Fig. 10.6*b* as point P is pulled outward, shows that $\mathbf{r}''$ becomes parallel to $\mathbf{r}$. Therefore, it will under the conditions of Eq. 10.53 be valid to assume that

$$\begin{aligned} &\text{(a)} \qquad \vartheta'' \cong \vartheta \\ &\text{(b)} \qquad \sin \vartheta'' \cong \sin \vartheta \end{aligned} \tag{10.54}$$

This result will immediately permit us to remove from under the integral sign in Eq. 10.51 the term containing $\sin \vartheta''$.

We now introduce the second assumption that *the largest dimension of the physical radiator is much smaller than a wave length.* More precisely, we shall assume that

$$\sqrt{l^2 + 4r_0^2} \ll \frac{\lambda}{2\pi} \tag{10.55}$$

It is important to observe that Eqs. 10.53 and 10.55 represent approximations based on two different distance scales. In Eq. 10.53 we are restricting the point P at which the field is being determined to lie far away from the physical radiator as compared to its own largest dimension. In Eq. 10.55, however, we are assuming that the largest dimension of the radiator itself is much smaller than a wave length. This pair of conditions therefore does permit the distance of observation r to be comparable to a wave length; but this situation can only arise if the physical dimensions of the radiating element itself are really considerably smaller than a wave length.

The combination of Eqs. 10.52 and 10.53 would permit us to set r'' approximately equal to r in the algebraic terms in square brackets in Eq. 10.51. That is to say, insofar as the "geometric spreading" of the spherical coordinate system produces a weakening of the field strength, the difference between the distances r'' and r is negligible as long as the point of observation is far away compared to the physical dimensions of the radiator.

Matters are quite different, however, with respect to the term $\beta_0 r''$ in the exponential in Eq. 10.51. There, the difference between r and r'', *measured in wave lengths,* determines the relative phase of arrival at point P of the radiation from various current elements on the cylinder. In this connection the assumption of Eq. 10.55 is most important.

It amounts to the statement that there is negligible phase difference at the point of observation P between the field contributions from current elements on the cylinder that are separated by the maximum possible distance. Accordingly, under these conditions we may put

$$\beta_0 r'' \equiv \frac{2\pi r''}{\lambda} \cong \frac{2\pi r}{\lambda} \tag{10.56}$$

It follows that Eq. 10.51 may be simplified to read

$$H_\varphi(\mathbf{r}) \cong -\frac{\beta_0^2 \sin \vartheta}{4\pi}\left[\frac{1}{j\beta_0 r} + \frac{1}{(j\beta_0 r)^2}\right] e^{-j\beta_0 r} \int_0^{2\pi} \int_{-l/2}^{l/2} \mathrm{K}_s(z') r_0 \, d\varphi' \, dz' \tag{10.57}$$

But,

$$\int_0^{2\pi} \mathrm{K}_s(z') r_0 \, d\varphi' = 2\pi r_0 \mathrm{K}_s(z') = \mathrm{I}_z(z') \tag{10.58}$$

where $\mathrm{I}_z(z')$ is the total current on the circumference of the cylinder at the position z'. Similarly, the current supplied by the gap voltage V_0 is

$$\begin{aligned} \int_0^{2\pi} \mathrm{K}_s(0) r_0 \, d\varphi' &= \mathrm{I}_z(0) = 2\pi r_0 \mathrm{K}_s(0) \\ &= \text{terminal current} \equiv \mathrm{I}_0 \end{aligned} \tag{10.59}$$

In view of Eq. 10.58, we may write

$$\int_0^{2\pi} \int_{-l/2}^{l/2} \mathrm{K}_s(z') r_0 \, d\varphi' \, dz' = \int_{-l/2}^{l/2} \mathrm{I}_z(z') \, dz' \tag{10.60}$$

Since in the case of the physical antenna it is convenient to deal with the terminal current at the prescribed source, denoted by I_0 in Eq. 10.59, we may appropriately *define* an *effective height* d of the physical antenna by the relation

$$\int_{-l/2}^{l/2} \mathrm{I}_z(z') \, dz' \equiv \mathrm{I}_0 d \tag{10.61a}$$

In other words, if we knew the current distribution on the surface of the cylinder, we could compute the effective height d of the physical antenna from the relations

$$d \equiv \frac{\int_{-l/2}^{l/2} \mathrm{I}_z(z') \, dz'}{\mathrm{I}_z(0)} = \frac{\int_0^{2\pi} \int_{-l/2}^{l/2} \mathrm{K}_s(z') r_0 \, d\varphi' \, dz'}{\int_0^{2\pi} \mathrm{K}_s(0) r_0 \, d\varphi'} = \frac{\int_{-l/2}^{l/2} \mathrm{K}_s(z') \, dz'}{\mathrm{K}_s(0)} \tag{10.61b}$$

We must now deal with the effect of the currents on the cylinder ends in Fig. 10.6*a*. These ends have been redrawn in Fig. 10.6*c*. It is not difficult to see from the circular symmetry of the current distributions on these ends that, at any point P in space outside the cylinder, they also contribute only a magnetic field in the φ direction. Moreover, Eq. 10.55 certainly implies that both l and the diameter $2r_0$ are individually much less than $\lambda/2\pi$. There is negligibly little phase difference between the radiation arriving at a remote point P from current elements at a and b in Fig. 10.6*c*, because this phase difference is measured by the distance ae in that figure, a distance small compared to a wave length in view of $r_0 \ll \lambda$. Since the current elements in question, however, are oppositely directed (by symmetry), their net contribution at P is approximately zero. The same comment applies for the elements c and d in the figure. Even if these terms did not cancel each other in this manner, the elements at a and c would tend to cancel each other, as would the elements at e and d. The phase difference between such pairs, based respectively on the distances ch and dg in the figure, is also small compared to a wave length on account of the distance l having the same property.

Therefore Eq. 10.57 turns out to give the entire contribution of the physical radiator to the φ component of the magnetic field. In view of Eqs. 10.61, that result may now be rewritten in terms of terminal current I_0 and effective height d as follows:

$$\left.\begin{aligned}
&\text{(a)}\ \mathrm{E}_r \cong -\frac{\mathrm{I}_0 d\beta_0^2}{2\pi}\sqrt{\frac{\mu}{\epsilon}}\cos\vartheta\left[\frac{1}{(j\beta_0 r)^2}+\frac{1}{(j\beta_0 r)^3}\right]e^{-j\beta_0 r}\\
&\text{(b)}\ \mathrm{E}_\vartheta \cong -\frac{\mathrm{I}_0 d\beta_0^2}{4\pi}\sqrt{\frac{\mu}{\epsilon}}\sin\vartheta\left[\frac{1}{j\beta_0 r}+\frac{1}{(j\beta_0 r)^2}+\frac{1}{(j\beta_0 r)^3}\right]e^{-j\beta_0 r}\\
&\text{(c)}\ \mathrm{H}_\varphi \cong -\frac{\mathrm{I}_0 d\beta_0^2}{4\pi}\sin\vartheta\left[\frac{1}{j\beta_0 r}+\frac{1}{(j\beta_0 r)^2}\right]e^{-j\beta_0 r}
\end{aligned}\right\}\begin{array}{c} d \ll \lambda \\ r \gg d \\ (10.62)\end{array}$$

The electric field components in Eqs. 10.62 have been calculated from H_φ, using Maxwell's equations; or, more simply, the electric field has been written by analogy from Eqs. 10.41*a* and 10.41*b*, in view of the similarities between Eqs. 10.62*c* and 10.41*c*.

The form of Eqs. 10.62 shows that the effective height d of the physical dipole may be interpreted in two ways, each somewhat different from that in which it was originally defined. First, regarding the field at sufficiently great distances from a sufficiently small physical dipole, the effective height d may now be said to represent the *physical* length of an equivalent dipole whose current is uniform at the value I_0 (the terminal current of the actual dipole in question) along its entire length. Since in the actual case the current $|I(z)|$ is smaller at the "open ends" $z = \pm l/2$ than at the source $z = 0$, the effective height is smaller than the physical height ($d < l$).

Alternatively, of course, it should be clear that d also represents that quantity which, when multiplied by the terminal current I_0, gives the current moment of an *infinitesimal* point dipole source whose electromagnetic field at the point of observation is very nearly that actually observed from the short physical radiator itself. That is, seen from sufficient distances, a sufficiently short physical dipole "looks" electromagnetically like a point dipole source, whose dipole current moment is the actual terminal current multiplied by the *effective* height. Thus the effective height could be determined not only from its defining relation (Eq. 10.61), which gives it in terms of the detailed current distribution along the physical conductors, but also in terms of the field strength (Eq. 10.62) measured at a point sufficiently far away from the element in question. In this latter determination, however, we must know the terminal current I_0 which produces the measured field.

An important point about our work so far is the fact that we have not attempted to find the field at points very close to the physical radiator itself. It would be important to know this field if we were interested in the precise value of the impedance seen by the driving source V_0 in Fig. 10.6*a* or 10.6*b*. In general, this field is likely to be quite complicated. Fortunately, however, we can say a good deal about this impedance for the short electric dipole without actually going through all the details of the field solution.

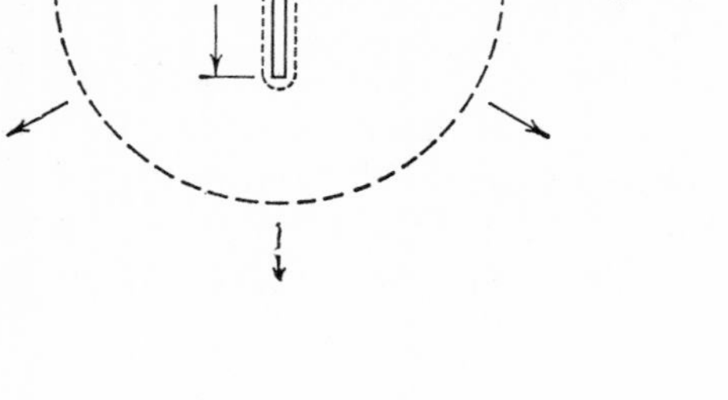

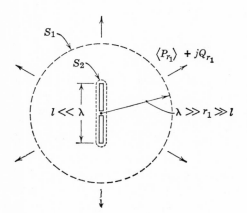

Fig. 10.7. For the discussion of the impedance of a short physical dipole.

In Fig. 10.7, consider a sphere of radius r_1 drawn about our physical dipole. This radius r_1 is chosen to be much larger than the physical dimension of the dipole, but much smaller than a wave length. Such a choice is possible, provided the physical dimensions of the radiator are much smaller than a wave length. Outside of the sphere S_1, defined by the radius r_1, we know that the electromagnetic field is very nearly given by Eq. 10.62, the field of an electric dipole. We have not computed the field in the region of space inside of S_1 and outside of an imaginary sheath S_2 which encloses tightly the physical radiating element, as shown in Fig. 10.7.

If the conductors of the physical dipole are perfect ones, we may write for the impedance seen by the driving gap source the relation

$$\mathrm{Z}_{\text{in}} \equiv \frac{\mathrm{V}_0}{\mathrm{I}_0} = \frac{2(\langle P_{\text{in}}\rangle + jQ_{\text{in}})}{|\mathrm{I}_0|^2} = \frac{2(\langle P_d\rangle + j2\omega\langle W_m - W_e\rangle)_{\text{outside } S_2}}{|\mathrm{I}_0|^2} \tag{10.63}$$

But,

$$(\langle P_d\rangle + j2\omega\langle W_m - W_e\rangle)_{\substack{\text{outside}\\ S_2}} = (\langle P_d\rangle + j2\omega\langle W_m - W_e\rangle)_{\substack{\text{between}\\ S_1 \text{ and } S_2}} + (\langle P_d\rangle + j2\omega\langle W_m - W_e\rangle)_{\substack{\text{outside}\\ S_1}} \tag{10.64}$$

Now in view of Eqs. 10.47a and 10.62, we know that

$$(\langle P_d\rangle + j2\omega\langle W_m - W_e\rangle)_{\text{outside } S_1} = \langle P_r\rangle + jQ_r = \frac{1}{2}|\mathrm{I}_0|^2\left[\frac{\sqrt{\mu/\epsilon}}{6\pi}(\beta_0 d)^2 + \frac{\sqrt{(\mu/\epsilon)}d^2}{6\pi j\beta_0 r_1^{\,3}}\right] = \frac{1}{2}|\mathrm{I}_0|^2\left(R_r + \frac{1}{j\omega C''}\right) \tag{10.65}$$

in which we have defined

(a) $$R_r = \frac{\sqrt{\mu/\epsilon}}{6\pi}(\beta_0 d)^2$$

(b) $$C'' = 6\pi\left(\frac{r_1}{d}\right)^2 \epsilon r_1 \tag{10.66}$$

Inasmuch as not only the conductors themselves but also the medium surrounding them have been assumed to be free of dissipation, Eqs. 10.64 through 10.66 yield

$$\mathrm{Z}_{\text{in}} = \frac{(4j\omega\langle W_m - W_e\rangle)_{\substack{\text{between}\\ S_1 \text{ and } S_2}}}{|\mathrm{I}_0|^2} + R_r + \frac{1}{j\omega C''} \tag{10.67}$$

But we can say something about the nature of the average stored energy in the space close to the physical dipole. Reviewing for a moment our technique illustrated in Figs. 10.6*a* and 10.6*b* of adding up the field contributions at a point P produced by all of the elementary point-source dipoles on the surface of the cylinder, and examining the complete field of such elementary point sources as given in Eq. 10.41, we notice that the dominant terms in the field will be those arising from the electric rather than the magnetic contributions. This dominance arises from the fact that the electric field varies as the inverse cube of the normalized distance $\beta_0 r$ whereas the magnetic field varies only as the inverse square. Moreover, we have already pointed out that the relative phase shift between the contributions to the field at an external point from current elements at various points on the surface of the cylinder is negligible because of the small dimensions l and r_0. That is to say, the factor $e^{-j\beta_0 r''}$ may be taken at the value 1 for all points on the cylinder when P, the point of observation, is electrically close to the dipole. In short, it is quite clear that the field within the sphere S_1 of Fig. 10.7 is that which corresponds essentially to the field of a *static* electric dipole, accompanied by a small quasi-static magnetic field contribution; i.e.,

$$\frac{(4j\omega\langle W_m - W_e\rangle)_{\substack{\text{between}\\ S_1 \text{ and } S_2}}}{|\mathrm{I}_0|^2} = \frac{1}{j\omega C'} \tag{10.68}$$

in which C' is very nearly the electrostatic capacitance of the two conductors that form the arms of the physical dipole in Fig. 10.6. Actually, since $r_1 \gg l$, it is to be expected that the largest portion of the electrostatic energy of this structure will be contained within, rather than outside, the sphere S_1, so that C' may be expected to be much greater than C'' in Eq. 10.67. In any case, we may certainly say that

$$\mathrm{Z}_{\text{in}} = R_r + \frac{1}{j\omega(C' + C'')} = R_r + \frac{1}{j\omega C} \tag{10.69}$$

in which R_r is what we call the *radiation resistance* (given by Eq. 10.66*a*), and C is capacitance which, although usually difficult to compute, is measurable reasonably easily as the low-frequency (static or quasi-static) capacitance of the physical dipole structure. In vacuum, the intrinsic impedance of free space is approximately 120π, as indicated in Eq. 10.47*b*. Therefore the important parameters of the short electric dipole for such a medium are

$$\begin{aligned} &\text{(a)} \qquad R_r = 20(\beta_0 d)^2 = 8 \times 10^2 (d/\lambda)^2 \\ &\text{(b)} \qquad C \cong \text{quasi-static capacitance of physical dipole} \end{aligned} \tag{10.70}$$

It is important to notice that the radiation resistance, first calculated essentially in Eq. 10.47, comes only from those terms of E_ϑ and H_φ that vary inversely with r. These terms are the ones that persist at large distances from the physical source, all other terms in the field components varying at least as fast as r^{-2}. This means that the real time-average power radiated may always be computed from the *distant* field, although this power must, as remarked following Eq. 10.41, be constant through all spheres surrounding the source, and, indeed, must be supplied directly at the driving terminals. The assumed lossless nature of both the radiating structure and the surrounding space means that the power radiated at great distances from the source must show directly in the form of the radiation resistance at the terminals of the radiator. The radiation resistance of a radiating structure is therefore somewhat easier to calculate than the input reactance, because the former depends only on the form of the distant field while the latter depends upon the details of the field in the immediate neighborhood of the radiator.

It is also worth noting from Eq. 10.70 that the radiation resistance may not be very large for physical structures which meet the approximations we have had to make in connection with the electrically short dipole. For example, if d/λ is equal to 1/10, R_r is equal to about 8 ohms. The approximations involved do not permit very much larger values of d/λ, so the 8-ohm figure given here is actually rather optimistic. A value more in keeping with our approximations would be $d = \lambda/100$. In that case, the radiation resistance would only be 80 milliohms. Should there be any loss in the conductors comprising the dipole arms and/or the dielectric material in the neighborhood of the conductor, the input impedance of the dipole would have to be modified accordingly, and it becomes clear that the achievement of a useful radiation efficiency from a practical dipole, which is small compared to $\lambda/2\pi$, may easily become a very serious problem. It is also quite clear that, in view of the dependence of the radiation resistance upon the square of the effective height d, any methods are definitely worth while that can make more uniform the current distribution along a given electrically short antenna without increasing its actual height, but nevertheless thereby increasing its effective height. Considerations of efficiency and the achievement of relatively uniform current distributions (or large effective heights) are often of paramount importance in the design of broadcast antennas for transmitting in the A-M standard broadcast band of frequencies or at lower frequencies. This situation arises because, with wave lengths of order of 1000 feet or more, it is not always a simple matter to construct antennas the electrical length of which is great.

In connection with the impedance of a physical antenna, there is one question over which we passed rather quickly in our discussion of the short physical electric dipole. We have assumed that d is a real number. If we refer back to Eqs. 10.60 and 10.61, it is not at all obvious that the integrals defining the effective height d are, in fact, real numbers. The point is that, unless the current density distribution $K_s(z')$ has the same phase ($\pm 180°$) at all points on the surface, there is indeed no assurance that the effective height is real.

This question can be answered, however, in terms of Eq. 10.51, which defines the magnetic field at a point of observation in terms of the surface current density $K_s(z')$. If the point of observation P is *on the surface* of the cylinder, the φ component of the magnetic field must be equal to the value of the surface current density K_s at that point. This is a consequence of the perfectly conducting nature of the cylinder, and is a boundary condition equivalent to the vanishing of the tangential component of the electric field on the boundary. For such points on the boundary, Eq. 10.51 becomes an integral equation in the surface current density K_s. But we have previously pointed out that, over the largest distances involved in the physical dimensions of the dipole, the exponential factor controlling the relative phases of the contributions at any point produced by infinitesimal current elements at other points has a variation which is completely negligible. Hence this exponential factor may effectively be eliminated from the integrand. Similarly, because of the short dimensions of the dipole compared with an air wave length, the term in $(r'')^{-2}$ will dominate the one containing only $(r'')^{-1}$, as was the case implied in our discussion relating to the evaluation of the stored energy term (Eq. 10.68) at points in the immediate neighborhood of the dipole. With these two consequences of the shortness of the dipole dimensions in terms of a wave length, the integrand in Eq. 10.51 contains only real quantities multiplying the unknown $K_s(z')$, which means that the current density solution of the integral equation has no phase variation with position.

Such reasoning as that just employed however becomes invalid in case the dimensions of the physical dipole are appreciable compared to a wave length. Under such circumstances, we can (among other things) expect the reactance to differ quite radically from its quasi-static value. Although it is not simple to see from Eq. 10.51 how the current distribution will change under these circumstances, some elements of the situation may be appreciated in terms of the transmission-line principles we have studied previously. In particular, with reference to Fig. 10.6, it is expected that the two conductors of the physical dipole will be carrying symmetrically equal and oppositely directed currents with respect to the gap source at the center. Moreover, the charge distributions on these conductors will also be of equal magnitude and opposite sign at corresponding points on the upper and lower arms. The reason for the variation of current along the conductor, and, accordingly, for the variation of charge density on the surface is, of course, the displacement current that travels through space between the opposite charges on the two arms. There will also be a displacement current connecting the two oppositely charged end faces of the conducting cylinders.

It is now not very difficult to understand, qualitatively at least, that the situation is quite similar to the pair of wires of a two-wire transmission line terminated in a capacitance. The only difference is that the separation between the conductors increases with position along them, so that the effective capacitance per unit length decreases with distance. Similarly, the inductance per unit length increases with distance. In some ways, then, the two conductors of the physical dipole are behaving like a tapered transmission line, in which the phase velocity tends, at least, to remain constant (as the inductance per unit length rises and the capacitance per unit length drops with distance along the system). Similarly, the characteristic impedance increases with distance. The terminating capacitance, represented by the end faces, completes the picture, and makes it quite clear why, at very low frequencies for which all of the dimensions are short compared to a wave length, the input impedance exhibits capacitive reactance.

But we can now see a good deal more about what happens as the length of the structure is increased. In terms of a transmission line terminated in a capacitance, it is easily appreciated that, when the distance $l/2$ from the gap source at the center to the end face of one arm becomes slightly *less* than a quarter of a wave length, we should expect a series resonance at the input terminals and a more or less perfect standing wave of current distribution along the conductor. Similarly, as the length of one arm approaches one-half wave length, we should expect a shunt resonance at the center driving terminals, etc.

We must point out, however, that the lossless transmission-line discussion is neglecting the very radiation loss for maximization of which the structure is primarily designed. To the extent that the radiated power is small compared to the stored energy, one would expect these arguments to become increasingly accurate. It should not be a surprise, therefore, that with very thin conductors, which tend to store a lot of energy in the strong fields existing over the surface of small radius, the current distribution along the conductor tends to be very nearly sinusoidal and the input impedance very much like that of a transmission line with small losses. In fact, in that case, with such a very small conductor radius, the effective capacitance of the remotely separated end faces is almost negligibly small, and the antenna behaves with respect to the gap source at the center very much like a piece of lossless transmission line open-circuited at the far end. When however the radius of the conductors becomes comparable to their length, one can reasonably understand that the stored energy per volt applied becomes less; the current distribution becomes more uniform along the length, because of the now rather large end-face area; and the effective Q of the antenna decreases considerably. That is to say, the more uniform current distribution increases the effective height of the dipole for a given physical length, thus increasing the radiation resistance while the increased radius decreases the reactance. Under these circumstances, we should not be surprised to learn that the current distribution along the conductor is not very close to sinusoidal, and the input impedance does not vary with frequency as violently from series to shunt resonance as is the case for very thin conductors.

We do not intend to pursue here any further the complicated details of antenna impedance, as related to physical structure, or the properties of electrically long dipoles. Instead we shall devote ourselves to an introductory discussion of transmitting, receiving, and array problems in which the individual radiators are taken to be short electric dipoles.

10.7 Radiation Characteristics

The first terms in E_ϑ and H_φ in Eq. 10.62 are usually called the *radiation field* of the short electric dipole, since the other terms become negligible when $\beta_0 r$ is large. For $\beta_0 r \gg 1$, we have

$$\begin{aligned} \mathrm{E}_\vartheta &\cong j\sqrt{\frac{\mu}{\epsilon}}\,\frac{\mathrm{I}_0 d}{2\lambda r}\sin\vartheta\, e^{-j\beta_0 r} \\ \mathrm{H}_\varphi &\cong j\,\frac{\mathrm{I}_0 d}{2\lambda r}\sin\vartheta\, e^{-j\beta_0 r} \qquad (10.71) \\ &\cong \sqrt{\frac{\epsilon}{\mu}}\,\mathrm{E}_\vartheta \end{aligned}$$

and the (real) radial power density:

$$S_r \cong \frac{1}{2}\sqrt{\frac{\mu}{\epsilon}}\left|\frac{\mathrm{I}_0 d}{2\lambda r}\right|^2 \sin^2\vartheta$$

It is the simplest form of radiation field associated with an electromagnetic radiator. The field is independent of the angle φ, varies sinusoidally with angle ϑ, and inversely as the distance from the radiator.

A plot of the relative angular distribution of either field intensity or power density S_r is called the *radiation pattern in field intensity or power respectively*. Thus the polar radiation pattern, in field, in the plane of the dipole is a horizontal figure eight (∞)—with the two circles tangent to the axis of the dipole—and that in a plane bisecting the dipole is a circle. In the latter plane the radiation is maximum.

Since the radiation is not uniform over all spherical angles, *the ratio of radiation power density in a given direction to the average density is defined as the "gain" in that direction.* If we define an isotropic radiator as one that radiates uniformly in all directions, the above "gain" is then relative to an isotropic radiator which is radiating the same total power. Mathematically, we have in general for the gain $G(\vartheta, \varphi)$

$$G(\vartheta, \varphi) \equiv \frac{S_r(r, \vartheta, \varphi)}{\dfrac{1}{4\pi}\displaystyle\int_0^{2\pi} d\varphi \int_0^{\pi} \sin\vartheta \, d\vartheta S_r(r, \vartheta, \varphi)} \tag{10.72}$$

where $S_r(r, \vartheta, \varphi)$ is the (real) radial component of Poynting's vector at large distances, and $\sin\vartheta \, d\vartheta \, d\varphi$ is the incremental solid angle. For an electric dipole, the gain is easily shown to be

$$G(\vartheta, \varphi) = \tfrac{3}{2} \sin^2 \vartheta \tag{10.73}$$

which has a maximum value 1.5. Note that, by definition,

$$\frac{1}{4\pi}\int_0^{2\pi} d\varphi \int_0^{\pi} \sin\vartheta \, d\vartheta G(\vartheta, \varphi) \equiv 1$$

The gain is, of course, a concept applicable to any antenna, using Eq. 10.72.

10.8 Coupled Dipoles

We have so far covered some fundamental aspects of *one* dipole. We shall now investigate the behavior of a pair of dipoles. This problem is more than twice as complicated as that of a single dipole because of the presence of the mutual coupling. One of the practical aspects of the phenomenon is the detection or reception of the radiation. We are also interested in the effect of one radiator upon another.

Consider two electrically short physical dipoles of the type shown in

Fig. 10.6, oriented arbitrarily in space but separated by a distance large compared to their own dimensions. This separation need not, of course, necessarily be large compared to a wave length. Let the terminal currents at the gap sources of the dipoles be I_1 and I_2 respectively when both are being driven simultaneously. Let the effective heights of these dipoles be d_1 and d_2—with these being defined, however, when each is radiating separately in the other's complete absence.

With terminal currents I_1 and I_2 present, the field ($\mathbf{E}$, $\mathbf{H}$) surrounding the two dipoles can be considered to be the result of the currents actually flowing over the conducting surfaces, when the actual conductors themselves are removed. This is true because the correct current distributions are, by very nature, precisely those that flow to make the field satisfy the boundary conditions on the conductors. Consequently, we may take advantage of the linearity of the system to resolve ($\mathbf{E}$, $\mathbf{H}$) at every point in space surrounding the dipoles into two field contributions

$$\begin{aligned} \mathbf{E} &= \mathbf{E}_1 + \mathbf{E}_2 \\ \mathbf{H} &= \mathbf{H}_1 + \mathbf{H}_2 \end{aligned} \tag{10.74}$$

where ($\mathbf{E}_1$, $\mathbf{H}_1$) is the field that would be produced by *all* the currents on dipole 1, *when all the currents on dipole 2 are absent* (*zero*), and ($\mathbf{E}_2$, $\mathbf{H}_2$) is the field produced by all currents on dipole 2 when those of dipole 1 are absent.

There are two properties of a partial field like ($\mathbf{E}_1$, $\mathbf{H}_1$) that are very important to appreciate as matters of principle, especially because we are immediately going to make approximations concerning them. First, for a given value of I_1, the current distribution defining ($\mathbf{E}_1$, $\mathbf{H}_1$) is *not* the same as would occur on dipole 1 radiating alone in free space. The presence of dipole 2 and its driving current I_2 modify by a "proximity effect" the current distribution on dipole 1. That is, the "effective height" and the symmetry giving rise to ($\mathbf{E}_1$, $\mathbf{H}_1$) are not such as to yield equations like Eq. 10.62 with I_1d_1 in place of I_0d. Nevertheless, because of the assumed large spacing between the two dipoles, compared to their dimensions, *we expect their mutual coupling to be weak.* Accordingly, we shall *assume* that ($\mathbf{E}_1$, $\mathbf{H}_1$) *is* the field of dipole 1 in free space, driven by terminal current I_1. Thus, Eq. 10.62 with I_1d_1 for I_0d gives correctly this field at appropriately distant points in space, and the close field defines a capacitance C_1 analogous to that in Eq. 10.70. Of course, analogous comments apply to ($\mathbf{E}_2$, $\mathbf{H}_2$).

The second property of importance is that the condition of dipole 2 when ($\mathbf{E}_1$, $\mathbf{H}_1$) is defined may *not* be arrived at simply by open-circuiting its gap terminals. Such an operation forces $I_2 = 0$, but it does *not* force *all* the surface currents on dipole 2 to vanish. Nevertheless, with the center gap *really open* ($I_2 = 0$ means no conduction or displacement current inside it), and with the far end faces of the conductors small and facing away from each other, each conductor of the dipole is forced to have very nearly zero current at its far end as well as exactly zero current at its gap end. With the assumed dipole dimensions much less than $\lambda/2\pi$, there just is not enough space to permit much current to build up on the portion of the conductor between these ends. Thus we shall assume that calculations made using $(\mathbf{E}_1, \mathbf{H}_1)$ in the space surrounding the dipoles do apply to the situation $I_2 = 0$, without additional conditions.

To investigate the mutual effects between the two dipoles considered above, we shall first investigate the power flow in the neighborhood of the first dipole. In terms of the partial fields defined by Eq. 10.74, the resultant complex Poynting vector is

$$\mathbf{S} = \tfrac{1}{2}(\mathbf{E}_1 + \mathbf{E}_2) \times (\mathbf{H}_1{}^* + \mathbf{H}_2{}^*)$$

$$= \tfrac{1}{2}(\mathbf{E}_1 \times \mathbf{H}_1{}^* + \mathbf{E}_1 \times \mathbf{H}_2{}^* + \mathbf{E}_2 \times \mathbf{H}_1{}^* + \mathbf{E}_2 \times \mathbf{H}_2{}^*) \quad (10.75)$$

Power densities, like energy densities, are not generally additive quantities because they depend upon products of fields. The cross terms in the above equation do not always vanish. We cannot study the divergence of the Poynting vector right at the first dipole, because of the mathematical singularity of $\mathbf{E}_1$ and $\mathbf{H}_1$ near the gap source. We shall, therefore, calculate the surface integral of the Poynting vector through a sheath surrounding the dipole, like surface S_2 of Fig. 10.7. The complex power flowing *out* of such a closed surface (S_1) surrounding the first dipole is

$$P = \int_{S_1} \mathbf{S} \cdot d\boldsymbol{a}$$

$$= \frac{1}{2} \int_{S_1} d\boldsymbol{a} \cdot (\mathbf{E}_1 \times \mathbf{H}_1{}^* + \mathbf{E}_1 \times \mathbf{H}_2{}^* + \mathbf{E}_2 \times \mathbf{H}_1{}^* + \mathbf{E}_2 \times \mathbf{H}_2{}^*) \quad (10.76)$$

The integral (Eq. 10.76) contains four terms. The first one is, on the basis of the approximations discussed above, the surface integral of the Poynting vector around the first dipole in the absence of the second one. Thus from Eqs. 10.69 and 10.70

$$P_{11} = \frac{1}{2} |I_1|^2 \left[20(\beta_0 d_1)^2 + \frac{1}{j\omega C_1} \right] \quad (10.77)$$

The fourth term is by strict definition the surface integral of the Poynting vector due to the second dipole, around the first dipole, in the absence of the first dipole. Since obviously no time-average power is emitted or absorbed by the first dipole under the conditions defining $(\mathbf{E}_2, \mathbf{H}_2)$, the real part of the integral must vanish anyway. The imaginary part will also vanish if the dimensions of the volume enclosed by the surface (i.e., the volume of dipole 1) are small compared to λ. We have already assumed such dimensions.

For the second integral, the magnetic field of the second dipole is continuous through the volume of the integral, provided the first dipole is electrically small. Since the surface of integration is a cylinder coaxial with the first dipole, the tangential component of $\mathbf{E}_1$ is also continuous. Thus the second integral will vanish.

The only integral remaining to be evaluated is the third one.

$$\mathrm{P}_{21} = \frac{1}{2}\int_{S_1} d\boldsymbol{a}\cdot\mathbf{E}_2 \times \mathbf{H}_1{}^* \tag{10.78a}$$

For this integral, if the dimensions of the cylinder S_1 are small (compared to λ), $\mathbf{E}_2$ can be considered to be constant over the whole volume enclosed. The integral then becomes

$$\begin{aligned}\mathrm{P}_{21} &= \frac{1}{2}\int_{S_1} \mathbf{E}_2\cdot\mathbf{H}_1{}^* \times d\boldsymbol{a}\\ &= \frac{1}{2}\mathbf{E}_2\cdot\int_{S_1} \mathbf{H}_1{}^* \times d\boldsymbol{a}\\ &= -\frac{1}{2}\mathbf{E}_2\cdot\boldsymbol{i}_1\int_z \mathrm{I}_1{}^*\, dz\\ &= -\tfrac{1}{2}\mathbf{E}_2\cdot\boldsymbol{i}_1\mathrm{I}_1{}^*d_1\end{aligned} \tag{10.78b}$$

The unit vector $\boldsymbol{i}_1$ is in the direction of the current on the first dipole. The last form of P_{21} can be interpreted as one half of the product of the conjugate of the "current moment" and the component of the electric field parallel to it. This is the coupling term.

The complete integral (Eq. 10.76) is then given by

$$\mathrm{P}_1 = \tfrac{1}{2}|\mathrm{I}_1|^2\mathrm{Z}_{11}' - \tfrac{1}{2}\mathrm{I}_1{}^*\mathbf{E}_2\cdot\boldsymbol{i}_1 d_1 \tag{10.79a}$$

Similarly, the complex power radiated from the second dipole can be shown to be:

$$\mathrm{P}_2 = -\tfrac{1}{2}\mathrm{I}_2{}^*\mathbf{E}_1\cdot\boldsymbol{i}_2 d_2 + \tfrac{1}{2}|\mathrm{I}_2|^2\mathrm{Z}_{22}' \tag{10.79b}$$

where Z_{11}' and Z_{22}' are by our approximations the input impedances, given by expressions like Eq. 10.77, of the first and second dipoles respectively in the absence of the other.

Now, because the whole system between the two dipoles is linear, the terminal input voltages V_1 and V_2 of the dipoles must be related to the terminal currents by equations of the form

$$\begin{aligned} &\text{(a)} \qquad V_1 = I_1 Z_{11} + I_2 Z_{12} \\ &\text{(b)} \qquad V_2 = I_1 Z_{21} + I_2 Z_{22} \end{aligned} \qquad (10.80)$$

If the medium surrounding the two dipoles is isotropic (as well as linear), the reciprocity theorem [1] applies to the circuit containing the space around the dipoles, and we have in Eqs. 10.80

$$Z_{12} = Z_{21} \qquad (10.81)$$

Note that Z_{ij} measures voltage at i produced by *terminal* current at j, with terminals i open (*terminal* current $I_i = 0$). From Eqs. 10.80 and 10.81, we have for the complex power

$$\begin{aligned} &\text{(a)} \qquad P_1 = \tfrac{1}{2}|I_1|^2 Z_{11} + \tfrac{1}{2} I_2 I_1^* Z_{12} \\ &\text{(b)} \qquad P_2 = \tfrac{1}{2} I_1 I_2^* Z_{21} + \tfrac{1}{2}|I_2|^2 Z_{22} \end{aligned} \qquad (10.82)$$

To facilitate comparison between Eqs. 10.79 and 10.82, note that, since $\mathbf{E}_1$ is proportional to I_1 and $\mathbf{E}_2$ is proportional to I_2, $\mathbf{E}_1/I_1$ and $\mathbf{E}_2/I_2$ are functions of coordinates only. We can then rearrange Eqs. 10.79 in the form

$$\begin{aligned} &\text{(a)} \qquad P_1 = \frac{1}{2}|I_1|^2 Z_{11}' - \frac{1}{2} I_2 I_1^* \left(\frac{\mathbf{E}_2}{I_2}\right)_{\text{at dipole 1}} \cdot \boldsymbol{i}_1 d_1 \\ &\text{(b)} \qquad P_2 = -\frac{1}{2} I_1 I_2^* \left(\frac{\mathbf{E}_1}{I_1}\right)_{\text{at dipole 2}} \cdot \boldsymbol{i}_2 d_2 + \frac{1}{2}|I_2|^2 Z_{22}' \end{aligned} \qquad (10.83)$$

It follows that we may set

$$\left.\begin{aligned} &\text{(a)} \qquad Z_{11} \cong Z_{11}' \\ &\text{(b)} \qquad Z_{22} \cong Z_{22}' \\ &\text{(c)} \qquad Z_{12} = Z_{21} \cong -\left(\frac{\mathbf{E}_2}{I_2}\right)_{\text{at dipole 1}} \cdot \boldsymbol{i}_1 d_1 \\ &\qquad\qquad \cong -\left(\frac{\mathbf{E}_1}{I_1}\right)_{\text{at dipole 2}} \cdot \boldsymbol{i}_2 d_2 \end{aligned}\right\} \qquad (10.84)$$

The difference between the strict conditions defining Z_{ij} and those involved in precise definitions of Z_{11}', Z_{22}', $\mathbf{E}_1$, and $\mathbf{E}_2$ have been elimi-

[1] See for example the proof given in, Robert M. Fano, Lan Jen Chu, Richard B. Adler, *loc. cit.*

nated only through our assumptions of small dipoles and comparatively large spacing.

We have through Eqs. 10.84 reduced the problem of two weakly coupled dipoles to a two-terminal-pair network problem. The effect of any source or load connected to the two-terminal-pair system can now be analyzed on the basis of conventional circuit theory.

The above theory can easily be extended to cover the general case of N small dipoles with weak mutual coupling.

10.9 The Receiving Properties of a Dipole

If the second dipole discussed in the last section is far, far away from the first dipole, the field due to the second dipole at the first one is essentially a uniform plane wave. Let $\mathbf{E}_0$ be the complex electric field of the uniform plane wave at the first dipole when the latter is absent. The voltage across the input terminal of the first dipole is then given by Eqs. 10.80 and 10.84 as

$$\mathrm{V}_1 = \mathrm{I}_1\mathrm{Z}_1 - \mathbf{E}_0 \cdot \boldsymbol{i}_1 d_1 \qquad (10.85)$$

The second term $\mathbf{E}_0 \cdot \boldsymbol{i}_1 d_1$ is the open-circuit voltage observed at the terminals. The first term can be interpreted as the voltage drop across the input impedance Z_1 of the dipole. Looking into the terminals of a dipole in the presence of an external field, we see an equivalent circuit as shown in Fig. 10.8; namely, an open-circuit voltage in series with the impedance Z_1.

Let E_0 be the complex magnitude of the *linearly polarized* electric field of a uniform plane wave incident upon the dipole, and let ψ be the angle between the electric field vector $\boldsymbol{E}_0$ and the axis of the dipole. The open-circuit voltage then has the complex magnitude $\mathrm{E}_0 d_1 \cos \psi$. To visualize the relationship between the angle ψ and the direction of propagation of the incident wave, let us choose the latter to coincide with the z-axis, the electric field vector along the x-axis, and the dipole along an arbitrary radial direction (ϑ, φ), as shown in Fig. 10.9. Then

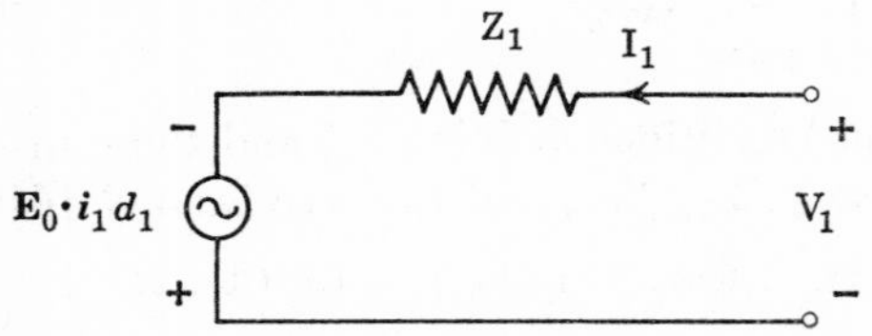

Fig. 10.8. Equivalent circuit of a short receiving electric dipole.

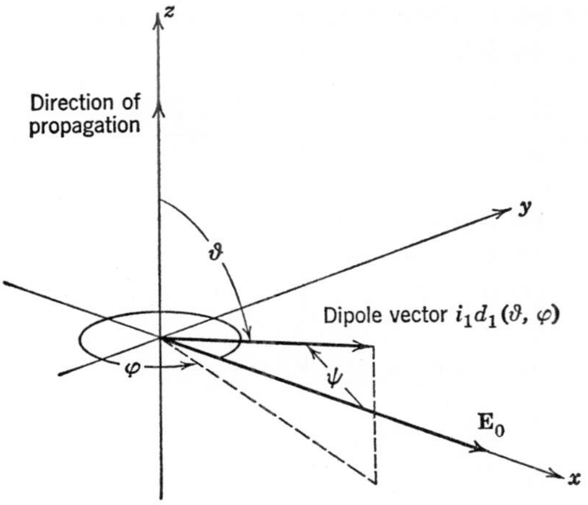

Fig. 10.9. Geometry of induced voltage for a receiving dipole.

$\cos\psi$ is the direction cosine of the dipole vector with respect to the x-axis.

$$\cos\psi = \sin\vartheta\cos\varphi$$

The plane containing the direction of propagation and the dipole is usually defined as the plane of incidence. The angle φ is measured from the electric vector to the plane of incidence (not the normal to the plane). For a given ϑ, maximum open-circuit voltage occurs when the electric field is in the plane of incidence ($\varphi = 0$). Under this condition, the open-circuit voltage is proportional to $\sin\vartheta$, identical with the angular variation of the radiation field of the dipole. If the dipole can be oriented freely in all directions, it will have a maximum open-circuit voltage when the dipole is parallel to the electric field of the incident wave.

It will be interesting to know the maximum power a load connected to the dipole can absorb from the plane wave. The power *absorbed* is given as the real part of $-\mathrm{P}_1$, since P_1 was defined as the complex power flowing *out* of the dipole. To evaluate P_1, we must have some knowledge of the load connected to the dipole. Let the load impedance be Z_L. From the equivalent circuit in Fig. 10.8, we find

$$\mathrm{I} = \frac{\mathbf{E}_0 \cdot \boldsymbol{i}_1 d}{\mathrm{Z}_1 + \mathrm{Z}_L} \tag{10.86}$$

The power absorbed by the load will then be $|\mathrm{I}_1|^2 R_L/2$, where R_L is the load resistance. The real power absorbed will have a maximum, with respect to variation of load impedance, of

$$\langle P\rangle_{\text{max}} = \frac{|\mathrm{E}_0 d \cos \psi|^2}{(2)(4)20\beta_0{}^2 d^2} \tag{10.87}$$

achieved when the load impedance is the *conjugate* of the dipole impedance. We observe that the maximum power is *independent of the "effective height" of the dipole.* This maximum power can be written

$$\langle P\rangle_{\text{max}} = A(\vartheta, \varphi)S_0 = \left(\frac{3\lambda^2}{8\pi} \cos^2 \psi\right) S_0 \tag{10.88a}$$

where

$$S_0 = \frac{1}{2}\sqrt{\frac{\epsilon}{\mu}}\,|\mathrm{E}_0|^2$$

is the magnitude of the power density of the incident wave and

$$A(\vartheta, \varphi) = \frac{3\lambda^2}{8\pi} \cos^2 \psi = \frac{3\lambda^2}{8\pi} \sin^2 \vartheta \cos^2 \varphi \tag{10.88b}$$

has the dimensions of an area. This factor $A(\vartheta, \varphi)$ is called the *receiving cross section* of the electric dipole. It has a maximum value $3\lambda^2/8\pi$ when the (linearly polarized) electric field is parallel to the dipole.

The definition of the receiving cross section is

$$A(\vartheta, \varphi) = \frac{\langle P\rangle_{\text{max}}}{S_0} \tag{10.89}$$

As the ratio of the maximum (with respect to load impedance) of the power received from a linearly-polarized plane wave to the incident wave power density, for a given antenna orientation, it is useful in general—not just for electric dipoles.

It should be emphasized that, when a dipole is used for receiving and is providing appreciable current to a load, some current also flows along the dipole. It produces a field in space associated with the induced current element. This field is usually called a *scattered field.* It reacts with the transmitting dipole through the mutual impedance Z_{12}, and may modify the input impedance of the transmitter appreciably if the two are close enough.

10.10 Radiation from Two or More Dipoles

We now wish to discuss the properties of two or more dipoles that are close together electrically, but not in terms of their dimensions. We shall consider how these dipoles radiate.

The mutual effect of two dipoles can be represented by a two-terminal-pair network with prescribed self- and mutual impedances, as we have discussed before. If a source is connected to one dipole, and a load is connected to the other, we can visualize a limited range of possible complex values of $(I_2d_2)/(I_1d_1)$ achieved by varying the load impedance. If, however, both dipoles are connected to the same source, through separate networks, the choice of $(I_2d_2)/(I_1d_1)$ is practically unlimited. To determine the relative currents is a conventional circuit problem. We shall therefore restrict our discussion to the external field, assuming that the dipole terminal currents are known. In particular, we are interested in the field at electrically great distances from the dipoles—the "radiation" field.

The radiation field of a single current element parallel to the z-axis and located at the origin can be rewritten as

$$E_\vartheta \cong j\,\frac{60\pi I_0 d}{\lambda r}\, e^{-j\beta_0 r} \sin\vartheta \qquad (10.90)$$

$$H_\varphi \cong E_\vartheta / 120\pi$$

We notice that over any sphere with the current element at the center, the time phases of E_ϑ and H_φ remain constant. In optical terms, the sphere is called the wave front of the point source. The total phase delay or retardation from the current to the field at a distance r away is $\beta_0 r - \pi/2$. When we have several current elements distributed in space, the waves from them will arrive at a distant point with different retardations. This difference varies with the spherical angles of the point of observation with respect to the current elements. We have, then, an interference phenomenon similar to that discussed for oblique incidence of plane waves in Chapter 7. Although the interference phenomenon is generally undesirable, it can be used to advantage to concentrate the radiated power in chosen directions when the relative phase and magnitude of the currents in the dipole elements can be controlled accurately. In the following, we will therefore be discussing what may be called the principles of simple antenna arrays.

Let the current element I_1d_1 be located at a point (x_1, y_1, z_1) near the origin. For the radiation field at a distance sufficiently far away

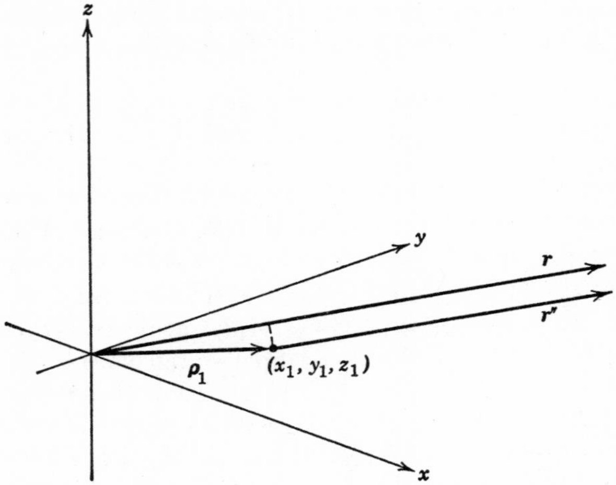

Fig. 10.10. Geometry of an array of dipoles.

from the origin, Eq. 10.90 is valid, provided the variable r in the exponent is considered to be the distance measured *from* the dipole *to* the point of observation. This distance will be designated as r''.

In general, the difference between the spherical angle (ϑ, φ) to the point of observation, as seen from the origin, and (ϑ'', φ'') as seen from the dipole, is known as *parallax*. See Fig. 10.6*b* for a similar case. The parallax approaches zero as the distance r'' approaches infinity. The variable r'' in the amplitude factor in Eq. 10.90 can be approximated as r under the same condition. However, the distance r'' in the exponent is measured in terms of the wave length, and controls the phase of the field in space. This distance r'' can be evaluated from Fig. 10.10 in terms of (r, ϑ, φ). When the point of observation (r, ϑ, φ) is far away, the vectors r and r'' are practically parallel to each other. The difference in length is the projection of the vector $\boldsymbol{\rho}_1$ (from the origin to (x_1, y_1, z_1)) on the vector $\boldsymbol{r}$. Thus for large r''

$$\begin{aligned} r - r'' &= \boldsymbol{a}_r \cdot \boldsymbol{\rho}_1 \\ &= \boldsymbol{a}_r \cdot (\boldsymbol{a}_x x_1 + \boldsymbol{a}_y y_1 + \boldsymbol{a}_z z_1) \end{aligned} \tag{10.91}$$

The dot products of the unit vectors are the direction cosines with respect to the three Cartesian axes, so

$$r'' = r - x_1 \sin \vartheta \cos \varphi - y_1 \sin \vartheta \sin \varphi - z_1 \cos \vartheta \tag{10.92}$$

If there are N current elements, *all oriented parallel to the z-axis* and radiating at the same frequency, the resultant radiation field is evidently

$$\mathrm{E}_\vartheta \cong j \frac{60\pi}{\lambda r} e^{-j\beta_0 r} \sin \vartheta \sum^{N} \mathrm{I}_n d_n e^{j\beta_0 \mathbf{a}_r \cdot \boldsymbol{\rho}_n} \tag{10.93}$$

The coefficient in front of the summation is the complex radiation field of a unit current element located at the origin. It has an angular distribution $\sin \vartheta$, which is characteristic of each of the radiating elements. The whole coefficient is called the *element factor* (E.F.) of the entire space pattern of the field because it contains all the features of the field of each element of the array. The summation in Eq. 10.93 is independent of the radiation characteristics of the elements but depends upon the geometry of the arrangement as well as on the relative phases and amplitudes of the excitation. It is called the *array factor* (A.F.). In radio engineering, the judicious choice of excitation and physical arrangement of an array provides a convenient means of directing the radiation in desired directions.

As a simple example, consider the two dipoles oriented along the z-axis and placed on it. One is located at the origin, and the other at a distance $z = c$ above the origin. Let the current moments be of the same magnitude, but of different phases, given by Id and $Ide^{-j\psi}$ respectively. Then the array factor is

$$\text{A.F.} = Id(1 + e^{j\beta_0 c \cos \vartheta - j\psi}) \tag{10.94}$$

The exponent of the second term is the total phase difference between contributions of the two dipoles to the field at a distant point. This difference is due jointly to the difference in the paths of transmission and the difference in the phase of excitation. The factor can be rewritten as

$$\text{A.F.} = 2Ide^{j(\pi c/\lambda) \cos \vartheta - j(\psi/2)} \cos \left[\left(\frac{\pi c}{\lambda} \cos \vartheta \right) - \frac{\psi}{2} \right] \tag{10.95}$$

Because of the bad choice of the origin, the phase angle of the array factor is no longer constant over a sphere with a center at the origin. We note that the phase variation around the sphere is one half that of the second dipole over the same sphere. Thus, if we shift the origin up along the z-axis by a distance $c/2$, the phase variation of the A.F. around the new sphere will vanish.

The array factor has a maximum value of $2Id$ when

$$\left(\frac{\pi c}{\lambda} \cos \vartheta \right) - \frac{\psi}{2} = n\pi \qquad n = 0, \pm 1, \pm 2, \cdots$$

This occurs wherever the phase difference of the radiation fields from the two elements is zero or an integer multiple of 2π:

$$(\beta_0 c \cos \vartheta) - \psi = 2n\pi \qquad n = 0, \pm 1, \pm 2, \cdots \tag{10.96}$$

Since $0 \leq \vartheta \leq \pi$, the value of $\beta_0 c \cos \vartheta$ is limited to the range $-\beta_0 c$ to $\beta_0 c$. The spacing c can therefore be chosen to restrict the number of maxima within the physical angle ϑ.

The A.F. is zero wherever

$$\left(\frac{\pi c}{\lambda} \cos \vartheta\right) - \frac{\psi}{2} = \pm(2n + 1)\frac{\pi}{2} \qquad n = 0, 1, 2, \cdots$$

or

$$(\beta_0 c \cos \vartheta) - \psi = \pm(2n + 1)\pi \qquad n = 0, 1, 2, \cdots$$

which is an odd multiple of π. The waves from the two dipoles then arrive at the point of observation with opposite phases. Between the maximum and minimum, the array factor varies cosinusoidally with $\cos \vartheta$. With two parameters, $\beta_0 c$ and ψ, we can specify angles for maximum and minimum radiation arbitrarily between 0 and π.

For the case $\psi = 0$, the two current moments along the z-axis are identical. The array factor is always maximum in the equatorial plane for any spacing c. When the spacing c is equal to λ, the path difference from the elements to any point along the $\pm z$-axis is a full wave length, and we have constructive interference along these directions. The radiation field, however, is actually zero for $\vartheta = 0$ or π, on account of the element factor. When the spacing is 2λ or more, constructive interference will occur at many values of ϑ, with destructive interference in between. The situation is similar to that of passing light through two narrow slits. After passing through the slits, the light from a monochromatic source appears on a screen as a number of lines. Since the line spacing is fixed when measured in wave lengths, the phenomenon can be used to determine the frequencies of optical radiation.

The case $\psi = 0$ discussed above has a trivial limit when the spacing becomes zero ($c \to 0$). The two dipoles (now each infinitesimal, necessarily) combine to form a single current element of current moment

$$I'd' = I_1 d + I_2 d = 2Id$$

and the theory of a single dipole applies.

A different situation occurs, however, when the two current elements are equal in amplitude but opposite in phase. We shall see that as $c \to 0$ in this case, the radiation approaches zero everywhere (for a finite value of Id). Thus, when $\psi = \pi$ and c/λ is small, the array factor becomes

$$\text{A.F.} = -j2Id \sin\left(\frac{\pi c}{\lambda}\cos\vartheta\right)$$

$$\cong -j\frac{2Id\pi c}{\lambda}\cos\vartheta \tag{10.97}$$

and the radiation field E_ϑ becomes

$$E_\vartheta \cong \frac{60\pi^2 Idc}{\lambda^2 r} e^{-j\beta_0 r}\sin 2\vartheta \tag{10.98}$$

This is the radiation field of a *collinear quadrupole,* which has a more complicated radiation pattern than that of a dipole. For a given quadrupole "current moment" Idc, the radiation field is inversely proportional to the square of the wave length, instead of just the wave length as is the case for a dipole. On account of the condition $c \to 0$, the quadrupole radiation is a second-order quantity as compared with the radiation from a dipole. Note that $\sin 2\vartheta$ can be recognized as a solution to Eq. 10.19 with $C = -6$.

For an arbitrary complex ratio of $I_1 d/I_2 d$, the excitations of the two closely spaced current elements considered above can be resolved into symmetrical and antisymmetrical phasor components. The symmetrical parts contribute the dipole field in space, and the antisymmetrical parts contribute the collinear quadrupole field. We have, therefore, covered all cases comprised of two collinear dipoles.

An extensive generalization of the foregoing limiting process to include other orientations and more dipoles would lead us eventually to a complete set of "spherical harmonics" (solutions of Eq. 10.19) expressing the field caused by complicated "point sources" near the origin.

We wish next to consider power relations for the two collinear dipoles, taking account of the mutual effects. Including the element factor, the appropriate radiation field of the collinear dipoles and the corresponding radial component of the Poynting vector may be written from our previous expressions, as follows:

$$E_\vartheta = j\frac{120\pi}{\lambda r} Ide^{-j\beta_0 r}\sin\vartheta\cos\left[\left(\frac{\pi c}{\lambda}\cos\vartheta\right) - \frac{\psi}{2}\right] \tag{10.99}$$

$$S_r = \frac{1}{2}\sqrt{\frac{\epsilon}{\mu}}|E_\vartheta|^2 = \frac{|E_\vartheta|^2}{240\pi} \tag{10.100}$$

where the origin has been placed at $z = c/2$, between the dipoles. The total power radiated by the two dipoles is the surface integral of S_r over a complete sphere. Thus

$$P = 60\pi \left(\frac{Id}{\lambda}\right)^2 \int_0^{2\pi} d\varphi \int_0^{\pi} \sin^3 \vartheta \cos^2 \left[\left(\frac{\pi c}{\lambda} \cos \vartheta\right) - \frac{\psi}{2}\right] d\vartheta$$

Let $\cos \vartheta = x$. Then

$$-\sin \vartheta \, d\vartheta = dx$$

$$\sin^2 \vartheta = 1 - x^2$$

and the ϑ integral becomes

$$\int_{-1}^{1} (1 - x^2) \cos^2 \left[\left(\frac{\pi c}{\lambda} x\right) - \frac{\psi}{2}\right] dx$$

Through successive integrations by parts, we have

$$P = 80 \left(\frac{\pi Id}{\lambda}\right)^2 \left\{1 + 3 \cos \psi \left[-\frac{\cos \beta_0 c}{(\beta_0 c)^2} + \frac{\sin \beta_0 c}{(\beta_0 c)^3}\right]\right\} \qquad (10.101)$$

The power radiated from *each* dipole can be computed directly from the input conditions. The mutual impedances of the two dipoles are, according to Eqs. 10.84 and 10.41*a*,

$$Z_{21} = Z_{12} = \frac{\beta_0{}^2 d^2}{2\pi} \sqrt{\frac{\mu}{\epsilon}}\, e^{-j\beta_0 c}[(j\beta_0 c)^{-2} + (j\beta_0 c)^{-3}]$$

$$= \frac{\beta_0{}^2 d^2}{2\pi} \sqrt{\frac{\mu}{\epsilon}} \left[-\frac{\cos \beta_0 c}{(\beta_0 c)^2} + \frac{\sin \beta_0 c}{(\beta_0 c)^3} + j\frac{\sin \beta_0 c}{(\beta_0 c)^2} + j\frac{\cos \beta_0 c}{(\beta_0 c)^3}\right] \qquad (10.102)$$

In evaluating Z_{12} above, we located the origin of the spherical coordinates at the second dipole. The coordinates of the first dipole are then (c, π, φ). The dot product of the unit vectors $\boldsymbol{a}_r$ and $\boldsymbol{i}_1$ $(=\boldsymbol{a}_z)$ is -1, and the value of $\cos \vartheta$ in Eq. 10.41*a* is also -1. The other mutual impedance Z_{21} can be evaluated independently in a similar way. The two can be shown to be identical, confirming the reciprocity theorem.

The complex power at the inputs of the dipoles are, respectively,

$$\text{(a)} \qquad P_1 = \tfrac{1}{2}|I_1|^2 Z_{11} + \tfrac{1}{2} I_1{}^* I_2 Z_{12}$$

$$\text{(b)} \qquad P_2 = \tfrac{1}{2}|I_2|^2 Z_{22} + \tfrac{1}{2} I_1 I_2{}^* Z_{12} \qquad (10.103)$$

For the present case,

$$\begin{aligned} \mathrm{I}_1 &= I \\ \mathrm{I}_2 &= Ie^{-j\psi} \\ \mathrm{I}_1{}^*\mathrm{I}_2 &= I^2e^{-j\psi} \\ \mathrm{I}_1\mathrm{I}_2{}^* &= I^2e^{+j\psi} \end{aligned}$$

The real parts of the input powers then become

$$\langle P_{1,2}\rangle = \frac{1}{2} I^2 20(\beta_0 d)^2 \left\{1 + 3\cos\psi\left[-\frac{\cos\beta_0 c}{(\beta_0 c)^2} + \frac{\sin\beta_0 c}{(\beta_0 c)^3}\right] \pm 3\sin\psi\left[\frac{\sin\beta_0 c}{(\beta_0 c)^2} + \frac{\cos\beta_0 c}{(\beta_0 c)^3}\right]\right\} \qquad (10.104)$$

where the (+) and (−) signs refer respectively to $\langle P_1\rangle$ and $\langle P_2\rangle$. Because of the phase difference ψ in excitations, the powers radiated from the two dipoles are not identical. The sum of the two is, however, quite properly equal to the total power radiated, as given by Eq. 10.101. The coefficients of $I^2/2$ in $\langle P_1\rangle$ and $\langle P_2\rangle$ are the total input resistances of each of the dipoles, including the mutual effects.

We have two special cases of interest, mentioned previously. One is the in-phase condition $\psi = 0$, and the other is the out-of-phase condition $\psi = \pi$. For both cases, $\sin\psi = 0$ and $\beta_0 c \to 0$ as a limit. The important term in Eq. 10.104 becomes

$$-\frac{\cos\beta_0 c}{(\beta_0 c)^2} + \frac{\sin\beta_0 c}{(\beta_0 c)^3} = \frac{2}{3!} - \frac{4}{5!}(\beta_0 c)^2 + \cdots$$

When the two dipoles are excited in phase, $\cos\psi = 1$. The radiation resistance of each, including the mutual effects, is $40(\beta_0 d)^2$, according to Eq. 10.104, and the radiation resistance corresponding to the sum of the two powers is $20(\beta_0 2d)^2$. In this case, the two dipoles behave like a single dipole with an effective length $2d$, which we also found from the array factor (Eq. 10.95).

The other case, $\cos\psi = -1$, relates to the radiation from a collinear electric quadrupole. With this phase opposition, the radiation resistance of each dipole is now

$$20(\beta_0 d)^2\left[\frac{(\beta_0 c)^2}{10}\right]$$

a quantity proportional to the fourth power of frequency.

For the general case, $\mathrm{I}_1 d \neq \pm\mathrm{I}_2 d$, when we resolve the current element excitations into symmetrical and antisymmetrical phasor components, the symmetrical components will radiate like a dipole, and the antisymmetrical components will radiate like a collinear quadrupole. The radiation field is primarily that of the dipole, with its $\sin \vartheta$ distribution. The quadrupole radiation is hardly detectable on account of its low radiation resistance.

PROBLEMS

Problem 10.1. The earth receives from the sun radiation of intensity 0.15 w/cm^2. Given that the distance between the sun and the earth is 149×10^6 km, find: (a) The total power output of the sun, assuming that it is an isotropic source. (b) The rms electric field intensity at the earth due to the sun's radiation.

Problem 10.2. Discuss the rélative merits and disadvantages concerning a *long* transmission line connecting the transmitter of a TV station to its antenna, if: (a) The line is matched at both ends. (b) The line is matched at the transmitter only. (c) The line is matched at the antenna only. (d) The line is mismatched at both ends.

Problem 10.3. Discuss the fading of an A-M broadcast-band signal received by an automobile radio, when the car enters a tunnel or passes under a highway overpass bridge. Make some rough calculations of the magnitude of the effect, and see if they agree with your experience. Consider factors which might produce a discrepancy.

Problem 10.4. Find approximately the maximum power that may be fed into a very short dipole at a wave length of 10 m without requiring an FCC license. Assume the peak permitted value of the radiation field in the direction of maximum intensity is specified at $15\sqrt{2}$ μv/m, at a distance of $\lambda/2\pi$.

Problem 10.5. A certain electric dipole has an effective length $\lambda/10$ in air. It is to be used at the same frequency under fresh lake water, which for the purposes of this problem may be treated as an infinite, lossless, dielectric with $\epsilon = 9\epsilon_0$. (a) Estimate the radiation resistance of the antenna under water, stating carefully any additional assumptions you make. (b) Estimate the ratio of the input reactance under water to that in air, giving reasons for your conclusions. (c) Suppose the original dipole were of effective length $\lambda/3$ in air. Point out briefly any new features of the problem which would have to be taken into account to answer (a) and (b).

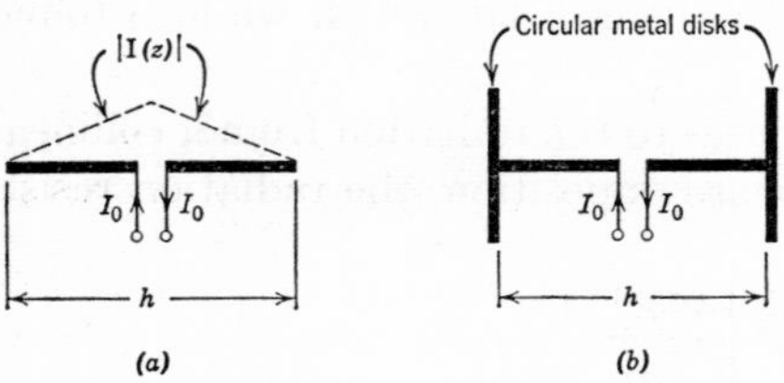

Fig. 10.11. Problem 10.6.

Problem 10.6. A short antenna of actual length $h(<\lambda/4)$ has the current distribution shown in Fig. 10.11*a*. (a) Find the "effective height" d of this antenna as an electric dipole. (b) Metal disks of small diameter ($<h/2$) are added at the ends of the antenna as shown in Fig. 10.11*b*. Indicate the approximate effect of these disks upon the current distri-

bution and discuss briefly the major changes, if any, they produce in: (i) the antenna radiation resistance; (ii) the antenna reactance; (iii) the radiation pattern.

Problem 10.7. Two solid metal cones, each of altitude $\lambda/10$ and base diameter $\lambda/20$, are to be used as arms for a center-fed electric dipole antenna. (a) Would you arrange them as in Fig. 10.12*a* or 10.12*b*? Explain precisely. (b) Find the approximate radiation resistance you expect for your choice of (a).

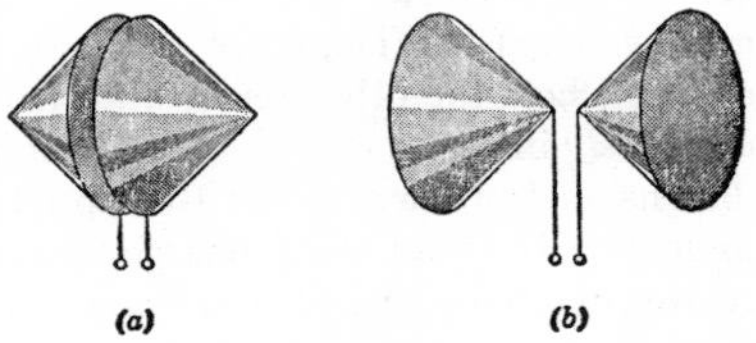

Fig. 10.12. Problem 10.7.

Problem 10.8. A radio ham wishes to have a meter length of No. 10 copper wire (diameter 100 mills) radiate energy at a frequency of 1 mc. Discuss with appropriate calculations the essential problem he faces.

Problem 10.9. (a) Find the radiation field from a half-wave dipole, fed at the center, assuming a sinusoidal current distribution along the dipole. (b) Plot the field radiation pattern in the plane of the antenna. (c) Evaluate the radiation resistance of the half-wave dipole. See Jahnke and Emde, *Tables of Functions*, for helpful integrals.

Problem 10.10. Consider the class of antennas whose radiation field is expressible in the form:

$$\mathrm{H}_\varphi = \frac{\mathrm{I}_0}{2\pi r} F\mathrm{f}_\varphi(\vartheta, \varphi) e^{-j\beta r} e^{j\xi}$$

$$\mathrm{E}_\vartheta = \sqrt{\frac{\mu}{\epsilon}}\, \mathrm{H}_\varphi$$

where f_φ is the field pattern, whose maximum value is unity in some direction, F is a function of geometry and frequency, and ξ is a fixed phase. (a) Show that for the above class of antennas $\pi R_r G_m = \sqrt{(\mu/\epsilon)}\, F^2$, where R_r is the antenna radiation resistance and G_m is its maximum gain (i.e., the same integral is involved in the calculation of both R_r and G_m). (b) Show that both the short dipole and the half-wave antennas belong to the above class by evaluating the respective F's. See Prob. 10.9.

Problem 10.11. Two center-fed $\lambda/2$ antennas are designed for use at the same frequency. One has a much larger wire diameter than the other. Which one has: (a) Greater physical length? (b) Higher radiation resistance? (c) Higher efficiency for radiation? (d) Higher Q? Explain your answers.

Problem 10.12. An infinitesimal electric dipole radiator is at the center of a spherical cavity of radius R defined by a perfectly conducting wall. (a) Find the field produced by the dipole. (b) Discuss the impedance of the dipole in this environment as a function of R and the frequency. (c) In view of the presence of receiving antennas, hills, aeroplanes, buildings, trees, etc., why are we still justified in using only the "outgoing" wave solution for a broadcast or transmitting antenna? State a general criterion for neglecting the incoming wave in such a way that the criterion is satisfied by your answer to this part, and by your answer to (a) above.

Problem 10.13. (a) Expand the electric dipole solution (Eq. 10.41) into the form $E_r = E_{r,0} \sum_{n=0}^{\infty} a_n(j\beta_0 r)^n$, $E_\vartheta = E_{\vartheta,0} \sum_{n=0}^{\infty} b_n(j\beta_0 r)^n$, $H_\varphi = H_{\varphi,1} \sum_{n=0}^{\infty} c_n(j\beta_0 r)^n$, where $E_{r,0}$ and $E_{\vartheta,0}$ are the components of the electrostatic field $\boldsymbol{E}_0$ produced by a static charge moment p, and $H_{\varphi,1}$ is the corresponding quasi-static magnetic field. Be sure to *derive* $H_{\varphi,1}$ from $\boldsymbol{E}_0$ by quasi-static methods, explaining very carefully any boundary conditions needed. (b) Explain the fact that some terms are missing from the expansions of (a) (i.e., some of the a_n, b_n, or c_n are zero), and show, qualitatively at least, how the other terms arise from each other. Is this expansion expressible *completely* in terms of $\beta_0 r$, as has been the case in our frequent expansions for one-dimensional problems? (c) Re-express $E_{r,0}$, $E_{\vartheta,0}$, and $H_{\varphi,1}$ in terms of the current moment Id instead of the charge moment p. Considering zero-frequency behavior, reinterpret the physical significance of these fields. Can you *derive* $E_{r,0}$, and $E_{\vartheta,0}$ from $H_{\varphi,1}$ by quasi-static methods? How are they related? (d) Repeat (b) with the viewpoint of (c). (e) Develop all the differential equations for a complete solution of the electric dipole problem by power-series expansions in frequency. It is convenient to choose as field variables $r\mathrm{H}_\varphi$, $r\mathrm{E}_\vartheta$, E_r, and to start from the field of an electrostatic dipole moment p. Proceed with the actual solution of these equations until you reach a problem at least as complicated to solve as Eq. 10.16. Do not solve that problem, but show at least that your answer to (a) is a solution. (f) At a given frequency, how well do the expansions discussed above converge for various values of distance r?

Problem 10.14. (a) Noting a relationship similar to duality between Eqs. 10.15 and 10.14, find quickly an *outgoing radiation* solution of Eqs. 10.15 for the case $\partial/\partial\varphi \equiv 0$ and $\mathrm{E}_\varphi \sim \sin\vartheta$. This solution corresponds to a "magnetic" dipole whose strength will appear in an arbitrary constant (like A of Eq. 10.30). (b) Write the approximate form of the solution (a) when $\beta_0 r \ll 1$.

Problem 10.15. (a) Find the static magnetic field produced by a current I_0 flowing around a small loop of radius R (area $A = \pi R^2$). Let the loop normal be the z-axis, and use spherical coordinates. Carry out the solution for $r \gg R$ only. (*Hint:* Use the magnetostatic vector potential.) (b) Find the quasi-static electric field produced by this loop. (c) Show that, by proper interpretation and proper choice of the arbitrary constant in it, your solution to Prob. 10.14 can represent the field of a small ($R \ll \lambda/2\pi$) current loop, at large distances from it.

Problem 10.16. (a) Using the results of Prob. 10.15, find the radiation resistance, transmitting gain, and receiving cross section of a small loop antenna (magnetic dipole). Compare with the electric dipole. (b) Check your work in Probs. 10.14 and (a) as follows: Compute separately the Z_{12} and Z_{21} of a network comprising an electric dipole and a small loop antenna, widely separated. For simplicity, orient them to achieve maximum coupling. You should find out that $\mathrm{Z}_{12} = \mathrm{Z}_{21}$ in agreement with the reciprocity theorem.

Problem 10.17. Derive the field of a small loop antenna by integration of the electric dipole contributions (Eq. 10.41) around a ring of small diameter. Make use of symmetry to find the simplest field first by direct integration. The rest of the fields can then be found from Maxwell's equations directly.

Problem 10.18. Using the results of Probs. 10.14 to 10.16, give a discussion of the impedance of a loop antenna, as a function of its size, similar to that contained in Sec. 10.6 for the electric dipole.

Problem 10.19. An electric dipole of current moment Id is mounted at a height h above a perfectly conducting ground plane. (a) It is desired to replace the effect of the plane upon the space above it by that of an "image" dipole. For what position, orientation, and complex moment of the image dipole are the boundary conditions satisfied at the ground plane, if the original dipole is oriented: (i) horizontal? (ii) vertical? (iii) 45° off the vertical? Make sketches showing your answers, and demonstrate satisfaction of boundary conditions. (b) If the dipole is vertical, and driven by a source between ground and its lower end (with $h \to 0$), what is the radiation resistance? (c) Answer (b) if the dipole is still center-driven and $h \to 0$. (d) If the dipole is horizontal, and $h \gg \lambda$, determine and sketch versus position the electric field strength *on* the ground plane along a line passing directly under the dipole and perpendicular to it. (e) Repeat the pertinent parts of (d) for a vertically oriented dipole. (f) If the dipole is horizontal, and $h \gg \lambda$, sketch the approximate field radiation pattern at distances much larger than h in a vertical plane through the dipole and perpendicular to it. Repeat for the vertical plane through the dipole and parallel to it. (g) For both cases in (f) calculate the approximate vertical angular position and the magnitude of the first minimum and the first maximum of the pattern above the ground plane. Indicate the polarization in each case. (h) Repeat the pertinent parts of (f) and (g) for a vertically oriented dipole.

Problem 10.20. Two z-directed electric dipoles located on the y-axis carry currents of equal *magnitude.* (a) Find *all* combinations of spacing D and relative time phase α which result in *one and only one* direction of zero radiation (for the distant field) in the x, y-plane. Note, for example, that north and south are regarded as two different directions. (b) In your results for (a), how does the phase of the radiation vary over a large sphere whose center lies on the y-axis midway between the two dipoles? (c) Among your solutions to (a), find *all* cases that have in addition *one and only one* direction of largest radiation in the x, y-plane. (d) Express the maximum value of the radial component of the Poynting vector in the solutions of (c), as a function of the magnitude of the current, the dipole moment, and the electrical spacing ($\beta_0 D$), over the range of $\beta_0 D$ to which the solution applies. Find especially an approximate expression for this power density correct to terms in $(\beta_0 D)^2$ when $\beta_0 D$ is small.

Problem 10.21. Two identical very short z-directed electric dipoles, each of effective height d, are parallel and spaced a distance D apart along the y-axis. (a) If the reactance of each dipole alone (with the other absent) is X: Find expressions for Z_{11}, Z_{22}, and Z_{12} which describe the network behavior between terminal pairs 1–1 and 2–2 of the two dipoles. (b) If $I_2 = -I_1$, find the *real part* of the impedance seen by the source driving terminal pair 1–1 and that seen by the source driving terminal pair 2–2. (c) Compare the total time-average power radiated with that which would be radiated by one of the dipoles *alone* carrying current I_1. (d) Discuss the results of (c) if $\beta_0 D \ll 1$, and then as $\beta_0 D \to 0$. (e) Suppose $I_2 = I_1 e^{j(\pi - \beta_0 D)}$. Find $\langle P_r \rangle$, the total time-average power radiated, and plot versus $\beta_0 D$ in the range $0 \leq \beta_0 D \leq \pi/2$. Obtain particularly an expression for $\langle P_r \rangle$ correct to terms in $(\beta_0 D)^2$ for small $\beta_0 D$. (f) Using the results of (e), along with those of Prob. 10.20 *d*, determine which of the situations in Prob. 10.20*c* leads to the largest maximum antenna gain. Evaluate this gain and compare it with that of a single electric dipole.

Problem 10.22. Apply the results of Probs. 10.19–10.21 to determine the principal radiation patterns, radiation impedance, and maximum gain of a short electric dipole oriented parallel to an infinite, perfectly conducting plane. Plot the radiation resistance and maximum gain versus distance from the plane in wave lengths.

Problem 10.23. A short open-wire transmission line extending in the z-direction may be considered as composed of two parallel dipoles. Phase angle between currents in the two wires is 180°, and spacing between wires is $\lambda/100$. Calculate with suitable approximations the radiated power, compared with that from a single wire carrying the same current.

Problem 10.24. How should two dipoles spaced one-half wave length apart be oriented, and what should the relative phase of the driving currents be (if their magnitudes are equal) for optimum propagation to two towns 120° apart?

Problem 10.25. Three dipole antennas with moments in the $+z$-direction are spaced a half wave length apart along the x-axis. The antennas are identical, and the exciting currents have the same magnitude. Maximum radiation is to be obtained in the positive x-direction. (a) What must be the phase angles between exciting currents in the three antennas? (b) Compare the Poynting vector, obtained at a distant point along the x-axis from this array, with the Poynting vector at the same point, obtained from a single antenna. (c) Determine whether or not it is possible with this spacing ($\lambda/2$) to choose the current phases so maximum radiation will occur in only *one* direction.

Problem 10.26. Two short current elements carry currents of equal magnitude but 90° out of the time phase. The elements form a cross at the origin, one polarized along the $+z$-axis and the other along the $+x$-axis. (a) On the surface of a sphere of large radius ($\beta_0 r \gg 1$), what is the nature of the polarization of the electric field at the points: (i) $\vartheta = 0$; (ii) $\vartheta = \pi/2$, $\varphi = 0$; (iii) $\vartheta = \pi/2$, $\varphi = \pi/2$; (iv) $\vartheta = \pi/4$, $\varphi = 0$. (b) What is the locus of all points on the sphere at which the polarization is linear? Circular? (c) What fraction of the total power radiated is supplied by each antenna? (d) Plot the far-field radiation pattern in power for the x, z-plane. (e) Repeat (a) through (d) if the two elements are fed in time phase.

Problem 10.27. Three $+z$-directed dipoles of equal moment magnitudes are located with space positions and time phases described by the notation (x, y, z; $\alpha°$), as follows: #1(0, 0, 0; 0); #2($\lambda/2\sqrt{2}$, 0, 0; $-90°$); #3(0, $\lambda/2\sqrt{2}$, 0; $-90°$). (a) Derive an expression for $\|\mathbf{E}\|$ at distant points in the x, y-plane. (b) Find the direction in which $\|\mathbf{E}\|_{\max}$ occurs for this array and the polarization and magnitude of this maximum field strength.

Problem 10.28. Four identical $+z$-directed dipoles carry currents of equal magnitude. Their space positions and time phases are described by the notation of Prob. 10.27, as follows: #1$\left(-\frac{\lambda}{4}, 0, 0; 0°\right)$; #2$\left(0, +\frac{\lambda}{4}, 0; +90°\right)$; #3$\left(+\frac{\lambda}{4}, 0, 0; 0°\right)$; #4$\left(0, -\frac{\lambda}{4}, 0; -90°\right)$. (a) Find the direction of maximum radiation in the x, y-plane. (b) Sketch the field radiation pattern in the x, y-plane.

INDEX

DATE DUE

ELECTROMAGNETIC ENERGY TRANSMISSION AND RADIATION

RICHARD B. ADLER

LAN JEN CHU

ROBERT M. FANO

THE M.I.T. PRESS
Cambridge, Massachusetts, and London, England

First MIT Press edition, June 1969
Second printing, March 1973

Library of Congress catalog card number: 60-10305
ISBN 0 262 01024 0

Printed in the United States of America